Human Impact

Researchers combined 17 sets of data on direct and indirect human influence to create this map of the estimated human impact on marine ecosystems. The map includes data on shipping, fishing, pollution, invasive species, temperature changes, ultraviolet light changes, and ocean acidification.

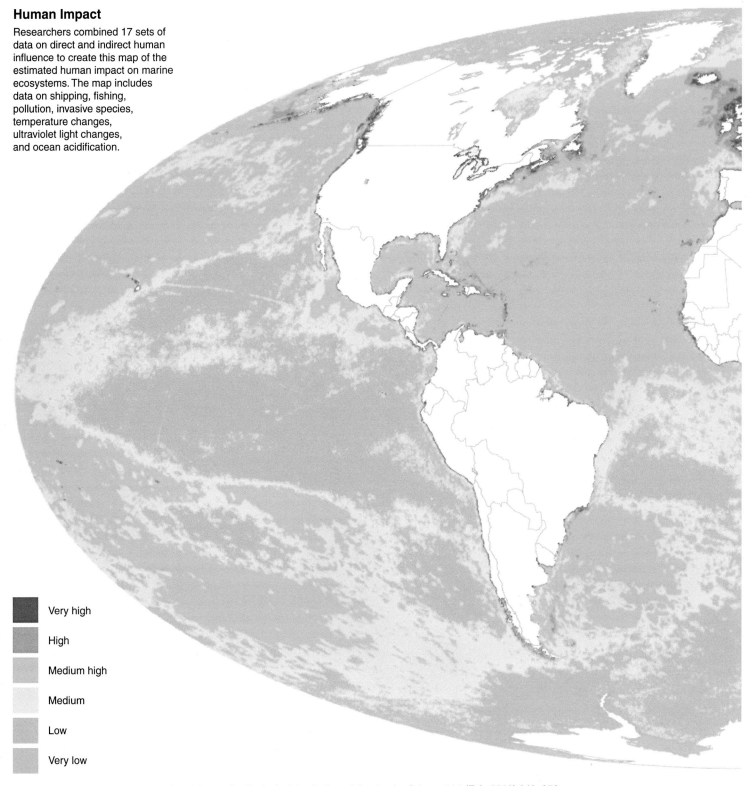

Very high

High

Medium high

Medium

Low

Very low

Sources: Benjamin S. Halpern, National Center for Ecological Analysis and Synthesis; *Science* 319 (Feb. 2008):948–952.

You can learn more about the impact humans have on marine ecosystems in each chapter of the text. You can learn more about the effects of fishing, pollution, invasive species, and temperature change in Part 3, Marine Ecosystems, and Part 4, Humans and the Sea.

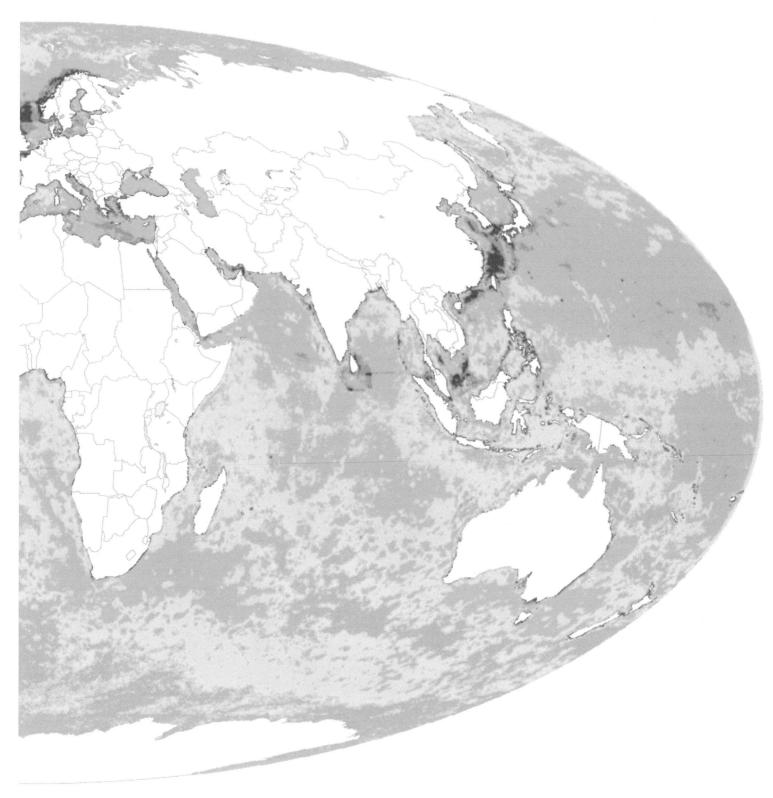

INTRODUCTION TO
Marine Biology

FOURTH EDITION

George Karleskint, Jr.
St. Louis Community College–Meramec

Richard Turner
Florida Institute of Technology

James W. Small, Jr.
Rollins College

CENGAGE

Australia • Brazil • Canada • Mexico • Singapore • United Kingdom • United States

CENGAGE

Introduction to Marine Biology,
Fourth Edition
George Karleskint, Jr.; Richard Turner;
James W. Small, Jr.

Publisher: Yolanda Cossio

Acquisitions Editor: Peggy Williams

Developmental Editor: Alexis Glubka

Assistant Editor: Shannon Holt

Editorial Assistant: Sean Cronin

Media Editor: Lauren Oliveira

Marketing Manager: Tom Ziolkowski

Marketing Assistant: Jing Hu

Marketing Communications Manager: Darlene Macanan

Senior Content Project Manager: Carol Samet

Creative Director: Rob Hugel

Art Director: John Walker

Print Buyer: Karen Hunt

Rights Acquisition Specialist: Dean Dauphinais

Production Service: Edward Dionne, MPS Limited

Text Designer: Riezebos/Holzbaur Group

Text Researcher: Tracy Metivier

Photo Researcher: PreMedia Global

Copy Editor: Michael Ryder

Cover Designer: John Walker

Cover Image: Mauricio Handler/ National Geographics

Compositor: MPS Limited

For product information and technology assistance, contact us at
**Cengage Customer & Sales Support, 1-800-354-9706
or support.cengage.com.**

For permission to use material from this text or product, submit all requests online at **www.cengage.com/permissions.**

Library of Congress Control Number: 2012933733

Student Edition:
ISBN-13: 978-0-357-67096-5
ISBN-10: 0-357-67096-5

Cengage
200 Pier 4 Boulevard
Boston, MA 02210
USA

Cengage is a leading provider of customized learning solutions with employees residing in nearly 40 different countries and sales in more than 125 countries around the world. Find your local representative at: **www.cengage.com.**

To learn more about Cengage platforms and services, register or access your online learning solution, or purchase materials for your course, visit **www.cengage.com.**

Printed at CLDPC, USA, 12-20

Contents Overview

Jon L. Hawker

Amanda Rohde/istockphoto.com

Contents

Red Barn Studio/istockphoto.com

Part 3: Marine Ecosystems

rusm/istockphoto.com

George Peters/istockphoto.com

Preface

Introduction to Marine Biology is intended for undergraduate students majoring in a variety of disciplines and for biology students interested in learning more about the marine environment. Many such students may already be interested in the field of marine biology. Others, however, are studying marine biology to fulfill a general education requirement, and they may have a fear of science courses. Like many, these students are generally intrigued by the marine environment and especially marine organisms. Having grown up with television programs, movies, and Internet sites dealing with the ocean and marine organisms, they have a natural interest in this subject area. This text strives to use this interest as a starting point for teaching biological science as it applies to marine organisms and the ecology of the marine environment. Each of the authors has been teaching marine biology for more than 30 years to both majors and nonmajors. In their lectures and field courses they stress an ecological approach to the study of marine organisms, and they have used this same approach in preparing this text.

The main focus of the text is the *ecology of the marine environment*—that is, the ways in which marine organisms interact with each other and with their physical environment. The authors believe that there is no better way to teach about the delicate balance of natural systems than within the context of marine ecosystems. This text also strives to educate students about the importance of marine ecosystems to terrestrial ecosystems and to humankind. The authors hope that by studying marine biology, students will be able to make well-informed decisions when voting and when they engage in activities that have an impact on the natural world, especially with respect to the ocean. The authors hope that this text will provide the factual foundation necessary for making such important decisions.

New to This Edition

Our overall goal with the fourth edition was to write a reader-friendly book that helped students develop a deeper appreciation of the diversity and beauty of life in the ocean while also developing a critical understanding of human impact on the sea and on sea life. Specifically, we have:

- Strengthened the coverage of ecology by introducing a new boxed feature, the "Impact Bubble," that presents current events and topics regarding the human impact on the marine environment.

- Organized the fourth edition to help students connect concepts by numbering the A-heads throughout the text. These numbered A-heads now tie sections to features, such as the "Have You Wondered?" chapter opening questions and the end-of-chapter Key Concepts.

- Expanded learning tools for students. The "Have You Wondered?" chapter opening questions, which draw students into the topics, have been updated and are now revisited in the end of chapter material in a new section. A running glossary now defines all glossary terms where they appear in the text. The "In Summary"

boxes are now bulleted to make the key points of each section even clearer, and boxes now also include references to the "Have You Wondered?" questions if the answers are covered in that section. In the "Questions for Review" section, new matching questions have been added to further help students prepare for exams.

- A new text design enhances the learning experiences. Over 60 photos and illustrations are new or redesigned, enhancing the student learning experience. A spectacular photo opens each chapter, piquing the motivation and attention of students.

In this edition, we have revised every chapter to include recent discoveries and to emphasize the impact that humans have on marine creatures and the marine environment. We have incorporated appropriate, timely material throughout, as the following examples illustrate: Chapter 2 contains a revised Population section; Chapter 3 has a new boxed feature, "Ecology and the Marine Environment: Seamount Communities" and a reorganized section on continental drift; Chapter 4 includes new content and an illustration of the ocean conveyor; Chapter 8 contains a revised section on sponges, additional material on cubozoans, clarification of cubozoans as a separate cnidarian class, and an updated phylogenetic tree; Chapter 9 includes changes that reflect current taxonomy, such as coverage on Sipunculids being moved to the section on Annelids; Chapter 12 presents new content on whales, a revised section on echolocation, and an added section on dolphin communication and behavior; Chapter 13 presents a completely revised sandy intertidal section; Chapter 15 has been reorganized to clarify the structure of the topics in the chapter; Chapter 19 has updated fishery statistics and revised material on aquaculture, especially as it applies to salmon farming; and Chapter 20 contains a reorganized section on pollution, added coverage of the Deepwater Horizon disaster, and the section on global warming was renamed "Climate Change" and the coverage was updated.

Learning Aids

This edition now contains the following learning aids to help students master the text material and use their knowledge.

1. "Have You Wondered?" questions begin each chapter to catch students' interest and motivate them to read the chapter to find answers to these questions.

2. New numbered A-heads throughout the text. These numbered A-heads are now tying sections to features, such as the "Have You Wondered?" chapter opening questions and the end of chapter Key Concepts.

3. Boldface terms: Throughout the text, important terms appear in boldface. All boldfaced terms are defined in the running glossary and in the glossary at the end of the text.

4. New running glossary defines every boldfaced term in the margins for easy student reference.

5. Pronunciation guide: Terms in the text that might pose pronunciation problems for beginning students are accompanied by phonetic transcriptions.

6. New and updated boxed readings: Boxed readings are organized into three categories: "Ecology and the Marine Environment," "Marine Biology and the Human Connection," and "Impact Bubble." The first two types provide students with an interesting and engaging focus on important aspects of marine biology. The third type is a new boxed reading that presents current events and topics regarding human impact on the marine environment.

7. Updated "In Summary" boxes are now bulleted to make key points of each major section even clearer, and they now also include references to the "Have You Wondered?" questions when the answers are covered in that section.

8. "In Perspective" summary tables: Each of the chapters devoted to marine organisms contains a summary table that reinforces classification and important biological and ecological aspects of the organisms discussed.

9. Phylogenetic trees at the beginning of Chapters 6–12 help students see evolutionary relationships among organisms discussed in each chapter.

10. Key Concepts: These appear at the end of each chapter and identify its main points.

11. The new "Are You Still Wondering?" section revisits the chapter opening "Have You Wondered?" questions, so the students can confirm the answers to these questions.

12. Updated Questions for Review: There are four levels of these end-of-chapter questions. The first level, Multiple Choice, comprises objective questions that deal primarily with vocabulary, terminology, and some basic concepts. The second level, Matching, is new to this edition and has been added to further help students prepare for exams by asking them to match terms/objects with the appropriate definitions/descriptions. The third level, Short Answer, consists of essay questions that test students' recall of facts and their ability to synthesize information. The last level, Thinking Critically, involves higher-order thinking skills and problem solving. The answers to the first- and second-level Questions for Review are provided in the Appendix of the text. Suggested responses to the other questions are found in the Instructor's Manual.

13. Suggestions for Further Reading: At the end of each chapter is a short list of articles and books that supplement the chapter's content. These articles and books are written at a level appropriate for students using this text. A few articles from more technical scientific literature are also included for those students and instructors looking to find in-depth coverage of some of the chapter topics.

14. Separate Glossary: Located at the end of the text, the Glossary defines all the boldfaced terms that appear in the text.

Organization

Introduction to Marine Biology is arranged in four parts that cover the environment, organisms, and ecosystems of the ocean, and humans' relationship to the ocean.

PART 1: THE OCEAN ENVIRONMENT

Chapter 1 introduces the science of marine biology, presents the scientific method, and orients the student to the rest of the text. Chapter 2 presents the basic principles of ecology as they apply to marine systems. This information is presented early in the text to support the main theme of ecology of the marine environment. Chapters 3 and 4 introduce the physical aspects of the marine environment and discuss their importance to the organisms that live in the ocean.

PART 2: MARINE ORGANISMS

Chapter 5 introduces the student to basic biological concepts such as the chemical basis of life, cell structure and function, energy transfer in biological systems, evolution, and biological classification. Chapters 6 through 12 survey all of the major groups of marine organisms and examine their interrelationships. These chapters are organized on the basis of feeding relationships, proceeding from organisms that produce their own food to those that rely on other organisms for food. Descriptions of animals are presented in a traditional format, beginning with invertebrates and working upward through the vertebrate classes to mammals. The focus of the individual chapters is the role that each group of organisms plays in the overall web of marine life.

PART 3: MARINE ECOSYSTEMS

Chapters 13 to 18 cover the major marine ecosystems. Each chapter in this part examines how the interactions of the physical and biological environment make each ecosystem unique and how these factors influence the number and kinds of marine organisms that inhabit a given area. The survey begins at the coastline with the intertidal zone (Chapter 13) and estuaries (Chapter 14) and then progresses to offshore ecosystems, coral reefs (Chapter 15), and the continental shelves and neritic zone (Chapter 16). Chapter 17 deals with the pelagic ecosystem of the open sea, and Chapter 18 addresses life in the ocean's depths.

PART 4: HUMANS AND THE SEA

Part 4 of the text examines the impact that humans have had and continue to have on the marine environment. Chapter 19 addresses fisheries management and the consequences of overfishing as well as problems associated with the extraction of nonliving products from the seas. The chapter also deals with how the use of these natural resources has changed over the past century, and it discusses the effects of this change on the marine environment. Chapter 20 examines marine pollution, climate change, introduced species, and habitat destruction and presents ways for students to help stem and possibly reverse environmental damage.

SUPPLEMENTS

Introduction to Marine Biology is accompanied by a supplement package that has been designed to aid students in learning and instructors in teaching, including helpful classroom preparation material such as the Instructor's Manual, Test Bank, ExamView computerized testing, and transparencies. In addition, the PowerLecture, a complete all-in-one reference for instructors, contains PowerPoint slides of images from the text, stepped art from the text, zoomable images, bonus photos, JoinIn clicker content, test bank, and Instructor's Manual.

Students will find a robust book-specific website at *www.cengage.com/biology*. This complementary site features chapter-by-chapter online tutorial quizzes, a final exam, chapter outlines, chapter review, chapter-by-chapter Web links, flash cards, and more.

A Lab Manual specially designed to help students learn more about marine life forms and their habits is also available for purchase with this edition.

ACKNOWLEDGMENTS

The production of a text such as this one involves the collaborative efforts and creative talents of many individuals. Without the help of editors, reviewers, and other professionals, this book would never have been published. The authors would like to acknowledge the highly professional and supportive staff at Cengage Learning.

The authors would also like to express their gratitude to their colleagues at St. Louis Community College–Meramec, Florida Institute of Technology, and Rollins College for their support and input and to their students, who have provided them with valuable feedback.

The authors would like to dedicate this book to the following:

To my wife, Shaun, and my children, Brad, Kristy, Tim, and Samantha, who share my enthusiasm for the ocean and its organisms. To my

grandsons, Zachary and Maxwell, who are beginning to discover the wonders of the sea.

—GK

To my grandchildren, Adam, Marnie, Marion, Lincoln, Haven, and Pollyanna, some of whom already show interest not only in charismatic marine megafauna but also the shells, worms, and crabs of the sea.

—RT

To my wife, Carolyn, who has put up with my being away with students on marine biology field trips for over 30 years. To my students, past and present, who have made those trips the great experiences they were.

—JS

REVIEWERS

The authors appreciate the diligent work done by the following individuals, who took time from their busy schedules to read and make suggestions concerning the manuscript. Their help and input was invaluable in creating the final text.

Rachel Venn Beecham, *Mississippi Valley State University*

Shari Bookstaff, *Skyline College*

W. Randy Brooks, *Florida Atlantic University*

Linda Fergusson-Kolmes, *Portland Community College*

Louis F. Gainey, Jr., *University of Southern Maine*

Richard S. Grippo, *Arkansas State University*

John Gunderson, *Tennessee Technological University*

Fiona M. Harper, *Rollins College*

Claudia Ibarra, *Irvine Valley College*

Gerard Loisel, *Nova Southeastern University*

Pamela Lynch, *SUNY–Suffolk County*

Carol Mankiewicz, *Beloit College*

Will Patterson, *University of West Florida*

Robert Reavis, *Glendale Community College*

Kristin M. Scheible, *Massaponax High School*

Jyotsna Sharma, *University of Texas at San Antonio*

Jeff Wooters, *Pensacola Junior College*

The authors are committed to providing the best possible text for teaching and learning about marine biology and welcome user feedback. Comments or suggestions for how we can improve this text in future editions are appreciated and can be sent to the authors in care of the publisher.

Have You Wondered?

Science and Marine Biology

THERE IS PERHAPS no better way to learn about the delicate balance of natural systems than to study marine organisms and the communities they form. This pursuit is not only fascinating in itself but also teaches us how important the ocean and its inhabitants are to humans and why we should try to preserve and conserve these resources. The diversity of marine organisms has attracted naturalists and scientists from the earliest days of human history, and the rich history of marine biology has strongly influenced modern science. This chapter will introduce you to the importance of oceans and their inhabitants and the developments that have formed the modern science of marine biology.

1-1 Importance of the Ocean and Marine Organisms

The world ocean is the principal physical feature of our planet. It covers nearly 71% of the earth's surface and represents the last great expanse on this planet to be charted and explored. The physical characteristics of this great body of water directly and indirectly affect our everyday lives, and the living organisms that inhabit them are an important source of food and natural products.

The ocean acts as an enormous solar-powered engine that drives the various weather patterns affecting terrestrial environments **(Figure 1-1a)**. Phenomena such as El Niño Southern Oscillation (ENSO), which can cause droughts in Peru, flooding in Texas, and relatively mild winters in the Midwest, are the result of changes originating in the Pacific Ocean. The actions of waves and tides change the contours of continents and affect the lives of people who inhabit coastal areas.

Ocean productivity—the amount of food marine organisms can produce and the number of organisms the ocean can support—is a leading area of research in marine ecology. The sea has provided, and still provides, a substantial amount of the world's food supply **(Figure 1-1b)**. The United Nations reports that more than 80 million metric tons (1 metric ton = 1.1 U.S. tons) of marine fish and shellfish (molluscs and crustaceans) are harvested annually.

> **Ocean productivity** is the amount of food produced by marine organisms, and the number of organisms the ocean can support.

Marine organisms also provide us with important materials for industry and medicine. The commercial harvesting and processing of these organisms and their products provide jobs for millions of people worldwide.

Marine organisms play another significant role in scientific research: Biologists have found that many species inhabiting the sea and seashores are ideally suited to the study of such varied fields as ecology, physiology, biochemistry, biogeography, behavior, genetics, and evolution. Experiments using marine organisms have

FIGURE 1-1 THE IMPORTANCE OF OCEANS. (a) The exchange of heat energy between the atmosphere and the oceans is responsible for creating the weather patterns that affect terrestrial habitats. The light area in this photo is a tropical storm developing in the Pacific Ocean. **(b)** The ocean supplies a significant amount of food in the form of fish, shellfish, and seaweeds.

(a)

(b)

provided biologists with considerable information not only about the organisms studied but also the workings of biology in general. Discoveries based on experiments with marine organisms have greatly advanced our understanding of biology.

Only by learning more about the sea and its inhabitants will we be able to realize the full potential of the world ocean.

IN SUMMARY

- The world ocean covers nearly 71% of our planet's surface.
- The world ocean affects weather patterns.
- The world ocean provides food and vital resources for human populations.

1-2 Study of the Sea and Its Inhabitants

Oceanography is the study of the oceans and their phenomena, such as waves, currents, and tides. The science of oceanography draws from many different disciplines, including chemistry, physics, geology, geography, meteorology, and even biology. The study of the living organisms that inhabit the seas and their interactions with each other and their environment is **marine biology (Figure 1-2)**. These two areas of study are not completely distinct from each other; they frequently overlap, and it is necessary to combine elements from both fields to form a complete picture of the ocean and its inhabitants.

With this in mind, the theme of this book is the ecology of marine organisms—that is, the interplay and interdependence of

Oceanography is the study of the physical characteristics of the ocean.

Marine biology is the study of marine organisms and their interactions.

organisms with each other and their environment. There is tremendous diversity among the organisms in the ocean, and we know very little about many of them. Some organisms may be important new sources of food, useful commercial products, or chemicals that could be used as medicines **(Figure 1-3)**. Millions of visitors are drawn yearly to seaside resorts to admire the beauty and variety of marine organisms. This is a key factor in industries such as ecotourism and sports such as scuba diving. Unless we learn more about marine organisms, we will never discover all of the substantial contributions that they could make to our lives, now and in the future.

We hear a great deal today about the impact humans have on the environment, including the sea. There are even some who claim that the sea is doomed and that we cannot reverse the damage that has been done. Although the media play a primary role in informing us about

FIGURE 1-2 MARINE BIOLOGY. The marine organisms in this tidal pool interact with and are influenced by each other and their physical environment. The study of marine organisms and these interactions is the science of marine biology.

FIGURE 1-3 MARINE ORGANISMS AND MEDICINE. The cartilage that makes up the skeletons of sharks is an important source of antiangiogenesis factor, a chemical that prevents tissues from establishing a blood supply. This chemical may be useful in the fight against cancer by depriving tumors of blood, thus killing them.

human impact on the sea, they interpret much of this information for us as well. The attitudes that people take toward such topics as the dumping of trash, disposal of radioactive and industrial wastes, oil spills, climate change, and overfishing are greatly influenced by the media. To make intelligent decisions about the ocean, as well as the rest of the environment, responsible citizens need a foundation of facts when analyzing these complex issues. This is perhaps the most important reason for a concerned citizen to learn about marine biology.

IMPACT BUBBLE

ONE OF THE MOST important reasons for studying marine biology is to be an informed citizen when voting on issues that could impact the sea and its inhabitants. Consider the issue of opening up more coastal areas for offshore oil drilling. With the ever-increasing demands for energy and a move to become less dependent on foreign oil, there is greater pressure on the federal government to allow more offshore drilling for oil. Although there are proven oil reserves in coastal regions, the extraction of the oil must be balanced against the potential ecological damage. Construction of offshore rigs destroys kelp beds, reefs, and coastal wetlands. These are important habitats for marine organisms, and their destruction also poses a threat to fisheries and ocean productivity. Pollution in the form of drilling fluid and metal cuttings are dumped into the ocean during the drilling process. In addition, over its lifetime an offshore rig will dump into the water toxic metals such as lead and mercury and cancer-causing chemicals such as toluene and benzene. These materials can cause health problems for marine organisms, interfere with their ability to reproduce, and increase mortality. Offshore drilling operations expose marine life to the possible threat of oil spills that could severely damage or even devastate whole populations. Your vote for politicians who support the environment or on referendums for offshore drilling could have an important impact on the future health of our oceans.

IN SUMMARY

- Oceanography is the study of the oceans and their phenomena.
- Marine biology is the study of marine organisms and their interactions with each other and their environment. (Have You Wondered? #1)
- Knowledge of marine biology helps us understand how marine organisms relate to us.
- Knowledge of marine biology helps us understand how human activities affect the marine environment. (Have You Wondered? #2)
- Knowledge of marine biology helps conscientious citizens make prudent decisions regarding activities that involve and affect the sea. (Have You Wondered? #2)

1-3 Marine Biology: A History of Changing Perspectives

Human interest in the sea probably dates back to the time we first set eyes on it, fished its waters, and sailed across it. The great expanse of water with its variety of strange and wonderful creatures has inspired awe, wonder, curiosity, myth, and at times, fear. This interest in the sea and its creatures laid the foundation for the sciences of oceanography and marine biology.

Like living organisms, the science of marine biology has evolved over time. Early investigations of the seas' creatures centered on simple observations of marine organisms that were easily accessible from the shore. As ships and equipment became available, humans set out to conquer the sea and attempted to control its awesome power. In their quest for discoveries, explorers and merchants alike eagerly set about emptying the sea of its contents, often overexploiting its resources. Improvements in shipbuilding and navigation opened up new frontiers in marine exploration. These same technical improvements also allowed the science of marine biology to expand as a body of knowledge. More recently, advances in submersibles, robotics, computers, and other technology have further broadened our view of the marine environment. This new knowledge has led marine biology to a more global perspective of the interrelatedness of ocean habitats and the interactions between ocean and terrestrial habitats.

EARLY STUDIES OF MARINE ORGANISMS

Early attempts to study the seas' creatures can be traced back to the ancient Greeks and Romans. The Greek philosopher Aristotle, who was an accomplished naturalist, was one of the first to develop a scheme of classification, which he called the "ladder of life." His writings contain descriptions of more than 500 species, almost one-third of which are marine. He also studied fish gills, proposing that they functioned in gas exchange, and made detailed observations of the anatomy of the cuttlefish (*Sepia*).

Pliny the Elder was the foremost of the Roman naturalists. His only surviving work is the 37-volume *Natural History*, which contains mostly information about terrestrial animals, but it does include references to marine fishes, clams, and mussels. During the Middle Ages, the

Dead Zones

Even people who do not live near the ocean can have an impact on marine communities and ecosystems. Burning of fossil fuels, for instance, increases the amount of greenhouse gases in the atmosphere, leading to climate change such as global warming, rising sea levels, and increased acidity of the ocean. Another way in which inland populations can affect the marine environment is by contributing nutrients to the ocean. Nitrogen is a major nutrient that supports the growth of algae in aquatic ecosystems. When nitrogen-containing chemicals from terrestrial sources reach the ocean they support an enormous increase in the growth of algae. When the algae die, the decomposition of their remains robs the water of oxygen. Marine organisms that can swim away, such as fishes, migrate to better water while those that cannot, such as clams and worms, die from lack of oxygen. The decomposition of their bodies removes more oxygen from the water, making a bad situation even worse. The result of this excessive decomposition is an area of ocean water that is oxygen depleted. Because so little marine life can survive in such an area, it is referred to as a **dead zone**.

A major source of the nitrogen-containing chemicals entering the ocean is agricultural runoff that flows into rivers that drain into the sea. Although farmers are not the only culprits in the nitrogen overload, they contribute the most. Studies show that 51% of the nitrogen entering the ocean from land comes from commercial fertilizer, 30% from animal manure, 5% from sewage treatment, and the remainder from other sources.

The problem of dead zones is worldwide. The U.N. Environment Program lists more than 150 dead zones around the world. These dead zones range from under 2.5 square kilometers (1 square mile) to several thousand square kilometers. The earliest recorded dead zones were in Chesapeake Bay, the Baltic Sea, Scandinavia's Kattegat Strait, the Black Sea, and the northern Adriatic Sea near Italy. Others have appeared off South America, China, Japan, southeast Australia, and New Zealand. The most infamous, however, is in the northern Gulf of Mexico **(Figure 1-A)**.

The decline in Gulf oxygen levels was first noticed in 1983. Levels remained relatively constant until 1993, when the Mississippi River's 100-year flood dumped enough fertilizer into the Gulf to double the size of the dead zone in 1 year. Since the mid-1990s, scientists have been monitoring conditions in the Gulf of Mexico off the coast of Louisiana, where a vast area of ocean (over 15,625 square kilometers, or 6,120 square miles) is an oxygen-depleted dead zone. Researchers studying the dead zone have linked the problem to farm chemicals that flow down the Mississippi River from Midwest farmlands. Farmers in the Midwest use millions of tons of nitrogen fertilizer for their crops of corn and soybeans. A large amount of this fertilizer washes into the Mississippi River following rainstorms and snowmelt and is eventually dumped into the Gulf of Mexico.

The dead zone in the Gulf of Mexico may well be the world's largest, and fixing this environmental problem will be most difficult. The problem pits Midwest farmers against Gulf fishers who have lost much of their livelihood as a result of the dead zone. Farmers are not receptive to talk of regulation, contending that fertilizers are necessary for their crops; and as demand increases for corn to produce ethanol for fuel, this problem is likely to worsen. Elected officials are not eager to take on the formidable farm lobby or attempt to divide the costs of Gulf cleanup among the 31 states in the Mississippi River drainage. Voters may eventually have the opportunity to make a difference by voting for senators and representatives who are willing to attack the fundamental problems and help restore the dead zone to life.

Catholic Church became the primary overseer of scholastic pursuits but gave most of its attention to matters of theology and philosophy. The study of natural history still depended on the works of the early Greek and Roman naturalists. Arabian philosophers of the Middle Ages involved themselves with interpreting and explaining the works of the ancient naturalists, rather than engaging in their own studies and observations. It would be the late 18th century before biologists would again conduct studies of the marine environment based on original observations.

> A **dead zone** is an area of ocean that has become oxygen depleted.
>
> An **atoll** is a ring of coral reef that encloses a lagoon.

RENEWED INTEREST IN MARINE ORGANISMS

During the late 18th and early 19th centuries, biology expanded into several disciplines. This was a time of great discovery, fueled in part by the development of better sailing ships and navigational instruments.

The exploration of new lands and new sea routes provided information for the various branches of science, including biology. Scientists such as the French naturalist Jean-Baptiste Lamarck and the anatomist Georges Cuvier, studied and described many marine organisms during this period.

In December 1831, the vessel HMS *Beagle* set sail on a 5-year voyage of exploration that would take it around the globe. Among the members of the expedition was Charles Darwin. During the voyage, Darwin, father of the theory of evolution by natural selection and an early marine biologist, was able to observe marine life firsthand and collect many specimens of marine organisms. On the basis of his observations of **atolls** (ring-shaped coral reefs that enclose a lagoon) in the Pacific, he proposed an explanation of atoll development that is still widely accepted. It was during this voyage that he began to formulate what would eventually become his theory for the process of evolution. In 1859, Charles Darwin published his landmark work, *On the Origin of Species by Means of Natural Selection*. This work stimulated many scientists to investigate the causes of adaptations observed in marine organisms. It also sparked study of the interrelationships among

According to the Pew Oceans Commission report, nitrate runoff pollution from the northern and midwestern Mississippi River watershed is one of eight major threats to ocean wildlife, polluted beaches, and collapse of commercial fishing in U.S. waters. Such pollution has contributed to a growing "dead zone" in the Gulf of Mexico where the Mississippi meets the ocean. Pollution drains oxygen from the ocean ecosystem, causing fish kills and depletion of ocean wildlife.

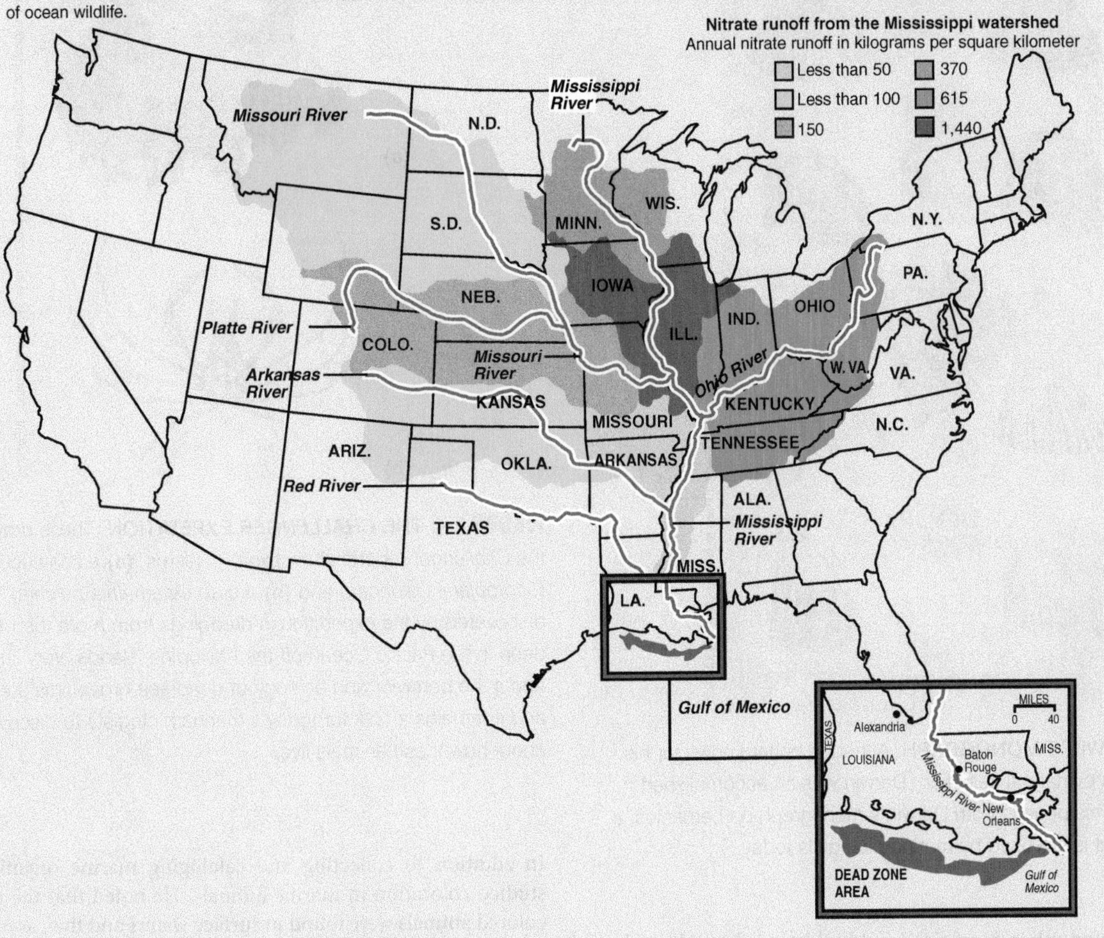

Nitrate runoff from the Mississippi watershed
Annual nitrate runoff in kilograms per square kilometer

Less than 50 | 370
Less than 100 | 615
150 | 1,440

Reprinted with permission from the St. Louis Post Dispatch, © 1997.

FIGURE 1-A HYPOXIC ZONE. Agricultural nutrients draining into the Gulf of Mexico have contributed to the depletion of oxygen in an area off Louisiana referred to as the "dead zone."

marine organisms and between marine organisms and their environment. In the years following the voyage of the *Beagle*, Darwin engaged in a detailed study of the barnacles that inhabit the rocky coasts of England and produced a monograph on the subject that is still used today **(Figure 1-4)**.

In the early part of the 19th century, it was generally agreed that living organisms could not survive in the cold and darkness of the ocean depths, an idea proposed by the English naturalist Edward Forbes. Evidence to the contrary was produced when a transatlantic telegraph cable linking the United States and England failed shortly after it was laid in 1858 and had to be retrieved. The cable was located at a depth of approximately 1.7 kilometers (1 mile) in the northern Atlantic Ocean. As the repair ship retrieved the cable, it was found to be covered with all sorts of marine organisms that had never been seen before. These organisms apparently flourished in the depths of the ocean, and their discovery sparked investigations by several countries into life on the ocean floor. Subsequent dredging expeditions recovered animals from as deep as 4.42 kilometers (14,500 feet, or more than 2¾ miles).

BEGINNINGS OF MODERN MARINE SCIENCE

By the end of the 19th century, many countries, including the United States, were spending significant amounts of time and money to learn more about the marine environment and its creatures.

Challenger Expedition

As a result of increasing interest in the marine environment, the British Admiralty organized the *Challenger* expedition, which lasted 3½ years. The expedition was named after the research vessel HMS *Challenger*, a ship containing state-of-the-art equipment and research facilities. When the *Challenger* returned to England in May 1876, it had crisscrossed the major oceans of the world and brought back enough information to fill 50 volumes of scientific reports. During this expedition, more than 4,700 new species of marine organisms were collected and described. Many of these new species were dredged from great depths **(Figure 1-5)**.

The *Challenger* expedition gave birth to the modern sciences of marine biology and oceanography, and even today marine scientists continue to refer to the *Challenger* reports in their research. Later expeditions

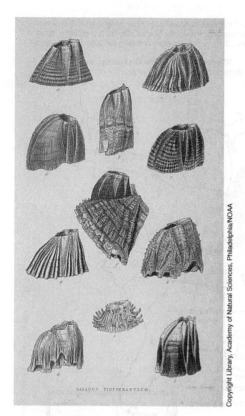

FIGURE 1-4 DARWIN'S MONOGRAPH. Although better known for his theory of evolution by natural selection, Darwin was an accomplished marine biologist. This page is from Darwin's monograph on barnacles, a reference work that is still used by marine biologists today.

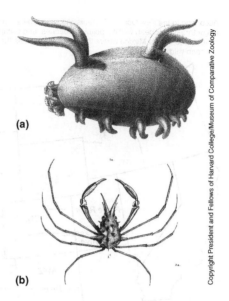

FIGURE 1-5 THE *CHALLENGER* EXPEDITION. These drawings from the *Challenger* reports show two organisms, **(a)** a sea cucumber (*Scotoplanes globosa*) and **(b)** a crab (*Anamathia pulchra*), that were first discovered by the expedition in dredgings from more than 1,000 meters deep in the Pacific Ocean off the Philippine Islands. Very little is known about the behavior and ecology of deep-sea organisms such as these, and it remains a task for today's marine biologists to discover more about how these animals live.

In addition to collecting and cataloging marine organisms, Agassiz studied coloration in marine animals. He noted that the most brightly colored animals were found in surface waters and that, as one proceeded deeper, brilliant colors gave way to blues and greens and ultimately reds and blacks. He theorized that the colors were related to the absorption of different wavelengths of light at different depths, a theory later

by Great Britain and other countries would add to and build on the broad base of information accumulated by this groundbreaking expedition.

Charles Wyville Thomson, the driving force behind the *Challenger* expedition and its chief scientist, collected samples of the microscopic organisms floating in the water **(Figure 1-6)**. Previously, biologists had paid little attention to these organisms, but now they were coming under close scrutiny. In 1887, Victor Hensen coined the term **plankton** to describe all of the different organisms that float or drift in the sea's currents. These tiny organisms are at the base of the ocean's complex food webs, and only in the last 50 years have marine biologists started to understand the specific roles that these organisms play.

Marine Studies in the United States

American scientists were also contributing to the ever-growing body of knowledge that was marine biology. Through expeditions and the founding of marine laboratories, the United States became a leader in the emerging field of marine studies.

Expeditions of Alexander Agassiz In 1877, the American naturalist Alexander Agassiz began the first of several expeditions he would direct to investigate the organisms of the sea. Agassiz collected samples of animals from hundreds of locations and dredged animals from depths of 180 to 4,240 meters (600 to 14,000 feet). The amount of material he collected rivaled that brought back from the *Challenger* expedition **(Figure 1-7)**.

Plankton are organisms that drift in ocean currents.

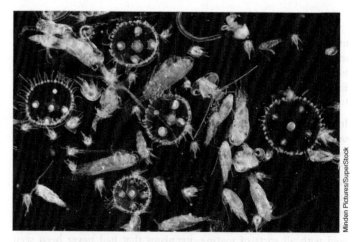

FIGURE 1-6 PLANKTON. Some examples of marine plankton, organisms that float or drift in the sea's currents. Charles Wyville Thomson, the chief scientist of the *Challenger* expedition, was one of the first scientists to seriously investigate the role of plankton in marine communities.

FIGURE 1-7 ALEXANDER AGASSIZ. Alexander Agassiz was one of the foremost U.S. marine biologist of the 19th century. He is pictured here in his laboratory with jars containing some of the marine specimens that he collected during his many expeditions.

FIGURE 1-8 MARINE BIOLOGICAL LABORATORY AT WOODS HOLE. This Massachusetts facility is one of many institutions worldwide whose goal is to learn more about the organisms that inhabit the oceans.

proved correct. He also noted a great deal of similarity in the deepwater organisms on the east and west coasts of Central America and hypothesized that the Pacific and the Caribbean were at one time connected. Agassiz spent much of the latter part of his life studying the structure and formation of coral reefs.

Founding of the First Marine Biological Laboratory Alexander's father, Louis Agassiz, founded the Museum of Comparative Zoology at Harvard University. He also founded the first marine biology laboratory in the United States in July 1873. Originally located on Penikese Island off the Massachusetts coast, the Anderson Summer School of Natural History was founded to help teachers at all levels improve their methods of teaching natural history. Agassiz was a firm believer in learning through observation and hands-on acquisition of knowledge. It was Agassiz's hope that the Summer School would allow him the opportunity to teach his methods to others using marine organisms as subjects of study. A primary reason for locating his school by the sea was that all the major groups of organisms are represented in the ocean, and many groups are found *only* in the ocean. The animals would be within easy reach and could be studied in their natural habitats. Agassiz's goal of teaching biology through direct observation is still a major objective of courses offered at marine laboratories around the world.

Agassiz's school was the inspiration and predecessor of the Marine Biological Laboratory at Woods Hole **(Figure 1-8)**, which was founded in 1888 in a little fishing village on Cape Cod, Massachusetts. In 1922, the Woods Hole Oceanographic Institution was constructed down the street from the Marine Biological Laboratory. Their proximity allowed a great deal of interchange between the two institutions. Today, the Marine Biological Laboratory is a major, internationally recognized research institution. Louis Agassiz's influence on the field of marine

biology can still be felt at the Marine Biological Laboratory at Woods Hole. One of his favorite sayings, "Study nature, not books," is prominently posted in the library to remind students and researchers alike of this important concept.

Other U.S. Marine Laboratories During the 20th century, other notable research institutions were founded, such as the Scripps Institution of Oceanography in California, the University of Miami's Rosenstiel School of Marine and Atmospheric Science in Florida, the Harbor Branch Oceanographic Institution in Florida, the Friday Harbor Laboratories of the University of Washington, and Duke University Marine Laboratory in North Carolina, to name a few. Understanding commercial fishery production was a driving force in the establishment of many marine laboratories and continues to be a chief goal. In addition to research on ocean productivity and the interrelationships of marine organisms, modern marine laboratories focus on the use of marine organisms to solve fundamental problems in biology, such as the control of cell division, the process of embryological development, the functioning of the nervous system, and the applications of biotechnology.

MARINE BIOLOGY IN THE 20TH CENTURY

Early in the 20th century, expeditions were mounted to study the Arctic and Antarctic seas. Individuals such as the Norwegian Fridtjof Nansen and the Englishman Sir Alistair Hardy **(Figure 1-9)** led expeditions that collected information and organisms from these two areas. Nansen was interested in reaching the magnetic North Pole as well as charting the waters around the Pole. He was not successful in his attempt to reach the Pole, but his expedition to gather information about the polar seas was quite successful. Hardy's interest in the biology and commercial exploitation of whales spurred his expeditions to the Antarctic Sea. In addition to making new observations concerning whales, he also increased the amount of information available concerning the Antarctic Sea. His book *The Open Sea: Its Natural History* remains a classic in the field of marine biology.

Since the latter part of the 19th century, advances in the field of marine biology have given us both a broader outlook and a deeper view into the sea itself. With this new perspective comes a fuller understanding about

FIGURE 1-9 SIR ALISTAIR HARDY. The Englishman Sir Alistair Hardy led expeditions to the Antarctic Sea to study whales.

FIGURE 1-10 THE SUBMERSIBLE *ALVIN*. Submersibles like *Alvin* allow marine scientists to investigate life in the ocean's deepest recesses.

global ecology. For years, our technology for exploiting the sea has outpaced our understanding of marine biology, but advances in the science have led to new attitudes about conservation and resource management. Starting in the last century, some investigators began to center their attention not only on the relationships of marine organisms to each other but also on the impact of human activities, such as fishing and pollution, on the marine environment.

MARINE BIOLOGY TODAY

We live in the Information Age. Each day, new findings that increase our comprehension of the living world that surrounds us are added to the world's data banks. Knowledge of the inner workings of the sea as well as its inhabitants is increasing at breakneck speed, aided by many advances in technology. Today, deep-sea submersibles such as *Alvin* **(Figure 1-10)** can take marine scientists to the very floor of the ocean to view and collect organisms that live in the deepest recesses of the sea. Researchers can live in underwater habitats while they observe the activities of these organisms. The advent of the Internet and the "information superhighway" allows scientists and nonscientists around the world to share enormous amounts of information collected on a variety of subjects, including marine biology. With the aid of these new tools, marine biologists are discovering the multifaceted interrelationships among marine organisms and marine systems and their ties to terrestrial environments.

In the following chapters, you will learn more about the sea and the organisms that inhabit it. You will learn how both physical and biological factors interact to produce the complex ecosystems of the marine environment. You will be introduced to the methods of science as they apply to the study of marine biology. And you will learn what impact humans have had on this truly wondrous realm.

Science is an endeavor or study.

Hypotheses are explanations that can be tested by experiments.

Inductive reasoning is the process of reasoning whereby a general explanation is derived from a series of observations.

IN SUMMARY

- The science of marine biology has advanced over the years as new technologies have been developed.

- Initial studies of marine organisms can be traced back to the ancient Greeks and Romans.

- The modern sciences of oceanography and marine biology originated with the *Challenger* expedition in the 19th century.

- The first marine laboratory in the United States was founded by Louis Agassiz in 1873 for the purpose of helping teachers improve their methods of teaching natural history.

- In the early 20th century, expeditions were mounted to study to the Arctic and Antarctic seas and their inhabitants.

- Marine laboratories play a vital role in both marine and basic biological research. (Have You Wondered? #3)

1-4 Process of Science

A particular endeavor or study becomes a **science** when the principles on which it is based can be presented as **hypotheses**, explanations that can be tested by experiments. A good hypothesis can explain past events and predict the outcome of current or future experiments. All branches of science seek to organize observations so that hypotheses can be formed that suggest relationships between the observations. These hypotheses are then tested by experiments. The data gathered from these experiments are evaluated, and logical conclusions are drawn from the information at hand.

SCIENTIFIC METHOD

Not all researchers follow the same approach to doing science. Some believe that science should be done inductively. **Inductive reasoning** involves looking at individual observations and proposing a general explanation for them. For instance, a marine biologist might observe that the octopus and squid–animals classified as cephalopods–have arms with suckers and conclude, using inductive reasoning, that all cephalopods have arms with suckers. Other researchers believe that good science should be done deductively. In **deductive reasoning**, observations suggest some general principle or idea from which specific statements can be derived. Thus a marine biologist might note that all cephalopods have arms with suckers, and because a cuttlefish is a cephalopod, it must have arms with suckers. Most scientists feel free to use either method and sometimes the best science is accomplished by using both methods. For example, if a body of facts suggests some general principle (inductive reasoning), then specific points can be drawn from that principle (deductive reasoning), and those points can then be tested.

Scientists use an organized, commonsense approach to their work that begins with observations of the natural world. On the basis of these observations, the scientist asks questions, proposes hypotheses, designs experiments to test the hypotheses, gathers results, and draws conclusions. This orderly pattern of gathering and analyzing information is called the **scientific method**. The scientific method represents what might be called a scientific approach to interpreting the natural world. It should be emphasized that the steps of the scientific method are simply an idealized approach to problem solving in science. Individual researchers tend to bring their own variations to the plan as they construct and conduct their experiments.

Step 1: Making Observations

Generally, the first step in the process of science is to gather observations that stir the curiosity of scientists, causing them to ask questions concerning the subject that is being studied. Making observations is a substantial aspect of the scientific process, but observations alone are not very helpful without some thread or pattern that ties them together. It is not enough just to note facts. A good researcher is not only a good observer but also a creative thinker. For instance, like many visitors to a rocky shore, a biologist named J. H. Connell observed that barnacles grew on rocks in groups forming definite zones. Connell also noticed that two different types of barnacle occupied distinctly different zones on the rocks. These observations set him wondering about the possible causes for the separation. Were the two types separated because of some characteristic of their environment, such as type of rock or amount of moisture? Or was it due to some biological interaction, such as competition? To answer these questions, Connell would have to form hypotheses and test them experimentally. In another case, researchers found that night-feeding fishes, such as moray eels, ocean catfish, and some shark species, are unable to locate their food by sight. These observations led the researchers to ask how these fishes were finding their prey. Again, finding the answer would involve forming a hypothesis and testing it experimentally.

Step 2: Using Inductive Reasoning to Form a Hypothesis

After noticing a recurring pattern or a relationship between observations, a researcher uses inductive reasoning to make an educated guess as to the probable answer to the question asked. This process is called *formulating a hypothesis*. On the basis of his observations of the barnacles, Connell hypothesized that biological interactions between species were responsible for the pattern he observed. In the example of the night-feeding fishes, researchers hypothesized that the fishes rely on their sense of smell to locate their prey. A hypothesis may be as simple as proposing a cause-and-effect relationship between observed events, or it may be more complex and propose a model for the way a particular process may work. There can be more than one method of formulating a hypothesis. Insights can arise from accident and intuition as well as from methodical observations.

Step 3: Using Deductive Reasoning to Design Experiments

Regardless of how the hypothesis is stated, to be valid it must be testable. It is this step of testing the hypothesis that sets science apart from other disciplines and enables scientists to produce accurate explanations of natural phenomena. Once the hypothesis has been formed, it can be used to predict what the consequences might be if the hypothesis is correct. This process of reasoning from a general statement to predicting consequences involves the use of deductive reasoning (the if/then process). For instance, Connell hypothesized that the two types of barnacle were restricted to their particular zones by biological interaction. If that were true, then removing one of the types of barnacle would allow the other to move into its zone. Connell could then design experiments, such as scraping one type of barnacle off the rocks, to test his hypothesis. In the example of the night-feeding fishes, researchers hypothesized that the fishes used their sense of smell to find their prey. On the basis of this hypothesis, if the fishes could not smell their prey, they would not be able to locate it. Again, researchers could then design experiments to test this hypothesis, such as plugging the fishes' nostrils.

Hypotheses are tested by performing experiments or sometimes, in fieldwork, by making systematic and detailed observations. For this step to be effective, scientists must design experiments that will disprove incorrect hypotheses. A well-designed experiment involves running two trials at the same time. In these two trials only one factor, the **experimental variable**, is altered. The trial that contains the experimental variable is called the **experimental set**. The trial without the experimental variable is known as the **control set**.

Experiments sometimes produce results that have occurred purely by chance and cannot be duplicated. Therefore, before accepting a hypothesis, researchers repeat the experiment several times to ensure that the results are valid. Another scientist working independently must be able to repeat the same experiments and get essentially the same results for a hypothesis to be valid and accepted. Alternative experiments should also be performed to ensure that other variables have not been overlooked. Only if many different approaches fail to disprove a hypothesis will it be considered accurate; even then, the acceptance of the hypothesis is subject to change if new information comes to light. Thus the aim of a scientist is not to prove something but to test it many times to achieve a more accurate explanation of the phenomenon being studied. Whenever possible, the data that are collected in an experiment are quantitative, or numerical. Numerical data are easier to compile and subject to statistical analysis.

Deductive reasoning is the process of reasoning whereby observations suggest a general principle from which a specific statement can be derived.

The **scientific method** is an orderly pattern of gathering and analyzing information.

The **experimental variable** is the factor that is altered in an experiment.

The **experimental set** is the experiment that contains the experimental variable.

The **control set** is the experiment that does not contain the experimental variable.

Steps 4 and 5: Gathering Results and Drawing Conclusions

After all of the results of the experiments are gathered, the researcher draws conclusions and either accepts or rejects the hypothesis. In the research with the night-feeding fishes, researchers found that the fishes could not find their prey with their nostrils plugged. They concluded that these fishes locate their prey with their sense of smell and accepted their hypothesis.

If the results of an experiment do not support the hypothesis, the investigator will then alter or modify the hypothesis in light of the new information and begin a new round of testing. This process may be repeated several times before a researcher arrives at a hypothesis that is consistently supported by experimentation. Connell found that removing one type of barnacle from the rocks did not always allow the other type to move into the newly vacated space. Through a series of several experiments, Connell was able to show that the location of a particular type of barnacle on the rocks was due to a combination of physical and biological factors. Over time, some hypotheses are expanded upon and eventually become generally accepted as theories. A scientific **theory** represents a body of observations and experimental support that have stood the test of time.

> A **theory** is a body of observations that have stood the test of time.
>
> **Observational science** is an approach to doing science when controlled experiments are not feasible.

PLANT GROWTH IN A SALT MARSH: A CASE STUDY OF THE SCIENTIFIC METHOD

Ivan Valiela is an ecologist who has spent many years studying salt marshes. While conducting studies on salt marshes in New England, he observed that marsh grasses that grew in areas receiving higher levels of nutrients seemed to grow taller than marsh grasses in areas where nutrient input appeared to be limited **(Figure 1-11)**. Analysis of soil samples from different areas suggested that the levels of nitrogen, an essential nutrient in plant growth, were higher where the plants were taller. These observations led him to hypothesize that the growth of marsh grass was limited by the availability of nitrogen. On the basis of this hypothesis, he predicted that if nitrogen were added to the soil, then marsh grass would grow larger or faster or both.

To test his hypothesis, Valiela picked several plots of salt marsh that were as identical as possible in several respects. In each plot the types of plants, plant density, soil type, freshwater input, and height above the average tide level were as similar as possible. To his experimental plots, Valiela added nitrogen-containing fertilizer. The control plots received no nitrogen-containing fertilizer but otherwise were subjected to the same environmental conditions as the experimental plots.

All through the growing season, Valiela monitored the plots, measuring the growth rate of the plants, chemical composition of the leaves, and rate at which the leaves decayed when they were removed from the plant. He analyzed his data statistically and, among other things, found that several well-known species of marsh grasses grew taller and larger than the controls when they were supplemented with nitrogen-containing fertilizer. Further tests of the hypothesis continued to support these findings, thus leading to the conclusion that a major limiting factor in the growth of salt marsh plants is the availability of nitrogen.

ALTERNATIVE METHODS OF SCIENCE

Although the process of forming hypotheses and testing them with experimentation is central to scientific study, it is not the only approach that scientists use. There are times when scientists must depend on observation alone. For instance, certain behaviors of humpback whales may occur more frequently while mating, leading to the hypothesis that these behaviors are important for attracting a mate. In this situation, it is not possible to conduct controlled experiments to test the hypothesis, so the weight of observations alone must support or deny the hypothesis. This approach to science is sometimes referred to as **observational science**.

IN SUMMARY

- Scientists use an orderly method of investigation known as the *scientific method* to learn about the natural world. (Have You Wondered? #4)
- The scientific method begins with observations of natural phenomena.
- Hypotheses are formed from observations using inductive reasoning.
- Deductive reasoning is used to design experiments to test hypotheses.
- Conclusions drawn from experiments lead scientists to either accept or reject their hypotheses.
- Observational science is an approach to doing science when controlled experiments are not feasible.

KEY CONCEPTS

1. Marine and terrestrial environments are interrelated, interactive, and interdependent. (1-1)

2. The ocean is an important source of food and other resources for humans. (1-1)

3. Marine biology is the study of the ocean's diverse inhabitants and their relationships to each other and their environment. (1-2)

4. The history of marine biology is one of changing perspectives that have shaped the modern science and its applications. (1-3)

5. Marine laboratories play an important role in education, conservation, and biological research. (1-3)

6. It is important to study marine biology in order to make informed decisions about how the oceans and their resources should be used and managed. (1-2)

7. Scientists use an organized approach called the scientific method to investigate natural phenomena. (1-4)

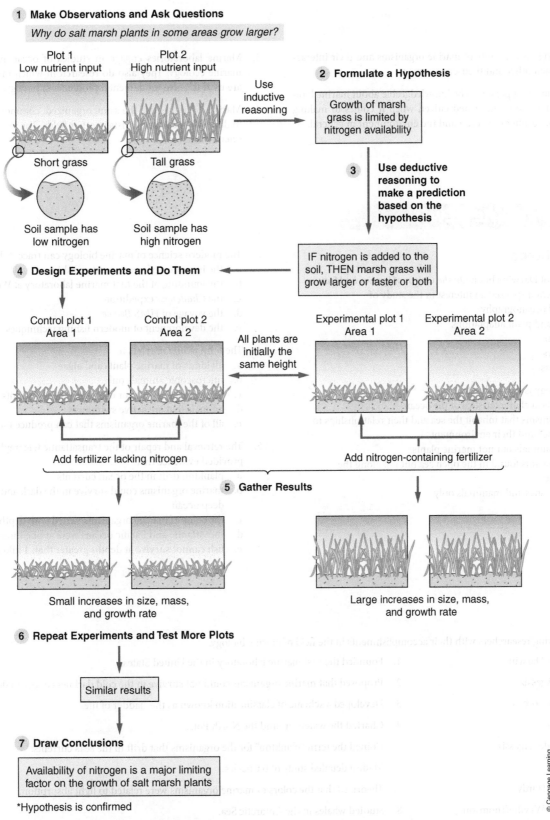

1 **Make Observations and Ask Questions**

Why do salt marsh plants in some areas grow larger?

Plot 1
Low nutrient input

Plot 2
High nutrient input

Use inductive reasoning

2 **Formulate a Hypothesis**

Growth of marsh grass is limited by nitrogen availability

Short grass

Tall grass

3 **Use deductive reasoning to make a prediction based on the hypothesis**

Soil sample has low nitrogen

Soil sample has high nitrogen

IF nitrogen is added to the soil, THEN marsh grass will grow larger or faster or both

4 **Design Experiments and Do Them**

Control plot 1
Area 1

Control plot 2
Area 2

All plants are initially the same height

Experimental plot 1
Area 1

Experimental plot 2
Area 2

Add fertilizer lacking nitrogen

Add nitrogen-containing fertilizer

5 **Gather Results**

Small increases in size, mass, and growth rate

Large increases in size, mass, and growth rate

6 **Repeat Experiments and Test More Plots**

Similar results

7 **Draw Conclusions**

Availability of nitrogen is a major limiting factor on the growth of salt marsh plants

*Hypothesis is confirmed

© Cengage Learning

FIGURE 1-11 THE SCIENTIFIC METHOD. The scientific method is an orderly pattern of gathering and analyzing information in science. It is the approach that many researchers use in doing science. Shown here are the steps of this method as they apply to Ivan Valiela's experiments on salt marsh plants.

ARE YOU STILL WONDERING?

1. Marine biology is the study of marine organisms and their interactions with each other and their environment.

2. It is important for a person to be knowledgeable about marine biology in order to be an informed citizen when it comes to making decisions concerning the ocean and the environment in general.

3. Marine laboratories engage in studies of ocean productivity and marine ecology. They also do research in which marine organisms are used to solve fundamental problems in biology.

4. Marine biologists use the same organized, commonsense approach to doing science as scientists in other fields. The process is called the scientific method.

QUESTIONS FOR REVIEW

Multiple Choice

1. Publication of Darwin's book *On the Origin of Species by Means of Natural Selection* sparked an interest in the study of
   a. physical oceanography
   b. animal and plant adaptations
   c. plankton
   d. barnacles
   e. polar seas

2. Marine biology is the study of
   a. the physical characteristics of the ocean
   b. the organisms that inhabit the sea and their relationships to each other and their environment
   c. marine animals but not marine plants
   d. the organisms found in the open sea but not along the shoreline
   e. marine fishes and mammals only

3. The modern science of marine biology can trace its beginnings to
   a. the invention of steam-powered ships
   b. the founding of the first marine laboratory at Woods Hole
   c. the *Challenger* expedition
   d. the voyage of HMS *Beagle*
   e. the development of modern fishing techniques

4. The term *plankton* refers to
   a. all kinds of marine plants and algae
   b. microscopic animals only
   c. organisms that float or drift in the sea's currents
   d. animals that are active swimmers
   e. all of the marine organisms that can produce their own food

5. The retrieval and repair of the transatlantic telegraph cable in 1858 provided evidence that
   a. plankton drift in the ocean currents
   b. marine organisms could survive in the dark and cold of the deep ocean
   c. the color of marine organisms varied with depth
   d. the Atlantic and Pacific oceans were at one time connected
   e. fish cannot survive at depths greater than 1 kilometer

Matching

Match the following researchers with their accomplishments in the field of marine biology.

a. Charles Darwin

b. Louis Agassiz

c. Edward Forbes

d. Aristotle

e. Alexander Agassiz

f. Pliny

g. Alistair Hardy

h. Charles Wyville Thomson

i. Victor Hensen

j. Fridtjof Nansen

1. Founded the first marine laboratory in the United States.

2. Proposed that marine organisms could not survive in the cold darkness of ocean depths.

3. Developed a scheme of classification known as the "ladder of life."

4. Charted the waters around the North Pole.

5. Coined the term "plankton" for the organisms that drift in the sea's currents.

6. Made a detailed study of barnacles.

7. Theorized that the colors of marine organisms were related to light absorption.

8. Studied whales in the Antarctic Sea.

9. Wrote the 37-volume *Natural History*, which includes references to marine animals.

10. Was the chief scientist of the *Challenger* expedition.

Short Answer

1. What was Louis Agassiz's goal in founding the first marine laboratory in the United States?

2. Explain the significance of the *Challenger* expedition.

3. Describe how the focus of marine biology has changed from early times to the present day.

4. Describe the kinds of research performed at marine biology laboratories today.

5. Compare inductive and deductive reasoning.

Thinking Critically

1. Apply the scientific process to a scenario of your own (real or imagined). Develop a hypothesis to explain something you have observed, and design an experiment to test your hypothesis. Try to relate the steps in your process to the generalized steps of the scientific method presented in the chapter.

SUGGESTIONS FOR FURTHER READING

Dunn, R. J. 2001. William Francis Thompson (1888–1965) and the Dawn of Marine Fisheries Research in California, *Marine Fisheries Review* 63(2).

Dybas, C. L. 2005. Dead Zones Spreading in World Oceans, *Bioscience* 55(7):552–557.

Mee, L. 2006. Reviving Dead Zones, *Scientific American* 295(5):79–87.

Stix, G. 2009. Darwin's Living Legacy, *Scientific American* 300(1):38–43.

Valiela, I., and J. Teal. 1979. The Nitrogen Budget of a Salt Marsh Ecosystem, *Nature* 280:652–656.

Have You Wondered?

1. What factors determine where marine organisms live? (2-2)
2. How the physical environment affects marine organisms? (2-2)
3. How the growth of marine populations is regulated? (2-3)
4. What determines the characteristics of marine communities? (2-4)
5. How energy flow affects the function of an ecosystem? (2-5)

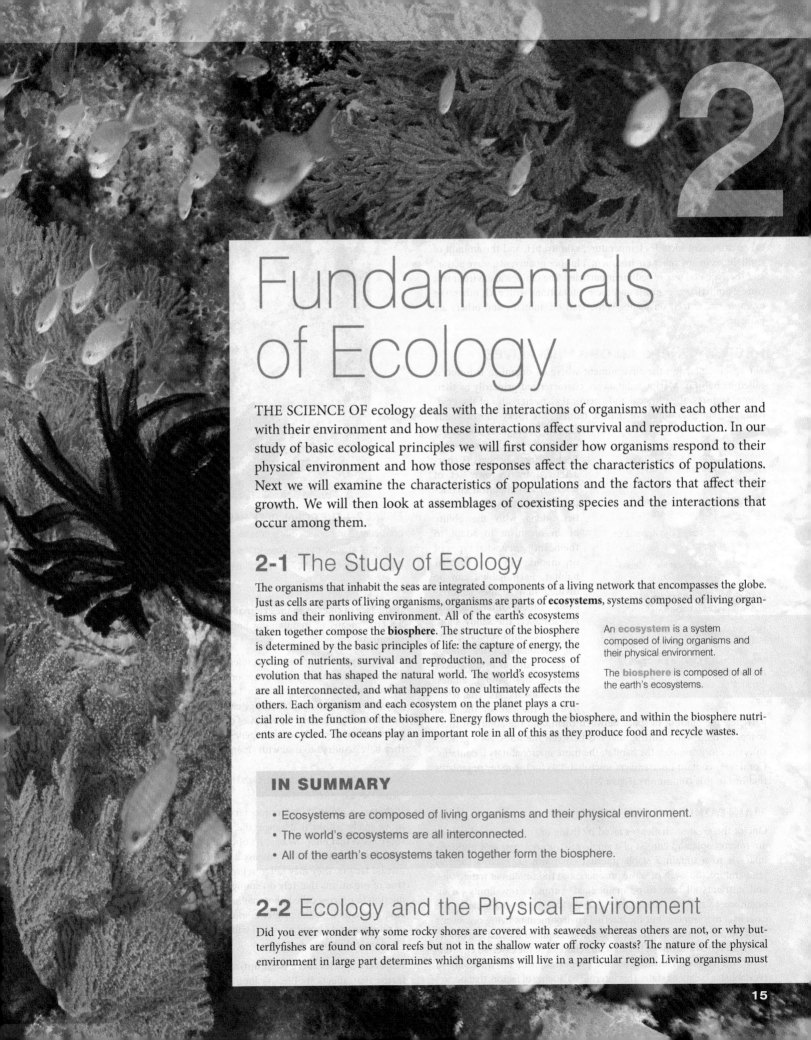

Fundamentals of Ecology

THE SCIENCE OF ecology deals with the interactions of organisms with each other and with their environment and how these interactions affect survival and reproduction. In our study of basic ecological principles we will first consider how organisms respond to their physical environment and how those responses affect the characteristics of populations. Next we will examine the characteristics of populations and the factors that affect their growth. We will then look at assemblages of coexisting species and the interactions that occur among them.

2-1 The Study of Ecology

The organisms that inhabit the seas are integrated components of a living network that encompasses the globe. Just as cells are parts of living organisms, organisms are parts of **ecosystems**, systems composed of living organisms and their nonliving environment. All of the earth's ecosystems taken together compose the **biosphere**. The structure of the biosphere is determined by the basic principles of life: the capture of energy, the cycling of nutrients, survival and reproduction, and the process of evolution that has shaped the natural world. The world's ecosystems are all interconnected, and what happens to one ultimately affects the others. Each organism and each ecosystem on the planet plays a crucial role in the function of the biosphere. Energy flows through the biosphere, and within the biosphere nutrients are cycled. The oceans play an important role in all of this as they produce food and recycle wastes.

> An **ecosystem** is a system composed of living organisms and their physical environment.
>
> The **biosphere** is composed of all of the earth's ecosystems.

IN SUMMARY

- Ecosystems are composed of living organisms and their physical environment.
- The world's ecosystems are all interconnected.
- All of the earth's ecosystems taken together form the biosphere.

2-2 Ecology and the Physical Environment

Did you ever wonder why some rocky shores are covered with seaweeds whereas others are not, or why butterflyfishes are found on coral reefs but not in the shallow water off rocky coasts? The nature of the physical environment in large part determines which organisms will live in a particular region. Living organisms must expend energy to survive and reproduce. The demands of their environment determine how much energy is necessary to survive and whether there will be enough energy to reproduce. This in turn influences the distribution of organisms in the marine environment.

THE ENVIRONMENT

An organism's **environment** consists of all the external factors acting on that organism. These factors can be physical (**abiotic factors**) or biological (**biotic factors**). The physical environment consists of the nonliving aspects of an organism's surroundings. For marine organisms, the physical environment includes temperature, salinity, pH, and the amount of sunlight, as in any other environment, plus ocean currents, wave action, and the type and size of sediment particles. The biological environment consists of living organisms and their interactions with each other. For example, some feed on plants, some feed on animals, and others are parasites.

HABITAT: WHERE AN ORGANISM LIVES

The specific place in the environment where an organism is found is called its **habitat**. Marine habitats are characterized primarily by their abiotic features, the physical and chemical characteristics of the environment. Some examples of marine habitats are rocky shores, sandy shores, mangrove swamps, coral reefs, and deep-sea vents. Each of these habitats is characterized by its own set of physical and chemical characteristics, and these characteristics, along with the ability of an organism to adapt to them, influence what types of organisms can live in that habitat. Each habitat can be divided into smaller subdivisions called **microhabitats**. For instance, the sandy shore habitat contains several different microhabitats for microscopic organisms in the spaces between the sand granules. These habitats are characterized by the size of the sand particles, the amount of space between them, and the ability of these spaces to hold water during intervals between the tides. As a general rule, the more complex the habitat, the more microhabitats it contains. Coral reefs contain thousands of microhabitats for the many organisms that live in this community **(Figure 2-1)**.

The **environment** consists of all external factors acting on an organism.

Abiotic factors are nonliving aspects of the environment.

Biotic factors are living aspects of the environment.

The **habitat** is the specific location where an organism is found.

Microhabitats are smaller subdivisions of a habitat.

Homeostasis is the internal steady state of a cell or organism.

The **optimal range** is the range of environmental factors to which an organism is best adapted.

MAINTAINING HOMEOSTASIS

One of the greatest challenges faced by living organism—whether they are microscopic and consist of a single cell, or are larger and multicellular—is to maintain a stable internal environment. Factors such as temperature, the levels of waste products, and the amount of water, salts, and nutrients all have to be maintained within narrow limits for an organism to survive. These factors can be constantly changing in the external environment, but the internal environment of living organisms must be maintained within the boundaries required for that organism's survival. When any one of these factors changes in the external environment, the organism must make the proper adjustments internally to reestablish a balanced state. This internal balancing of factors that occurs

© Ralph A. Clevenger/CORBIS

FIGURE 2-1 THE CORAL REEF HABITAT. Large habitats, such as the coral reef, can contain many smaller microhabitats. Microhabitats in the coral reef include the crevices in the coral, the sediments surrounding the coral stands, and even the tissues of the organisms themselves.

in the face of changes in the external environment is called **homeostasis**, and the means of maintaining homeostasis is vital to the life of all organisms.

Different types of organisms can tolerate different levels of fluctuation in environmental factors, and some organisms may be able to maintain homeostasis with respect to some factors but not others. An organism's response to environmental change can involve changes in its physiology or behavior. Ultimately, the ability of organisms to survive in their natural environment depends on their genes and the evolutionary adaptations they have acquired to deal with changing environmental conditions.

Homeostasis and the Distribution of Marine Organisms

If there is a range of environmental conditions and animals are able to move freely within that range, the animals will preferentially occupy those areas that offer the best set of conditions. If, however, animals are forced to occupy habitats that have a less than optimal range of environmental factors, they may fail to reproduce or may even die. The same is true of organisms that rely on sunlight to provide energy for food production (a process called *photosynthesis*). These organisms will thrive in environments with the proper amounts of sunlight and nutrients, such as nitrogen and phosphorus, but will fail to thrive if the environment is too deficient in these factors.

For every species there is an **optimal range** for each environmental factor that affects its life. As long as the factors remain within the

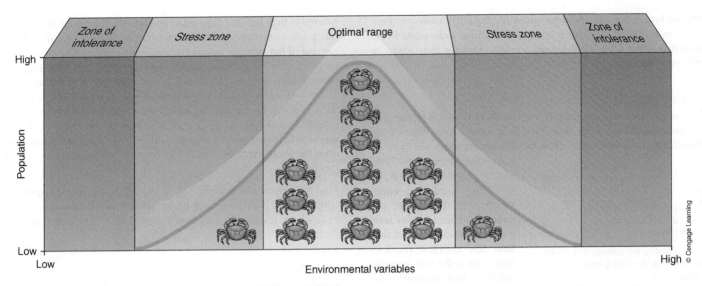

FIGURE 2-2 OPTIMAL RANGES. An organism survives and reproduces best when environmental factors affecting it fall within an optimal range. Although organisms can live outside of their optimal ranges, they expend more energy maintaining homeostasis, leaving less energy available for reproduction.

optimal range, the organism should be able to survive and reproduce. When one or more of the environmental factors is outside the optimal range, an organism's chances of survival decrease **(Figure 2-2)**. **Zones of stress** are regions above or below the optimal range of an environmental variable. An organism may be able to exist in a stress zone if the stress is not too great. If the stress is high, the organism may have to expend so much energy maintaining homeostasis that it will not have enough left to reproduce. Beyond the stress zones are zones where an environmental variable is so far from the optimal range that the organism cannot survive. These areas are called **zones of intolerance**. Consider, for example, the crabs in **Figure 2-2**. For a particular species, the optimal environmental temperature might be 20°C to 30°C, and this is where most of the crabs are found. The crabs may be able to tolerate temperatures slightly warmer (30°C to 35°C) or cooler (15°C to 20°C), but in these stress zones the amount of energy that the crab has to expend to cool down or warm up is so great that very little remains for reproduction. Consequently, these animals have fewer offspring and the numbers of crabs in these areas are fewer. Temperatures that are even higher (above 35°C) or lower (below 15°C) constitute zones of intolerance. At these temperatures the crabs would have to expend so much energy trying to maintain a reasonable internal temperature (maintain homeostasis) that there would not be enough left over for other life processes, much less reproduction. Thus the crabs could not exist in water with these temperatures. The same principle applies to all environmental factors that affect an organism.

CHARACTERISTICS OF THE ABIOTIC ENVIRONMENT

A marine organism's abiotic environment includes such factors as sunlight, temperature, salinity, pressure, nutrients, and wastes. The ability of an organism to tolerate changes in these and other environmental factors plays a major role in determining that organism's distribution in the marine environment. If an organism must expend too much energy to maintain homeostasis when any of these environmental factors changes, it will not be able to survive in that area of the sea.

Sunlight

Sunlight plays an essential role in the marine environment. It powers the process of photosynthesis that, directly or indirectly, provides energy to nearly all forms of life on earth. The largest group of photosynthetic organisms in marine environments is the **phytoplankton**, mostly microscopic, plantlike organisms and bacteria that float in ocean currents. Phytoplankton, together with seaweeds and plants, are the primary sources of nutrients and energy for marine animals. The distribution of these leading food producers is determined by the available sunlight and nutrients. In cloudy coastal waters, as in those of some North Atlantic bays and estuaries, phytoplankton can survive only in the shallowest areas because sunlight penetrates to a depth of less than 1 meter (3.3 feet). In the very clear water of the South Pacific, there may be enough sunlight for photosynthesis at depths of 200 meters (660 feet). Some phytoplankton migrate vertically within the water to gain maximum exposure to the sun's radiant energy to power their photosynthesis.

A **stress zone** is a region above or below the optimal range in which the organism must expend more energy than normal to maintain homeostasis.

A **zone of intolerance** is a region so far removed from an organism's optimal range that it cannot survive.

Phytoplankton are photosynthetic organisms that float in the ocean's currents.

Sunlight is also necessary for vision. Many animals rely on their vision to capture prey, avoid predators, and communicate with each other. The distribution of these animals is affected by the depth to which sunlight can penetrate and allow for accurate vision. Fishes and other animals that live in areas of the ocean that do not receive enough light for good visibility must rely on other senses, such as taste and smell, to find food, avoid predators, and find mates.

Excessive sunlight can be a problem as well. Organisms that live in the harsh environments along shorelines that are periodically covered with water and then exposed by the changing tides can be subject to the intense heat of the sun and **desiccation** (drying out) that results from overexposure to sunlight (solar energy). Many algae suffer pigment destruction when exposed to intense sunlight, limiting their ability to photosynthesize.

Temperature

The majority of marine animals are **ectotherms**, which means they obtain most of their body heat from their surroundings **(Figure 2-3a)**. Ectotherms become sluggish when the temperature drops and more active when temperatures rise. Marine mammals and birds, on the other hand, are **endotherms**. An endotherm can maintain a constant body temperature because its **metabolism** (the chemical reactions within its cells) can generate sufficient heat internally. Because these animals maintain temperatures that are usually higher than those of their surroundings, they have to be very well insulated so that they do not lose too much heat to the water that surrounds them **(Figure 2-3b)**.

Temperature often influences the distribution of organisms in shallow water and in the intertidal zone, the region that is covered at high tide and exposed at low tide. Shallow water, like that in tide pools, can experience large temperature changes in a short time because of hot summer sun or freezing winter nights. Organisms that survive in this habitat must be able to adapt quickly to a wide range of temperatures. The same is true of organisms that live in intertidal zones. The temperatures in these environments change quickly and substantially, and only the hardiest organisms are able to survive in this type of habitat. Large bodies of water such as oceans do not change temperature rapidly; consequently, organisms that live in the open ocean away from the shore experience relatively constant temperatures at a given depth.

Most organisms can tolerate only a specific range of environmental temperatures. Temperatures above or below this critical range disrupt metabolism, resulting in decreased ability to reproduce, injury, or even death.

Salinity

Salinity is a measure of the concentration of dissolved inorganic salts in the water. Substances that are dissolved in water are generally referred to as **solutes**. The membranes of living cells are almost always permeable to water, but they are not always permeable to the substances that are dissolved or suspended in water. All organisms must maintain a proper balance of water and solutes in their bodies (maintain homeostasis) to keep their cells alive. When a solute cannot move across the cell membrane to reach a balanced state on both sides (equilibrium), water moves instead to achieve the balance. The movement of water across a membrane in response to differences in solute concentration is called **osmosis (Figure 2-4)**. The process of osmosis is vital to the life of a cell. If a cell loses too much water, it will become dehydrated and die. On the other hand, if a cell takes in too much water, it will swell and, in the case of

Desiccation is the process of drying out.

Ectotherms are animals that obtain most of their body heat from the environment.

Endotherms are animals that obtain most of their body heat from metabolism.

Metabolism is the sum of all of the chemical reactions that occur within cells.

Salinity is a measure of the salts dissolved in water.

Solutes are substances that are dissolved in water.

Osmosis is movement of water across a semipermeable membrane from an area of low solute concentration to an area of high solute concentration.

(a)   (b)

FIGURE 2-3 TEMPERATURE. (a) Ectotherms such as this crab obtain most of their body heat from their surroundings. If the environmental temperature rises or falls, so will their body temperature. **(b)** Endotherms such as this penguin can maintain a constant body temperature by generating heat internally through metabolism. Endotherms are well insulated to prevent excess heat loss.

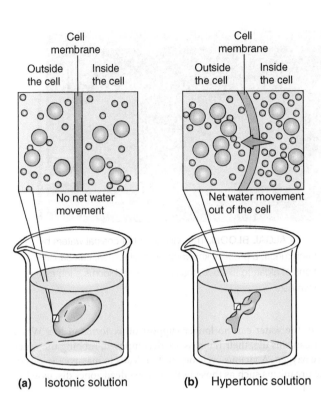

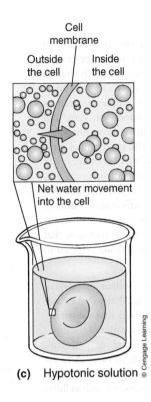

| (a) Isotonic solution | (b) Hypertonic solution | (c) Hypotonic solution |

FIGURE 2-4 OSMOSIS. Water tends to move from areas of lower solute concentrations to areas of higher solute concentrations. **(a)** An isotonic solution contains the same concentration of solute molecules (green) and water molecules (blue) as a cell. Cells placed in isotonic solutions do not change because there is no net movement of water. **(b)** A hypertonic solution contains a higher concentration of solute than a cell. A cell placed in a hypertonic solution will shrink as water moves out of the cell to the surrounding solution by osmosis. **(c)** A hypotonic solution contains a lower solute concentration than a cell. A cell placed in a hypotonic solution will swell and possibly rupture as water moves by osmosis from the environment into the cell.

© Cengage Learning

animal cells, possibly burst. Living organisms invest both time and energy in maintaining the proper amount of salts and water in their cells and the fluid that surrounds them. This is a particularly important challenge for many marine organisms, because they live in water that has a high concentration of solutes. Some marine animals, such as the spider crab (*Macrocheira*) and other inhabitants of the deep sea, cannot regulate the salt concentration of their body fluids. Because their body surfaces are permeable to both salts and water, the concentration of solutes in their body fluids rises and falls with changes in the concentration of solutes in seawater. This is not a problem for these animals because the deep sea is a very stable environment, and they are seldom exposed to fluctuations in salt and water content. They do, however, have very little ability to withstand changes in salinity, and if they are moved to a less-stable environment, such as an estuary, they will die since the crab would now be in an hypotonic environment and too much water would enter its cells (see **Figure 2-4**).

Along coastal areas, the concentration of salts in seawater can vary greatly, especially in bays, estuaries, and tide pools. In these areas, pockets of water can lose moisture to evaporation, thus concentrating the salt content, or gain moisture through rain and freshwater runoff, thus diluting the normal salt concentration of the seawater. Animals that thrive in this type of environment, such as the fiddler crab (*Uca pugnax*), must be able to adjust the salt content of their body tissues by regulating salt and water retention. As we will see in the following chapters, salinity is critical in determining the distribution and types of organisms in many marine habitats.

Pressure

The pressure at sea level is 760 mm Hg, or 1 atmosphere (14.7 pounds per square inch). Because water is so much denser than air, for every 10 meters (33 feet) below sea level in the ocean, the pressure increases by 1 atmosphere. For instance, the pressure at an average ocean depth of 3,700 meters is 370 atmospheres (2.7 tons per square inch). A styrofoam

wig head lowered to a depth of 4,000 meters, where the pressure is 400 atmospheres, is compressed to approximately one third of its original size **(Figure 2-5)**. As you might imagine, very few surface-dwelling organisms are able to survive at these depths (although pressure may not be the only factor limiting the number of organisms at great depths). The pressure of the water not only affects organisms that inhabit the deep regions of the seas but also poses problems for animals that sometimes frequent these depths, such as whales and some species of deep-sea fishes. These animals possess specialized adaptations that allow them to survive at great depths.

NOAA

FIGURE 2-5 PRESSURE. To demonstrate the pressure in the sea's depths, this Styrofoam wig head, originally the size of a human head, was lowered to a depth of 4,000 meters (13,200 feet). The pressure at this depth is so great that it compressed the Styrofoam to the size you see in this photograph.

Metabolic Requirements

The availability of nutrients strongly influences the distribution of organisms in the marine environment. The term **nutrient** refers not just to food but to all of the organic and inorganic materials that an organism needs to metabolize, grow, and reproduce. The chemical composition of seawater supplies many mineral nutrients, such as nitrogen and phosphorus, required by phytoplankton, seaweeds, and plants, which in turn supply other marine organisms with nutrients. For instance, the mineral calcium, essential for the synthesis of mollusc shells, coral skeletons, and the external covering of crustaceans, is readily available in seawater. The availability of a key nutrient may play a significant role in the distribution of organisms in the marine environment. For instance, marine plants may have sufficient amounts of sunlight, but if they do not have enough nitrogen in their environment, they will not be able to thrive. Nutrients, such as nitrogen, that limit the number or distribution of marine organisms are referred to as **limiting nutrients**.

A particularly important requirement for metabolism is oxygen. Oxygen is produced as a by-product of the photosynthesis that occurs in phytoplankton, seaweeds, and plants. Life evolved in an environment that lacked free oxygen, and when free oxygen entered the environment, it was probably the first harmful chemical, or pollutant, released into the earth's atmosphere. Although toxic to most early life forms, the presence of free oxygen revolutionized life on earth by producing an environment that would allow the evolution of multicellular organisms.

Oxygen in seawater, coming from photosynthesis or the atmosphere, dissolves at or near the surface. The ability of water to dissolve oxygen depends on its temperature and salinity. Cooler less-salty water of the open sea contains more oxygen than warm water in a tide pool with a high salt content. Not all organisms require oxygen for life. Some bacteria are **anaerobic organisms** (**anaerobes**), meaning that they can survive and even thrive in the absence of oxygen. These organisms are often found in areas where the amount of oxygen is very limited, such as the ocean's depths, salt marshes, sand and mud flats, and between sediment particles, and they are responsible for the familiar odor of rotten eggs associated with many of these areas. **Aerobic organisms** (**aerobes**), such as plants, algae, animals, and the majority of marine microbes, require oxygen for their survival and are limited to regions of the ocean that contain sufficient quantities of oxygen.

Although nutrients are necessary for life, too much nutrient material in seawater can be a problem. Water runoff from land can greatly increase the nutrient levels in some coastal waters. This process of nutrient enrichment is known as **eutrophication**. Eutrophication leads to population explosions of certain types of photosynthetic plankton, an event referred to as an **algal bloom** (**Figure 2-6**). When these excess plankton die, they are decomposed by bacteria. The process of decomposition robs the water of oxygen. As the oxygen is

Nutrients are substances that an organism needs to grow, metabolize, and reproduce.

Limiting nutrients are nutrients that limit the number or distribution of organisms in a particular area.

Anaerobic organisms (**anaerobes**) are organisms that can survive in an environment that lacks oxygen.

Aerobic organisms (**aerobes**) are organisms that require oxygen for survival.

Eutrophication is the process whereby a body of water becomes enriched with nutrients.

An **algal bloom** is a population explosion of phytoplankton.

2007 Mark Conlin/V&W/Image Quest Marine

FIGURE 2-6 ALGAL BLOOM. The enrichment of coastal waters by nutrient-rich runoff, a process known as eutrophication, can cause populations of algae to explode, resulting in algal blooms such as the one shown here.

depleted, the water can no longer support other forms of life. When these organisms die, their remains are decomposed, robbing the water of more oxygen. A vicious cycle begins that leads to massive die-offs of marine organisms such as those in the ocean's dead zones discussed in Chapter 1.

Metabolic Wastes

All organisms produce waste products when they metabolize. Most living organisms release carbon dioxide as a product of respiration. Animals excrete nitrogen-rich waste products, and plants release oxygen when they photosynthesize. Most of the time, waste products of metabolism are either removed from the environment or broken down and recycled by a variety of organisms, especially bacteria. In some environments, waste products can accumulate to toxic levels, and only organisms with adaptations to deal with such inhospitable environments will survive. Certain small tide pools and coastal marsh areas are especially susceptible to the problems of accumulating metabolic wastes because the exchange of waste-laden water with new, uncontaminated water is limited.

IN SUMMARY

- An organism's environment consists of all the external factors acting on that organism.

- The specific place in the environment where an organism is found is called its habitat.

- Organisms expend energy to maintain homeostasis, a relatively constant environment for their cells. (Have You Wondered? #1)

- If the amount of energy expended in maintaining homeostasis is too great, organisms may not be able to reproduce or survive and will not be found in such habitats.

- Characteristics of the abiotic environment, such as sunlight, temperature, salinity, pressure and metabolic requirements, determine the amount of energy an organism must expend to maintain homeostasis. (Have You Wondered? #2)

2-3 Populations

To a biologist, a **population** is a group of the same species that occupies a specified area. Members of a population interact with each other and are able to breed with each other. They rely on the same resources and are influenced by the same environmental factors. The characteristics of populations are determined by the interactions between individuals and their environment. In nature, populations are separated from one another by barriers that prevent organisms from interacting or breeding. The population, rather than the individual, is the basic unit that many ecologists study.

POPULATION RANGE AND SIZE

Every population has geographical boundaries. The **geographical range** of a population is the geographical area within which it lives. When studying populations, biologists begin by defining the area in which the organisms will be studied. One researcher may study the population of hermit crabs in a single salt marsh, while another may study the population of tuna in a region of the South Pacific.

Marine biologists are not only interested in the geographical range of a population but also the size of the population within its range. Sometimes it is possible to determine population size by simply counting the number of individuals within the geographical range. For instance, we could count the number of barnacles on a particular rock or the number of sea anemones in a given tide pool. It may even be possible to count the number of manatees or whales in a given area from an airplane or helicopter. In many cases, however, it is not possible to directly count all of the individuals in a population, so biologist use techniques called **sampling methods** to estimate population size.

One common sampling method involves counting the number of individuals in a representative area or plot within the range. The number of these plots in a given range is then determined, and the number of individuals per plot is multiplied by the number of plots to estimate the population's size (**Figure 2-7a**). Another method would be to look for evidence of an organism's presence, such as burrows or the telltale trails that some organisms leave in the sand, and count these to get an idea of the population size.

In a procedure known as the **mark-recapture method**, animals are captured and tagged or marked before being released. After waiting a sufficient amount of time for these animals to randomly mix back into the population, a sample of animals is taken again. The ratio of marked to unmarked animals can then be used to give an estimate of population size (**Figure 2-7b**). For instance, say we capture and tag 10 nurse sharks and release them back into their range. Several weeks later, we catch 10 nurse sharks again from this range, and 2 of these 10, or 20%, are tagged. From this we could estimate that 20% of the nurse shark population is tagged. Since we initially tagged 10 sharks, they would represent 20% of the entire population in the area, so the size of the population would be approximately 50 sharks (20% of 50 = 10). Of course this assumes that any tagged individual has an equal chance of getting caught again, which may not be a correct assumption.

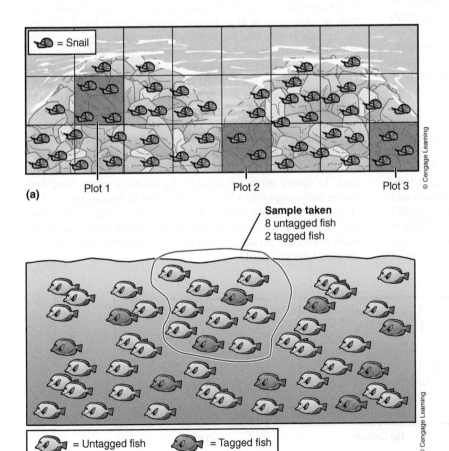

A **population** is a group of the same species inhabiting a specific area.

The **geographical range** is the geographical area where a population lives.

Sampling methods are procedures used to estimate the size of a population.

The **mark-recapture method** estimates population size by capturing and marking or tagging individuals, then releasing them back into the wild and counting the number recaptured at a later time.

FIGURE 2-7 SAMPLING TECHNIQUES. (a) We can estimate the number of snails on these coastal rocks by dividing the area into smaller divisions (plots) and then counting the number in several representative plots. We could then find the average number per plot and multiply by the number of plots in the area. **(b)** The number of fish in this range can be estimated by capturing a sample from the population and marking them with tags. The tagged individuals are then released and allowed to mix with the population. Sometime later another sample is taken and the ratio of tagged to untagged individuals is determined. We assume that this represents the ratio of tagged to untagged individuals in the population. Since we know how many tagged individuals we released into the population, we can estimate the total population size.

DISTRIBUTION OF ORGANISMS IN A POPULATION

Population density is the number of individuals per unit area or volume, for instance, the number of barnacles on a square meter of rock or the number of plankton in a cubic centimeter of surface water. Population density can frequently vary within a large range because not every part of the range will provide suitable habitat. **Dispersion** refers to the pattern of spacing among individuals within the range. Dispersion patterns range from **clumped**, in which individuals are densely packed in patches, to **uniform**, if they are evenly spaced, to **random**, if the spacing varies in an unpredictable pattern (**Figure 2-8**).

Clumping may be caused by variations in the organism's abiotic environment. Barnacles and other rock dwellers may grow only on certain rocks that are exposed the least amount of time during low tide. Herbivorous snails are likely to be clumped in areas where there is sufficient algae to eat. Uniform spacing is often the result of competition among species. In seaweeds, uniform spacing may reflect competition for sunlight. Some sessile organisms, such as corals, may kill competitors to prevent them from growing too close. Some animals establish a territory of a certain size and defend it against the intrusion of others. Random spacing, which occurs when there is a lack of strong interactions among individuals, is not very common in the marine environment.

Population density is the number of individuals per unit area or volume.

Dispersion is the pattern of spacing among individuals within a range.

In a **clumped pattern** individuals are densely packed in patches.

In a **uniform pattern** individuals are evenly spaced.

In a **random pattern** spacing among individuals varies in an unpredictable pattern.

Generation time is the average time between an individual's birth and the birth of its offspring.

Survivorship is how long, on average, an individual of a given age could be expected to live.

CHANGES IN POPULATION SIZE

Populations change in size as individuals are added through reproduction and immigration and individuals are eliminated by death and emigration. Typically, rates of reproduction and death vary among different age groups and sexes within a population. If we look at the individuals in each age group within a population, we would probably find that each age group has a characteristic birth and death rate. Typically, the very young and old individuals in a population are more likely to die than individuals of intermediate age who have youthful strength and the survival skills that come with maturity. Also, individuals of intermediate age are usually more successful at reproduction. For example, a given number of 10-year-old female fur seals produce about twice as many pups during a breeding season as an equal number of 5-year-old or 18-year-old females. In fish, however, it is usually the older, larger females that produce the most offspring. In general, a population of individuals of prime reproductive age will grow faster than a population with proportionately more younger or older individuals.

Another important factor affecting population growth is the **generation time**, the average time between an individual's birth and the birth of its offspring. Other factors being equal, a shorter generation time will usually result in a faster-growing population.

The sex ratio also affects population growth. The number of young that can be produced is usually related to the number of females in a population. The number of males may be less important because in many species—seals and sea lions, for example—a single male can mate with several females. In these populations, a decline in males will have little impact on population size. In species where individuals form mating pairs, such as some marine birds, a reduction in males would be more likely to affect the number of young produced.

Survivorship

Survivorship is another factor that affects population size. **Survivorship** refers to how long, on average, an individual of a given age could be expected to live. This information is frequently displayed by a graph known as a **survivorship curve (Figure 2-9a)**. A survivorship curve shows the number of individuals in a population still alive at each age from reproduction to death. Survivorship curves can be classified into three general types. A **Type I curve** shows relatively low death rates during early and middle life, with death rates increasing among older individuals. Large marine animals, such as whales, that produce few offspring and provide them with good care often exhibit this pattern. The opposite pattern is a **Type III curve**, which reflects high mortality rates for the young but a flat rate for those individuals that survive to a certain critical age. This pattern is associated with many marine organisms that produce very large numbers of offspring but provide little or no parental care, such as fishes and bivalves. **Type II curves** fall in the middle, showing more or less constant mortality rates over time. Many species fall in between these basic patterns or show more complex patterns. In marine birds, for example, the mortality rate is often high among the young (as in a Type III curve) but fairly constant in adults (Type II curve) **(Figure 2-9b)**. Crabs, on the other hand, show a "stair-step" type of curve, with brief periods of increased mortality rate during molts followed by periods of lower mortality rate when their protective exterior skeleton is hard.

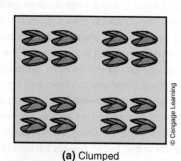

(a) Clumped

(b) Uniform

(c) Random

FIGURE 2-8 DISPERSION PATTERNS. The pattern of spacing of individuals in a population is known as dispersion. Possible patterns are **(a)** clumped, **(b)** uniform, and **(c)** random.

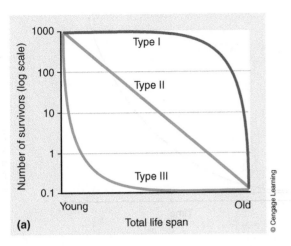

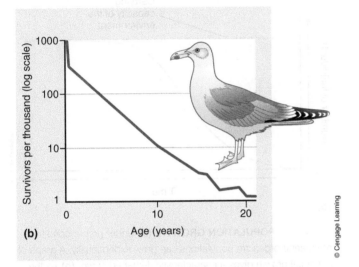

FIGURE 2-9 SURVIVORSHIP CURVES. (a) Survivorship curves indicate the number of individuals still living at each age from birth to death. Type I curves indicate increased death rates as the organisms age. Type II curves indicate relative constant death rates over time. Type III curves indicate higher death rates among the young but increased survival for those organisms that survive past a certain age. **(b)** Not all organisms have life histories that conveniently fit one of the three curves in part (a). For instance, herring gulls, *Larus argentatus*, have Type III survivorship curves as chicks and Type II survivorship curves as adults.

Life History

The **life history** of any organism can be divided into three phases: birth, reproduction, and death. Three aspects of life history affect the number of offspring a female will produce in her lifetime: clutch size, number of reproductive events, and age at first reproduction.

Clutch Size The number of offspring produced each time an organism reproduces is called the **clutch size**. Generally, organisms with a large clutch size produce small eggs or offspring; thus each offspring has little energy to start life on its own. Large clutch size is typical of organisms with a Type III survivorship pattern. Offspring from small clutches are generally larger and have a better chance of surviving and are more typical of organisms with Type I or Type II survivorship curves. In many organisms, such as some fishes, clutch size increases with body size. In other organisms clutch size can vary seasonally, with clutches being larger during seasons when environmental factors are more favorable for the young.

Number of Reproductive Events Some organisms, such as Pacific salmon and the octopus, reproduce only once in their lifetime, whereas others can reproduce repeatedly. The advantage of a single reproductive event is an organism's ability to invest all of its energy budget into the production of offspring. In contrast, organisms that reproduce repeatedly over their life span must divide their energy among maintenance, growth, and reproduction.

Age at First Reproduction For organisms that reproduce repeatedly, the timing of reproduction can have a large effect on a female's lifetime reproductive output. If a female uses some of her energy to reproduce at a younger age, she will have less energy available for maintenance and growth, reducing her potential to produce larger clutches in the future. On the other hand, a female that delays reproduction invests more energy in maintenance and growth. Although this strategy might increase her potential to reproduce in the future, she could die before producing any offspring. Observations based on experimental models suggest that if older females produce much larger clutches than younger ones, and if there is a good chance of surviving to an older age, then female reproductive output is maximized when first reproduction is delayed.

An organism's **biological fitness** is measured by how many of its offspring survive to produce their own offspring. As a result, the inherited characteristics of life history that result in the most offspring being produced over time will become more common in a population. Of course, all of the variables that can affect life history cannot be maximized simultaneously because all organisms have a limited energy budget that demands trade-offs. For example, the production of many offspring with little chance of survival may result in fewer descendants than the production of a few well-cared-for offspring that can compete effectively for limited resources.

A **survivorship curve** is a graph that shows the number of individuals in a population still alive at each age from reproduction to death.

A **Type I survivorship curve** shows relatively low death rates during early and middle life, with increasing mortality among older individuals.

A **Type II survivorship curve** shows more or less constant mortality rates over time.

A **Type III survivorship curve** shows high mortality rates for the young but a flat rate for individuals that survive to a certain critical age.

Life history comprises the three phases of an organism's life: birth, reproduction, and death.

Clutch size is the number of offspring produced each time an organism reproduces.

Biological fitness is a measure of how many of an organism's offspring survive to produce their own offspring.

POPULATION GROWTH

There are many different ways by which a population can increase in size. The number of new individuals added through reproduction can increase while the death rate remains constant; the rate at which individuals die can decrease while the birth rate remains constant; or the birth rate can rise while the death rate declines. Populations can grow as the result of immigration of new individuals from other populations, or it can decline because of emigration of members to neighboring

Recruitment is the addition of new members to a population through reproduction or immigration.

Larval settlement is the rate at which aquatic larvae leave the water column and settle to the bottom.

Exponential (or **logarithmic**) **growth** is growth that is initially slow but accelerates with time.

Logistic growth is growth that is exponential at first but then levels off.

The **carrying capacity** is the point at which the environment contains as many organisms as it can sustain for an extended period.

Density-dependent factors are population-regulating factors that have a greater effect as the population size increases.

populations. The addition of new members to a population through reproduction or immigration is frequently termed **recruitment**. The majority of marine organisms living on rocks and bottom sediments have planktonic larval stages. The size of these populations can be affected by the rate at which the larvae leave the water column and settle to the bottom, a process called **larval settlement**.

When populations of organisms have sufficient food or nutrients and are not affected greatly by predators or disease, they will grow rapidly. For example, phytoplankton will undergo large population increases when there is an increase in available nutrients. In a short time, the phytoplankton reproduce so quickly that they exploit all of the available nutrients or space; they then begin to die back. If we were to graph the growth of the phytoplankton population as a function of time, it would be a J-shaped curve (**Figure 2-10a**). Initial growth would be slow, represented by the bottom of the J, but the doubling of the population again and again by cell division, represented by the nearly vertical top of the J, would produce ever-bigger numbers. This pattern of growth is called **exponential**, or **logarithmic, growth** and is characteristic of rapidly growing populations. Although some populations can grow exponentially for short periods, none can sustain this type of growth for long. Eventually, some resource needed for growth or survival will become scarce, and the growth rate will begin to decline.

A more typical growth pattern for a population is **logistic growth,** in which the original growth is exponential, but then the rate of population increase reaches an equilibrium point and population growth falls to zero (**Figure 2-10b**). A population reaches its equilibrium point when it contains as many organisms as its environment can sustain for an extended period. That point is called the **carrying capacity** of the environment. Growth rate slows as a population reaches its carrying capacity for any of three reasons: the birth rate drops, the death rate rises, or birth rates and death rates change at the same time.

The carrying capacity inhibits the tendency of populations to grow exponentially. Any number of environmental factors can limit population size. Insufficient amounts of sunlight can limit the number of photosynthetic organisms, or insufficient numbers of prey can limit the number of predators. Although it may appear that the carrying capacity of a specific environment for a particular species is constant, it often is not. The abiotic and biotic factors that determine the carrying capacity can vary, for instance, with time of year, and thus the carrying capacity will vary as well. The typical pattern shown in **Figure 2-10b** predicts the growth of laboratory populations and some populations in nature, but most natural populations exhibit many different growth patterns that are variations of the typical pattern.

Population Regulation

What factors determine the carrying capacity of an environment? Biologists have attempted to divide these factors into two categories. Factors

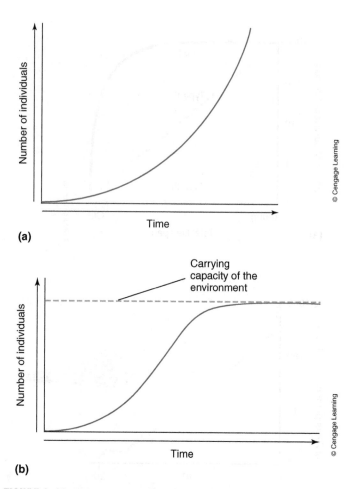

(a)

(b)

FIGURE 2-10 POPULATION GROWTH. (a) Under proper conditions, some marine organism populations can grow exponentially. A graph of exponential growth gives a characteristic J-shaped curve. **(b)** As the growth of a population approaches its carrying capacity, the graph flattens out. As a result, the logistic growth curve shown here has somewhat of an S-shape.

that vary with population size are called density-dependent factors. Factors that do not vary with population density are called density-independent factors.

A **density-dependent factor** is one that has greater effect as the population increases in size. Density-dependent factors reduce the population growth rate by decreasing reproduction and by increasing mortality rate in a crowded population. Decreased availability of resources, such as food supplies, often limits reproduction in crowded populations. Population density also influences individual organisms' health and survivorship. Plants and seaweeds that grow under crowded conditions tend to be smaller, and smaller organisms are less likely to survive and produce offspring. Animals often experience higher mortality rates at high population densities, too. For instance, predation can be an important density-dependent factor. A predator will almost certainly encounter and capture more prey when the prey population increases.

For some animal species, internal rather than external factors appear to regulate population size. For instance, high densities and overcrowding can cause stress, which results in physiological changes that delay sexual maturation and cause reproductive organs to shrink, thus inhibiting reproduction. The mechanisms underlying these responses have yet to be discovered. Other factors, including the abiotic environment and population interactions, can affect population size. Although populations in the field appear to fluctuate around carrying capacity, this does not necessarily mean that the phenomenon is solely the result of density-dependent factors at work.

Density-independent factors are not related to population size. They affect the same percentage of individuals regardless of how large or small the population is. The most common and important density-independent factors are related to weather and climate. Sudden, extreme, or unpredictable changes in environmental conditions, such as hurricanes, can periodically wipe out large numbers of organisms. Such events may decimate certain populations in an area without regard to their size or density.

Over the long term, many populations remain fairly stable in size and are thought to be close to a carrying capacity that is determined by density-dependent factors. This general stability, however, is sometimes disrupted by density-independent factors. In some cases, density-dependent and density-independent factors act together to regulate a population. The relative importance of density-dependent and density-independent factors may also vary seasonally.

Organisms whose populations are regularly decimated tend to reproduce early and produce large numbers of offspring. Because this strategy maximizes their intrinsic rate of reproduction (r), they are called **r-strategists** and their growth curves resemble **Figure 2-10a**. Species whose populations are more often in equilibrium, on the other hand, are often called **K-strategists** because they seem to maximize their carrying capacity, or K. Their growth curves resemble **Figure 2-10b**. The majority of marine organisms, however, are neither pure r-strategists nor K-strategists but lie somewhere in a continuum between these two extremes.

IN SUMMARY

- A population is a group of the same species that occupies a specific area.

- Within the geographical range of a population, organisms are dispersed in patterns that are clumped, uniform, or random.

- Factors that affect reproduction and mortality rate, such as survivorship and life histories, have a significant effect on the size of populations. (Have You Wondered? #3)

- Initially, populations grow quickly, a process known as exponential growth, but such growth cannot be maintained indefinitely.

- Characteristics of the environment, such as space and available food, limit the number of organisms an area can support. This limit is called the carrying capacity of the environment.

- The carrying capacity is set by density-dependent factors that decrease reproduction or increase mortality rate as a population grows.

- Population growth can also be limited by density-independent factors.

2-4 Communities

A biological **community** is composed of interacting populations of different species that occupy one habitat at the same time. The species that make up a community are linked to some degree by competitive relationships, predator–prey relationships, and symbiotic relationships. For instance, we can talk about the populations of anemones, mussels, seaweeds, sea stars, and snails that inhabit a rocky shore on the Pacific coast. This assemblage of interacting populations would make up a rocky shore community **(Figure 2-11)**. Communities can be large or small, depending on the area that is being discussed. Communities in nature are rarely isolated from each other, resulting in interactions among different communities. As indicated earlier in the chapter, biological communities both influence and are influenced by their abiotic environment.

> **Density-independent factors** are population-regulating factors that are not related to population size.
>
> An **r-strategist** is an organism that reproduces early and produces large numbers of offspring.
>
> A **K-strategist** is an organism that seems to maximize its carrying capacity.
>
> A **community** is a group of interacting populations that inhabit a specific area.
>
> A **niche** is the role a species plays in a community.

NICHE: AN ORGANISM'S ENVIRONMENTAL ROLE

What an organism does in its environment (in a sense, its occupation) is its **niche**. For example, mussels stick to rocks and filter seawater for food; crabs scavenge; and some worms burrow into bottom sediments, extracting organic material as they do so. A full description of an organism's niche would include the range of environmental and biological factors that affect its ability to survive and reproduce. Because a niche is so complex and involves so many different factors, it is not possible to show a picture of a niche as you could a habitat, but it is possible to examine different aspects of the niche separately to see how each affects an organism **(Figure 2-12)**.

Cary Kalscheuer/Shutterstock

FIGURE 2-11 COMMUNITY. A community is composed of populations of different species interacting with each other. On this rocky coast populations of limpets, sea anemones, seaweeds, crustaceans, sea stars, and snails all interact with each other in a number of ways to form a balanced community.

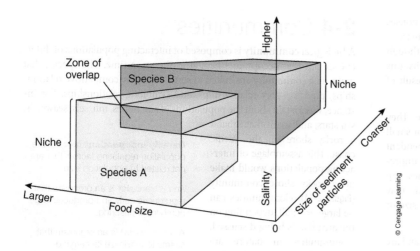

FIGURE 2-12 A NICHE. An organism's niche is determined by a variety of abiotic and biotic factors acting together on the organism. This three-dimensional graph shows how several factors (food size, salinity, and size of sediment particles) interact to form niches for two species of burrowing worm. On the basis of this graph, we can see that species A prefers to burrow in substratum composed of smaller sediment particles where the salinity of the water is low and prefers to feed on medium- to large-sized food items. Species B, on the other hand, prefers coarser sediments where the salinity of the water is higher and prefers smaller food items. The zone of overlap indicates the combination of sediments, salinity, and food that would meet the requirements of both organisms.

If we were to examine the organisms on a rocky shore, we would notice that their distribution from high tide line to low tide line is determined by factors such as available moisture, the length of time exposed to air during low tide, their ability to withstand the force of waves, and the characteristics of the rock itself. This would be an abiotic view of this particular niche.

Biotic factors that describe a niche include predator–prey relationships, parasitism, competition for the same resources, and organisms that provide shelter for other organisms. If we return to the example of a rocky shore, we would find that some organisms such as blue mussels (*Mytilus edulis*) are distributed in their particular zone because of the zone's abiotic characteristics. Predators, such as sea stars, are found in an overlapping zone because of the abundance of prey, namely the mussels. The seaweed that grows on the rocks provides food and shelter for a variety of small crustaceans, and the distribution of snails is determined by the amount and distribution of seaweed.

A **fundamental niche** is the broadest possible niche that a population can potentially occupy.

A **realized niche** is the portion of a fundamental niche that a population actually occupies.

An organism's behavior also plays an important role in defining its niche. Behavioral factors, such as when and where an organism feeds, how it mates, where it bears its young, and social behaviors, influence an organism's niche. For instance, two very similar species that require the same kind of food can coexist if one feeds at night and the other during the day, provided that the amount of available food will support both organisms. This is because each species' niche is defined by multiple factors that regulate the species' distribution in the environment

The broadest possible niche that an organism can potentially occupy is called its **fundamental niche**. The fundamental niche takes into account the range of resources that a population is theoretically capable of using under ideal circumstances. In reality, interactions with other organisms often restrict a species to a small part of its fundamental niche, called the **realized niche**. The realized niche takes into account the resources that a population is actually using.

Connell's Barnacles

As far back as Darwin, biologists have recognized that the organisms that live on rocky coasts are arranged in horizontal layers. These layers are arranged in an orderly and predictable fashion from the splash zone, which receives moisture from waves that crash against the rocks, to the subtidal zone. The biologist J. H. Connell, from the University of California, proposed an explanation for this arrangement of organisms in 1961.

Connell performed experiments on the rocky coasts of Scotland, where he noticed that two different genera of barnacles, *Semibalanus* and *Chthamalus*, live next to each other in separate layers on the rocks **(Figure 2-13)**. Members of the genus *Chthamalus* live in the layer of organisms just above the high-tide line. In the zone just beneath this, populations of the genus *Semibalanus* replace populations of *Chthamalus*. Barnacles are sedentary animals, and the species that Connell worked with are relatively small and live in large, dense populations. Because of these characteristics, Connell could perform experiments that would have been impossible with larger, more mobile organisms.

On one section of rocky shore, Connell observed the normal growth, reproduction, and daily activities of the two types of barnacle. This section of shore acted as the control for his experiment. On other sections he performed experiments. In one, he transplanted barnacles to different zones on the rocks. In another, he completely cleared the barnacles from areas of rock to see which species would colonize. Connell determined from these experiments that individuals of *Chthamalus* were better adapted to living above the high-tide mark. Members of this genus had a much greater tolerance for the higher temperature and desiccation that occurs in this zone compared with the members of the genus *Semibalanus*. The members of the genus *Semibalanus*, however, grow faster than those of *Chthamalus*. When larvae of both species settle on bare rock in the lower zone, the members of the genus *Semibalanus* compete better for the vital resource of space. The *Semibalanus* barnacles would either overgrow the *Chthamalus* barnacles, causing them to starve, or undercut the slower-growing *Chthamalus* barnacles, literally prying them off the rocks. As a result, the *Chthamalus* barnacles did not have the opportunity to become established in this zone. *Chthamalus* barnacles live in the upper zone because they are better adapted to the harsh environment of this zone. *Semibalanus* lives in the lower zone where both species can survive but where *Semibalanus* is a superior competitor.

Although the fundamental niche of both barnacles extended over relatively large areas of the habitat, abiotic and biotic factors limited them to a smaller part of their fundamental niche (their realized niche). Experiments such as these show the importance of the interplay between the physical environment and biological factors, such as competition, in determining the distribution of organisms in an ecosystem.

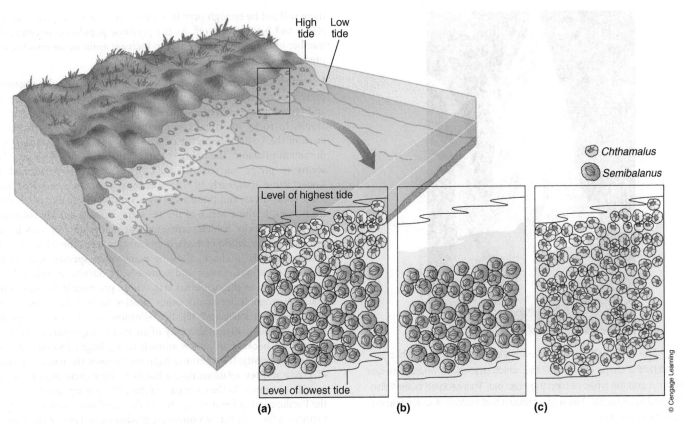

High tide   Low tide

Level of highest tide

Level of lowest tide

(a)   (b)   (c)

Chthamalus
Semibalanus

© Cengage Learning

FIGURE 2-13 FUNDAMENTAL vs. REALIZED NICHE. **(a)** J. H. Connell noted that two species of barnacle in the genera *Chthamalus* and *Semibalanus* live in separate layers on the rocky shores of Scotland. *Chthamalus stellatus* live just above the high tide line and is replaced by *Semibalanus balanoides* beneath this point. **(b)** When the population of *Chthamalus* was experimentally removed, the population of *Semibalanus* did not colonize the open space. **(c)** When the *Semibalanus* population was removed, however, *Chthamalus* quickly colonized the open space.

BIOLOGICAL INTERACTIONS

An organism's biotic environment is composed of all of the different species with which the organism comes in contact and the ways that the organism interacts with those species. Some of the more notable species interactions are competition, predator–prey relationships, and symbioses.

Competition

When organisms require the same limited resource, such as food, living space, or mates, competition occurs. Competition may occur between different species (**interspecific competition,** or interspecies competition) or between members of a single species (**intraspecific competition,** or intraspecies competition), and it prevents two groups of organisms from occupying the same niche. In other words, usually no two groups of organisms can use exactly the same resources in exactly the same place at exactly the same time. If this were to occur, chances are good that one group would be more efficient at survival and reproduction. Possible results of competition are local extinction of the less-successful competitor (a process that ecologists call **competitive exclusion**), displacement of the less-successful competitor to another area where there is less or no competition, or selection for specializations that would lessen the competition.

Individuals that are successful at using resources not in demand by other individuals experience less competition and usually produce more off-spring. For example, several species of angelfish (*Holocanthus*) feed almost

exclusively on sponges, a food source utilized by few other species. Many times, two groups of organisms competing for the same resource have developed anatomical and behavioral specializations that enable them to use that resource more efficiently. If a variety of similar, yet distinct, resources are available, a single niche can be subdivided into two or more smaller niches that have only minimal overlap. This process that allows organisms to share a resource is called **resource partitioning.** For instance, plankton feeders on coral reefs divide their food species in a number of different ways. Some plankton-feeding fishes, such as sea bass (*Anthias*), feed close to the reef, whereas damselfish (*Chromis*) go on short forays away from the reef to feed on plankton. The sea bass and damselfishes swarm after plankton during the day, whereas soldierfishes (Holocentridae) hunt for plankton at night.

Many similar species can occupy similar small niches if the anatomy, feeding behavior, and preferred territory of each species differ just a little. Many butterflyfishes (Chaetodontidae) have elongated jaws that allow them to reach deep into cracks and crevices for small invertebrates. Some species, however, have a blunt snout and must feed on the surface

Interspecific competition is competition between different species.

Intraspecific competition is competition among members of the same species.

Competitive exclusion is local extinction of the less-successful competitor in a competitive situation.

Resource partitioning is a process that allows organisms to share a resource.

FIGURE 2-14 COMPETITION. Competition among butterflyfishes is limited by the shape of their mouths, which determines where they can find food and the types of food they can eat. This saddled butterflyfish (*Chaetodon ulietensis*) has a blunt mouth that restricts it to feeding on the surface of corals.

of coral heads, where they suck food-laden mucus from the coral animals **(Figure 2-14)**. These different characteristics allow several similar species to thrive in the same coral reef community.

Predator–Prey Relationships

Another important interaction between species occurs between predators and their prey. The number of **herbivores** (plant-eating animals) that a given area can support is relative to the amount of vegetation that is available for food. The number of herbivores, in turn, supports a certain number of predators that use the herbivores as food. If the number of herbivores in an area increases so that there is not enough food for them all, some will starve and the number of individuals will decline. As the number of individuals declines, the vegetation may be able to grow back, and as a result of more food, the population of herbivores may increase again.

A similar situation can exist between **carnivores** (meat-eating animals) and their prey. As a prey population increases, predators may feed preferentially on that species, consuming a higher percentage of individuals. Some predators will concentrate their efforts on prey that are more abundant because it conserves energy to do so. Predators will feed on the available prey as long as they are available and then switch as the prey numbers decline. This can cause density-dependent regulation of the prey population. If the number of predators increases too quickly, however,

Herbivores are animals that eat vegetation.

Carnivores are animals that eat other animals.

A **keystone species** is a species that has a greater effect on community structure than its numbers might suggest.

there will not be enough prey to support them, and the predators will starve and begin to die off. As the predator population declines, the number of prey species may start to increase again as the result of less predation.

In some habitats, the presence and activities of a particular organism prevent one or a few aggressively colonizing species from multiplying and crowding out others and thus dominating a community. For example, in rocky intertidal communities along the northern Pacific coast of the United States, the ochre sea star (*Pisaster ochraceus*) is a dominant predator. The ochre sea star feeds on a variety of prey but seems to prefer mussels (*Mytilus*). In experiments performed by Robert Paine in the early 1970s, the ochre sea stars were removed from some rocky areas, and those areas were kept relatively free of sea stars for 5 years. In the absence of the sea stars, mussels quickly colonized more of the rocky habitat, crowding out many other species in the process. The predation of mussels by sea stars keeps these highly competitive animals in check, allowing many other species, such as sea anemones, chitons, snails, and seaweeds, to survive in this habitat. Animals, such as the ochre sea star, whose presence in a community makes it possible for many other species to live there, are called **keystone species (Figure 2-15)**. By definition, the effect of a keystone species on the biological diversity of an area is disproportionate to its abundance, because only a few animals have a large effect on the community's diversity. If something happens to upset the natural balance of predator–prey relationships, a boom-or-bust cycle frequently occurs. For instance, in the early part of the 19th century, sea otters along the Pacific coast (a keystone species in this case) were hunted nearly to extinction for their fur. Sea otters are predators, and one of their favorite foods is the sea urchin, which feeds on a variety of seaweeds. As the number of sea otters decreased, the number of sea urchins increased, and they began to overgraze forests of kelp, a giant seaweed (brown alga) that dominates their habitat. In localized areas, the kelp was almost eradicated. Because many fish species relied on the kelp for cover, its loss led to a reduction in the local fish populations, which in turn depressed local populations of eagles which relied on the fish for food. Fortunately, sea otters were eventually protected by the Marine Mammal Protection Act, and the sea urchin population was thus again brought under control. Populations of sea urchins still thrive, however,

FIGURE 2-15 KEYSTONE SPECIES. This ochre sea star (*Pisaster ochraceus*) limits the size of the mussel population in this community. This prevents the mussels from crowding out other species of rock dwellers.

in sewage-polluted areas where sea otters cannot survive and in some coastal areas of the North Pacific where killer whales feed on sea otters. Human intervention can also disrupt the delicate interactions between predators and their prey, as in southern California, where commercial harvest of sea urchins for food threatens the balance of this community, once again by reducing the size of the sea urchin population on which sea otters depend.

Symbiosis: Living Together

Some organisms have developed very close relationships with each other, to the extent that one frequently depends on the other to survive. Any change in environmental factors that affects one partner will invariably affect the other as well. This arrangement is called **symbiosis**, a term that means "living together." Types of symbiotic relationships are distinguished by the nature of the relationship between the two organisms. There are three main types: mutualism, commensalism, and parasitism. Although in theory each of these symbiotic relationships is clearly defined, determining where one ends and the other begins can be difficult. In nature the distinctions between commensalism and mutualism, and commensalism and parasitism, are frequently vague and open to interpretation.

Mutualism In **mutualism**, both organisms benefit from the relationship. Sometimes, as in the case of coral animals (called polyps) and their zooxanthellae (a type of single-celled, photosynthetic organism), the two organisms are so dependent that they appear and function as a single organism. The polyp provides the zooxanthellae with nitrogen, phosphate, and carbon dioxide—nutrients needed for photosynthesis. In return, the zooxanthellae provide the polyps with food in the form of carbohydrates. Also, the removal of carbon dioxide from the polyps by the photosynthesizing zooxanthellae makes it easier for the polyp to form its stony skeleton.

This type of mutualistic relationship is at one extreme of the spectrum. Not all mutualistic relationships are equally beneficial to both organisms involved. In some mutualistic relationships, two organisms can be separated but at the risk that one or both may not thrive or may die, as in the case of Pacific clownfishes (*Amphiprion*) and their mutualistic anemones **(Figure 2-16a)**. The body of the clownfish is coated with a special mucus that protects it from the anemone's toxic stings. The fish acclimates the anemone to its presence by rubbing against it so that the fish's mucous covering picks up the anemone's scent. Thus the anemone doesn't recognize the fish as food or foe. The clownfish gains protection by living within the stinging tentacles of the anemone. In return, the clownfish defends the anemone from other fishes that might eat the anemone's tentacles, rendering it defenseless and unable to feed. In this case, both organisms benefit from the relationship, but the benefit is greater for the clownfish. Although the anemone can often survive without the clownfish, the clownfish cannot survive very long without its symbiotic partner.

Commensalism In **commensalism**, one organism benefits from the relationship, whereas the other partner is neither harmed nor benefited. An example of commensalism is the relationship between the remora fish (*Echeneis*) and some species of shark and ray **(Figure 2-16b)**. Remoras have a flattened suction cup on top of their heads that allows them to attach to the shark's body. In addition to getting a free ride on the shark, they eat the scraps of food left over when a shark feeds. Remoras are also less likely to be attacked by predators because they are attached to the shark. The shark, on the other hand, does not appear to benefit from the relationship, but is not harmed either, although a certain amount of hydrodynamic drag presumably results from the attached remoras.

Parasitism In **parasitism**, one organism, the **parasite**, lives off another organism, the **host**. The parasite benefits from the relationship, whereas the host is harmed **(Figure 2-16c)**. Many fishes and marine mammals are infected by parasitic worms, such as nematodes. Nematodes live in the body of their hosts and derive their nourishment from them. As a result, the host is weakened and becomes more vulnerable to disease and predation.

Symbiosis is a situation in which two different organisms live together in close association.

Mutualism is a symbiotic relationship in which both partners benefit.

Commensalism is a symbiotic relationship in which one partner benefits and the other does not but is not harmed.

Parasitism is a symbiotic relationship in which one partner benefits at the expense of the other.

A **parasite** is the member of a parasitic relationship that benefits by living off of its partner.

A **host** is the member of a parasitic relationship that supports the parasite and is harmed in the relationship.

(a)

(b)

(c)

FIGURE 2-16 **SYMBIOTIC RELATIONSHIPS. (a)** Mutualism: clownfish take refuge in the tentacles of a sea anemone. **(b)** Commensalism: Remora fishes attached to this shark gain protection from predators. **(c)** Parasitism: Nematode worms in the swimbladder of a marine eel derive nutrition from the organ's blood supply and weaken the host.

ECOLOGY & THE MARINE ENVIRONMENT

The Amphipod and Sea Butterfly

Not all symbiotic relationships fit neatly into one of the three categories of mutualism, commensalism, or parasitism. For instance, marine biologists working in the ocean around Antarctica observed sea butterflies (*Clione antarctica*) being carried around on the backs of amphipods (*Hyperiella dilatata;* **Figure 2-A**). The observations started them wondering about the relationship and raised several questions, such as which organism was doing the carrying and how the association was formed. Both field and laboratory observations clearly confirmed that the amphipods were doing the swimming while the sea butterflies were attached to their backs. When solitary amphipods and sea butterflies were placed in a laboratory aquarium, the amphipods were observed to swim quickly to a sea butterfly and grab it with their anterior appendages. They then used their sixth and seventh appendages to position the sea butterfly on their backs and hold it there.

From the standpoint of traditional symbiosis, the relationship did not make good sense in that both partners seemed to be negatively affected. Studies revealed that amphipods with sea butterflies attached could swim only half as fast as they could without their passenger. Presumably this would make them easier prey for their predators and make it more difficult for them to capture their prey. Other studies suggested that while attached to the amphipod the sea butterflies were unable to feed. The relationship appeared to be a disadvantage for both partners, but if that were true, why did it form? This question prompted researchers to ask if one or the other of these organisms derived a benefit from the relationship. Because the amphipods purposely "kidnapped" the sea butterflies, it was hypothesized that the advantages for amphipods outweighed the disadvantages.

Analysis of stomach contents from several predatory fishes revealed that the plankton-feeding fish *Pagothenia*, as well as at least five other species, were major predators of the amphipods. This observation led researchers to wonder if the presence of the sea butterfly somehow deterred predators. To answer this question, researchers offered captive *Pagothenia* codfish muscle to determine whether the fish were hungry. They then presented the fish with amphipods, and the amphipods were quickly eaten. Another group of hungry *Pagothenia* were given sea butterflies to eat. The fish began to feed on the sea butterflies, but immediately spit them out with a violent head shake. The same response was observed when hungry *Pagothenia* were offered amphipods carrying sea butterflies.

Having established that fish rejected both sea butterflies and amphipods with attached sea butterflies, researchers wanted to know if the fishes' response was due to a chemical cue or a visual one. To find out, researchers homogenized some sea butterflies and added the extract to fish meal to form food pellets. Control fish were fed food pellets without the homogenate, and the experimental group was fed fish pellets with the homogenate. The control *Pagothenia* readily ate their pellets without the homogenate, but the experimental fish rejected the pellets with the homogenate. The results of this experiment suggested that some chemical from the sea butterfly was responsible for the avoidance response.

To identify the specific chemical, researchers homogenized more than 4,000 sea butterflies and separated the extracted compounds into two groups: fat soluble and water soluble. Food pellets were made containing each of the two extracts and fed to hungry *Pagothenia*. The fish readily ate pellets made with the water-soluble extract but rejected pellets containing the fat-soluble extract. This suggested that one or more chemicals in the fat-soluble extract were distasteful to the fish. The fat-soluble fraction was further processed, and five pure organic compounds were isolated. Each of these compounds was tested, and one was determined to be the compound of interest. Using several chemical techniques, the compound was identified as a previously unknown chemical that investigators named pteroenone. Apparently the production of this compound by sea butterflies deters predators. The results of these experiments revealed that this may be a unique case of symbiosis in which one species "kidnaps" another as a way of protecting itself against predators. Biologists are tentatively referring to this as antagonistic symbiosis.

James McClintock, Bill Baker, American Scientist

FIGURE 2-A AMPHIPOD AND SEA BUTTERFLY. The amphipod and sea butterfly exhibit an interesting symbiotic relationship in which the sea butterfly is "kidnapped" by the amphipod and provides protection against fish predators.

IN SUMMARY

- A community is composed of interacting populations of organisms that occupy the same habitat at the same time. (Have You Wondered? #4)

- The role an organism plays in its environment, in a sense its "profession," is its niche.

- The broadest niche an organism can occupy is its fundamental niche.

- Interactions with other organisms and the physical environment, however, limit it to a smaller part of the niche called its realized niche.

- The biotic environment of an organism includes interactions with other species, such as competition, predator–prey relationships, and symbioses.

- Symbiosis occurs when two different organisms develop an intimate living association.

- The major categories of symbiotic relationships are mutualism, commensalism, and parasitism.

2-5 Ecosystems: Basic Units of the Biosphere

Assemblages of organisms interact with the physical environment to produce a relatively stable system, the ecosystem. Energy flows through the ecosystem, maintaining the life of the organisms that make up its communities. Important nutrients cycle from one part of the ecosystem to another and from living organisms to the abiotic environment. These processes maintain the stability of the ecosystems.

ENERGY FLOW THROUGH ECOSYSTEMS

All living organisms require energy to live, grow, and reproduce. The source of this energy for practically all life on earth is the sun. Organisms that are capable of photosynthesis convert the radiant energy of the sun into the chemical energy of food molecules. These molecules in turn serve as a source of nutrition for both the photosynthesizers and the organisms that feed on them. As each organism in turn feeds on another, energy is funneled through levels of the ecosystem.

Producers

Some organisms contain special pigment molecules, such as chlorophyll, that capture the sun's energy. This energy is then stored in organic molecules that can serve as a source of energy for the organisms that produce them or as food for other organisms. The process whereby the energy of sunlight is captured and stored in organic molecules is called **photosynthesis (Figure 2-17)**. In photosynthesis, light energy is used to combine relatively low-energy molecules of carbon dioxide and water to form high-energy carbohydrate molecules such as the sugar glucose. In advanced photosynthetic organisms, such as plants, oxygen gas is also released in the process. In the marine environment, the primary photosynthetic organisms are phytoplankton, seaweeds, and plants. Because these organisms are able to produce their own food, as well as food for other organisms, they are called **autotrophs** (from *auto*, meaning "self," and *troph*, meaning "feed"), or **producers**. More than one half of the photosynthesis (measured in kilocalories per year) on earth occurs in the oceans.

In the marine environment, not all producers are photosynthetic. Some are chemosynthetic, using the energy from chemical reactions, rather than sunlight, to form organic molecules from carbon dioxide and other compounds. For example, bacteria that inhabit regions of the ocean floor where water heated by the earth's core seeps through (deep-sea vents) and no light is available produce their food by chemosynthesis.

Measuring Primary Productivity

The amount of food being produced in an area of the ocean can be determined by measuring primary productivity. **Primary productivity** refers to the rate at which energy-rich food molecules (organic compounds) are being produced from inorganic materials. You will notice from the equation for photosynthesis in **Figure 2-17** that it would be possible to determine the amount of photosynthesis that is taking place by measuring either how much carbon dioxide and water are being used in the process or how much carbohydrate (glucose) and oxygen are being produced. This is the basis for methods used to measure primary production.

The traditional method employed by marine biologists is the **light-dark-bottle method**. In this procedure, two bottles that are identical in all respects except that one is clear (light can penetrate) and the other opaque (light cannot penetrate) are used. Equal volumes of ocean water taken from the same area and depth are placed in each bottle. The water sample in each bottle contains the

> **Photosynthesis** is the process whereby the energy of sunlight is captured and stored in organic molecules.
>
> An **autotroph** is an organism that can produce its own food and is also known as a **producer**.
>
> **Primary productivity** is the rate at which energy-rich food molecules are being produced from inorganic materials.
>
> The **light-dark-bottle method** is an experimental method for determining primary production.

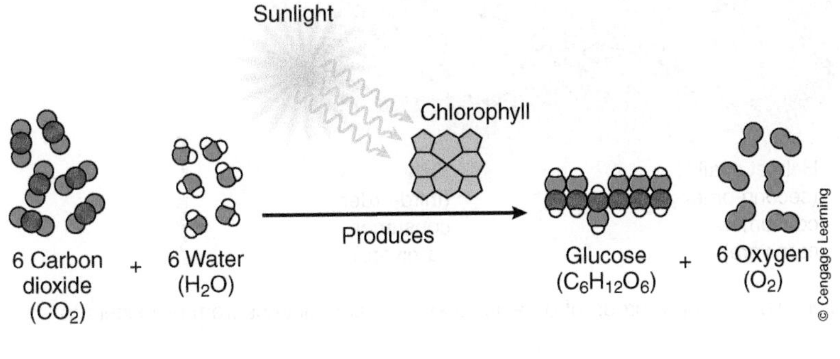

FIGURE 2-17 PHOTOSYNTHESIS. In the process of photosynthesis, carbon dioxide and water combine to form a sugar called glucose. Oxygen is a by-product of the reaction. The energy for the process is supplied by sunlight. Special molecules, such as the green pigment chlorophyll, absorb light energy and make it available to power the photosynthetic process. The glucose produced by photosynthesis can be used by the photosynthetic organism as food or to make other important molecules.

Sunlight

Chlorophyll

6 Carbon dioxide (CO_2) + 6 Water (H_2O) → Produces → Glucose ($C_6H_{12}O_6$) + 6 Oxygen (O_2)

© Cengage Learning

A **heterotroph** is an organism that relies on other organisms for food and is also known as a **consumer**.

First-order consumers (also known as **primary consumers**) are animals that feed directly on producers.

Second-order consumers (also known as **secondary consumers**) are carnivores that feed on herbivores.

Third-order consumers (also known as **tertiary consumers**) are carnivores that feed on other carnivores.

Omnivores are consumers that feed on both producers and other consumers.

Detritivores are organisms that feed on detritus.

plankton present in the water source. Another sample of seawater taken from the same area is analyzed separately to determine the water's initial concentration of oxygen. The light and dark bottles are stoppered and lowered on a line to the depth from which the water sample was taken. The sample in the clear bottle is able to receive sunlight, and the producers photosynthesize, releasing oxygen, and they and the zooplankton respire, consuming oxygen. The sample in the opaque bottle does not receive any sunlight, so there is no photosynthesis, but the organisms respire using the oxygen in the sample. After a time determined by the researcher, the samples are retrieved and the oxygen content of each bottle is measured. It is assumed that the same amount of respiration occurred in both bottles, so that value is subtracted from the oxygen content in the bottle in which photosynthesis occurred. The oxygen remaining is assumed to have come from photosynthesis and is thus an estimate of the primary production in the water sample. A problem with this method is that the technique for measuring oxygen is not sensitive enough to pick up changes in water samples containing small amounts of phytoplankton.

Another method for measuring primary production involves measuring the amount of carbon incorporated into the organic products of photosynthesis. A radioactive form of carbon, ^{14}C, is used in this measurement. The ^{14}C is introduced into the water in the form of bicarbonate ion (HCO_3^-). After allowing the phytoplankton in the sample to photosynthesize for a known time, they are filtered out of the water and

a radiation counter is used to measure the amount of radioactive carbon in the sample. The amount of radioactive carbon present in the sample represents carbon that has been incorporated into food molecules and is thus proportional to the amount (rate) of primary production. To correct for the possible uptake of ^{14}C by nonphotosynthetic organisms, a dark-bottle control may also be used. This method tends to give a low estimate of productivity, especially in water that is low in nutrients or has a low density of phytoplankton.

Other methods for determining primary production are being developed that use satellite images to estimate the amount of chlorophyll in the water, and then, on the basis of estimates of how much radiant energy is being absorbed by the chlorophyll, primary productivity can be estimated.

Consumers

Organisms that rely on other organisms for food are collectively called **heterotrophs** (from *hetero*, meaning "other," and *troph*, meaning "feed"), or **consumers**. **First-order consumers** (also known as **primary consumers**) are those that feed directly on producers. These organisms are also called herbivores. **Second-order consumers** (also known as **secondary consumers**) are carnivores that feed on herbivores. **Third-order consumers** (also known as **tertiary consumers**) are carnivores that feed on other carnivores, and so on. **Omnivores** are consumers that feed on both producers and other consumers. **Detritivores** are organisms that feed on **detritus**, organic matter such as animal wastes and bits of decaying tissue. **Decomposers** are organisms that break down the tissue of dead organisms and help to recycle nutrients. Detritivores and decomposers are also considered consumers because they cannot produce their own food and rely on other organisms for their organic nutrients.

Food Chains and Food Webs

In every ecosystem, producers and consumers are linked by feeding relationships called **food chains (Figure 2-18)**. For instance, in tropical regions, sea urchins feed on seagrass and seaweeds, and helmet snails feed on the urchins. Humans, fishes, and marine mammals, in turn, prey on the urchins and helmet snails. This is an example of a food

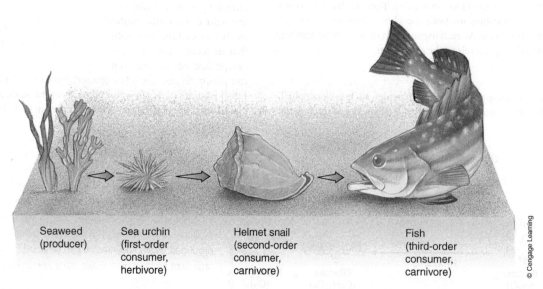

| Seaweed (producer) | Sea urchin (first-order consumer, herbivore) | Helmet snail (second-order consumer, carnivore) | Fish (third-order consumer, carnivore) |

© Cengage Learning

FIGURE 2-18 A FOOD CHAIN. Food chains depict the feeding relationships among a group of organisms as a linear sequence from producers to higher-level consumers.

chain. The concept of food chains, although helpful in establishing feeding relationships, is too simplified for what actually occurs in an ecosystem. If we were to study the dynamic feeding relationships in a particular community or ecosystem, we would notice that the relationships are interconnecting and much more complex than the simple food chain implies. These complex feeding networks are referred to as **food webs**. **Figure 2-19** shows an example of a simplified marine food web.

Other Energy Pathways

Not all energy pathways in marine ecosystems involve one organism consuming another. For instance, it is difficult for one-celled organisms to trap small molecules within their cells, and some inevitably leak into the environment. In addition, phytoplankton may release some of their photosynthetic products into the surrounding seawater. These organic molecules that are lost or released into the water column are referred to as **dissolved organic matter (DOM)**. These energy-rich organic molecules can be taken in by plankton and other small organisms and used as a nutrient source. These smaller organisms are eaten by larger animals, and thus the energy

of the organic compounds is funneled through the oceanic food web.

Detritus represents an enormous supply of energy for marine organisms. The major sources of detritus are decaying plant and algal matter that is not consumed by grazing herbivores, animal wastes, and bits and pieces of animal tissue. All of this material contains energy trapped in the organic compounds that compose them. As the detritus rains down through the water column, it serves as an important food source for planktonic bacteria and some zooplankton. When detritus settles to the bottom, it is consumed by animals, such as worms or clams, which in

Detritus is organic matter such as animal wastes and bits of decaying tissue.

Decomposers are organisms that break down the tissue of dead organisms.

A **food chain** is a linear feeding relationship that links producers and consumers.

A **food web** is a complex network of feeding relationships among producers and consumers.

Dissolved organic matter (DOM) is the organic material that is lost or released into the water column from single-cell organisms.

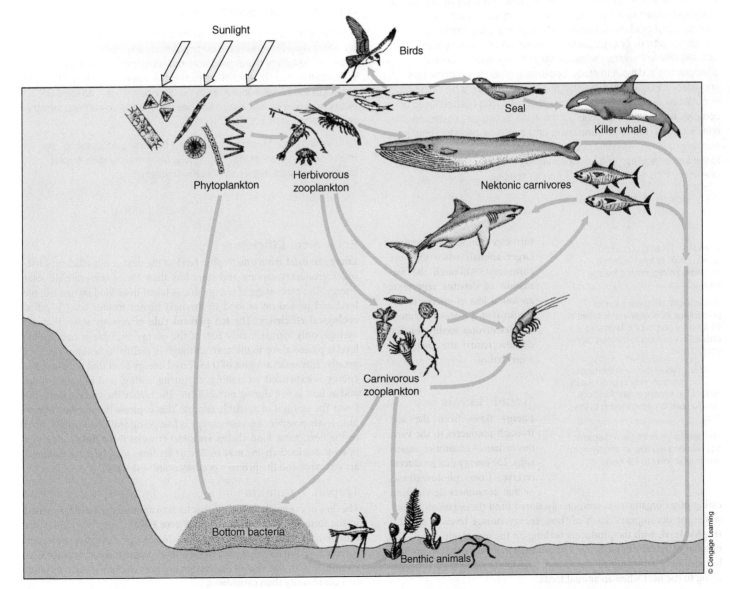

FIGURE 2-19 A FOOD WEB. Food webs show the complex, interconnecting feeding relationships among members of a community or ecosystem.

IMPACT BUBBLE

OUR PLANET IS CURRENTLY in the early to middle stages of a mass extinction. The International Union for the Conservation of Nature has estimated that 800 plant and animal species have gone extinct in the last 500 years and that another 16,000 or more are currently at risk of extinction. Unlike mass extinctions of the past, this one is largely the result of human activity and is characterized by the loss of larger-bodied animals in general and top order consumers in particular. Of all the ways that humans interfere with nature, the elimination of top-order consumers, such as sharks and tuna, may have the greatest impact because it occurs on a global scale. Also, the changes are permanent, whereas other environmental changes caused by humans are potentially reversible over time. The process of removing large top-order consumers from nature is referred to as trophic downgrading, because changes in the abundance and distribution of top consumers can cause major changes in many levels of an ecosystem. The removal of top-order consumers impacts their prey, and the effects continue downward through food webs. The loss of top consumers can reduce the length of food chains and alter the amount of grazing by herbivores, a process known as a trophic cascade. This in turn affects the abundance of producers. These types of changes are often abrupt, sometimes difficult or impossible to reverse, and commonly lead to radically different energy pathways within an ecosystem. Trophic downgrading is difficult to study, because in order to determine how one species interacts with another, the interactions must be disturbed. Populations of large top-order consumers have long been reduced or completely removed from much of the world, so we cannot observe the effects of large top-order consumers until they have been lost from an ecosystem. Even when this occurs in nature, ecosystem changes due to the loss or addition of a species may require many years to observe due to the long generation times of some species.

The loss of top-order consumers frequently results in dramatic changes in the abundance and species composition of prey and producers, leading to ecosystem changes. Recall the effects of removing the ochre seastar from rocky coastal areas in the Pacific Northwest. Another example can be found on many of the world's coral reefs, where overfishing of top predators, such as large reef fishes, alters the patterns of predation and grazing in coral reef ecosystems. The loss of these top-order consumers causes shifts in the delicate balance between corals and coralline algae and other benthic organisms. The loss of large predators results in a decrease in the competitive advantage of corals and coralline algae, resulting in dramatic changes in the benthic landscape.

The loss of top-order consumers has indirect effects on ecosystems as well. For example, overfishing of the spiny lobster (*Panulirus interruptus*) in southern California led to an increase in the local populations of sea urchins, the prey of the spiny lobster. The increase in the population density of sea urchins allowed for the rapid spread of sea urchin wasting disease, collapsing the urchin population and letting many producer organisms grow unchecked.

A common feature of many successful invasive species is that, freed from their natural predators, their populations are no longer controlled. Likewise the loss of native predators makes an ecosystem more vulnerable to invasion as it offers niches that need to be filled. The rapid spread of the invasive lionfishes in the Caribbean Sea is an excellent example of how an introduced species can quickly spread and alter the invaded ecosystems.

To have any hope of understanding and managing ecosystems, we must become aware of how our actions affect the complex trophic levels and delicate balances of natural systems.

A **trophic level** is a position in a food chain or food web that indicates an organism's feeding relationships.

Ecological efficiency is the percentage of energy that is taken in as food by one trophic level and passed on as food to the next higher trophic level.

The **ten percent rule** is the tenet that, on average, only approximately 10% of the energy available at one trophic level is passed along to the next.

An **energy pyramid** is a diagram representing the flow of energy from one trophic level to the next.

turn can channel the energy to larger animals when they are consumed. Although the formation of detritus represents an initial loss of energy to organisms in the water column, the detritivores feeding on the detritus return the energy to food chains.

Trophic Levels

Energy flows from the sun through producers to the various orders of consumer organisms. The energy that producers receive from photosynthesis or that consumers derive from eating other organisms is temporarily stored until the organism is consumed or decomposes. Each of these energy storage levels is called a **trophic level**, with the producers making up the first trophic level. The number of actual trophic levels that can exist in an ecosystem is limited, because only a fraction of the energy available on one level is passed along to the next when an animal feeds.

Ecological Efficiency

Energy transfer from one trophic level to the next is not efficient. First-order producers capture and store less than 1% of the available solar energy. The percentage of energy that is taken in as food by one trophic level and passed on as food to the next higher trophic level is called **ecological efficiency**. The **ten percent rule** of ecology states that, on average, only approximately 10% of the energy available at one trophic level is passed along to the next, although in reality the amount can vary greatly. This small amount of transferred energy is all that remains after energy is expended on finding, capturing, eating, and digesting food, and as heat is lost during metabolism. The higher the trophic level, the lower the amount of available energy. This controls the number of trophic levels possible. Because energy is lost going from one trophic level to the next, most food chains are short (two or three links), although open-water food chains may be five or six links long because nutrients are so scarce and the primary producers are very small.

Trophic Pyramids

The flow of energy from one trophic level to another can be diagrammed in the form of an **energy pyramid (Figure 2-20a)**. Energy pyramids not only indicate that the amount of available energy decreases with each trophic level but also that each trophic level supports fewer organisms as a result. It is for this reason that herbivores are generally more numerous in a community than carnivores.

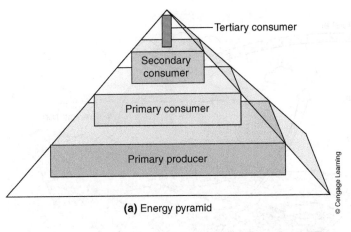

(a) Energy pyramid

Tertiary consumer

Secondary consumer

Primary consumer

Primary producer

© Cengage Learning

FIGURE 2-20 ECOLOGICAL EFFICIENCY. **(a)** The flow of energy from one trophic level to another can be represented as an energy pyramid, emphasizing the decrease in available energy from one level to the next. This diagram also represents a pyramid of numbers. The size of each box indicates the relative number of organisms that can be supported by each level. **(b)** Typically, a pyramid of biomass resembles a pyramid of numbers. A pyramid of biomass indicates the mass (kilograms) of living organisms that each level can support. For instance, 10,000 kilograms of phytoplankton are required to support 1,000 kilograms of herbivorous zooplankton and, when channeled through several layers of consumers, would be enough to support 0.1 kilogram (100 grams) of tuna or about enough to make one tuna sandwich. (The numbers have been rounded to illustrate the general principle.)

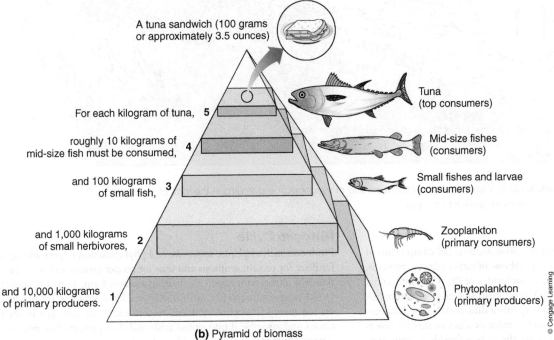

A tuna sandwich (100 grams or approximately 3.5 ounces)

For each kilogram of tuna, **5**

roughly 10 kilograms of mid-size fish must be consumed, **4**

and 100 kilograms of small fish, **3**

and 1,000 kilograms of small herbivores, **2**

and 10,000 kilograms of primary producers. **1**

Tuna (top consumers)

Mid-size fishes (consumers)

Small fishes and larvae (consumers)

Zooplankton (primary consumers)

Phytoplankton (primary producers)

© Cengage Learning

(b) Pyramid of biomass

These patterns are only generalizations; the actual energy relationships in ecosystems are much more complicated. Although energy pyramids give the best overall picture of community structure, ecologists also use ecological **pyramids of biomass** (the amount of all living tissue in an area) and **pyramids of numbers** to indicate relationships among trophic levels **(Figure 2-20b)**.

BIOGEOCHEMICAL CYCLES

The nutrients that are necessary for life are available in limited supply and are recycled. The cycling of these various nutrients involves biological, physical, and chemical processes. For this reason they are frequently referred to as **biogeochemical cycles**. A cycle exists for each of the nutrients needed for life, such as water, carbon, nitrogen, and phosphorus. In this section we will examine a few of the more important cycles.

Hydrologic Cycle

The most abundant compound in all living organisms is water. Obviously, the cycling of water is of primary importance to all living things, as well as to the marine environment **(Figure 2-21)**. Because the regions of the earth around the equator receive the most heat energy and sunlight, this is the area of the ocean that supplies the greatest

amount of water to the atmosphere via evaporation.

Water vapor is carried north and south from the equator and from west to east within each hemisphere. As the air masses rise and cool, the water falls to earth as precipitation. Although most sea salt remains in the ocean, some enters the air owing to waves crashing on the beach and similar action. Sea salt acts as **precipitation nuclei**, airborne particulates that attract water droplets. When these get heavy enough, they fall back to earth as precipitation. Precipitation over the oceans returns a great deal of water to the ocean. Precipitation that falls on land collects in the many rivers and streams that carry water back to the sea. On the way, minerals and organic substances dissolve in the water. Thus freshwater that enters the marine environment returns not only water but also large amounts of organic and inorganic nutrients.

A **pyramid of biomass** is a diagram representing the amount of all living tissue at each trophic level.

A **pyramid of numbers** is a diagram indicating the relative number of organisms at each trophic level.

A **biogeochemical cycle** is the combination of the biological, physical, and chemical processes that are involved in recycling nutrients in an ecosystem.

Precipitation nuclei are airborne particulates that attract water droplets.

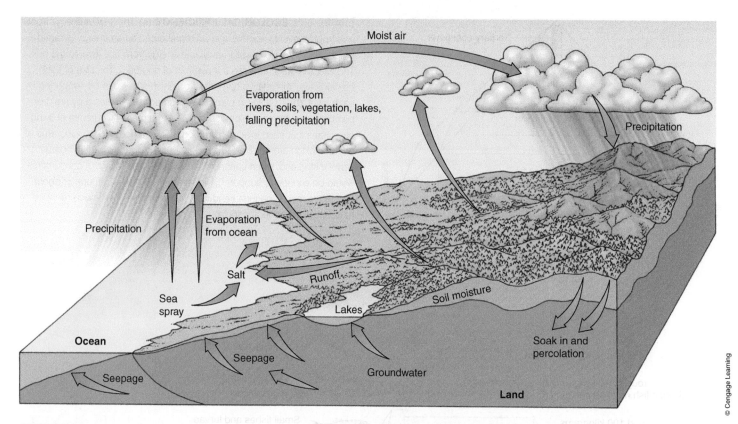

FIGURE 2-21 THE HYDROLOGIC CYCLE. Water leaves the oceans by way of evaporation and returns in the form of precipitation. Rivers and streams collect the precipitation that falls on land and return it to the sea.

Carbon Cycle

Compounds containing the element carbon are essential components of all living organisms. Carbon is the backbone of carbohydrates, proteins, lipids, and nucleic acids—the basic molecules of life. Carbon cycling, then, is vital to the scheme of life. Living organisms produce carbon dioxide when they respire. When an organism dies, a variety of decay organisms break down its tissues, and a major product of this process is carbon dioxide. Marine producers use the carbon dioxide in photosynthesis to make carbohydrates, which in turn can be used to make other organic molecules **(Figure 2-22)**.

Carbon dioxide (CO_2) reacts with seawater to form carbonic acid (H_2CO_3), which in turn forms hydrogen ions (H^+) and bicarbonate ions (HCO_3^-), as shown in the following equation:

$$CO_2 + H_2O \rightarrow H_2CO_3 \rightarrow H^+ + HCO_3^-$$

The bicarbonate ions can then be taken up by some marine organisms and combined with calcium to form the calcium carbonate needed for their shells and skeletons. The calcium carbonate from shells, corals, animal skeletons, and coralline algae eventually collects in bottom sediments and becomes limestone. Geological upheavals sometimes bring the limestone to the surface, where it is weathered by wind, rain, and other physical forces. The calcium carbonate released from the limestone then washes back into the sea by way of rivers and streams. Another source of calcium carbonate is the recycling of dead shells and skeletons by organisms, such as boring sponges.

When some marine plants and animals die, their tissues are trapped in sediments and decay, ultimately forming deposits of fossil fuel. Most of the world's oil reserves were formed in this way. When these fuels are burned, carbon dioxide is released. The carbon dioxide can then be recycled via photosynthesis.

Nitrogen Cycle

Producers such as plants, seaweeds, and phytoplankton require nitrogen fertilizer for protein synthesis and thus for proper growth and reproduction. The nitrogen they need is usually in a simple form, such as ammonia (NH_3), ammonium (NH_4^+), nitrite (NO_2^-), or nitrate (NO_3^-). The producers use energy from photosynthesis to concentrate the nitrogen in their tissues and assemble it into amino acids and then proteins. The nitrogen is passed in the form of amino acids and proteins to consumers when they eat producers or other consumers. Animals process the amino acids and proteins and excrete nitrogen in the form of ammonia, urea, and uric acid. Certain bacteria convert some of the ammonia into nitrites and nitrates that can be used again by the producers to make more amino acids and proteins **(Figure 2-23** on page 38). When an organism dies, decomposers release nitrogen into the environment from the tissue, and various groups of bacteria then convert the nitrogen into forms used by producers.

The atmosphere consists of almost 79% nitrogen and represents a major reservoir for this critical nutrient. Electrical discharges during thunderstorms produce nitrates, which enter the oceans during precipitation. Some microorganisms are capable of converting atmospheric nitrogen into forms that are usable by producers in a process called nitrogen fixation. The major nitrogen-fixing organisms in the marine environment are cyanobacteria, and we will examine the role they play in this process more closely in Chapter 6.

Runoff from the land may contain nitrogen from fertilizers, sewage, and dead plants and animals, as well as from animal wastes. This nitrogen can collect in shallow coastal waters and support a large amount of phytoplankton growth. In the open ocean, dead organisms and the nutrients they contain sink from the sunlit surface water and are unavailable for the producers in those areas. Thus large quantities of nutrient-laden material end up on the ocean floor. In certain areas, such as the coasts of California

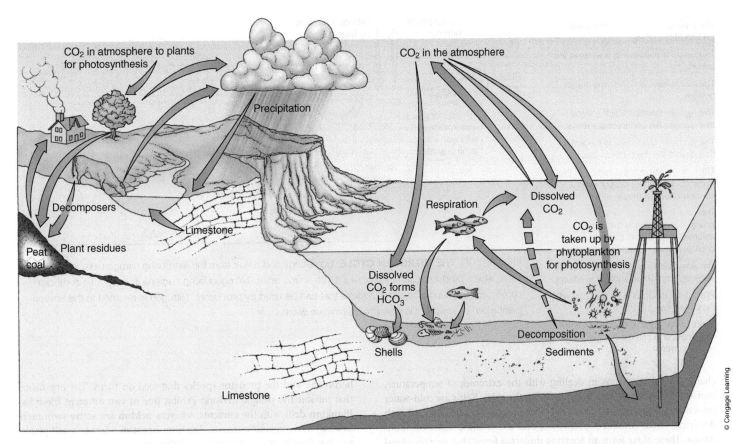

FIGURE 2-22 THE CARBON CYCLE. Carbon dioxide from the atmosphere that is dissolved in seawater is used by producers to make food through the process of photosynthesis. When the food is metabolized in respiration, the carbon dioxide is returned to the environment. Some carbon dioxide is converted into bicarbonate ions, and the bicarbonate ions into carbonate ions that are incorporated into the shells of marine organisms. When these organisms die, their shells sink to the bottom. Some of the shells are compressed to form limestone, some of which is moved by geological processes to the surface, where erosion will return the carbon to the sea.

and Peru, the combination of winds and ocean currents brings large quantities of this nutrient-laden material from the bottom back into the sunlit surface water, where it can support the growth of phytoplankton.

IN SUMMARY

- Energy constantly flows through ecosystems.
- The energy for most life on earth comes from sunlight.
- Producers capture the energy of sunlight in the chemical bonds of organic molecules. (Have You Wondered? #5)
- Consumer organisms rely on the energy-rich molecules from producers as a source of food, because they cannot synthesize their own.
- In every ecosystem, producers and consumers are linked by feeding relationships called food chains.
- In reality, the interactions among most living organisms are more complex than simple food chains. These complex feeding networks are known as food webs.
- A trophic level is a position in a food chain or food web that indicates an organism's feeding relationships and acts as an energy storage level.

- The ten percent rule of ecology states that the average amount of energy passed from one trophic level to the next is approximately 10%.
- Nutrients are constantly recycled from one generation to the next through biogeochemical cycles.

2-6 The Biosphere

The biosphere includes all of earth's communities and ecosystems. Communities and ecosystems in nature are rarely isolated from each other, resulting in interactions among different systems. Some examples of marine communities and ecosystems are estuaries, salt marshes, mangrove swamps, rocky shores, sandy shores, kelp forests, coral reefs, and open ocean.

Estuaries occur where partially enclosed areas of the sea receive freshwater runoff to produce an area of mixed salinities. Because estuaries are such changeable environments, they are home to opportunistic species that can cope with environmental variation.

The intertidal zone is the area of shore defined by the high-tide and low-tide mark. In the northeastern United States and along much of the Pacific coast, this area is rocky shoreline. In the southeastern part of the country, sandy shores predominate. These areas offer particular

The **pelagic division** is the water portion of the ocean.

The **water column** is the water in the ocean.

The **benthic division** is the ocean bottom.

The **neritic province** is the water that overlies the continental shelves.

The **oceanic province** is the water that covers the deep ocean basins.

The **photic zone** is the region of the water column where sunlight can support photosynthesis.

The **disphotic zone**, or **twilight zone**, is the region of the water column where light is dim and photosynthesis cannot take place.

The **aphotic zone** is the region of the water column where sunlight is absent.

Plankton are organisms that drift with the ocean currents.

Nekton are organisms that are active swimmers and can move against the ocean currents.

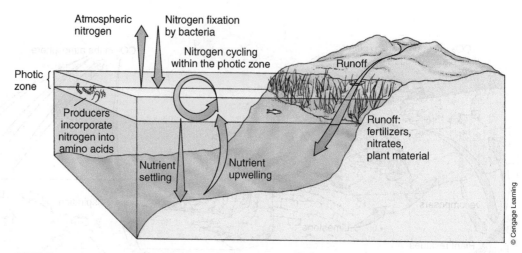

FIGURE 2-23 **THE NITROGEN CYCLE.** Upwellings and runoff from the land bring nitrogen into the photic zone, where producers can incorporate it into amino acids. Nitrogen-fixing bacteria in the photic zone can convert atmospheric nitrogen into forms that can be used by producers. Nitrogen is returned to the environment when organisms die or animals eliminate wastes.

challenges to organisms in dealing with the extremes of temperature, moisture, and salinity that occur between the tides. Kelps are cold-water brown algae found off the coasts of North America, Japan, Siberia, South America, Great Britain, Scandinavia, and the Atlantic coast of South Africa. These algae form an amazing undersea forest that provides food and shelter for many marine organisms. Coral reefs are complex communities formed by living organisms, including coral animals and coralline algae. They are found in subtropical and tropical waters and are home to thousands of species. The great diversity of life on coral reefs reflects the thousands of niches that are available. We will study the characteristics of these communities and ecosystems in more detail in Part 3 of this book.

DISTRIBUTION OF MARINE COMMUNITIES AND ECOSYSTEMS

Ecologists frequently divide the marine environment into two major divisions: the **pelagic division**, composed of the ocean's water (the **water column**), and the **benthic division**, the ocean bottom. These divisions can be subdivided into zones on the basis of three characteristics: distance from land, light availability, and depth **(Figure 2-24)**. We will examine the physical characteristics of these zones in more detail in later chapters.

Pelagic Division

The pelagic division is subdivided into the **neritic province** and the **oceanic province**. The neritic province is composed of the water that overlies the continental shelves. The larger oceanic province consists of the water that covers the deep ocean basins. The pelagic division can also be divided into the **photic zone**, where sunlight is present to support photosynthesis; the **disphotic zone**, or **twilight zone**, where light is dim and photosynthesis cannot take place; and the **aphotic zone**, where sunlight is absent. The photic zone contains not only the greatest number of photosynthetic organisms but also the largest number of animals. As we saw earlier, in typical predator–prey relationships, large populations of photosynthesizers can support correspondingly large populations of

herbivores and the predator species that feed on them. The organisms that inhabit the pelagic division exhibit one of two different lifestyles. **Plankton** drift with the currents, whereas **nekton** are active swimmers that can move against currents. The term **neuston** refers to small plankton that float at or near the surface of the ocean.

Benthic Division

The benthic division begins at the shore with the **intertidal zone**. This region of ocean bottom is covered with water only during high tide. During low tide it is exposed to air. The **shelf zone** extends from the line of lowest tide to the edge of the continental shelf. From the edge of the continental shelf to a depth of 4,000 meters (13,200 feet) is the **bathyal zone**. The **abyssal zone** extends from 4,000 to 6,000 meters. At an ocean bottom depth greater than 6,000 meters (19,800 feet) lies the **hadal zone**.

Benthic organisms live primarily in or on the bottom sediments. Benthic organisms that live on the bottom such as sponges and coral are called **epifauna**; those that live in the bottom sediments such as clams and worms are called **infauna**.

IN SUMMARY

- The biosphere contains all of the earth's communities and ecosystems.

- Estuaries, rocky coasts, sandy shores, salt marshes, mangrove swamps, kelp forests, and coral reefs are some examples of marine communities and ecosystems.

- Ecologists divide the marine environment into two major divisions: the pelagic division, composed of the ocean's water (the water column), and the benthic division, the ocean bottom.

- The divisions of the marine environment can be subdivided into zones on the basis of distance from land, light availability, and depth.

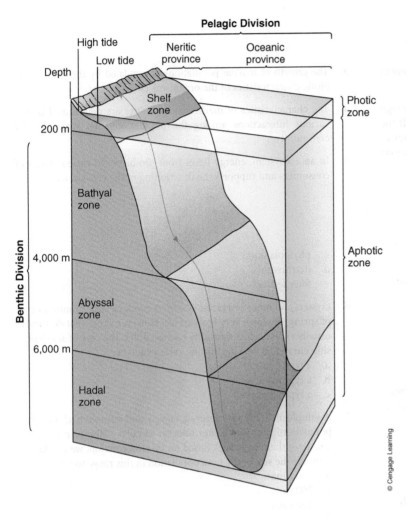

Pelagic Division

High tide

Low tide

Neritic province — Oceanic province

Depth

Shelf zone

Photic zone

200 m

Bathyal zone

Aphotic zone

4,000 m

Abyssal zone

6,000 m

Hadal zone

Benthic Division

© Cengage Learning

FIGURE 2-24 OCEAN DIVISIONS AND ZONES. Ecologists frequently divide the ocean into two major divisions: the pelagic division, consisting of the water column, and the benthic division, consisting of the sea bottom. The pelagic division can be subdivided based on the availability of sunlight (photic zone and aphotic zone) or distance from the shore (neritic province and oceanic province). The benthic division can be subdivided on the basis of depth (intertidal zone, shelf zone, bathyal zone, abyssal zone, and hadal zone).

Neuston are small plankton that float at or near the surface of the ocean.

The **intertidal zone** is the region of ocean bottom that is covered with water only during high tide.

The **shelf zone** is the region of ocean bottom that extends from the line of lowest tide to the edge of the continental shelf.

The **bathyal zone** is the region of ocean bottom that extends from the edge of the continental shelf to a depth of 4,000 meters.

The **abyssal zone** is the region of ocean bottom that extends from 4,000 to 6,000 meters.

The **hadal zone** is the region of ocean bottom that lies at a depth greater than 6,000 meters.

Epifauna are benthic organisms live primarily on the bottom sediments.

Infauna are benthic organisms that live in the bottom sediments.

KEY CONCEPTS

1. Ecology is the study of relationships among organisms and the interactions of organisms with their environment. (2-1)

2. An organism's environment consists of biotic (biological interactions) and abiotic (physical characteristics of the environment) factors. (2-2)

3. An organism's habitat is where it lives.

4. All organisms expend energy to maintain homeostasis. (2-2)

5. Abiotic factors of the environment, such as sunlight, temperature, salinity, exposure, and pressure, along with biotic factors, will influence where marine organisms can live. (2-2)

6. The basic unit studied by most ecologists is the population. (2-3)

7. Population growth is influenced by generation time, survivorship, and life history. (2-3)

8. Most populations initially grow at an exponential rate, but as they approach the carrying capacity of the environment, the growth rate levels off. (2-3)

9. Density-dependent factors play a major role in determining the carrying capacity. (2-3)

10. An organism's niche is the role the organism plays in its community. (2-4)

11. Biological interactions among species in a community include competition, predator–prey relationships, and symbiosis. (2-4)

12. Marine ecosystems consist of interacting communities and their physical environments. (2-5)

13. Energy in ecosystems flows from producers to and through consumers. (2-5)

14. The average amount of energy passed from one trophic level to the next is approximately 10%, and this ultimately regulates and limits the number and biomass of organisms at different trophic levels. (2-5)

15. With the exception of energy, everything that is required for life is recycled. (2-5)

16. The marine environment is divided into two major divisions, the pelagic division and benthic division. (2-6)

1. The physical and biological aspects of an organism's environment determine where a marine organism can live.

2. An organism's physical environment determines how much energy an organism will need to expend to maintain homeostasis. If the amount of energy needed to maintain homeostasis is too great, there may not be enough left for reproduction or the organism may die and thus be excluded from that particular area.

3. The growth of marine populations is regulated by the abiotic and biotic characteristics of the organism's environment.

4. The characteristics of marine communities are determined by the various interactions among the populations that make up the community.

5. In an ecosystem, energy flows from producers to various levels of consumers and supports the functioning of the ecosystem.

QUESTIONS FOR REVIEW

Multiple Choice

1. According to **Figure 2-12**, species B prefers or tolerates _____ better than species A.
   a. higher salinity, coarse sediments, and larger food
   b. lower salinity, fine sediments, and smaller food
   c. higher salinity, fine sediments, and larger food
   d. lower salinity, coarse sediments, and smaller food
   e. higher salinity, coarse sediments, and smaller food

2. The upside-down jellyfish depends on algae in its body for certain nutrients. The algae are protected by the jellyfish and supplied with nutrients. This relationship would be an example of
   a. mutualism
   b. commensalism
   c. parasitism

3. According to the ten percent rule, how many kilograms of phytoplankton would be needed to produce 10 kilograms of fish that were second-order consumers?
   a. 1 kilogram
   b. 10 kilograms
   c. 100 kilograms
   d. 1,000 kilograms
   e. 10,000 kilograms

4. The ultimate source of energy for most life in the ocean is
   a. photosynthesis
   b. the sun
   c. thermal vents
   d. predation
   e. phytoplankton

5. The most important primary producers in marine ecosystems are
   a. seaweeds
   b. plants
   c. phytoplankton
   d. detritivores
   e. filter feeders

6. Oysters and other broadcast spawners produce large numbers of offspring, of which very few survive. However, those that do survive usually exhibit a low mortality rate as adults. The type of survivorship curve that best fits this life cycle would be
   a. a type I curve
   b. a type II curve
   c. a type III curve

7. A sample of 50 tuna is captured, tagged, and released back into the population. Four weeks later, another sample of 50 tuna is taken and 10 of them have tags. Based on this information, we would estimate the size of the tuna population in this range to be
   a. 100 tuna
   b. 200 tuna
   c. 250 tuna
   d. 500 tuna
   e. 1,000 tuna

8. The dispersion pattern that frequently results when there is competition among species is
   a. clumping
   b. uniform
   c. random

9. An organism's fitness is measured in terms of
   a. how long it lives
   b. how many times it reproduces
   c. how many of its offspring survive to produce offspring
   d. clutch size
   e. the range of habitats it can occupy

Matching

Match the following terms with the appropriate definition:

   a. community   1. The location where an organism can be found.

   b. habitat   2. A system composed of living organisms and their physical environment.

   c. niche   3. A group of interacting populations occupying the same area at the same time.

   d. ecosystem   4. A position in a food chain or food web that indicates an organism's feeding relationships.

   e. trophic level   5. The role an organism plays in its environment.

Match each of the following zones with the appropriate description:

a. benthic zone

b. photic zone

c. neritic province

d. abyssal zone

e. disphotic zone

6. The area composed of the water that lies over the continental shelf.

7. The region of the water column that receives sunlight but not enough to power photosynthesis.

8. The ocean bottom.

9. The region of the water column that receives enough sunlight to power photosynthesis.

10. The region of ocean bottom at depths of 4,000 to 6,000 meters.

Match each of the following feeding types with the appropriate description:

a. herbivore

b. carnivore

c. omnivore

d. detritivore

e. decomposer

11. An animal that eats both vegetation and other animals.

12. An organism that breaks down the dead tissue of other organisms.

13. An animal that eats only other animals.

14. An organism that feeds on bits of decaying material and organic wastes.

15. An animal that eats only vegetation.

Short Answer

1. What are six abiotic factors that affect the distribution of organisms in an ecosystem?

2. What would probably happen to the natural balance in a community if the population of predators dramatically decreased or was wiped out?

3. What is the difference between a community and an ecosystem?

4. Describe the three main types of symbiotic relationships found in nature.

5. Describe the marine nitrogen cycle.

6. Why are there fewer marine organisms in the ocean' depths?

7. Why are so many marine organisms ectotherms?

8. Why is energy transfer between trophic levels inefficient?

9. How can groups of similar species avoid competition?

10. How can primary production be measured?

11. Why can't most populations continually grow at an exponential rate?

12. Describe a detritus-based food chain.

13. Describe two methods that could be used to determine the size of a population in the wild.

14. What is the difference between an organism's fundamental niche and its realized niche?

Thinking Critically

1. Does the area of overlap in **Figure 2-12** indicate that species A and species B have overlapping niches and will be in direct competition under those conditions? Why or why not?

2. Why does it make good ecological sense for whales to feed on plankton?

3. Although the open ocean receives plenty of radiant energy and has a larger area, it is not nearly as productive as the shallow coastal seas. Why?

4. Why do organisms that live in tide pools have to be more tolerant of changes in salinity than organisms that live in the open sea?

5. Which type of competition would you think would be more intense: interspecific or intraspecific? Why?

6. Would it be possible to have a pyramid of numbers where the trophic level above was larger than the trophic level below? Explain using an example with marine organisms.

7. Under what conditions would it be advantageous for a female to delay reproduction until later in life?

SUGGESTIONS FOR FURTHER READING

Appenzeller, T. 2004. The Case of the Missing Carbon, *National Geographic* 205(2): 88–117.

Bertness, M. D., S. D. Gaines, and M. E. Hay. 2001. *Marine Community Ecology*. Sunderland, Mass.: Sinauer Publishing.

Kaiser, M. J., M. J. Attrill, S. Jennings, D. N. Thomas, D. K. A. Barnes, A. S. Brierley, N. V. C. Polunin, D. G. Raffaelli, and P. J. le B. Williams. 2005. *Marine Ecology: Processes, Systems, and Impacts*. Oxford, England: Oxford University Press.

Myers, R. A., J. K. Baum, T. D. Shepherd, S. P. Powers, and C. H. Peterson. 2007. Cascading Effects of the Loss of Apex Predatory Sharks from a Coastal Ocean, *Science* 315(5820): 1846–1850.

Paine, R. T. 1974. Intertidal Community Structure: Experimental Studies on the Relationship between a Dominant Competitor and Its Principal Predator *Oecologica* 15: 93–120.

Zimmer, C. 2000. Do Parasites Rule the World?, *Discover* 21(8): 80–85.

Have You Wondered?

1. **How the ocean originally formed? (3-1)**
2. **What the difference is between an ocean and a sea? (3-1)**
3. **What the seafloor looks like? (3-3)**
4. **How earthquakes occur in the ocean? (3-2)**
5. **Why the ability to navigate the ocean is useful to marine biologists? (3-5)**

Geology of the Ocean

THE PHYSICAL CHARACTERISTICS of the environment play an important role in determining the kinds of organisms that can live in a given area and the traits that they will exhibit. Before we begin our study of the oceans' inhabitants and their interactions with each other and their environment, we need to gain a basic understanding of the physical characteristics of the ocean itself. In this chapter you will learn how the ocean was formed and about the forces that shaped the physical features of the ocean in the past and continue to shape them today. You will be introduced to the methods scientists and navigators use to locate their position when they are at sea with no landmarks for reference points and how this information is used to produce charts that accurately represent the geography of the ocean. We will also examine the role that living organisms play in forming some of the physical features of the ocean environment.

3-1 World Ocean

Current theories hold that our solar system was formed more than 5 billion years ago, and that the planet earth formed about 400 million years later. The primitive earth and its surrounding atmosphere were quite different in the beginning than they are now. At that time, the surface of the earth was so hot that water could not remain there. There was no free oxygen in the atmosphere, as there is today. In this environment, the ocean began to form about 4.2 billion years ago and continued to form for the next 200 million years. Life first evolved in this newly formed ocean. Today, the world ocean covers 70.8% of the planet's surface and is the most visible physical feature of our planet when viewed from space.

PRIMITIVE EARTH AND FORMATION OF THE OCEAN

It is generally believed that for the first billion years of its existence, the earth was mainly composed of silicon compounds, iron, magnesium oxide, and small amounts of other elements. Geologists hypothesize that originally the earth was composed of cold matter and that, over time, several factors–such as energy from space and the decay of radioactive elements–contributed to raising its temperature.

The process of heating continued for several hundred million years, until the temperature at the center of the earth was high enough to melt iron and nickel. As these elements melted, they moved to the earth's core, displacing lighter, less dense elements, and eventually raised the core temperature to approximately 2,000°C. Molten material from the earth's core moved to the surface and spread out, creating some of the features of the early earth's landscape. It is probable that the melting and solidifying happened repeatedly, ultimately separating the lighter elements of the earth's crust from the deeper, denser elements. This separation of elements eventually produced the various layers of the planet, which will be discussed in a later section.

In these early times, any water present on the planet was probably locked up in the earth's minerals. As the cycles of heating and cooling took place, water, in the form of water vapor, was carried to the surface, where it cooled, condensed, and formed the ocean.

OCEAN AND THE ORIGIN OF LIFE

The earth's early atmosphere is thought to have been formed by some of the hot gases escaping from deep within the planet. Initially, the atmosphere contained gases such as carbon monoxide, carbon dioxide, nitrogen, methane, and ammonia. Because it is so chemically active, any free oxygen originally present would have combined with other elements to form compounds called oxides. Gaseous oxygen did not begin to accumulate in the atmosphere until the evolution of modern photosynthesis.

The **world ocean** is a continuous mass of water that covers most of the planet.

An **ocean basin** is a portion of the deep ocean floor.

An **ocean** is the body of water in an ocean basin.

The **Southern** or **Antarctic Ocean** is the water surrounding the continent of Antarctica.

In the early 20th century, biologists such as the Russian A. I. Oparin and the Englishman J. B. S. Haldane theorized that heat, lightning, and radiation, forms of energy that were prevalent in the early days of earth, may have provided enough energy to produce the organic molecules needed for life from the gases present in the primitive atmosphere. In 1953, while a graduate student at the University of Chicago working under Harold Urey, Stanley Miller did some experiments in which he tested this hypothesis. Using an apparatus similar to the one shown in **Figure 3-1**, he

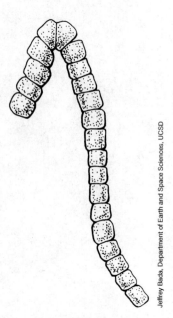

FIGURE 3-2 **OLDEST KNOWN FOSSILS.** These fossils of marine bacteria are between 3.4 and 3.5 billion years old and represent some of the earth's earliest life forms.

attempted to replicate the conditions present on the primitive earth. The results of Miller's experiments showed that organic molecules could form from the gases present in the primitive atmosphere, including some of the basic molecules necessary for life. Since Miller's initial experiments, other scientists have done similar experiments and achieved similar results. Biologists theorize that as these molecules formed, they accumulated in the ocean, and that over time the ocean became a highly concentrated nutrient soup. In this environment, molecules became organized and the first cells evolved. Because there was no free oxygen in the atmosphere, the first cells must have been anaerobic–most likely heterotrophs that gained nutrition from the organic molecules in the surrounding water. It is not surprising, then, that the oldest known fossils are of marine bacteria **(Figure 3-2)**. The fossils were found in northwestern Australia and are between 3.4 and 3.5 billion years old.

THE OCEAN TODAY

Today, 1.37 billion cubic kilometers (approximately 362×10^{18} gallons) of water covers the earth's surface and forms the ocean, the largest habitat on the planet. The ocean, which is frequently referred to as the **world ocean**, is a continuous mass of water that covers most of the planet. Traditionally, humans have divided the world ocean into artificial compartments called oceans and seas using the boundaries of continents and imaginary lines such as the equator. In reality, there are few natural divisions that can be used to separate the world ocean into smaller bodies of water. Beneath the surface of the world ocean are four main basins (regions of ocean floor), and water flows freely among them. These major **ocean basins** are the Pacific, Atlantic, Indian, and Arctic basins. The body of water in each ocean basin is referred to as an **ocean (Figure 3-3)**. The Pacific Ocean is the largest, and the Arctic Ocean is the smallest. Each ocean has its own characteristic surface area, volume, and depth, as shown in **Table 3-1**. Oceanographers also refer to the water surrounding the continent of Antarctica as the **Southern** or **Antarctic Ocean**.

Other divisions of the world ocean, such as seas and gulfs, are so named for our convenience and are in reality only temporary features of the single

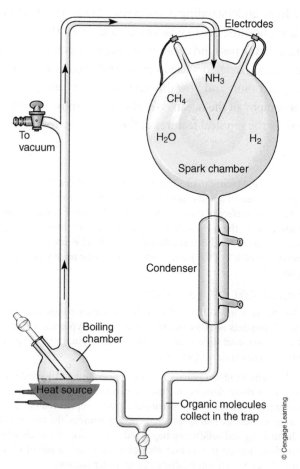

FIGURE 3-1 **MILLER'S APPARATUS.** Using an apparatus similar to the one shown here, Stanley Miller was able to demonstrate that simple organic compounds, including some necessary for life, could have formed under the conditions found on the primitive earth.

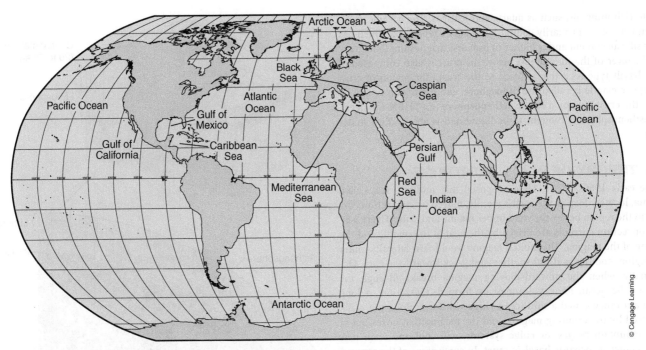

FIGURE 3-3 THE WORLD OCEAN. There are four main ocean basins, and the body of water in each basin is traditionally referred to as an ocean. These are the Pacific, Atlantic, Indian, and Arctic Oceans. Other common smaller divisions of the world ocean, such as seas and gulfs, are temporary features that are named for convenience.

TABLE 3-1 CHARACTERISTICS OF MAJOR OCEAN BASINS

Ocean	Surface Area (km²)	Volume (km³)	Average Depth (m)
Atlantic	82,400,000	323,600,000	3,575
Pacific	165,200,000	707,600,000	4,282
Indian	73,400,000	291,000,000	3,962
Arctic	12,257,000	13,702,000	1,117

© Cengage Learning

world ocean. A **sea** is a body of salt water that is smaller than an ocean and is more or less landlocked. The Mediterranean Sea, Red Sea, and Caribbean Sea are a few examples. A **gulf** is a smaller body of water that is mostly cut off from the larger ocean or sea by land formations. Some examples are the Gulf of California, Gulf of Mexico, and Persian Gulf.

IN SUMMARY

- The world ocean is believed to have formed approximately 4.2 billion years ago when water vapor escaping from minerals in the earth cooled and condensed on the earth's surface. (Have You Wondered? #1)

- The first cells evolved in the early ocean.

- Today the world ocean covers nearly 71% of the planet's surface.

- The body of water in each of the four major ocean basins is referred to as an ocean. These are the Atlantic, Pacific, Indian, and Arctic Oceans.

- Smaller subdivisions of the oceans are seas and gulfs. (Have You Wondered? #2)

3-2 The Changing Seafloor

The features of the earth are constantly changing, and the seafloor is no exception. New seafloor is continually being added as old seafloor is removed. This process reshapes the features of the ocean beneath the surface of the water and influences marine organisms, especially those that live in and on the bottom.

LAYERS OF THE EARTH

The earth is made up of several layers **(Figure 3-4)**. At the center of the planet is an inner core, which is solid (because of the extremely high pressure), very dense, very hot, and rich in iron and nickel. Surrounding the inner core is an outer core that consists of a transition zone and a thick layer of liquid material. The liquid material has the same composition as the inner core, but it is under less pressure and thus is cooler. The next layer, the **mantle**, is the thickest layer and contains the greatest mass of material. The mantle is mainly composed of magnesium-iron silicates. Although the outer region of the mantle is thought to be rigid, the inner layer is able to flow slowly. The outermost layer of the earth, the **crust**, is the thinnest and coolest of all.

Because the crust is the most easily accessible portion of the earth, more is known about it than the other layers. The continental crust is thicker and slightly less dense than the oceanic crust and is mainly composed of granite, which contains mostly lightweight

A **sea** is a body of salt water that is smaller than an ocean and is more or less landlocked.

A **gulf** is a body of water smaller than a sea that is mostly cut off from the larger ocean or sea by land formations.

The **mantle** is the thickest layer of the earth lying just beneath the crust.

The **crust** is the outermost layer of the earth.

silicate-rich minerals, such as quartz and feldspar. The oceanic crust, by comparison, is primarily composed of basalt-type rock, which has lower silicate content and is higher in iron and magnesium than granite. The layer of the mantle just below the crust is also composed of rigid, basalt-type rock that is fused to the crust. The region of crust and upper mantle is called the **lithosphere**, and the region of mantle below the crust is known as the **asthenosphere** (as-THEN-uh-sfeer). The asthenosphere is thought to be liquid and is able to flow under stress.

THE THEORY OF PLATE TECTONICS

In the early 1960s, H. H. Hess proposed that molten rock, called **magma**, located deep in the earth's mantle, moved by convection currents to the region below the solid upper mantle and crust **(Figure 3-5)**. The convection currents are driven by the heat of the mantle and the cooling of the magma. The molten magma would flow laterally under this region, cooling as it did so, then sink back toward the core. Occasionally, when the upward-moving magma breaks through the crust of the ocean floor, volcanoes are formed. Over time, a long mountain range called a **midocean ridge** will form along the crack produced by the erupting magma, and the magma that oozes out of these mountain ranges, or **ridge systems**, will cool and form new crust, known as **oceanic basaltic crust**. In some areas of the range, a **rift valley** runs along the length of a portion of the mountain crests. These are areas of high volcanic activity. Steep-sided **fracture zones**, linear regions of unusually irregular ocean bottom, may also occur, running perpendicular to the ridges and rises and separating sections of the range.

Because new crust is being formed and the earth is not growing larger, there must be some area where old crust is removed. In regions called **subduction zones**, such as the deep recesses of ocean trenches, old crust at the bottom of the trenches sinks and eventually reaches the mantle, where it is liquefied and recycled by convection currents into the earth's core **(Figure 3-6)**.

The **lithosphere** is the region of crust and upper mantle.

The **asthenosphere** is the region of mantle below the crust.

Magma is molten rock.

A **midocean ridge** (or **ridge system**) is a long mountain range that forms along the crack produced by erupting magma.

Oceanic basaltic crust is new crust that forms when magma oozing out of a ridge system cools.

A **rift valley** is an area of high volcanic activity that runs along the length of a midocean ridge.

A **fracture zone** is a steep-sided, linear region of unusually irregular ocean bottom that runs perpendicular to a ridge system.

A **subduction zone** is an area where old crust sinks and eventually reaches the mantle.

Most of the magma that rises from the deep mantle is blocked by the rigid lithosphere and moves horizontally under it, ultimately descending along the sides of ascending convection currents and carrying pieces of the lithosphere with it. This activity produces a movement of the crust called **seafloor spreading**.

All of the information concerning the movement of the earth's crust is combined to form the **theory of plate tectonics**. This theory views the lithosphere as a series of rigid plates **(Figure 3-7a)** separated by the earthquake belts of the world—that is, the trenches, ridges, and faults. There are seven major lithospheric plates: the Pacific, Eurasian, African, Australian, North American,

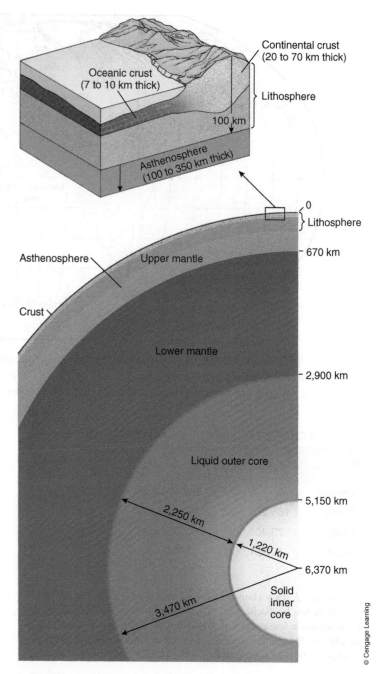

FIGURE 3-4 COMPOSITION OF THE EARTH. The earth is composed of several layers. The surface layer is the crust, which is fused to the lithosphere, the outermost portion of the mantle just beneath it. The asthenosphere lies beneath the lithosphere and is able to flow under stress.

South American, and Antarctic plates. Each plate is composed of continental or oceanic crust, or both.

At the midocean ridges, where plate boundaries move apart as new lithosphere is formed, **divergent plate boundaries** occur. **Convergent plate boundaries** occur at trenches, where plates move toward each other and old lithosphere is destroyed. The plates move past each other at regions known as **faults** (see **Figure 3-7b**).

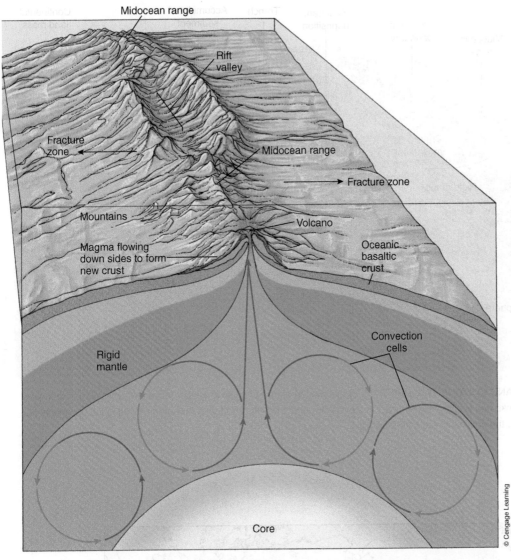

Seafloor spreading is the movement of the earth's crust produced by magma moving horizontally beneath the lithosphere.

The **theory of plate tectonics** is the theory that the movement of continents is the result of the movement of the rigid plates on which they rest.

Divergent plate boundaries are regions at the midocean ridges where plate boundaries move apart as new lithosphere is formed.

Convergent plate boundaries are regions where plates move toward each other and old lithosphere is destroyed.

Faults are regions where plates move past each other.

A **transform fault** is a special kind of fault in which each side is formed by a different plate, and the plates scrape against each other as they move in opposite directions.

An **escarpment** is a nearly continuous line of cliffs with sharp vertical drops formed by the movement of lithospheric plates along a fault.

A **rift zone** is a region where the lithosphere splits, separates, and moves apart as new crust is formed.

FIGURE 3-5 FORMATION OF OCEANIC CRUST AND MOUNTAINS. Molten material called *magma* rises from the earth's core to the upper mantle. The hot magma (red arrows) rises because it is less dense than the surrounding material. As magma reaches the mantle it cools (blue arrows), becomes more dense, and sinks back toward the core. This cycling of magma from the core to the mantle and back that results from changes in temperature and density is called *convection*. Occasionally, the heated magma breaks through the earth's crust, forming volcanoes. Lines of volcanoes form mountain ranges called *midocean ridges*.

A **transform fault** is a special kind of fault found in sections of the midocean ridge. Each side of a transform fault is formed by a different plate, and the plates scrape against each other as they move in opposite directions. The fault zone produced by this movement is the site of frequent earthquakes. The motion of the plates along these faults produces a sharp demarcation and often a nearly continuous line of cliffs with sharp vertical drops, known as **escarpments**. In these regions, there are sudden changes in ocean depth.

Regions where the lithosphere splits, separates, and moves apart as new crust is formed are called **rift zones**. The midocean ridge and rise systems represent the major rift zones. It is generally thought that rifting occurs when rising magma stretches the overlying crust, creating a sunken rift zone. As this process continues, the rift spreads and the fault deepens and cracks, allowing the magma to seep through and eventually form a ridge. When this happens to the lithosphere under the continents, the rift zone can eventually fill with seawater. The Red Sea is an example of this type of formation.

As a plate moves away from the rift zone, it cools and thickens. At the rift, thinning of the crustal plate and increased flow of magma into the rift cause the land mass to separate. A low-lying region of oceanic basalt is formed, and a new ocean basin and ridge system forms as well.

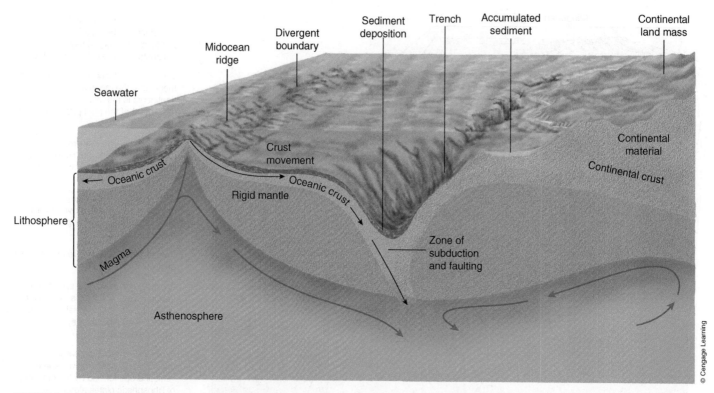

FIGURE 3-6 **SEAFLOOR SPREADING AND CONTINENTAL DRIFT.** Rising magma forms new oceanic crust that moves away from the midocean ridges. At subduction zones, old crust sinks and is ultimately returned to the mantle, where it melts and forms new magma. Since the continents rest on the basaltic crust, as the crust moves, the continents are carried along.

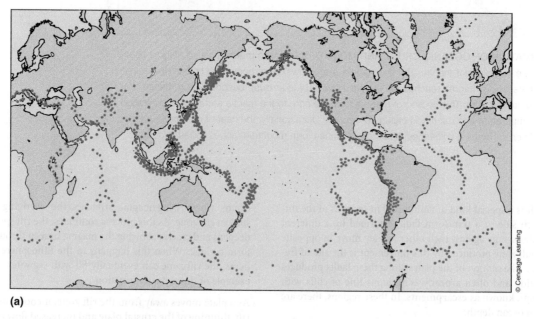

(a)

FIGURE 3-7 **EARTHQUAKE ZONES AND TECTONIC PLATES.** **(a)** This map shows the earth's earthquake zones. Each red dot represents an earthquake epicenter (the area of the earth's surface directly above the origin of an earthquake). Notice that the earthquake zones correspond closely to the location of the major plate boundaries (see part b). The distribution of volcanoes also follows along these lines.

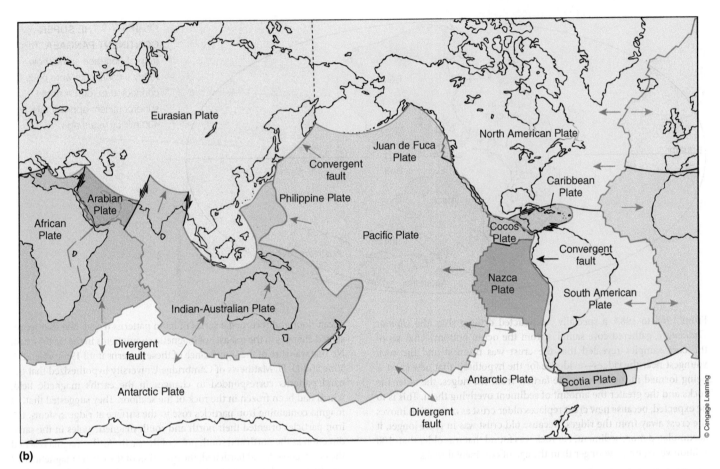

(b)

FIGURE 3-7 (b) This map shows the location of the major tectonic plates and their general direction of movement. Compare the plate boundaries to the earthquake zones shown in (a).

MOVING CONTINENTS

With the advent of more accurate world maps in the later 17th century, several scientists became intrigued with the observation that the continents fit together like pieces of a jigsaw puzzle **(Figure 3-8)**. For instance, the eastward bulge of South America would fit well with the west coast of the African continent. Also, the comparative study of rock formations, the placement of mountain ranges, and the distribution of fossils all suggested that the continents were at one time connected.

This idea was not new. As early as the 1600s, the English scholar Sir Francis Bacon suggested that the continents may have once been connected to each other. In the latter part of the 19th century, the Austrian geologist Edward Suess even proposed a name, **Gondwanaland**, for the fusion of southern continents. In 1915, the German meteorologist Alfred Wegener proposed that at one time there was only one supercontinent, which he named **Pangaea**. He suggested that forces associated with the earth's rotation were responsible for breaking the supercontinent into a northern portion, **Laurasia**, composed of Europe, Asia, and North America, and a southern portion, Gondwanaland, composed of India, Africa, South America, Australia, and Antarctica. Following World War II, there was renewed interest in the theory of moving continents. With more sophisticated equipment, studies were made during the 1950s to determine what forces might be acting to produce the continental movement.

The theory of plate tectonics put forth an explanation of continental movement or **continental drift**: The process of seafloor spreading causes movement of tectonic plates, and because many of those plates have attached continents, the continents move, much in the same way that boxes move on a conveyor belt.

EVIDENCE FOR SEAFLOOR MOVEMENT

Evidence that supports the theory of plate tectonics includes observations on the distribution of earthquakes, the temperature of the sea bottom, the age of rock samples from the seafloor, the analysis of core samples drilled through the ocean sediments, and changes in magnetic fields. Earthquakes are known to occur around the globe in narrow zones that correspond to areas along ridges and trenches, the most active areas of crustal movement **(Figure 3-7a)**. The highest seafloor temperatures occur in the regions of the ridges; temperatures decrease with distance from the ridge. These observations are consistent with heated magma oozing from the ridges and cooling as the seafloor spreads away from the ridges.

Gondwanaland is the name given to fused southern continents of the ancient supercontinent.

Pangaea is the name given to the ancient supercontinent.

Laurasia is the northern portion of the ancient supercontinent.

Continental drift is the movement of continents due to the movement of the seafloor.

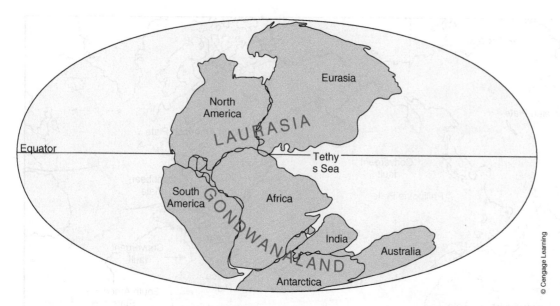

FIGURE 3-8 THE SUPER-CONTINENT PANGAEA. This map of Pangaea shows how today's continents were once connected to form a single supercontinent approximately 400 million years ago.

From 1968 to 1983, a specially constructed drilling ship, the *Glomar Challenger*, gathered core samples from the ocean bottom. Analysis of the core samples revealed that the crust was thinnest and the rock youngest near the ridges—evidence for the hypothesis that new crust is being formed in these areas. The farther from the ridges, the older the rocks and the greater the amount of sediment overlying them. This is to be expected, because new crust replaces older crust as convection moves the crust away from the ridges. Because old crust was in place longer, it accumulated more sediment. No rock was found that was older than 180 million years, much younger than the age of continental rocks.

Over the past 76 million years, there have been 170 reversals of the earth's magnetic field, when its north–south polarity switched. No one knows what causes these reversals, but their existence can be demonstrated by measuring the magnetic poles of magnetized rocks that were formed at different times. For example, when iron particles in a magnetic field are heated, they become magnetized as they cool, forming miniature compass needles. The north and south magnetic poles of these particles line up with the direction of the north and south magnetic poles of the earth when they form. Examination of the north and south poles of magnetized iron found in rocks of various ages in the earth's crust indicates that reversals in the earth's magnetic field have occurred in the past.

In 1960, oceanographers from the Scripps Institution of Oceanography towed magnetometers (devices that measure magnetic fields) over the ocean floor and recorded a series of band patterns **(Figure 3-9)** that represented changes in the polarity of magnetic components in the earth's crust. No one was sure of the significance of these patterns until 1963, when F. J. Vine and D. H. Matthews of Cambridge University hypothesized that the band patterns corresponded to changes in the earth's magnetic field, which had been frozen in the rock of the seafloor. They suggested that, as magma containing iron particles rose to the surface at ridge systems, the iron particles oriented their north and south magnetic poles in the same direction as the north and south poles of the earth. As the magma reached the crust, cooled, and hardened, the direction of the earth's magnetic field was frozen in the magnetized iron particles in the rocks. As new crust formed and the plates moved apart, these magnetized rocks were carried away from the ridge system. Over time, this process would produce strips of crust parallel to the ridges that recorded alternating reversals in the polarity of the magnetic field, accounting for the band patterns observed by the research team from the Scripps Institution.

Vine and Matthews deduced that if their hypothesis were correct, there would be symmetrical, mirror-image band patterns on the two sides of a ridge and the oldest rocks would be found farthest away from the ridge. Furthermore, the rocks' polarity should be the same as that in similarly dated rocks found on land.

To test their hypothesis, they measured the polarity and age of rock samples taken from various sites relative to ridges on the ocean bottom.

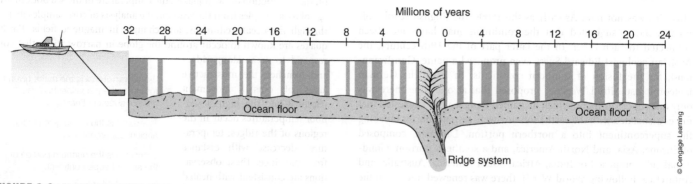

FIGURE 3-9 MAGNETOMETER DATA. This pattern of bands was produced when a magnetometer was towed across the floor of the ocean. The alternation of dark and light bands indicates reversals in the earth's magnetic field over millions of years. Notice that the older rock is farther away from the site of crust formation and that the newer rock is closer to it.

WHOI/D. Foster/Visuals Unlimited.

FIGURE 3-10 VENT COMMUNITY. These vestimentiferan worms are members of a thriving community found in the Galápagos Rift. Because sunlight does not penetrate to this depth, these organisms rely on chemosynthetic bacteria for food.

The results confirmed their hypothesis: The band patterns on either side of the ocean ridge were mirror images of each other, and the ages of the rocks matched their predictions. The conclusions drawn–that new crust was being formed on the seafloor and was spreading–support the theory of plate tectonics.

RIFT COMMUNITIES

Rift communities, also known as **deep-sea vent communities**, are thriving communities of marine organisms that depend on the specialized environments found at divergence zones in the ocean floor. The first of these communities was discovered in 1977 by Dr. Robert Ballard and J. F. Grassle of Woods Hole Oceanographic Institution. It was found at a depth of 2,500 meters (8,250 feet) in the Galápagos Rift, which lies between the East Pacific Rise and the coast of Ecuador. Rift communities consist of a variety of animals, such as clams, worms, and crabs, some of which are relatively large (**Figure 3-10**). These organisms depend on the chemosynthetic activity of bacteria for their nutrients. Rift communities are unique because they represent food webs that exist in the absence of sunlight. We will examine the biology and ecology of deep-sea vent communities in detail in Chapter 18.

IN SUMMARY

- The seafloor is constantly changing as new seafloor is formed and old seafloor is removed.
- The theory of plate tectonics holds that the lithosphere is composed of a series of rigid plates that are separated by the earthquake belts of the world.
- The crustal plates move horizontally when molten magma from the earth's core moves to the crust and breaks through, pushing the plates apart and forming new crust. (Have You Wondered? #4)

- Old crust is removed in the deep trenches and other subduction zones of the ocean.
- The continental masses are not locked into position but are constantly moving at a very slow rate.
- Geologists have determined that in the past there was only one large continent that fragmented one or more times to form the continents that we know today.
- Communities of marine organisms known as deep-sea vent communities are found in some rift zones.

3-3 Ocean Bottom

The ocean bottom has geological features similar to those found on land. Mountain ranges, canyons, valleys, and great expanses are all part of the vast underwater landscape (**Figure 3-11**). These physical features of the ocean bottom are called **bathygraphic features**, and unlike their counterpart topographical features on land, they change relatively slowly. Erosion is slow in the relatively calm recesses of the ocean, and changes mainly involve sedimentation, uplift, and subsidence. We can divide the ocean bottom into two major regions. The region that lies beneath the neritic zone is called the **continental margin**. Beyond the continental margin lies the ocean basin.

CONTINENTAL MARGINS

The continental margin is composed of the continental shelf and the continental slope. **Continental shelves** are shallow, submerged extensions of continents. They are composed of continental crust, mainly granite, that is covered by sediments, and they have physical features similar to the edge of the nearby continent. Continental shelves are generally flat areas, averaging 68 kilometers (40 miles) wide and 130 meters (430 feet) deep, that slope gently toward the bottom of the ocean basin. The width of a continental shelf is frequently related to the slope of the land it borders. Mountainous coasts such as the western coast of the United States usually have a narrow continental shelf, whereas low-lying land such as the eastern coast of the United States usually has a wide one.

The features of continental shelves can be modified by erosive processes, such as waves, and processes that deposit sediments. The greatest modification has occurred at times of low sea level, such as during the ice age, when the continental shelves were exposed.

The transition between the continental shelf (continental crust) and the deep floor of the ocean (oceanic crust) is called the **continental slope**. At the point where the continental shelf ends and the continental slope begins there is an abrupt change in the landscape called the **shelf break**. From this point there is a slight increase in the

Rift communities (or **deep-sea vent communities**) are communities of marine organisms that depend on the specialized environments found at divergence zones.

Bathygraphic features are physical features of the ocean bottom.

The **continental margin** is the ocean bottom that lies beneath the neritic zone.

Continental shelves are the shallow, submerged extensions of continents.

The **continental slope** is the transition between the continental shelf and the deep floor of the ocean.

The **shelf break** is the point where the continental shelf ends and the continental slope begins.

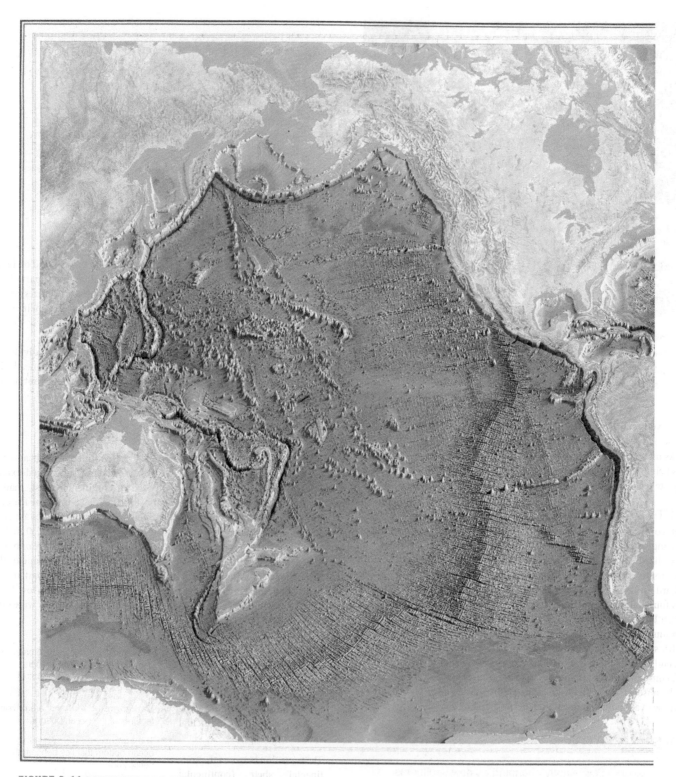

FIGURE 3-11 BATHYGRAPHIC CHART. Bathygraphic charts such as this one (*World Ocean Floor Panorama* by Bruce C. Heezen and Marie Tharp) depict the physical features of the ocean bottom.

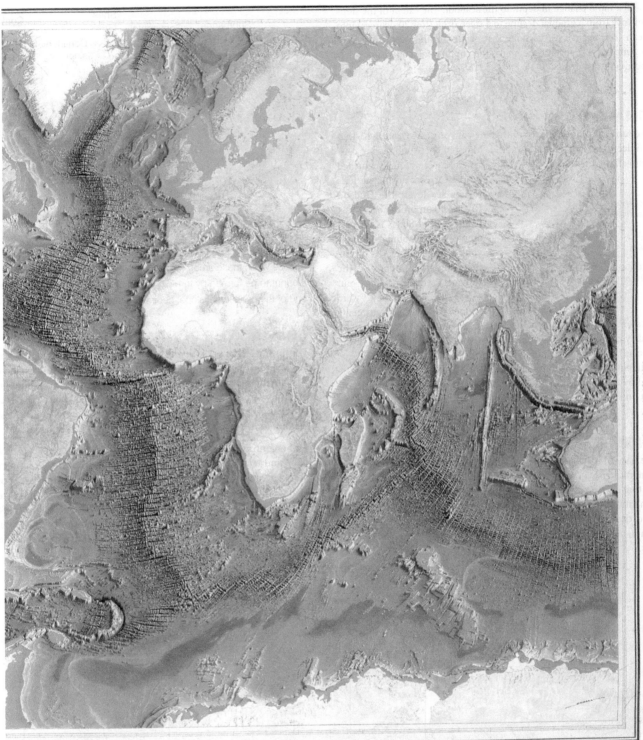

slope and a rapid change in depth to the seafloor. The extent of the sloping can vary from a gradual decline to a steep drop (low 20 degrees) into
an ocean trench, such as the slope that occurs off the western coast of
South America. Because of the angle, the continental slope usually has
less sediment.

SUBMARINE CANYONS

Some continental slopes have **submarine canyons** (see **Figure 3-12**) that
are similar to canyons found on land. Many of these submarine canyons are aligned with river systems on land and were probably formed
by these rivers during periods
of low sea level. The Hudson
River canyon on the east coast
of the United States is one example. Other submarine canyons have ripple marks on the
floor, and at the ends of the
canyons sediments fan out,
suggesting that they were
formed by moving sediments
and water. Oceanographers believe that these canyons were formed by
turbidity currents, swift avalanches of sediment and water that erode
a slope as they sweep down and pick up speed. At the end of the slope,
the current slows and the sediments fan out. Turbidity currents can be
caused by earthquakes or the collapse of large accumulations of sediments on steep slopes.

Submarine canyons are canyons in
the ocean bottom.

Turbidity currents are swift
avalanches of sediment and water.

A **continental rise** is a gentle slope
at the base of a steep continental
slope.

CONTINENTAL RISE

At the base of a steep continental slope there may be a gentle slope, a
continental rise (see **Figure 3-12**), produced by processes such as landslides that carry sediments to the bottom of the continental slope. Most
continental rises are located in the Atlantic and Indian Oceans and
around the continent of Antarctica. In the Pacific Ocean, it is more
common to find trenches located at the bottom of the continental
slopes.

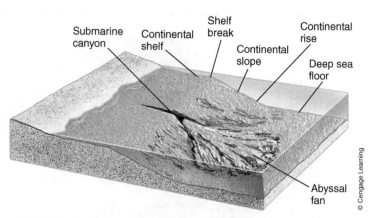

FIGURE 3-12 CONTINENTAL SHELF. Continental shelves are generally
flat areas that extend from the continental land mass into the sea. Many
physical features of continental shelves were formed during glacial
periods when these areas were exposed. Submarine canyons are
thought to be formed by turbidity currents. As the current slows at the
end of the continental slope, sediments fan out to form an abyssal fan.

SHAPING THE CONTINENTAL SHELVES

During the glacial periods, much of the ocean water was incorporated into
the polar ice sheets, and the continental shelves were at times uncovered.
Erosion carved valleys in the shelves, waves altered their contours, and rivers deposited sediments. When the glaciers melted and the sea level rose,
the continental shelves were again covered by water, retaining the surface
features that were established during the time they were uncovered.

Some continental shelves still experience changes today. The Mississippi and Amazon Rivers, for instance, carry mud, silt, and sand out to
sea to be deposited on the continental shelves. These thick deposits of
sediment have a significant effect on the productivity of these areas, as
we shall see in Chapter 16. However, not all shelves have this thick layer
of sediment. For instance, the continental shelf around the eastern tip
of Florida is bare of sediment. The swift Florida current sweeps the sediments northward to deeper water, leaving the shelf sediment free.

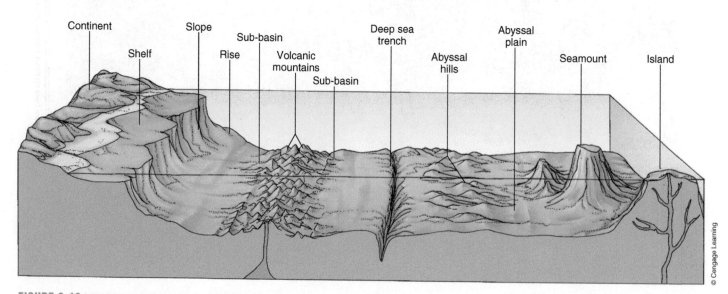

FIGURE 3-13 LANDSCAPE OF THE OCEAN FLOOR. Like the surface of the continents, the ocean floor displays a variety of physical features,
including mountains, hills, trenches, and expansive plains.

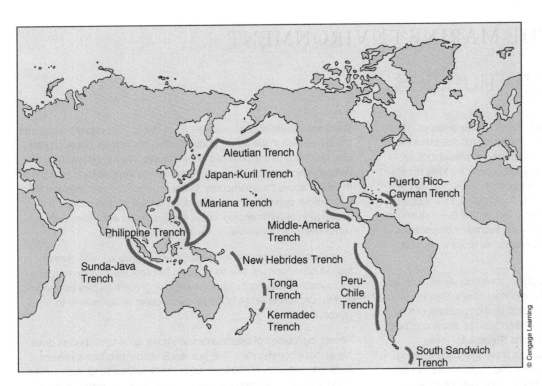

© Cengage Learning

FIGURE 3-14 OCEAN TRENCHES. This map shows the location of the major ocean trenches.

The **abyssal plain** is a flat expanse at the bottom of an ocean basin.

Abyssal hills are hills found on an abyssal plain.

A **seamount** is a steep-sided formation that rises sharply from the ocean bottom.

Island arcs are chains of volcanic islands.

OCEAN BASIN

The floor of the ocean, referred to as the ocean basin, covers slightly more of the earth's surface than the continents. As noted earlier in this chapter, there are four main ocean basins: the Pacific, Atlantic, Indian, and Arctic basins. Ocean basins are composed of basaltic rock that is covered by a thick blanket of sediments that have either washed down from the continental shelves or settled there from the surface waters. Some of the planet's most spectacular geological formations are found in the ocean basins.

Abyssal Plains and Hills

The bottom of many ocean basins is a flat expanse called the **abyssal plain (Figure 3-13)**. These plains extend from the seaward side of the continental slope and are formed by sediments deposited by turbidity currents, as well as sediments falling from above. Dotting the ocean floor are **abyssal hills** rising as high as 1,000 meters (3,300 feet). Abyssal hills cover as much as 50% of the Atlantic seafloor and as much as 80% of the seafloor of the Pacific and Indian Oceans. They are formed by volcanic action and the movements of the seafloor described earlier.

Seamounts

Another feature of ocean basins is the **seamount**, a steep-sided formation that rises sharply from the bottom. All seamounts are formed from underwater volcanoes and are most prevalent in the Pacific Ocean. Some seamounts show evidence of coral reefs and surface erosion, suggesting that at one time they were above the surface. Plate tectonic movement of the ocean floor, the natural process of compaction that volcanic material undergoes, subsidence due to cooling of the ocean floor, erosion, and the increased weight of sediments at the top may be the reasons for the sinking of these structures.

Ridges and Rises

A continuous series of large, underwater, volcanic mountains runs through every ocean of the world and stretches some 68,000 kilometers (40,000 miles) around the earth. The ridges and rises of these mountain ranges separate the ocean basins into a series of smaller, deepwater sub-basins.

Trenches

Other impressive features of the ocean floor are deepwater trenches. These are more common and more spectacular in the Pacific Ocean and are usually associated with chains of volcanic islands called **island arcs**. For instance, the deepest of all ocean trenches, the Mariana Trench, is associated with the Mariana Islands chain **(Figure 3-14)**. A portion of the Mariana Trench, the Challenger Deep, is 11,020 meters (6.85 miles) deep, the deepest spot on earth. The Peru-Chile Trench extends for more than 6,120 kilometers (3,600 miles) along the coast of South America, making it the longest of the ocean trenches. By comparison, the Atlantic has only two, relatively short trenches, the South Sandwich Trench and the Puerto Rico–Cayman Trench. **Table 3-2** lists the major ocean trenches and their depths.

TABLE 3-2 DEPTH OF MAJOR OCEAN TRENCHES

Trench	Depth in Meters	Depth in Feet
Puerto Rico–Cayman Trench	8,605	28,232
Peru-Chile Trench	8,065	26,460
Middle America Trench	6,669	21,880
South Sandwich Trench	7,235	23,736
Aleutian Trench	7,635	25,194
Kuril Trench	10,542	34,587
Japan Trench	8,513	27,929
Mariana Trench	11,020	36,366
Tonga Trench	6,000	19,800
Sunda-Java Trench	7,450	24,440
Kermadec Trench	10,047	32,962
Philippine Trench	10,190	33,431
New Hebrides Trench	8,100	26,730

© Cengage Learning

ECOLOGY & THE MARINE ENVIRONMENT

Seamount Communities

Seamounts are undersea mountains with extremely steep slopes. They are found throughout the world's oceans; current estimates place their number at around 100,000. Seamounts often have exposed rock surfaces and many intricate surface features that provide a variety of habitats for marine organisms. Because seamounts are subject to very vigorous currents and are associated with hard substrates, they support distinctive biological communities. Physical factors that affect seamount communities include the characteristics of the seamounts themselves as well as depth from the sea surface, temperature of the water, and availability of nutrients.

Shallow seamounts that reach into the photic zone host algae, such as kelp and calcareous green algae, and sometimes corals. Deeper seamounts are dominated by organisms that feed on particles of food suspended in the water, including hard and soft corals, sea anemones, hydroids, sponges, sea squirts, and crinoids **(Figure 3-A)**. These organisms usually require a hard surface for attachment and a strong flow of water to supply food, remove waste, and disperse their larvae. The densest populations of these organisms are usually near the seamount summit, where locally induced currents are stronger. Also living on the hard substrate are crabs, lobsters, sea stars, sea urchins, sea cucumbers, worms, and snails.

Sediments are common towards the base of seamounts or on the tops of flat-topped seamounts known as **guyots**. The sediments provide habitat for a wide variety of infauna and epifauna, including worms, crustaceans, snails, clams, and sea squirts. In areas where currents are strong there are few burrowing organisms, due to the fact that strong currents remove lighter sediments leaving behind the coarser sediments that make a poor habitat for burrowing. Larger animals living on the sediments include sponges, stalked barnacles, gorgonians, sea anemones, and crinoids. Giant protozoans known as xenophyophores are particularly common. Many of these organisms are suspension feeders and favor areas exposed to strong currents.

Corals may form cold, deep-water reefs on seamounts and provide a large variety of microhabitats, including spaces between the corals, coral surface, dead coral pieces, and holes bored into coral by other organisms. The diversity of organisms associated with these cold-water reefs is comparable to that of tropical reefs. Associated with these reefs are many suspension feeders and animals such as worms and sponges that bore into the coral. Mobile predators such as fish, crabs, sea stars, and snails are common. These reefs may also provide shelter for juvenile species of deep-sea fishes. Gorgonians and whip corals sometimes form extensive meadows. These soft corals are associated with enhanced densities of other organisms such as fish and may be important foraging areas for large predators, such as Hawaiian monk seals (*Monachus schauinslandi*).

There are reports of large populations of lantern fish, mysid shrimp, and squid that feed in the waters above seamounts at night but during the day stay close to the sides of the seamounts. They appear to feed on vertically migrating plankton that may become trapped by currents in the water above the seamounts. These organisms in turn are food for commercial species such as sharks, rays, tuna, and swordfish. The association of commercially valuable fish species with seamounts is well known, and these waters are intensely fished.

Seamounts are important as habitat, feeding grounds, and sites of reproduction for deep sea and pelagic fish species. Orange roughy (*Hoplostethus atlanticus*) form spawning aggregations over seamounts, and the Japanese eel (*Anguilla japonica*) come to seamounts to reproduce.

Recent exploration of seamounts has shown quite high species diversity in these communities. The Tasman Seamounts off southeastern Australia contain over 850 species. In the New Caledonia region, over 2,000 species have been identified, half of which are new to science. Many species are endemic (not found anywhere else) to specific seamounts. The fauna of seamounts is highly unique and may have a very limited distribution restricted to a single geographic region, seamount chain, or even single seamount.

FIGURE 3-A **SEAMOUNT COMMUNITIES.** This seamount off the coast of Alaska provides a habitat for a thriving community that includes sponges, soft corals, tunicates and barnacles.

LIFE ON THE OCEAN FLOOR

The water above the continental shelf is comparatively shallow, and sunlight can sometimes reach the bottom. Because the coastal seas are next to the continental land masses, they receive high levels of nutrients washed from the land. The combination of plentiful sunlight and high nutrient levels makes these areas very productive, and marine life on the continental shelf is abundant.

Although abyssal plains may contain nutrients, they do not receive any sunlight. Thus there is no photosynthesis, and food production is limited to the chemosynthetic activity of bacteria in vent communities. Organisms that live on the ocean floor rely heavily on food delivered from the sunlit surface waters. Because primary production is so limited, life is not as abundant in this environment compared with the continental shelf.

IN SUMMARY

- The seafloor can be divided into the continental margins, composed of the continental shelf and the continental slope, and ocean basins.
- Continental shelves are the exposed extensions of continents composed of continental crust.
- The features of continental shelves are shaped by erosive forces and the deposition of sediments. The transition between the continental shelf and the deep seafloor is the continental slope.
- At the bottom of the continental slope is the ocean basin.
- Some features of the ocean basin are the abyssal plain, mountain ranges, ridges, valleys, trenches, and seamounts. (Have You Wondered? #3)

3-4 Composition of the Seafloor

The seafloor is composed of basaltic rock that is covered by a coating of sediment. Sediment is composed of loose particles of inorganic and organic material. The particles that make up these sediments may come from living organisms, the land, the atmosphere, or even the sea itself and accumulate over time on the ocean floor. The amount and type of sediments that are found on continental shelves, continental slopes, and abyssal plains are particularly important to the organisms that inhabit these areas. The sediments provide a habitat for many organisms and a source of nutrients for others.

The many different types of sediments in the ocean can be classified on the basis of particle size or according to their source. A scheme of classifying sediments on the basis of particle size was devised in the late 19th century. In this scheme of classification, the finest particles are clay measuring less than 0.004 mm (0.00016 inch) and the largest are boulders measuring greater than 256 mm (10 inches). **Table 3-3** lists the different types of sediment particles found in the ocean and along its shores classified according to particle size.

Sediments can also be classified according to their origin. Sediments formed by the precipitation of dissolved minerals from seawater are called **hydrogenous sediments**. Sediments formed from the hard parts of dead organisms are called **biogenous sediments**. **Terrigenous sediments** come from land and are washed into the sea. **Cosmogenous sediments** are formed from dust and debris from outer space.

TABLE 3-3 SEDIMENT TYPES

Sediment	Size of Sediment Particle (diameter)
Clay	Less than 0.004 mm (1.6×10^{-4} inch)
Silt	0.004 to 0.062 mm (1.6×10^{-4} to 2.5×10^{-3} inch)
Sand	0.062 to 2 mm (0.0025 to 0.08 inch)
Granule	2 to 4 mm (0.08 to 0.16 inch)
Pebble	4 to 64 mm (0.16 to 2.5 inches)
Cobble	64 to 256 mm (2.5 to 10 inches)
Boulder	Greater than 256 mm (10 inches)

© Cengage Learning

HYDROGENOUS SEDIMENTS

Some sediments form from seawater as a result of a variety of chemical processes. Carbonates, phosphorites, and manganese nodules are examples of sediments that may form in salt water and accumulate on the seafloor. In shallow areas, calcium salts, sulfates, and rock salt can form on the bottom of pools when enough water evaporates to cause these minerals to precipitate.

A **guyot** is a seamount with a flat top.

Hydrogenous sediments are sediments formed by the precipitation of dissolved minerals from seawater.

Biogenous sediments are sediments formed from the hard parts of dead organisms.

Terrigenous sediments are sediments washed from land into the sea.

Cosmogenous sediments are sediments formed from dust and debris from outer space.

BIOGENOUS SEDIMENTS

Sediments from living organisms (biogenous sediments) are mostly particles of corals, mollusc shells, and the shells of microscopic, planktonic organisms. In the deep sea almost all of the biogenous sediments are composed of the remains of single-celled organisms, such as diatoms, radiolarians, foraminiferans, and coccolithophores, and of molluscs known as pteropods, or sea butterflies **(Figure 3-15)**. These organisms produce shells of calcium carbonate (foraminiferans, coccolithophores, and molluscs) or silica (diatoms and radiolarians).

Dr. Leslie Sautter/Ocean Explorer/NOAA

FIGURE 3-15 BIOGENOUS SEDIMENTS. Biogenous sediments are primarily composed of the shells of organisms such as foraminiferans, molluscs, and corals.

ECOLOGY & THE MARINE ENVIRONMENT

Animal Sculptors of the Seafloor

In the late 1970s, while surveying the floor of the Bering Sea for geological hazards to offshore oil platforms, C. Hans Nelson discovered pits and furrows that could not be attributed to any known geological process. Nelson was aware that many marine mammals either live in or frequent these relatively shallow waters between Alaska and Siberia, and he wondered whether any of these animals might be responsible for the disturbances he discovered. Investigating the disturbances, Nelson and his colleagues discovered that California gray whales (*Eschrichtius gibbosus*) produce the pits and that Pacific walruses (*Odobenus rosmarus*) produce the furrows. Both animals alter the seafloor in the course of feeding on the continental shelf and, in the process, introduce more sediment into the water of the region than the Yukon River, which annually dumps 60 million metric tons of sediment into the sea.

The gray whales are not permanent residents of this region. They leave their breeding grounds off Baja California in March and migrate up the Pacific coast to feed in the waters of the northeastern Bering Sea. When the winter ice melts, the water is relatively calm and teems with life. The whales spend May to November feeding in these rich waters before returning to their breeding grounds, and while feeding, they dig up huge amounts of sediment searching for the small, bottom-feeding crustaceans known as amphipods **(Figure 3-B)**, which are their preferred food. Initially, the evidence for gray whales disturbing the seafloor was circumstantial. As far back as the 19th century, biologists had suspected that whales feed on animals in the sea bottom. Whalers frequently reported whales coming to the surface with muddy water streaming from their mouths, and when the stomachs of captured animals were opened, they were full of bottom-dwelling animals that have since been identified as amphipod crustaceans. When the whale feeds, it rolls to one side so that its mouth is parallel to the seafloor. It then retracts its tongue, creating a suction that draws a large amount of food-laden sediment into the mouth, where a series of fibrous plates, called baleen, growing from the jaw, filter out the water.

Two other findings suggested that whales were responsible for the seafloor disturbances. In 1979, aerial observers who were tracking feeding gray whales by following emitted mud plumes found them to be concentrated in the Chirikov Basin, the area where the bottom disturbances were found. Nelson had sampled bottom sediments from several sites on the continental shelf of the Bering Sea and found that a sheet of sand covering the bottom of the Chirikov Basin was inhabited by the amphipods that were the whales' favorite food. In addition, the extent of this sand sheet matched the area where the whales were observed to be feeding. Kirk Johnson and his colleagues confirmed that the pits

Nelson found had been caused by the gray whales. Using direct measurements of the pits and data obtained from sonograms (impressions of the seafloor derived from sound waves), they were able to show that the pits could indeed have been formed by whales and that the area where the pits are located corresponded to the whales' feeding grounds.

It was later realized that walruses were also involved in changing the shape of the seafloor in the region. Pacific walruses disturb the bottom as they forage for clams and other types of bottom-dwelling prey, producing grooves or furrows as they feed. The walruses are year-round inhabitants of the area. In 1972, Samuel Stoker of the University of Alaska saw the furrows from a submarine and noticed walruses feeding nearby. He suggested that the furrows were the feeding tracks of the walruses. Ten years later, John Oliver of Moss Landing Marine Laboratories showed that clam shells that appeared to have recently been excavated and emptied were found along the furrows. These observations confirmed the findings of sand, gravel, and clam shells in the stomachs of walruses by Eskimos, who concluded that they were bottom feeders. Researchers noted that the furrows were generally in clam beds and that the width of the furrows was approximately the same as a walrus snout. This finding was consistent with Oliver's suggestion that walruses unearth their food with their lips and not with their tusks, as was previously assumed.

The feeding activities of whales and walruses seem to be beneficial to the ecosystem and enhance the area's productivity. The feeding activity of the whales separates mud deposited by the Yukon River from the sand, thus preserving the sandy bottom as an ideal habitat for amphipods. During feeding, only adult amphipods are trapped on the baleen, whereas the smaller juveniles escape to replenish the supply. The feeding activity also releases nutrients from the bottom sediment and moves them into the water column, where they stimulate the growth of plankton.

FIGURE 3-B AN AMPELISCID AMPHIPOD. Crustaceans such as this are a favorite food of gray whales. As the whales filter the amphipods from the sea bed they alter the contours of the seafloor.

Les Watling, Darling Marine Center, University of Maine

When they die, their remains settle on the sea bottom. If more than 30% of an area's sediment is made up of these fine biogenous particles, the sediment is termed **ooze**: calcareous ooze (calcium) or siliceous ooze (glass), depending on which of the two shell types is dominant. The amount of biogenous sediment in an area depends on the mass of organisms contributing to the sediment and the ability of the water to dissolve the minerals in the shells.

TERRIGENOUS SEDIMENTS

Some sediments are produced from continental rocks by the various actions of wind, water, freezing, and thawing. These particles are carried from the land by water, wind, ice, and gravity and are deposited primarily on the continental shelves. Mud is found in ocean basins throughout the world wherever marine life is too scarce to form biogenous sediments. Mud is composed of clay and silt. Clay is composed of fine powdered rock, and its color depends on the type of minerals that it contains. For instance, red clay contains large amounts of iron compounds. It is believed that this fine powder is blown from land by the wind or rinsed by rain from the atmosphere. The particles are so fine that they may remain suspended in the water for many years before settling.

COSMOGENOUS SEDIMENTS

Iron-rich particles from outer space strike the surface of the ocean and slowly drift to the seafloor. Many dissolve before reaching the bottom. Although not as numerous as other sediments, small amounts are found scattered on the bottom of all seas.

IN SUMMARY

- The seafloor is covered with a variety of sediments that can be classified according to size or on the basis of their origin.
- Sediment particles can be formed by the action of physical processes on rocks, from the seawater itself, and from space.
- A substantial source of ocean sediment comes from small organisms that have shells of silica or calcium carbonate that form biogenous sediments when the organisms die.

3-5 Finding Your Way around the Sea

Sometimes it is necessary for marine biologists to return to the same point in the ocean to take samples or to make observations. This can be quite challenging on the open sea, out of the sight of land, because everything looks the same. There are no obvious features or landmarks that can be used to determine one's exact location. Under these circumstances, marine biologists must have an accurate knowledge of navigational techniques. Navigators use maps and charts, as well as sophisticated modern technology, to locate specific areas of the ocean.

> **Ooze** is a type of sediment composed of more than 30% of fine biogenous particles.
>
> A **map** is a two-dimensional representation of the land features of the earth.
>
> A **chart** is a two-dimensional representation of the oceans and their features.
>
> A **chart projection** is a projection of the earth's surface, together with reference lines, onto a surface.

MAPS AND CHARTS

A two-dimensional representation of the earth's three-dimensional surface is called a **map** or **chart**. Maps usually show the land features of the earth, whereas charts display the oceans and their features. Maps and charts are made by projecting the features of the earth's surface together with reference lines onto a surface to produce a **chart projection (Figure 3-16)**. Imagine a clear glass globe

IMPACT BUBBLE

A FISHING TECHNIQUE KNOWN as deep-sea trawling is devastating corals and pristine marine habitats, such as seamounts, that have gone untouched since the last ice age. During visits to the seafloor and seamounts, researchers have discovered thriving coral reefs and many species that are new to science. Over the past five years, however, these surveys have also revealed that deep-sea habitats are suffering severe impacts from bottom trawling. Where an area has been trawled there's not much living on the surface of the ocean floor.

Trawling was originally developed for use in shallow waters with smooth seafloors, but as fish populations declined and technology improved, fishing fleets began trawling in much deeper water. Trawl nets are giant, heavy-duty nets that are dragged over the seafloor at depths of more than 1 kilometer (0.6 mile), and each trawler drags a trawl net over an area of ocean around 33 square kilometers (11.9 square miles). The nets are fitted with rubber rollers called "rock hoppers" that disrupt the sea bottom and destroy corals that provide habitats for fish and other marine organisms. Anything that's standing up from the bottom, such as a sponge or coral, is usually bent over, broken, or removed.

If the sea bottom is muddy, the nets dig into it. Since animals in the muddy bottom mostly live in the upper 3-4 cm (1 ¼ to 1 ½ inches), the nets damage their burrows and tubes. Some animals may recover from this disruption by digging new burrows or making new tubes, but many animals invest large amounts of energy in making their burrows and tubes. In fact, some marine worms are unable to reform their protective tubes because of the energy cost. Some animals raise their young in these burrows and tubes. If they are destroyed, the young may be trapped, unable to burrow their way out of the mud, and they die.

Fishers will return again and again to trawl a favorite area, ultimately creating an entirely new bottom community. Organisms that reproduce quickly can frequently colonize a damaged area, but animals with a longer, more stable lifestyle are lost, especially if an area has been trawled repeatedly. When researchers examine heavily trawled areas, they find the whole bottom community has changed to species that are capable of quickly colonizing disturbed habitats. For some fish species this may not be bad. For instance, in the North Sea, areas that are trawled repeatedly show increased numbers of flatfish species but fewer of other species that would normally inhabit the area. Although the damage to bottom communities may be reversible, the time scale for recovery for the deeper sea bottom is long. Some areas require a century or two to recover.

Among the most threatened sites are cold-water coral reefs in temperate regions, some of which have just recently been discovered. The Norwegian government has banned deep-sea trawling over the Røst reef, the largest cold-water reef in the world, which was first discovered in 2002. The 3-kilometer-wide (1.8 mile) reef is teeming with life and stretches nearly 40 kilometers (24 miles) at a depth of 450 meters (1,426 feet). Similar bans are in place at a number of other sites around the world, but what is urgently needed is a network of protected areas where any type of fishing that involves dragging equipment across the seabed is banned. About half of the continental shelves get trawled each year, although that number fluctuates as patterns of fish distribution change. Some researchers have proposed that we protect 20-30% of the continental shelf area. Others feel we should protect most of the sea bottom and severely restrict trawling. What do you think?

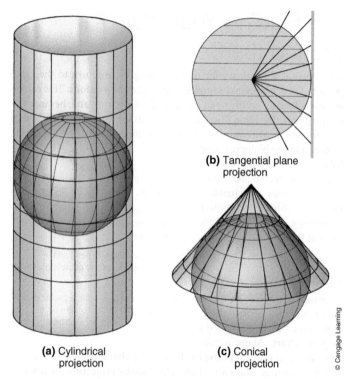

(a) Cylindrical projection

(b) Tangential plane projection

(c) Conical projection

© Cengage Learning

FIGURE 3-16 MAP PROJECTIONS. If a light were placed inside a transparent globe, the pattern of light and shadow falling on a piece of paper placed in a particular position would form a map. The three most common arrangements for producing a map are **(a)** the cylindrical projection, **(b)** the tangential plane projection, and **(c)** the conical projection.

with reference lines drawn on the surface and a light bulb placed inside. As the light shines through the globe, it projects an image onto a piece of paper that is placed in a particular position with reference to the globe. Depending on the position of the paper, one could make three basic types of chart projections: cylindrical, conical, or tangential plane.

The most familiar maps and charts, such as those found in texts (including this one) and in classrooms, are modifications of the cylindrical type called a **Mercator projection** (for examples see **Figures 3-3** and **3-7**). Even though the pole regions are greatly distorted and the poles themselves cannot be shown with this projection, it has the advantage that a straight line drawn on it is a line of true direction (constant compass heading), and as such is useful for navigation. Charts of the ocean that show lines connecting points of similar depth are called **bathymetric charts** (**Figure 3-17a**); these are similar to topographical maps showing land elevations. A chart that uses perspective drawing (picturing the sea as it appears to the eye with reference to a relative depth), coloring, or shading

A **Mercator projection** is a modification of a cylindrical chart projection.

Bathymetric charts are charts that show lines connecting points of similar depth in the ocean.

A **physiographic chart** is a chart that uses perspective drawing, coloring, or shading to show depth.

Latitude is a reference line drawn as a circle around the surface of the earth that is parallel to the equator.

Longitude is a line drawn at a right angle to latitude lines.

The **equator** is a circle drawn around the center of the earth perpendicular to its axis of rotation.

A **parallel** is the same as a line of latitude.

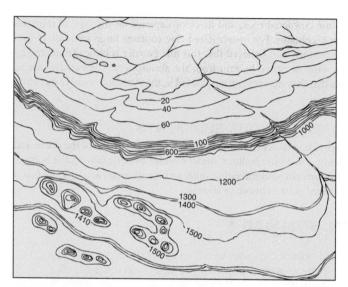

(a) Bathymetric chart

(b) Physiographic chart

© Cengage Learning

FIGURE 3-17 BATHYMETRIC AND PHYSIOGRAPHIC CHARTS. **(a)** A bathymetric chart indicates variations in ocean depth by lines that connect areas of similar depth. **(b)** A physiographic chart shows the same information as a bathymetric chart, but it uses coloring or shading rather than lines to indicate areas of similar depth.

instead of lines to show the varying depths of the ocean is called a **physiographic chart** (**Figure 3-17b**).

REFERENCE LINES

To locate a specific spot on the surface of the earth or navigate from one place to another, we need a frame of reference. For this we use a grid composed of two sets of lines, **latitude** and **longitude**. The grid is superimposed on the earth's surface and divides it into sections.

Latitude

The earth is essentially a sphere. The circle drawn around the center of the earth perpendicular to its axis of rotation is the **equator** (**Figure 3-18a and b**), designated 0 degrees latitude. Each half of the earth is then

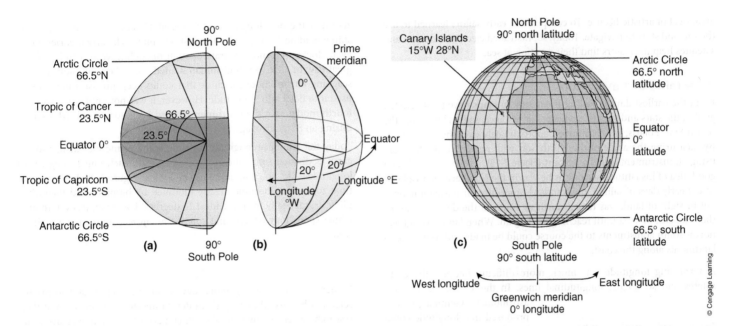

FIGURE 3-18 **LATITUDE AND LONGITUDE. (a)** The point of reference for latitude is the equator (0 degrees latitude). Latitude lines are then drawn at equal intervals from the equator to each pole (90 degrees latitude). **(b)** The prime, or Greenwich, meridian is the reference point for lines of longitude (0 degrees longitude). Other lines of longitude are designated by the angle they form with respect to the prime meridian and by direction (east or west). **(c)** Together, lines of latitude and longitude form a grid that, when superimposed on the earth's surface, can be used to locate the precise position of any point on the globe.

divided again by other latitude lines at equal intervals ending at the poles. The North Pole is 90 degrees north with reference to the equator, and the South Pole is 90 degrees south. Notice that each line of latitude forms a progressively smaller circle, and that it is necessary to state whether the latitude is north (N) or south (S) of the equator. Latitude lines are sometimes referred to as **parallels** because they are parallel to the equator as well as each other.

In addition to the equator, other well-known parallels include the **Tropic of Cancer** (23.5 degrees N) and the **Tropic of Capricorn** (23.5 degrees S), which set the boundaries of what is called the tropics, or tropical zone, and the **Arctic Circle** (66.5 degrees N) and the **Antarctic Circle** (66.5 degrees S). The odd latitudes (numbers that are not multiples of ten) for these parallels are determined by the position of the sun and radiant energy during the June and December solstices (times of the year when the overhead position of the sun is farthest from the celestial equator). The Tropic of Cancer lies at the point where the sun shines directly overhead during the summer (June) solstice. Likewise, the Tropic of Capricorn lies where the sun shines directly overhead during the winter solstice (December). The Arctic Circle marks the northernmost limit of sunlight at the December solstice. In the area north of the Arctic Circle the sun does not rise for one or more days on approximately December 21 and does not set for one or more days on approximately June 21. The Antarctic Circle marks the southernmost limit of sunlight at the June solstice. The area south of the Antarctic Circle experiences constant darkness for one or more days on approximately June 21 and constant sunlight for one or more days on approximately December 21.

Longitude

Lines of longitude, also known as **meridians**, are at right angles to the latitude lines, extending from the North Pole to the South Pole. The primary line of longitude is an arbitrary one that passes through the Royal Naval Observatory in Greenwich, England. This line is known as the **prime**, or **Greenwich, meridian** and is designated 0 degrees longitude. Directly opposite the prime meridian at 180 degrees is a line that approximates the **international date line**. Longitudinal lines are identified by the angle that they form with respect to the prime meridian and their direction from it, east (E) or west (W). Unlike latitude lines, longitude lines all form circles of the same size, because each passes through the earth's poles and equator.

Divisions of Latitude and Longitude

Using latitude and longitude, one can exactly locate any point on the earth's surface. For instance, the Canary Islands lie at 15 degrees W and 28 degrees N **(Figure 3-18c)**. A degree of latitude or longitude covers a large area, so each degree is subdivided into 60 minutes and each minute into 60 seconds for giving a more precise location. A common unit of distance used in navigation is the **nautical mile**, which is equal to 1 minute of latitude (1.85 kilometers, or 1.15 land miles).

NAVIGATING THE OCEAN

Early navigators had a difficult time accurately determining their position. Charts and maps were not particularly accurate and exhibited a

The **Tropic of Cancer** is the latitude that lies at the point where the sun shines directly overhead during the summer (June) solstice (23.5 degrees N).

The **Tropic of Capricorn** is the latitude that lies at the point where the sun shines directly overhead during the winter solstice (December) (23.5 degrees S).

The **Arctic Circle** is the northernmost limit of sunlight at the December solstice (66.5 degrees N).

The **Antarctic Circle** is the southernmost limit of sunlight at the June solstice (66.5 degrees S).

A **meridian** is the same as a line of longitude.

The **prime meridian** is the primary line of longitude that passes through the Royal Naval Observatory in Greenwich, England.

great deal of artistic license. To compensate, early sailors learned to use the sun and stars to navigate. Today, sophisticated electronic devices and satellites help navigators find their position at sea.

Principles of Navigation

From the earliest days of sailing the ocean, humans have used celestial guides, the stars and sun, to navigate. For instance, sailors knew that the North Star, Polaris, was positioned approximately over the North Pole. By measuring the angle of the North Star with respect to the horizon, using an instrument called a **sextant (Figure 3-19)**, a seaman could get a good idea of his latitude, as long as he was in the Northern Hemisphere. In the early days of sailing, the standard procedure for navigation, once out of sight of land, was to sail north or south to the desired latitude, then sail east or west until reaching land again. When land was sighted, north–south adjustments to the course could be made with reference to landmarks along the coast.

Determining longitude was much more difficult, because the earth rotates and so do the longitudinal lines. In the 1500s, the Flemish astronomer Gemma Frisius proposed that longitude could be determined by setting a clock to exactly noon when the sun was at its zenith (highest point above a reference longitude), then moving to a different longitude and taking the time when the sun reached its zenith at the new position. Because the earth rotates 15 degrees each hour, one could determine the new longitude by finding the difference in time: If the zenith at the new longitude occurred at 1 p.m. (one hour later), the ship would be 15 degrees to the west. If the

> The **Greenwich meridian** is the same as the prime meridian.
>
> The **international date line** is a line directly opposite the prime meridian at 180 degrees longitude.
>
> A **nautical mile** is equal to 1.85 kilometers or 1.15 land miles.
>
> A **sextant** is an instrument used to measure the angle of a star with respect to the horizon.
>
> A **chronometer** is a clock that keeps precise time.
>
> The **global positioning system (GPS)** is a technology that utilizes a system of satellites to locate an exact position anywhere on the earth.

zenith at the new longitude occurred at 11 a.m. (23 hours later), the ship would be 15 degrees to the east. Despite such early theories, however, clocks were not accurate enough on sailing vessels to be of any great use. In 1735, an Englishman named John Harrison built the first seagoing **chronometer**, a clock that could keep precise time, and he used it to determine longitude. However, it was not until 1761, with his fourth model, that the instrument proved itself, losing only 51 seconds during an 81-day voyage.

Today the reference longitude for time is the prime meridian, and the earth is divided into time zones **(Figure 3-20)**, each one 15 degrees of longitude wide. You will notice as you look at Figure 3-20, however, that time zones have become greatly modified by humans, and they often differ dramatically from the ideal 15 degrees. For example, even though China spans four actual time zones, the entire country is on a single time.

Global Positioning System

Modern electronic navigational devices are very important to marine scientists because they help them determine the exact position of their research vessels in any type of weather, permitting them to return again and again to the same position for more sampling or study. The most sophisticated of these modern systems is the **global positioning system**, or **GPS**. GPS utilizes a system of satellites that can be used to find an exact position anywhere on the earth at any time and in any weather. There are three parts to GPS: the space component, the user component, and the control component. The space component consists of 24 satellites, each in its own orbit 11,000 nautical miles above the earth. The satellites are spaced in such a way that at any time at least four of them will be above the horizon from any point on earth. The user component consists of receivers, which can be held in your hand or mounted in a vessel (or other vehicle). The control component consists of five ground stations located around the world that assure the satellites are working properly.

Each GPS satellite takes 12 hours to orbit the earth and is equipped with an atomic clock, a radio transmitter, and a computer. The user-component unit receives the satellite signal and determines the amount of time it took the signal to reach the receiver. The difference between the time the signal was sent and the time it was received enables a computer in the receiver to calculate the distance to the satellite. By calculating this information from at least three different satellites, the receiver can calculate a position (latitude and longitude) that is accurate to less than 1 meter (3.3 feet), depending on the quality of the receiver being used.

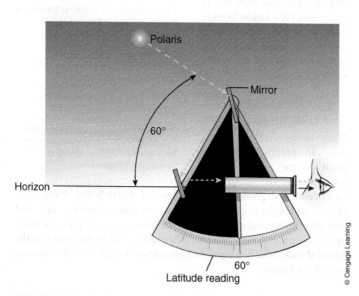

FIGURE 3-19 A SEXTANT. Sextants such as this one are used to measure the angle of the North Star with respect to the horizon. This information can then be used to determine a ship's latitude.

© Cengage Learning

IN SUMMARY

- Navigators use charts and maps with lines of latitude and longitude to locate their position while at sea.
- Lines of latitude run perpendicular to the long axis of the earth, which extends from pole to pole.
- Lines of longitude run perpendicular to lines of latitude.
- Modern navigational techniques use sophisticated electronic equipment and satellites to help marine biologists find their location accurately anytime, anywhere, and in any weather. (Have You Wondered? #5)

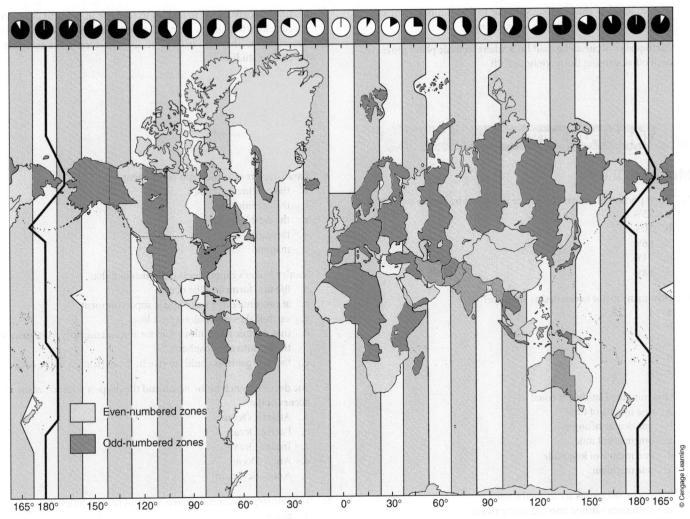

FIGURE 3-20 THE EARTH'S TIME ZONES. The reference longitude for time is the prime meridian. Because the earth rotates 15 degrees each hour, each of the world's time zones is 15 degrees wide. Yellow bands represent even-numbered hours and pink bands represent odd-numbered hours. Countries shown in yellow are in the closest yellow time zone, whereas countries shown in red are in the closest pink time zone. Notice that some countries (e.g., China) span more than one time zone, but the country is on one time. Countries in blue or purple set their own time independent of the time zones.

KEY CONCEPTS

1. The world ocean has four main basins: the Atlantic, Pacific, Indian, and Arctic. (3-1)

2. Life first evolved in the ocean. (3-1)

3. The earth's crust is composed of moving plates. (3-2)

4. New seafloor is produced at ocean ridges, and old seafloor is removed at ocean trenches. (3-2)

5. The ocean floor has topographical features similar to those found on continents. (3-3)

6. The seafloor is composed of sediments derived from both living and nonliving sources. (3-4)

7. Latitude and longitude determinations are particularly necessary for precisely locating positions in the open sea, where there are no features at the surface. (3-5)

ARE YOU STILL WONDERING?

1. Early in earth's history, the planet went through many cycles of heating and cooling. During the heat cycle, water trapped in the earth's minerals was liberated and evaporated. As the water vapor rose in the atmosphere and cooled, it formed liquid water that rained back upon the earth and formed the oceans.

2. The term ocean is applied to a body of water that occupies one of the major ocean basins. A sea is a body of saltwater that is smaller than an ocean and is more or less land-locked.

3. The seafloor has characteristics similar to the continents, including canyons, mountains, valleys, and vast plains.

4. Earthquakes occur along fault lines where tectonic plates move past each other, causing the seafloor to shift.

5. Marine biologists need to be able to navigate in order to document specifically where in the ocean they are taking samples or making observations. They also need to be able to return to the same place for further studies.

QUESTIONS FOR REVIEW

Multiple Choice

1. What portion of the earth's surface is covered by oceans today?
   a. 25%
   b. 50%
   c. 66%
   d. 71%
   e. 88%

2. How many major ocean basins are there?
   a. 1
   b. 2
   c. 3
   d. 4
   e. 5

3. One minute of latitude is equal to
   a. one minute of time
   b. one degree of angle
   c. one nautical mile
   d. one minute of longitude
   e. one meridian

4. Biogenous sediments are formed from
   a. sediments washed into the sea by rivers
   b. chemical reactions between elements in seawater
   c. meteor particles showering down on oceans
   d. the shells of plankton and small animals
   e. clay dust blown from the land by winds

5. The theory of plate tectonics helps to explain
   a. the formation of volcanoes
   b. the presence of rift communities
   c. the origins of biogenous sediments
   d. the movement of continents known as continental drift
   e. the physical features of continental shelves

6. Basalt-type rock is the main component of
   a. the continental crust
   b. the oceanic crust
   c. the earth's mantle
   d. the earth's core
   e. magma

7. Stanley Miller's experiments demonstrated that
   a. life first formed in the ocean
   b. at one time there was a single supercontinent
   c. earthquakes occur along fault lines
   d. that organic molecules could form spontaneously from gases in the primitive atmosphere
   e. living organisms could survive in the dark abyss of the ocean

8. The deepest trench in the ocean and the deepest point on earth occurs in the
   a. Atlantic Ocean
   b. Pacific Ocean
   c. Indian Ocean
   d. Arctic Ocean
   e. Antarctic Ocean

9. The primary producers in deep-sea vent communities are
   a. algae
   b. phytoplankton
   c. chemosynthetic bacteria
   d. photosynthetic bacteria
   e. seaweeds

10. A sextant is used to
    a. determine latitude
    b. determine longitude
    c. determine sea depth
    d. determine time at sea
    e. determine satellite position

Matching

Match the following terms with the appropriate definition:

a. prime meridian

b. Tropic of Cancer

c. Tropic of Capricorn

d. equator

e. international date line

1. The latitude 23.5 degrees north of the equator.

2. The longitude line that lies approximately 180 degrees from the Greenwich Meridian.

3. The primary line of longitude.

4. The latitude 23.5 degrees south of the equator.

5. A circle drawn around the center of the earth perpendicular to its axis of rotation.

Match each of the following terms with the appropriate description:

a.	continental shelf	6.	A gentle slope at the base of a steep continental shelf.
b.	continental rise	7.	Region of the ocean floor where old crust sinks to the earth's core and is recycled.
c.	ridge system	8.	The submerged edge of a continent.
d.	rift valley	9.	A long mountain range that forms along cracks produced by erupting magma.
e.	subduction zone	10.	An area of high volcanic activity that runs along the length of a midocean ridge.

Match each of the following terms with the appropriate description:

a.	seamount	11.	A region where the lithosphere splits, separates, and moves apart as new crust is formed.
b.	abyssal plain	12.	A steep-sided formation that rises sharply from the ocean bottom.
c.	rift zone	13.	A linear region of unusually irregular ocean bottom that runs perpendicular to ocean ridges.
d.	convergent plate boundary	14.	A flat expanse at the bottom of an ocean basin.
e.	fracture zone	15.	A region where plates move toward each other and old lithosphere is destroyed.

Match the following terms with the correct definition:

a.	mantle	16.	Molten rock.
b.	lithosphere	17.	The region of mantle that lies below the earth's crust.
c.	asthenosphere	18.	The thickest layer of the earth and the one that contains the most material.
d.	crust	19.	The part of the earth composed of the crust and upper mantle.
e.	magma	20.	The outermost layer of the earth.

Short Answer

1. What was the earth's early atmosphere like, and how did it influence the evolution of cells?

2. What evidence supports the theory of plate tectonics?

3. What processes are responsible for the formation and sculpting of the continental shelves?

4. What is a seamount, and how is it formed?

5. Explain how the oceans were originally formed.

6. Explain how time can be used to help determine a ship's longitude.

7. Explain how biogenous sediments are formed.

8. Describe how the continents are thought to move apart.

9. What is the global positioning system?

Thinking Critically

1. How would you expect continental drift to affect the distribution of bottom-dwelling marine organisms?

2. Would you expect to find more actively swimming fish species above the continental shelves or in the large expanses of open sea? Explain.

3. While doing research along a coastal area, you discover that the bottom sediments are predominantly calcareous ooze. What does this imply about the local conditions?

SUGGESTIONS FOR FURTHER READING

Clift, P., 2004. Moving Earth and Heaven, *Oceanus* 42(2): 91–94.

Detrick, R., 2004. Motion in the Mantle, *Oceanus* 42(2): 6–12.

Dick, H. J., 2004. Earth's Complex Composition, *Oceanus* 42(2): 36–39.

Glatzmater, G. A., and P. Olson. 2005. Probing the Geodynamo, *Scientific American* 292(4): 50–57.

Gurnis, M., 2001. Sculpting the Earth from Inside Out, *Scientific American* 284 (3): 40–47.

Keleman, P. B., 2004. Unraveling the Tapestry of the Ocean Crust, *Oceanus* 42(2): 40–43.

Keleman, P. B., 2009. The Origin of the Land Under the Sea, *Scientific American* 300(2): 42–47.

Have You Wondered?

1. How the physical characteristics of ocean water affect marine organisms? (4-1)

2. What role ocean currents play in the distribution of marine organisms? (4-4)

3. How waves are formed? (4-6)

4. Why the western coast of South America is such a productive fishing area? (4-5)

5. What causes tides? (4-7)

Water, Waves, and Tides

LIVING CELLS CONTAIN about two-thirds water by mass and have a concentration of salts similar to that of seawater. The similarity between the salt concentration in cells and seawater reflects the fact that life evolved in the ocean. From the beginning, water has played a central role in the evolution of living organisms on this planet and continues to play a vital role in the maintenance of life. Marine organisms are constantly influenced by water and the forces associated with it, such as waves and tides. In this chapter you will learn about the characteristics of water, especially seawater, that make it unique and vital to life. You will also learn about currents, waves, and tides so that you can better understand the role these forces play in the biological communities that we will study in later chapters.

4-1 Nature of Water

Not only is water necessary for life, but it is also the most abundant component of living things. For instance, most marine organisms are between 70% and 80% water by mass. By comparison, terrestrial organisms are approximately 66% water by mass. The importance of water to living organisms is related to its unique physical and chemical properties (Table 4-1).

PHYSICAL PROPERTIES OF WATER

Water is an excellent **solvent** (medium for dissolving other substances). It has a high boiling point and freezing point compared with similar chemical compounds. It is denser in its liquid form than it is as a solid. Water helps to support marine organisms by buoyancy and provides a medium for the different chemical reactions that are necessary for life. Many of these physical properties are related to the structure of the water molecule itself.

Structure of a Water Molecule

Water molecules are composed of two atoms of hydrogen bonded to one atom of oxygen. Chemists refer to the water molecule as being **polar** because its different parts have different electrical charges. Although the water molecule as a whole is electrically neutral, the electrons that form the chemical bond between the hydrogen and oxygen atoms are attracted more toward the large, positively charged oxygen atom. This causes the oxygen end of the molecule to have a slightly different charge than the two hydrogen ends (**Figure 4-1a**). The oxygen atom carries a slight negative electrical charge, and, by the same token, the hydrogen atoms carry a slight positive electrical charge.

A **solvent** is a medium for dissolving other substances.

A **polar** molecule is a molecule in which different parts of the molecule have different electrical charges.

Freezing Point and Boiling Point

Because water molecules are polar, they have a tendency to come together and form hydrogen bonds with each other. **Hydrogen bonds** are weak attractive forces that occur between the slightly positive hydrogen atoms of one molecule and the slightly negative oxygen ends of a neighboring molecule **(Figure 4-1b)**. The ability of water molecules to form hydrogen bonds accounts for many of water's special properties. For instance, the relatively high boiling point of water (100°C) reflects the substantial amount of energy that is required to overcome the attractive forces of the hydrogen bonds so that individual water molecules can separate from each other and attain a gaseous state. The comparatively high freezing point of water (0°C) is also due to the presence of hydrogen bonds. Because the molecules have a natural attraction to each

Hydrogen bonds are weak attractive forces that occur between the slightly positive hydrogen atoms of one molecule and slightly negative atoms of another molecule.

TABLE 4-1 PHYSICAL PROPERTIES OF WATER

Boiling point	100°C
Freezing point	0°C
Heat capacity	1.00 cal/g/°C
Density (at 4°C)	1.00 g/cm³
Latent heat of fusion	80 cal
Latent heat of vaporization	540 cal

© Cengage Learning

other, less energy is required to fix them into position so that they will form a solid. When water does freeze, the molecules move away from each other because of repulsive electrical forces between electrons of neighboring atoms. As a result, solid water (ice) is less dense than liquid water, and it floats **(Figure 4-1c)**. This characteristic is significant

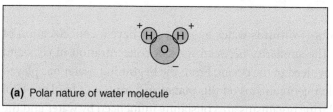

(a) Polar nature of water molecule

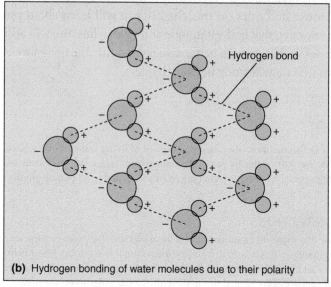

Hydrogen bond

(b) Hydrogen bonding of water molecules due to their polarity

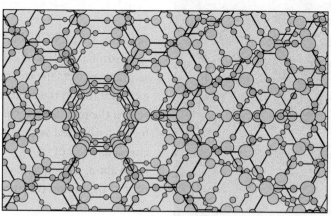

(c) Structure of water molecules in a solid state (ice)

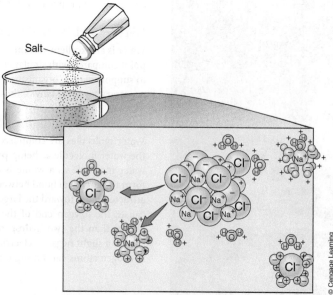

Salt

(d) Salt crystals dissolving in water

© Cengage Learning

FIGURE 4-1 WATER MOLECULES. **(a)** A water molecule contains two atoms of hydrogen and one atom of oxygen. Water molecules are polar because the hydrogen atoms have a slight positive charge and the oxygen has a slight negative charge. **(b)** The polar nature of water molecules causes them to be attracted to each other and form hydrogen bonds. The hydrogen bonding in water accounts for its relatively high freezing point and boiling point. **(c)** Because of the polar nature of water molecules, when water freezes the molecules spread out. As a result, solid water (ice) is less dense than liquid water and floats. **(d)** The ions in salts are attracted to polar water molecules and readily move into solution. In solution, the ions are surrounded by water molecules, so there is less chance of their being attracted to each other and reforming crystals.

because a body of water freezes at its surface, forming an insulating layer, and the water below stays liquid. This keeps the oceans from freezing solid and allows living organisms to survive, even when the ocean surface is frozen.

Water As a Solvent

The polar nature of water molecules also accounts for water's remarkable solvent properties. Other polar substances, such as salts, readily dissolve in water. Salt crystals are made up of individually charged particles called **ions**, which are attracted to the polar water molecules. As ions separate from the crystal, the crystals dissolve in the water. The attraction of polar water molecules for charged ions then helps to keep the ions in solution once the crystals have dissolved **(Figure 4-1d)**. Water is not able to dissolve molecules that are nonpolar, such as oil and petroleum products.

Cohesion, Adhesion, and Capillary Action

The hydrogen bonds in water cause water molecules to be very cohesive; that is, they stick together. This cohesiveness gives water a high **surface tension**. The water molecules at the surface of the liquid have a greater attraction for other water molecules than they do for molecules in the air. As a result, the molecules at the surface form a tight, closely packed layer held by the hydrogen bonds of the water molecules beneath the surface. The high surface tension contributes to the high boiling point of water, because water must take in enough energy (the **latent heat of evaporation**) to overcome the surface tension before the molecules move into the vapor state. Many small organisms such as water striders (*Halobates*) take advantage of water's surface tension to provide them support **(Figure 4-2)**.

Water is also attracted to the surface of objects that carry electrical charges. This property is called **adhesion** and accounts for the ability of water to make things wet. It also accounts for the ability of water to rise in narrow spaces, a property known as **capillary action**. Water adheres to the surface of sediment particles and can become trapped in the spaces between them. The smaller the sediment particles, the smaller the spaces between them, which increases the movement of water by capillary action. Microscopic organisms that live in the spaces between sediment particles rely on adhesion and capillary action to supply the water necessary to support life **(Figure 4-3)**. Without adhesion and capillary action, sediments could not trap or move sufficient amounts of water for these organisms to survive.

Specific Heat

Because the presence of hydrogen bonds greatly restricts the movement of water molecules, it takes significantly more energy to raise the temperature of 1 gram of water 1°C than other common liquids. The amount of heat energy required to change the temperature of 1 gram of a substance 1°C is called the **specific heat**, or **thermal capacity**. Large bodies of water such as oceans maintain a more or less constant temperature because of the relatively large amount of energy they can absorb without changing temperature. Because the temperature of the ocean is relatively stable, marine organisms, with the exception of intertidal plants and animals, have not evolved mechanisms for adapting to rapid fluctuations in temperature, as have terrestrial plants and animals.

Water and Light

The ability of light to penetrate water has important consequences for marine life. As you learned in Chapter 2, light (solar radiation) powers the process of photosynthesis, which produces food for most life in the

> An **ion** is a particle that carries an electrical charge.
>
> **Surface tension** is the cohesive force that holds molecules together at the surface of a liquid.
>
> The **latent heat of evaporation** is the amount of energy required to move a substance from the liquid state to the vapor state.
>
> **Adhesion** is the property of water that causes it to be attracted to the surface of objects that carry electrical charges.
>
> **Capillary action** is the ability of water to rise in narrow spaces.
>
> **Specific heat** is the amount of heat energy required to change the temperature of 1 gram of a substance 1°C.
>
> **Thermal capacity** is a synonym for specific heat.

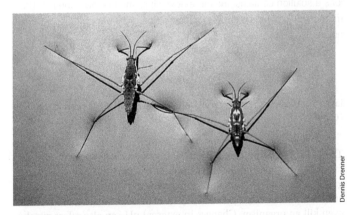

FIGURE 4-2 SURFACE TENSION. Since water molecules have a greater attraction for each other than for molecules in the air, they form a tight layer at the surface that is able to support small organisms, such as this water strider.

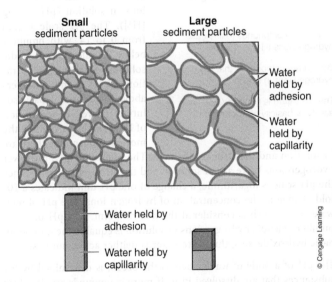

FIGURE 4-3 ADHESION AND CAPILLARY ACTION. Water moves through the tiny spaces between sediment particles and adheres to the surface of the particles, supplying moisture for numerous organisms that live in bottom sediments. The smaller the particles, the smaller the spaces between them and the more capillary water they can hold.

sea and is required for vision. However, only a small portion of the light that reaches the sea drives photosynthesis. Much of the solar radiation is reflected by the water into the atmosphere, and wave action increases the amount of light that is reflected.

The ability of water to selectively absorb certain wavelengths of light is another notable physical property of water. Low-energy red, orange, and yellow light is quickly absorbed, whereas high-energy blue, green, and violet light can penetrate deeper. This explains why objects appear more blue and green when viewed underwater. Clear water, such as the water around tropical coral reefs, appears blue because blue light predominates. Coastal seas appear greener because the water contains more sediments and plankton that absorb more of the blue, allowing the wavelengths of green light to dominate. In the turbid water of the Atlantic Ocean, almost 65% of the light that enters the water is absorbed within the first meter (3.3 feet). In clear ocean water, such as the tropical South Pacific Ocean, long-wavelength, low-frequency infrared light is absorbed within the first few centimeters, whereas short-wavelength, high-energy ultraviolet light can penetrate as deep as 200 meters (660 feet), the limit of the photic zone. In most clear water, however, less than 1% of the light entering the ocean can penetrate deeper than 100 meters (330 feet). The absorbed light energy is converted to heat that helps to warm the water. Heat energy, however, cannot power photosynthesis, so photosynthetic organisms are restricted to the upper, sunlit surface waters where there is sufficient radiant energy.

CHEMICAL PROPERTIES OF WATER

Chemical compounds are classified as acids or bases on the basis of how they ionize in water. **Acids** are compounds that release hydrogen ions (H^+) when they are added to water, whereas **bases** can bind hydrogen ions and remove them from a solution. To indicate whether a solution is acidic or basic, biologists use a shorthand notation called the **pH scale (Figure 4-4)**. The pH, an indicator of the number (concentration) of hydrogen ions in a volume of a solution, is equal to the negative logarithm of the concentration of hydrogen ions in solution (pH = –log [H^+]). The pH scale ranges from 0 to 14, with numbers below 7 indicating an acidic solution (higher concentration of H^+) and numbers above 7 indicating a basic solution (lower concentration of H^+). The lower the pH, the more hydrogen ions there are in solution and the stronger the acid. The higher the pH, the fewer hydrogen ions there are in solution and the stronger the base. Since the pH scale is logarithmic, a change of one pH unit indicates a 10-fold change in the concentration of hydrogen ions. The pH of pure water is 7, which is considered the neutral point. At a pH of 7, the number of acidic hydrogen ions in solution is equal to the number of basic hydroxide ions; thus pure water is neither acidic nor basic.

The pH of a body of water, such as the ocean, is determined by the substances that are dissolved in it. If more compounds are dissolved that release hydrogen ions, the water will be acidic. On the other hand, if there are more ions that can bind to hydrogen ions and remove them from the solution, the water will be basic. Seawater is slightly basic, with an average pH of 8. The reason for this is that seawater contains a

Acids are compounds that release hydrogen ions (H^+) in solution.

Bases are compounds that bind hydrogen ions in solution.

The **pH scale** is a measure of the acidity or basicity of a solution.

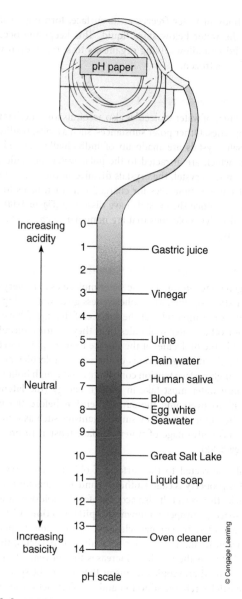

FIGURE 4-4 pH SCALE. The pH scale is a convenient way of expressing the acidity of a solution. A pH of 7 is neutral. The more hydrogen ions a solution contains, the more acidic it is and the lower the pH. The more hydroxide ions a solution contains, the more basic it is and the higher the pH.

large amount of ions, mainly bicarbonate (HCO_3^-) and carbonate (CO_3^{2-}), that can bind hydrogen ions and remove them from solution.

The pH of an organism's internal and external environment is a vital factor in determining the distribution of marine life. Changes in an organism's internal pH can affect the function of vital molecules, such as enzymes, and interfere with metabolism. Such a change could injure or even kill an organism. Changes in external pH can also affect metabolism and interfere with growth. For example, hard corals cannot grow in water that is acidic because the low pH inhibits their ability to form external skeletons.

MARINE BIOLOGY & THE HUMAN CONNECTION

Carbon Dioxide and Ocean pH

The increased amount of greenhouse gases, such as carbon dioxide, in the atmosphere is affecting more than just ocean temperature, climate, and weather patterns; it is also affecting the oceans pH. Historically, the ocean absorbs about one-third of the carbon dioxide that enters the atmosphere from human activity. With the increase in the atmospheric levels of carbon dioxide and other greenhouse gases, more of these gases are dissolving in ocean water, causing the ocean to become more acidic. Current estimates set the amount of carbon dioxide being absorbed by the ocean at 22 tons per day. This has caused the ocean to become 30% more acidic than it was at the beginning of the Industrial Revolution, and some scientists believe it could be more than 150% more acidic by the end of this century. Dr. Steven Emerson at the University of Washington even thinks that it is possible that

the pH of the ocean could be lowered permanently.

The change in ocean pH affects all marine organisms, from plankton to whales. The increased acidity of the water can erode the shells of animals such as oysters, clams, and pteropods, and interfere with these animals' ability to grow and repair their shells. The low pH can also weaken the external skeletons of important crustaceans such as crabs, shrimp, and krill. Without their protective coverings most of these animals would die. As seawater becomes more acidic, colonies of coral have a more difficult time forming their protective skeletons and may completely disappear, causing the collapse of the marine environment's most productive community. Animals such as squid are very sensitive to lower pH. The increased acidity affects the circulation of blood in their bodies and their ability to

exchange gases with the water, leading to increased mortality rate.

Krill and pteropods are a major source of food for juvenile fishes, such as salmon, herring, pollock, cod, and mackerel. They are also an important food source for many whales. Penguins and some seals feed on krill, and many marine carnivores feed on squid. By causing the death of organisms that are important links in food chains, changes in ocean pH have the potential to disrupt entire food webs.

Marine organisms would not be the only ones affected by these disastrous changes in ocean pH. Worldwide, 500 million to 1 billion people depend on fish and shellfish for their survival. The collapse of oceanic food chains and food webs could thus lead to starvation and death in human populations as well.

IN SUMMARY

- The unique physical and chemical properties of water make it a critical component of all living cells. (Have You Wondered? #1)

- Many of water's properties are due to the polar nature of the water molecules and their ability to form hydrogen bonds.

- Since large bodies of water such as the ocean can absorb large quantities of heat, the temperature of the ocean environment is relatively stable.

- Water reflects some solar radiation and selectively filters out certain wavelengths of light.

- Biologists use a chemical shorthand called the pH scale to indicate the acidity of a solution.

- The ocean's pH is about 8 because it contains large amounts of bicarbonate and carbonate ions.

4-2 Salt Water

Ocean water is referred to as salt water because of the high quantity of dissolved salts compared with freshwater. As you learned in Chapter 2, the salinity of ocean water is an environmental factor that strongly influences marine organisms as they work to maintain appropriate levels of salts and water in their bodies. The variety of salts and other substances found dissolved in seawater is also important in determining the kinds and distribution of organisms in the marine environment.

COMPOSITION OF SEAWATER

Most of the salts that are present in seawater are in their ionic form (Table 4-2). A total of six ions are responsible for 99% of the dissolved salts in the ocean: sodium (Na^+), magnesium (Mg^{2+}), calcium (Ca^{2+}), potassium (K^+), chloride (Cl^-), and sulfate (SO_4^{2-}). Other elements that are dissolved in seawater but present in concentrations of less than one part per million are called **trace elements**. The

Trace elements are elements present in seawater at concentrations of less than one part per million.

proportions of these major salts in seawater are relatively constant, and even though salinities can change, the relative proportions do not over much of the world ocean. This helps marine organisms because it makes this aspect of their environment predictable.

TABLE 4-2 MAJOR IONS IN SEAWATER

Ion	g/kg of Seawater	Percentage by Weight
Chloride (Cl^-)	19.35	55.07
Sodium (Na^+)	10.76	30.62
Sulfate (SO_4^{2-})	2.71	7.72
Magnesium (Mg^{2+})	1.29	3.68
Calcium (Ca^{2+})	0.41	1.17
Potassium (K^+)	0.39	1.10
Bicarbonate (HCO_3^-)	0.14	0.40
Total		99.76

© Cengage Learning

The ions and minerals in seawater are vital to the lives of marine organisms. Snails, clams, corals, and various crustaceans, for instance, require calcium to form their shells. Even minerals that are present in trace quantities are necessary to many marine organisms. Several marine organisms can concentrate trace minerals in their tissues at much higher levels than exist in seawater. For example, kelps (*Macrocystis*) and shellfish (shrimp, oysters, and so on) are able to concentrate high levels of iodine in their tissues. Kelps have even been commercially harvested as a source of this element, an essential nutrient in the human diet.

SALINITY

Seawater is approximately 3.5% salt and 96.5% water by mass. Sodium chloride (NaCl), or table salt, is the most common salt present. Since salinity (the concentration of salt in a given volume of water) is usually expressed in either grams of salt per kilogram of water or parts per thousand (ppt, symbolically represented as ‰), seawater would have an average salinity of 35 grams per kilogram, or 35 parts per thousand. The salinity of surface seawater varies with latitude and the topographical features of an area. These variations are the result of evaporation, precipitation, freezing, thawing, and freshwater runoff from nearby land masses. As you might expect, areas of ocean that are close to the mouths of rivers (freshwater) have comparatively low salinity, whereas surface waters in subtropical areas of the world where evaporation is high and precipitation is low have comparatively high surface salinities (**Figure 4-5**).

Evaporites are salt deposits that form when saltwater becomes cut off from the ocean and the water evaporates.

The processes of evaporation and precipitation contribute heavily to the salinity of surface water in the midocean. Between 10 degrees N and 10 degrees S of the equator, rainfall is heavy and the surface water has a relatively low salinity. The regions of ocean at approximately 30 degrees N and S, the latitudes that correspond with many of the world's deserts, are characterized by evaporation exceeding precipitation. As a result,

they have relatively high salinity. Farther north and south, from 50 degrees latitude, precipitation is again heavy and the surface water is relatively less salty. Near the poles, freezing removes water from the sea, leaving the salt behind. Thus the water beneath the ice has relatively high salinity.

CYCLING OF SEA SALTS

The original sources of sea salts were rocks and other constituents of the earth's crust and interior. It is currently thought that these salts entered the ocean dissolved in water that seeped up through the ocean floor. Several processes continue to contribute salts to the ocean. Rocks previously formed on the seafloor release their ions as they are broken down by physical and chemical processes. Volcanic eruptions produce gases, such as hydrogen sulfide and chlorine, which then dissolve in rainwater and enter the ocean with precipitation. River water carries large amounts of ions, formed by the weathering of rocks, into the sea.

Geologists believe that the composition of the ocean has been the same for about the last 1.5 billion years, despite runoff from the land annually adding 2.5×10^{12} kilograms (about 250 million dump-truck loads) of salt to the sea. For the salinity of the ocean to remain the same over time, the amount of salt added to the oceans by runoff must be balanced by salt removal. Salt is removed from seawater in several ways. Some salt ions can react with each other to form insoluble complexes that precipitate to the ocean floor. Alternatively, when waves strike a beach, the sea spray is carried off by air currents and forms a layer of salt on surrounding rocks, land, buildings, and vehicles.

During the development of the earth, shallow extensions of the ocean became cut off from the sea, and the water evaporated. This process produced salt deposits called **evaporites** (the salt flats of southern California are an example). Similar situations exist today when small, shallow bodies of water become separated from the ocean and the water evaporates, leaving the salt behind. Biological processes also play a role in salt cycling. Certain ions absorbed by marine organisms are later returned to the environment when the organisms excrete them or die and decompose. Salts concentrated in the body tissues of marine organisms

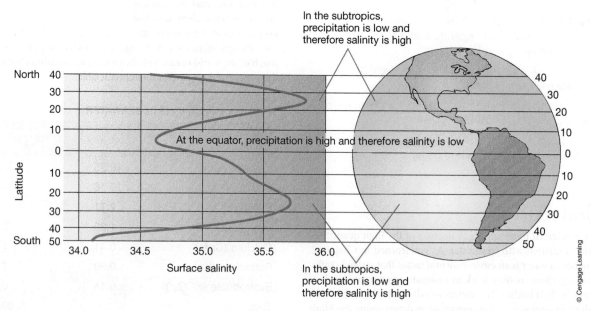

FIGURE 4-5 OCEAN SALINITY. The salinity of the surface water of the ocean varies with latitude as a result of regional differences in evaporation and precipitation.

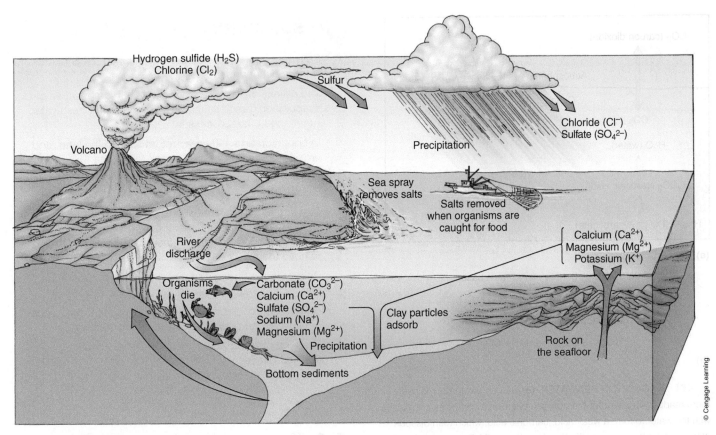

FIGURE 4-6 SEA SALT CYCLING. Salts are constantly being added to and removed from seawater by a variety of processes.

are permanently removed if the organisms are harvested for food. The process responsible for removing the largest amount of salt, however, involves ions sticking to the surface of fine particles, a process known as **adsorption**. These particles then settle to the bottom and become trapped in the sediment. Clay particles formed from weathered rock and carried to the oceans by rivers attract ions and attach them to their surfaces by the process of adsorption. Ions and minerals deposited in this manner in the ocean's sediments can later be moved by geological processes to areas well above sea level, where weathering can remove the minerals and return them to the oceans **(Figure 4-6)**.

GASES IN SEAWATER

The gases of the atmosphere, primarily oxygen (O_2), carbon dioxide (CO_2), and nitrogen (N_2), are also found in seawater **(Table 4-3)**. These

TABLE 4-3 GASES FOUND IN SEAWATER

Gas	Percentage by Volume in Atmosphere	Percentage by Volume in Surface Seawater	Percentage by Volume in Ocean Total
Nitrogen (N_2)	78.08	48	11
Oxygen (O_2)	20.99	36	6
Carbon dioxide (CO_2)	0.03	15	83
Other gases	0.95	1	

© Cengage Learning

gases dissolve at the surface of the sea from the atmosphere and are introduced into the water by biological processes. All of the gases are distributed throughout the water by mixing processes and currents.

Gases from Biological Processes

Some bacteria, phytoplankton, algae, and plants use carbon dioxide in the process of photosynthesis, producing oxygen as a by-product. Most living organisms require oxygen and produce carbon dioxide. The process of decomposition, which requires respiration by bacteria or other organisms, also uses oxygen and produces carbon dioxide. Oxygen as a by-product of photosynthesis is added to seawater only near the surface. Just below the sunlit surface waters is the **oxygen-minimum zone**, an area in which oxygen is depleted by resident animal life but not replaced by photosynthesis. Carbon dioxide, on the other hand, is added at all depths by the processes of decomposition and respiration. The levels of these two gases in seawater have a profound effect on the kinds of organisms that can live in a given area and the number of organisms a region can support.

Adsorption is the process whereby ions stick to the surface of fine particles.

The **oxygen-minimum zone** is the area just below the sunlit surface waters in which oxygen is depleted by the resident animal life but not replaced by photosynthesis.

Solubility of Gases in Seawater

The proportion of oxygen and carbon dioxide in surface seawater is greater than it is in the atmosphere. The amount of nitrogen in seawater is less. Oxygen is far more soluble in seawater than nitrogen, and carbon

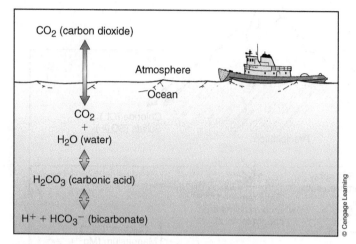

(a)

In acid, bicarbonate can act as a base.

$$H^+ + HCO_3^- \longrightarrow H_2CO_3 \longrightarrow H_2O + CO_2$$

In base, bicarbonate can act as an acid.

$$HCO_3^- + OH^- \longrightarrow H_2O + CO_3^{2-}$$

(b)

FIGURE 4-7 GASES IN SEAWATER. (a) Carbon dioxide combines more readily with seawater than oxygen because it combines chemically with the water to form a weak acid, carbonic acid, which then dissociates into a hydrogen ion and a bicarbonate ion. **(b)** Bicarbonate ions act as a buffer to help maintain the pH of seawater. If there is excess acid, it can combine with the hydrogen ion to form carbonic acid, which will break down to carbon dioxide and water. If the water is too basic, it can donate a hydrogen ion to form a carbonate ion and water.

dioxide is more soluble than oxygen. Carbon dioxide dissolves in seawater more readily than oxygen because it combines chemically with the water to form a weak acid called carbonic acid, which then dissociates into hydrogen ions and bicarbonate ions **(Figure 4-7a)**. The chemical combination of carbon dioxide with water allows the seawater to hold much more carbon dioxide than the other two gases, which do not chemically combine. The amount of gas that water can hold depends on the temperature, salinity, and pressure of the water. For instance, cold water holds more gas than warm water, and more gas will dissolve if the salinity of the water is low or the gas pressure is high. For these reasons, warm tropical waters contain less oxygen than the cold water of polar seas.

Role of Bicarbonate as a Buffer

The bicarbonate that is formed from the solution of carbon dioxide plays the vital role of a **buffer**, a substance that can maintain the pH of a solution at a relatively constant point. Bicarbonate ions are buffers because they remove excess hydrogen ions if the water becomes too acidic or release hydrogen ions if the water becomes too basic **(Figure 4-7b)**. By counteracting changes in the hydrogen ion concentration, the bicarbonate helps to maintain the pH of the water at a more or less constant value, maintaining a stable environment for marine organisms. The large amount of bicarbonate ions in the ocean is the main reason seawater has a slightly basic pH.

A **buffer** is a substance that can maintain the pH of a solution at a relatively constant point.

IN SUMMARY

- Seawater contains a number of salts; sodium chloride is the most abundant.
- The salinity of seawater varies with latitude because of evaporation, precipitation, freezing, thawing, freshwater runoff, and other processes.
- Evaporation and surf spray return salts to the land, and erosion processes return the salt to the sea.
- Seawater also contains gases, such as oxygen and carbon dioxide.
- Oxygen levels are highest and carbon dioxide levels lowest in the upper regions of the ocean, where physical processes and photosynthesis add and remove the two gases.
- Deeper in the ocean, there is more carbon dioxide as the result of respiration and decay processes.
- Carbon dioxide is more soluble in seawater than oxygen because it can chemically combine with the water.
- The bicarbonate ion that forms from the chemical reaction between carbon dioxide and water is a buffer and accounts for seawater's slightly basic pH.

4-3 Ocean Heating and Cooling

As mentioned earlier in this chapter, it takes a large amount of heat energy to change the temperature of a body of water as big as an ocean. The ability of the ocean to store and release heat has substantial implications for life on land as well as life in the sea.

EARTH'S ENERGY BUDGET

The earth and the atmosphere continually receive energy from space, primarily in the form of radiant energy. They return energy to space in the form of heat. For the earth to maintain a relatively constant annual average temperature, the amount of heat gained has to be roughly equal to the amount lost. An imbalance in these processes can produce global warming or global cooling.

Energy Input

The sun's radiant energy is responsible for warming the earth's surface. The latitudes between the Tropic of Cancer and the Tropic of Capricorn receive the greatest amount of radiant energy, the middle latitudes receive moderate amounts, and the poles receive the least. This difference in the amount of radiant energy striking different parts of the earth's surface is caused by the earth's spherical shape and its atmosphere **(Figure 4-8)**. The greatest amount of radiant energy is received where the sun's rays strike the earth at a right angle, around the equator. As you move to the north or the south, the angle of the sun's rays relative to the earth's surface increases because of the earth's curved surface. Consequently, at the middle latitudes and the poles, the same amount of sunlight falls on a larger area of the earth. Thus these regions receive proportionately less radiant energy. The earth's atmosphere also absorbs some of the radiant energy before it strikes the earth. Atmospheric

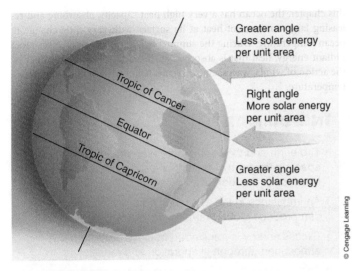

FIGURE 4-8 DISTRIBUTION OF SOLAR ENERGY. Because of the earth's curved surface, more radiant energy reaches the tropics than the temperate regions or the regions surrounding the poles.

absorption causes the amount of radiant energy reaching the earth's surface to decrease with increasing latitude. The decrease is due to the longer path through the atmosphere that the sun's rays must travel to reach the earth's surface at higher latitudes. The combined effect of the earth's curved surface and atmosphere is a larger amount of radiant energy being received and transformed to heat around the equator, less in the temperate regions, and least at the poles.

Energy Output

Because the earth's temperature is not constantly increasing over the short term, heat must be lost as well as gained. Some incoming solar radiation is reflected into space by the atmosphere and the surface of the earth and plays no role in heating the earth's surface. Of the remaining energy, most is absorbed by the earth's surface, and the rest is absorbed by the atmosphere. Ultimately, an equivalent amount of heat energy is reradiated back into space. The atmosphere loses more energy back to space than it gains from the sun, whereas the earth loses less energy back to space. In this way the atmosphere cools, and the surface of the earth warms.

To maintain the earth's temperature, the excess energy must be transferred from the earth to the atmosphere, and this is accomplished by evaporation and radiation **(Figure 4-9)**. Energy from the earth's surface produces water vapor from liquid water by evaporation. This energy is then transferred to the atmosphere when the vapor condenses to form rain or freezes to form hail or snow. The heat that radiates from the earth's surface is absorbed by the water content of the atmosphere. Approximately two-thirds of the heat transferred to the atmosphere is transferred by evaporation, and the other one third is transferred by radiation and other processes. The accumulation in the atmosphere of **greenhouse gases** such as carbon dioxide, methane, and chlorofluorocarbons (CFCs) can prevent heat energy from radiating back to space. Like the glass in a greenhouse, these gases tend to block the escape of heat energy, causing an increase in the earth's average temperature known as global warming. The increase in average sea temperature that has occurred because of global warming is pushing some marine organisms into their stress zones, discussed in Chapter 2.

Greenhouse gases are gases that accumulate in the atmosphere and prevent heat energy from radiating back to space.

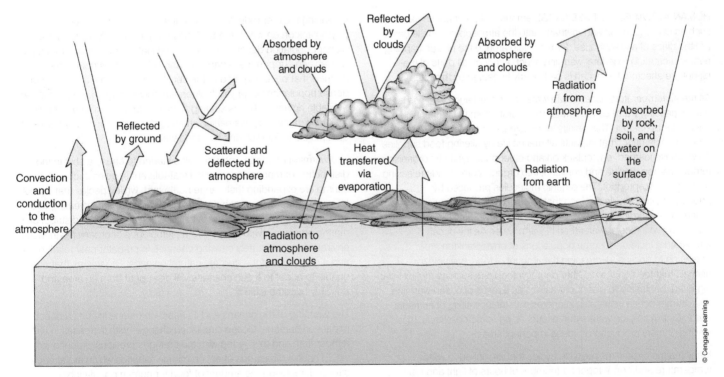

FIGURE 4-9 THE EARTH'S ENERGY BUDGET. Only about 47% of incoming solar radiation is absorbed by the earth's surface. The remainder is absorbed by the atmosphere or reflected into space. Heat is lost from the surface of the earth by the processes of evaporation, radiation, conduction, and convection. The width of the arrows in the diagram indicates the relative amount of energy being transferred.

SEA TEMPERATURE

When considering the amount of heat energy in an area of ocean, we must take into account several factors, such as the amount of energy absorbed at the surface, the loss of heat energy by evaporation, the transfer of energy into or out of the area by ocean currents, the warming or cooling of the overlying atmosphere by heat from the surface, and heat lost from the ocean back to space by radiation. Since all of these processes change with time, regions of the ocean show both daily and seasonal variations in temperature **(Figure 4-10)**. More heat is gained than lost in the region around the equator, whereas more heat is lost than gained at the higher latitudes. Winds and ocean currents are responsible for removing the excess heat from the tropics and moving it to the higher latitudes, thereby maintaining the overall pattern of surface temperatures across the earth.

The amount of solar radiation reaching the earth also undergoes an annual cycle of seasonal variations. These variations are most pronounced between the latitudes 40 degrees and 60 degrees N and 40 degrees and 60 degrees S because the angle of the sun's rays changes so dramatically at these latitudes with changes in season. This seasonal change in radiant energy produces seasonal changes in the ocean's surface temperature. Between the Tropic of Cancer and the Tropic of Capricorn, the amount of solar radiation remains relatively constant, and the surface temperature of the ocean also remains fairly uniform. As the sun makes its annual migration between the Tropic of Cancer and the Tropic of Capricorn, its rays cross this area twice, producing only a small, semiannual variation in the intensity of solar radiation. As mentioned previously in this chapter, the ocean has a very high heat capacity, absorbing and releasing large amounts of heat at its surface with very little change in ocean temperature. During the summer, when the ocean receives more radiant energy, heat that is absorbed at the surface is moved deeper by the action of winds, waves, and currents, helping to stabilize the surface temperature.

IN SUMMARY

- The earth and the atmosphere receive energy from space in the form of radiant energy and return energy to space in the form of heat.

- An imbalance in the amount of heat gained and lost can produce global warming or cooling.

- Excess energy is transferred from the earth to the atmosphere through evaporation and radiation.

- The accumulation of greenhouse gases can prevent heat from radiating back to space.

- Regions of the ocean show both daily and seasonal variations in temperature.

- Winds and ocean currents are responsible for removing the excess heat from the tropics and moving it to the higher latitudes, so that the overall pattern of surface temperatures across the earth is maintained.

IMPACT BUBBLE

HUMAN ACTIVITIES RELEASE LARGE amounts of greenhouse gases, such as carbon dioxide and methane, into the atmosphere. These gases, like the glass of a greenhouse, form a shield around the planet, trapping heat and creating a general warming effect. Although all parts of the planet are affected, the ocean is particularly impacted by this change.

As the air temperature rises, ocean water is warmed and becomes less dense, causing warm surface water to separate from the nutrient-filled cold water beneath it. This results in a decrease in primary production and a domino effect that impacts all marine life by altering food supplies. As you have learned, phytoplankton and other photosynthetic organisms remove carbon dioxide and convert it into organic compounds releasing oxygen as a by-product. The organic molecules produced by photosynthesis provide the food that feeds almost every marine organism. According to a recent NASA study, phytoplankton is more likely to thrive in cooler water than in warm water. As the ocean temperature increases, some populations of phytoplankton are beginning to decline, decreasing the amount of food available to animals that feed on them. Other photosynthesizers such as multicellular algae are beginning to disappear as well as a result of ocean warming. The warmer ocean water also interferes with the upwelling of nutrients into the sunlit surface waters that occurs along some coastlines, thus limiting primary production in these important areas.

Many marine organisms have life cycles that are linked to water temperature and/or photoperiod (number of hours of light and dark). Some phytoplankton, for instance, are starting their annual reproductive cycles earlier in the season due to the warming water. Other marine organisms whose life cycles are linked to photoperiod, however, are still beginning their reproductive cycles at their usual time. Since the phytoplankton on which many of these organisms rely for food reach higher population densities earlier than usual, there is not as much food available to support the organisms whose life cycle is not altered. Animals that once migrated to the sunlit surface waters to feed on the dense populations of phytoplankton are now finding an area lacking in available nutrients. This is starting to drive some organisms to begin their reproductive cycles earlier, creating an asynchronous pattern that disrupts entire food chains and food webs.

The warming of ocean water has also started to cause a shift in the distribution of marine organisms. Heat-tolerant species such as some shrimp are expanding their range northward, while species that do not tolerate warmer water such as clams and flounder are retreating to a smaller range where the water is still cool. This shift in the distribution of marine organisms will lead to new assemblages of organisms in these environments, ultimately causing changes in predator-prey relationships and a disruption of the delicate natural balance. Organisms that are unable to adapt to these changes will disappear from an area and possibly become extinct.

The warming of the oceans and its effect on marine life has a direct impact on humans. Ocean circulation changes with increases in air temperature and this, along with the changes previously mentioned, would have a disastrous effect on marine fisheries which are already stressed, reducing the amount of food for human populations. Decreasing the amount of greenhouse gases released into the atmosphere will go a long way in protecting the ocean, marine life, and humans from these potentially drastic changes.

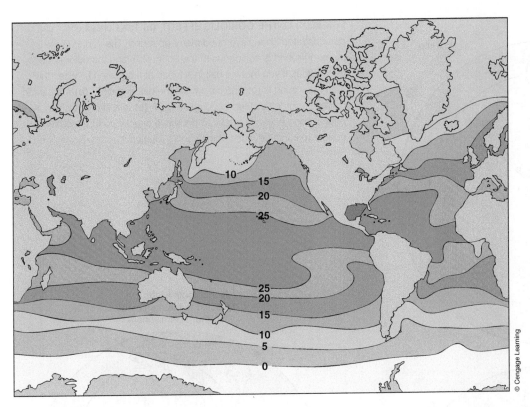

FIGURE 4-10 AVERAGE SURFACE TEMPERATURES OF THE WORLD OCEAN. The surface temperature of the sea varies from one region to the next as the result of differences in the amount of solar radiation received, loss of heat via evaporation, and heat energy transferred by ocean currents and the atmosphere.

4-4 Winds and Currents

Approximately 10% of ocean water is involved in surface currents. Driven by winds, the currents move water in the upper 400 meters (1,320 feet) of the ocean. These currents carry plankton in predictable paths through the surface waters, changing the distribution of planktonic food and dispersing planktonic larvae. Surface currents also influence the distribution of nekton, because predators follow the plankton and plankton feeders on which they prey.

WINDS

Winds result from horizontal air movements that are caused by factors such as temperature and density. As air heats, its density decreases and it rises; as it cools, its density increases and it falls toward the earth. The density of air is determined by its temperature, water vapor content, and atmospheric pressure. Air density decreases when it is warmed, when its vapor content increases, or when the atmospheric pressure decreases. Air becomes denser when it is cooled, its vapor content decreases, or the atmospheric pressure increases. These forces produce two sets of winds that circulate in opposite directions: winds at the earth's surface and upper winds such as the jet stream.

Wind Patterns

As noted previously, the region of the earth around the equator receives large amounts of the solar energy that heats the air and water. This combination of warm air with a high water vapor content produces less dense air that rises. At the poles, on the other hand, the air is cooling, moisture has been removed by precipitation, and there is increased atmospheric pressure. This combination of factors causes the air to become denser and sink. The denser air from the poles is then sucked toward the equator to replace the less dense air that has moved away.

These processes establish a basic pattern of upper air flow (winds) from the equator toward the north in the Northern Hemisphere and from the equator toward the south in the Southern Hemisphere (**Figure 4-11**).

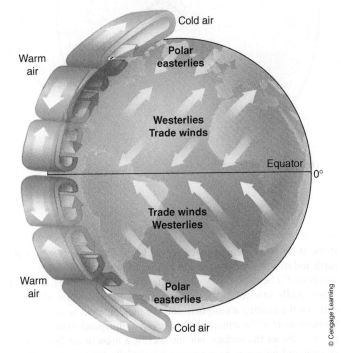

FIGURE 4-11 NORTH–SOUTH AIR FLOW. As warm air from the equator moves north and south, it displaces colder, denser polar air that then flows toward the equator. This produces the basic pattern of upper air currents.

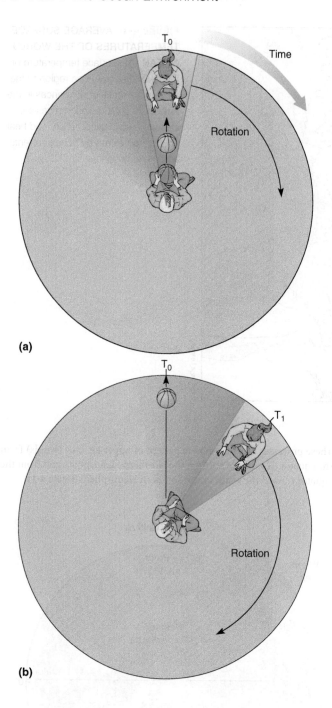

(a)

(b)

FIGURE 4-12 THE CORIOLIS EFFECT. (a) The Coriolis effect can be demonstrated on a playground merry-go-round. The center of the merry-go-round is analogous to the South Pole and the outer edge to the equator. If you throw a ball to a friend at time zero (T_0) when the merry-go-round is rotating clockwise, the ball at time T^1 (b) appears to curve to your left instead of going straight. (c) In a similar fashion, wind patterns and ocean currents do not move directly north or south but appear to curve because of the earth's rotation.

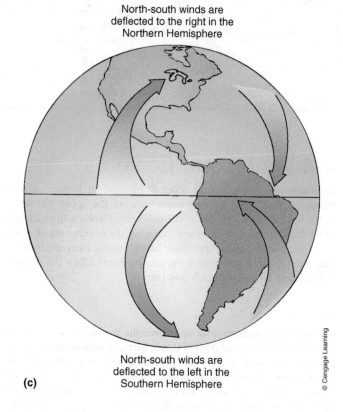

North-south winds are deflected to the right in the Northern Hemisphere

North-south winds are deflected to the left in the Southern Hemisphere

(c)

Coriolis Effect

If the atmosphere were attached tightly to the earth's surface, and the earth and its atmosphere moved in unison, these north and south movements would be the primary air movements. The atmosphere, however, is not rigidly attached to the earth's surface, and the amount of friction between the moving atmosphere and the moving earth is so small that the movement of the atmosphere is somewhat independent of the movement of the earth's surface. For instance, if a mass of air lies over the equator and appears to be stationary, it is actually moving eastward with the rotating earth.

As the earth rotates on its axis, a point at the equator moves faster than a point at a higher latitude. Both points make one full revolution per day, but the point at the equator has to travel a longer distance; therefore, it must be moving faster. For this reason, air moving north or south away from the equator toward the poles moves progressively faster than the earth beneath it. This causes the track the air mass follows to appear to curve relative to the earth's surface. In the Northern Hemisphere this effect causes the air to deflect to the right of the direction of the air movement, and in the Southern Hemisphere the deflection of the air is to the left of the direction of air flow.

This apparent deflection of the path of the air is called the **Coriolis effect** (Figure 4-12a and b). The Coriolis effect is named after Gaspard Coriolis, who developed a mathematical model for deflection in frictionless motion when the motion is referenced to a rotating body. The Coriolis effect is applicable to any rotating body, and describes not just air movements but also movement of surface water and currents.

Surface Wind Patterns

Because of the Coriolis effect, air that rises at the equator does not move straight to the north or the south but is deflected as it moves **(Figure 4-12c)**. At 30 degrees N and S, the equatorial air sinks and either moves back along the surface of the water to the equator or to 60 degrees N or S. The air that reaches the poles cools and sinks and begins to move back toward the equator. As the air flows back toward the equator, it warms and picks up water vapor, and by the time it reaches 60 degrees N or S, it rises again. The net result of these processes is the division of the earth's surface into three convection cells in each hemisphere.

Between the equator and 30 degrees N, the surface winds are deflected to the right, producing the **northeast trade winds**. In the Southern Hemisphere, the winds are deflected to the left, producing the **southeast trade winds**. (Winds are designated with reference to the direction from which they are coming, not to which they are blowing.) Between 30 degrees N and 60 degrees N, the deflected surface air produces winds that blow from the south and west, producing the **westerlies**. In the Southern Hemisphere, the westerlies blow from the north and the west. Between 60 degrees N and the North Pole, the winds blow from the north and east, forming the **polar easterlies**. Between 60 degrees S and the South Pole, the easterlies blow from the south and east.

Low-density air rises at 0 degrees and 60 degrees N and S, whereas high-density air descends at 30 and 90 degrees N and S. Areas where the air rises are zones of low atmospheric pressure and usually exhibit clouds and rain. Areas where the air descends are associated with high atmospheric pressure, clear skies, and low precipitation.

Surface winds move from areas of high pressure to areas of low pressure **(Figure 4-13)**. In the areas of vertical movement between the wind belts, wind movement is unsteady and unreliable. In the early days of sailing ships, these areas were a cause of great concern to sailors because they could be stranded for days without a breeze to move them. The area of rising air at the equator is known as the **doldrums**, and the areas of descending air at 30 degrees N and S are known as the **horse latitudes**. It is said that the horse latitudes received their name from early sailing ships carrying horses. If a ship became becalmed for a long time, there would eventually not be enough drinking water to support both the crew and the horses, so the horses were thrown overboard.

The **Coriolis effect** is the apparent deflection of the paths of winds and ocean currents that results from the rotation of the earth.

Northeast trade winds are surface winds that occur between the equator and 30° north latitude.

Southeast trade winds are surface winds that occur between the equator and 30° south latitude.

Westerlies are surface winds that blow from the west.

Polar easterlies are winds that blow from the east between 60° north or south and the poles.

The **doldrums** are areas of rising air at the equator.

The **horse latitudes** are the areas of descending air at 30 degrees north and south.

A **gyre** is a circular pattern of water flow that occurs within an ocean basin.

OCEAN CURRENTS

Ocean currents are produced when wind blowing across the surface of the ocean pushes and pulls the water, causing it to move. Like rivers on land, the ocean currents move water in predictable patterns within the ocean basins.

Surface Currents

Winds transfer energy from the air to water by friction. This action produces waves that transfer energy to the water, causing a mass of water beneath the moving air to flow and form a surface current. The main driving forces for surface currents are the trade winds (easterlies and westerlies) in each hemisphere. As the water moves, it begins to accumulate in the direction that the wind is blowing. Gravity acts on the mass of accumulating water and tries to pull it back against the pressure gradient, but the water continues moving according to the Coriolis effect in a circular pattern.

Coriolis Effect Just as with air, the friction between the moving water and the earth is very small, and the water's motion is described by the Coriolis effect. Since water moves more slowly than air, it takes the water longer to move the same distance. During this longer time, the earth rotates farther out from under the water than it would from under air. Therefore the slower-moving water is deflected to a greater degree than air. In the Northern Hemisphere, the current is deflected to the right of the prevailing wind direction, and in the Southern Hemisphere, the deflection is to the left. In the open sea, this deflection can be as much as a 45-degree angle from the wind direction.

Gyres The position of the continents and the features of the ocean basins interfere with the continuous flow of water and contribute to the deflection of currents, so that water flows in a circular pattern around the edge of an ocean basin. This type of flow pattern is called a **gyre**. There are five major gyres in the world ocean, three in the Southern

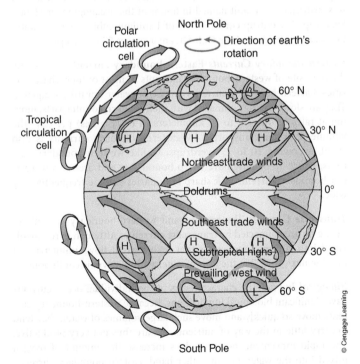

FIGURE 4-13 SURFACE WIND PATTERNS. Differences in air temperature and density are responsible for producing global wind patterns. This figure shows the planet's major wind patterns.

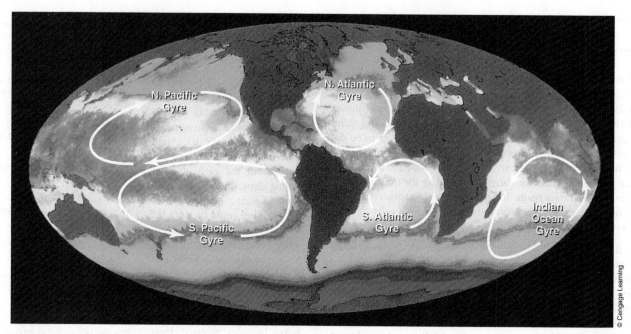

FIGURE 4-14 **GYRES.** A combination of surface winds, the Coriolis effect, position of continental land masses, and the shape of ocean basins causes surface currents to form circular currents of water called gyres. This map shows the location of the five major gyres.

Hemisphere and two in the Northern Hemisphere **(Figure 4-14)**, and each for the most part is independent of the others. A sixth major surface current is the **Antarctic Circumpolar Current**, which is driven by the strong westerly winds. This is the largest of the ocean currents and flows continuously eastward around the continent of Antarctica. Since it is not deflected by a continental mass, it is technically not considered a gyre.

Classification of Currents

Although the water within a gyre flows continuously, oceanographers divide each gyre into interconnected currents. Each current can be distinguished on the basis of distinct temperature and flow characteristics. The currents can be classified by position as eastern-boundary currents, western-boundary currents, or transverse currents. The water in a current can move long distances along a well-defined path, and frequently the flow of water in these currents is compared to the flow of water in a river. With few exceptions these currents have a nearly constant pattern **(Figure 4-15)**.

Western-Boundary Currents The fastest and deepest currents are the **western-boundary currents** found along the western boundaries of ocean basins. They move warm water toward the poles in each gyre. The **Gulf Stream** is the largest of these western-boundary currents, moving at an average speed of 2 meters per second (5 miles per hour). At this rate the Gulf Stream can travel more than 160 kilometers (100 miles) per day. As you might imagine, the amount of water that a western-boundary current can move is quite large. The volume of water transported in an ocean current is measured in **sverdrups** (sv), a unit named in honor of the oceanographer Harald Sverdrup. A sverdrup equals a flow of 1 million cubic meters of water per second. The flow of the Gulf Stream averages 55 sverdrups.

Eastern-Boundary Currents **Eastern-boundary currents** are exactly the opposite of western-boundary currents in most of their characteristics. Eastern-boundary currents carry cold water toward the equator. They are slow moving compared with their western counterparts, moving at rates of only 10 to 20 kilometers (6 to 12 miles) per day. The volume of water transported by eastern-boundary currents is also small, averaging 10 to 15 sverdrups. Unlike western-boundary currents, which frequently have sharp boundaries and are narrow, eastern-boundary currents lack sharp boundaries and are frequently very wide.

Transverse Currents The eastern- and western-boundary currents in each gyre are connected by **transverse currents**. Although gyres are divided into currents, remember that each current flows uninterrupted into the next, such that the flow of water within a gyre is continuous.

Biological Impact The characteristics of the individual ocean currents have significant biological implications. Because western-boundary currents move so quickly and move such large volumes of water, they tend to carry little in the way of nutrients and are thus not very productive. The rapid movement, however, does increase the amount of oxygen mixed into the water. On the other hand, eastern-boundary currents, because they move less water and move more slowly, tend to be more productive and contribute to the mixing of nutrients into surface waters that is discussed later in this chapter.

The **Antarctic Circumpolar Current** is an ocean current that flows continuously eastward around the continent of Antarctica.

Western-boundary currents are ocean currents that occur along the western boundaries of ocean basins.

The **Gulf Stream** is the largest of the western-boundary currents.

A **sverdrup (sv)** is a measure of the volume of water moved in an ocean current and is equal to a flow of 1 million cubic meters of water per second.

Eastern-boundary currents are ocean currents that occur along the eastern boundaries of ocean basins.

Transverse currents connect eastern and western boundary currents within a gyre.

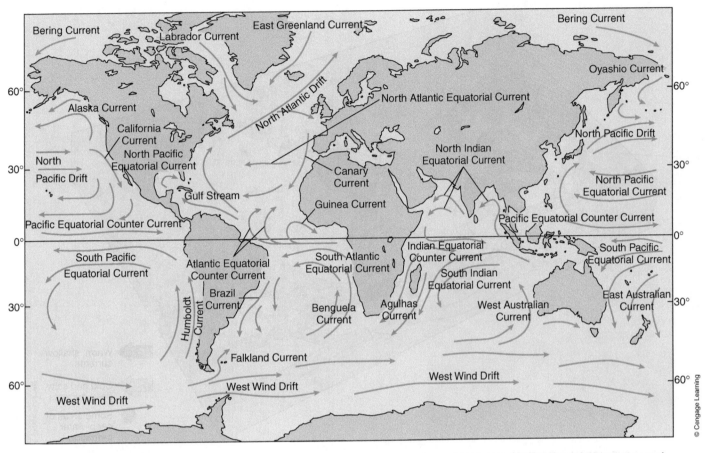

FIGURE 4-15 MAJOR OCEAN CURRENTS. Ocean currents are produced by the driving force of wind. Notice the similarities in the patterns of ocean currents and the surface winds shown in Figure 4-13.

The Ocean Conveyor

Water not only moves within ocean basins but also between basins. This occurs when warm tropical water moves toward the poles, cools, and sinks to become deep water that flows into neighboring ocean basins. Recent models propose that this flow of deep water is connected to surface circulation. One popular model compares water flow in the ocean to a huge conveyor belt driven for the most part by processes that occur in the North Atlantic Ocean. In this model, warm water from the Pacific and Indian Oceans moves along the coast of Africa and North America to enter the North Atlantic. As this warm water is carried north toward the polar sea, it loses heat to the atmosphere by radiation and its density increases. As a result, the water sinks and moves into other basins as deepwater flow. The deep water flows south toward Antarctica and mixes with deep water formed in the Southern Ocean. This mass of combined deep water then moves around the continent of Antarctica and flows north into the Pacific and Indian Oceans. As the water enters the Pacific Ocean in the vicinity of Indonesia and Australia, it is driven by winds and currents, causing it to rise where it is warmed by contact with the atmosphere. The water then travels to the Indian Ocean and, driven by the currents flowing around the southern tip of Africa, joins the Gulf Stream where it is carried back toward the North Atlantic. This is a slow process that requires one thousand years or more to make a complete circuit. This slow circulation moves water between the hemispheres and is interconnected with the more rapid flow of water in surface gyres **(Figure 4-16)**.

The slow, steady flow of water in the conveyor belt distributes dissolved gases and nutrients. Where water masses sink, they carry dissolved oxygen to the ocean's depths. Where water masses rise, they return nutrients to the surface. This process helps maintain high levels of primary production. The ocean conveyor also plays an important role in transporting plankton and larval stages of marine organisms among the ocean basins.

Ekman Transport

The energy transferred from winds to water at the surface continues down to other layers of water. The movement of water at the surface causes a layer of water below it to move as a result of friction. The movement of this deeper layer of water is deflected to the right in the Northern Hemisphere and to the left in the Southern Hemisphere as a result of the Coriolis effect. This process continues to affect successively deeper layers of water to a depth of about 100 meters (330 feet). Each layer slides horizontally over the layer beneath, moving at an angle to the layer above and moving slower than the layer above because of frictional loss of energy **(Figure 4-17)**. The spiral flow of water that results is called the **Ekman spiral** after the Swedish oceanographer who developed a mathematical model to describe the process. The net movement of water to the 100-meter depth is called **Ekman transport**.

The **Ekman spiral** is the spiral flow of water from the surface of the ocean to deeper water.

Ekman transport is the net movement of water by the Ekman spiral from the surface to a depth of 100 meters.

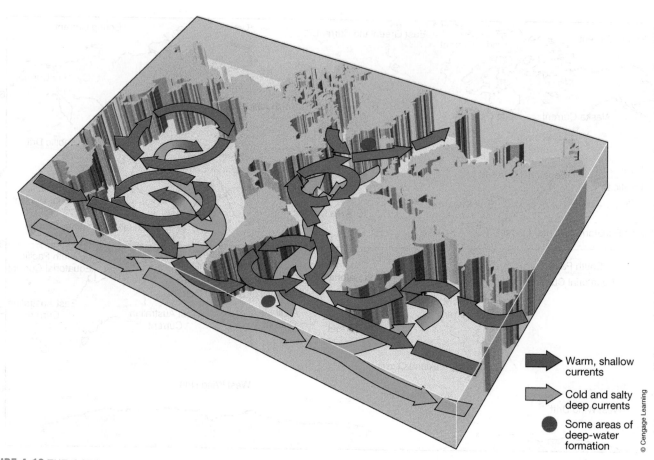

Warm, shallow
currents

Cold and salty
deep currents

Some areas of
deep-water
formation

© Cengage Learning

FIGURE 4-16 THE OCEAN CONVEYOR. Water moves among the ocean basins beginning in the North Atlantic, where deep water flows south to Antarctica, where it mixes with more deep water and then moves to the Pacific and Indian Oceans, where the water warms, rises, and flows around the tip of Africa and back to the North Atlantic.

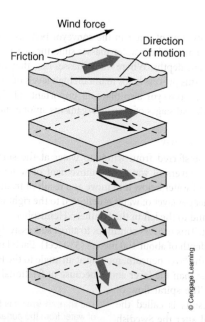

Wind force

Friction

Direction
of motion

© Cengage Learning

FIGURE 4-17 EKMAN SPIRAL. The movement of surface water is passed along to deeper layers of water, causing them to move in a characteristic pattern known as the Ekman spiral. As you go deeper, each successive layer of water moves at a slower speed and an angle to the layer above, producing the spiral movement.

IN SUMMARY

- Winds are produced by differences in the density of air.

- Warmer air at the equator rises and moves toward the poles, whereas colder air at the poles sinks and returns to the equator.

- The apparent deflection of air masses in the two hemispheres is called the Coriolis effect.

- Wind patterns are responsible for driving ocean currents.

- Like air masses, ocean currents are affected by the Coriolis effect, causing them to deflect.

- The combination of wind, gravity, and the Coriolis effect produces gyres, circular patterns of water flow at the ocean's surface.

- The ocean conveyor slowly moves water from one hemisphere to another.

- The characteristics of ocean currents affect life in the sea. (Have You Wondered? #2)

- Water also flows from the surface to a depth of about 100 meters, a process known as Ekman transport.

4-5 Ocean Layers and Ocean Mixing

Density is a physical property of a substance: the mass of the substance in a given volume. It is generally measured in grams per cubic centimeter (g/cm^3). For instance, pure water has a density of $1\ g/cm^3$. Because seawater contains dissolved salts, it has a higher density than pure water ($1.0270\ g/cm^3$). Variations in the surface temperature and salinity of the ocean control the water's density. Density increases when salinity increases or temperature decreases (recall that water reaches its maximum density at a temperature of 4°C). Density decreases when salinity decreases or temperature increases. Salinity increases when evaporation occurs or when seawater comes out of solution to form ice. Precipitation, influx of river water, melting ice, or a combination of factors all can contribute to a decrease in salinity. Pressure also can affect the density of seawater, but pressure effects at the surface are minor.

In the open ocean, surface temperature is more decisive than salinity in determining the water's density. For instance, surface water in the tropics has the highest salinity in the open ocean, but the water is so warm that it remains less dense than the water below it and does not sink. On the other hand, water in the North Atlantic has a lower salinity but is much colder, and thus the surface water is denser and sinks. Salinity becomes a more important factor in determining the density of water close to shore. This is especially evident in semi-enclosed bays that receive a large amount of freshwater runoff. Wave action contributes to the density and mixing of ocean water close to shore, where the water is relatively shallow. The combination of different surface temperatures and salinities produces regions of ocean with different densities **(Figure 4-18)**. Denser water sinks until it reaches water of similar density, whereas less dense water rises. This situation is responsible for layered water in the ocean.

CHARACTERISTICS OF OCEAN LAYERS

The characteristics of temperature, salinity, and density change with depth. The surface layer of the ocean extends down to about 100 meters (330 feet), is warmed by solar heating, and is well mixed by a variety of processes. From 100 meters to 1,000 meters (3,300 feet), temperatures decrease, creating layers of water of increasing density **(Figure 4-19a)**. This zone of rapid temperature change is called a **thermocline**. Similarly, below the surface waters in the temperate zone, salinity increases with depth to about 1,000 meters **(Figure 4-19b)**. This zone is called a **halocline**. The changes in temperature and salinity in the

> **Density** is a physical property of matter equal to mass divided by volume.
>
> A **thermocline** is a zone of rapid temperature change.
>
> A **halocline** is a zone of rapid change in salinity.

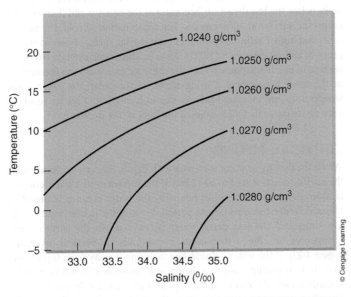

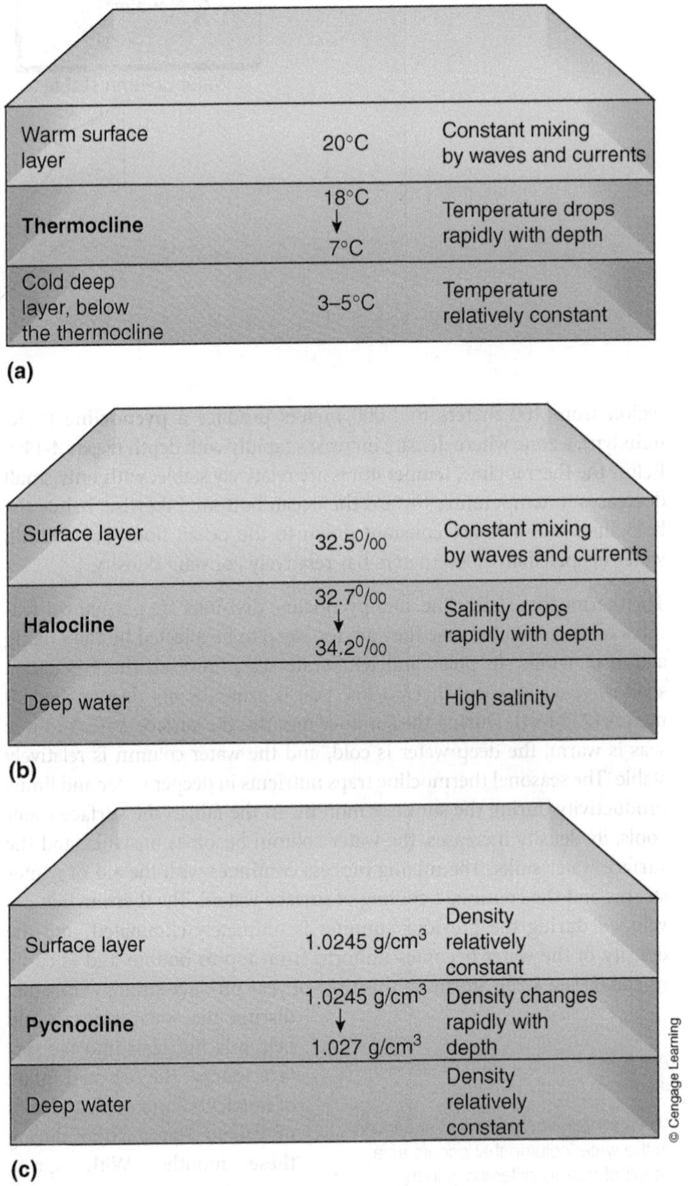

FIGURE 4-18 EFFECTS OF TEMPERATURE AND SALINITY ON THE DENSITY OF SEAWATER. Increases in salinity or decreases in water temperature result in an increase in the density of seawater. This graph shows the combined effects of salinity and temperature on seawater densities.

Warm surface layer	20°C	Constant mixing by waves and currents
Thermocline	18°C ↓ 7°C	Temperature drops rapidly with depth
Cold deep layer, below the thermocline	3–5°C	Temperature relatively constant

(a)

Surface layer	32.5⁰/₀₀	Constant mixing by waves and currents
Halocline	32.7⁰/₀₀ ↓ 34.2⁰/₀₀	Salinity drops rapidly with depth
Deep water		High salinity

(b)

Surface layer	1.0245 g/cm^3	Density relatively constant
Pycnocline	1.0245 g/cm^3 ↓ 1.027 g/cm^3	Density changes rapidly with depth
Deep water		Density relatively constant

(c)

FIGURE 4-19 CHANGES IN TEMPERATURE, SALINITY, AND DENSITY OF SEAWATER WITH DEPTH. (a) Changes in water temperature with depth. **(b)** Changes in salinity with depth. **(c)** Changes in density with depth.

© Cengage Learning

FIGURE 4-20 SEASONAL
CHANGES AND VERTICAL
MIXING. Seasonal changes in the
temperature and salinity of surface
waters of temperate and polar seas
produce a mixing effect that brings
nutrient-rich bottom water closer to
the surface and oxygen-rich surface
water closer to the bottom.

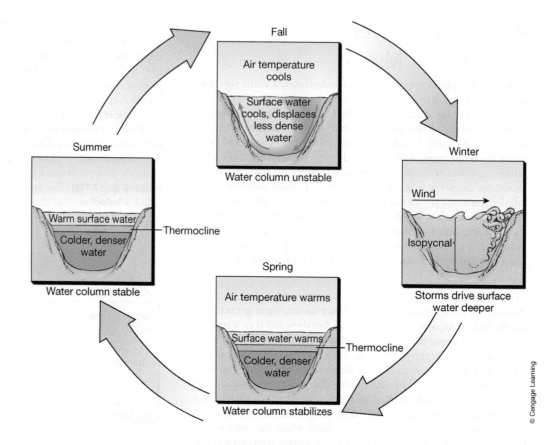

region from 100 meters to 1,000 meters produce a **pycnocline** (PIK-nuhklyn), a zone where density increases rapidly with depth **(Figure 4-19c)**. Below the thermocline, temperatures are relatively stable, with only small decreases in temperature toward the ocean bottom. Likewise, below the halocline, salinities are constant down to the ocean floor. As a result, water deeper than 1,000 meters has relatively constant density.

The thermocline, halocline, and pycnocline divisions are permanent features of the ocean because they are too deep to be affected by the mixing action of winds. In polar and temperate seas, however, there is also a shallow-water seasonal thermocline that is generally not deeper than 70 meters (231 feet). During the summer months, the surface water in these seas is warm, the deep water is cold, and the water column is relatively stable. The seasonal thermocline traps nutrients in deeper water and limits productivity during the summer months. In the fall, as the surface water cools, its density increases, the water column becomes unstable, and the surface water sinks. The mixing process continues with the aid of winter storms and the continued cooling of surface waters. The thermocline developed during the previous summer is ultimately eliminated, and the density of the water becomes uniform from top to bottom and is easily mixed **(Figure 4-20)**. Storms at this time of year produce strong winds that disrupt the water column and help mix nutrients into the surface waters. The seasonal influx of nutrients supports high levels of primary production during these months. With spring comes warmer temperatures, and the thermocline begins to reestablish itself, stabilizing the water column, a process that continues through the summer.

A **pycnocline** is a zone where density increases rapidly with depth.

Vertical overturn is the changeover in the water column that occurs as a result of density differences in the water.

Isopycnal means the water column has the same density from top to bottom.

HORIZONTAL MIXING

Seawater that is not very dense, such as the warm low-salinity water around the equator (1.0230 g/cm³), remains at the surface. Surface water at 30 degrees N and 30 degrees S is also warm, but it has a higher salinity and is denser (1.0267 g/cm³) than the equatorial waters. This difference in density causes the water at 30 degrees N to form a curved layer that sinks below the less dense surface water at the equator and then rises to rejoin the surface at 30 degrees S. The combination of colder temperatures and higher salinities produces even denser surface water at 60 degrees N and 60 degrees S (1.0276 g/cm³). This water extends from the surface in one hemisphere below the other surface waters to the surface of the other hemisphere. During winter at the poles, the lower water temperature and increased salinity due to formation of sea ice result in very dense water that sinks toward the floor of the ocean.

VERTICAL MIXING

When the density of water increases with depth, the water column from the surface down is said to be stable. If the top water in a water column is more dense than the water below it, the water column is unstable. Unstable water columns do not persist, because the denser water at the top sinks and the less dense water below rises to the surface. This changeover in the water column produces a **vertical overturn** **(Figure 4-21)**. If the water column has the same density from top to bottom, it is neutrally stable, or **isopycnal** (eye-soh-PIK-nuhl). Neutral stability means there is no tendency for water in the column to sink or rise. A water column that is neutrally stable can be easily vertically mixed by wind, wave action, or currents.

Any process that increases the density of surface water will cause vertical movement of the water, or vertical mixing. Vertical mixing is one of the principal processes for the exchange of water from top to bottom

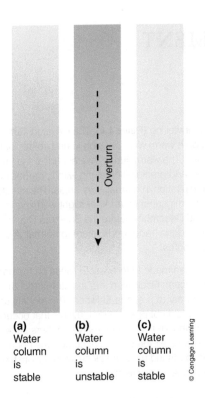

FIGURE 4-21 VERTICAL OVERTURN. **(a)** In a stable water column, water density increases with depth. **(b)** In an unstable water column, the water at the top of the column is denser than the water beneath. As a result, the denser surface water sinks, and the less dense water beneath is displaced to the surface. This process is called vertical overturn. **(c)** An isopycnal water column has the same density from top to bottom and is stable.

(a) Water column is stable

(b) Water column is unstable

(c) Water column is stable

Overturn

© Cengage Learning

throughout the world ocean. Since bottom water usually contains abundant nutrients from the settling of organic matter from above and the process of decomposition, vertical mixing provides a means of exchanging this nutrient-rich bottom water with oxygen-rich surface water. Since density is generally controlled by temperature and salinity, this type of circulation is also known as **thermohaline circulation**.

UPWELLING AND DOWNWELLING

Sometimes the horizontal movement of water that is produced by the winds can cause surface water to move vertically. This phenomenon is known as **wind-induced vertical circulation**. Since the quantity of water in the oceans is essentially fixed, any movement of water from one place to another would cause a similar but opposite movement to replace the water that leaves an area. For instance, when dense surface water sinks to a depth at which it is no longer denser than the water beneath it, it stops sinking. At this point, the water begins to move horizontally, making room for denser water that is sinking behind it. Eventually, this water rises back to the surface, replacing the surface water that has been sinking. At the same time, surface water is also moving horizontally into areas where the surface water is sinking. Vertical movement upward is called an **upwelling**, and downward vertical movement is called a **downwelling**. Upwellings and downwellings mix usually cold, nutrient-rich bottom water with the warmer, sunlit surface waters.

Equatorial Upwelling

In most tropical seas, productivity is low because of the pycnocline that separates the nutrient-rich bottom water from the sunlit surface waters.

The tropical waters along the equator are by contrast quite productive regions. This is due to water from currents on either side of the equator deflecting toward the poles, pulling surface water away that is replaced by deeper water. This movement of water is known as the **equatorial upwelling**. The equatorial upwelling brings nutrients to the surface and supports a high level of productivity all along the equator, which in turn provides food for large schools of fish, such as tuna.

Coastal Upwelling

Upwellings also occur along coastal areas **(Figure 4-22a)**. Winds blowing parallel to the coast produce surface currents that deflect according to the Coriolis effect. The Ekman transport that results causes the water to move offshore. As the water moves offshore, deeper water along the shore rises to take its place. The deeper water is rich in nutrients from sedimentation and decay processes, and it supplies the needs of photosynthetic organisms and the zooplankton that are vital parts of oceanic food chains. Along coastlines where prevailing winds blow more or less continuously, such as the west coast of South America, the constant upwelling supports high levels of biological productivity. In other areas, upwellings are seasonal. In the areas where upwelling occurs, phytoplankton thrive and rapidly reproduce. This forms the basis for productive food webs that may include anchovies, herring, lobster, cod, hake, bluefish, tuna, and other commercial species. As much as 50% of the world's commercial fish catch comes from upwelling areas that amount to only about 0.1% of the ocean's surface.

> **Thermohaline circulation** is movement of water in the water column due to changes in water temperature and salinity.
>
> **Wind-induced vertical circulation** is horizontal movement of water produced by winds that cause surface water to move vertically.
>
> An **upwelling** is movement of bottom water to the surface to replace water that is sinking.
>
> A **downwelling** is horizontal movement of water into areas where surface water is sinking.
>
> The **equatorial upwelling** is movement of deeper water to the surface to replace water that is being moved away from either side of the equator.

Coastal Downwelling

When winds blow water toward a coastline, it tends to be forced downward and then returns to sea along the continental shelf **(Figure 4-22b)**. Downwelling does not directly affect ocean productivity, but it does force gases and nutrients from the surface into deeper waters and plays a significant role in the distribution of marine organisms.

Because downwelling carries oxygen-rich surface water to deeper areas, many organisms can live in the deep water where downwellings occur. This process is especially crucial for organisms that live below the photic zone.

DEEPWATER CIRCULATION

The currents that we have discussed so far have primarily been in the surface waters, but there are also horizontal and vertical currents in deep ocean water below the pycnocline. In the ocean depths, differences in density cause the water to move, rather than wind energy. As previously mentioned, water density is most affected by salinity and temperature. The densest water in the ocean is Antarctic Bottom Water (1.0279 g/cm³). The majority of Antarctic Bottom Water forms in the winter in an area known as the Weddell Sea. As the water freezes, most salt remains in solution, producing water that is extremely dense because of the cold temperature and high salinity. This dense water sinks and mixes with water from the Antarctic Circumpolar Current. The mixture of water

ECOLOGY & THE MARINE ENVIRONMENT

El Niño Southern Oscillation

Not only do oceans supply the water that allows life to exist on land, they also influence other aspects of terrestrial climate. In 1983, it appeared that dramatic changes were occurring in the earth's climate, manifested in a variety of ways. Some areas such as California and the coasts of South America were drenched with rains that caused severe flooding. Other areas such as Australia and Indonesia, suffered terrible droughts, whereas parts of Polynesia were hammered by severe typhoons. The effects of these changes were also seen in marine organisms. Off the coast of South America, large numbers of fishes and sea birds died for no apparent reason, and many marine mammals, such as whales, dolphins, and seals, disappeared from their usual feeding grounds.

Scientists studying these events found that they were the result of changes in wind patterns and ocean currents that occur occasionally in winter. Farmers and fishermen in Peru had been aware of the phenomenon for years and had named it El Niño ("the Child"), in honor of the Christ child, since the phenomenon usually occurred around Christmas. An El Niño is usually a short-term, local phenomenon, but in 1983 it was particularly severe.

The coastal waters of Peru and Ecuador are highly productive fishery areas because of the nearly constant upwelling of deep, nutrient-rich water. The upwelling is driven by the Pacific trade winds, which move large volumes of water westward along the ocean's surface. Upwelling cold, nutrient-rich water along the coast replaces the water moving out to sea. Sometimes the trade winds lessen, typically around Christmas, and warm tropical water moves eastward across the Pacific to accumulate along the coasts of North and South America.

The warm water kills cold-water organisms that are the basis of food chains for many marine fishes, mammals, and birds. The loss of food causes the death of these animals. In severe cases the number of decaying organisms is so great that the surface water contains large quantities of hydrogen sulfide. Although the increased temperature of the coastal waters usually ends by April, sometimes it may persist for as long as a year. Severe El Niños have occurred frequently since 1953, with the two most severe occurring in 1982–1983 and 1997–1998.

No one is certain of the exact cause of El Niños, but certain processes have been correlated with the phenomenon's appearance. One process is the Southern Ocean Oscillation, in which the atmospheric pressure on one side of the Pacific increases while the pressure on the opposite side decreases. The pressure changes then reverse. Normally, there is a high-pressure system over Easter Island in the eastern Pacific and a low-pressure system over Indonesia in the western Pacific. Under these conditions the east-to-west trade winds are strong and constant, and upwellings occur along the coasts of Peru and Ecuador. When the normal pressure system reverses, the trade winds break down and warm water from the western Pacific moves eastward, ultimately warming the surface waters along the coast of South America and

depressing the upwelling **(Figure 4-A)**. The elevated surface temperature is the result of both warm water moving in and colder, denser water not being able to rise. The warm surface water carries the low-pressure zone of rising air and precipitation along with it, producing large amounts of precipitation in normally dry areas. The effects of El Niño usually lessen by midsummer. Another slight warming is often observed in November and December before the Southern Ocean Oscillation reverses again and the trade winds return to normal. A severe El Niño can last for 15 months.

Attempts have been made to forecast El Niños on the basis of changes in the Southern Ocean Oscillation and the strength of the trade winds, both of which appear to precede El Niños. The models used to study previous El Niños did not predict some events that occurred and predicted others that did not occur. As researchers seek more clues that might increase their forecasting accuracy, the phenomenon of El Niño continues to remind us of the close interrelationship between marine and terrestrial ecosystems.

G. Jacobs, Stennis Space Center/SPL/Photo Researchers, Inc.

FIGURE 4-A EL NIÑO SOUTHERN OCEAN OSCILLATION. The orange and yellow band in the lower right of this picture is warm water of the 1991 El Niño. The orange and yellow band further north is a sea temperature change caused by the 1982–1983 El Niño. El Niños are thought to be the result of changes in atmospheric pressure over the eastern and western Pacific Ocean that produce changes in the trade winds, causing a mass of warm water to move east across the Pacific from Asia. This changes the position of the atmospheric jet streams and results in climate shifts across the globe.

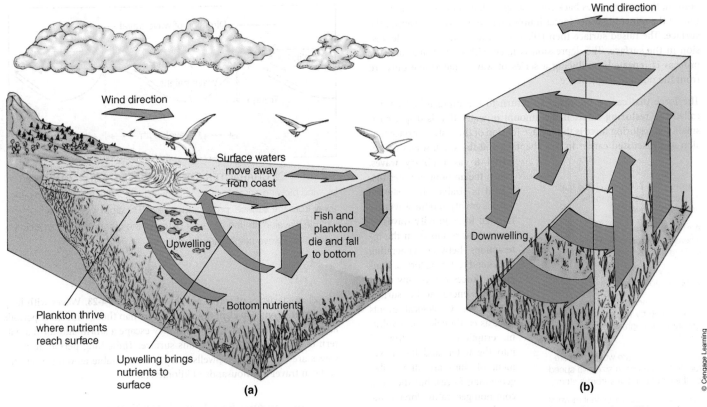

FIGURE 4-22 UPWELLING AND DOWNWELLING ZONES. **(a)** Upwelling zones are regions where bottom water is rising to the surface. This can be the result of changes in density or divergent surface currents. **(b)** Downwelling zones are areas where surface water is sinking. Downwelling can result from increases in density of the surface water in a region or the action of wind-driven surface currents that produce convergence.

continues to sink to the bottom and then begins to spread north. The movement of Antarctic Bottom Water is extremely slow. In the Pacific basin it may take as long as 1,600 years to reach the Arctic Sea. Since the Atlantic basin is much smaller, the water moves faster and may reach the North Atlantic in 800 years. Dense water forms and sinks to the bottom in the Arctic Ocean as well, but the characteristics of the ocean basin prevent most of it from moving southward. A deep channel east of Greenland does allow some North Atlantic Deep Water to flow into the North Atlantic.

Another deepwater mass forms in the Mediterranean Sea. Here evaporation exceeds the input of freshwater, producing very dense water that sinks. The density is most pronounced in the winter when the water salinity reaches 38 parts per thousand. The Mediterranean Deep Water flows through the Strait of Gibraltar and into the Atlantic Ocean, where it is found along the bottom of most of the central Atlantic. Some Mediterranean Deep Water has also been found as far south as the Antarctic. The salinity of Mediterranean deep water is higher than other deep water but it is warmer and so tends to sit atop of other deep water.

IN SUMMARY

- The density of seawater is primarily determined by temperature and salinity.
- Wave action and currents play leading roles in mixing deep and surface waters.
- Changes in density contribute to vertical mixing.

- The nutrients brought to the surface by upwellings help support very productive food chains. (Have You Wondered? #4)
- Downwellings bring oxygen to deeper water.
- Differences in water density contribute to bottom currents and deepwater circulation.

4-6 Waves

Waves are the result of forces acting on the surface of the water. A wave in the ocean does not represent a flow of water but a flow of energy or motion. The energy that is added to the water to cause the wave is either dissipated at sea or transferred to a beach or structures on the beach when the wave strikes.

WAVE FORMATION

A force that disturbs the water's surface, such as a stone dropped into the water or wind blowing across the water's surface, is called a **generating force**. The disturbance produced by the generating force moves outward, away from the point of the disturbance. Consider, for instance, water into which a rock is dropped. As the rock hits, the surface water is pushed aside. Then, as the rock sinks to the

A **generating force** is a force that disturbs the water's surface.

© Cengage Learning

bottom, the water moves back into the space left behind. The momentum of the returning water forces it upward so that it is raised above the surface. The raised surface then falls back down and creates a depression in the surface. The depression is filled with more water, and the process is repeated, producing a series of waves that ripple outward from the point of disturbance.

The force that causes the water to return to the undisturbed level is called the **restoring force**. If the amount of water that is displaced is small, the restoring force is the surface tension of the water (surface tension was discussed earlier in this chapter), and the small waves are referred to as **capillary waves**. When the amount of water displaced is sizable, the restoring force is gravity and the waves are referred to as **gravity waves**. A wave, then, results from the interactions between generating forces and restoring forces. Generating forces can be any event that adds energy to the surface of the sea. Geological events such as earthquakes and volcanic eruptions, objects dropped into the water, and the movement of ships are all possible generating forces, but the most common generating force is the wind.

As wind blows across the surface of still water, it creates drag (friction) that lifts some water away from the surface **(Figure 4-23)**. If the amount of water displaced is very small, the surface tension of the water pulls it back to restore a smooth surface, and a series of ripples are formed. If the force of the air is greater than a small breeze, more friction is created. As the water is stretched by the wind, more of it is pulled away from the surface, and a combination of surface tension and gravity acts to pull the water back to the surface. As the surface becomes rougher, it becomes easier for the wind to add more energy. The frictional drag between the air and the water is increased, and the waves become progressively larger.

> A **restoring force** is a force that causes water to return to the undisturbed level.
>
> A **capillary wave** is a small wave where the restoring force is the surface tension of water.
>
> A **gravity wave** is a large wave whose restoring force is gravity.
>
> A **progressive wave** is a wave that is generated by wind, restored by gravity, and progresses in a particular direction.
>
> **Forced waves** are waves that are forced to increase in size and speed by the input of energy from a storm.
>
> **Free waves** are waves that move at speeds determined by the wave's length and period.
>
> The **period** of a wave is the time required for one wavelength to pass a fixed point.
>
> A **swell** is a long-period, uniform wave.
>
> **Deepwater waves** are waves that occur in water that is deeper than one half of a wave's wavelength.
>
> The **fetch** is the distance along the water over which the wind blows.

TYPES OF WAVES

Most ocean waves are generated by wind and restored by gravity, and they progress in a particular direction. This type of wave is called a **progressive wave**. Progressive waves can be formed by local storm centers or by the prevailing winds of the wind belts, such as the trade winds or westerlies. Waves produced by storms at sea move outward from the storm center in all directions. As wind waves are formed by the storm, they are forced to increase in size and speed by the input of energy from the storm; for this reason they are also known as **forced waves**. When the energy from a storm or other generating force no longer has an effect on the waves, they become **free waves**, moving at speeds that are determined by the wave's length and **period**, the time required for one wavelength to pass a fixed point (a wave's speed is calculated by dividing its length by its period).

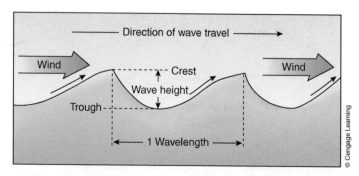

FIGURE 4-23 **CHARACTERISTICS OF WAVES.** The drag produced by wind blowing across the surface of still water lifts some of the water away from the surface, producing a wave. Principal wave characteristics include wavelength (the distance between the crests of two consecutive waves) and wave height (amplitude).

The characteristics of waves are shown in **Figure 4-23**. Waves with long periods and long wavelengths move faster than those with short periods and short wavelengths. Eventually, they escape a storm and form a pattern of wave crests on the ocean's surface. These long-period, uniform waves are called **swells**. Swells carry a considerable amount of energy and can travel for thousands of kilometers.

Deepwater and Shallow-Water Waves

Waves that occur in water that is deeper than one half of a wave's wavelength are called **deepwater waves**. The height of a deepwater wave increases with wind speed, the duration that the wind blows, and the **fetch**, or distance over the water that the wind blows. If the wind speed is low, the waves will be small, regardless of duration or fetch. A strong wind blowing for a short time does not produce large waves even if the fetch is large. If a strong wind blows for a long time over a short fetch, the waves are again small. Large waves are produced only when all three factors (wind speed, wind duration, and fetch) are of high magnitude.

Breakers

When a deepwater wave enters shallow water it becomes a **shallow-water wave** (the depth of the water is less than half the wavelength). The **surf zone** is the area along a coast where waves slow down, become steeper, break, and disappear **(Figure 4-24)**. As waves approach the shore, the decreasing depth begins to affect their shape and speed, and their crests become flatter. The friction from the lower part of the wave dragging on the bottom slows the forward movement of the wave. Whereas the length and speed of a deepwater wave are determined by the wave period, the length and speed of a shallow-water wave are determined by the depth of the water.

Breakers form in the surf zone when the bottom of the wave slows but its crest continues moving toward the shore at a speed faster than that of the wave. As a result, the crest overtakes the base of the wave in front of it and eventually falls into the preceding trough and breaks up. The two most common types of breakers are **plungers** and **spillers**. Plunging breakers form when the beach slope is steep. The crest curls and curves over and outruns the rest of the wave. It then breaks with a sudden loss of energy and a splash. Spilling breakers are more common and are found on flatter beaches, where the energy is dissipated more gradually as the wave moves over the shallow bottom. This action produces waves

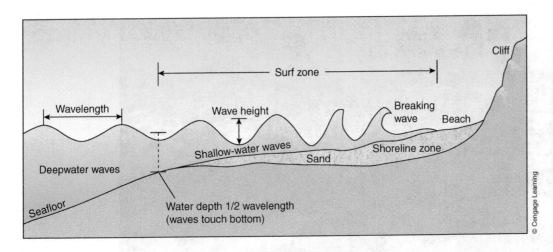

FIGURE 4-24 BREAKERS. As a wave approaches shallow water, the lower part of the wave may become slowed by friction with the bottom, although the top of the wave is not affected much. The result is a wave known as a breaker.

that are less spectacular and consist of turbulent water and bubbles that flow down the face of the wave. Spilling breakers last longer than plungers because they lose energy more gradually. If you were a surfing enthusiast, you would look for spillers to get a longer ride and plungers to get a more exhilarating ride.

As waves enter shallow water they almost always approach the shore at an angle. The portion of the wave that reaches shallow water first slows, but the portion still in deeper water continues at the original speed. The wave bends, a process called **wave refraction**.

Tsunamis

Sudden movements of the earth's crust produce earthquakes, which may produce large **seismic sea waves**, or **tsunamis**. These waves are sometimes called **tidal waves**, but oceanographers use that term to describe a totally unrelated event. If a large area of the seafloor rises or falls, it will cause a proportional movement in the surface of the sea. The disruption of the surface forms waves with long wavelengths, long periods, and low height. Because tsunamis are such long waves, they behave like shallow-water waves when they move out from the point of the seismic disturbance. As the wave approaches a coast or an island, the wave energy is compressed into a smaller volume of water as the depth decreases. The sudden increase in energy causes the height of the wave to increase dramatically, and the energy rapidly dissipates as the water races over the land mass. The wave's surge over land can cause mass destruction, wrecking buildings and docks and depositing vessels high on dry land **(Figure 4-25)**.

IN SUMMARY

- Waves are produced by forces acting at the surface of the water, most commonly wind. (Have You Wondered? #3)
- The force raises the water, and capillary action or gravity restores the water to its original position.
- The energy transferred to the water remains with the wave until it can be dissipated, as when it crashes against the shore.
- Large seismic sea waves, or tsunamis, are caused by earthquakes in the seafloor.

4-7 Tides

The periodic changes in water level that occur along coastlines are called **tides**. Tides play a considerable role in the life of many marine organisms, especially those that live in the area exposed at low tide and covered by high tide (the intertidal zone). The adaptations that allow intertidal organisms to survive and thrive in this potentially hostile environment will be explored later in the text.

WHY TIDES OCCUR

Tides are the result of the gravitational pull exerted on the ocean by the moon and the sun. The easiest way to explain the effects of the sun and the moon that result in tides is to imagine the earth as being completely covered with water and the tides behaving as ideal waves, ones that are not influenced by friction. In this model, the moon is held in orbit around the earth by the earth's gravitational force. There is also an apparent force called centrifugal force acting in the opposite direction and pulling the moon away from the earth **(Figure 4-26)**. Together, the earth and moon are held in their orbit around the sun by the gravitational attraction between the sun and the center of mass of the earth–moon system. An opposing centrifugal force pulls the center of mass of the earth–moon system away from the sun. For the earth–moon system to remain in orbit around the sun, the gravitational force must equal the centrifugal force. Because the moon also exerts a gravitational pull on the earth, there must be an opposing centrifugal force to prevent the earth from moving toward the moon. This centrifugal force is

Shallow-water waves are waves that occur in water that is less deep than half the wave's wavelength.

The **surf zone** is the area along a coast where waves slow down, become steeper, break, and disappear.

Breakers are waves that form in the surf zone when the bottom of the wave slows but its crest continues moving toward the shore at a speed faster than that of the wave.

Plungers are breakers with crests that curl, curve over, and outrun the rest of the wave.

Spillers are breakers that dissipate energy gradually as the wave moves over shallow bottom.

Wave refraction is the bending of a wave as it approaches the shore.

Seismic sea waves are large waves produced by earthquakes in the sea.

Tsunami is a synonym for seismic sea wave.

Tidal wave is a term sometimes used as a synonym for tsunami.

Tides are periodic changes in water level that occur along coastlines.

FIGURE 4-25 TSUNAMI. This damage to the city of Banda Aceh in Indonesia was caused by a tsunami that struck following an earthquake centered off of Sumatra, Indonesia, on December 26, 2004. Tsunamis form when activity along fault lines causes large areas of seafloor to rise or fall, producing proportional movement at the surface of the ocean.

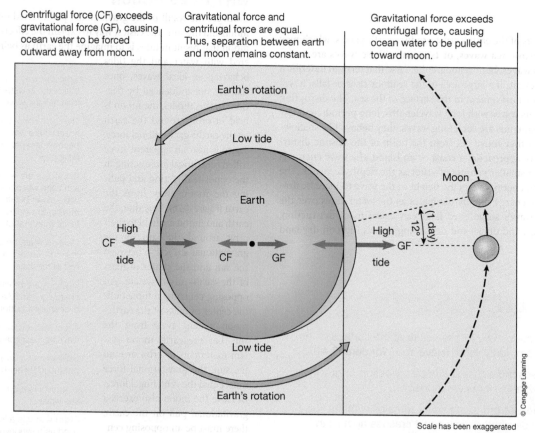

Centrifugal force (CF) exceeds gravitational force (GF), causing ocean water to be forced outward away from moon.

Gravitational force and centrifugal force are equal. Thus, separation between earth and moon remains constant.

Gravitational force exceeds centrifugal force, causing ocean water to be pulled toward moon.

Earth's rotation

Low tide

Moon

Earth

High
CF
tide

CF   GF

High
GF
tide

(1 day)
12°

Low tide

Earth's rotation

Scale has been exaggerated

FIGURE 4-26 TIDES. Tides occur when the gravitational force of the moon pulls ocean water toward it, while the centrifugal force of the earth–moon system forces a mass of ocean water to move in the opposite direction. The earth rotates beneath the water such that when a point on the earth's surface is under a bulge the tide is high, and when a point on the surface is not under a bulge the tide is low.

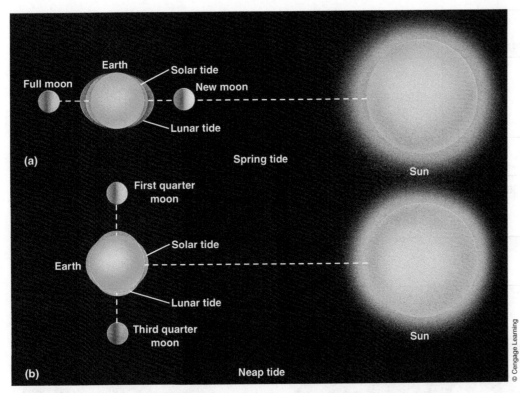

Spring tides are the highest and lowest tides in a tidal cycle.

Neap tides are tides with the smallest change between high and low tide.

A **diurnal tide** is the condition of having only one high tide and one low tide each day.

A **semidiurnal tide** is the condition of having two high tides and two low tides each day.

A **mixed semidiurnal tide** occurs when high and low tides are at different levels.

High water is the greatest point to which the high tide rises.

Low water is the lowest point to which the tide falls.

FIGURE 4-27 SPRING AND NEAP TIDES. **(a)** Spring tides occur when the moon and the sun are in alignment and the gravitational pull of the two are working together. **(b)** Neap tides occur when the sun is at a right angle to the position of the moon. The relatively smaller (due to its much greater distance) gravitational pull of the sun cancels some of the moon's gravitational pull, producing smaller tides.

created by the earth's center of mass rotating around the center of mass of the earth–moon system. Calculations using Newton's law of gravity, which describes the gravitational force between two or more bodies, indicate that the opposing forces are in balance at the earth's center, but at the earth's surface, the forces are not in balance. Although the mass of the moon is much less than the mass of the sun, it is much closer to the earth. When the forces are calculated for the effects of the sun and the moon on the water at the earth's surface, the moon's gravitational pull is found to have the greater effect.

The water on the side of the earth facing the moon is acted on by a gravitational pull from the moon that is larger than the opposing centrifugal force. Because water is fluid, it moves to a point directly under the moon and produces a bulge. On the opposite side of the earth, the centrifugal force acting on the water is larger than the moon's gravitational pull, producing another bulge on the side of the earth opposite the moon. At the same time that the two bulges form, areas of low water form between them. Because the earth rotates on its axis, a location on the earth's surface would experience a high tide when it is under one of the bulges and a low tide when it is not under one. Therefore, most locations experience two high tides and two low tides each day. Although one rotation of the earth takes 24 hours, the moon also moves slightly in its orbit each day, so that a location on the earth's surface needs an extra 50 minutes to come into alignment with the moon again. For this reason a full tidal cycle takes 24 hours and 50 minutes.

SPRING AND NEAP TIDES

Although the moon plays the greater role in producing tides, the sun also participates. Even though the sun's mass is much larger than that of the moon, it is so far away from the earth that its tide-producing

gravitational force is only about 46% that of the moon. The bulges that are produced by the sun are smaller and usually masked by the moon's.

Twice a month, at the full moon and new moon, the earth, moon, and sun are all in a straight line. As a result, the gravitational pull of the sun is added to that of the moon. These are the times of the highest and lowest tides, referred to as **spring tides**, not for the season of the year but for the water "springing up" at the shore. During the first and last quarter of the moon, the sun and moon are at right angles to each other. The pull of the sun cancels some of the moon's pull and produces **neap** (from an Anglo-Saxon word meaning "napping") **tides**, which have the smallest change between the high and low tide (**Figure 4-27**).

TIDAL RANGE

Tidal range is greatly influenced by geographical factors such as the contours of shorelines and the depth of coastal waters. Remember that the model for tides assumes the water is moving as an ideal wave. As the earth rotates, continents' positions interfere with tidal crests, altering their size and movement. The shape of ocean basins can also affect tidal patterns and height. For these reasons, some coastal areas have only one high tide and one low tide each day. This condition is referred to as a **diurnal tide**. Most areas experience two high tides and two low tides each day, a condition called a **semidiurnal tide**. In the typical semidiurnal tide, the two high tides are the same height and the two low tides are at about the same level. If the high and low tides are at different levels, the tide is referred to as a **mixed semidiurnal tide** (**Figure 4-28**).

In a tidal system, the greatest height to which the high tide rises is called **high water** and the lowest point is called **low water**. Observations of

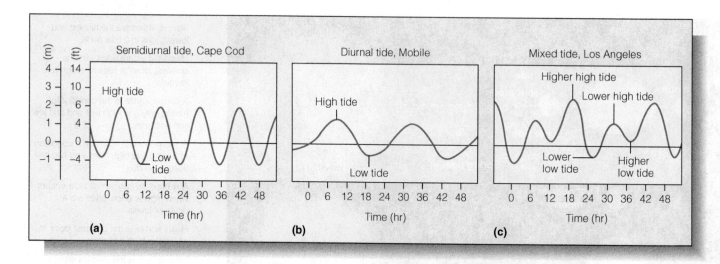

(a) Semidiurnal tide, Cape Cod

(b) Diurnal tide, Mobile

(c) Mixed tide, Los Angeles

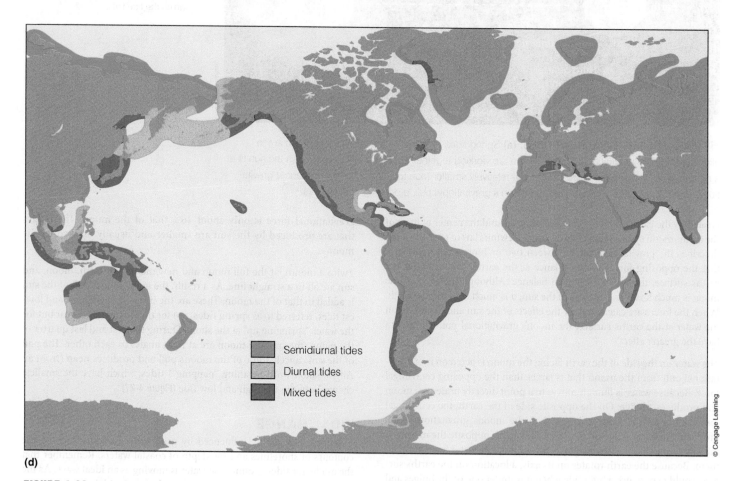

(d)

FIGURE 4-28 TIDAL PATTERNS. Tidal patterns are influenced by geographical factors, resulting in three main types of tide: **(a)** semidiurnal tide, **(b)** diurnal tide, and **(c)** mixed tide. The map **(d)** shows the geographical distribution of the three types of tide.

A **flood tide** is a rising tide.

An **ebb tide** is a falling tide.

Tidal currents are currents associated with tidal movements.

Slack water is the period of time during which tidal currents slow down and then reverse.

tides over a long period are used to determine the average tide. A rising tide is called a **flood tide**, because the water floods the coast, and a falling tide is referred to as an **ebb tide**. **Tidal currents**, associated with the rising and falling of the tides, can be very swift and sometimes dangerous as they move water onshore and offshore. During the change of tide from high to low or low to high is a period called **slack water**, when tidal currents slow down and then reverse.

IN SUMMARY

- The periodic rise and fall in the level of coastal waters is known as a tide.

- Tides are the result of the gravitational pull of the moon and the sun acting on the ocean. (Have You Wondered? #5)

KEY CONCEPTS

1. The polar nature of water accounts for many of its physical properties. (4-1)

2. Seawater contains a number of salts, the most abundant being sodium chloride. (4-2)

3. Salts are constantly being added to and removed from the oceans. (4-2)

4. The earth and the atmosphere receive energy from space in the form of radiant energy and return energy to space in the form of heat. (4-3)

5. The exchange of energy between oceans and the atmosphere produces winds that drive ocean currents and weather patterns. (4-4)

6. The density of seawater is mainly determined by temperature and salinity. (4-5)

7. Vertical mixing of seawater carries oxygen to the deep and nutrients to the surface. (4-5)

8. Waves are the result of forces acting on the surface of the water. (4-6)

9. The gravitational pull of the moon and the sun and the centrifugal forces of the earth–moon system on the ocean produce tides. (4-7)

ARE YOU STILL WONDERING?

1. Since water has a high thermal capacity, large bodies of water like the ocean have fairly stable temperatures, making it easier for marine organisms to maintain a favorable body temperature. Very cold liquid water is more dense than solid water; therefore, ice floats and water freezes at the surface–not from the bottom up–thus preventing marine organisms from being trapped in ice. Capillary action helps water move in the spaces between sediment particles, providing moisture for infauna. These are just a few of the ways in which the ocean's physical characteristics affect marine organisms.

2. Ocean currents carry the plankton that is food for many marine animals. Large consumers that feed on plankton or on plankton-feeding animals follow the currents to gain a meal.

3. Waves are formed when energy, such as that from the wind, is transferred to the water.

4. The upwelling off the western coast of South America brings nutrient-rich bottom water to the sunlit surface water. This combination of nutrients and sunlight results in large populations of phytoplankton. The phytoplankton form the basis of a very productive food web that includes large fish populations.

5. Tides are the result of the gravitational pull of the moon and the sun, and the centrifugal forces of the earth–moon system, pulling on the ocean.

QUESTIONS FOR REVIEW

Multiple Choice

1. The relatively high boiling and freezing points of water are due to
   a. the size of the water molecules
   b. the ability of water molecules to form hydrogen bonds
   c. the low thermal capacity of water
   d. the neutral pH of water
   e. the shape of water molecules

2. Compounds that release hydrogen ions when they dissolve in water are called
   a. acids
   b. bases
   c. salts
   d. neutral compounds
   e. organic compounds

3. Carbon dioxide gas is produced by living organisms during the process of
   a. photosynthesis
   b. digestion
   c. respiration
   d. chemosynthesis
   e. anabolism

4. Which of the following statements concerning the salinity of seawater is true?
   a. The salinity of coastal waters increases following a hurricane.
   b. The salinity of ocean water around the equator is low due to evaporation.
   c. A tidepool in the tropics would usually have water of low salinity.
   d. The salinity in the polar seas is high due to the formation of polar ice.
   e. The higher the salinity of seawater, the less dense it is.

5. The Coriolis effect refers to
   a. the process by which ocean currents are formed
   b. the movement of nutrient-rich bottom water to the surface in exchange for oxygen-rich surface water
   c. the apparent deflection of the path of air currents and ocean currents that results from the earth, its atmosphere, and ocean water all moving at different speeds
   d. the vertical exchange of water due to differences in density
   e. the horizontal exchange of water between the poles and the equator that occurs as the result of changes in the temperature and salinity of the water

6. Ocean currents are produced by
   a. earthquakes
   b. winds
   c. differences in water density
   d. upwellings and downwellings
   e. tides

7. The restoring force for large waves is
   a. wind
   b. capillary action
   c. adhesion
   d. friction
   e. gravity

8. When the sun, moon, and earth are all in alignment, _____ tides occur.
   a. ebb
   b. neap
   c. spring
   d. seasonal
   e. no

Matching

Match the following terms with the appropriate definition:

a. pycnocline
b. halocline
c. thermocline
d. vertical overturn
e. thermohaline circulation

1. An area of rapid temperature change with change in depth.
2. The change that occurs in an unstable water column as denser surface water sinks.
3. A change in salinity with a change in depth.
4. The flow of ocean water that is driven by density changes.
5. A zone of water where density increases rapidly with depth.

Match the following terms with the appropriate definition:

a. plunger
b. forced wave
c. fetch
d. swell
e. tsunami

6. The length of time a wind blows.
7. Large waves produced by earthquakes in the ocean.
8. A type of wave whose crest curls over and outruns the rest of the wave.
9. A wave produced by storm winds.
10. Long-period, uniform waves.

Short Answer

1. What roles do photosynthesis and respiration play in the distribution of gases in seawater?

2. What factors are responsible for the earth's prevailing wind patterns?

3. What combination of factors produces neap tides?

4. Explain how the polar nature of water molecules influences water's physical characteristics.

5. Explain how salt from the sea is returned to the land.

6. Explain how vertical mixing of seawater occurs.

7. Describe how winds are produced.

8. Explain how waves are formed.

9. Explain how breakers are formed.

10. Describe how upwellings occur.

11. What is the biological importance of upwelling and downwelling zones?

12. What are gyres and how are they formed?

13. Why is carbon dioxide more soluble in seawater than oxygen?

14. List three different processes that circulate water in the oceans.

15. Why is the shallow-water thermocline in polar and temperate seas seasonal?

Thinking Critically

1. Would it be easier for a planktonic organism to float in water with high salinity or low salinity? Explain.

2. Why do upwelling zones and downwelling zones support more biomass than areas of the open sea where these zones don't exist?

3. Why do coastal cities usually experience cooler summers and warmer winters than cities of the same latitude that are inland?

4. Sometimes the discharge of organic wastes from land into coastal waters results in a phytoplankton bloom. How would this affect organisms that live in the water below the surface and on the bottom?

5. What effect do you think boundary currents have on coastal communities of organisms?

SUGGESTIONS FOR FURTHER READING

Clark, P. U., N. G. Pisins, T. F. Stocker, and A. J. Weaver 2002. The Role of the Thermohaline Circulation in Abrupt Climate Change, *Nature* 415: 863–869.

Detrick, R. 2004. The Engine that Drives the Earth, *Oceanus* 42(2): 6–12.

Garrison, T. 2002. *Oceanography*, 6th ed. Pacific Grove, Calif.: Thomson Learning.

Geist, E. L., V. V. Titov, and C. E. Synolakis, 2006. Tsunami: Wave of Change, *Scientific American* 294(1): 56–63.

Krajik, K. 2001. Message in a Bottle, *Smithsonian* 32(4):36–47.

Mundy, C. J., S. M. Kathmann, and G. K. Schenter, 2007. A Special Brew. *Natural History* 116(9): 32–36.

Stutz, B. 2004. Rogue Waves, *Discover* 25(7): 48–55.

Have You Wondered?

1. What the definition of a species is? (5-3)

2. What natural selection is? (5-3)

3. How new species evolve? (5-3)

4. How marine organisms are named and classified? (5-4)

5

Biological Concepts

LIVING ORGANISMS are composed of cells. These basic units of life contain all of the chemicals necessary to support life and pass along hereditary information. To understand the wide variety of organisms that inhabit the sea, it is necessary to know something about the cells and the chemicals that make up cells. Through the process of cell division, cells reproduce and pass along genetic information that allows their offspring to continue functioning in the marine environment. If you spend time observing marine organisms, you will surely notice that all of their activity seems to be directed toward two objectives: survival and reproduction. In animals, for instance, behaviors and body structures are specialized to gather food, produce reproductive cells, and use those reproductive cells to produce new individuals. Other behaviors, such as defense against predators, enhance an organism's ability to survive so that it can continue to feed and reproduce. All organisms that are alive today descended from ancestors that were well adapted to their environments and survived long enough to reproduce successfully.

5-1 Building Blocks of Life

To understand living organisms, one must have a basic understanding of the variety of compounds from which organisms are built. Some of the most important chemicals in living organisms are large molecules called **macromolecules**. There are four major classes of macromolecules: carbohydrates, lipids, proteins, and nucleic acids.

CARBOHYDRATES

The molecules known as **carbohydrates** contain the elements carbon, hydrogen, and oxygen, frequently in a ratio of 1:2:1, or CH_2O, thus the name *carbohydrate* (carbon water). The most common carbohydrates in living organisms are sugars and polysaccharides.

Sugars

The most common sugars in nature are monosaccharides and disaccharides. **Monosaccharides**, or simple sugars, are small molecules usually containing five or six carbon atoms. The five-carbon sugars ribose and deoxyribose are essential to the structure of nucleic acids (discussed later), as well as other molecules in cells. The six-carbon sugar **glucose** is the basic fuel molecule for living cells (**Figure 5-1a**).

Macromolecules are large molecules.

Carbohydrates are molecules that contain the elements carbon, hydrogen, and oxygen, frequently in the ratio of 1:2:1.

Monosaccharides are simple sugars.

Glucose is a monosaccharide that serves as the basic fuel molecule for living cells.

Disaccharides are composed of two monosaccharides held together by a chemical bond. **Sucrose**, common table sugar, is a disaccharide that contains a molecule of glucose and a molecule of fructose (another six-carbon sugar). Marine plants and algae use sucrose to transport sugars within their bodies **(Figure 5-1b)**. Another disaccharide is lactose. **Lactose** is known as milk sugar because it is found in mammalian milk. Lactose supplies much of the energy that newborn mammals require.

Disaccharides are carbohydrates composed of two monosaccharides.

Sucrose is a disaccharide that contains a molecule of glucose and a molecule of fructose.

Lactose is the sugar found in mammalian milk.

Polysaccharides are large molecules made up of chains of monosaccharides.

A **polymer** is a large molecule that consists of the same basic units linked together.

Starches are polysaccharides composed of chains of glucose molecules that function in energy storage.

Glycogen, also known as animal starch, functions in energy storage.

Cellulose is a structural carbohydrate found in the cell walls of plants, algae, and some microorganisms.

Polysaccharides

The carbohydrates known as **polysaccharides** are made up of chains of monosaccharides and are examples of polymers. A **polymer** is a large molecule that consists of the same basic units linked together. **Starches** are polysaccharides composed of chains of glucose molecules. The glucose molecules are bonded together in a way that makes them easily separated for use as a ready source of energy or to build other molecules. They are produced by plants, seaweeds, and some microorganisms and are an efficient way for cells and organisms to store large amounts of glucose for future use. Starches are also important food sources for many animals. A polysaccharide similar to plant starch, **glycogen**, or animal starch, is produced by animals and some microorganisms to store glucose for future use.

Other polysaccharides are used to support cells and multicellular bodies structurally rather than for storage. **Cellulose,** an example of a structural polysaccharide, is found in the cell walls of plants, seaweeds, and some microorganisms **(Figure 5-2a)**. Like starches and glycogen, cellulose is a polymer of glucose molecules. Cellulose differs from starch and

(a)

(b)

FIGURE 5-1 SUGARS. **(a)** The sugar glucose supplies energy for the cells of marine organisms, such as this barracuda. **(b)** Marine plants, such as this mangrove, use the disaccharide sucrose to transport sugars from one part of the plant to another.

(a)

(b)

FIGURE 5-2 POLYSACCHARIDES. **(a)** The leaves, stems, and roots of this marsh grass are composed of the polysaccharide cellulose. **(b)** This blue crab's hard, protective shell contains the polysaccharide chitin.

(a)

(b)

(c)

FIGURE 5-3 LIPIDS. (a) These whales are insulated from the cold by a thick layer of blubber that is composed of triglyceride fat. **(b)** Steroids, such as the hormone testosterone, control this fish's reproduction. **(c)** Lipids known as waxes waterproof the cells of this alga.

glycogen in the way the glucose molecules are arranged, which accounts for its strength and durability. Because the glucose molecules in cellulose are highly resistant to disassembly, materials such as wood, rope, and paper may last for decades or centuries. Another structural polysaccharide, **chitin** (KY-tin), a polymer composed of modified glucose molecules, is also strong and durable. Chitin is in the cell walls of fungi and the hard exterior skeletons of some marine animals such as crabs and lobsters **(Figure 5-2b)**.

LIPIDS

Fats, oils, and waxes are all examples of **lipids**, macromolecules composed primarily of carbon and hydrogen. Marine organisms use simple fats, or **triglycerides**, to store energy, to cushion vital organs, and

FIGURE 5-4 PROTEINS. The muscles that allow this bird to fly are composed of protein.

to increase buoyancy. Homeothermic animals use triglycerides as insulation to trap heat **(Figure 5-3a)**. Another kind of lipid, **phospholipid**, is similar to triglycerides and is the major structural component of membranes that surround cells and some of the internal components of cells.

Steroids are lipids that function as chemical messengers within the bodies of animals **(Figure 5-3b)**. Waxes coat the exposed surfaces of some marine plants and seaweeds and act as a water barrier **(Figure 5-3c)**. Waxes are also found in the body coverings of some marine animals and in the ear openings of some marine mammals.

PROTEINS

Proteins are polymers made up of basic units called **amino acids**. Twenty different amino acids make up the various proteins found in living organisms. Within cells, individual amino acids are assembled into chains called **polypeptides**. The polypeptide chains are then coiled and folded into complex, three-dimensional protein molecules. Their complex structures allow protein molecules to serve several functions. For instance, the primary structural components of animals, muscles and connective tissues, are composed of protein **(Figure 5-4)**. Proteins known as

Chitin is a polysaccharide found in the cell walls of fungi and the external skeletons of arthropods.

Lipids are macromolecules composed primarily of carbon and hydrogen.

Triglycerides are simple fats.

Phospholipid is the major structural component of cell membranes.

Steroids are lipids that function as chemical messengers in animals.

Proteins are polymers of amino acids.

Amino acids are the basic building blocks of proteins.

Polypeptides are chains of amino acids.

TABLE 5-1 PROTEINS AND THEIR FUNCTIONS

Types of Proteins	Function
Enzymes	Biological catalysts that speed up the rate of chemical reactions in cells
Structural Proteins	Make up body parts of animals, such as hair, skin, scales, tendons, cartilage
Contractile Proteins	Make up muscle
Messenger Proteins	Send signals from one cell to another and from organ to organ
Transport Proteins	Transport important substances, such as oxygen and fatty acids
Storage Proteins	Store important materials in cells, such as iron
Antibodies	Protect animals from foreign proteins and disease-causing microbes
Toxins	Help to capture prey and protect animals from predators

© Cengage Learning

enzymes are essential for life. **Enzymes** are biological catalysts that speed up the rate of chemical reactions, allowing metabolism to function efficiently. Some proteins, such as hemoglobin, transport chemicals within organisms, whereas others store chemicals. **Table 5-1** lists the many functions of proteins and some examples.

NUCLEIC ACIDS

Nucleic acids are polymers of molecules called nucleotides. Each **nucleotide** is composed of a five-carbon sugar, a nitrogen-containing base, and a phosphate group. Two types of nucleic acids are found in living organisms: **deoxyribonucleic acid (DNA)** and **ribonucleic acid (RNA)**.

DNA

DNA is a large molecule with the shape of a helix **(Figure 5-5)**. Each nucleotide found in DNA contains the sugar deoxyribose, a phosphate group, and one of the following four nitrogen-containing bases: adenine, guanine, cytosine, or thymine. The nucleotides are linked by the phosphate groups to form long chains, or strands, of DNA. In the DNA molecule two strands of nucleotides are wound around each other to form the characteristic double helix (except for some viruses that have a single strand of DNA). DNA contains an organism's genetic material, or **genes**, and is capable of copying itself so that the genes can be passed from one generation to the next. Encoded in the genes are instructions for synthesizing the various proteins that an organism needs. The proteins are responsible for an organism's appearance and

Enzymes are biological catalysts.

Nucleic acids are polymers of nucleotides.

A **nucleotide** is a molecule composed of a five-carbon sugar, a nitrogen-containing base, and a phosphate group.

Deoxyribonucleic acid (DNA) is the molecule that makes up an organism's genetic material

Ribonucleic acid (RNA) is a molecule that functions in the synthesis of protein.

A **gene** is a unit of hereditary information located on a molecule of DNA.

Protein synthesis is the process whereby cells manufacture proteins.

Messenger RNA (mRNA) is a molecule transcribed from DNA that carries instructions for making a protein at a ribosome.

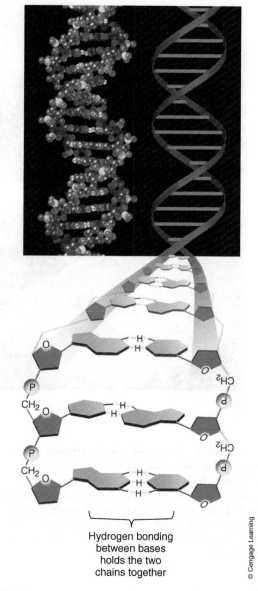

Hydrogen bonding between bases holds the two chains together

© Cengage Learning

FIGURE 5-5 NUCLEIC ACIDS. The nucleic acid DNA is composed of units called nucleotides. Each nucleotide contains the five-carbon sugar deoxyribose (red pentagons), a nitrogen-containing base (blue structures), and a phosphate group (yellow spheres). The nucleotides are linked together by the phosphate groups and form long chains. Two chains are wound together to form the characteristic double helix.

control its function, growth, and reproduction. In many cells, most of the DNA is located in a specialized structure called the nucleus. In addition, smaller amounts of DNA are found in other parts of the cell such as mitochondria and chloroplasts (described later).

RNA

RNA is composed of nucleotides that contain the sugar ribose and the same bases found in DNA, with the exception that in RNA the base uracil takes the place of thymine. RNA molecules are usually single-stranded (although some viruses have double-stranded RNA) and can come in different forms that have evolved for specific roles in the cell. RNA functions in **protein synthesis.** In this process, the information carried by individual genes is copied, or transcribed, from DNA onto a molecule of RNA known as **messenger RNA (mRNA)**.

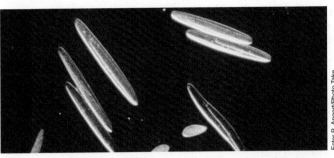

(a)

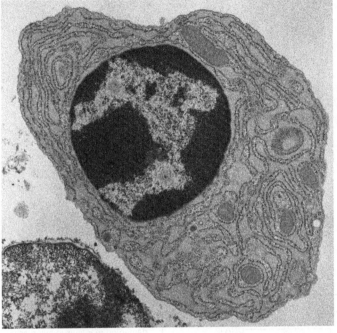

(b)

FIGURE 5-6 TYPES OF CELLS. **(a)** Prokaryotic cells, such as these bacterial cells, do not have a nucleus or membrane-bound organelles. **(b)** Eukaryotic cells, such as this animal cell, have a well-defined nucleus and many membrane-bound organelles.

The message carried by the mRNA is then decoded, or translated, by a cellular structure known as a ribosome, composed of protein and another type of RNA called **ribosomal RNA (rRNA)**. The ribosome synthesizes protein by connecting the appropriate amino acids that are brought to the ribosome by a third type of RNA called **transfer RNA (tRNA)**. The proteins produced by this process make new cell parts and enzymes that give the cells their characteristics.

IN SUMMARY

- Four main groups of large molecules are necessary for life.
- Carbohydrates supply energy, and some contribute to the structure of plants, fungi, seaweeds, some microorganisms, and some invertebrates.

- Lipids function in the storage of energy, in insulation, for buoyancy, and as chemical messengers.
- Proteins provide structural materials for making many body parts and act as chemical messengers to coordinate the activities of many cells.
- Enzymes are proteins that function as catalysts, regulating the metabolism of cells.
- Nucleic acids are molecules that carry genetic information and direct the synthesis of proteins.

5-2 Cells

All living organisms are composed of basic units called **cells**, and every cell is capable of the basic processes of life: metabolism, growth, and reproduction. Each cell is surrounded by a **cell membrane** that separates the contents of the cell from the external environment. The cell membrane is a selective barrier, regulating what goes in and out of the cell. Some cells also have a **cell wall**, an external nonliving structure, that gives them protection and support. The fluid content of a cell is called **cytosol**. Cytosol is a complex mixture of chemicals but consists mostly of water. In addition to cytosol, some cells contain internal structures called **organelles** that perform specific jobs within the cell. The combination of cytosol and organelles is called **cytoplasm**.

Ribosomal RNA (rRNA) is a nucleic acid found in organelles known as ribosomes.

Transfer RNA (tRNA) carries amino acids to ribosomes so that they can be incorporated into protein.

Cells are the basic units of which all living things are composed.

The **cell membrane** separates the contents of the cell from the external environment.

A **cell wall** is an external nonliving structure that gives some cells support.

Cytosol is the fluid content of a cell.

Organelles are structures found within cells that perform specific functions.

Cytoplasm is the combination of cytosol and organelles.

Prokaryotic cells are cells that lack a nucleus.

Prokaryotes are organisms composed of prokaryotic cells.

Eukaryotic cells are cells that have a well-defined nucleus and many membrane-bound organelles.

Eukaryotes are organisms composed of eukaryotic cells.

TYPES OF CELLS

There are two major types of cells in nature: prokaryotic cells and eukaryotic cells. **Prokaryotic cells (Figure 5-6a)** lack a nucleus and do not have any membrane-bound organelles. Organisms with prokaryotic cells are called **prokaryotes**, and they are always unicellular. Examples of prokaryotes are the many different types of marine bacteria and archaeons. **Eukaryotic cells (Figure 5-6b)** have a well-defined nucleus and many membrane-bound organelles. Organisms with eukaryotic cells, or **eukaryotes**, can be either unicellular or multicellular. Seaweeds, fungi, plants, and animals are examples of marine eukaryotes.

ORGANELLES

Eukaryotic cells contain several different organelles. Just as organs in a multicellular organism specialize in certain functions, organelles have specific functions within a cell **(Figure 5-7)**.

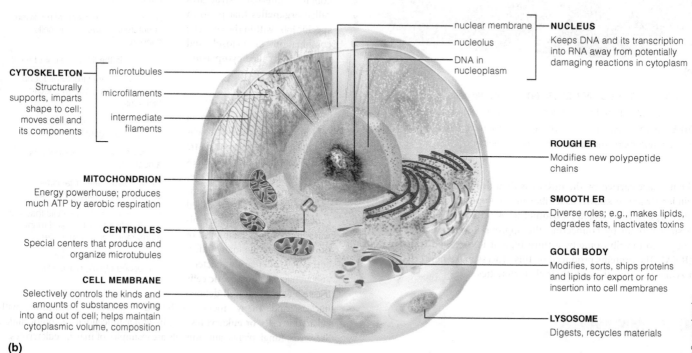

CELL WALL
Protects, structurally supports cell

CHLOROPLAST
Specializes in photosynthesis

CENTRAL VACUOLE
Increases cell surface area, stores metabolic wastes

nuclear membrane
nucleolus
DNA in nucleoplasm

NUCLEUS
Keeps DNA and its transcription into RNA away from potentially damaging reactions in cytoplasm

CYTOSKELETON
Structurally supports, imparts shape to cell; moves cell and its components

microtubules
microfilaments
intermediate filaments (not shown)

MITOCHONDRION
Energy powerhouse; produces much ATP by aerobic respiration

ROUGH ER
Modifies new polypeptide chains

SMOOTH ER
Diverse roles; e.g., makes lipids, degrades fats, inactivates toxins

GOLGI BODY
Modifies, sorts, ships proteins and lipids for export or for insertion into cell membranes

CELL MEMBRANE
Selectively controls the kinds and amounts of substances moving into and out of cell; helps maintain cytoplasmic volume, composition

LYSOSOME-LIKE VESICLE
Digests, recycles materials

(a)

CYTOSKELETON
Structurally supports, imparts shape to cell; moves cell and its components

microtubules
microfilaments
intermediate filaments

nuclear membrane
nucleolus
DNA in nucleoplasm

NUCLEUS
Keeps DNA and its transcription into RNA away from potentially damaging reactions in cytoplasm

MITOCHONDRION
Energy powerhouse; produces much ATP by aerobic respiration

CENTRIOLES
Special centers that produce and organize microtubules

ROUGH ER
Modifies new polypeptide chains

SMOOTH ER
Diverse roles; e.g., makes lipids, degrades fats, inactivates toxins

GOLGI BODY
Modifies, sorts, ships proteins and lipids for export or for insertion into cell membranes

CELL MEMBRANE
Selectively controls the kinds and amounts of substances moving into and out of cell; helps maintain cytoplasmic volume, composition

LYSOSOME
Digests, recycles materials

(b)

FIGURE 5-7 CELL STRUCTURE. (a) A diagram of a generalized plant cell showing the various organelles. **(b)** A diagram of a generalized animal cell showing the various organelles.

Nucleus and Ribosomes

The most obvious organelle in a eukaryotic cell is the **nucleus**, a large structure surrounded by a nuclear membrane. It contains the cell's DNA and acts as the cell's control center. The DNA in the nucleus is combined with protein to form structures called **chromosomes**, which contain an organism's genes. Eukaryotic cells contain one or two copies of each different chromosome. In addition to being the repository of a cell's genes, the nucleus is where ribosomes are formed. In eukaryotes, ribosomes are assembled in a specific area of the nucleus called the **nucleolus**. As we noted earlier, ribosomes are organelles that function in the synthesis of proteins. Ribosomes are not surrounded by a membrane and are found in both prokaryotic and eukaryotic cells.

Organelles Involved in Synthesis, Processing, and Storage

In eukaryotic cells a series of membranes called **endoplasmic reticulum (ER)** wind through the cytoplasm. Some endoplasmic reticulum has ribosomes attached to its surface (**rough endoplasmic reticulum**, or **RER**), and some does not (**smooth endoplasmic reticulum**, or **SER**). Rough endoplasmic reticulum functions in the modification of proteins as they are synthesized. Smooth endoplasmic reticulum functions in the synthesis of lipids and carbohydrates and the detoxification of harmful substances. Organelles called **Golgi apparatuses** function in the modification of molecules and place membranes around them. These packages of chemicals may be stored for later use or released from the cell. Lysosomes are an example of the membrane-bound sacs produced by the Golgi apparatus. **Lysosomes** contain enzymes that function in digestion. **Vacuoles** are structures surrounded by a membrane and may contain food, wastes, gas, or water.

Organelles Involved in Energy Conversion

The most complex organelles found in eukaryotic cells are those involved in energy conversions. **Chloroplasts** convert the radiant energy of light into chemical energy. They are found in photosynthetic organisms such as phytoplankton, seaweeds, and marine plants. **Mitochondria** transfer the chemical energy in food to molecules of **adenosine triphosphate**, or **ATP**. ATP supplies the energy for most of the metabolism and other activities that occur in living cells, such as transport of materials across cell membranes and cellular movement. All eukaryotic cells contain mitochondria, even if they are photosynthetic. Unlike other cellular organelles, chloroplasts and mitochondria are not synthesized by the cell but reproduce themselves.

Organelles of Movement

Some cells have organelles called **flagella** (singular, flagellum) and **cilia** (singular, cilium) that move the cell. Although the basic structures of flagella and cilia are the same, there are some differences. Cells with flagella usually have one to four of these long hair-like structures, which they use to propel themselves through their watery environment **(Figure 5-8a)**.

The **nucleus** is an organelle that contains the cell's genetic material

Chromosomes are structures composed of DNA and protein that contain an organism's genes.

The **nucleolus** is an area of the nucleus where ribosomes are formed.

The **endoplasmic reticulum (ER)** is a series of membranes that winds through the cytoplasm.

Rough endoplasmic reticulum (RER) is endoplasmic reticulum that has ribosomes attached to its surface.

Smooth endoplasmic reticulum (SER) is endoplasmic reticulum that does not have ribosomes attached to its surface.

Golgi apparatuses are organelles that function in the modification of molecules and packaging molecules in membranous sacs.

Lysosomes are membrane-bound sacs containing digestive enzymes.

Vacuoles are membrane-bound structures that may contain food, wastes, gas, or water.

Chloroplasts are organelles that convert the radiant energy of light into chemical energy.

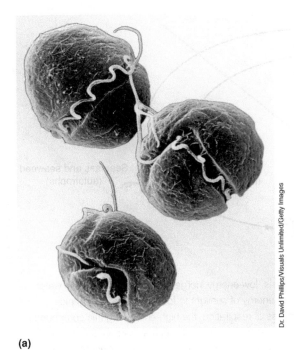

Dr. David Phillips/Visuals Unlimited/Getty Images

(a)

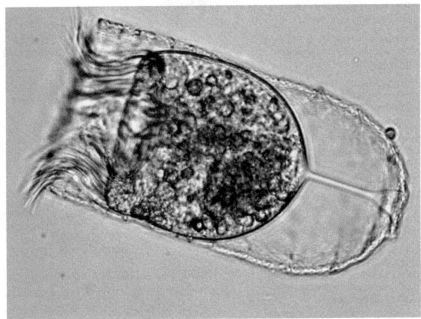

Copyright Nicole J. Huber

(b)

FIGURE 5-8 FLAGELLA AND CILIA. **(a)** This dinoflagellate has a pair of flagella that it uses to move in the water column. **(b)** The cilia on this tintinnid allow it to capture plankton for food.

Cilia are short hair-like structures that are numerous enough to cover large areas of a cell's surface **(Figure 5-8b)**. Cilia are used by single cells to move through their aqueous environment. In multicellular animals, cilia may cover the surface of some cells and move material along a cell's surface.

ENERGY TRANSFER IN CELLS

As we noted in Chapter 2, energy is necessary to power all of life's activities; without energy, life would be impossible. Autotrophs use energy from their environment to synthesize food molecules, which can then be used as a source of energy to form the ATP molecules that will be used to power the cell's various activities. Heterotrophs rely on other organisms for energy. The food they take in is broken down, and some of the energy is transferred to ATP molecules for the organism's use. The rest is released as heat energy **(Figure 5-9)**.

Photosynthesis

You learned in Chapter 2 that in the process of photosynthesis low-energy molecules such as carbon dioxide and water combine to form high-energy food molecules such as carbohydrates. In photosynthetic prokaryotes this occurs in areas of the cell where the membrane folds in to form a surface to which the molecules and enzymes required for this vital process attach. In eukaryotic cells photosynthesis occurs in organelles called chloroplasts, small oval structures surrounded by two membranes. Within the chloroplasts are membrane-bound discs called **thylakoids (Figure 5-10a)** that contain pigment molecules such as chlorophyll that are needed to trap the radiant energy of light. Surrounding the thylakoids is a fluid material called the **stroma** that contains the enzymes necessary to convert the carbon in carbon dioxide into an organic form, a process called **carbon fixation**.

Cellular Respiration

Cellular respiration releases energy stored in food molecules. Most of this energy-conversion process occurs in mitochondria. Like chloroplasts, mitochondria are complicated organelles surrounded by two membranes. The inner membrane is folded many times **(Figure 5-10b)**, increasing the total surface area available for the attachment of molecules that play key roles in energy conversion. In a stepwise series of complex reactions that begin in the cell's cytoplasm, food molecules are broken down and some of the energy released is used to synthesize molecules of ATP. During cellular respiration the carbon from the food molecules combines with oxygen, producing carbon dioxide as a waste product. Organisms exchange gases with the surrounding seawater to supply oxygen for respiration and remove the waste carbon dioxide.

CELLULAR REPRODUCTION

Cells reproduce by cell division. Cell division in prokaryotic cells is relatively simple. In eukaryotic cells, however, the process is more complex and the nucleus divides before the cell.

Cell Division in Prokaryotes

Prokaryotic cells often have only a single, circular chromosome. The chromosome must be duplicated before the cell divides so that each new

Mitochondria transfer the chemical energy from food to a form of chemical energy used by cells.

Adenosine triphosphate (ATP) is a molecule used by cells to power cellular processes.

Flagella are long, hair-like structures found on some cells that function in locomotion.

Cilia are short, hair-like structures that cover the surface of some cells and function in locomotion and the movement of materials across the cell surface.

Thylakoids are membrane-bound disks found within chloroplasts that contain the pigment molecules necessary for photosynthesis.

Stroma is the fluid material surrounding thylakoids in chloroplasts that contains the enzymes necessary to produce organic molecules from carbon in carbon dioxide.

Carbon fixation is the process of forming high-energy organic compounds from carbon dioxide.

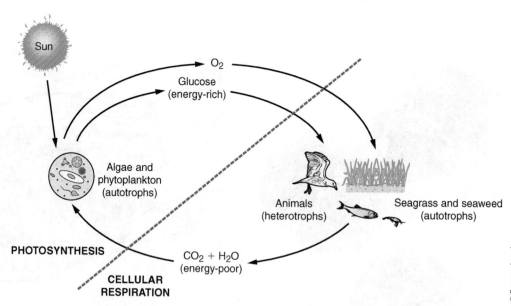

© Cengage Learning

FIGURE 5-9 ENERGY TRANSFER. In photosynthesis, low-energy inorganic compounds such as water and carbon dioxide are combined using the radiant energy of sunlight to form high-energy organic compounds that can be used for food. In the process of respiration, the high-energy organic compounds produced by photosynthesis are broken down to low-energy compounds such as carbon dioxide and water. Some of the energy released by the process is transferred to adenosine triphosphate (ATP) and used to power cellular processes.

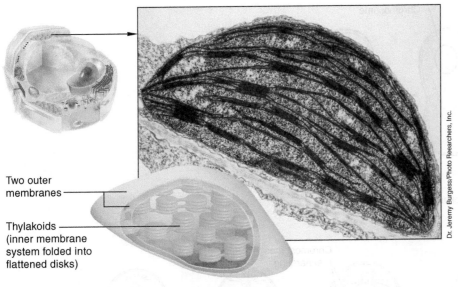

FIGURE 5-10 CHLOROPLASTS AND MITOCHONDRIA. (a) Photosynthesis occurs in the chloroplasts of eukaryotic cells. Pigments, such as chlorophyll, carotenoids, and xanthophylls, in the thylakoid membranes trap the light energy that drives the process. **(b)** In eukaryotes, respiration occurs in an organelle called the mitochondrion. The highly folded internal membrane provides a large surface area for the attachment of enzymes that are necessary for the process of breaking down food molecules.

Two outer membranes

Thylakoids (inner membrane system folded into flattened disks)

(a)

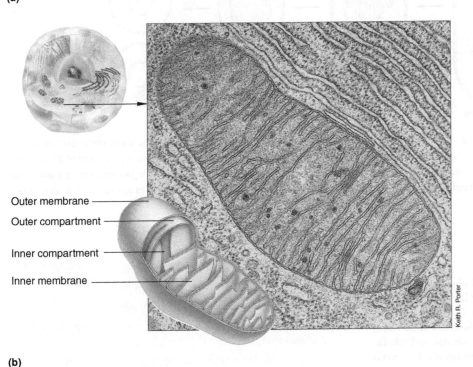

Outer membrane
Outer compartment
Inner compartment
Inner membrane

(b)

daughter cell will receive a complete copy of the necessary genetic information. Once duplicated, the two chromosomes migrate to opposite sides of the cell. The cell then splits in two, forming two new daughter cells **(Figure 5-11a)**. This is known as **binary fission**.

Cell Division in Eukaryotes

Before a eukaryotic cell can divide, its nucleus divides, a process called **mitosis**. Before the nucleus can divide, all of its chromosomes must be duplicated. During mitosis the nuclear membrane disappears and the two sets of chromosomes separate. A new nuclear membrane then forms around each of the two identical sets of chromosomes.

Mitosis ensures that each new cell will receive a complete set of chromosomes containing the necessary genetic information. Once the nucleus has divided, the cell divides. In cells with cell walls, a new wall forms between the two nuclei, producing two new cells. In animal cells, which lack a cell wall, the cytoplasm pinches off between the two nuclei, forming two new cells (see **Figure 5-11b**). The division of the cell into two daughter cells is called **cytokinesis**.

Binary fission is the process of cell division in prokaryotes.

Mitosis is the process of nuclear division that precedes cell division in eukaryotes.

Cytokinesis is the process of cell division in eukaryotes.

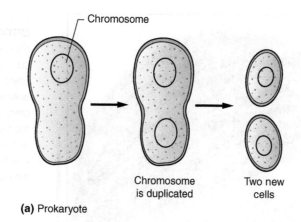

Chromosome

Chromosome
is duplicated

Two new
cells

(a) Prokaryote

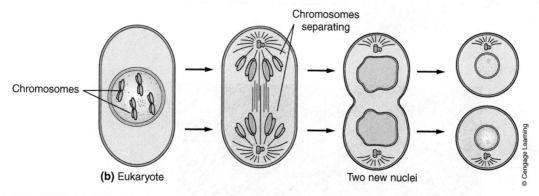

Chromosomes
separating

Chromosomes

(b) Eukaryote

Two new nuclei

© Cengage Learning

FIGURE 5-11 CELL DIVISION. (a) Cell division in prokaryotes is called binary fission. In this process the cell duplicates its single chromosome, the two chromosomes are separated, and the cell splits into two. **(b)** In eukaryotes cell division (cytokinesis) is preceded by mitosis, division of the nucleus. Once the nucleus has been duplicated and the two new nuclei separated to opposite sides of the cell, the cell divides. In organisms with cell walls, cell division is accomplished by forming a new cell wall between the two nuclei. In animals, cell division involves the formation of furrows that deepen until the cell is divided into two.

LEVELS OF ORGANIZATION

Although many marine organisms are composed of only a single cell, others are made up of many cells that work together at various levels of organization to form an organism **(Figure 5-12)**. Cells that serve one particular function are grouped into types of **tissues**. For instance, many muscle cells join to form muscle tissue in animals, and tubular cells make up vascular tissue to carry liquids through plants. Several different tissues can combine to form more complex structures called **organs**. Organs also perform specialized functions. The stomach is an animal organ whose function is digestion: It is composed of tissues that line the inside and outside of the organ and nerve and muscle tissue. Groups of organs can combine to form **organ systems**. For instance, an animal digestive system can be made up of an oral cavity, pharynx, stomach, and intestines, all working together to fully digest food. The most complex organisms are made up of several organ systems all working together to maintain the organism's life.

Tissues are groups of cells that perform a specific function.

Organs are structures composed of different tissues that perform a specific function.

An **organ system** is a group of organs that perform a specific function.

IN SUMMARY

- All living things are composed of cells.
- Prokaryotic cells lack a nucleus and membrane-bound organelles.
- Eukaryotic cells have a nucleus and many membrane-bound organelles.
- The organelles found in eukaryotic cells perform specific functions within the cell, much like organs within a multicellular organism.
- The molecule ATP supplies energy for metabolism and other cellular activities.
- All cells reproduce by a process of cell division, giving rise to new cells.
- In multicellular organisms, groups of similar cells form tissues.
- Different tissues combine to form organs, and organs can combine to form organ systems.

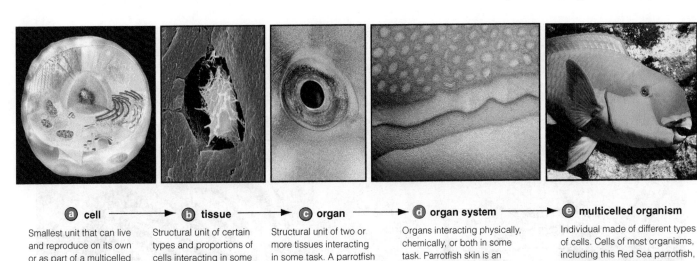

a cell → **b tissue** → **c organ** → **d organ system** → **e multicelled organism**

Smallest unit that can live and reproduce on its own or as part of a multicelled organism. It has an outer membrane, DNA, and other components.

Structural unit of certain types and proportions of cells interacting in some task. Many cells (*white*) made this bone tissue from their own secretions.

Structural unit of two or more tissues interacting in some task. A parrotfish eye is a sensory organ used in vision.

Organs interacting physically, chemically, or both in some task. Parrotfish skin is an integumentary system with tissue layers, organs such as glands, and other parts.

Individual made of different types of cells. Cells of most organisms, including this Red Sea parrotfish, are organized as tissues, organs, and organ systems.

FIGURE 5-12 LEVELS OF ORGANIZATION. Individual cells can group together as a unit and form tissues that perform a single function. Several types of tissue come together to form organs, and groups of organs performing the same function make up an organ system. Organ systems working together form an organism. *(Lisa Starr with PDB file courtesy of Dr. Christina A. Bailey, Department of Chemistry and Biochemistry, California Polytechnic State University, San Luis Obispo, CA.)*

5-3 Evolution and Natural Selection

The sea is home to large numbers of living organisms, each one seemingly well suited to survive and reproduce in its particular habitat. How and when did these organisms evolve, and what role does the environment play in determining the characteristics of organisms that can live in any given area? The answers to these and other questions lie in **evolution**, the process by which populations of organisms change over time.

DARWIN AND THE THEORY FOR EVOLUTION

In 1831, Charles Darwin set sail aboard HMS *Beagle* for a voyage of discovery around the world. During the next five years, he observed and collected specimens from many parts of the globe and began to formulate his ideas on the mechanism by which evolution operates.

Voyage of Discovery

In the years before Darwin's great voyage, geologists were finding that the earth was much older than most people believed and that layers of rock of different ages contained fossils of life forms that were different from but similar to living species **(Figure 5-13)**. Observations such as these laid the groundwork for the theory of evolution by natural selection that Darwin would eventually propose. Darwin was aware of these fossil findings as he set sail on the *Beagle*. As he traveled, he read books by the geologist Charles Lyell in which Lyell outlined his hypotheses for geological change. Lyell proposed that physical features of the earth such as mountains and valleys were formed over long periods by the same slow geological processes, such as uplifting and erosion, that occur today. Darwin was greatly influenced by two conclusions drawn from the observations of geologists such as Lyell. First, if geological change is slow and continuous, then the earth must be very old. Second, slow and subtle changes that occur over centuries and millennia can produce substantial changes. Darwin realized that the age of the earth allowed ample time for gradual changes to occur in different populations of organisms and for new forms of organisms to arise.

As he observed and collected specimens from around the world, Darwin was amazed at the diversity of living organisms. He marveled at how each organism he observed seemed to have what he referred to as a "perfection of structure" for doing whatever was necessary to survive and produce offspring. He wondered what natural processes might be responsible for the evolution of the diverse life forms and their beneficial characteristics **(Figure 5-14)**.

Formulating a Theory for Evolution

Following his return to England, Darwin began to document his ideas on the process of evolution. In 1838, he read a 40-year-old essay by the English mathematician Thomas Malthus. Malthus had observed that the human population was continuing to grow and, if unchecked, would eventually run out of space and food. Malthus believed that the consequences of uncontrolled population growth would be famine, disease, and war, and that these consequences were external controls that would keep the human population from growing too large. If this were true of humans, who have relatively few offspring, Darwin reasoned, it must be even more true of plants and animals that produce numerous offspring. After years of work, Darwin came up with a scientific explanation for why populations generally do not exhibit unchecked growth and how they can change over time. He called his hypothesis "evolution by natural selection."

> **Evolution** is the process whereby populations of organisms change over time.

For many years, Darwin discussed his ideas with only a few scientific colleagues. It was not until 1858, when he read a paper by the naturalist Alfred Russell Wallace, that he was encouraged to present his views.

John Cancalosi/Photolibrary/Getty Images

FIGURE 5-13 FOSSILS. This fossil crinoid is nearly identical to species living today in the ocean depths.

Wallace had been working in the country now known as Malaysia and had formulated a theory for evolution similar to the one Darwin had been working on for years. Wallace's paper, and a paper by Darwin were presented at a meeting of the Linnaean Society in 1858. A year later, Darwin published his classic work, *On the Origin of Species by Means of Natural Selection*.

Theory of Evolution by Natural Selection

Darwin was quite familiar with the process known today as **artificial selection**, which is practiced by farmers and animal breeders. In artificial selection, only animals or plants with certain desirable traits are selected for breeding in an effort to produce more animals or plants with the same traits. Darwin believed that a similar process was occurring in nature as the result of what he called **natural selection**, a process that favors the survival and reproduction of those organisms that possess variations best suited to their environment.

Unlike artificial selection, natural selection acts without the purpose of a farmer or a breeder. Note that natural selection does not have the ability to cause variations that are better suited than others. The variations that occur are due to chance mutations. Only after the variations appear can they be affected by natural selective forces–the physical and biological

Artificial selection is the practice of allowing only animals or plants with certain desirable traits to reproduce.

Natural selection is the process that favors the survival and reproduction of organisms that possess variations best suited to their environment.

characteristics of the environment (such as temperature, salinity, predation, availability of food, and so on) that favor the survival of one organism over another. In fact, the process of evolution operates without any ultimate goal, selecting those forms of organisms that are best able to survive and reproduce under the environmental conditions in which

Frederick R. McConnaughey/Photo Researchers, Inc.

FIGURE 5-14 BENEFICIAL ADAPTATIONS. This triggerfish is adapted for the aquatic life on a coral reef. Fins help the fish maneuver in water, and the laterally compressed body allows the fish to easily navigate among the corals. Powerful jaws allow the triggerfish to feed on small animals with protective body coverings that are found on the reef.

they live. According to Darwin's theory, selection occurs over many generations as organisms become better adapted by a process that can involve all aspects of an organism, such as anatomy, physiology, and behavior.

Darwin's theory of evolution by natural selection contains four basic premises:

1. All organisms produce more offspring than will survive to reproduce.

2. There is a great deal of variation in traits among individuals in natural populations. Many of these variations can be inherited.

3. The amount of resources necessary for survival (food, light, living space, and so on) is limited. Therefore organisms must compete with each other for these resources.

4. Those organisms that inherit traits that make them better adapted to their environment are more successful in the competition for resources. They are more likely to survive and produce more offspring. The offspring inherit their parents' traits, and they continue to reproduce, increasing the number of individuals in a population with the adaptations necessary for survival.

Some modern evolutionary biologists believe that many traits become established in a population without necessarily being advantageous. Some traits may remain not because they are beneficial, but because they confer no disadvantage. Others may be linked to traits that are beneficial, and the linked trait is "extra baggage" that is not harmful. To put it another way, the combination of traits exhibited by a successful organism represents a balance among several selective forces. **Table 5-2** summarizes some key points concerning the evolutionary process.

TABLE 5-2 EVOLUTION: WHAT, WHO, HOW

	Random Genetic Changes	Natural Selection	Evolution
What Is It?	The raw material acted upon by natural selection	Difference in reproductive success between genetically different individuals of a population	Genetic change in a population from generation to generation
Whom Does It Affect?	Only individuals	Individuals and populations	Only populations
How Does It Work?	Produces an organism with an individual genetic makeup that: 1. is unique 2. may be inherited by the organism's offspring	Produces an organism with genetic differences that: 1. are advantageous under the existing environmental conditions 2. make it more likely to survive and reproduce, passing these advantages on to offspring	Natural selection affects generations and produces a population that is better adapted to existing environmental conditions
		Wrong: "Individuals can evolve." Right: "Individuals do not evolve." The genes that determine the anatomical features of the mouth cannot be modified in an individual. Therefore, this seahorse has no way to change its mouth to bite or chew prey if the small organisms that it feeds on were to become scarce or disappear completely.	
		Wrong: "Evolution produces perfection." Right: "Organisms are not perfectly adapted." Almost all adaptations represent a compromise in which the adaptation has advantages and disadvantages. For example, a turtle's shell provides good protection against sharks and other predators, but this same shell slows locomotion, decreases maneuverability, and makes reproduction challenging.	
		Wrong: "Evolution has a purpose." Right: "Evolution involves chance." Selective factors are unpredictable, as is the course of evolution. For example, when shoreline rocks are continuously submerged, this rock anemone survives and reproduces. If geological changes or human intervention causes these same rocks to be raised above the water line, the same genes that allowed the anemone to survive previously will not help it survive in the dry environment, and the anemone will die.	
Bacterium magnified 30,000,000 times		Wrong: "More evolved equals better." Right: "New species are not better than older species." Any species alive today is successful (so far), whether it originated 200 years ago or 200 million years ago. Is a bacterium that evolved a billion years ago better than a humpback whale? It is less beautiful to the human eye, perhaps, but the bacterium is just as successful. Indeed, bacteria may very well outsurvive the humpback whale and all of its relatives.	

GENES AND NATURAL SELECTION

At the time Darwin proposed his theory of evolution by natural selection, cell division, genes, and chromosomes had not yet been discovered. Since that time, the discovery of DNA and its role in heredity has revolutionized the field of biology. Modern discoveries in the fields of **genetics** (the study of heredity) and **molecular biology** (the study of the structure and function of nucleic acids such as DNA) have not contradicted Darwin's theory but have expanded and deepened our understanding of the evolutionary process.

Modern Evolutionary Theory

Today's view of how evolution occurs, the **modern synthetic theory of evolution**, is essentially Darwin's 1858 idea as it has been refined by modern genetics. Recall from our previous discussion that a gene is a unit of hereditary information. It contains the chemical instructions for making proteins, and these proteins are largely responsible for the structure and function of organisms–in other words, their traits. Genes produce traits, such as body color, longer fins, or better vision, when the genetic information is translated into proteins by protein synthesis. The protein molecules can then form parts of the organism (structural proteins), direct the organism's biological activities (chemical messengers), or aid in the production of other molecules necessary for the organism's structure and function (enzymes).

Genetics is the branch of biology that involves the study of heredity.

Molecular biology is the branch of biology that involves the study of the structure and function of nucleic acids.

The **modern synthetic theory of evolution** is an explanation for evolution that combines natural selection with modern genetics.

Alleles are different forms of a gene.

Asexual reproduction is the process whereby offspring are produced from a single parent without the fusion of sex cells.

Sexual reproduction is the process whereby offspring are produced by the fusion of sex cells produced by each of two parents.

The **haploid number** is the number of different chromosomes in an organism.

The **diploid number** is twice the haploid number.

Gametes are sex cells.

Genes can exist in different forms called **alleles**. We now know that the variation in natural populations is largely the result of the alleles present in individual organisms. For instance, a snail can have a gene for shell color (the trait) and there can be several alleles (alternative forms) of the gene, so that some snails would have red shells, others yellow or brown **(Figure 5-15)**. Each snail would have two of each kind of chromosome in its cells (one from each parent) and thus would have two alleles for each gene. The alleles could be the same, for instance, two yellow alleles, or different, one yellow and one red. There would be many possible combinations and variations in shell color from the same genes. Because chromosomes from the two parents are randomly mixed in their offspring, several offspring from the same parents can receive different combinations of alleles. This accounts for the variations among offspring of the same parents.

Role of Reproduction

Genes are transmitted from one generation to another when DNA in the form of chromosomes is passed from parents to offspring in the process of reproduction. Reproduction can involve a single parent (asexual reproduction) or two parents (sexual reproduction).

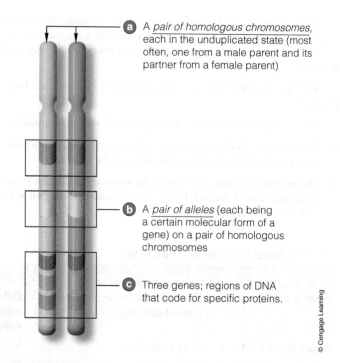

ⓐ A *pair of homologous chromosomes*, each in the unduplicated state (most often, one from a male parent and its partner from a female parent)

ⓑ A *pair of alleles* (each being a certain molecular form of a gene) on a pair of homologous chromosomes

ⓒ Three genes; regions of DNA that code for specific proteins.

© Cengage Learning

FIGURE 5-15 GENES. Chromosomes contain an organism's genes. Each gene can exist in more than one form, and each form of a gene is called an allele.

Asexual Reproduction In **asexual reproduction (Figure 5-16a)**, offspring receive their genes from only one parent, and as a result, they are identical to the parent in all of their characteristics (clones). This type of reproduction introduces no variety into the population. The only source of variation is mutation. Asexual reproduction is common in organisms that live in stable environments where there is a benefit to reproducing many individuals in a short time. Asexual reproduction is very efficient because every member of the population can reproduce and mating is not required. Populations of microorganisms can achieve high population densities (e.g., blooms) rapidly through asexual reproduction.

Sexual Reproduction In **sexual reproduction** chromosomes or parts of chromosomes from two parents are combined, often by fertilization **(Figure 5-16b)**. The resulting recombination of genes forms a genetically different individual, whereas asexual reproduction produces clones of identical individuals. Sexual reproduction results in greater variation on which natural selection can act, but it is not as efficient as asexual reproduction because large amounts of resources and energy are wasted in forming cells that will not be fertilized or survive. Also, only those organisms successful in having a mate are able to reproduce; this limitation is itself a situation on which natural selection acts. Although the process of sexual reproduction takes many forms in prokaryotic and eukaryotic organisms, the typical situation in eukaryotes will be described here.

Recall that eukaryotic cells may have either one or two copies of each of the organism's different chromosomes. Organisms with one copy are said to have a **haploid number**, represented by N, where N is the number of different chromosomes. Organisms with two copies have a **diploid number** ($2N$). Sexual reproduction in eukaryotic organisms occurs when haploid cells, usually called **gametes**, fuse, producing one

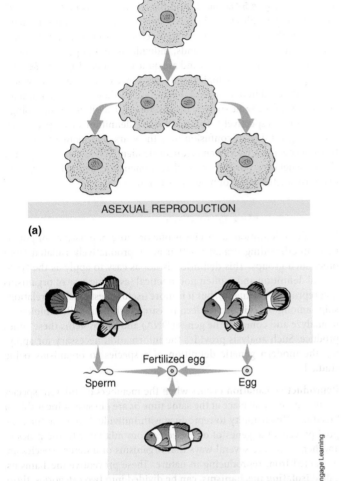

ASEXUAL REPRODUCTION

(a)

Sperm Fertilized egg Egg

SEXUAL REPRODUCTION

(b)

© Cengage Learning

FIGURE 5-16 ASEXUAL AND SEXUAL REPRODUCTION. **(a)** Asexual reproduction requires only one parent and produces offspring that are genetically identical to the parent (clones). **(b)** Sexual reproduction requires two parents to contribute genes to the next generation. This results in offspring that have new combinations of genes (half from each parent) and are not identical to their parents.

diploid cell, the **zygote**, by fertilization. Life cycles of sexually reproducing eukaryotes include at least an alternation between individuals or stages that are haploid and diploid. Life cycles occur in amazing variety, and they might include periods of asexual reproduction. In some organisms, gametes are almost indistinguishable from each other, whereas in others they are clearly distinct as eggs and sperm. Haploid and diploid stages can vary widely in life span, and they may be single celled or multicellular. In some organisms there might be additional haploid and diploid stages that differ in their contribution to feeding, dispersal, and reproduction. In all variants of the life cycle one problem is clear: At some point, a diploid (2N) stage must become a haploid (N) stage. Otherwise, if the number of chromosomes were not reduced to one copy of each, one generation would have cells with twice as many chromosomes as the previous generation. The potential problem of chromosome doubling with each generation is solved by the process of meiosis.

Meiosis Haploid cells are formed by a special kind of cell division called **meiosis**, or reduction division, from diploid cells. Meiosis is called reduction division because it produces cells with nuclei containing only one half the number of chromosomes (N) as the parent cell (2N). The main difference between meiosis and mitosis (described earlier in the chapter) is that chromosomes are duplicated once in meiosis, but the cell divides twice **(Figure 5-17)**. Meiosis produces four cells, each with one of each kind of chromosome. The resulting haploid cells are called gametes if they soon fuse in sexual reproduction and **spores** if they later undergo mitosis in asexual reproduction. A series of generations of spores or other haploid stages might eventually produce cells asexually that become gametes, which then form the diploid zygote by fertilization.

Meiosis not only reduces the number of chromosomes in the cell's nucleus but also increases the amount of variation in populations. During the initial phase of meiosis, all like chromosomes connect with each other for a short time. During this phase, alleles of the same gene on neighboring chromosomes can switch from one chromosome to another. This **crossing over** produces chromosomes with combinations of alleles that did not exist before the meiotic division. This recombination of alleles increases the amount of variation within populations. As meiosis continues, more variety is added as the chromosomes are randomly assorted into new combinations in the gametes or spores that did not exist in the cells of the previous generation. During fertilization, a male gamete fuses with a female gamete to produce a new individual that has inherited half its chromosomes from one parent and half from the other.

A **zygote** is a cell formed by the fusion of two gametes during fertilization.

Meiosis is a form of cell division in which haploid cells are produced from diploid cells.

Spores are haploid cells that undergo mitosis in asexual reproduction.

Crossing over is a process that occurs during meiosis in which alleles of the same gene on neighboring chromosomes switch from one chromosome to another.

A **gene pool** is all of the alleles in a given population that exist at a given time.

As you can see, the sexual reproduction that occurs in each generation produces a constant reshuffling of alleles in the species' **gene pool**, all of the alleles in a given population that exist at a given time. This random mixing of alleles by sexual reproduction, along with the random sorting of alleles during gamete production and other genetic processes such as mutation, creates variations among different individuals of a species. The variations are then acted on by natural selection.

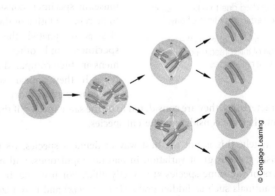

© Cengage Learning

FIGURE 5-17 MEIOSIS. In sexually reproducing organisms, sex cells are formed by meiosis. In meiosis, the number of chromosomes in a cell is reduced by one half.

Population Genetics

For a population of organisms to survive and reproduce under changing environmental conditions, it must be able to adapt to them. A population's ability to adapt is limited by the genes and alleles of those genes carried by the individuals in the population. Only those individuals that have combinations of genes and alleles that allow adaptations to their particular surroundings are likely to survive and leave offspring. An organism's biological success, or **fitness**, is measured by the number of its own genes present in the next generation. The more offspring an individual contributes to the population, the more biologically successful that individual is.

Fitness is an organism's biological success as measured by the number of its own genes present in the next generation.

A **species** (older definition) is an organism with a definable set of characteristics that is visibly different from other similar organisms.

A **species** (modern definition) is one or more populations of potentially interbreeding organisms that are reproductively isolated from other such groups.

Morphology is the appearance of an organism.

A **type specimen** is a museum specimen considered to be representative of the species it represents.

Sexual dimorphism is the condition of males and females having different morphology.

Isolating mechanisms are mechanisms that prevent different species from interbreeding in nature.

Habitat isolation is the prevention of interbreeding due to populations occupying different habitats.

Anatomical isolation is the prevention of interbreeding due to members of different species having incompatible copulatory structures.

Behavioral isolation is the prevention of interbreeding due to differences in mating behaviors.

Temporal isolation is the prevention of interbreeding due to differences in breeding seasons or time of day.

Biochemical isolation is the prevention of interbreeding due to biochemical differences.

EVOLUTION OF NEW SPECIES

Each different kind of organism is a different **species**. What exactly, though, is a species? The definition of species has changed over the years as our understanding of evolution and molecular biology has improved.

Typological Definition of Species

In the past, a *species* was defined as an organism with a definable set of characteristics that is visibly different from other similar organisms. Two fishes the same size and shape but having a different color pattern would be considered different species in the historical sense. This definition of species is called typological and is based on **morphology**, or the structure and appearance of the organism. For the purpose of identification, a museum specimen considered to be representative of the species is designated the **type specimen**, and other specimens are then compared to the type. If they appear similar based on certain morphological characteristics, they are considered to be the same species; if they are different, they are considered different species.

Unfortunately, this is not the best way to define a species. As noted previously, there is great variation in natural populations, and not all members of the same species are exactly alike. For instance, in some marine animals such as fiddler crabs (*Uca pugnax*) and rosy razorfish (*Hemipteronotus martinicensis*), the males **(Figure 5-18a)** and females **(Figure 5-18b)** look quite different—a condition known as **sexual**

dimorphism. In other species such as the bluehead wrasse (*Thalassoma bifasciatum*; **Figure 5-18c and d**) and the spotted hogfish (*Bodianus pulchellus*), the juveniles look distinctly different from the adults. In many marine snails, such as the communal nerite (*Nerita communis*), which is found in the Philippine Islands, individuals in a population show a variety of colors and patterns and are quite dissimilar **(Figure 5-18e)**. According to the strict typological concept of species, these animals would be considered different species, when indeed they are really the male and female of the same species, the juvenile and adult forms, or ecological variants, respectively. The typological definition of a species has caused a great deal of confusion over the years, and many of the same types of organism have been erroneously identified as separate species. Today, whenever possible, several specimens are defined as types to demonstrate the range of natural variation.

Modern Species Definition

By modern definition, a species is one or more populations of potentially interbreeding organisms that are reproductively isolated from other such groups. This definition is not as easy to apply as the typological definition and is often not practical (as in the case of organisms that reproduce asexually), but it is more useful in establishing relationships among organisms. Modern research techniques allow biologists to analyze and compare the genes (DNA) and the proteins these genes produce. Such analysis provides the information necessary for applying the modern genetic definition of a species to organisms being studied.

Reproductive isolation occurs when the members of different species are not in the same place at the same time or are physically incapable of breeding. This inability to reproduce with individuals outside the gene pool prevents the genes of one species from mixing with the genes of another. There are several ways that organisms of different species are prevented from reproducing in nature. These preventive mechanisms, called **isolating mechanisms**, can be divided into two categories: those that prevent fertilization from ever occurring and those that prevent successful reproduction following fertilization.

Isolating Mechanisms that Prevent Fertilization Isolating mechanisms that prevent fertilization include differences in habitat, breeding time, anatomy, and behavior. For instance, if one species of snail lives on sandy beaches in the Caribbean, and a similar species lives in mangrove swamps, they will not be able to interbreed because they never encounter each other. This mechanism is referred to as **habitat isolation**.

In invertebrate animals different species frequently have incompatible copulatory organs that prevent similar species from reproducing with each other. This is an example of **anatomical isolation**.

Some animals exhibit special behaviors during the breeding season, and only members of the same species recognize the behavior as courtship. This constitutes **behavioral isolation**. The male common cuttlefish (*Sepia officinalis*), for instance, presents a striped color pattern that identifies him to a female *Sepia officinalis* as a prospective mate. Only female *Sepia officinalis* recognize the color pattern as a mating display. Members of different species do not recognize or respond to this courtship display and thus do not attempt to mate with the displaying individual.

Most organisms reproduce only at certain times of the year or under certain conditions (temperature, salinity, hours of light). If the time that members of one species are ready to reproduce does not coincide with

(a)

(b)

(c)

(d)

(e)

FIGURE 5-18 MORPHOLOGICAL DIVERSITY WITHIN A SPECIES. Some species, such as this rosy razorfish *(Hemipteronotus martinicensis)*, exhibit sexual dimorphism, that is, the **(a)** male and **(b)** female look distinctively different. **(c)** The juvenile bluehead wrasse *(Thalassoma bifasciatum)* does not look the same as **(d)** the adult of the species. **(e)** Some marine organisms, such as these snails *(Nerita communis)*, exhibit a large range of variation as adults.

the time members of a related species are reproducing, the two species will be separated by **temporal isolation**.

Members of closely related species that succeed in copulating generally do not produce offspring because biochemical differences between their gametes prevent successful fertilization. This is referred to as **biochemical isolation**.

Isolating Mechanisms that Prevent Successful Reproduction Following Fertilization Incompatible genes can prevent a fertilized egg from developing further. In some instances, fertilization and development may be successful, but the offspring, called **hybrids**, may be infertile (unable to reproduce) or ecologically weak (unable to successfully compete with the well-adapted parental species) and thus quickly die. These mechanisms prevent successful reproduction even if mating and fertilization do take place.

Process of Speciation

Speciation refers to the mechanisms by which new species arise. Speciation frequently begins when two or more populations of the same species become geographically isolated from each other or when segments of a single large population become geographically isolated from other segments. This type of speciation is called **allopatric** ("different homeland") **speciation**. Although allopatric speciation is not the only type of speciation, it is thought to be responsible for the formation of the greatest

Hybrids are the offspring of two different but closely related species.

Speciation is the mechanism whereby new species arise.

Allopatric speciation is the process whereby new species are formed when populations become geographically isolated from each other.

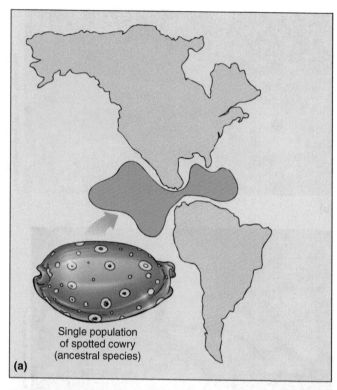

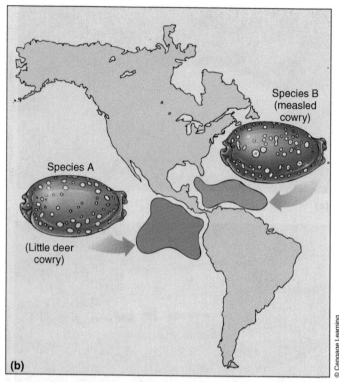

FIGURE 5-19 **ALLOPATRIC SPECIATION.** **(a)** More than 4 million years ago, the Pacific Ocean and the Caribbean Sea were connected and home to many populations, such as the ancestral population of spotted cowries. **(b)** When the land mass known as Central America was formed between 3 and 4 million years ago, many animal populations, including the spotted cowry population, became geographically isolated from each other, resulting in the evolution of similar yet distinct species. The measled cowry (*Cypraea zebra*) is found in the Gulf of Mexico, and the little deer cowry (*Cypraea cervinetta*) is found along the Pacific coast of Central America. These two species are quite similar and very likely share a common ancestor. When brought together today in an aquarium, they can no longer mate successfully.

number of species. Geological changes such as earthquakes, mountain building, and the formation of islands can separate populations. This geographic isolation because of geological change is often the first step in evolution. Geographic isolation is followed by the appearance of biological isolating mechanisms because, once two populations have become physically isolated, there is no longer any gene flow between them. From that point on, natural selection will operate on each population independently. This is especially true when the separated populations are subject to different environmental conditions.

For example, more than 4 million years ago, the Caribbean Sea and Pacific Ocean were connected between North and South America. The geological changes that formed Central America between 3 and 4 million years ago separated many species of fish, invertebrates, and seaweeds into Pacific and Caribbean populations. These isolated populations evolved independently as they adapted to the different temperature, salinity, sediments, food availability, predators, and other conditions in the two seas. Although some species from the two sides of Central America are morphologically similar, others have diverged in such traits as color, habitat preference, breeding behavior, and breeding season. These differences act as reproductive isolating mechanisms, and the populations would no longer interbreed, even if the physical barrier between them is removed or if they jointly colonize a common area. These populations, formerly the same species, have become separate species **(Figure 5-19)**.

IN SUMMARY

- Charles Darwin introduced the theory of evolution by natural selection in 1859 in his book *On the Origin of Species by Means of Natural Selection.* (Have You Wondered? #2)

- Natural selection functions by favoring the survival of individuals with the combination of traits best suited to their particular place and time.

- Over long periods, new characteristics accumulate in populations, and new, better-adapted species gradually replace older ones that are not as well adapted.

- The modern synthetic theory of evolution is basically Darwin's idea refined by modern genetics.

- Individuals that possess genes for traits that best adapt them to their environment are the most likely to survive and reproduce, thus contributing more of their genes to the population's gene pool.

- Biological success is measured in terms of the number of genes an organism contributes to the population.

- In the past, a species was defined as a group of organisms with an observable set of characteristics that was different from other, similar organisms. (Have You Wondered? #1)

- Today a species is defined as one or more populations of potentially interbreeding organisms that are reproductively isolated from other such groups. (Have You Wondered? #1)

- New species may form when environmental or genetic changes cause portions of a population to become reproductively isolated from the rest of the group and natural selection acts independently on them. (Have You Wondered? #3)

5-4 Classification: Bringing Order to Diversity

We use several schemes to classify the organisms we study in marine biology. The terms *plankton, nekton,* and *benthos* classify organisms based on where and how they live. The terms *herbivore, carnivore,* and *decomposer* classify them according to ecological roles. To deal with the problems of classifying the large number of organisms that inhabit the earth, biologists have developed systems that use physical and molecular characteristics to organize groups of organisms into domains and their subdivisions to reflect their evolutionary relationships.

While studying living organisms, biologists find it helpful, and indeed necessary, to use a universal system of naming the more than 1 million known species. Naming organisms enables scientists all over the world to exchange information about them. Most organisms do not have non-scientific names, and those that do have names that vary from one country to another and even among geographical regions within a country. Biologists need a system of naming things so that every researcher in the field calls the same organism by the same name.

LINNAEUS AND THE BINOMIAL SYSTEM OF NAMING

The system that biologists now use for naming organisms was introduced by Carl Linnaeus, a Swedish botanist, in the mid-1700s. During Linnaeus's time, organisms were classified on the basis of a lengthy description. Linnaeus found this method too cumbersome and developed a standardized method that he refined over the years. In 1753, he introduced the idea of **binomial nomenclature**, a system that uses two names to identify an organism. The first name of a proper scientific nomenclature is the **genus** (plural, genera), and Linnaeus placed all organisms that shared certain physical characteristics in the same genus. For instance, almost all butterflyfishes belong to the genus *Chaetodon* **(Figure 5-20)**. The second name is the **species epithet**, which identifies a particular kind, or species (plural is also species), of organism in the genus. For instance, *Chaetodon longirostris* is the longnose butterflyfish, whereas *Chaetodon ocellata* is the spotfin butterflyfish. The first letter of the genus name is always capitalized and the species epithet is always in lowercase. When the scientific name appears in print, it is set in italics or underlined.

> **Binomial nomenclature** is a system of nomenclature that uses two names to identify an organism.
>
> The **genus** is the first name of a proper scientific nomenclature.
>
> The **species epithet** is the second part of the two-part scientific name.

When Linnaeus developed the system of binomial nomenclature, he placed organisms into genera on the basis of similar morphology. He then grouped organisms into larger categories called classes and orders. These larger groups were made up of organisms that bore a superficial resemblance but were not in all cases related. Today, we still use the binomial system of nomenclature but organisms are now classified on the basis of evolutionary relationships, not solely on the basis of structural similarities. If two species share a genus name, it means they not only may appear similar, but there is evidence that they also share a common ancestor. Common ancestry is determined by the number of shared characteristics and molecular similarities.

(a) Jon Hawker

(b) Fred McConnaughey 1985/Photo Researchers

FIGURE 5-20 BUTTERFLYFISHES. Two different species of butterflyfish, **(a)** *Chaetodon longirostris* (long-nose butterflyfish), and **(b)** *Chaetodon ocellata* (spotfin butterflyfish).

TAXONOMY: THE SCIENCE OF CLASSIFICATION

Taxonomy is the branch of biological science that deals with classification. Several categories are currently used to show the complex evolutionary relationships among related organisms. The major categories, from most inclusive to least inclusive (most specific), are domain, kingdom, phylum, class, order, family, genus, and species. Students should always keep in mind that classification is a human endeavor, and that the important thing is not the memorization of these or other categories but a basic understanding of why each organism is classified as it is.

Early Schemes of Classification

During Linnaeus's time, and until the 1960s, biologists grouped all living organisms into one of two kingdoms: Plantae and Animalia. The kingdom Animalia contained the organisms that fed on other organisms (heterotrophs) and were capable of independent movement. All other organisms (autotrophs and saprotrophs, or decomposers) were classified in the kingdom Plantae.

With the invention of the electron microscope and the accumulation of more information from all fields of biology, it became apparent that many prokaryotes were more closely related to each other than they were to eukaryotes and that many unicellular organisms were not closely related to multicellular organisms. There were also major differences among organisms within the groups of multicellular organisms. This led to the creation of more kingdoms.

Modern Classification

With the advent of modern molecular techniques, our knowledge of the relationships among organisms has increased dramatically. Most biologists agree that the old system of classification does not adequately reflect the actual evolutionary relationships among organisms. Continuing studies at the molecular level have led to new categories, and classification schemes continue to change as evolutionary biologists attempt to accurately represent evolutionary relationships among organisms.

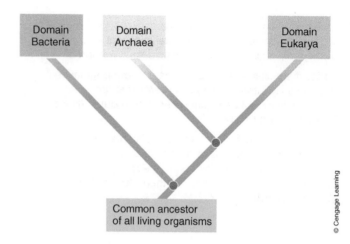

FIGURE 5-21 DOMAINS. Biologists currently recognize three domains of living organisms. The domains Eubacteria and Archaea contain the prokaryotes. All of the eukaryotes are assigned to the domain Eukarya.

Domains During the 1980s, biologists began to recognize that not all prokaryotes were closely related. Molecular studies resulted in a proposal for a new category higher than kingdom, the **domain**. The domains **Archaea** and **Eubacteria** both contain single-cell, prokaryotic organisms that differ from each other in several ways at the biochemical level. For instance, members of the domain Eubacteria have molecules containing amino acids and glucose, called peptidoglycans, in their cell walls while members of the domain Archaea do not. All eukaryotic organisms were placed in the domain **Eukarya**. Today, most biologists recognize the three-domain system of classification **(Figure 5-21)**.

Kingdoms The domains Eubacteria and Archaea each contain a single kingdom **(Figure 5-22)**. The domain Eukarya contains three well-defined

FIGURE 5-22 KINGDOMS. Domains are subdivided into kingdoms. The domains Eubacteria and Archaea each contain a single kingdom. The domain Eukarya contains several kingdoms. Note that some groups of protists are more closely related to plants, whereas other protists are more closely related to animals and fungi.

Taxonomy is the branch of biological science that deals with classification.

Domains are taxonomic categories higher than kingdoms.

Archaea is the domain that contains prokaryotes that lack peptidoglycans in their cell walls.

Eubacteria is the domain that contains prokaryotes that contain peptidoglycans in their cell walls.

Eukarya is the domain that contains all eukaryotic organisms.

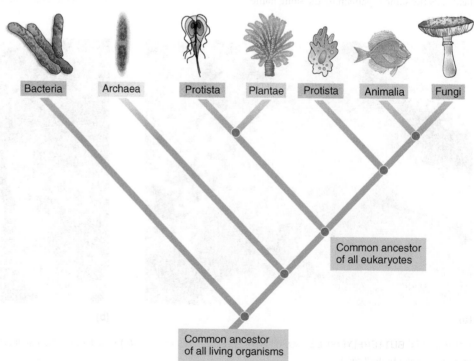

kingdoms and many phyla that are clearly not related by ancestry. The three kingdoms in the domain Eukarya are Fungi, Plantae (plants), and Animalia (animals). **Fungi** are eukaryotic organisms that are not capable of photosynthesis and whose cell walls contain the polysaccharide chitin. Some fungi are unicellular whereas others are multicellular. They are primarily saprotrophs, although there are some parasitic species. Yeasts, molds, and mushrooms are examples of fungi. **Plants** have cells with cell walls containing the polysaccharide cellulose. All plants are multicellular and capable of photosynthesis. **Animals** are all multicellular. Their cells lack cell walls and they are ingestive heterotrophs, meaning that they rely on other organisms for food, which they take into their bodies to digest.

Protists The domain Eukarya also contains the organisms generically referred to as **protists**. Protists are eukaryotic organisms that do not fit the definition of animal, plant, or fungus. Some examples of protists are algae, protozoans, and slime molds. Although the majority of protists are unicellular, there are also multicellular forms. Their cells may or may not have cell walls, and the composition of the cell walls can be quite variable. Nutritionally they can be autotrophs, heterotrophs, or saprotrophs, and some can even be combinations of two nutritional types. Some biologists group all protists into a separate

kingdom, Protista, but as more and more molecular studies are done, it has become clear that grouping all of these organisms together is artificial. It makes more sense to deal with these organisms as separate phyla in the domain Eukarya. Their relationships to each other and to other eukaryotes are complex, and biologists are just beginning to unravel them with the aid of modern molecular techniques. One of many possible schemes for classifying eukaryotes is shown in **Figure 5-23**.

Tracing Relationships

The evolutionary history (a record of the descendants of an ancestor) of a species or group of related species is called **phylogeny**. These relationships

Fungi are eukaryotic organisms that are not photosynthetic and have cell walls composed of chitin.

Plants are multicellular eukaryotic organisms that are photosynthetic and have cell walls composed of cellulose.

Animals are multicellular eukaryotic organisms that are heterotrophic and whose cells lack cell walls.

Protists are eukaryotic organisms that do not fit the definition of animal, plant, or fungus.

Phylogeny is the evolutionary history of a species or group of related species.

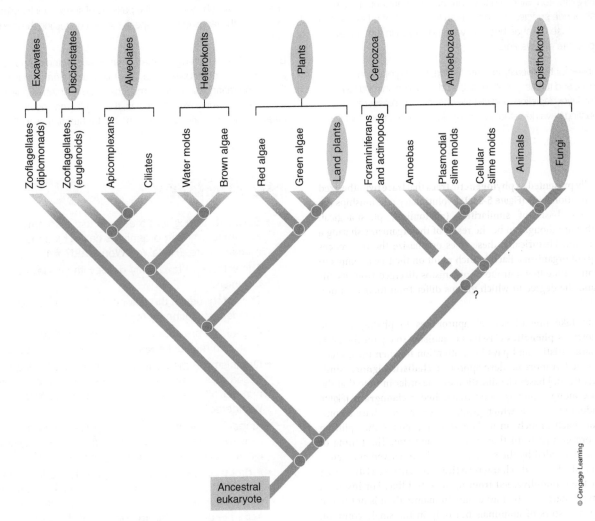

FIGURE 5-23 CLASSIFICATION OF EUKARYOTES. This diagram shows one of many possible classification schemes for the eukaryotes. Many taxonomists classify eukaryotes into eight major groups (indicated by the colored ovals above the brackets). The colored ovals below the brackets indicate the three well-defined eukaryotic kingdoms. All of the other groups indicated are protists.

© Cengage Learning

IMPACT BUBBLE

LIFE IN THE OCEAN is extremely diverse. In fact, 80% of all phyla contain marine organisms. Because it is difficult to study directly the marine life in much of the ocean, the full extent of marine biodiversity is unknown. Consequently, estimates of the number of marine species varies widely from hundreds of thousands to millions. We know little regarding the range of many species and the size of their populations. Even for marine species that are relatively easy to monitor, such as those living along shorelines, we have little data on biodiversity. Since historical data on marine biodiversity are virtually nonexistent, we do not know to what extent biodiversity has changed over the last several centuries. We need to be concerned with biodiversity because it maintains the health of ecosystems. It also provides economic and recreational value for human populations.

Changes in the size of populations can impact biodiversity. When the size of a population decreases, the genetic variation in that population is reduced. This loss of genetic diversity hampers a species' ability to adapt to environmental changes and stresses, such as global climate change. The inability of a species to adapt to environmental changes can lead to its extinction. Since species are not independent of each other, the loss of one species can often lead to the loss of others. When biodiversity decreases, niches become vacated, which in turn threatens the stability of entire ecosystems.

The major causes for biodiversity loss are fishing, hunting of marine mammals, birds, and turtles, pollution, habitat destruction, introduced species, ozone layer depletion, and climate change. The pressures of fishing and hunting have depleted some populations to the point that it is no longer economically feasible to harvest them. While not extinct, these species are no longer successfully filling their niche in their respective ecosystems. Nutrient and toxic chemical pollution are invariably associated with a reduction in biodiversity due to the toxic effects of pollutants. Some species can adapt to or even thrive under conditions that would be too stressful for other marine organisms, allowing them to invade niches that were previously occupied by other organisms. In some polluted areas, these organisms can come to dominate a biological community, thus changing its entire nature and function. This may lead to an even greater loss of biodiversity. Fishing operations, such as trawling, can destroy bottom habitats and decrease species populations. Coastal habitats are subject to a number of physical alterations, such as coastal development for homes and resorts and changes in the coastal landscape to support aquaculture. The introduction of non-native species typically results in the loss of native species as the introduced species thrive in the absence of natural predators or outcompete native organisms. Depletion of the ozone layer allows more potentially lethal radiation to reach the sea, damaging or killing planktonic organisms that are at the base of most oceanic food chains. Climate changes alter the patterns of species distribution, expanding the range of some species while contracting the range of others.

Biodiversity is threatened in many marine ecosystems, and concern about the loss of biodiversity is becoming an important public issue. Protecting marine biodiversity must become an international priority. If we do not conserve extant species, the results may not just be catastrophic for marine ecosystems but for humans as well.

---

are traditionally presented as **phylogenetic trees** that trace hypothesized evolutionary relationships **(Figure 5-24a)**. Evolutionary relationships are decided on the basis of similarities (anatomical, physiological, behavioral) that are thought to be the result of the organisms sharing a common ancestor. Historically, these trees emphasize the differences among groups of organisms. Each branch point on the tree attempts to show the relative time that a group of organisms diverged from its ancestral line and the degree to which groups differ from their common ancestor.

Biologists can take one of several approaches to phylogeny. The approach known as **phenetics** classifies organisms solely on the basis of similar characteristics and pays little attention to when these characteristics evolved. A more modern approach, **cladistics**, ignores similarity of structure and bases classification on the order in time that the branches arise along a phylogenetic tree called a **cladogram (Figure 5-24b)**. A cladogram tells us which groups of organisms share a common ancestor. Each branch on a cladogram represents the splitting of two or more groups from their common ancestor. The timing of branch points is decided by the sequence in which derived characteristics originated. **Derived characteristics** are characteristics that evolved after the group diverged from its ancestral line. For instance, hair is a derived characteristic found only in mammals; it is not found in the earlier ancestors of mammals but only in the single common ancestor—the first mammal. So it must have evolved when the common ancestor of mammals split from the existing animal line of reptiles at the time.

IN SUMMARY

- Biologists throughout the world use a common system for naming organisms called binomial nomenclature. (Have You Wondered? #4)
- The science of taxonomy deals with classifying organisms.
- The categories in the taxonomic scheme are organized to reflect evolutionary relationships.
- Organisms are subdivided into one of three domains based on molecular and physical characteristics.
- Organisms in the domains are placed into kingdoms based on several characteristics, such as cell type, number of cells, and feeding type. (Have You Wondered? #4)
- Kingdoms are divided into progressively less inclusive taxonomic categories as follows: phylum, class, order, family, genus, and species.
- A group of eukaryotes called protists are frequently placed into their own kingdom, but these organisms are not related evolutionarily and are better treated as separate phyla within the domain Eukarya.
- Biologist use diagrams such as phylogenetic trees and cladograms to reflect the evolutionary history of organisms.

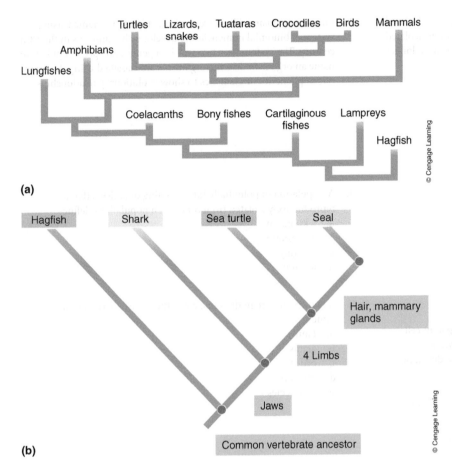

(a)

(b)

FIGURE 5-24 PHYLOGENY. **(a)** A phylogenetic tree of vertebrates. This traditional representation divides vertebrates into different groups based on physical features inherited from a common ancestor. **(b)** A cladogram showing the evolutionary relationships of some vertebrates. Each branch on this diagram represents the time that a group of organisms split from their common ancestor. The timing of these branch points was determined by the sequence in which derived characteristics originated.

Phylogenetic trees are diagrams that trace hypothesized evolutionary relationships among organisms.

Phenetics is a process for classifying organisms based solely on similar characteristics.

Cladistics is a process for classifying organisms based on the order in time that branches arise along a phylogenetic tree.

A **cladogram** is a diagram consisting of a series of branches, where each branch point represents novel characteristics that are unique to all species on the branch.

Derived characteristics are characteristics that evolved after a group diverged from its ancestral line.

KEY CONCEPTS

1. To understand living organisms, one must have a basic understanding of the variety of compounds from which organisms are built. (5-1)

2. Four groups of macromolecules are necessary for life: carbohydrates, lipids, proteins, and nucleic acids. (5-1)

3. All living organisms are composed of cells. (5-2)

4. Cells can be either prokaryotic or eukaryotic. (5-2)

5. Cells produce new cells by the process of cell division. (5-2)

6. Evolution is the process by which the genetic composition of populations of organisms changes over time. (5-3)

7. Natural selection favors the survival and reproduction of those organisms that possess variations that are best suited to their environment. (5-3)

8. A species is a group of physically similar, potentially interbreeding organisms that share a gene pool, are reproductively iso-

lated from other such groups, and are able to produce viable offspring. (5-3)

9. The binomial system of nomenclature uses two words, the genus and the species epithet, to identify an organism. (5-4)

10. Most biologists classify organisms into taxonomic categories that reflect theories about evolutionary relationships. (5-4)

11. Phylogenetic trees and cladograms indicate evolutionary relationships among groups of organisms. (5-4)

ARE YOU STILL WONDERING?

1. The modern definition defines a species as one or more populations of potentially interbreeding organisms that are reproductively isolated from other such groups.

2. Natural selection is the process by which nature, through physical and biological factors, selects organisms that possess characteristics best suited to survive better and have greater reproductive success in their particular environment than other less well adapted organisms.

3. New species evolve when members of a population become isolated from each other and natural selection acts separately on the isolated groups until they undergo sufficient change that they can no longer reproduce with each other.

4. Marine organisms, like terrestrial organisms, are named using the system of binomial nomenclature developed by Linnaeus in the 18th century. The system uses two words, a genus and species epithet, to name an organism. Marine organisms are classified using the methods of phenetics or cladistics to show evolutionary relationships.

QUESTIONS FOR REVIEW

Multiple Choice

1. If a cell lacked ribosomes, it would not be able to
   a. synthesize lipids
   b. digest food
   c. synthesize proteins
   d. form membranes
   e. signal to other cells

2. During the process of meiosis
   a. two new cells are formed that are identical to the parent cell
   b. chromosomes are not duplicated before division occurs
   c. the number of chromosomes in a cell's nucleus is reduced by one half
   d. two gametes fuse to form a new individual
   e. two cells are formed that contain twice as many chromosomes as the parent cell

3. Darwin proposed that evolution occurs as the result of
   a. cosmic forces
   b. human intervention
   c. artificial selection
   d. natural selection
   e. inherent need

4. A population of potentially interbreeding organisms that is reproductively isolated from other such populations defines
   a. a kingdom
   b. a community
   c. a family
   d. a genus
   e. a species

5. If two species are in the same class, they must also be in the same
   a. family
   b. genus
   c. order
   d. phylum
   e. subspecies

Matching

Match the following molecules with their appropriate function:

a. DNA	1.	The building block of protein.
b. ATP	2.	The molecule that contains an organism's genetic information.
c. amino acid	3.	The molecule used by homeotherms for insulation.
d. cellulose	4.	The molecule that provides the energy for most cellular processes.
e. lipid	5.	A structural molecule found in the cell walls of plants.

Match each of the following characteristics with the appropriate kingdom:

a. Bacteria	6.	Multicellular, heterotrophic organisms with cells lacking cell walls.
b. Protista	7.	Prokaryotic cells.
c. Plantae	8.	Heterotrophic organisms with cells having cell walls containing chitin.
d. Animalia	9.	Mostly unicellular eukaryotes that are not closely related from an evolutionary standpoint.
e. Fungi	10.	Multicellular, photosynthetic organisms with cells having cell walls containing cellulose.

Short Answer

1. How can you distinguish between flagella and cilia?

2. How can you distinguish prokaryotic cells from eukaryotic cells?

3. What are the four points of Darwin's theory of evolution by natural selection?

4. What are the higher categories in modern classification?

5. Describe how new species are formed.

6. Describe how populations can become reproductively isolated.

7. Describe the characteristics of organisms in each of the three domains.

8. Compare asexual and sexual reproduction.

9. Why does it make more sense not to place protists into one kingdom?

10. How does the phenetic approach to taxonomy differ from the cladistic approach?

Thinking Critically

1. While studying plankton samples, a friend asks you how you can distinguish protozoans from small animals such as crustaceans. What would you answer?

2. While on a field trip, you collect two snails that look very similar. How would you determine if they were the same or different species using the methods available in 1858? How would you determine it today?

3. While working in the field you collect several similar specimens of marine fishes from populations in two different localities. What experiments would you perform to determine if these fishes were the same or different species? What results might you expect from these experiments, and what might the results prove?

SUGGESTIONS FOR FURTHER READING

Carroll, S. B., B. Prudhomme, and N. Gompel 2008. Regulating Evolution, *Scientific American* 298 (5): 61–67.

Conniff, R. 2008. On the Origin of a Theory, *Smithsonian,* 39 (3): 86–93.

Edgar, G. J., C. R. Samson, and N. S. Barrett 2005. Species Extinction in the Marine Environment: Tasmania as a Regional Example of Overlooked Losses in Biodiversity, *Conservation Biology* 19 (4): 1294–1300.

Gould, S. J. 1999. Branching through a Wormhole, *Natural History* 108(2).

Quammen, D. 2007. A Passion for Order, *National Geographic* 211(6).

Stoeckle, M. 2003. Taxonomy, DNA, and the Bar Code of Life, *Bioscience* 53(9).

Vermeij, G. J. 2002. Why Are There No Lobsters on Land or Bats at Sea? The Answer Appears to Be a Biological Version of Beginner's Luck, *Natural History* 111(1).

Zimmer, C. 2008. What Is a Species? *Scientific American* 298 (6): 72.

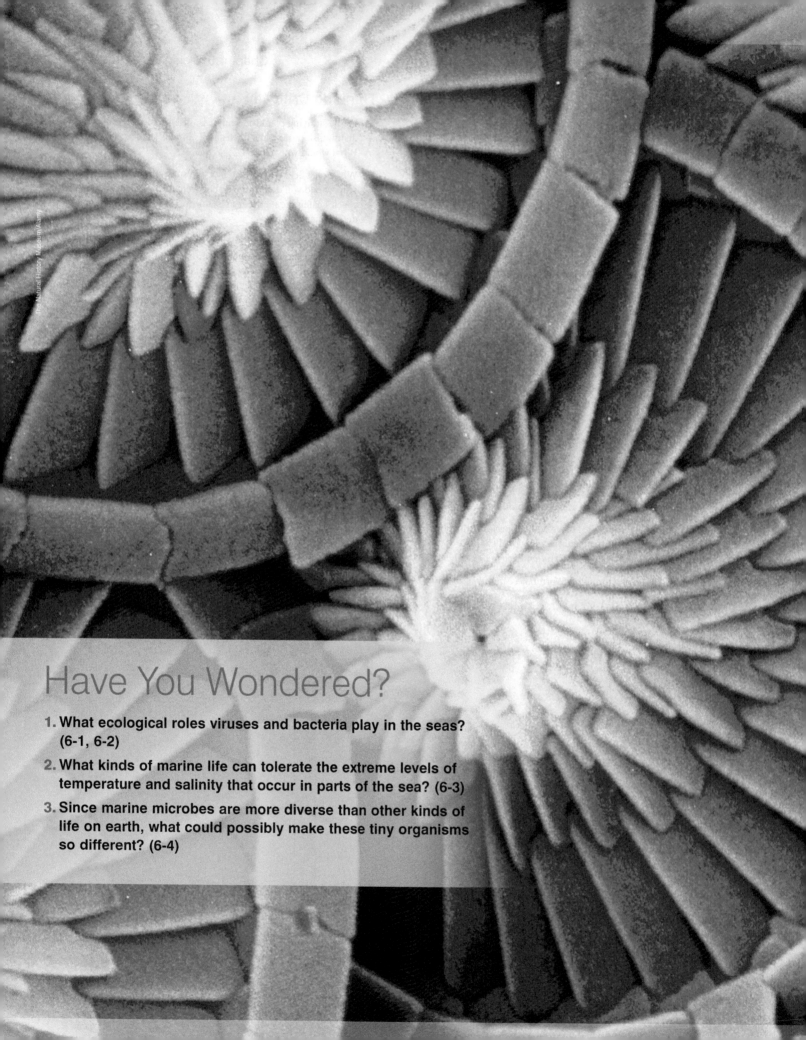

Have You Wondered?

1. **What ecological roles viruses and bacteria play in the seas? (6-1, 6-2)**

2. **What kinds of marine life can tolerate the extreme levels of temperature and salinity that occur in parts of the sea? (6-3)**

3. **Since marine microbes are more diverse than other kinds of life on earth, what could possibly make these tiny organisms so different? (6-4)**

6

Marine Microbes

MICROBES ARE ORGANISMS that are invisible to the naked eye. They include viruses, one-celled organisms, and occasionally fungi. Here we will deal with viruses and one-celled organisms and will include the fungi because most marine fungi are one-celled organisms or at least microscopic. These microbes belong to all three domains of life (Eubacteria, Archaea, Eukarya; **Figure 6-1**) and play many roles in marine ecosystems, such as producer, consumer, decomposer, mutualist, and parasite. Although unseen by people, they are the most numerous organisms in the sea.

6-1 Marine Viruses

As in so many biological disciplines, **virology**, the study of viruses, had a late start in the marine sciences. Viral diseases among humans and their domesticated crops and animals were known to ancient civilizations, and Edward Jenner's vaccinations against smallpox were reported more than 300 years ago. It was not until the mid-1900s, shortly after the chemical nature of viruses was becoming understood, that the first marine virus, one that infected bacteria, was discovered. Viruses of marine eukaryotic hosts were reported in the 1970s. Reliable counts of viral abundance made in the 1980s led to a better understanding of the ecological roles of marine viruses in the 1990s. The first decade of the new millennium was a period of molecular studies, with emphasis on genome sequencing and the diversity of marine viruses. Genomic studies continue to be an important focus in marine virology. We examine marine viruses in this chapter because of their abundance—greater than any other organism in the sea—diversity, and significance in marine food webs, population biology, and disease.

A **microbe** is a living organism too small to examine without using a microscope.

Virology is the study of viruses.

Viral replication is the manufacture and assembly of new viruses within a cell under the control of the infecting virus.

A **virologist** is a scientist who studies viruses.

VIRAL CHARACTERISTICS

Most authorities do not consider viruses to be alive because they are little more than stores of genetic information, bits of DNA or RNA surrounded by protein. Unlike living organisms, viruses have no metabolism. They rely entirely on host cells for energy, material, and organelles for duplicating themselves, a process called **viral replication**, and they can reproduce only inside a host cell. Despite this, **virologists** (scientists who study viruses) use terms such as "life cycle" and "live virus" that sound as if viruses are alive, and the two prevailing hypotheses about the origins of viruses postulate that they come from living cells. One hypothesis states that viruses are highly reduced prokaryotic parasites, whereas the other views them as renegade genes.

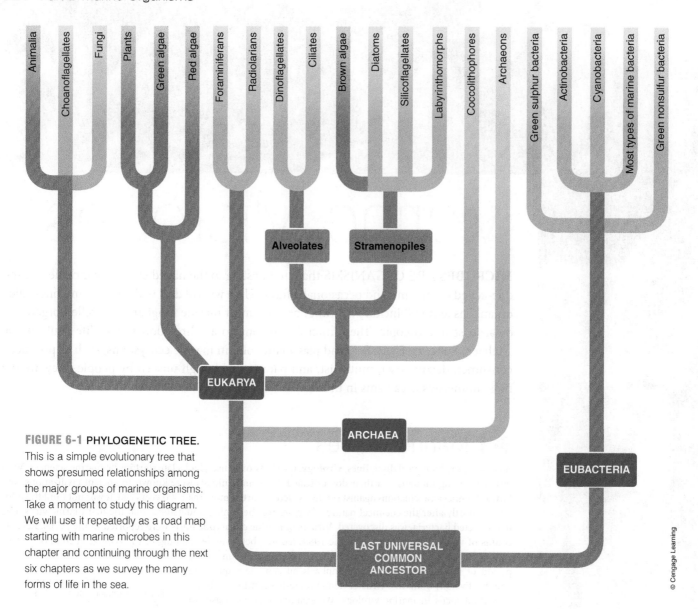

Animalia · Choanoflagellates · Fungi · Plants · Green algae · Red algae · Foraminiferans · Radiolarians · Dinoflagellates · Ciliates · Brown algae · Diatoms · Silicoflagellates · Labyrinthomorphs · Coccolithophores · Archaeons · Green sulphur bacteria · Actinobacteria · Cyanobacteria · Most types of marine bacteria · Green nonsulfur bacteria

Alveolates · Stramenopiles

EUKARYA

ARCHAEA

EUBACTERIA

LAST UNIVERSAL COMMON ANCESTOR

© Cengage Learning

FIGURE 6-1 PHYLOGENETIC TREE.
This is a simple evolutionary tree that shows presumed relationships among the major groups of marine organisms. Take a moment to study this diagram. We will use it repeatedly as a road map starting with marine microbes in this chapter and continuing through the next six chapters as we survey the many forms of life in the sea.

Viruses **(Figure 6-2)** seem to infect all groups of living organisms. They are **pathogens**, agents of disease and death. Individual viruses are usually host specific, but families of viruses vary widely in the groups they infect. For example, the Papillomaviridae family of viruses, which includes human papillomavirus, is restricted to vertebrates, but the Rhabdoviridae family of viruses infects vertebrates, insects, and plants. The many families of DNA viruses that are specific for bacteria are generally called **bacteriophages** (or simply **phages**), which translates as "eaters of bacteria." Some DNA viruses, such as the recently described Phycodnaviridae, infect eukaryotes. Most RNA viruses in the sea seem to infect eukaryotic organisms, particularly those that are photosynthetic microbes.

Viral Structure

Viruses are subcellular particles that vary in size from 10 to 400 nanometers, barely overlapping the lower size range of bacteria. Outside the host cell, a virus particle is called a **virion**. A virion is composed of a nucleic acid core surrounded by a coat of protein called a **capsid** **(Figure 6-3)**. The nucleic acid core can be either DNA or RNA and contains a small number of genes (from 3 to almost 300) that are unable to replicate outside a host cell. The combination of the virus's genetic material and protein is called a **nucleocapsid**, and sometimes it is coated with an **envelope**, a membrane derived from the host's nuclear membrane or cell membrane. The capsid protects the virus from the environment.

Viruses occur commonly in three shapes: icosahedral, helical, and binal (see **Figure 6-3**). **Icosahedral viruses** have a capsid with 20 triangular

A **pathogen** is a microbe that causes disease or mortality.

A **bacteriophage** is a virus that infects a bacterium.

The term **phage** is a shortened form of *bacteriophage*.

A **virion** is the infective viral particle released by a host cell.

The **capsid** is the outer protein coating of a virion.

The **nucleocapsid** is the combined capsid and the core of nucleic acids of a virion.

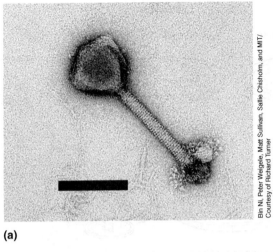

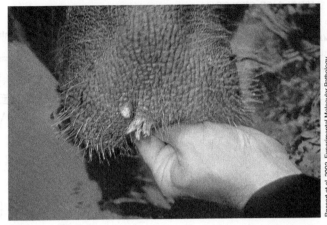

(a)   **(b)**

Bin Ni, Peter Weigele, Matt Sullivan, Sallie Chisholm, and MIT/ Courtesy of Richard Turner

Bossart et al. 2002. Experimental Molecular Pathology 72(1):37–48/Gregory D. Bossart

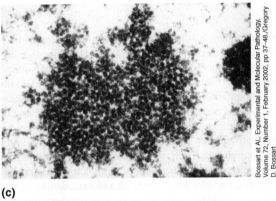

(c)

Bossart et Al., Experimental and Molecular Pathology, Volume 72, Number 1, February 2002, pp 37–48./Gregory D. Bossart

FIGURE 6-2 EXAMPLES OF MARINE VIRUSES. Marine viruses infect organisms from bacteria to mammals, depending on the kind of virus. **(a)** This bacteriophage, which belongs to the family Myoviridae, infects *Prochlorococcus*, a marine planktonic cyanobacterium that is the most abundant photosynthetic cell on earth. Myoviruses are binal DNA viruses with long contractile tails, and they include the phage of the human intestinal bacterium *E. coli*. The scale bar is 100 nm. **(b)** Manatees are one of several marine mammals that develop fibropapillomas. This animal has several on its upper lip. **(c)** The papilloma is caused by viruses in the skin cells. Papillomaviruses have invaded and replicated in this cell, and they appear as these clusters of tiny polygons.

faces composed of protein subunits. In **helical viruses**, the protein subunits of the capsid spiral around the central core of nucleic acid. **Binal viruses** are those with icosahedral heads and helical tails. Some more complex virions have filaments and other parts that are used for attachment to and infection of the host cell.

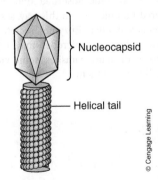

Nucleocapsid

Helical tail

© Cengage Learning

FIGURE 6-3 STRUCTURE OF MARINE VIRUSES. Mature virions of many marine viruses consist of an icosahedral (20-sided) head and a helical tail. Such viruses, as in the example here and in Figure 6-2(a), are called binal viruses, having two parts. The head is a nucleocapsid, consisting of a protein coat (capsid) protecting the genetic material (not shown) inside. Icosahedral viruses consist only of the 20-sided head. Helical viruses have a nucleocapsid that looks like a tail without the head.

Viral Life Cycles

Two broad categories of life cycles are recognized in viruses **(Figure 6-4)**. They differ in the length of time the virus remains in its host. The **lytic cycle** is a rapid one of infection, replication of viral nucleic acids and proteins, assembly of virions, and release of virions by rupture **(lysis)** of the cell. In the **lysogenic cycle**, the viral nucleic acid is inserted into the host genome and may reside there through multiple cell divisions before becoming lytic.

BIODIVERSITY AND DISTRIBUTION OF MARINE VIRUSES

Marine viruses are 10 times more abundant than marine prokaryotes. Densities may reach 10^{10} virions per liter (approximately 10^{10} per quart) in surface waters (epipelagic zone) and 10^{13} per kilogram (2×10^{13} per pound) of sediment. Recent studies have found an incredible diversity of viruses in the oceans. Samples of ocean water and sediments are estimated to contain 100 to 10,000 different viral genotypes, any one of which makes up no more than 2% of

The **envelope** is an outer covering of a virion derived from the cell or nuclear membrane of the host.

An **icosahedral virus** has a capsid with 20 sides.

A **helical virus** has a capsid of spirally arranged proteins.

A **binal virus** has an icosahedral head and a helical tail.

In a **lytic cycle**, a virus has no dormant phase in the host before initiating viral replication.

Lysis is the rupture of a host cell and release of its contents.

In a **lysogenic cycle**, a virus remains dormant in the host cell awhile before initiating viral replication.

FIGURE 6-4 VIRAL LIFE CYCLES. Many viruses undergo a rapid cycle of infection, direction of the host genome to replicate viral particles, and release of the mature virions by rupture (lysis) of the host cell after self-assembly of the viral parts. Other viruses add to this life cycle a resting phase in which the viral genome becomes inserted into the host genome, possibly undergoing replication as cells divide through several generations. The virus then becomes lytic and infects other cells. Such a virus is said to be lysogenic.

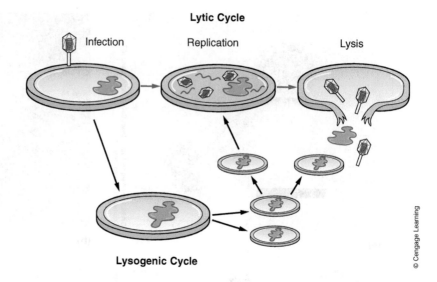

the viral community. Many marine viruses, both DNA viruses and RNA viruses, are genetically distinct from other known viruses, but much work is needed before we can characterize them genetically or understand their relationships with other viruses. A giant lytic virus that infects a marine eukaryotic microbe has recently been discovered. The 544 genes characterized so far in its genome make it the second largest virus known to science.

We now know that virions in plankton samples are not inactive contaminants but infective agents. Most marine planktonic viruses are icosahedral or binal bacteriophages with DNA and lytic life cycles. Members of the Podoviridae, a family of viruses with short tails, seem especially abundant. The bacteriophages include those that infect cyanobacteria (cyanophages), as well as the abundant heterotrophic bacteria. Eukaryotic phytoplankton also host marine viruses (phycoviruses), as do marine animals. Some animal viruses have lysogenic life cycles (e.g., green-turtle fibropapilloma retrovirus, sea-slug retrovirus). Viruses in marine sediments typically are helical and lysogenic.

Seston are particles, living or dead, that are suspended in seawater.

ECOLOGY OF MARINE VIRUSES

The ecological roles of marine viruses are largely based on lysis of their hosts. The immediate outcome of lysis is death of the cell, and the longer-term effect is some level of population control of bacteria and other microbes in plankton communities. Plankton diversity may be maintained or restored by the suppression or decline of blooms (e.g., red or brown tides) caused by viral infection. This occurs in addition to or instead of limitation by predation and lack of nutrients. Viruses can alter biogeochemical cycles and planktonic food webs by bacterial lysis. This reduces the amount of food available for eukaryotic microbes and releases nutrients into the seawater, thus increasing community respiration through elevated bacterial metabolism. It also facilitates sedimentation of particles (dead cells) to the seafloor, also by bacterial activity of consolidating particles until they reach a size and density on which gravity takes effect. Because lysis of some photosynthetic microbes might release the greenhouse gas dimethyl sulfide, viruses are thought to contribute to global climate change.

Marine viral populations probably are controlled by several abiotic and biotic factors. Near the surface, wavelengths of visible and ultraviolet light penetrating the ocean can damage or alter viral nucleic acids, causing their elimination. Because they are little more than complex macromolecules, viruses can become adsorbed onto suspended particles (**seston**) and rendered noninfective; in this way, seston are comparable in function to a charcoal filter. Enzymes secreted by bacteria may destroy virions, and eukaryotic microbes may ingest them without harm. As is the case with many parasites, some virions will attach to nonhost cells and fail to infect them.

IMPACT BUBBLE

IN ADDITION TO THEIR impact on microbial populations, viruses are responsible for chronic infection as well as mass mortality in populations of marine animals. Diseases include infections of commercially harvested fish (herring, salmon) and invertebrates in mariculture operations (shrimp, oysters), papillomas in green turtles and manatees, morbillivirus (related to distemper, measles, and mumps) in seals and whales, and poultry viruses in seabirds (penguins). Scientists attribute the increased incidence of mass mortality and the emergence of new viral diseases to environmental stress on immune systems of marine animals and to exotic species introduced by human activities.

IN SUMMARY

- Viruses are more abundant than other microbes in the sea.

- Marine planktonic viruses are icosahedral and lytic and are responsible for the death of many bacteria and phytoplankton in the epipelagic zone.

- Through this process, viruses play a significant role in marine food chains and in the cycling of mineral nutrients in the sea. (Have You Wondered? #1)

- Many emerging diseases of marine animals are caused by viruses.

6-2 Marine Bacteria

The domain Eubacteria includes the original inhabitants of the sea, dating back several billion years. Over that time, they have evolved a diversity of metabolic pathways, accomplishing biological tasks in ways not found in members of other domains today. According to the most widely accepted hypothesis on evolution of the Eukarya, bacteria have repeatedly formed symbiotic associations with their own descendents to give rise to much of the microbial diversity covered in the rest of this chapter. Their unique metabolism, widespread presence, and abundance in the sea make bacteria vital components of marine ecosystems. Marine bacteria are primary producers, decomposers, agents in biogeochemical cycles, food for other marine inhabitants, modifiers of marine sediments, and symbionts and pathogens.

GENERAL CHARACTERISTICS

Bacteria have cells with a simple, prokaryotic organization, a general feature shared with the domain Archaea. Recall that prokaryotic cells lack nuclei and other membrane-bound organelles such as mitochondria and chloroplasts. Bacterial cells have a single chromosome of DNA that is circular and contains relatively few genes, and most are surrounded by a nonliving cell wall that is made from a special combination of sugars and amino acids and that gives support and protection. Bacteria reproduce asexually by **binary fission**. In this process, a cell's genetic material (DNA) is duplicated, a membrane forms between the duplicated DNA molecules (chromosomes), and the cell splits into two daughter cells of roughly equal size. Each daughter cell then grows rapidly and repeats the process. Locomotion in bacteria most often involves a flagellum, with a structure and function quite different from that of eukaryotes, a bent filament that turns like a corkscrew and is driven by a protein engine fueled by protons.

Marine bacteria come in a wide range of shapes and sizes. Most are rod shaped and known as the **bacillus (Figure 6-5a)**. The **coccus** has a spherical shape (see **Figure 6-5a**). The **spirillum (Figure 6-5b)** looks like a corkscrew and is least-often encountered in marine habitats. In addition to these three standard bacterial shapes, marine actinobacteria, which live in marine sediments, look more like fungi **(Figure 6-5c)**. Marine bacteria

> In **binary fission**, one cell splits into two after the original cell has duplicated its genetic material.
>
> A **bacillus** is a rod-shaped bacterium.
>
> A **coccus** is a spherical bacterium.
>
> A **spirillum** is a bacterium shaped like a corkscrew.

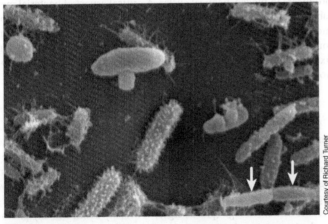

(a)

Courtesy of Richard Turner

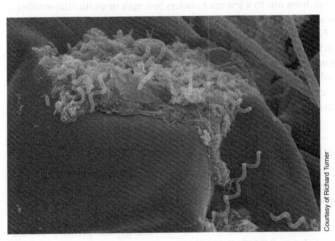

(b)

Courtesy of Richard Turner

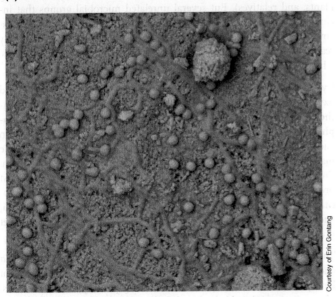

(c)

Courtesy of Erin Gontang

FIGURE 6-5 BACTERIAL SHAPES. Marine bacteria come in three standard shapes: rods (bacilli), spheres (cocci), and spirals (spirilla). But other shapes occur. **(a)** Bacilli are the elongate cells that are most numerous in this electron micrograph of the underside of a juvenile marine mite that lives on beach hoppers. Two daughter bacilli (arrows) near the bottom right corner have just formed through binary fission. A coccus is the one spherical cell near the upper left corner. **(b)** This spirillum lives on the surface of an adult marine mite. **(c)** *Salinospora* is a genus of marine actinobacteria that lives in sediments. It forms long strands and spherical reproductive structures that look like the mycelium and ascospores of the eukaryotic fungi, covered later in the chapter. The strands form by chromosome replication without complete binary fission.

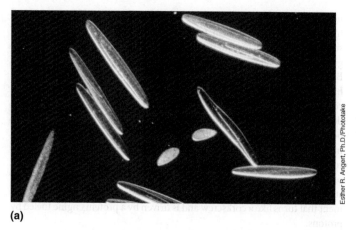

(a)

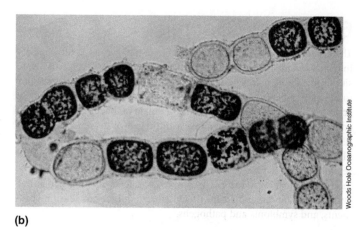

(b)

FIGURE 6-6 **THE LARGEST BACTERIA ON EARTH.** In the late 1980s to late 1990s, two marine bacteria in succession were proposed to be the largest prokaryotes on earth. **(a)** *Epulopiscium fishelsoni*, described in 1988, is a bacillus that lives in the gut of the brown surgeonfish. Note the enormous size of these prokaryotic cells compared to the two much smaller nearby paramecia, which are eukaryotes. **(b)** *Thiomargarita namibiensis*, described in 1999, is a coccus that lives in sediments off the coast of southwest Africa.

range from one to a few micrometers in length or width. The smallest is about 200 nanometers long, and the largest is 750 micrometers (about 1/32 inch) in diameter. In fact, the two largest bacteria are marine: the bacillus *Epulopiscium fishelsoni* (80 micrometers by 600 micrometers) from the gut of the brown surgeonfish (*Acanthurus nigrofuscus*) from the Red Sea **(Figure 6-6a)**, and the coccus *Thiomargarita namibiensis* (750 micrometers in diameter) from nutrient-rich sediments off southwest Africa **(Figure 6-6b)**.

NUTRITIONAL TYPES

Bacteria acquire nutrients in a variety of ways and play multiple roles in marine ecosystems **(Figure 6-7)**. Some marine bacteria are primary producers (autotrophs), manufacturing organic molecules from inorganic molecules. The source of energy for primary production in photosynthesis is the sun, and in chemosynthesis it is simple and abundant chemicals. Other bacteria are heterotrophs, requiring organic molecules for their nutrition and using them to produce additional bacterial biomass. Heterotrophic marine bacteria obtain external organic matter by absorption across the cell wall and membrane, a style of nutrition called **osmotrophy**. When osmotrophic bacteria encounter organic molecules too large to absorb, they break them down into smaller ones with digestive enzymes they secrete (**exoenzymes**). In this way, heterotrophic bacteria are decomposers, recycling dead organic substances as living bacterial biomass

and releasing inorganic molecules that become available to primary producers in biogeochemical cycles. A critical aspect of bacterial metabolism is the ability of some groups to break the strong bond of molecular nitrogen (N_2) and convert the nitrogen into ammonium ion, a form usable by other living organisms. Without such nitrogen fixation, there would be insufficient nitrogen in many habitats to support life in the sea.

Cyanobacteria

Among the almost two dozen phyla of bacteria, the most familiar are the photosynthetic marine bacteria called **cyanobacteria** (blue-green bacteria, phylum Cyanobacteria; **Figure 6-8**). Another name, "blue-green algae," is used less commonly now because our understanding of how organisms are related has rendered the word *alga* almost useless in its broader sense. An **alga** (plural, algae) is broadly defined as any photosynthetic organism that is not a plant (mosses, ferns, conifers, flowering plants, and relatives). But several unrelated microbial groups that include algae also include nonphotosynthetic organisms, as you will discover later in this chapter. Use of the term *alga* has become increasingly restricted to the three phyla of seaweeds and their unicellular relatives, which we will treat in the next chapter.

Cyanobacteria are found in environments that are high in dissolved oxygen, and they produce free oxygen as a product of their photosynthesis **(Figure 6-9a)**. Excess photosynthetic products are stored as **cyanophycean starch** and oils. The photosynthetic activity of blue-green bacteria accounts for a major proportion of the production of organic matter and oxygen in the seas. In particular, species of two genera, *Prochlorococcus* and *Synechococcus*, are among the smallest and most productive bacteria in plankton communities. Of the two genera, *Prochlorococcus* is believed to be the most abundant life form in the sea, and it is well adapted to the nutrient-poor open ocean of the tropics. Yet it was discovered only in 1988.

The primary photosynthetic pigments in cyanobacteria are **chlorophyll *a*** and, in some, **chlorophyll *b***. Both pigments are also characteristic of land plants, but chlorophyll *b* is rare among photosynthetic microbes. These pigments absorb wavelengths of light in the red and blue parts of the spectrum, and they reflect green wavelengths, which then tend to

Osmotrophy is the absorption of small organic molecules from the external medium across the cell membrane.

An **exoenzyme** is an enzyme released by osmotrophic microbes for external digestion.

Cyanobacteria are photosynthetic prokaryotes that have chlorophyll *a* and release oxygen as a by-product of their photosynthesis.

An **alga** is any photosynthetic organism, unicellular or multicellular, in which all cells are photosynthetic.

Cyanophycean starch is the kind of starch stored in cyanobacteria.

Chlorophyll *a*, the most common photosynthetic pigment of autotrophs, absorbs primarily violet and red light.

Chlorophyll *b* is a primary photosynthetic pigment found in few microbes, in green algae, and in all plants; it absorbs primarily blue and red light.

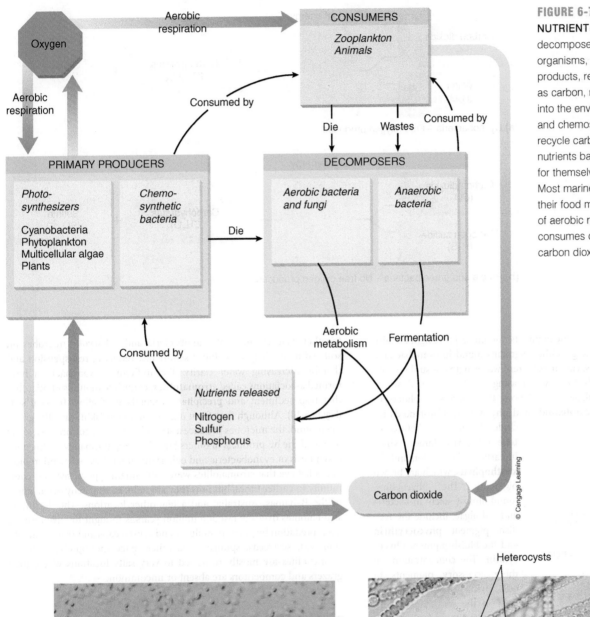

© Cengage Learning

FIGURE 6-7 BACTERIA AND NUTRIENT CYCLING. Bacteria decompose the bodies of dead organisms, as well as animal waste products, releasing nutrients such as carbon, nitrogen, and sulfur back into the environment. Photosynthetic and chemosynthetic bacteria can recycle carbon dioxide and other nutrients back into food molecules for themselves and consumers. Most marine organisms metabolize their food molecules by the process of aerobic respiration, which consumes oxygen and returns carbon dioxide to the environment.

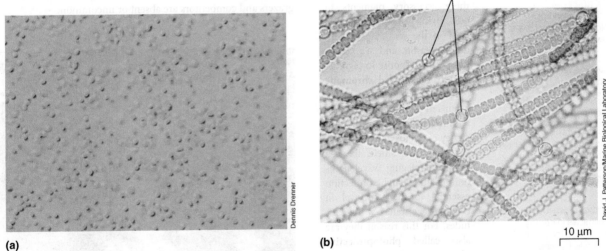

FIGURE 6-8 BLUE-GREEN BACTERIA. Phylum Cyanobacteria includes unicellular and filamentous blue-green bacteria. **(a)** Probably the most common cells in the ocean are those of species of the planktonic blue-green *Prochlorococcus*. **(b)** The blue-green color seen in cells of this filamentous blue-green bacterium is the result of the green pigment chlorophyll *a* and the blue pigment phycocyanin. The enlarged cells are heterocysts, within which nitrogen is fixed for incorporation into amino acids, nucleic acids, and other nitrogen-containing organic compounds.

FIGURE 6-9 BACTERIAL PHOTOSYNTHESIS. (a) Blue-green bacteria carry out photosynthesis in the same way that green plants do, using water as a source of electrons and hydrogen atoms to make carbohydrates from carbon dioxide and water and releasing oxygen in the process. **(b)** Purple bacteria and green bacteria use a variety of compounds (such as hydrogen sulfide) rather than water and therefore do not release oxygen. In the reaction shown here, sulfate is released by oxidation of sulfide.

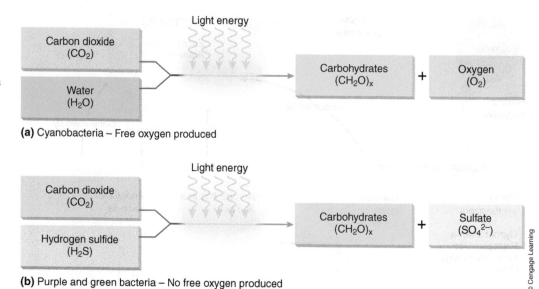

define the color of the cell. These pigments not only capture light but also convert it to chemical energy. Other pigments found in cyanobacteria, called accessory pigments, capture different wavelengths of sunlight and pass it to the chlorophylls, thereby increasing the efficiency with which the bacteria harvest available light. Among these accessory photosynthetic pigments are the **carotenoids** and **phycobilins**. Carotenoids include the familiar **beta-carotene** (of carrot fame), which imparts a yellow color, and the **xanthophylls**, which are brown pigments. The phycobilins, found almost solely in bacteria and red algae, include the reddish pigment **phycoerythrin** and the bluish pigment **phycocyanin**. The concentrations of these accessory pigments in cells change as the quality of light in the sea changes with depth, season, and sea conditions. This response in pigment composition, called **chromatic adaptation**, occurs in many groups of photosynthesizers. In addition to increasing the efficiency of photosynthesis, accessory pigments shield the cell against damaging wavelengths or intensities of light, protection essential to their survival in clear surface waters at low latitudes. For this reason they are also called **photoprotective pigments**.

Cyanobacteria may exist as isolated cells or form colonies of dense mats or long filaments (see **Figure 6-8**). Their presence as tufts on rocks may make walking in the intertidal zone treacherous because they secrete **mucilage**, a slippery biological glue that holds cells of the filaments together. Felt-like mats of cyanobacteria and eukaryotic microbes on soft sediments help to stabilize sediments, reducing resuspension and thereby increasing water clarity. For millennia, cyanobacteria have formed associations called **stromatolites**, coral-like mounds of microbes that trap sediment and precipitate minerals in shallow tropical seas **(Figure 6-10)**. Although they form only a thin living "skin" over the growing mound, the microbes produce a strongly stratified community, some living off the by-products of others but all relying ultimately on the photosynthesis of cyanobacteria and eukaryotic microbes. Some microbiologists believe that stromatolites were a dominant type of seafloor community in earlier times, but now they are reduced to patchy occurrences in the Bahamas, Australia, and a few other locations. The decline of stromatolites over the last 500 million years is thought to have resulted from predation by grazing molluscs and crustaceans and from competition with seaweeds, sponges, and other space monopolizers. Today, stromatolites are mostly restricted to very salty locations where their grazers and competitors are absent or uncommon.

Carotenoids are a class of accessory pigments that absorb blue light and protect chlorophylls from damage.

Phycobilins are a class of accessory pigments that capture wavelengths less used by chlorophylls and transfer energy to them.

Beta-carotene is a yellow or orange carotenoid pigment.

Xanthophylls are a kind of carotenoid pigment that confer a yellow or brown hue to some organisms.

Phycoerythrin is a red phycobilin that absorbs green light.

Phycocyanin is a blue phycobilin that absorbs orange light.

In **chromatic adaptation**, a photosynthetic organism can alter the kind and quantity of photosynthetic pigments in response to changes in the wavelengths and intensity of sunlight.

Photoprotective pigments consist of complex chemicals in primary producers that reduce the potential harm to chlorophylls by light.

FIGURE 6-10 STROMATOLITES. Stromatolites are formed by dense mats of blue-green bacteria and other microbes. Some fossil stromatolites date back 3.4 billion years.

Other Photosynthetic Bacteria

Other groups of photosynthetic bacteria, such as the green and purple sulfur and nonsulfur bacteria, offer a contrast to cyanobacteria because they are anaerobic and do not produce oxygen as a by-product of photosynthesis. These organisms cannot tolerate the high levels of oxygen found in many marine habitats. Sulfur bacteria, for instance, are **obligate anaerobes**, tolerating no oxygen, whereas nonsulfur bacteria are **facultative anaerobes**, respiring when in low oxygen or in the dark and photosynthesizing anaerobically when in the presence of light. The primary photosynthetic pigments in these bacteria are **bacteriochlorophylls**.

Sulfur bacteria use the hydrogen sulfide, sulfur, and hydrogen available in anaerobic habitats rather than water and thus do not produce oxygen. For example, the splitting of hydrogen sulfide (H_2S) releases elemental sulfur or sulfate rather than oxygen (see **Figure 6-9b**). Nonsulfur bacteria use small organic molecules more than they use hydrogen or sulfur compounds. Purple sulfur and purple nonsulfur bacteria are found in surface sediments and in locally anoxic waters that get enough sunlight to support photosynthesis. These bacteria, rather than cyanobacteria, are the photosynthetic component of microbial communities that form in mats of drifting seaweeds and seagrasses that accumulate along the shores of quiet lagoons and bays. As objectionable as these putrefying mats are to humans, microbial decay makes them important to the ecology of the seashore.

Chemosynthetic Bacteria

Light is absent from most ocean waters and sediments, but bacteria still can be autotrophic in these habitats, using inorganic chemicals rather than light as a source of energy. Such bacteria are called **chemosynthetic bacteria**. Chemosynthetic bacteria use energy derived from chemical reactions that involve substances such as ammonium ion (NH_4^+), sulfides (S^{-2}) and elemental sulfur (S), nitrites (NO_2^-), hydrogen (H_2), and ferrous ion (Fe^{+2}). The energy is used to manufacture organic food molecules, usually with carbon dioxide (CO_2) as the carbon source. Chemosynthesis is far less efficient than photosynthesis, and if large quantities of inorganic compounds are not available, rates of cell growth and division slow. Therefore chemosynthetic bacteria are found in areas where the inorganic substances they require are in abundance, and they usually are anaerobic.

Some chemosynthetic bacteria live around deep-sea hydrothermal vents **(Figure 6-11)**. Many occur in dense concentrations in overlying waters or as microbial mats within or near warm (8°C to 23°C) vents, where suspension feeders and benthic grazers consume them as food. These bacteria convert the sulfide ions that spew from the vents to sulfur and sulfate and use the energy released to produce food. These chemosynthetic bacteria form the base of a productive food chain that consists of a diverse assemblage of organisms, including worms, clams, and crabs, that make up the hydrothermal vent community. Others live symbiotically with benthic invertebrates.

Some chemosynthetic bacteria occur in shallower habitats. The largest bacterium, *Thiomargarita namibiensis* ("sulfur pearl of Namibia"; see **Figure 6-6b**), lives in sediments at 100 meters (330 feet) depth off the southwestern coast of Africa. It uses nitrate ions to convert the high concentration of sulfides in its habitat into granules of elemental sulfur within its cell. This bacterium and others like it off Chile and Peru are characteristic of upwelling areas where the high productivity of plankton communities results in nutrient-rich sediments that are high in hydrogen sulfide.

Heterotrophic Bacteria

Heterotrophic bacteria are decomposers. They use available organic matter in their surroundings to obtain energy and material for synthesis of their own compounds and for general metabolism. Their respiration and fermentation return to the environment many inorganic chemicals that become available to primary producers (see **Figure 6-7**). These bacteria release exoenzymes with the capacity to digest cellulose, lignin, chitin, keratin, and other natural molecules that are otherwise resistant to decay. In fact, we now know that enzymes evolved by bacteria are able to degrade plastics and other artificial polymers in the sea. Without these heterotrophic bacteria, many kinds of complex substances would accumulate in the oceans and remain unavailable for millennia as nutrients for other organisms. Organic substances in the sea usually occur as particles suspended in the water or accumulated on the ocean floor. Soon after formation, these particles quickly become populated with bacteria. The release of exoenzymes presents the risk of dilution in the surrounding environment. To combat this risk, the bacteria secrete mucilage that glues them to the particles. The mucilage keeps the exoenzymes close to the source of food and prevents the digestive by-products from washing away from the bacterium. In these microbial communities, the different kinds of bacteria on the same food particle provide a variety of enzymes capable of digesting a wide range of nutrients, thus producing a broader "menu" of by-products to absorb.

The association of bacteria with particles in the water column aids in three processes: consolidation, lithification, and sedimentation. Sticky surfaces, resulting from the secretion of mucilage and the alteration of the electric charge of particles, cause adjacent particles to adhere to each other (**consolidation**). The metabolic activity of the bacteria usually alters the pH of the area immediately around the particle, often causing minerals to precipitate and form a cement between particles (**lithification**). Consolidation and lithification increase the aggregation of particles. As aggregate size increases, the ability of the water to hold the particles in suspension declines, and the particles begin to settle (**sedimentation**). Particles might accumulate on the seafloor, at places where layers of water of different density meet (such as at thermoclines), or in large, cobwebby, drifting structures called **marine snow** formed by mucus secreted by many kinds of plankton. As particles accumulate, they become available to suspension and deposit feeders. These consumers survive on the large microbial communities growing on the particles. In addition to these attached bacteria, free bacteria have a prominent part in marine communities,

Mucilage is a gelatinous secretion of algal cells for attachment of cells and for protection.

A **stromatolite** is a coral-like community of microbes that form a thin layer of living cells and filaments over an accumulated mass of dead stony material.

An **obligate anaerobe** thrives only in the absence of oxygen.

A **facultative anaerobe** thrives in the presence or absence of oxygen.

Bacteriochlorophylls are a class of primary photosynthetic pigments that do not release oxygen.

Chemosynthetic bacteria can form organic molecules from inorganic molecules using other chemicals rather than sunlight as a source of energy.

Consolidation is the aggregation of suspended particles in seawater by bacteria.

Lithification is the conversion of consolidated particles into a solid mass with mineral cement.

Sedimentation is the settlement of particles from suspension in water.

Marine snow is a loose network of suspended live and dead particles arising from webs of mucus released by many kinds of microbes and zooplankton.

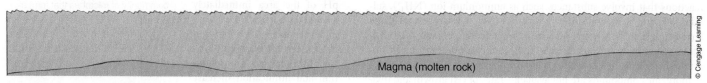

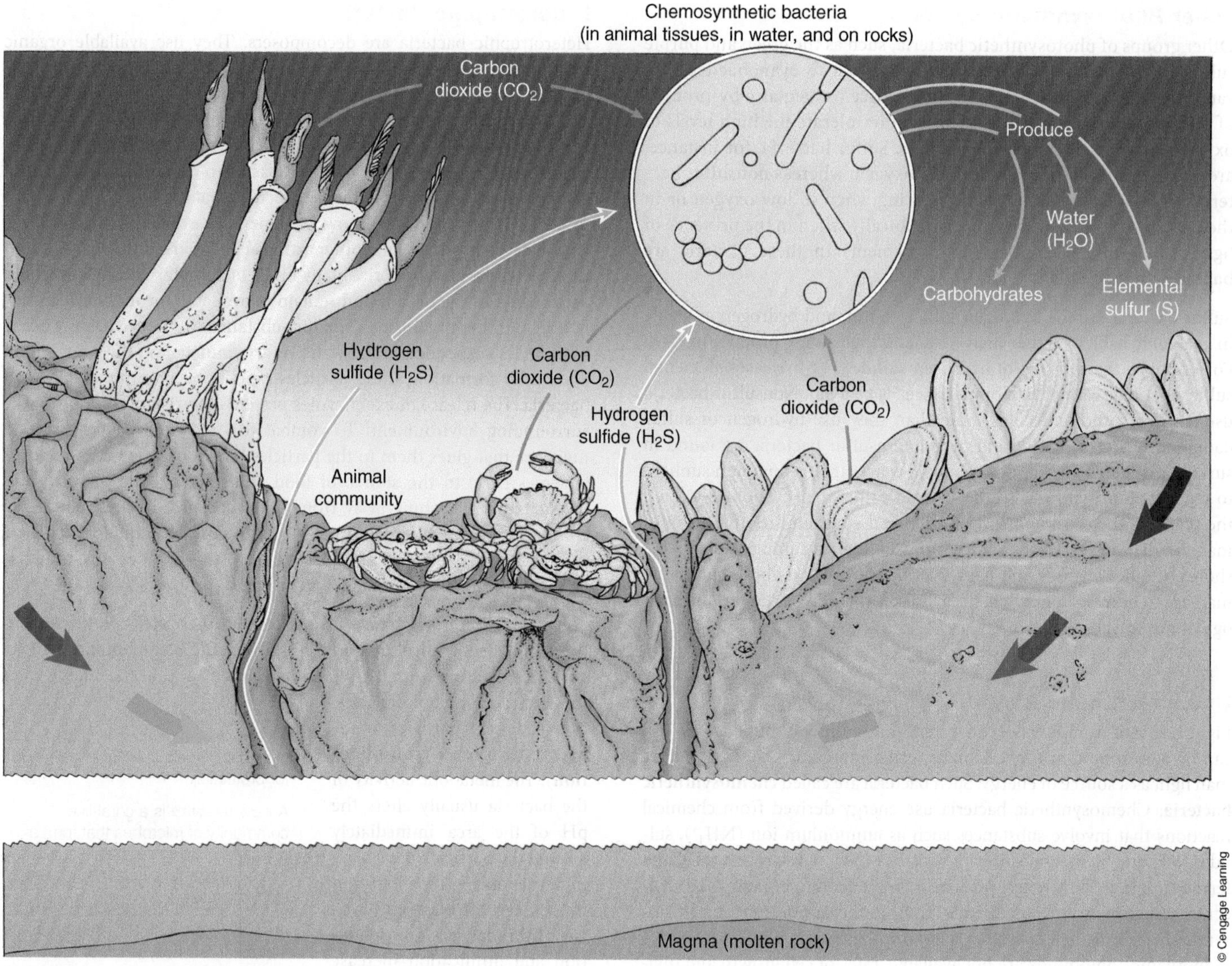

Chemosynthetic bacteria
(in animal tissues, in water, and on rocks)

Carbon
dioxide (CO$_2$)

Produce

Water
(H$_2$O)

Carbohydrates

Elemental
sulfur (S)

Hydrogen
sulfide (H$_2$S)

Carbon
dioxide (CO$_2$)

Carbon
dioxide (CO$_2$)

Hydrogen
sulfide (H$_2$S)

Animal
community

Magma (molten rock)

© Cengage Learning

FIGURE 6-11 CHEMOSYNTHESIS. Bacteria and archaeons that live in deep-sea hydrothermal vent communities can produce food molecules from carbon dioxide and hydrogen sulfide, using the energy derived from chemical reactions rather than from sunlight. Seawater in the earth's crust warms and expands from proximity to the magma. The water picks up dissolved reduced minerals and rises up through the crust, entering channels that exit at the vents. On exposure to cold ocean bottom water, the vent waters cool rapidly. Some of the minerals precipitate, forming chimneys at the vents, and some minerals remain dissolved and provide resources for chemosynthetic microbes.

usually in food webs involving several of the microbial groups that we will discuss later in this chapter.

NITROGEN FIXATION AND NITRIFICATION

The major process that adds new usable nitrogen to the sea is **nitrogen fixation**. This process converts the abundant molecular nitrogen dissolved in seawater to ammonium ion **(Figure 6-12)**. Nitrogen often is scarce in marine ecosystems. It is required to form amino acids, nucleotides, and other compounds. Nitrogen is recycled to autotrophic organisms when it

Nitrogen fixation is the conversion of atmospheric nitrogen into ammonium ion.

Nitrogenase is the enzyme used by bacteria to fix nitrogen.

is released as metabolic waste from living organisms and when decomposers such as bacteria and fungi break down dead organic matter to ammonium ions. This rapid uptake by primary producers and loss of nitrogen to settling particles may leave a deficit of nitrogen in surface waters. One adaptation to nitrogen limitation in the open sea is small cell size and low number of genes, because proteins and nucleic acids contain much nitrogen. The planktonic *Pelagibacter ubique* has the smallest genome of free-living cells, and the genomes of *Prochlorococcus* and *Synechococcus* are only slightly larger. Nitrogen might also be limiting in marine sediments, as in seagrass beds and mangrove swamps.

Only some cyanobacteria (and a few archaeons, described later) are capable of fixing nitrogen. These organisms have an enzyme, **nitrogenase**, that is capable of breaking the strong molecular bond between the two

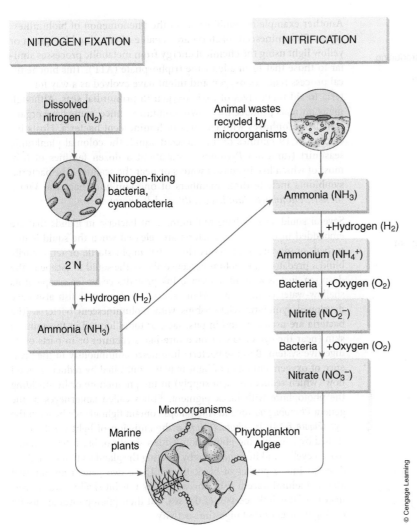

NITROGEN FIXATION

NITRIFICATION

FIGURE 6-12 NITROGEN FIXATION AND NITRIFICATION. Nitrogen dissolved in seawater is converted by blue-green bacteria and some archaeons to ammonium ion; this process is called nitrogen fixation. In nitrification, bacteria convert ammonium to nitrites and nitrates, forms of nitrogen that are more readily used by primary producers.

A **heterocyst** is a specialized cell in which conditions favorable for nitrogen fixation are maintained.

Nitrification is the conversion of ammonia from animal wastes and dead tissue into nitrate ions.

Endosymbiotic theory is the body of evidence supporting the idea that some one-celled organisms have evolved by the incorporation of other one-celled organisms or their organelles into the host cell.

A **plastid** is an organelle in a eukaryotic cell derived originally by endosymbiosis with a cyanobacterium.

An **endosymbiont** is a guest organism or organelle that lives within a host organism or cell.

The **hydrogen hypothesis** is a concept on the origin of eukaryotic cells that postulates an endosymbiotic relationship between a host archaeon that needed hydrogen for chemosynthesis and a guest bacterium that released hydrogen and became a mitochondrion.

atoms of nitrogen that make up the molecules of nitrogen gas. Because nitrogenase is sensitive to oxygen, nitrogen fixation must be accomplished in an anaerobic environment. This is, therefore, a difficult task for a group of aerobic autotrophs that produce oxygen as they photosynthesize. In many filamentous cyanobacteria, nitrogen fixation occurs in a **heterocyst**, a thick-walled cell in which photosynthesis is altered to prevent the release of oxygen (see **Figure 6-8b**). In addition to the thick cell wall, diffusion of oxygen into the heterocyst is further reduced by a coating of mucilage and by adjacent cells that consume oxygen. What little oxygen diffuses across the cell wall is used rapidly within the heterocyst by respiration. Other cyanobacteria accomplish nitrogen fixation under environmental conditions that do not require the formation of heterocysts. Any ammonium ion (NH_4^+) that diffuses into the surrounding seawater becomes available for other primary producers, but they usually require bacterial conversion of the ammonium to nitrite (NO_2^-) and nitrate (NO_3^-) ions, a process called **nitrification**, before they are able to use the nitrogen.

SYMBIOTIC BACTERIA

Symbiosis is the biological phenomenon in which two different organisms form a very close association with each other. Many bacteria

have evolved symbiotic relationships with a variety of marine organisms. The symbiotic relationships that probably have the longest history are those now represented by the mitochondria and chloroplasts of members of the domain Eukarya. Other more recently evolved symbioses involve associations in which the bacterial guest is clearly distinguishable from its eukaryotic host.

Endosymbiotic theory is a body of evidence that supports various ideas that mitochondria, **plastids** (of which chloroplasts are one kind), and hydrogenosomes (a third kind of "organelle" not discussed here) evolved as symbionts within other cells, that is, that these intracellular structures are **endosymbionts** rather than organelles. This theory has its origin in the early 1900s. Since the 1960s, revival of the theory has resulted in several hotly contested views on details of how eukaryotes and their endosymbionts evolved. One view, the **hydrogen hypothesis (Figure 6-13)**, has much support from scientific evidence. It proposes that the endosymbiont was a hydrogen-releasing anaerobic eubacterium (which became the mitochondrion) within a hydrogen-requiring chemosynthetic archaeon. Formation of the nucleus (and therefore the eukaryotic condition) was favored as a way to separate different stages in the process of converting genetic information in DNA into proteins.

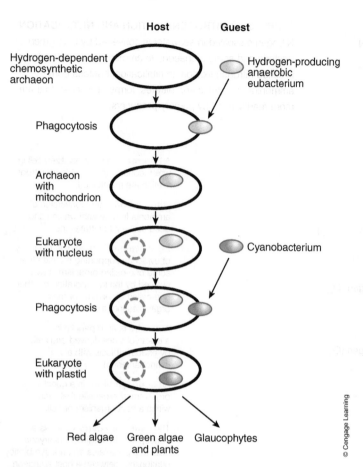

Host   **Guest**

Hydrogen-dependent chemosynthetic archaeon

Hydrogen-producing anaerobic eubacterium

Phagocytosis

Archaeon with mitochondrion

Eukaryote with nucleus   Cyanobacterium

Phagocytosis

Eukaryote with plastid

Red algae   Green algae and plants   Glaucophytes

© Cengage Learning

FIGURE 6-13 PRIMARY PLASTID ENDOSYMBIOSIS. The hydrogen hypothesis on the evolution of eukaryotes with mitochondria and plastids involves several steps. This is one of a few strongly competing hypotheses in current endosymbiotic theory.

Another example of symbiosis is in the phenomenon of bioluminescence. Bioluminescent bacteria are capable of emitting blue-green or yellow light using the chemical energy from metabolic processes similar to those that form adenosine triphosphate (ATP). This biochemical process requires oxygen and might have evolved as a way for bacteria to rid themselves of toxic oxygen in primordial seas. Although many animals produce their own bioluminescence, some have organs called **photophores** with cultures of luminescent bacteria. Hosts include 2 of 19 families of luminescent squid, the colonial planktonic seasquirt (tunicate) *Pyrosoma*, and about a dozen families of fish, most of which live in coastal waters or near the seafloor. The bacterial symbionts include three members of one family of bacteria: *Vibrio fischeri*, *Photobacterium leiognathi*, and *P. phosphoreum*.

Several squid species have bioluminescent bacteria in glands that are embedded in the ink sac. Bacteria are released when the squid is disturbed, producing a luminous cloud that might startle or temporarily blind a predator. The bioluminescence allows the squid to escape in the deep sea, just as a cloud of ejected ink provides cover for escape of its shallow-water counterparts. Many species of deepwater fish also have developed symbiotic relationships with bioluminescent bacteria. The bacteria are usually found in pits, sacs, or tubes located in the animal's skin (near the eye or jaw or on a lure-like structure) or in parts of its digestive system. Because bacteria luminesce continuously in the presence of oxygen, emission of light may be controlled by reducing blood flow (which reduces oxygen supply) to the photophore or by shielding the photophore with black pigment. Fishes called lanterneyes in the genera *Photoblepharon* and *Anomalops* contain light glands beneath the eye **(Figure 6-14)**. In *Photoblepharon* the emission of light can be controlled by covering the photophore with a sheath of black tissue, much like an eyelid, and in *Anomalops* by rotating the gland so that the light is directed into a pocket of black tissue. These two genera are unusual among bioluminescent fishes because they inhabit shallow-water reefs instead of the dark recesses of the sea, but their photophores are useful during their period of nighttime activity.

Plastid endosymbiosis is the evolutionary process by which a heterotrophic host cell gains the ability to photosynthesize from a photoautotrophic guest cell.

A **deep-sea vent community** is a community of marine organisms that depend on the specialized environment found at divergence zones in the ocean floor.

A **photophore** is an organ in some organisms that produces bioluminescence.

Plastids arose later as endosymbiotic cyanobacteria, a process called primary **plastid endosymbiosis**. Molecular and other data indicate that primary plastid symbiosis occurred once, forming the ancestor common to red algae, to green algae and plants, and to a third group called glaucophytes. The relationship between the cyanobacterial guest and its archaeal or eukaryotic host became increasingly interdependent as genes in the plastid were transferred to the chromosome of the host.

In **deep-sea vent communities** (see **Figure 6-11**), chemosynthetic bacteria live within the tissues of marine tube worms and clams. These bacteria supply their hosts with organic food molecules through chemosynthesis, whereas the hosts supply the bacteria with carbon dioxide and other essential nutrients, particularly sulfides. Since the discovery of bacterial symbioses in vent communities, similar relationships have been found in shallow-water habitats.

John B. Corliss

FIGURE 6-14 BACTERIAL BIOLUMINESCENCE. In addition to light production in free bacteria, the bioluminescence of many animals is due to symbiotic associations of bacteria in organs called photophores. In lanterneyes, the photophore occurs below the eye, as shown here in *Photoblepharon palpebratum*. This fish is unusual among luminescent fishes by its habitation of shallow-water reefs. Light emission during its nighttime period of activity can be blocked by raising a black partition like an eyelid. By "blinking," the fish can lure prey, confuse predators, and communicate with other lanterneyes.

high salinities, low pH, and high pressure. Although their adaptations to extreme habitats typify archaeons, we now recognize them also as members of the plankton.

NUTRITIONAL TYPES

The domain Archaea includes photosynthetic, chemosynthetic, and heterotrophic organisms. Some can fix nitrogen, and most are anaerobes. Although many aspects of their nutrition are similar to those of bacteria, other aspects are unusual. Much is yet to be learned about their metabolism.

Most archaeons are **methanogens**, anaerobic organisms that live in environments rich in organic matter. The organic matter is metabolized for energy, and methane gas is a waste product of this metabolism. Some methanogens are chemosynthetic, using hydrogen as a source of energy. Indeed, most archaeons that are primary producers are chemosynthetic.

In contrast to the production of methane as waste by marine archaeons, some areas of research and engineering attempt to convert what we call "waste" *into* methane as a source of energy. *Methanocaldococcus jannaschi* and *Methanopyrus kandleri*, two of the many marine methanogens, both live at deep-sea hydrothermal vents. The first complete genome sequenced for an archaeon was that of *Methanocaldococcus jannaschi*—a study supported in part by the U.S. Department of Energy because of the potential of this microbe to give clues to the industrial production of methane. The scientific world was astounded to learn that 56% of the genes of *Methanocaldococcus jannaschi* were unique, giving further support to the three-domain system of life. Aside from its potential commercial use, methane is a greenhouse gas that may contribute to global warming. Methane production in anoxic habitats in the sea is less likely to affect global warming because much of the methane may be oxidized as it rises into overlying oxygenated waters.

Archaeons differ from bacteria in their biochemistry, genetic makeup, ability to produce methane, and tolerance of extreme environmental conditions.

A **methanogen** is an archaeon that produces methane gas in its metabolism.

Halobacteria are archaeons that require high concentrations of salt where they live.

A **halophile** is an organism that grows and reproduces best in the presence of salt.

Bacteriorhodopsin is a light-capturing protein that produces ATP in halobacteria.

One group of archaeons, the **Halobacteria**, are aerobic heterotrophs but also can capture light for ATP production. Unlike photoautotrophs, light capture does not result in production of organic compounds as food. Halobacteria belong to a group of archaeons known as **halophiles**, or "salt lovers," so-called because they were first discovered to grow best in extremely salty environments such as salt lakes and salt evaporation ponds. In fact, if the salt concentration of their environment drops below five times the concentration of normal seawater, such halobacteria will die. We now know that other halobacteria are abundant in plankton communities. Because warm salty water holds very little oxygen, halobacteria frequently experience a shortage of this important gas, and the amount of ATP that these bacteria can synthesize is greatly decreased. When oxygen levels are low and sunlight is plentiful, halobacteria employ a backup system using light energy to produce enough ATP to survive. Light is captured by purple proteins, **bacteriorhodopsins**, which are chemically similar to the

---

IN SUMMARY

- Bacteria have cells with a prokaryotic organization.

- Chemosynthetic and photosynthetic bacteria extract inorganic nutrients such as nitrogen, phosphorus, and carbon dioxide from the environment and incorporate them into organic molecules. (Have You Wondered? #2)

- Chemosynthetic bacteria use energy derived from chemical reactions (often involving compounds of sulfur) to produce their food molecules, whereas photosynthetic ones use the radiant energy from the sun.

- Some producers, such as blue-green bacteria, release oxygen during photosynthesis.

- Such primary producers, as well as heterotrophic bacteria, form the base of marine food webs. (Have You Wondered? #2)

- In addition, marine bacteria play a critical role in nitrogen fixation and nitrification. (Have You Wondered? #2)

- As decomposers, bacteria return dead organic matter to biogeochemical cycles as inorganic matter that primary producers can incorporate into living biomass.

- The chemosynthetic bacteria of hydrothermal vent communities and the bioluminescent bacteria found in association with deep-sea organisms are examples of recently evolved symbiotic relationships.

- The oldest symbiotic relationships of bacteria gave rise to mitochondria and chloroplasts, which are endosymbionts in eukaryotic cells.

6-3 Archaea

The domain Archaea includes microbes formerly considered to be bacteria. Molecular and ultrastructural analyses have revealed the Archaea to be quite distinct from the Eubacteria and Eukarya domains. Surprisingly, they share a number of features with each of the other two domains and might have a closer evolutionary relationship with the Eukarya, as indicated in the hydrogen hypothesis of endosymbiosis described previously.

GENERAL CHARACTERISTICS

Archaeons are small (0.1 to 15 micrometers) and have a much narrower range of sizes than do bacteria. Like bacteria, archaeons are prokaryotes, but they differ from bacteria in other ways. Most have a cell wall, which lacks the special sugar-amino acid compounds of bacterial cell walls. Some archaeons have only a layer of proteins or of proteins complexed with carbohydrates covering their cell membrane. The lipids that occur in the cell membranes of archaeons are quite different from those found in the membranes of bacteria. The differences are related in part to the need to stabilize the membranes under the extreme environmental conditions that characterize the habitats of many Archaea. The professional journal *Extremophiles*, first published in 1997, is dedicated to the biology of archaeons and other microbes that thrive in high and low temperatures,

visual pigment rhodopsin in the eyes of vertebrate animals. Patches of bacteriorhodopsin may occupy as much as 50% of the surface of the cell membrane, and the pigment molecules extend through the membrane such that a portion of the molecule is exposed at both surfaces. When light strikes the bacteriorhodopsin, the molecule undergoes a series of reactions in which a hydrogen ion is pumped from the inside to the outside of the cell. The flow of hydrogen ions back into the cell provides the energy for ATP synthesis. The existence of this mechanism suggests that primitive prokaryotic cells may have evolved systems to trap sunlight and make ATP long before the evolution of photosynthesis. The ability to use the energy of sunlight certainly would have had a tremendous effect on the evolution of early life.

HYPERTHERMOPHILES

Many archaeons tolerate very warm temperatures, but some, categorized as **hyperthermophiles**, are known to survive in temperatures exceeding 100°C, such as occur at deep-sea hydrothermal hot vents. These microbes often have names that refer to heat, fire, and smoke. For example, *Pyrolobus fumarii* translates as "fire lobe of the chimney," a name given to a hyperthermophilic primary producer collected from a black smoker chimney on the Mid-Atlantic Ridge at 3,650 meters (12,045 feet) depth. *Pyrolobus fumarii* cannot use organic matter as food but is chemosynthetic, using carbon dioxide and hydrogen to produce organic compounds. It accomplishes this and also divides at rates of up to once per hour at an optimal temperature of 106°C. It will grow at temperatures between 90°C and 113°C and tolerates brief exposure to 121°C. Recently an archaeon called "strain 121" from a hydrothermal vent in the Pacific Ocean better tolerated a temperature of 121°C in the laboratory than did *Pyrolobus fumarii*, and it also survived 2 hours of incubation at 130°C. The heat tolerance of hyperthermophilic archaeons is truly remarkable. This group of microbes holds much potential for exciting revelations of biological adaptation and for future biomedical or industrial applications.

A **hyperthermophile** is a microbe that grows and reproduces best at temperatures exceeding 100° C.

In **phagocytosis**, a cell engulfs a particle by inward folding and separation of the cell membrane to form a vacuole.

A **mycologist** is a scientist who studies fungi.

Aspergillosis is a fungal disease caused by the genus *Aspergillus*.

Mycology is the study of fungi.

A **yeast** is a single-celled fungus.

A **filamentous fungus** consists of long thread-like masses.

A **hypha** is a filament that makes up a section of a filamentous fungus.

IN SUMMARY

- Like bacteria, archaeons are prokaryotes.
- Most archaeons are methanogens—anaerobic organisms that live in environments that are rich in organic matter— and they make significant contributions to methane production.
- Archaeons have an unsurpassed ability in the natural world to tolerate extreme environmental conditions. (Have You Wondered? #3)

6-4 Eukarya

The domain Eukarya contains all organisms with eukaryotic cells, including plants, animals, and fungi, as well as photosynthetic microbes and single-celled animal-like protozoans. In addition to cell structure, eukaryotes are distinctive in their ability to ingest particles by **phagocytosis**, a process of inward budding of the cell membrane to form a vacuole that encloses a particle. Not only does phagocytosis give eukaryotes an additional mode of nutrition, it also allows the development of endosymbiotic relationships, as in the formation of mitochondria and plastids. In addition, eukaryotes evolved sexual life cycles that involve diploid genomes, meiosis, and fertilization.

FUNGI

We are familiar with many kinds of terrestrial fungi because they grow in our lawns and on trees (for example, puffballs and bracket fungi), cause diseases in plants (rusts, powdery mildews) and humans (ringworm, athlete's foot), are used in preparation of food and drink (yeasts), produce antibiotics (*Penicillium*), are on our menus (mushrooms, morels, truffles), and spoil our food (bread molds, ergot). Marine fungi are much less obvious to us because of their microscopic size, and they affect our lives much less directly. Less than 1% of known species of fungi are marine, but marine **mycologists** (biologists who study fungi) are finding that, although fungi are not as diverse as some other marine organisms, they are important in marine ecosystems as decomposers, prey, pathogens, and symbionts. Marine fungi are economically important because they decompose wooden structures such as pilings and boat hulls, and they cause diseases of a number of commercially harvested marine animals. The recent discovery of **aspergillosis** in soft corals (sea fans) is an area of concern. The fungal genus *Aspergillus*, a common fungus in terrestrial soils, causes spoilage of fruits and vegetables, produces dangerous chemicals called aflatoxins, and infects the human lung. The emergence of new marine fungal pathogens from terrestrial sources might cause a surge in marine mycology over the next decade.

History of Marine Mycology

The first marine fungus was discovered in 1849, but it was not until the mid-1900s that mycologists recognized that there are communities of fungi unique to the seas. Early progress in marine **mycology** (the study of fungi) was impeded because contaminant spores from land and freshwater habitats could not be distinguished from those that were truly marine. In much of the second half of the 1900s, marine mycology consisted of cataloguing new species or focused on descriptive studies of the roles of marine fungi. The difficulties of quantifying fungal biomass make it hard for mycologists to evaluate critically the contributions of fungi to the ecology of the sea. Application of recently developed techniques is likely to yield better estimates of fungal productivity in the future.

General Features of Fungi

Members of the kingdom Fungi are eukaryotes with cell walls of chitin, a carbohydrate more commonly associated with crustaceans and other arthropods. Many fungi are unicellular **yeasts** or have a unicellular yeast-like stage in their life cycle. **Filamentous fungi**, on the other hand, grow vegetatively into long filaments called **hyphae (Figure 6-15a)** composed of many cells that may or may not be divided from each other by cell walls. The branching of hyphae produces a

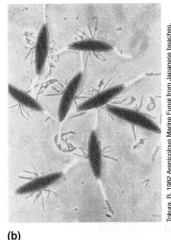

(a)   **(b)**

FIGURE 6-15 ARENICOLOUS MARINE FUNGUS. **(a)** Thread-like hyphae and large spherical ascocarps make up the mycelium of this sand-dwelling (**arenicolous**) marine fungus in the genus *Corollospora*. The very fine hyphae give a cobweb appearance over the surface of the grains of sand. Where adjacent hyphae have fused and undergone sexual reproduction, ascospores are aggregated within the large fruiting bodies, called ascocarps, comparable to terrestrial mushroom caps. **(b)** Within the ascocarps, ascospores mature and then are released. The spines and filaments on the ascospores aid their entrapment in sea foam, which transports them to distant shores.

tangled mass called the **mycelium**, such as the black hairy tufts that form bread mold.

Fungi are not photosynthetic. Like most bacteria, fungi are heterotrophic decomposers and recycle organic material. Their exoenzymes are capable of digesting a wide variety of biologically and synthetically produced molecules. Perhaps the most significant of these molecules is lignin. Destruction of the chemical lignin, a major component of wood, allows fungal enzymes called cellulases to reach the fibers of cellulose, which serves as the fungus's food source. Food reserves are stored as **glycogen**, often called "animal starch" because of its role as a storage carbohydrate in the animal kingdom. Unlike many heterotrophic bacteria, most fungi are strict aerobes, requiring high oxygen content in their habitats.

The kingdom Fungi is divided into four phyla: Chytridiomycota, with motile cells; Zygomycota, such as black bread mold; Basidiomycota, or club fungi, including the familiar mushrooms; and Ascomycota, or sac fungi. The last three phyla lack motile stages in their life cycles. The ascomycetes are the most diverse and abundant phylum of fungi in the sea, constituting 81% of the fewer than 500 known marine species of filamentous fungi. In addition, many of the nearly 200 marine yeasts are ascomycotes.

Ecology and Physiology of Marine Fungi

Some species of marine fungi are obligately marine, requiring ocean or brackish waters to grow and produce spores. Facultative marine fungi grow and may sporulate (form spores) in marine environments but are primarily of terrestrial or freshwater origin. Most terrestrial and freshwater fungi cannot grow in the sea because of growth-retardant chemicals in the seawater and their intolerance of the high salt content. Growth of hyphae requires taking in water across the cell membrane, and only obligate and facultative marine fungi are able to do this in seawater. In addition, the concentration of sodium ion in seawater is toxic to marine fungi, and much of the cellular activity of fungal

hyphae is directed toward ridding cells of excess sodium. For these and other reasons, some mycologists argue that marine fungi have evolved from terrestrial and freshwater fungi, probably multiple times.

In addition to their assignment to taxonomic groups, marine fungi can also be classified on the basis of the material on which they are found. By far, the most diverse group of marine fungi lives on or in wood of terrestrial origin (**Figure 6-16**). They decompose the logs swept by rivers into coastal environments and even into the sea. Logs on the seafloor in anoxic locations are devoid of fungi and are decomposed by bacteria. Second in diversity to marine fungi that decompose wood are those that inhabit coastal salt marshes of temperate zones. Grasses (*Spartina*) that grow in these marshes are intertidal. The fungi that live on them range from terrestrial species growing on the rarely submerged flower stalks to facultative marine species living on the higher parts of stems and leaves to obligate marine fungi growing on the lower plant parts frequently covered by seawater. Decomposition of plant matter in the anoxic sediments of salt marshes is the role of marine bacteria. The next most diverse are those fungi that live on marine algae worldwide and the mangroves of tropical coasts. Again, because mangroves are intertidal, fungi on mangrove trees range from those least adapted to marine life and found on branches, leaves, and upper parts of roots to those best adapted to marine life and found on lower but exposed parts of the root system. Among the fungi with lowest diversity are those that dwell on sand (see **Figure 6-15a**). Just as beaches are physically and physiologically stressful habitats for plants and animals, so they are for fungi. In the open sea, marine fungi in plankton communities decompose the chitinous remains of dead crustaceans that rain down to the seafloor or become caught in marine snow.

> A **mycelium** is the body of a fungus, whether a single cell or a large mass of hyphae.
>
> **Glycogen** is a polysaccharide found in some organisms that serves to store glucose reserves.

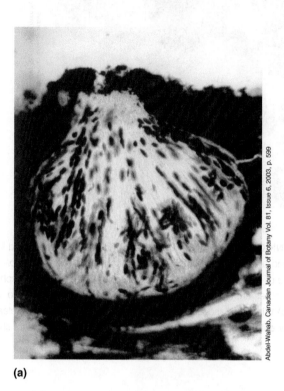

(a)

(b)

Abdel-Wahab, Canadian Journal of Botany Vol. 81, Issue 6, 2003, p. 599

Abdel-Wahab, Canadian Journal of Botany Vol. 81, Issue 6, 2003, p. 599

Budding is asexual reproduction in which two individuals are produced by unequal division of the adult.

A **conidiospore** is an asexually produced dispersal stage in the life cycle of a fungus.

A **fruiting body** is a sexually reproductive structure of a fungus.

An **ascocarp** is the fruiting body of an ascomycote fungus.

An **ascus** is one component of an ascocarp that produces four or eight spores.

Ascospores are haploid cells produced in an ascocarp.

A **lichen** is a mutualistic association between a fungus and an alga.

FIGURE 6-16 LIGNICOLOUS MARINE FUNGUS. **(a)** This ascocarp of a wood-inhabiting (**lignicolous**) fungus is composed of many asci embedded just below the surface of the timber. **(b)** Each ascus is a long tube-like structure within which an original cell divides by meiosis to produce four daughter cells, each of which then divides once by mitosis, resulting in a total of eight ascospores.

Reproduction of Marine Fungi

Marine yeasts reproduce asexually by **budding**, a kind of mitosis that produces daughter cells of unequal size. Filamentous marine fungi reproduce asexually, often by the production of spores called **conidiospores** that form at the tips of hyphae. The terminal spores on the end of a string of conidiospores are released and carried away by currents. On contacting an appropriate surface, the conidiospore germinates into a mycelium.

Sexual reproduction frequently involves a spore-forming structure called a **fruiting body**. Filamentous marine ascomycotes reproduce sexually by production of a fruiting body called an **ascocarp** (see **Figures 6-15a** and **6-16a**). Whereas terrestrial ascocarps often consist of a cluster of spore-bearing sacs, each called an **ascus** (plural, asci), marine species typically have ascocarps of one ascus. A zygote nucleus is formed when two haploid nuclei from different mycelia fuse. The zygote nucleus then develops by meiosis and mitosis into eight cells that mature into **ascospores** (see **Figures 6-15b** and **6-16b**). Released ascospores will become new adult fungi. A similar process occurs in marine basidiomycotes.

Mushrooms and fruiting bodies of other terrestrial fungi are elevated on long stalks above the ground and release their spores into air currents. Some terrestrial ascomycotes explosively shoot ascospores into the air to aid dispersal. In contrast to their terrestrial counterparts, marine fungi live in a denser medium. Long stalks of terrestrial species would easily break in water currents, and an explosive mechanism of dispersal would not be effective. Marine ascocarps are small

(maximally 2 to 4 millimeters [0.08 to 0.16 inches]) and have hard walls and short stalks. Ascocarps of sand-inhabiting fungi, which live in an abrasive environment, have the hardest walls of all marine fungi. Fungi on mangrove prop roots in the intertidal zone and in driftwood produce ascocarps that barely break the surface of the wood (see **Figure 6-16a**), reducing exposure of these delicate structures to the erosive effects of tidal currents. The ascocarp walls of these fungi dissolve and passively release ascospores into the surrounding seawater. Ascospores often have spines or threads that are sharp or sticky, improving their adherence to surfaces when they settle. Appendages also allow the ascospores of sand-dwelling fungi to become caught in sea foam as a transport mechanism to other beaches (see **Figure 6-15b**).

Maritime Lichens

Lichens are mutualistic associations between a fungus and an alga. The fungi most often are ascomycotes, and the algae are usually green algae or cyanobacteria. The fungus provides attachment, the general structure of the lichen, minerals, and moisture. The alga produces organic matter through photosynthesis. The association allows lichens to inhabit environments that are inhospitable to either member alone.

Several of the 3,600 species of lichen in North America are restricted to the seashore, where they most often occur on rocks in the supralittoral zone (**Figure 6-17**). Zonation of lichens on rocky shores is an outcome of their differing tolerances to submersion by tides, exposure to salt spray,

(a)

(b)   **(c)**

FIGURE 6-17 MARITIME LICHENS. (a) This rocky shore in British Columbia, Canada, has a broad white band of the volcano lichen *Coccotrema maritimum* **(b)** in the high supralittoral zone and a broad black band of sea tar *Verrucaria maura* in the splash zone **(c)**.

and abrasion by ice. Maritime lichens rarely occur on drift lumber, on coastal trees and shrubs, and in sand dunes. As prevalent as they are on some rocky shores, there seems to be little information on the role that maritime lichens play in ecological communities by the sea.

STRAMENOPILES

The **stramenopiles** (*stramen*, straw; *pilos*, hair) are a diverse group of eukaryotic organisms. One feature that unifies the group, but is not the sole characteristic, is the nature of the two flagella these cells possess **(Figure 6-18)**. One flagellum has a short simple form, usually with a light-sensing body at the base, and the second longer flagellum bears many hair-like filaments called **mastigonemes**, with a thickened base and a branching tip along the shaft. Whereas one flagellum senses light and pushes the cell as it undulates, the other is a more effective "feathery" swimming organelle that pulls the cell forward as it undulates. Members of some groups have flagellated stages that appear only very briefly in their life cycles. Other groups seem to have lost the simpler light-sensitive flagellum and retained the complex one. The term **heterokont** refers to the different (*heteros*) form of the two flagella (*kontos*, a pole). The closest relatives of the stramenopiles seem to be the dinoflagellates, which also have heterokont flagella.

Stramenopiles can be nonphotosynthetic or photosynthetic. Photosynthetic stramenopiles are usually golden brown and are referred to as **ochrophytes** (as in the color ochre). The dozen or so groups of ochrophytes evolved by **secondary plastid endosymbiosis** between a

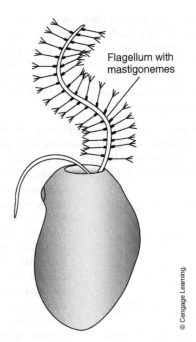

Flagellum with mastigonemes

FIGURE 6-18 STRAMENOPILES. Stramenopiles have two flagella. One is a shorter simple flagellum, and the other is longer and bears many hair-like mastigonemes.

heterotrophic stramenopile (or its ancestor) as the host and a red algal cell as the guest that provided the plastid. (Recall that red algae evolved by primary plastid endosymbiosis, with a cyanobacterium becoming the plastid.) Ochrophytes include diatoms, silicoflagellates, and brown algae (treated as a separate phylum in the next chapter). Most ochrophytes have chlorophyll *a*, chlorophyll *c* (molecularly unrelated to the true chlorophylls *a* and *b*), the yellow beta-carotene, and usually the brown pigment fucoxanthin. The end product of their photosynthesis is the complex carbohydrate **laminarin**.

Diatoms

Diatoms are among the most distinctive organisms in samples of marine phytoplankton, and they also are important members of benthic communities. In their marine planktonic species diversity (5,000 to 10,000 species), diatoms overshadow the remaining planktonic ochrophytes (20 to 30 species), planktonic green algae (100 species), dinoflagellates (more than 1,500 species), and other planktonic unicells (about 1,500 species). Diatoms are abundant in the oceanic province at high latitudes, whereas at low latitudes they are predominantly found in coastal waters or in areas of upwelling. Benthic habitats include soft sediments (as a part of microbial mats) or associations with living organisms. In addition, dense populations of diatoms may occur in sea ice.

Stramenopiles are a group of organisms that include diatoms and that have specialized flagella.

Mastigonemes are hair-like filaments that extend from the shaft of some flagella.

A **heterokont** cell bears two different flagella, one simple and one with mastigonemes.

Ochrophytes are photosynthetic stramenopiles.

Secondary plastid endosymbiosis is the condition of most eukaryotic photoautotrophs, in which an ancestor became host to a red algal cell that became the host's plastid.

Laminarin is the form of starch stored in ochrophytes.

Wherever they occur, diatoms often contribute a major portion of primary production. Many kinds of diatom can survive long periods of dim light or darkness if there is sufficient dissolved organic matter for heterotrophy. Diatoms typically store their food reserves as lipids (oils and fatty acids), as well as laminarin. The lipids help make diatoms more buoyant and provide a food source of high caloric value to herbivores such as copepods.

Diatom Structure A unique feature of diatoms is the cell wall **(Figure 6-19)**, called a **frustule**, composed of organic and mineral components. The organic part consists of the carbohydrate pectin, which is impregnated with a highly variable amount of silica (silicon dioxide). Silica, the substance of glass and many sands, may account for up to 95% of the diatom's weight; however, other diatoms have little silica in their frustules. Each half of the frustule is called a **valve**, one larger than the other and fitting over it like the cover of a shoebox. The surface of the frustule is ornamented with pores, knobs, grooves, spines, hooks, and other structures whose functions are only partly known. The intricate geometric patterns, however, are useful in identifying different species. As the diatom cell inside grows, the overlapping valves, which cannot grow, spread to accommodate the greater volume.

> A **frustule** is the two-part cell wall of diatoms.
>
> A **valve** is one part of the diatom frustule.
>
> A **centric diatom** has radially symmetrical valves.
>
> A **pennate diatom** has bilaterally symmetrical valves.
>
> The **raphe** is a slit along the valve by which some pennate diatoms move along surfaces.

Diatoms occur in two general shapes (see **Figure 6-19**): those with radially symmetrical valves (**centric diatoms**) and those with bilaterally symmetrical valves (**pennate diatoms**). Centric diatoms generally are planktonic, and the pennate diatoms are more typical of the benthos. The centric shape appeared early in diatom evolution, and the pennate shape evolved later several times independently. Diatoms may be as small as 2 micrometers in diameter and as long as 4 millimeters.

Locomotion in Diatoms Some benthic diatoms can move by secretion of mucilage from pores and slits, but the mechanism is still poorly understood. Movement is too slow to serve for avoidance of predators, but the ability to move might lessen the chance of depleting sources of nutrients on a microscopic scale. Only a few centric diatoms that are benthic can move, at least slowly, by release of mucilage through one to hundreds of specialized pores in the frustule. Many pennate diatoms can move, and those that have a long slit, or **raphe**, in the valve can move fastest. The secretion of mucilage also can elevate the diatom on a stalk for better exposure to the overlying water column, and it can bind sediment particles, increasing the stability of sediment where a surface layer of diatoms forms a mat over mud and sand.

Reproduction in Diatoms Diatoms reproduce asexually by fission. Diatoms do not share cell walls after fission, and most often the daughter cells separate and live independently. In some species, however, the frustules of adjacent diatoms remain connected to form filaments of many cells. Formation of chains of cells may reduce their rate of sinking in seawater and may reduce herbivory by plankton or small fishes whose mouths or filtration apparatuses cannot handle the larger chains.

FIGURE 6-19 DIATOMS. (a) This centric diatom of the genus *Thalassiosira* has radial symmetry, typical of planktonic diatoms. **(b)** The genus *Cocconeis* is a pennate diatom with biradial symmetry. This specimen was growing on the leaf of a seagrass.

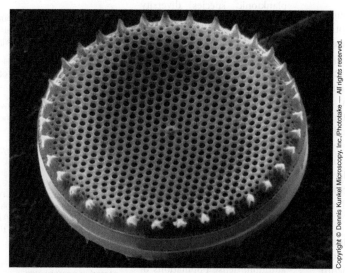

(a)

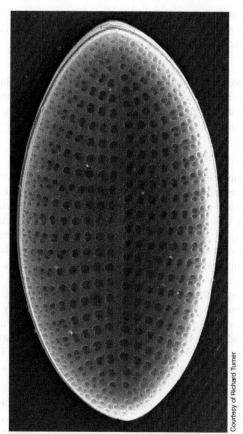

(b)

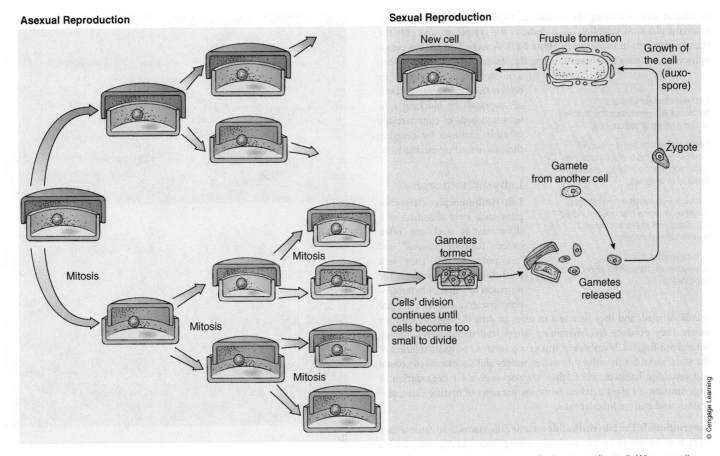

Asexual Reproduction

Mitosis

Mitosis

Mitosis

Mitosis

Cells' division continues until cells become too small to divide

Sexual Reproduction

New cell

Frustule formation

Growth of the cell (auxospore)

Zygote

Gamete from another cell

Gametes formed

Gametes released

© Cengage Learning

FIGURE 6-20 **DIATOM REPRODUCTION.** During asexual reproduction, diatoms divide by mitosis, each time producing a smaller cell. When a cell reaches a certain critically small size, it stops dividing, forms gametes, and sexual reproduction occurs. Regrowth of the zygote by auxospore formation restores the normal size of the cell.

When a diatom cell divides, each daughter cell inherits only one of the two valves of the parental frustule **(Figure 6-20)**. The cell is able to complete its frustule only by secreting a new, smaller valve within the inherited one. The daughter cell that inherits the larger valve has the potential to grow to the same size as the parent cell. The other daughter cell is limited in its growth by the size of the smaller parental valve it inherited. In each generation, half the cells are smaller than the parents, and the average cell size of the population decreases over successive generations. When the cell size reaches about 50% of the maximal cell size, the diatom forms an **auxospore**, which casts off the small frustule, increases the volume of its cell, and secretes a new frustule of normal dimensions. This period of growth is usually preceded by sexual reproduction. Diatoms smaller than one-third of the parental size cannot form auxospores, and they die.

Under poor environmental conditions, marine diatoms will form resting spores. During this process, the frustule thickens with additional silica, the cytoplasm becomes denser, stored organic reserves increase, and the cell sinks to the bottom sediments or becomes entangled in marine snow. Spores must be resuspended under favorable environmental conditions to germinate.

Diminishment of silica in the water is a major limiting factor in diatom populations, especially seasonally. Without silica, diatoms cannot reproduce or form spores.

Diatomaceous Sediments The silica part of diatom frustules cannot be decomposed by bacteria, and the frustules are almost insoluble in seawater, even under great pressure. When diatoms die, their frustules sink and accumulate on the seafloor as siliceous oozes, especially at high latitudes where diatoms are abundant. Eventually, some of these accumulations may form sedimentary rock. Over millions of years, geological events have moved some of these deposits to the surface. The deposits are called **diatomaceous earth**, and they are mined commercially. Because of the small size of the diatoms and the small pores in their frustules, diatomaceous earth makes an excellent material for filtering swimming pool water, beer, and champagne. It is a mild abrasive used in products such as silver polish and toothpaste. It is also used in soundproofing and insulation products. The nutrient reserves that are stored in diatoms as lipids also accumulate in the siliceous oozes. Most of the world's petroleum reserves were formed over the past 150 million years from diatom productivity and death.

> The **auxospore** is the stage in a diatom life cycle that restores the maximal size of the cell.
>
> **Diatomaceous earth** is an industrial material harvested from exposed deposits of dead diatom frustules.

Other Ochrophytes

Most other ochrophyte microbes belong to small and poorly studied groups. One group, the silicoflagellates, may be extremely abundant in cold marine waters. They have basket-shaped external skeletons composed of silica, as in diatoms. The cell wraps around the skeleton, which then only appears to be internal. Silicoflagellates also possess one or two flagella, at least one being the specialized flagellum with mastigonemes. Standard plankton sampling often destroys the cytoplasm, and only the skeleton remains in the samples **(Figure 6-21)**. A second group, the pelagophyceans, includes the bloom-forming alga *Aureococcus anophagefferens*, a nontoxic coastal species. Its brown tides can devastate beds of seagrasses by blocking sunlight and beds of commercially valuable molluscs by clogging their filter-feeding mechanisms.

Labyrinthomorphs are heterotrophic stramenopiles that are decomposers or pathogens.

Labyrinthulids are a group of labyrinthomorphs that includes the pathogen responsible for wasting disease in eelgrass.

Thraustochytrids are a group of labyrinthomorphs that are abundant decomposers in planktonic and benthic communities.

Haptophytes are a group of eukaryotic microbes that possess a haptonema.

Labyrinthomorphs

Labyrinthomorph stramenopiles once were classified with slime molds and are often called "fungoid protists" and "zoosporic fungi." They lack plastids and have osmotrophic nutrition like that of heterotrophic bacteria. Their cells are spindle shaped, and they secrete a mucous coating through which they move. They produce free-swimming stages, called zoospores, having heterokont flagella. Members of this group seem to be quite tolerant of the wide variation in salinity found in waters and sediments of coastal and estuarine habitats, where they are responsible for degradation of large amounts of plant detritus. Some are parasites of marine algae, seagrasses, and marine invertebrates.

Labyrinthulids The **labyrinthulids** include *Labyrinthula zosterae* **(Figure 6-22)**, now known to have caused the wasting disease of eelgrass (*Zostera marina*) in the North Atlantic in the early 1930s. As much as 90% of eelgrass acreage was lost in some geographical regions because of this disease. Recovery has been full in parts of its range, but eelgrass beds have not returned to other locations. The loss taught marine biologists a lesson about the critical role of seagrasses in the ecology of the seas. A related but unnamed species may have been responsible for a wasting disease of turtlegrass (*Thalassia testudinum*) in Florida Bay during 1987. Species of *Labyrinthula* seem to occur on other seagrasses as well as on

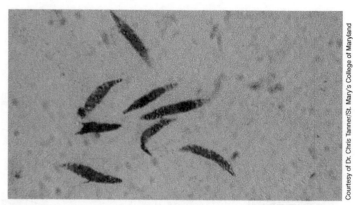

(a)

(b)

FIGURE 6-22 **LABYRINTHULIDS.** **(a)** These specimens of *Labyrinthula zosterae* were cultured from leaves of eelgrass. **(b)** Even healthy beds of seagrasses may show signs of labyrinthulid infection. Dark patches and streaks on leaves in this eelgrass bed show places where the microbe has destroyed tissue.

eelgrass and turtlegrass without causing disease. They are likely to be part of the normal seagrass flora, opportunistically invading living tissue only when the seagrass is under stress.

Thraustochytrids The **thraustochytrids** (thraw-stow-KY-trids) *Thraustochytrium* and *Schizochytrium* are planktonic and benthic decomposers. Some species are pathogens of marine organisms, including shellfish in both natural populations and aquaculture facilities. In addition to their beneficial role in marine environments and in contrast to their pathogenic impact, thraustochytrids are used commercially as a source of human dietary supplements. Oils extracted from *Schizochytrium* species are high in the polyunsaturated omega-3 fatty acid docosahexaenoic acid (DHA). DHA is known to be necessary in human neurological development and function, and it may reduce the concentration of undesirable lipids in blood.

HAPTOPHYTES

The phylum Haptophyta (also called prymnesiophytes) is closely related to the stramenopiles and the alveolates (a group discussed later in the chapter). It includes mostly photosynthetic organisms that were formerly grouped with the diatoms and silicoflagellates under an old name, Chrysophyta. The composition of their photosynthetic pigments is similar, but the **haptophytes** possess two similar simple flagella, both used

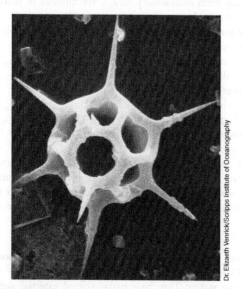

FIGURE 6-21 **SILICOFLAGELLATES.** Silicoflagellates are tiny phytoplankton related to diatoms. Their external shells are composed of silica.

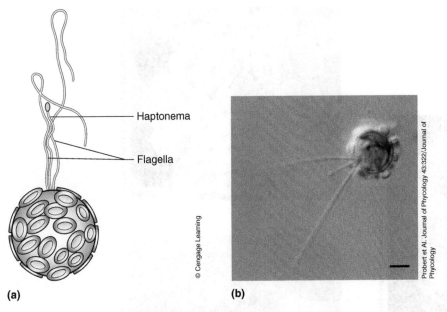

Haptonema

Flagella

© Cengage Learning

(a)

Probert et Al. Journal of Phycology 43:322/Journal of Phycology

(b)

FIGURE 6-23 HAPTOPHYTES. **(a)** Members of the phylum Haptophyta have a unique structure, the haptonema, that is used to collect food. **(b)** *Algirosphaera robusta* shows the short haptonema between the two longer flagella.

A **haptonema** is a rod-like organelle that projects between the two flagella of haptophytes and is used to capture prey.

Coccolithophores are photosynthetic haptophytes with the cell surface covered by calcareous scales.

A **coccolith** is one of numerous scales that cover the surface of coccolithophores.

Alveolates are a group of microbes with membranous sacs beneath the cell membrane.

An **alveolus** is a membranous sac beneath the cell membrane of an alveolate.

A **pellicle** is the complex of alveoli and the cell membrane in alveolates.

A **dinoflagellate** is an alveolate with cellulose in its alveoli and with two heterokont flagella for locomotion.

for locomotion, and a unique associated structure, the **haptonema** (*hapsis*, touch; *nema*, thread) **(Figure 6-23)**. The haptonema arises from the cell surface between the two flagella but does not undulate to propel the cell. The haptonema is very long in heterotrophic haptophytes, intermediate in length in haptophytes that mix photosynthesis and heterotrophy, and short or absent in photosynthetic haptophytes. The haptonema captures food on its sticky surface and then bends around to transfer the food to the posterior end of the cell, where ingestion occurs by phagocytosis. Haptophytes might have evolved from the same line as the ochrophytes or from a different event of secondary plastid endosymbiosis with a red alga. Most haptophytes are marine planktonic organisms.

Most of the 370 living species of haptophytes are **coccolithophores**, named after their surface coating of scales, or **coccoliths**, made of calcium carbonate **(Figure 6-24a, b)**. Coccoliths are produced internally within membranous vesicles and then are transported to the exterior of the cell. Production of coccoliths accounts for up to 40% of carbonate production in modern seas. Dead coccolithophores contribute heavily to formation of calcareous oozes on the seafloor and, where exposed by geological processes or changing sea levels, to chalk cliffs along the seashore. The White Cliffs of Dover and other formations along the English Channel, for example, are formed from coccolithophore ooze (see **Figure 6-24c**). Although some other haptophytes live in fresh water, coccolithophores are exclusively marine. They are significant members of the phytoplankton, particularly in the nutrient-poor open sea of the tropics, where they are heavily grazed by copepods and a variety of other zooplankton. It is believed that the high reflectance of the chalky coccoliths and their release of the greenhouse gas dimethyl sulfide may have a strong impact on global climate change, especially by bloom-forming species such as *Emiliania huxleyi*. But the 4,000 species of coccolithophores that have lived in the seas over the past 220 million years have had a significant impact on global carbon balance in addition to their contribution to the increase in atmospheric oxygen levels.

Because the shells of both diatoms and coccolithophores are relatively inert and do not decompose (although calcium carbonate dissolves at great depths), the amount of these sediments in the seafloor can be used to estimate what surface conditions were like in the past. Large drills are used to take core samples from the seafloor. These samples allow researchers to analyze the diatom and coccolith content and estimate the rate at which they accumulated in the bottom sediments. When surface waters were warm and carbon dioxide was plentiful in seawater for photosynthesis and calcification, there were large numbers of these organisms, and their remains rapidly accumulated on the bottom as they died. When climates were cool or carbon dioxide concentrations were lower, there were fewer of these organisms, and their remains accumulated more slowly.

Heterotrophic haptophytes are among the smallest (2 to 20 micrometers) eukaryotic plankton, giving them the ability to consume large numbers of yet smaller bacteria. In this way they transfer bacterial biomass to higher levels of marine food chains, largely through predation by marine ciliates, a group described in the next section.

ALVEOLATES

Scientists have recently regrouped several kinds of microbes into the phylum Alveolata on the basis of molecular biology and structure of their cells. **Alveolates** have membranous sacs (**alveoli**) beneath their cell membranes. If the combination of cell membrane and alveoli is complex, this structure at the cell surface is called a **pellicle (Figure 6-25)**. A pellicle differs from a cell wall because it exists within rather than outside the cell membrane. Marine alveolates include dinoflagellates and ciliates.

Dinoflagellates

Dinoflagellates are globular single-celled organisms (although a few species are colonial) with two flagella that lie in grooves on the cell

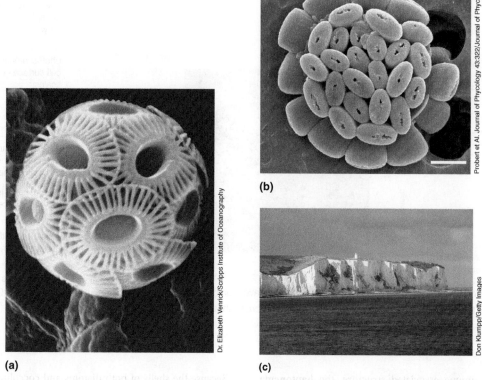

(a)

(b)

(c)

FIGURE 6-24 **COCCOLITHOPHORES.** Coccolithophores are tiny phytoplankton with cells covered by calcareous plates called coccoliths. **(a)** A coccolithophore with discoidal coccoliths. **(b)** *Algirosphaera robusta*, a coccolithophore with coccoliths shaped like petals. **(c)** Over a period of millions of years, coccolithophore tests accumulated in the bottom sediments of the seas around Great Britain. Geologic events ultimately exposed the deposits, forming chalk cliffs such as the White Cliffs of Dover.

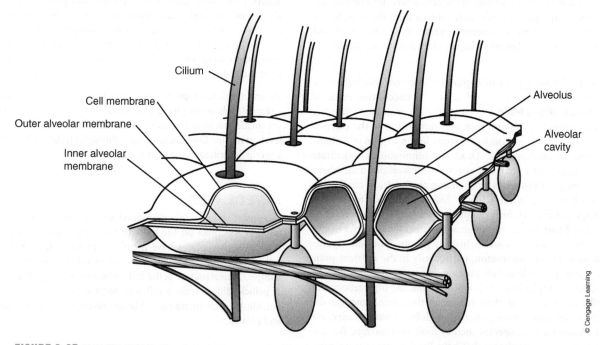

Cilium

Cell membrane

Outer alveolar membrane

Inner alveolar membrane

Alveolus

Alveolar cavity

FIGURE 6-25 **THE PELLICLE.** Alveolates have a complex layering of membranous sacs just inside the cell membrane. The example shown here is a ciliate, with many cilia, each surrounded at its base by donut-shaped alveoli. In dinoflagellates, the alveoli are filled with one or more layers of cellulose, forming the plates that strengthen the cell.

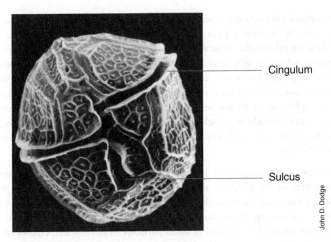

Cingulum

Sulcus

John D. Dodge

FIGURE 6-26 DINOFLAGELLATES. This scanning electron micrograph of the armored dinoflagellate *Gonyaulax* shows the intricate plates of cellulose within the pellicle. The two grooves that house the flagella (not shown) are the cingulum around the middle and the sulcus, which branches from the cingulum.

surface (**Figure 6-26**) in a unique arrangement. The pellicle often contains thin plates of cellulose within the alveoli. A unique decay-resistant chemical called **dinosporin** is associated with the cellulose plates, especially in dormant stages of the life cycle found in sediments. About 90% of known species of dinoflagellate are marine. Although most species are planktonic, some are benthic, some symbiotic, and others parasitic. Some dinoflagellates, like some bacteria, are bioluminescent and may produce spectacular nighttime displays in surface waters, giving rise to place names such as Bioluminescent Bay, Puerto Rico. Dinoflagellates range in size from 2 micrometers to 2 millimeters.

The relationship of dinoflagellates to other groups and to each other has been confused by their tendency to steal plastids from other phytoplankton over the millennia, as well as during their present-day feeding. Dinoflagellates are believed to have evolved by secondary plastid endosymbiosis with a red algal cell, giving rise to a plastid that is typical of autotrophic species. Since their origin, some lines of dinoflagellate evolution have lost the primitive plastid or replaced it by secondary or tertiary plastid endosymbiosis with green algae or haptophytes.

Dinoflagellate Structure The two flagella of dinoflagellates are heterokont, and both usually arise from pores on the side of the cell. A shorter ribbon-like flagellum with a single row of hair-like filaments encircles the cell in a horizontal groove, the **cingulum**, and produces a spinning motion to the dinoflagellate (*dinos*, whirling). A longer flagellum, with two rows of hairs, trails down a longitudinal groove, the **sulcus**, and powers most of the cell's forward motion. The thin cellulose plates in the pellicle are absent or few in **unarmored dinoflagellates**, and they occur in multiple layers in **armored dinoflagellates**. The plates of armored dinoflagellates are so thick that they are easily observed under the microscope, giving the impression of a soccer ball. The number, size, and shapes of plates are used by specialists to tell species apart. The surfaces of many armored dinoflagellates are ornamented with spines and other structures that reduce predation and the rate of sinking or that increase surface area for absorption of nutrients.

Dinoflagellate Nutrition Dinoflagellates obtain food by several means. The primitive plastid of photosynthetic dinoflagellates contains chlorophylls *a* and *c* and the accessory pigments beta-carotene and **peridinin**. The latter pigment, a carotenoid, gives dinoflagellates their typical golden-brown color. They store food as starch of the same composition as that of green plants. Many photosynthetic dinoflagellates are **mixotrophic**, which means that they supplement photosynthesis by either osmotrophy (absorbing dissolved nutrients) or **phagotrophy** (engulfing particles by phagocytosis). Only half the species of dinoflagellate are photosynthetic; the other half live mainly by a combination of phagotrophy and osmotrophy.

Because dinoflagellates are among the larger phytoplankton, their lower ratio of surface area to cell volume makes them less efficient in absorbing phosphorus and nitrates from seawater. These inorganic nutrients are largely obtained by eating other plankton (diatoms, other dinoflagellates, and many other kinds of microbes, as well as metazoan plankton, including copepods and fish eggs).

Reproduction in Dinoflagellates Dinoflagellates reproduce asexually by fission of the cell into two cells, at rates of up to one division per day. Each daughter cell of armored dinoflagellates must replace the cellulose plates of the missing half after fission. Sexual reproduction, known in a number of species, occurs by fusion of similar gametes. In most cases meiosis then occurs, resulting in adult dinoflagellates that are haploid. Life histories often include the production of **cysts**, dormant stages resistant to decay (by the presence of dinosporin) and to environmental stress. Cyst formation may follow seasonal sexual reproduction and is otherwise associated with the onset of unfavorable environmental conditions, such as reduced light, temperature, and nutrients.

Ecological Roles of Dinoflagellates Dinoflagellates play an important role in marine ecosystems. Together with diatoms and coccolithophores, they are a major component of the phytoplankton that provides food directly or indirectly to many marine animals. Their flagella, mixotrophic nutrition, ability to migrate vertically in the water column, and lack of dependence on silicon give dinoflagellates an ecological advantage over diatoms. For these reasons, dinoflagellates are more abundant than diatoms in tropical waters, particularly in the open sea where inorganic nutrients are in low concentration. On the other hand, aside from predation, to which diatoms are subjected as well, dinoflagellates are less tolerant of stormy seas and other turbulence. Their relatively large size and multiplated structure increase the chance of their cells being torn apart. Some dinoflagellates are parasitic and live in the intestines of marine crustaceans called copepods. Other species, collectively called **zooxanthellae** (ZOH-oh-zan-THEL-ee), lack the flagella

Dinosporin is a decay-resistant chemical in the cellulose plates of dinoflagellates.

The **cingulum** is the groove around the middle of a dinoflagellate.

The **sulcus** is the groove on the surface of a dinoflagellate that extends from the cingulum.

An **unarmored dinoflagellate** has a few thin layers of cellulose in its alveoli, giving the appearance of having no protective cell covering.

An **armored dinoflagellate** has many layers of cellulose in its alveoli, giving the appearance of having a protective cell covering.

Peridinin is a carotenoid pigment in dinoflagellates that gives them their golden-brown color.

A **mixotrophic** mode of nutrition is one that combines autotrophy and heterotrophy.

Phagotrophy is a mode of nutrition by which one organism eats another, either by phagocytosis or by taking into the mouth.

Zooxanthellae are dinoflagellates that are important symbionts of other organisms.

found in most other dinoflagellate species and are symbionts of jellyfish, corals, and molluscs. The zooxanthellae are photosynthetic and provide food for their host organisms, and the hosts provide carbon dioxide, other essential nutrients, and shelter.

Harmful Algal Blooms Some marine dinoflagellates are responsible for the phenomenon known as **harmful algal blooms** (HABs, formerly called "red tides"). HABs occur when photosynthetic dinoflagellates (or other primary producers) undergo a population explosion. The specific cause of these population explosions, or blooms, is not known. During a bloom, the number of organisms can be so great that they color the water red, orange, or brown. The species that cause HABs produce potent toxins that affect the nervous system, kidneys, muscles, or other parts of animals, either directly or through consumption of contaminated food. HAB toxins can kill fish, marine mammals, and humans. As bacteria decompose animals killed by the toxins, the oxygen content of the water may be depleted, leading to the death of other organisms otherwise unaffected by the toxins. Dead fish and other marine life can wash up on beaches and decay, creating sanitation problems and making the beaches unfit for recreational use.

Some dinoflagellate species produce toxins that are taken up by molluscs, in particular, clams and mussels. The molluscs are not harmed by the toxin, but it is concentrated in their tissues. Other animals that feed on these contaminated shellfish may be harmed or killed as a result. Humans who consume contaminated shellfish exhibit neurological problems such as loss of balance and coordination, a tingling sensation or numbness in the lips and extremities, slurred speech, and nausea. Death can occur if an amount of toxin sufficient to paralyze the respiratory muscles is ingested. This condition is known as **paralytic shellfish poisoning**, or PSP. Dinoflagellate toxins cannot be destroyed by cooking. Contamination of commercial shellfish by HABs results in a substantial loss of revenue to the shellfishing industry.

Ciliates

Ciliates are protozoans that bear cilia for locomotion and gathering food. The cilia may densely cover the body (i.e., the cell) in highly organized columns and rows, or they may be confined to certain patches on the cell. In many ciliates, adjacent cilia are fused in tufts or long rows called **membranelles (Figure 6-27)**. Membranelles often occur in the oral region, where an organelle called the **cytostome** serves as a permanent site for phagocytosis of food. Another distinctive feature of ciliates is their possession of two kinds of nucleus: a **micronucleus**, which carries one diploid set of chromosomes for inheritance by the next generation, and a **macronucleus**, which is much larger than the micronucleus and carries many copies of only those parts of the genome that are used for metabolism.

Marine ciliates range from 10 micrometers to 3 millimeters long and are members of the plankton and benthos, where they are major links in food chains and form symbiotic and parasitic relationships with many other marine organisms. Because suspension feeding by cilia is a highly efficient method for removing small particles from seawater, ciliates are significant components of microbial food webs. Marine ciliates reproduce asexually by binary fission and sexually by **conjugation** (the transfer of nuclei between fused cells). Fission may occur as often as once every 2 hours.

Types of Marine Ciliates Marine communities have a high diversity and abundance of ciliates, especially in sediments and coastal waters. Among the many groups of marine ciliates are three ecologically significant groups: the scuticociliates, oligotrichs, and tintinnids (see **Figure 6-27**). The scuticociliates have a dense and uniform distribution of cilia on their bodies. Oligotrichs have few cilia. Tintinnids usually lack body cilia and secrete an organic, loosely fitting shell called a **lorica**. The physical support and protection given by the lorica may be increased by cementing foreign particles to the organic membrane. The particles may be grains of silica or carbonate sand, or they may be leftovers of consumed food, such as coccoliths and diatom frustules.

Ecological Roles of Marine Ciliates Although some marine ciliates harbor autotrophic symbionts or plastids, most are heterotrophs. They are links in the transfer of production from heterotrophic and autotrophic (blue-green) bacteria to higher levels in the food chain. Ciliates less than 30 micrometers long typically graze on bacteria and heterotrophic nanoflagellates (flagellated plankton measuring 2 to 20 micrometers) and are a critical link in the bacterial loop of pelagic food webs. Many ciliates are herbivores; small ones take in diatoms, cyanobacteria, dinoflagellates, and autotrophic nanoflagellates; large ones eat other ciliates and small zooplankton. Ciliates in sediments capture bacteria, diatoms, and small animals that live among the grains. Most ciliates capture food by suspension feeding with their membranelles, but some actively chase prey. In turn, small ciliates are prey for larger ciliates, and ciliates in general are consumed by heterotrophic dinoflagellates, amoeboid "protozoans" (described in a later section), and animals, particularly small crustaceans, ciliated animals, and the larvae of many invertebrates. Planktonic ciliates reach their highest densities in nutrient-rich coastal waters, sometimes 5×10^5 cells per liter, largely consisting of small ciliates that eat bacteria.

CHOANOFLAGELLATES

Choanoflagellates are a phylum of marine and freshwater flagellated cells that look much like choanocytes, a type of cell found in sponges. Choanoflagellates are more closely related to animals than any other group of one-celled microbes, a conclusion based largely on molecular genetics. In fact, the entire genome of the marine choanoflagellate *Monosiga brevicollis* was sequenced in 2008. Choanoflagellates may

---

A **harmful algal bloom (HAB)** is formed by dense population of photosynthetic microbes and macroalgae that presents an environmental threat in natural and managed communities.

Paralytic shellfish poisoning is a human syndrome caused by ingestion of saxitoxin in seafood.

Ciliates are a group of alveolates that use cilia for locomotion and feeding.

Membranelles are ribbon-shaped or tufted arrangements of cilia that increase the effectiveness of locomotion and feeding.

A **cytostome** is an organelle in a ciliate where phagocytosis occurs.

The **micronucleus** is the smaller nucleus of a ciliate; it holds one diploid set of chromosomes for inheritance by the next generation.

The **macronucleus** is the larger nucleus of a ciliate; it holds many sets of chromosomes and plays a role in metabolism of the cell.

Conjugation is sexual reproduction that involves the exchange of nuclei between two fused cells.

A **lorica** is a loosely fitting external covering of a microbe.

Choanoflagellates are a group of microbes that filter suspended particles through a specialized collar surrounding the flagellum.

A **secondary metabolite** is a chemical made by cells for purposes other than routine metabolism.

MARINE BIOLOGY & THE HUMAN CONNECTION

Harmful Algal Blooms

Many kinds of marine algae undergo rapid population increases under conditions that are still poorly understood. Some are macroscopic algae such as the exotic *Caulerpa taxifolia* that has invaded the Mediterranean Sea and continues to expand its coverage of the seafloor. Microscopic algae often bloom as "red tides" in surface waters **(Figure 6-A)** and wreak havoc in coastal fisheries by killing commercially valuable species. Toxic species are especially notorious for their impact on populations of commercial fish and shellfish and the humans, birds, and marine mammals that consume them. In addition, toxin-bearing cells can become airborne as a result of rough sea-surface conditions and enter the lungs of marine mammals and humans when they inhale the mist, causing respiratory distress and even death. Blooms of nontoxic phytoplankton, including pelagophyceans and green algae, shade benthic communities by day and consume much oxygen while respiring at night. The pelagophyceans *Aureococcus* and *Aureoumbria* repeatedly cause brown tides off New England, the Mid-Atlantic states, and Texas, and the green alga *Resultor* caused significant loss of seagrass in the Atlantic lagoons of Florida in 2011. These harmful algal blooms (HABs) have become an area of significant interest to marine biologists. The journal *Harmful Algae*, launched in 2002, serves for scientific communication in this emerging discipline.

Sanford Berry/Visuals Unlimited, Inc.

FIGURE 6-A

HARMFUL ALGAL BLOOMS. "Red tides" are caused by population explosions of certain species of dinoflagellates, diatoms, and blue-green bacteria.

Toxic marine HABs are caused mostly by dinoflagellates. The toxins they produce are not part of their normal metabolism but are specially produced for protection. Such chemicals, called **secondary metabolites**, are highly complex end products of elaborate metabolic pathways that function under altered growth conditions. The best-known HAB toxins are the **saxitoxins**, a family of neurotoxins produced by dinoflagellates. The human syndrome caused by consumption of saxitoxins is called paralytic shellfish poisoning. Other human syndromes caused by other dinoflagellate toxins include **diarrhetic shellfish poisoning, neurotoxic shellfish poisoning**,

and **ciguatera fish poisoning**, the latter caused by benthic dinoflagellates. A fifth syndrome, **amnesic shellfish poisoning**, is caused by a toxin produced by blooms of diatoms.

Studying HABs is difficult for several reasons. HAB organisms are difficult to culture in the laboratory. Recreating the proper conditions for altered growth characteristics and inducing HAB organisms to produce toxins are also problematic. Solutions to these problems should ultimately help marine biologists better understand the causes of these harmful blooms and help them find ways to control or prevent them.

---

occur as single cells or in colonies, and the colonies may be stalked or embedded in a gelatinous mass **(Figure 6-28)**. The cell often is surrounded by a lorica composed of rods of silica. A single flagellum extends from the cell surface and is surrounded at its base by a funnel-shaped collar of tentacle-like projections called **microvilli**. The flagellum bears filaments like the mastigonemes of stramenopiles. Its movement creates a current that draws water and particles through the collar. Choanoflagellates are small organisms, about 10 micrometers or less. They are highly efficient consumers of bacteria, which become trapped between the microvilli and then are taken into food vacuoles by phagocytosis at the base of the collar. Although some are attached, free-swimming planktonic choanoflagellates in some locations may outnumber other groups of flagellated plankton, including dinoflagellates.

AMOEBOID PROTOZOANS

Amoeboid protozoans belong to several unrelated "protozoan" groups, once placed within the same phylum. They all have an organelle called a **pseudopod** that is an extension of the cell surface. Unlike the cilium and flagellum, pseudopods can change shape and are used for locomotion by benthic species and for food capture by benthic and pelagic ones. Amoeboid protozoans are heterotrophs, consuming bacteria and other small organisms that they find in the plankton,

Saxitoxins are a class of toxins produced by certain species of dinoflagellate that cause paralytic shellfish poisoning.

Diarrhetic shellfish poisoning is a human syndrome caused by consumption of shellfish that concentrate the toxic chemicals produced by blooms of certain species of dinoflagellate.

Neurotoxic shellfish poisoning is a human syndrome caused by consumption of shellfish that concentrate toxic chemicals produced by blooms of certain species of dinoflagellate.

Ciguatera fish poisoning is a human syndrome caused by consumption of fish that concentrate toxic chemicals produced by zooxanthellae.

Amnesic shellfish poisoning is a human syndrome caused by consumption of shellfish that concentrate the toxic chemicals produced by diatom blooms.

A **microvillus** is one of many short, hair-like, cellular extensions that form the collar of choanoflagellates.

A **pseudopod** is a finger-like projection of cytoplasm and membrane that functions in both locomotion and feeding in amoeboid protozoans.

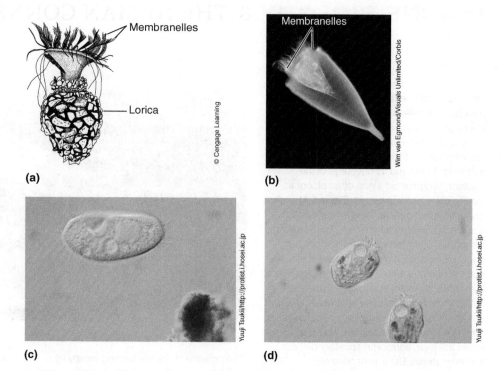

(a)

(b)

(c)

(d)

FIGURE 6-27 MARINE CILIATES. (a) Membranelles are tufts or long rows of fused cilia that function in locomotion and feeding. This marine ciliate is a tintinnid with its body partly encased in a lorica. **(b)** In this image of a tintinnid through a bright-field microscope, the membranelles at one end and the vase-shaped lorica are easily visible. There are no cilia on the body of the cell. **(c)** The marine ciliate *Pleuronema marina* has a uniform distribution of cilia typical of scuticociliates. Note how long the cilia extend from the cell. **(d)** *Strombidium armatum* is an oligotrich, with few cilia on the body and with membranelles at the tip of the cell.

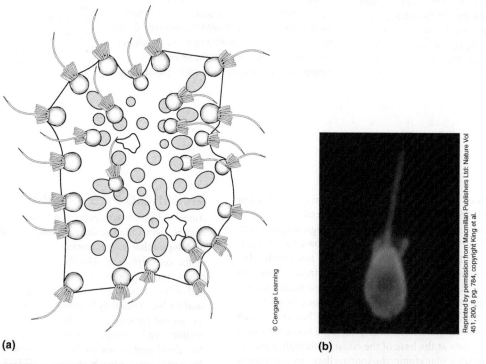

(a)

(b)

FIGURE 6-28 CHOANOFLAGELLATES. (a) A colony of the marine colonial choanoflagellate genus Proterospongia. The collar is composed of microvilli and surrounds a central flagellum. Cells are united in a gelatinous mass. **(b)** *Monosiga brevicollis* is a solitary choanoflagellate. The short faint collar is visible at the base of the long single flagellum. Its entire genome has recently been sequenced.

among grains of sediment, or on various surfaces. On contact, the prey becomes attached to the pseudopod. At some point along the pseudopod, the prey is then engulfed by phagocytosis and digested.

Some amoeboid protozoans have an unprotected cell membrane, although most have a shell, or test. The **test** is an externally secreted organic membrane that is often covered with foreign particles or strengthened by mineral secretions. In addition to a test, some amoeboid protozoans have an internal mineral skeleton. Pseudopods extend from a single opening or through multiple pores in the test.

This functionally similar but taxonomically mixed group of marine organisms includes two major phyla that are closely related: foraminiferans and actinopods. Their slender pseudopods give them the name filose amoebae. Foraminiferans are abundant and diverse on the seafloor and are perhaps the most ecologically significant planktonic protozoans. The actinopods include the radiolarians, acantharians, and heliozoans, of which the radiolarians predominate. Other groups of amoeboid protozoans have broad pseudopods and are rare in the sea. These lobose amoebae are very diverse as freshwater and parasitic microbes. They are more closely related to fungi, choanoflagellates, and animals than they are to the filose amoebae.

Foraminiferans

Foraminiferans, or **forams** (phylum Foraminifera), are amoeboid protozoans having branched pseudopods that form elaborate, net-like structures called **reticulopods**, used to snare prey. In addition, benthic forams use their reticulopods to crawl through sediments or over surfaces, and planktonic forams may use reticulopods to reduce the rate at which they sink through the water column. Forams consume large quantities of diatoms and bacteria, and some harbor symbiotic green and red algae and zooxanthellae (dinoflagellates). The symbionts provide nutrients to their hosts and, as in reef-building corals, might play a role in calcification, the process of depositing calcium carbonate in their tests. Some forams can attain sizes of 1 centimeter (⅜ inch) or more, and some fossil species reached 12 centimeters (5 inches) in diameter. Forams reproduce both sexually and asexually, and their life cycles are somewhat complex.

Foraminiferan Test Forams often produce elaborate multichambered tests of calcium carbonate **(Figure 6-29)**. As forams grow, they add more chambers, and the resulting test frequently resembles a microscopic snail shell. Benthic forams may have chambers in rows or spirals, and planktonic species typically have spherically symmetrical tests. The tests of planktonic species are usually thin and light and bear many spiny processes that help reduce the sinking rate. At depths greater than 2,000 meters (6,600 feet), benthic forams attach foreign particles to the organic membrane rather than deposit calcium carbonate because of calcium carbonate's solubility at great pressure and low temperature.

Although most foraminiferan species are bottom dwellers, a few species with enormous numbers of individuals are members of the zooplankton. The tests of dead planktonic forams are a major constituent of sediments in some areas of the deep ocean floor. These sediments are called **globigerina** (gloh-bij-eh-RYE-nuh) **ooze** because of the large number of tests of the genus *Globigerina*, a planktonic foram of low latitudes. In addition to foram tests, globigerina ooze may consist of as much as 25% coccoliths. Globigerina ooze is not found at depths exceeding 5,000 meters (16,500 feet) because the great pressure and cold temperature dissolve the carbonate tests faster than they can accumulate. Forams also contribute to the sand found on the beaches of many islands, including Tonga in the Pacific Ocean and the pink sands of Bermuda and the Bahamas.

Foraminiferans and Zooxanthellae Many forams from nutrient-poor waters maintain ecologically efficient relationships with zooxanthellae that live symbiotically within the foram's cytoplasm. By day, dinoflagellates migrate out along the reticulopods, gaining maximal exposure to sunlight. At night, they migrate deep into the central shell. Because it is difficult to keep planktonic foraminiferans alive in the laboratory, marine biologists have not yet been able to study the chemical basis of this relationship thoroughly. Experiments have demonstrated, however, that forams with zooxanthellae, when kept in normal sunlight, grow and reproduce more rapidly than those kept in the dark or those that have had the zooxanthellae chemically eliminated. Although the chemical evidence is lacking, the relationship seems to involve an exchange of materials between the forams and their symbionts similar to that occurring between corals and their zooxanthellae. The forams generate waste products such as carbon dioxide and ammonia, which are nutrients for the zooxanthellae, although toxic to the forams. The intracellular zooxanthellae are able to absorb the nutrients with little effort because they do not have to expend energy to reconcentrate them. This self-contained fertilizing and waste-removal system is extremely efficient for supporting life in nutrient-poor water.

Radiolarians

Radiolarians are a highly diverse class of zooplankton within the phylum Actinopoda **(Figure 6-30a)**, named for the **actinopods**, long needle-like pseudopods, that they bear. They have a central nuclear region surrounded by an external organic membrane, or **capsule**, perforated

---

A **test** is the external skeleton of an amoeboid protozoan.

A **foraminiferan** (or **foram**) is an amoeboid marine protozoan with specialized pseudopods and a calcareous test.

A **reticulopod** is a pseudopod of foraminiferans with branches that interconnect to form a net for the capture of particles.

Globigerina ooze is a seafloor sediment formed by the accumulated calcareous tests of foraminiferans.

A **radiolarian** is an amoeboid marine protozoan with specialized pseudopods and a skeleton of silica.

An **actinopod** is a needle-like pseudopod typical of radiolarians.

The **capsule** of radiolarians is an external organic layer that separates the inner nuclear region from the outer region.

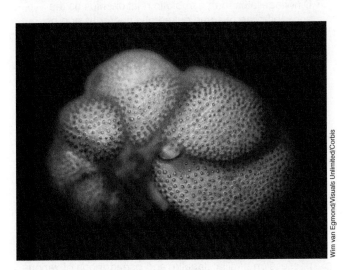

Wim van Egmond/Visuals Unlimited/Corbis

FIGURE 6-29 FORAMINIFERANS. Foraminiferans produce tests of calcium carbonate that contain chambers and resemble the shell of a microscopic snail. They use their pseudopods to capture prey.

Pseudopods

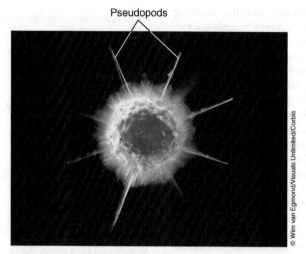

(a)

(b)

FIGURE 6-30 RADIOLARIANS. (a) Radiolarians consist of a central nuclear region surrounded by a frothy calymma. Long pseudopods extend from the cell in this live radiolarian to capture food. **(b)** The cell of the radiolarian is supported by a skeleton of silica, which often bears short radiating spines.

with one or many pores. Actinopods pass through the pores of the capsule and form a region called the **calymma**. The extended actinopods capture food and reduce sinking rates. In addition to actinopods, the cytoplasm in the calymma has large vacuoles to control buoyancy.

The cytoplasm of the calymma typically secretes an internal skeleton of silica. The skeleton may consist of numerous spines that radiate outward with the pseudopods, and in some species, the spines may be clustered, resembling children's jacks. Other radiolarian species have an intricate, fused skeleton that looks like a spherical glass ornament (see **Figure 6-30b**). Because silica is resistant to dissolving in seawater, even under great pressure, the skeletons of dead radiolarians accumulate on the seafloor as siliceous **radiolarian oozes** at depths beyond those at which foram tests dissolve. In contrast to diatomaceous oozes, which form at high latitudes, most radiolarian oozes form at low latitudes.

The **calymma** is the vacuolated outermost cytoplasm of a radiolarian.

Radiolarian ooze is a seafloor sediment formed by the accumulated siliceous skeletons of radiolarians.

Most radiolarians are planktonic and live in the photic zone, where they use their actinopods to capture a variety of phytoplankton and zooplankton. The larger radiolarians even capture copepods and other planktonic crustaceans. Zooxanthellae often are found in the calymma, and the radiolarians obtain nutrients from these symbiotic dinoflagellates.

Relatives of the radiolarians have similar construction and function, but they belong to different classes of actinopods. Acantharians secrete a skeleton of radiating spines made of strontium sulfate. Heliozoans, rare in the sea, lack an organic capsule between the nuclear region and calymma. Both groups are mainly planktonic, and many species harbor zooxanthellae.

IN SUMMARY

- Marine fungi, mostly sac fungi, are microscopic decomposers and pathogens.
- Fungi degrade cellulose, chitin, and other decay-resistant molecules that otherwise would accumulate in seafloor sediments.
- Marine fungi take advantage of water currents and sea foam for the transport of spores.
- Fungi form lichen associations with green algae and blue-green bacteria in maritime communities.
- Marine protists are highly diverse in their biochemistry, cell structure, cell coverings, and modes of nutrition, locomotion, and reproduction. (Have You Wondered? #3)
- Dinoflagellates, diatoms, coccolithophores, and silicoflagellates are photosynthetic producers and members of the phytoplankton.
- These photoautotrophs have evolved by secondary plastid endosymbiosis through phagocytosis of a red alga.
- Dinoflagellates form symbiotic relationships as the zooxanthellae of some eukaryotic microbes and of many invertebrates.
- Labyrinthomorphs are decomposers and pathogens.
- Heterotrophic consumers include ciliates (especially the tintinnids), choanoflagellates, foraminiferans, radiolarians, and many dinoflagellates.
- Among the consumers, a few groups are grazers of bacteria, allowing for the transfer of prokaryotic biomass to higher levels of marine food webs.
- Some eukaryotic microbes incorporate calcium carbonate and silicon dioxide into their cell walls or skeletons, adding to seafloor oozes on their death and sedimentation.
- The eukaryotic groups are distinguished by their cell coverings, the structure of their cell membranes, their possession of cilia, flagella, and pseudopods of various shapes for locomotion and prey capture, the chemistry of pigments and food-storage compounds, life-history characteristics, and many other features.

IN PERSPECTIVE

Taxonomic Group	Representative Organisms	Form and Function	Reproduction	Biochemistry	Ecological Roles
Viruses	Bacteriophages, cyanophages, phycoviruses, podoviruses, retroviruses, papillomaviruses, morbilliviruses	Icosahedral or helical as free virions, with protein capsid surrounding the genome; some with envelope from host cell membrane	Dependent on host mechanisms to make copies; either lytic or lysogenic	Genome of DNA or RNA, with capsid of protein subunits, and often with a few enzymes	Parasitic, major cause of mortality among plankton, possible control of bloom-forming microbes, important in alteration of pelagic food webs
Bacteria	Blue-green bacteria, green and purple sulfur and nonsulfur bacteria; *Prochlorococcus*, *Thiomargarita*, actinobacteria	Prokaryotic with cell wall containing aminated sugars; some flagellated; shapes mostly include rods and spheres, some mycelial	Binary fission	Widely varied modes of nutrition; photoautotrophs with bacteriochlorophyll or chlorophylls *a, b*	Chemoautotrophs and photoautotrophs, decomposers, nitrogen fixers, symbionts for light production or nutrition; form stromatolites with other microbes
Archaeons	*Methanocaldococcus*, *Pyrolobus*, halobacteria	Prokaryotic with cell wall lacking aminated sugars	Binary fission	Widely varied modes of nutrition; some autotrophs with bacteriorhodopsins; special lipids play role in stabilizing cell membranes under extreme environmental conditions	Chemoautotrophs and photoautotrophs, heterotrophs, methanogens; inhabitants of extreme environments
Fungi	Ascomycotes, including yeasts; lichens	Eukaryotic with cell wall of chitin; single-celled as yeasts or forming mycelium of hyphae	Asexual reproduction by budding or conidiospores; sexual reproduction with fruiting bodies and ascospores; passive dispersal of spores	Lignin digestion; glycogen storage; strict aerobes	Heterotrophs, decomposers; symbiotic with green algae or blue-green bacteria in lichens
Stramenopiles	Diatoms, silicoflagellates, *Aureococcus*, labyrinthulids, brown algae (discussed in Chapter 7)	Eukaryotic; heterokont flagella, one with hair-like filaments; diatoms with frustule of pectin and silica	Fission; sexual reproduction; auxospore formation	Chlorophylls *a, c*; lipid storage as oils and fatty acids; starch storage as laminarin	Photoautotrophs, some forming harmful blooms; heterotrophic decomposers; producers of seagrass wasting disease; form deep-sea deposits
Haptophytes	Coccolithophores, *Emiliania*	Eukaryotic; two simple flagella with haptonema at base; cell covering of calcium carbonate plates	—	Chlorophylls *a, c*	Photoautotrophs and phagocytic heterotrophs in plankton; form deep-sea deposits

Taxonomic Group	Representative Organisms	Form and Function	Reproduction	Biochemistry	Ecological Roles
Alveolates	Dinoflagellates; ciliates, including tintinnids	Eukaryotic; pellicle of membranous sacs; no cell wall but some with loosely fitting lorica; two heterokont flagella, or many cilia; cilia often fused as membranelles	Primarily by binary fission; ciliates with sexual reproduction by conjugation	Dinoflagellate pellicle with cellulose plates; autotrophic dinoflagellates with chlorophylls *a, c* and storage of simple starch	Photoautotrophs, osmotrophs, phagotrophs; planktonic and benthic; some symbiotic as zooxanthellae or forming harmful algal blooms
Choanoflagellates	Choanoflagellates, *Monosiga*	Eukaryotic; one flagellum with hair-like filaments, collar of microvilli; in gelatinous or branched colonies or as single cells with lorica	—	—	Planktonic and benthic filter feeders on bacteria
Foraminiferans	Forams, *Globigerina*	Eukaryotic; reticulopods; test of calcium carbonate or foreign particles	Asexual fission; some sexual reproduction with complex life cycles	—	Planktonic suspension feeders and benthic grazers; phagocytic; form deep-sea deposits
Actinopods	Radiolarians, acantharians, heliozoans	Eukaryotic; perforated organic capsule, actinopods; internal skeleton of silica or strontium sulfate	—	—	Planktonic suspension feeders and benthic grazers; phagocytic; form deep-sea deposits

Photos/Illustrations: **Viruses:** © Cengage Learning; **Bacteria:** David J. Patterson/Marine Biological Laboratory; **Fungi:** Tokura, R. 1982 Arenicolous Marine Fungi from Japanese beaches. Transaction of the Mycological Society of Japan 23:423–433/Kiyoto Kyoiku University; **Stramenopiles:** © Cengage Learning; **Haptophytes:** Dr. Elizabeth Venrick/Scripps Institute of Oceanography; **Alveolates:** John D. Dodge; **Choenoflagellates:** © Cengage Learning; **Foraminiferans:** Ed Reschke/Peter Arnold; **Actinopods:** Dr. Richard Kessel & Dr. Gene Shih/Visuals Unlimited.

KEY CONCEPTS

1. Microbial life in the sea is extremely diverse and includes members of all three domains of life, as well as viruses. (6-1, 6-2, 6-3, 6-4)

2. Marine virology is an emerging field of study, due to recognition of the critical role that viruses may play in population control of other microbes, in nutrient cycling, and in marine pathology. (6-1)

3. Photosynthetic and chemosynthetic bacteria and archaeons are important primary producers in marine ecosystems. (6-2, 6-3)

4. Primary and secondary plastid endosymbiosis accounts for the origin of all eukaryotic phytoplankton in the seas. (6-2, 6-4)

5. Heterotrophic bacteria, archaeons, and fungi play essential roles in recycling nutrients in the marine environment. (6-2, 6-3, 6-4)

6. Marine eukaryotic microbes are primary producers, decomposers, and consumers, and some contribute significantly to the accumulation of deep-sea sediments. (6-4)

7. Populations of several kinds of photosynthetic marine microbes may form harmful blooms that affect other marine and maritime organisms directly and indirectly. (6-4)

1. Although the roles that viruses and bacteria play in human disease might first come to mind, their roles in the sea are more diverse. Viruses control populations of other microbes, especially bloom formers. Through lysis of host cells, viruses influence marine food chains and alter biogeochemical cycles. Bacteria are important to other life in the sea on account of their wide metabolic diversity, their consumption by small eukaryotic grazers, their hastening of sedimentation, and their contribution to symbiotic relationships. In particular, the mitochondria and plastids of eukaryotes originated as marine bacteria.

2. Archaeons are known as extremophiles, organisms that can tolerate extreme environmental conditions. Two kinds of extremophile are hyperthermophiles (tolerant of high temperature) and halophiles (tolerant of high salinity).

3. Despite their small size, microbes are highly diverse in their biochemistry, cell structure, cell coverings, and modes of nutrition, locomotion, and reproduction.

QUESTIONS FOR REVIEW

Multiple Choice

1. Most marine viruses are
   a. alive
   b. bacteriophages
   c. human pathogens
   d. lysogenic
   e. noninfective

2. Which of the following is not a method of obtaining organic matter by marine microbes?
   a. chemosynthesis
   b. nitrogen fixation
   c. osmotrophy
   d. phagocytosis
   e. photosynthesis

3. Which of the following characteristics do bacteria and archaeons have in common?
   a. a prokaryotic cell structure
   b. chlorophyll *a*
   c. kinds of membrane lipids
   d. production of methane
   e. tolerance of very high temperatures

4. Progress in marine mycology has been slow because
   a. most of them are small yeasts
   b. of the presence of nonmarine contaminants in samples
   c. they do not reproduce sexually
   d. they rarely produce mushrooms
   e. they require anaerobic conditions

5. The auxospore is critical to the life history of diatoms because it
   a. can photosynthesize in the deep dark waters of the ocean
   b. helps disperse the population by flagellar locomotion
   c. is a resting stage that survives environmentally stressful periods
   d. is needed to reconstitute the original cell size
   e. produces many gametes by meiosis

6. Stramenopiles are united by their possession of
   a. different flagella
   b. many flagella
   c. no flagellum
   d. one flagellum
   e. two flagella

7. Which of the following is not an effect of harmful algal blooms?
   a. consumption of oxygen
   b. contamination of shellfish
   c. formation of deep-sea sediments
   d. release of toxins
   e. shading of benthic plants and seaweeds

8. Zooxanthellae are
   a. dinoflagellates
   b. flagellated
   c. foraminiferans
   d. parasites
   e. resting cysts in a life cycle

9. Globigerina ooze is formed by
   a. coccolithophores
   b. diatoms
   c. foraminiferans
   d. radiolarians
   e. tintinnids

10. Select the microbe that is properly paired with its organelle of locomotion.
    a. ascomycote, none
    b. choanoflagellate, collar
    c. ciliate, pellicle
    d. coccolithophore, coccolith
    e. phycovirus, flagellum

Matching

1. Match the following microbial groups with their modes of nutrition:

   a. archaeons
   b. choanoflagellates
   c. cyanobacteria
   d. labyrinthomorphs
   e. sulfur bacteria
   f. *Thiomargarita namibiensis*

   1. Aerobic photoautotrophs.
   2. Anaerobic photoautotrophs.
   3. Chemoautotrophs.
   4. Filterers of bacteria.
   5. Methanogenic heterotrophs.
   6. Osmotrophs.

2. Match the following microbial groups with their cell coverings or skeletons:

   a. bacteria
   b. coccolithophores
   c. diatoms
   d. dinoflagellates
   e. foraminiferans
   f. fungi
   g. radiolarians
   h. tintinnids

   1. Cell wall of chitin.
   2. Cell wall of pectin impregnated with silica.
   3. Cell wall of sugars and amino acids.
   4. Internal skeleton of silica.
   5. Multichambered test of calcium carbonate.
   6. Organic lorica, sometimes with foreign particles cemented to it.
   7. Pellicle of thin plates of cellulose.
   8. Surface scales of calcium carbonate.

Short Answer

1. What biotic and abiotic factors control viral activity in the sea?

2. Describe how marine bacteria hasten the fall of dead particles from surface waters.

3. Why are archaeons called "extremophiles"?

4. Explain why the calcareous skeletons of forams and coccolithophores do not accumulate at the greatest depths of the sea.

5. How do the cells of diatoms and dinoflagellates differ?

6. What significant role do heterotrophic bacteria, fungi, and labyrinthomorphs play in marine habitats?

7. Describe the methods of food capture by tintinnids, foraminiferans, and choanoflagellates.

Thinking Critically

1. If planktonic viruses lyse bacterial cells at a high rate, what effect might this have on food webs in the pelagic zone?

2. Many planktonic microbes have projections of some sort from their cells, cell walls, tests, and loricas. What are the possible advantages of these structures?

3. Marine microbes exhibit a wide range of sizes. If you wanted to study them, what methods would you use and what problems might you encounter in trying to collect them from seawater?

4. In what ways is *Prochlorococcus* better adapted for life in the open tropical seas than are diatoms?

SUGGESTIONS FOR FURTHER READING

Breitbart, M., L. R. Thompson, C. A. Suttle, and M. B. Sullivan. 2007. Exploring the Vast Diversity of Marine Viruses, *Oceanography* 20(2): 135–139.

Capriulo, G. M. 1990. *Ecology of Marine Protozoa*. New York: Oxford University Press.

DeLong, E. F. 2003. A Plenitude of Ocean Life, *Natural History* 112(4): 40–46.

Falkowski, P. G., and A. H. Knoll. 2007. *Evolution of Primary Producers in the Sea*. Boston: Elsevier Academic Press.

Fischer, M. G., M. J. Allen, W. H. Wilson, and C. A. Suttle. 2010. Giant Virus with a Remarkable Complement of Genes Infects Marine Zooplankton, *Proceedings of the National Academy of Sciences* 107: 19508–19513.

Hunter-Cevera, J., D. Karl, and M. Buckley. 2005. *Marine Microbial Diversity: The Key to Earth's Habitability*. Washington, D.C.: American Academy of Microbiology.

Hyde, K. D., and S. B. Pointing. 2000. *Marine Mycology: A Practical Approach*. Hong Kong: Fungal Diversity Press.

Kirchman, D. L. 2000. *Microbial Ecology of the Oceans*. New York: Wiley-Liss.

Lafferty, K. D., J. W. Porter, and S. E. Ford. 2004. Are Diseases Increasing in the Ocean?, *Annual Review of Ecology, Evolution, and Systematics* 35: 31–54.

Medina, M., A. G. Collins, J. W. Taylor, J. W. Valentine, J. H. Lipps, L. Amari-Zettler, and M. L. Soglin. 2003. Phylogeny of Opisthokonta and the Evolution of Multicellularity and Complexity in Fungi and Metazoa, *International Journal of Astrobiology* 2(3): 203–211.

Munn, C. B. 2011. *Marine Microbiology: Ecology & Applications*, 2nd ed. New York: Garland Science.

Poole, A. M., and D. Penny. 2006. Evaluating Hypotheses for the Origin of Eukaryotes, *Bioessays* 29(1): 74–84.

Porter, J. W. 2001. *The Ecology and Etiology of Newly Emerging Marine Diseases*. Dordrecht, The Netherlands: Kluwer Academic Publishers.

Rosenberg, E., O. Koren, L. Reshef, R. Efrony, and I. Zilber-Rosenberg. 2007. The Role of Microorganisms in Coral Health, Disease and Evolution, *Nature Reviews, Microbiology* 5(5): 355–362.

Suttle, C. A. 2007. Marine Viruses—Major Players in the Global Ecosystem, *Nature Reviews, Microbiology* 5(10): 801–812.

Valentine, D. L. 2007. Adaptations to Energy Stress Dictate the Ecology and Evolution of the Archaea, *Nature Reviews, Microbiology* 5(4): 316–323.

Zimmer, C. 2000. Sea Sickness, *Audubon* 102(3): 39–45.

Have You Wondered?

1. **What features distinguish the several groups of seaweeds? (7-1)**

2. **Aside from primary production, what roles seaweeds and plants play in the ocean? (7-1, 7-2)**

3. **How seaweeds and plants are distributed by depth, geography, and time? (7-1, 7-2)**

4. **What adaptations have allowed flowering plants, especially seagrasses, to invade the sea? (7-2)**

5. **How coastal populations historically and societies today have made use of seaweeds and marine plants? (7-1, 7-2)**

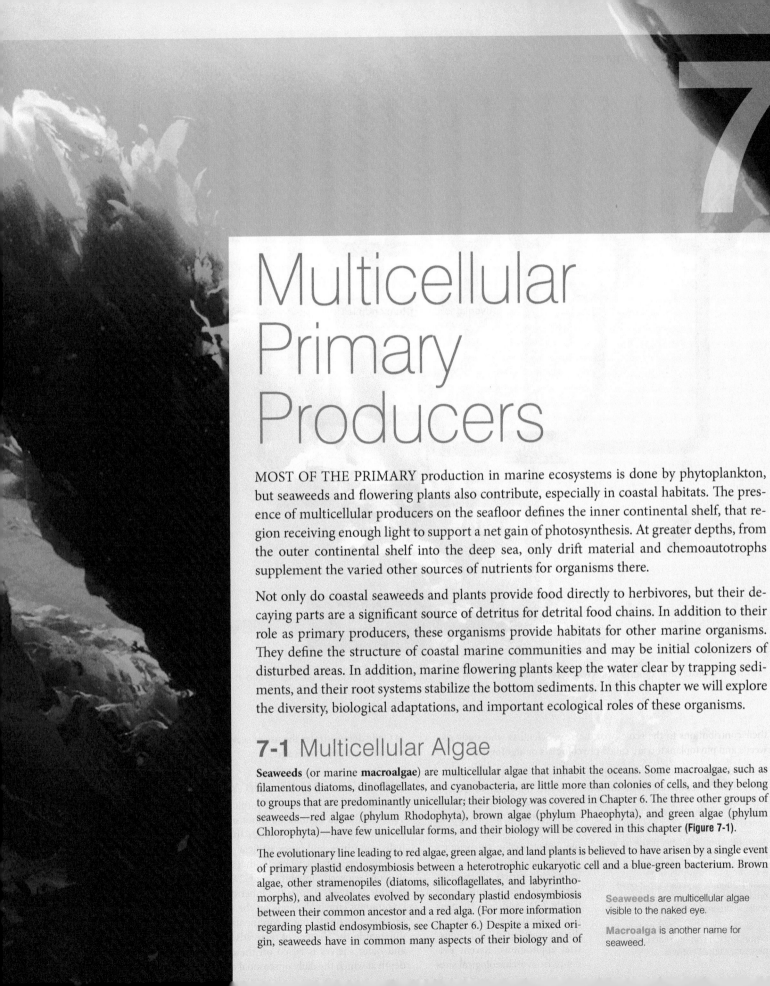

7

Multicellular Primary Producers

MOST OF THE PRIMARY production in marine ecosystems is done by phytoplankton, but seaweeds and flowering plants also contribute, especially in coastal habitats. The presence of multicellular producers on the seafloor defines the inner continental shelf, that region receiving enough light to support a net gain of photosynthesis. At greater depths, from the outer continental shelf into the deep sea, only drift material and chemoautotrophs supplement the varied other sources of nutrients for organisms there.

Not only do coastal seaweeds and plants provide food directly to herbivores, but their decaying parts are a significant source of detritus for detrital food chains. In addition to their role as primary producers, these organisms provide habitats for other marine organisms. They define the structure of coastal marine communities and may be initial colonizers of disturbed areas. In addition, marine flowering plants keep the water clear by trapping sediments, and their root systems stabilize the bottom sediments. In this chapter we will explore the diversity, biological adaptations, and important ecological roles of these organisms.

7-1 Multicellular Algae

Seaweeds (or marine **macroalgae**) are multicellular algae that inhabit the oceans. Some macroalgae, such as filamentous diatoms, dinoflagellates, and cyanobacteria, are little more than colonies of cells, and they belong to groups that are predominantly unicellular; their biology was covered in Chapter 6. The three other groups of seaweeds—red algae (phylum Rhodophyta), brown algae (phylum Phaeophyta), and green algae (phylum Chlorophyta)—have few unicellular forms, and their biology will be covered in this chapter (**Figure 7-1**).

The evolutionary line leading to red algae, green algae, and land plants is believed to have arisen by a single event of primary plastid endosymbiosis between a heterotrophic eukaryotic cell and a blue-green bacterium. Brown algae, other stramenopiles (diatoms, silicoflagellates, and labyrinthomorphs), and alveolates evolved by secondary plastid endosymbiosis between their common ancestor and a red alga. (For more information regarding plastid endosymbiosis, see Chapter 6.) Despite a mixed origin, seaweeds have in common many aspects of their biology and of

Seaweeds are multicellular algae visible to the naked eye.

Macroalga is another name for seaweed.

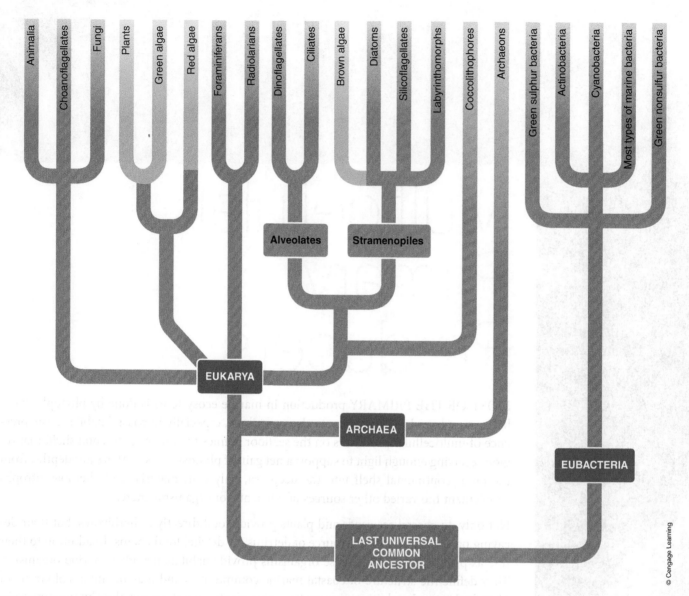

FIGURE 7-1 Taxonomic groups that are covered in this chapter on multicellular primary producers are highlighted in green.

their contributions to the ecology of the seas. Scientists who study sea-weeds and phytoplankton are called **phycologists** or **algologists**.

Seaweeds are important in the economy of coastal seas. They produce a three-dimensional structure to the otherwise two-dimensional space of the shallow seafloor, thereby creating habitat for many marine organisms. They contribute substantially to primary production and are consumed by a wide variety of animals, particularly sea urchins, snails **(Figure 7-2a)**, and fish. Humans use seaweeds for a variety of dietary, medicinal, and industrial applications. Recent evidence from archaeological sites

in Chile demonstrates the use of seaweeds by humans in the New World more than 14,000 years ago.

DISTRIBUTION OF SEAWEEDS

Most species of seaweed are benthic, growing on rock, sand, mud, and coral on the sea bottom, on other organisms, and as part of a **fouling community** (plants and animals that live on pilings, bulkheads, boat hulls, moorings, and other artificial surfaces). Some seaweeds attach to very specific surfaces, whereas other seaweeds are rather nonselective. In general, seaweeds inhabit about 2% of the seafloor. The presence of benthic seaweeds defines the inner continental shelf, where the marine community largely depends on the food and protection that seaweeds provide. Life on the outer continental shelf and in the deep sea is quite different in the absence of seaweeds. The distinction between the inner and outer shelves is based on the **compensation depth** of algae, the depth at which the daily or seasonal amount of light is sufficient to allow

A **phycologist** is one who studies seaweeds.

An **algologist** is another name for a phycologist.

A **fouling community** is an assemblage of organisms that grow on an intertidal or a submerged artificial structure.

Compensation depth is that depth beyond which primary producers generally cannot survive.

(a)

(b)

FIGURE 7-2 ZONATION OF SEAWEEDS. **(a)** Periwinkles and other gastropods graze on newly settled spores and early growth stages of seaweeds, sometimes eliminating them from zones along the shore. **(b)** Some of the clearest examples of seaweed zonation can be found in the intertidal zones of rocky shores. On this Scottish shore, brown algae grow in an upper zone and green algae in a lower zone. The growth of green algae here is partly due to nutrient enrichment from nearby seabird colonies. In general, the distribution of seaweeds by depth is not explained by their possession of different combinations of photosynthetic pigments (the chromatic adaptation hypothesis). Instead, zonation is explained by the responses of seaweeds to competition, herbivory, and physiological tolerances to many climatic factors.

enough photosynthesis to balance the metabolic needs of algae, with nothing left over to support growth.

The environmental factors that are most influential in governing the distribution of seaweeds are light and temperature. Some other abiotic factors critical in governing the distribution of seaweeds are duration of tidal exposure and desiccation (drying out), wave action and surge, salinity, and availability of mineral nutrients. The areas of the world most favorable to seaweed diversity include both sides of the North Pacific Ocean, Australia, southwestern Africa, and the Mediterranean Sea.

Effects of Light on Seaweed Distribution

The vertical and horizontal distributions of seaweeds are limited in part by the availability of sunlight and therefore vary by depth, latitude, sea conditions, and season. It was once thought that the vertical distribution of red, brown, and green algae could be explained by their accessory photosynthetic pigments, the presence of which gives the seaweeds their characteristic colors, a concept known as **chromatic adaptation**. Because green light penetrates deepest in coastal waters and the accessory pigments of red algae absorb mostly green wavelengths, red algae were thought to extend to the greatest depth of primary producers. It followed that green algae, which have pigments absorbing mostly blue and red wavelengths that diminish rapidly in seawater, should be found at the shallowest depths. Because accessory pigments of brown algae absorb intermediate wavelengths of light, brown algae would be expected to be most abundant at intermediate depths. Indeed, recent evidence seems to support the hypothesis of chromatic adaptation because the depth record (295 meters, or 973 feet) for seaweeds is held by a yet undescribed species of red algae from the Bahamas. However, the green alga *Rhipiliopsis profunda* at 268 meters (884 feet) is close behind this record.

The concept of chromatic adaptation was proposed in the 1800s and was accepted for about 100 years, until it was realized that such zonation did not occur and that the distribution of seaweeds depended more

on herbivory, competition, varying concentration of these specialized pigments, and the ability of seaweeds to alter their forms of growth (see **Figure 7-2b**).

Effects of Temperature on Seaweed Distribution

Temperature affects the distribution of seaweeds. The greatest diversity of algal species is in tropical waters. Farther north or south of the equator, the number of species decreases, and the species themselves are different. Many marine algae in colder latitudes are **perennials**, meaning that they live longer than 2 years. During the colder seasons only part of the alga remains alive; sometimes only a few cells but more often a mass of stem-like structures. When the temperature warms up in the spring, this body part initiates new growth. Temperature is not usually a limiting factor for algae that live in tropical and subtropical seas, although the temperature in intertidal areas may become too warm and contribute to seasonal mass mortality of many seaweeds and the animals they shelter. At high latitudes, freezing and scouring by ice may eliminate seaweeds from the intertidal and shallow subtidal zones.

> **Chromatic adaptation** is the alteration of kinds and quantities of photosynthetic pigments in response to changes in quantity and quality of light.
>
> A **perennial** is a plant or an alga that completes its life cycle in more than two growing seasons.
>
> A **thallus** is the body of an alga.
>
> **Vascular tissue** of higher plants provides physical support and transports water, minerals, and food.

STRUCTURE OF SEAWEEDS

The seaweed body is called the **thallus** (plural thalli) **(Figure 7-3a)**, and with few exceptions all cells of the thallus are photosynthetic. Because the thallus lacks the **vascular** (conductive) **tissue** of members of the kingdom Plantae, seaweeds do not have roots, stems, or leaves. The thallus can, however, occur in complex shapes, and some seaweeds are easily mistaken for plants. If most of the thallus is flattened, it may be called a

Reproductive structures

Air bladder

Blade

Stipe   Holdfast

(a)

© Cengage Learning

(b)

Courtesy of Richard Turner

(c)

Courtesy of Richard Turner

(d)

Courtesy of Richard Turner

(e)

L. Newman & A. Flowers/Photo Researchers, Inc.

FIGURE 7-3 VARIABLE SHAPES OF SEAWEEDS. (a) The thallus of a seaweed may consist of several distinct parts, as in rockweed (*Fucus*). But the shape can vary widely: **(b)** filamentous and branching as in the red alga *Spyridia filamentosa*, **(c)** tubular as in *Ulva intestinalis* (gutweed), **(d)** leaflike as in *Ulva lactuca* (sea lettuce), and **(e)** encrusting as in *Porolithon* (coralline algae).

MARINE BIOLOGY & THE HUMAN CONNECTION

Seaweeds and Medicine

Seaweeds have been used for centuries to treat a variety of illnesses. Asiatic cultures used seaweeds as long ago as 300 BC to treat glandular disorders, such as goiter. A goiter is an enlargement of the thyroid, a gland located in the neck. The thyroid gland requires iodine to produce its secretions. When an individual's diet is deficient in iodine, the gland frequently enlarges. Because many seaweeds concentrate iodine from seawater in their tissues, consumption of seaweed treats the condition.

The ancient Romans used seaweeds for treating burns and rashes and for healing wounds. The slimy mucilage that coats the blades of many seaweeds effectively blocks air and microorganisms from reaching an affected area. This relieves discomfort, helps prevent infection, and promotes healing. The red alga *Porphyra* contains vitamin C and was used by English sailors to prevent scurvy, a disease caused by vitamin C deficiency. *Porphyra* was more readily available than citrus fruits during ocean voyages and did not spoil as quickly. Several species of red algae were used to eliminate parasitic worms from the intestines of affected individuals. Kaenic acid, which is extracted from the red alga *Digenia*, is still used for this purpose.

In the past, phycocolloids that were isolated from red algae were used to treat a variety of intestinal ailments. Phycocolloids dissolve slowly and are not digested. They coat the lining of the stomach and intestines so that material in the digestive tract, such as acids, will not irritate it, thus alleviating some of the symptoms related to ulcers and stomachaches. Phycocolloids were used to treat constipation because they promote retention of fluid in the large intestine, which helps ease the movement of fecal material. Today, products from red algae such as agar and carrageenan are used in the treatment of ulcers. The pharmaceutical industry uses a variety of polysaccharides from red algae to coat pills and in the production of time-release capsules.

Probably the most important algal product used in medicine and research today is agar. Because agar can withstand high temperatures, it can be sterilized. Its porosity allows the movement of nutrients. It is solid at room temperature and resists decomposition by most microorganisms. These characteristics make agar an ideal medium for growing bacteria and fungi for study and research.

---

frond or **blade**. The structure attaching the thallus to a surface is the **holdfast**. Some seaweeds have a stem-like region, or **stipe**, between the holdfast and blade. Variation in the complexity of these parts of the thallus provides a rich diversity of form **(Figure 7-3)**.

BIOCHEMISTRY OF SEAWEEDS

The major distinctions among phyla of seaweeds are in their biochemistry: photosynthetic pigments, composition of cell walls, and the nature of their food reserves.

Photosynthetic Pigments

The color of the thallus is due to the wavelengths of light that are not absorbed by the combination of pigments in the seaweed. Chloroplasts in all seaweeds have chlorophyll *a*, which, along with chlorophylls *b* (green algae), *c* (brown algae), and *d* (red algae), is responsible for photosynthesis. Chlorophylls absorb blue and red wavelengths of light and pass green light. Accessory photosynthetic pigments, such as carotenes, xanthophylls, and phycobilins, absorb different wavelengths of light and pass the energy to chlorophylls for photosynthesis. In some cases the accessory pigments protect chlorophyll molecules from damage by light (providing a photoprotective function similar to a sunscreen).

Composition of Cell Walls

The cell walls of seaweeds are primarily composed of cellulose. In the case of calcareous algae, the cell wall may be impregnated with calcium carbonate. Many seaweeds secrete a slimy gelatinous mucilage, made of polymers of several sugars, that covers their cells. The mucilage can hold a great deal of water and may act as a protective covering that retards desiccation in intertidal algae exposed at low tide. It can also be sloughed off to remove organisms that may have become attached to the thallus. Some algae have a thick multilayered covering of protein called a **cuticle**. The cuticle is a protective layer and may be so thick that it gives the alga

an iridescent sheen. Other cell-wall components have commercial and medicinal value.

Nature of Food Reserves

When the amount of photosynthesis exceeds the immediate needs of the seaweed, the excess sugars are converted into polymers and stored in the cells as starches. The chemistry of these starch molecules differs among the groups of macroalgae. Unique sugars and alcohols may also be used as antifreezes by intertidal seaweeds at high latitudes where temperatures can drop below freezing.

REPRODUCTION IN SEAWEEDS

Seaweeds can reproduce both asexually and sexually. In some, asexual reproduction can occur when the thallus breaks into pieces, and each piece grows into a new alga, a type of asexual reproduction called **fragmentation**. In some coastal habitats, huge accumulations of seaweeds, called **drift algae**, form by fragmentation. Communities of drift algae can be physically complex and biotically diverse, but their high demand for oxygen may lead to rapid decline of the community, mass mortality, and decomposition. Some sargassum weeds reproduce entirely by fragmentation and form extensive floating mats covering hundreds of acres of sea surface.

A **frond** is a leaf-like part of a seaweed.

Blade is a synonym of frond.

The **holdfast** of a thallus anchors an alga to the seafloor.

The **stipe** is the stem-like part of a thallus connecting the frond to the holdfast.

A **cuticle** is the outermost nonliving layer of an organism.

Fragmentation is the production of new organisms from pieces of a parent organism.

Drift algae are seaweeds freed from attachment that can accumulate on windward shores.

A **sporangium** is the part of a seaweed that produces and releases asexual spores.

The **sporophyte** is the asexual, spore-producing stage in the life cycle.

Another type of asexual reproduction involves spore formation, typically by meiosis within a part of the thallus called the **sporangium**. Once created, the haploid spores are dispersed by water currents. Spores (and gametes) of green and brown algae usually have flagella that enable them to move in the water. If the spores settle in a suitable habitat, they germinate and form new thalli

(Figure 7-4). The stage of the life cycle that produces spores is the **sporophyte** ("plant that produces spores"), and it is diploid in chromosome number.

Sexual reproduction in seaweeds involves the formation of gametes that fuse to form a diploid, fertilized cell known as a zygote. The stage of the life cycle that produces gametes, the **gametophyte** ("plant that produces gametes"), is usually haploid. Gametes are typically produced by mitosis within structures in the gametophyte called **gametangia**. Gametophytes of some seaweeds are differentiated into

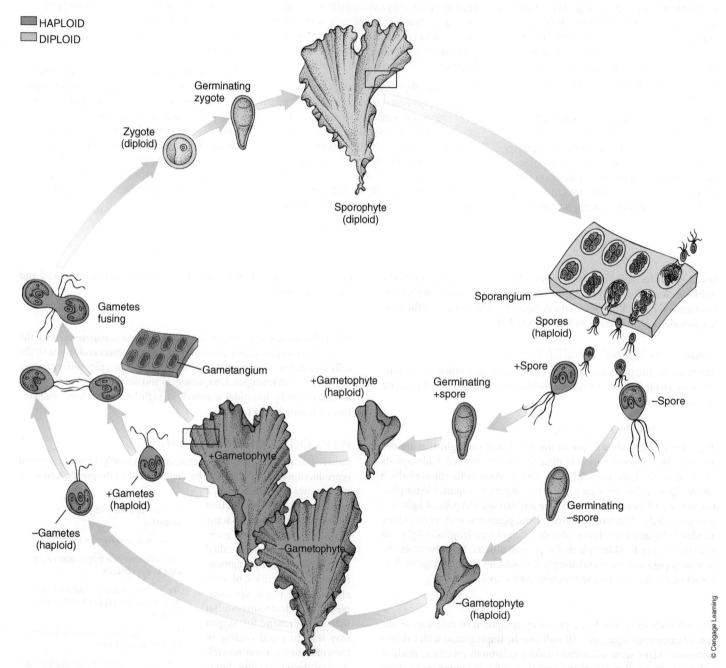

HAPLOID
DIPLOID

Germinating zygote

Zygote (diploid)

Sporophyte (diploid)

Sporangium

Spores (haploid)

+Spore

−Spore

Gametes fusing

Gametangium

+Gametophyte (haploid)

Germinating +spore

Germinating −spore

+Gametophyte

+Gametes (haploid)

−Gametes (haploid)

−Gametophyte

−Gametophyte (haploid)

© Cengage Learning

FIGURE 7-4 LIFE CYCLE OF THE SEA LETTUCE *ULVA*. In the sea lettuce *Ulva* (see Figure 7-3d), the diploid sporophyte produces by meiosis two different kinds of haploid motile spores, designated + and −. When the spores germinate (undergo mitosis), they grow into haploid gametophytes that release two kinds of motile gametes, also designated + and −. A + gamete fuses with a − gamete to produce a diploid zygote. The zygote settles on solid bottom, germinates, and forms a new sporophyte. The designations + and − are given because the spores, gametophytes, and gametes within each lineage cannot be distinguished by shape from the same stages in the other lineage.

male (sperm-producing) and female (egg-producing) gametophytes. After dispersal and settlement, the zygote grows by cell division (mitosis) into the sporophyte.

The possession of two or more separate multicellular stages (an asexual sporophyte and sexual gametophyte) in succession is called an **alternation of generations** (see **Figure 7-4**). Compared with a typical animal life cycle, most macroalgal life cycles are much more complex, sometimes involving as many as three multicellular generations of diploid sporophytes and haploid gametophytes, which may differ in size, shape, and longevity. The evolutionary advantage of an alternation of generations seems to lie in the differing adaptations of stages to herbivory, tolerances of environmental stress, ability to mask mutations in diploid cells, and lower nutrient requirements of haploid cells.

GREEN ALGAE

In contrast to life on land, life on the seafloor is not very green, but rather brown or red. Some relatives of land plants that add patches of green to the sea are green algae and seagrasses. **Green algae** (phylum Chlorophyta) are a diverse group of microbes and multicellular organisms that contain the same kinds of pigments found in vascular (land) plants: chlorophylls *a* and *b* and certain carotenoids. Two of the four classes of green algae are predominantly freshwater and terrestrial microbes, some the algal component of lichens. A third group, the prasinophyceans, probably represent the earliest green algae; prasinophyceans are poorly known marine phytoplankton. Seaweeds belong to the fourth class, the Ulvophyceae, which includes only about 1,100 species, or 13%, of green algae. Green algae are important as seasonal sources of food for marine animals. They also contribute to the formation of coral reefs. Green algae exhibit a rapid growth response to the presence of high levels of nutrients in polluted, near-shore waters and as such are members of fouling communities. Some green algae are introduced exotic species that are of special concern for marine conservation.

Structure of Green Algae

The majority of green algae are unicellular or are small multicellular filaments, tubes, or sheets. On the other hand, many marine green algae, especially those in the tropics, have a **coenocytic** (see-nuh-SIT-ik) thallus, which consists of a single giant cell or a few large cells containing more than one nucleus and surrounding a large vacuole **(Figure 7-5a)**. Coenocytic algae go through a developmental process in which the cell grows and the nucleus divides, but the cell does not divide. This process is repeated many times, resulting in a large multinucleate cell. The examples of this body plan shown in **Figure 7-5** demonstrate that a great diversity of form is possible even with this simplified organization of the cell. Some species appear to have a simple form, as in *Valonia* with its saclike thallus. Although the thallus at first appears quite simple, closer examination shows that it maintains its shape with an elaborate cell wall composed of cellulose. The cellulose is deposited in multiple layers for strength, similar to a multiple-ply automobile tire or plywood. The coenocytic alga *Caulerpa* forms elegant, feather-shaped thalli or structures that look like stems, leaves, and fruit. One of the most beautiful marine algae is *Acetabularia*. Although not technically coenocytic, as its relatives are, the uninucleate thallus is complex. *Acetabularia* has a slender stalk with a delicate, round, umbrella-like cap at the top. The cap functions in photosynthesis. This single-celled alga has been used by cell biologists to study the role of the nucleus in controlling cellular processes. Another interesting seaweed is *Codium*. This alga has a thallus composed of many long filaments wrapped together like a rope. *Codium* is of interest because of its healing response when damaged: A cut end will constrict to prevent loss of the cell sap of its coenocytic thallus. The green alga *Halimeda* has a coenocytic thallus that contains many segments, somewhat like a prickly pear and other cactuses, with cell walls containing deposits of calcium carbonate. Calcareous green algae such as *Halimeda* play an important role, along with coralline red algae and coral animals, in the formation of coral reefs. Deposits of *Halimeda* carbonate may exceed 50 meters (165 feet) depth in the Great Barrier Reef.

Response of Green Algae to Herbivory

Tolerance, avoidance, and deterrence are three ways that organisms adapt to predation, and the green algae provide good examples of each. Most green algae are small annuals. Rapid growth and the release of huge numbers of spores and zygotes prevent elimination of their populations by herbivores (**tolerance**). Their small size allows them to occupy crevices on rocky shores and reefs that thwart larger herbivores (**avoidance**). In addition to these survival tactics, coenocytic algae illustrate two adaptations to repel herbivores (**deterrence**). Deposits of calcium carbonate in the cell walls of *Halimeda*, *Acetabularia*, and other green algae require herbivores to have strong jaws, and they deter herbivory by filling the stomachs of grazing invertebrates and fish with minerals that have no nutritional value. Many coenocytic algae produce repulsive toxins that reduce herbivory on their otherwise nutritious thalli. Although few herbivores tolerate the toxins of coenocytic algae, one group of sea slugs (ascoglossans, molluscs related to nudibranchs) has evolved a piercing-and-sucking feeding style that takes advantage of the coenocytic thallus. A simple radula pierces the cell wall, and large quantities of the cell contents can be withdrawn from a single hole. In addition to getting a good meal, many ascoglossans harvest chloroplasts and retain the toxins. The chloroplasts—now stolen organelles—are maintained within cells of the slug's body as a source of sugars from photosynthesis, and the algal toxins protect the slug from predation.

Reproduction in Green Algae

A common marine form of green algae is sea lettuce, *Ulva* **(Figure 7-3d)**. Its life cycle is representative of the life cycles of green algae. The thallus of this alga resembles a large leaf of lettuce that may exceed 30 centimeters (12 inches) in length. It is found in intertidal and shallow coastal waters, where it attaches to the bottom by means of a tiny holdfast. The life cycle of *Ulva* demonstrates the basic alternation of generations of seaweeds **(Figure 7-4)**. The large leafy sporophyte and gametophyte stages are nearly identical. Their flagellated spores (released from the sporophyte) and gametes (released from the gametophyte) are also similar, but spores have four flagella and gametes two. Mating types are

---

A **gametophyte** is the stage in the life cycle of an alga or plant during which gametes are produced.

The **gametangium** is the part of a gametophyte where gametes are produced.

Alternation of generations describes the life cycle of seaweeds and plants that includes more than one multicellular stage, usually one sexual and one asexual.

Green algae are seaweeds that bear chlorophylls *a* and *b*.

A **coenocytic** body has one giant cell or a few large cells containing more than one nucleus.

Tolerance is a mechanism to compensate for loss of tissue to herbivory.

Avoidance is a mechanism to reduce herbivory by being in a different place or time than the herbivore.

Deterrence is a mechanism to reduce herbivory by repelling a predator.

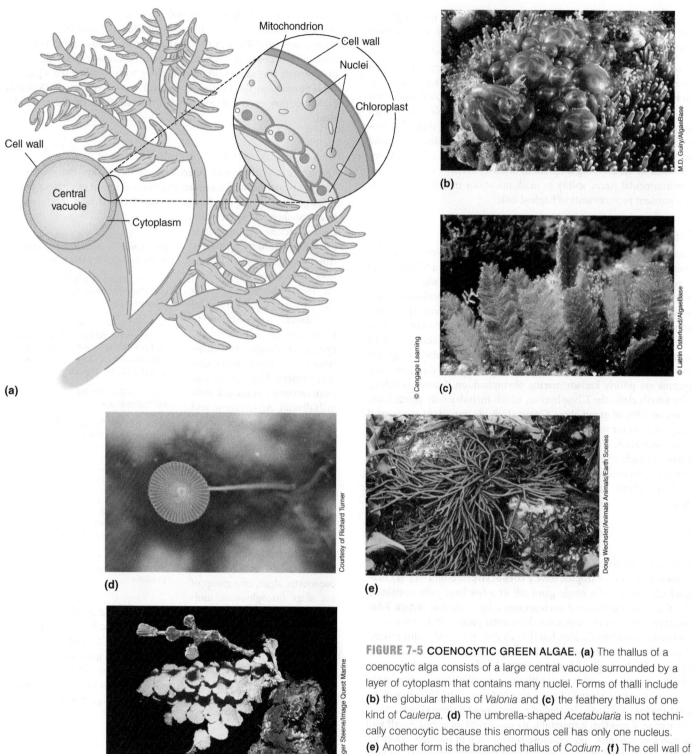

FIGURE 7-5 COENOCYTIC GREEN ALGAE. (a) The thallus of a coenocytic alga consists of a large central vacuole surrounded by a layer of cytoplasm that contains many nuclei. Forms of thalli include **(b)** the globular thallus of *Valonia* and **(c)** the feathery thallus of one kind of *Caulerpa*. **(d)** The umbrella-shaped *Acetabularia* is not technically coenocytic because this enormous cell has only one nucleus. **(e)** Another form is the branched thallus of *Codium*. **(f)** The cell wall of *Halimeda* contains calcium carbonate, which contributes to the formation of reef sediments.

designated + and −, a distinction recognizable as early as spore formation. Gametes of opposite mating types must fuse for fertilization to occur.

Human Uses of Green Algae

Over the course of human civilization, coastal populations have used seaweeds for many purposes, including food, fodder for livestock,

fertilizer or mulch in agriculture, wastewater treatment, and a source of chemicals. Mariculture has increasingly reduced our reliance on certain seaweeds that historically were harvested from the field. Among green algae, *Ulva* and other genera are used as food in Asian cuisines. Because their cell walls contain little more than cellulose, green algae, in contrast to red and brown algae, are not harvested for extraction of chemicals, as are red and brown algae.

RED ALGAE

Red algae (phylum Rhodophyta) are primarily marine organisms. About 98% of the 6,000 species are marine, which means that red algae have the highest diversity among the seaweeds. Although they are most diverse in tropical oceans, they can also be found as significant ecological components at higher latitudes. Red algae are mostly benthic in distribution. Some species can survive at depths as great as 200 meters (660 feet) in clear water. In addition to chlorophyll *a* and minor amounts of chlorophyll *d*, red algae contain the accessory pigments phycoerythrins and phycocyanins, reflecting the origin of their plastids from endosymbiosis with blue-green bacteria. The normally reddish color of these algae comes from the large amount of phycoerythrins. Phycoerythrin pigments are most effective at absorbing blue and green light, and they mask the green color of chlorophyll *a*. Red algae, however, are not always red. Their thalli can be a variety of colors, from yellow to black, because phycocyanin pigments absorb more red light than do phycoerythrins and the photoprotective carotenoid pigments, which absorb in several regions of the light spectrum.

Structure of Red Algae

Almost all red algae are multicellular and are less than 1 meter (3.3 feet) long. Their thalli vary widely in shape and organization **(Figure 7-6)**. Some species develop a large bladelike thallus similar to that of the green alga *Ulva*. Examples are the species of *Porphyra* that are widely cultivated as **nori** for human consumption in Asian cuisine. Many red algal thalli are composed of branching filaments that give the seaweed a bushy appearance, from delicate and fuzzy to robust and crispy. Others that are heavily calcified look like a variety of corals with flat encrusting plates, calcified nodules, erect branches, or leafy blades. Other red algae, along with greens and browns, form low, dense **algal turfs** of filamentous and branched thalli that carpet the seafloor over hard rock or loose sediment.

Red algae are seaweeds with chlorophylls *a* and *d* and with phycobilins.

Nori is a food made from the red alga *Porphyra*.

Algal turf is a thick low carpet of algae.

(a)

(b)

(c)

(d)

(e)

(f)

FIGURE 7-6 DIVERSITY OF RED ALGAE. **(a)** The edible sea laver, *Porphyra*, has broad flat blades. **(b)** The thallus of *Asparagopsis* is filamentous. **(c)** This *Gracilaria* has cylindrical branches. **(d)** The encrusting coralline alga *Porolithon* helps to cement loose calcareous reef sediments together. **(e)** *Amphiroa* is an example of a branching coralline alga. **(f)** Many small red algae, such as *Antithamnion*, are epiphytic on larger red algae and other seaweeds and marine plants.

Response of Red Algae to Herbivory

The annual red algae are a seasonal food source primarily for sea urchins, fish, molluscs, and crustaceans. Unlike the coenocytic green algae, red algae do not produce many toxins to deter herbivores. Their main defenses against herbivory are making their thalli less edible, changing their growth patterns, and evolving complex life cycles. Some red algae deter herbivores by incorporating calcium carbonate into the cell wall, making the thallus difficult to eat and digest and reducing its nutritional value. Others alter their growth patterns when subjected to predation, producing forms such as algal turfs that are more difficult for herbivores to graze. As a group, red algae are able to tolerate herbivory because of their rapid growth and annual and complex life cycles, which allow them rapidly to replace biomass lost to grazing. Many small red algae avoid large herbivores by growing in crevices.

Reproduction in Red Algae

Red algae exhibit a variety of life cycles, some of which can be quite complex **(Figure 7-7)**. Two unique features of their life cycle are the absence of flagella, even on spores and gametes, and the occurrence in most species of three multicellular stages: two sporophytes (carposporophyte and tetrasporophyte) in succession and one gametophyte (including male and female, usually as separate thalli). The life cycle can be described as follows: Sperm released by a male gametophyte are carried by water currents to a filament on a part of a female

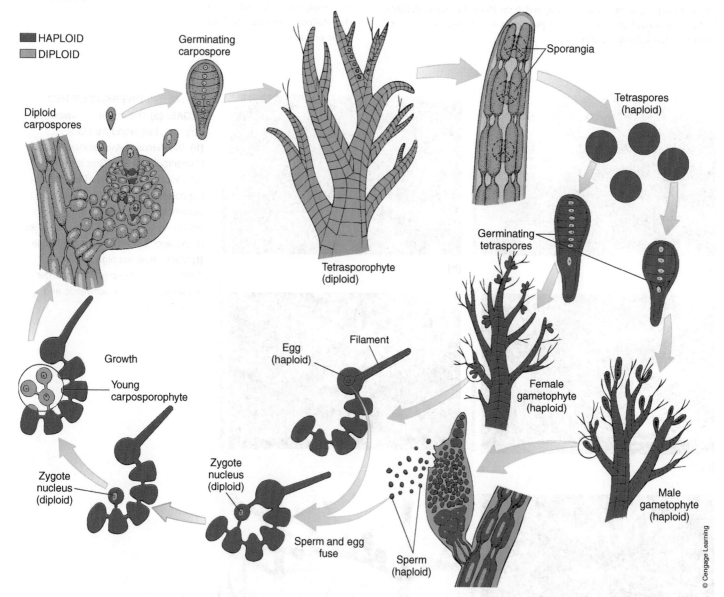

FIGURE 7-7 LIFE CYCLE OF A RED ALGA. Nonmotile haploid spores released by the tetrasporophyte germinate and grow into either male or female gametophytes. The sperm produced by the male gametophyte are nonmotile. They are carried by water currents to the female gametophyte and fertilize their stationary eggs. The zygote that results develops in place into a new generation, the diploid carposporophyte, which remains parasitic on the female gametophyte. Diploid carpospores released by the carposporophyte disperse, settle, and germinate into the tetrasporophyte.

gametophyte that contains an egg cell. After fertilization, the zygote divides while still attached to the female gametophyte, forming the **carposporophyte**, a stage unique to red algae. This "parasitic" stage produces nonmotile diploid spores called **carpospores**. When a carpospore settles on the seafloor or a host plant or animal, it germinates and forms a new adult alga, the **tetrasporophyte**. The tetrasporophyte is comparable to the sporophyte of green algae. The name tetrasporophyte serves to distinguish it from the carposporophyte and refers to the cluster of four ("tetra") haploid **tetraspores** that result from meiosis in this stage. The tetrasporophyte then releases nonmotile haploid spores, which disperse, settle, germinate, and form a new (haploid) gametophyte generation. The free-living sporophytes and gametophytes are, like *Ulva*, usually hard to tell apart, particularly in tropical red algae.

Ecological Relationships of Red Algae

A few of the smaller species of red algae are **epiphytes** (*epi* meaning "on," *phyte* meaning "plant"), meaning they grow on other algae or plants, or they are **epizoics** (*zoon* meaning "animal"), which grow on animal hosts **(Figure 7-8)**. Some red algae grow on a variety of hosts, whereas others grow only on a specific host. Some species of smaller red algae are parasitic on larger species of red algae, relying on nutrients from their hosts for their survival.

The cell wall of red algae consists of cellulose, a group of chemicals called phycocolloids (described in the following section) of commercial importance, and often a superficial coating of mucilage or a cuticle composed of protein. As noted previously, the cell wall may be impregnated with calcium carbonate, particularly in a group known as the **coralline algae** (see **Figure 7-6d–e**). Coralline algae precipitate calcium carbonate from seawater and help to cement bits and pieces of loose coral together. This process, called consolidation, is necessary to hold the many parts of a reef together and eventually forms large areas of nearly solid reef. More than half of the mass of some reefs is contributed by coralline algae. Like *Halimeda*, coralline algae also add to the accumulation of calcareous sediments on the reef.

Human Uses of Red Algae

Red algae deposit large amounts of **phycocolloids** (a group of polysaccharides) in their cell walls. Several phycocolloids are commercially valuable because of their gelling (stiffening) qualities. **Agar**, for example, forms thick gels at very low concentrations and resists degradation by most microbes. Its primary human use is in making solid culture media for growing bacteria in microbiology laboratories. It is also used in foods and pharmaceuticals as a thickening agent. Another phycocolloid from red algae, **carrageenan**, is also used as a thickening and binding agent in ice cream, pudding, and salad dressings and by the cosmetics industry for various creamy preparations.

The **carposporophyte** is a diploid multicellular stage in the life cycle of red algae that produces diploid spores.

A **carpospore** is the diploid, asexual, dispersal cell of red algae.

The **tetrasporophyte** is a diploid multicellular stage of red algae that produces spores by meiosis.

Tetraspores are haploid, asexual, dispersal cells of red algae.

An **epiphyte** is any organism that grows on a multicellular primary producer.

An **epizoic** is any organism that grows on an animal.

Coralline algae are a group of red algae that deposit calcium carbonate in their cell walls.

Phycocolloids are chemicals in cell walls that improve flexibility and strength and may be extracted for human use.

Agar is a phycocolloid of red algae that is used to make laboratory culture media.

Carrageenan is a phycocolloid of red algae that is used commercially as a thickening agent.

(a)

(b)

FIGURE 7-8 EPIPHYTIC AND EPIZOIC RED ALGAE. (a) The red alga *Polysiphonia lanosa* grows only as an epiphyte on a brown alga, knotted wrack (*Ascophyllum nodosum*), along the northeastern shores of North America. This obligate epiphyte is protected in the intertidal from desiccation by moisture provided by its larger host. **(b)** The red alga *Polysiphonia howei* has a wide distribution on hard intertidal and subtidal surfaces in the tropical western Atlantic Ocean. But it survives in the splash zone where it is facultatively epizoic on the Atlantic surf chiton, *Ceratozona squalida*.

Courtesy of Richard Turner

Brown algae are seaweeds that contain chlorophyll *c* and fucoxanthin.

Fucoxanthin is an accessory carotenoid of brown algae.

Bladders of seaweeds are gas-filled floats in the thallus.

Alginates are phycocolloids of brown algae that increase thallus flexibility and strength.

Some red algae are used by humans as a source of food. Irish moss (*Chondrus crispus*) is widespread on the shores of Ireland and many other parts of the North Atlantic Ocean. The Irish make a traditional pudding by boiling small amounts of Irish moss in milk. The dessert was adopted in New England, Canada, Great Britain, and France, where it is called blanc mange. Dulse (*Rhodymenia palmata*) has been used as a foodstuff in England for hundreds of years. In Japan and other parts of Asia, *Porphyra* species (laver, nori) are used in sushi, soups, and seasonings. In addition to being a source of human food, red algae are cultivated to produce animal feed and fertilizer in many parts of Asia.

BROWN ALGAE

Brown algae (phylum Phaeophyta) are a multicellular group of stramenopiles (a group introduced in Chapter 6). They include such familiar forms as rockweeds, kelps, and sargassum weed. The 1,500 species of brown algae are almost exclusively (99.7%) marine inhabitants. They have a higher species diversity in the sea than green algae and a lower diversity than red algae. With rare exception, such as some species of *Sargassum*, they are benthic. Brown algae range in size from microscopic, filamentous forms to the largest of all algae, the giant kelps **(Figure 7-9)**, which can attain lengths of 100 meters (330 feet). The characteristic olive-brown color of these algae is due to the carotenoid pigment **fucoxanthin**, which occurs also in diatoms and other ochrophytes and masks the green of chlorophylls *a* and *c*. Much of the familiar seaweed along the shore and in shallow water is brown algae.

FIGURE 7-9 KELP. Large kelp, such as these off the coast of California, form massive undersea forests.

FIGURE 7-10 SARGASSUM WEED. Some *Sargassum* species form extensive floating mats in an area of the North Atlantic Ocean known as the Sargasso Sea.

Distribution of Brown Algae

Brown algae for the most part are more diverse and abundant along the coastlines of high latitudes than in the tropics. In the western North Atlantic Ocean, along the northeastern coast of North America, for example, there are twice as many species of brown algae as in the Caribbean, Bahamas, and Gulf of Mexico. Brown algae are far less diverse than red and green algae in the tropics, but they rival the red algae in the northeast. Most of the large kelp or kelp-like brown algae are temperate forms. Tropical sargassum weeds **(Figure 7-10)** are notable exceptions, some species forming extensive free-floating mats in the subtropical open waters of the North Atlantic Ocean and occurring only sporadically along temperate shores when dispersed in the Gulf Stream. Another exception is the recent discovery of extensive, deep-water tropical forests of kelp near upwelling areas. Clear tropical surface waters, although poor in nutrients, allow the establishment of kelp at depths of 30 to 200 meters, where cold upwelling currents of deep ocean water provide the nutrients needed to maintain primary production.

Structure of Brown Algae

Most species of brown algae have thalli that are well differentiated into holdfast, stipe, and blade (see **Figure 7-3a**). Many develop large, flat, leaf-like blades, but the shapes of the blades vary widely **(Figure 7-11)**. The larger blades frequently have gas-filled structures, or **bladders**, that help to buoy the blade, allowing it to gain maximal exposure to sunlight. The gas bladders and blades are supported by a large stipe that is attached to the bottom by a branching holdfast. Some kelps, such as the sea palm *Postelsia*, have very thick and flexible stipes that hold the seaweed in heavy surf and surge (see **Figure 7-11b**). As you might suspect from their large size, many brown algae are perennials and offer a year-round source of food and habitat for marine animals.

The cell walls of brown algae are composed of cellulose and a group of phycocolloids called **alginates**, similar to those of red algae and of similar commercial importance. It is primarily the alginates that give the

(a)

(b)

(c)

(d)

(e)

(f)

FIGURE 7-11 DIVERSITY OF SHAPE IN THE BROWN ALGAE. **(a)** A dense mat of knotted wrack, *Ascophyllum,* in the intertidal zone. Gas bladders help buoy the alga at high tide. **(b)** The sea palm *Postelsia* has a tough stipe and blades that resist the heavy ocean surf of exposed Pacific coasts. **(c)** The flat blade of *Agarum* is perforated with holes, which might reduce resistance to waves and surge. **(d)** *Padina* looks like a cluster of potato chips or wood shavings. **(e)** *Dictyota* has a flat branching growth. **(f)** *Sphacelaria* is an example of a delicate and filamentous brown alga.

thallus the strength and flexibility to withstand the relentless waves, surge, and currents of the intertidal and subtidal zones of exposed coasts. Although they are not vascular plants, the larger kelps have specialized **trumpet cells** that conduct products of photosynthesis to deeper parts of the thallus. The lower stipes and holdfasts of some species cannot produce enough food for themselves by photosynthesis because the blades above absorb and block so much sunlight. In these algae, the trumpet cells carry food (primarily the alcohol **mannitol**) from the blade to the stipe and holdfast to support their metabolic needs.

Reproduction in Brown Algae

The life cycle of most brown algae consists of an alternation of generations between a sporophyte (often a perennial stage) and a gametophyte (usually an annual). The familiar giant kelp is an example of the diploid sporophyte stage. It reproduces by releasing haploid motile heterokont spores into the surrounding water. These spores, produced in extremely large quantities, are a food source for a variety of filter-feeding benthic animals and zooplankton. Spores that settle and germinate form microscopic male and female gametophytes. It is interesting that such a large alga as the kelp

has such a small gametophyte stage. The male gametophyte produces flagellated sperm that fertilize an attached egg on the female gametophyte. The zygote germinates in place, as in red algae, and the parasitic sporophyte on the short-lived female gametophyte eventually overgrows and obliterates it. In a specialized variation on the life cycle of brown algae, *Fucus*, or rockweed, and its relatives produce haploid sperm or egg cells by meiosis at the inflated reproductive tips, or **receptacles**, of the sporophyte in specialized chambers, or **conceptacles (Figure 7-12a)**. Receptacles might be confused with gas bladders, but they occur at the tips of the thallus and bladders do not. The gametophyte is eliminated from the life cycle. The eggs and sperm are then shed into the water, where fertilization takes place. The resulting zygote germinates as a new sporophyte. Thus only one multicellular stage exists in the life cycle of *Fucus* **(Figure 7-12b)**.

Trumpet cells of kelp carry food from fronds to deeper parts of the thallus.

Mannitol is an alcohol produced by brown algae as food.

Receptacles are swollen reproductive parts of a brown alga.

A **conceptacle** in brown algae is a chamber in a receptacle that holds the gamete-producing tissue.

FIGURE 7-12 LIFE CYCLE OF *FUCUS*. **(a)** In addition to gas bladders, the thallus of reproductive rockweed, *Fucus*, has inflated tips called receptacles. The small bumps on the surface of each receptacle are conceptacles, chambers within which the gametangia grow. **(b)** The life cycle of rockweed eliminates the independent gametophyte stage that is found in red and green algae, as well as in kelp and many other brown algae. The sporophyte is the only multicellular stage in the life cycle and is considered a sporophyte because it is diploid, as are sporophytes of other algae. Gametes form by meiosis in conceptacles embedded within receptacles at the tips of reproductive blades. Gametangia are released from the conceptacles and fall to the sea floor. Flagellated sperm and nonmotile eggs later are released into the seawater, where fertilization occurs. The zygote then settles to the bottom and develops into a new sporophyte.

Gas bladders

Receptacles

Courtesy of Richard Turner

(a)

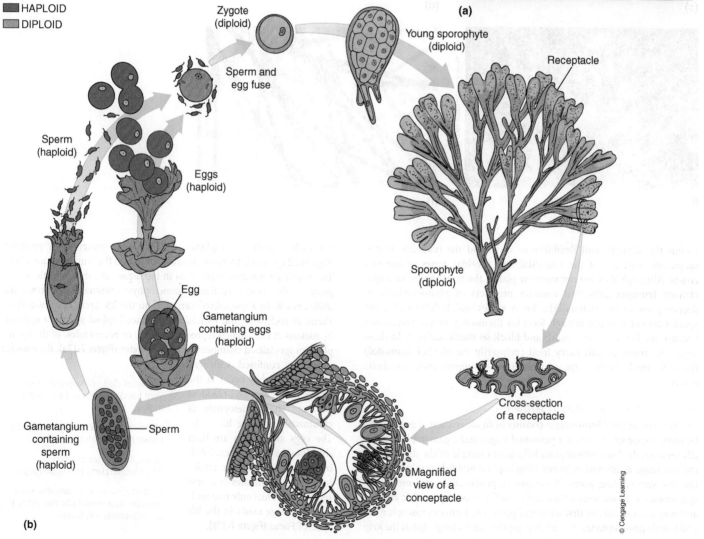

■ HAPLOID
■ DIPLOID

Zygote (diploid)

Sperm and egg fuse

Young sporophyte (diploid)

Receptacle

Sperm (haploid)

Eggs (haploid)

Sporophyte (diploid)

Egg

Gametangium containing eggs (haploid)

Cross-section of a receptacle

Gametangium containing sperm (haploid)

Sperm

Magnified view of a conceptacle

© Cengage Learning

(b)

In both rockweed and kelp, after the egg is fertilized it attaches to the bottom by mucilage and grows root-like structures called **rhizoids**. The holdfast eventually develops from the rhizoids and puts forth a shoot. The shoot grows larger and develops one or more zones from which further growth proceeds. In rockweed, these growth zones are located at the tips of the thallus. In kelp species, growth zones may be located farther back from the blade, often at the junction of the stipe and blade. As wave action breaks off the tips of kelp or herbivorous animals devour them, the central or basal growth area produces new tissue that is pushed upward and outward to take the place of lost tissue. This pattern of growth offers a renewable resource and is useful in commercial kelp harvest, in which only the canopy of the kelp forest is removed.

Brown Algae as Habitat

Although the rockweed *Fucus* is abundant on intertidal rocks, the majority of brown algae are found from the low tide line to a depth of about 10 meters (33 feet). The larger forms, the kelps, grow so profusely that they form offshore kelp forests. These forests are very efficient at capturing sunlight and are extremely productive. They are home to a diverse group of marine animals, including sea urchins, fishes, crustaceans, molluscs, sea lions, sea otters, and many more. An exceptional brown algal group is the sargassum weeds of the Sargasso Sea in the Atlantic Ocean. These free-floating clumps, buoyed by their gas bladders, form a complex three-dimensional habitat that is home to a variety of unique organisms **(Figure 7-13)**.

Human Uses of Brown Algae

The alginates (phycocolloids) of some brown algae are harvested for commercial use as thickening agents in the textile, dental, cosmetics, and food industries. Brown algae such as kelps concentrate iodine from seawater in their tissues. The concentration of iodine in some species of kelp may be 10 times that of the surrounding seawater. Before cheaper methods of obtaining iodine were developed, seaweeds were the main source of this trace nutrient, which is added to table salt to prevent goiter, a thyroid gland disorder. In some areas of the world, especially Asia, brown algae are used as food. Some coastal countries also use brown algae as cattle feed.

FIGURE 7-13 SARGASSUM WEED HABITAT. Free-floating clumps of sargassum weed form a complex three-dimensional habitat that is home to a variety of unique organisms such as this sargassum fish.

Image Quest Marine

IN SUMMARY

- Multicellular marine algae (seaweeds) are mostly benthic organisms. (Have You Wondered? #2)

- Their division into three major groups (red algae, brown algae, and green algae) is based on the photosynthetic and photoprotective pigments they contain. (Have You Wondered? #1)

- Algae have cell walls composed mainly of cellulose but also often containing various phycocolloids and calcium carbonate.

- Algal pigments are located within organelles called plastids, which originated from primary (red and green algae) or secondary endosymbiosis (brown algae).

- Many algae secrete a slimy covering of mucilage, which protects them from and retards desiccation.

- Algal life cycles tend to be complex, involving sexual and asexual, haploid and diploid, and motile and nonmotile stages of similar or dissimilar shapes and sizes.

- Red algae, the most diverse group of seaweeds, are widespread in the tropics. (Have You Wondered? #3)

- Brown algae are almost exclusively marine organisms and are most diverse at high latitudes. (Have You Wondered? #3)

- With the notable exception of sargassum weed, most brown algae can be found along continental shorelines. (Have You Wondered? #3)

- Green algae are found in shallow coastal waters. (Have You Wondered? #3)

- Many marine species of green algae exhibit a coenocytic body plan.

- Some red and green algae deposit calcium carbonate in their cell walls and are important contributors to the formation of coral reefs. (Have You Wondered? #2)

- Algae are a food source for many organisms, including humans, and they provide a habitat for many species. (Have You Wondered? #2)

- Algae also are a source of many commercial products. (Have You Wondered? #5)

7-2 Marine Flowering Plants

Kelp beds are defined by the presence of the kelp, which establish the structure of the bed and influence interactions among other inhabitants. Similarly, three other marine habitats are defined by their dominant plants: seagrass beds, salt marshes, and mangrove forests. Unlike kelp, which originated in the sea, plants along the shores and shallow ocean waters evolved from terrestrial or freshwater ancestors. In this section we will study the biology of seagrasses, marsh plants, and mangroves and their adaptations to life in the sea. The ecology of their habitats will be described in later chapters.

Rhizoids are root-like anchoring structures of a seaweed.

GENERAL CHARACTERISTICS OF FLOWERING PLANTS

The kingdom Plantae is divided into several major groups. The composition of the kingdom varies among authorities. Some recent views based on molecular studies emphasize the close evolutionary relationship among the green algae, a group called the charophyceans (**stoneworts**, once included with green algae), and plants in the strictest sense. Some stoneworts regularly occur in brackish waters around the world **(Figure 7-14)**. This group is structurally more complex than green algae and has structural features and a genome that indicate that ancestors of stoneworts gave rise to the land plants. Among the land plants, the presence of specialized vessels that carry water, minerals, and nutrients (**phloem**) and give structural support (**xylem**) separates the vascular plants such as ferns, conifers, and flowering plants from more primitive types such as mosses. The presence of phloem and xylem distinguishes the root, stem, and leaf of vascular plants from the holdfast, stipe, and blade of kelp and other seaweeds. The most advanced of the vascular plants reproduce by means of seeds and are called seed plants. The **seed** is a specialized structure containing the dormant embryonic plant (a new sporophyte) and a supply of nutrients surrounded by a protective outer layer produced by the microscopic female gametophyte and parts of the parent sporophyte. Thus the seed represents three generations. There are two groups of seed-bearing plants: conifers, such as pine trees, which bear seeds in structures known as cones, and **flowering plants** (phylum Anthophyta), which bear seeds in structures known as fruits. The **fruit**, produced by the parent sporophyte, is composed of several layers of tissues. It protects the seed and aids in its dispersal. Conifers are exclusively terrestrial plants, but a small number of salt-tolerant flowering plants known as **halophytes** are adapted to the marine environment.

A **stonewort** is a plant that is more complex than green algae and related to land plants.

Phloem is the vascular tissue of plants that carries food from leaves to other parts.

Xylem is the vascular tissue of plants that carries water from roots to other parts.

A **seed** is a dormant stage of vascular plants that bears the embryo within protective and nutritive layers.

Flowering plants are a group of vascular plants that produce seeds in a fruit.

Fruits are produced from the flower after pollination and contain the seeds.

Halophytes are plants that grow and reproduce best in the presence of salt.

Seagrasses are lily-like plants that live submerged in seawater.

A **hydrophyte** is a flowering plant that lives submerged under water.

INVASION OF THE SEA BY PLANTS

Flowering plants have adapted to maritime, estuarine, and fully marine habitats multiple times since they evolved on land about 150 million years ago. It must be kept in perspective that the ancestors of these plants had become adapted to a terrestrial existence over the preceding 330 million years. Gone were the delicate flagellated stages for dispersal by water, the independent gametophyte stages of the life cycle, the ability to metabolize in a low concentration of oxygen, the reliance on the buoyant effect of water, and the tendency for all parts of the body to absorb minerals from surrounding waters and to photosynthesize under diminished quantity and altered quality of light. Re-invasion of the sea required some major adaptations in form and

Courtesy of D. Scott Taylor

FIGURE 7-14 STONEWORTS. Some stoneworts, such as *Chara contraria,* grow in dense mats in brackish bays and ditches worldwide. Their structure is more complex than that of green algae, and molecular genetics demonstrates a close relationship with land plants.

function. On the other hand, the presence of flowering plants in marine communities has presented opportunities and challenges to the native species of the sea.

These flowering plants offer a new source of habitat and food, but they compete with the seaweeds for light and with other benthic organisms for space. Their bodies are composed of three polymers (cellulose, hemicellulose, and lignin) that most marine organisms cannot digest. Although cell walls of seaweeds contain cellulose, the more complex polymers hemicellulose and lignin bind together with cellulose to make a cell wall that is highly resistant to decay. In addition, parts of the plant body consist only of the walls of dead cells, whereas all cells in macroalgae are alive and contain digestible nutrients. Because of their nearly indigestible content, marine plants typically enter microbially based detrital food chains rather than the grazing food chains of herbivores. Marine plants have few competitors, and they tend to form extensive single-species stands on which other members of the community have come to depend. As in any dense population, disease and other disasters may wipe out its members. This potential was realized in the classic case of the wasting disease of eelgrass in the North Atlantic Ocean early in the 1900s. The massive die-off of eelgrass from an infection by labyrinthomorph microbes (described in Chapter 6) had far-reaching effects on both coastal marine and human communities.

SEAGRASSES

Seagrasses are **hydrophytes**, which means they generally live beneath the water. In contrast to marsh grasses and mangroves, which require periodic exposure by ebb tides to grow and reproduce, seagrasses can grow, flower, go to seed, and germinate while fully submerged. Among flowering plants they are best adapted to marine life and are the only ones that are truly marine. The 66 species of seagrasses represent about 0.02% of flowering plant species.

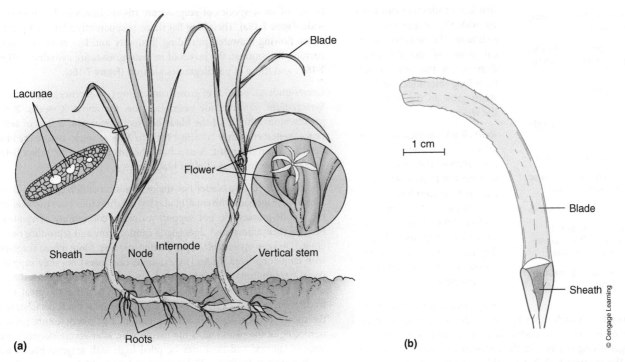

FIGURE 7-15 SEAGRASS ANATOMY. (a) The horizontal stems of seagrasses are called rhizomes; they grow within the sediment. Roots and vertical stems grow from the rhizomes. **(b)** Leaves, arising in this seagrass from the vertical stems, are divided into an upper green blade for photosynthesis and a lower pale sheath to wrap around the stem and support the blade. Lacunae in the aerenchyme of the blade carry oxygen to other plant parts and make the blade buoyant. The flowers of seagrasses are small and white.

Classification and Distribution of Seagrasses

Seagrasses are not true grasses but are related to lilies and a number of other groups of freshwater plants. It was formerly thought that they invaded the seas from shoreline shrubs, but recent molecular data firmly place seagrasses among freshwater relatives. There are 14 genera of seagrasses in 6 families, only 3 of which are entirely marine. These groups represent at least four independent invasions of the sea. One marine family includes the eelgrasses (*Zostera*) and surfgrasses (*Phyllospadix*). A second marine family includes the genus *Posidonia*, which forms extensive acreage in the Mediterranean Sea and along the cooler coastal regions of Australia. The third marine family contains the manateegrasses (*Syringodium*) and shoalgrasses (*Halodule*). A fourth family is mainly freshwater but includes three exclusively marine genera: paddle-grasses (*Halophila*), turtlegrasses (*Thalassia*), and the unusual *Enhalus*. There are 15 species of paddlegrass, almost 25% of total seagrass diversity. The two remaining families have many freshwater species but only one marine species each. The ditchgrasses (*Ruppia*) are taxonomically troublesome, and the marine species is yet to be officially described. An Australian species of *Lepilaena* is the only marine species of horned pondweed.

About half the species of seagrass inhabit the temperate zone and higher latitudes. The remaining species are tropical and subtropical in distribution. Seagrass diversity is highest in the Indo-West Pacific Ocean, particularly around Australia, which has many species that occur nowhere else. The Atlantic Ocean Basin, geologically younger than the Pacific and Indian Ocean Basins, has poor seagrass diversity, although diversity was higher before the Isthmus of Panama closed and changed the patterns of ocean circulation only 3 million years ago. A few species of seagrass form intertidal beds, but most occur subtidally to a maximum of 90 meters.

Structure of Seagrasses

Seagrass beds form primarily from **vegetative growth**. This means they grow in a fashion similar to many lawn grasses by extension and branching of horizontal stems (**rhizomes**) from which vertical stems and leaves arise. Seagrasses have the three basic parts of vascular plants: stems, roots, and leaves **(Figure 7-15)**.

Stems Stems look somewhat like bamboo, with cylindrical sections called **internodes** separated by rings (**nodes**). Rhizomes are horizontal stems, usually with long internodes. They have growth zones at their tips and periodically produce additional growth zones laterally for branching of more rhizomes or for production of vertical stems. Rhizomes generally lie in sand or mud. An exception is the rhizome of surfgrasses. Surfgrasses live on wave-beaten coasts, and their rhizomes are anchored to rocks by a dense tangle of branching roots. Many seagrasses store starch in their rhizomes for overwintering and to provide an early burst of growth in spring. The starchy deposits make seagrass meadows a food source for geese and other migratory birds.

In many genera of seagrass, vertical stems arise from the rhizomes. Vertical stems usually have short internodes and grow slowly upward toward the surface of the sediment. Their slow rate of growth ensures

Vegetative growth is a form of asexual reproduction in which growth regions produce additional units of the plant body of identical genetic makeup.

Rhizomes are underground horizontal stems.

An **internode** is a part of a stem between places where leaves, roots, and stems arise.

Nodes are joints at which internodes meet and where leaves, roots, and stems arise.

Root hairs are absorptive extensions of surface cells of a root.

Scale leaves are non-photosynthetic structures that protect the growing tip of a stem.

Foliage leaves produce a photosynthetic blade.

The **sheath** is the non-photosynthetic part of a leaf.

The **blade** is the green part of a leaf.

The **epidermis** is the outermost layer of cells of a multicellular organism.

Senescence is the process of aging.

Aerenchyme is a gas-containing tissue in vascular plants that consists of spaces between the cell walls.

Lacunae are the expanded, gas-filled spaces in aerenchyme.

that leaf production can keep up with the accumulation of sediment. The flowering vertical stems of the ditchgrass *Ruppia*, on the other hand, have long internodes, and the stems grow to the surface of the water.

Roots Roots, branched or not, arise from the nodes of stems and anchor the seagrass in sediment or to rock. The roots usually bear **root hairs**, cellular extensions from the root surface. Like their terrestrial ancestors, seagrasses largely depend on roots for absorption of mineral nutrients from the sediment.

Seagrasses and bacteria in the sediments interact with each other by way of the roots. Seagrass roots leak some oxygen into the surrounding sediment, which generally lacks oxygen. Nitrogen-fixing bacteria in the sediments thrive in the anoxic conditions of the seagrass bed, but the thin oxygen-containing region around the roots allows nitrifying bacteria to convert ammonia to nitrites and nitrates. These forms of nitrogen are easily absorbed and are used by seagrasses to make amino acids and other nitrogen-containing organic compounds. These physiological relationships between seagrasses and microbes point to the importance of sediment chemistry and to the potential problems from physical disturbance, such as dredging and propeller scarring of seagrass beds.

Leaves Leaves arise from the nodes of rhizomes or vertical stems. Short **scale leaves** of the rhizome protect the delicate growing tips. The long **foliage leaves** arising from vertical shoots have two parts: a lower protective **sheath** that bears no chlorophyll and an upper **blade** that accomplishes all photosynthesis in the plant. Unlike most land plants, the **epidermis** (surface layer of cells) of seagrass blades contains most of its chloroplasts. In addition to photosynthesis, cells in the leaves, and particularly in the sheath, control salt balance. Despite a large surface area, seagrass leaves absorb much less mineral nutrient from seawater than roots do from the sediment.

Leaves of most species of seagrass are ribbon-like, whether narrow or wide (**Figure 7-16a**). Their long flat shape is apparently adaptive for life in dense flowing seawater, providing flexibility and less resistance to the current. By contrast, the leaves of manateegrasses are cylindrical (**Figure 7-16b**), and those of paddlegrasses are oval (**Figure 7-16c**).

Leaves undergo periods of growth and aging (**senescence**). The leaves of *Syringodium filiforme*, for example, grow for about 2 weeks. Before growth stops, the tip of the blade begins to fragment, a sign of senescence. Senescence may continue for 2 to 3 more weeks before the blade detaches from the plant. A new blade begins growth 1 or 2 days before growth ceases in the next older blade.

The life cycle of the blades has major significance for epiphytes in the meadow. Seagrasses with small blades having short life cycles (for example, the paddle-grasses) do not support a diverse community of epiphytes, because their animal and algal guests cannot grow and reproduce rapidly enough to keep up with blade production and loss. More robust seagrasses with large long-lived blades (for example, eelgrasses) not only support a diverse community of epiphytes but also often harbor species that are unique to them as a habitat. If epiphytic growth becomes too great a load in currents, waves, or surge over the seagrass bed, the blade detaches from the sheath at their physically weak junction. This safety feature prevents uprooting of the plant and is the reason why, after severe storms, shorelines of lagoons and bays may be piled high with seagrass blades with rarely a stem to be found. Unless overburdened with incoming sediment during storms, seagrass beds can quickly recover from the damage.

Aerenchyme Internally, a most important gas-filled tissue in seagrasses is **aerenchyme** (AIR-en-kime). All parts of the plant have spaces called **lacunae** between the cells in this tissue (see **Figure 7-15a**). The lacunae provide a continuous system for gas transport through the plant. Because water carries much less oxygen than air and sediment is devoid of oxygen, the leaves must be a source of oxygen for the rest of the body. Daytime photosynthesis charges the aerenchyme with a high gas pressure. Sometimes during the day seagrass beds can be heard sizzling as oxygen bubbles are released from the surface of the blades. Oxygen diffuses through the lacunae to the stems and roots.

The aerenchyme provides buoyancy to the leaves, just as gas bladders do for brown algae, allowing the leaves to remain upright in the water column for full exposure to sunlight. Because seawater and aerenchyme buoy the leaves of seagrasses, vascular tissues in these plants are much reduced and not very rigid, and the cell walls contain less lignin, compared with their terrestrial relatives.

Courtesy of Dr. David Campbell

Florida Center for Environmental Studies

John Huisman/AlgaeBase

(a)   **(b)**   **(c)**

FIGURE 7-16 LEAF SHAPE IN SEAGRASSES. (a) A white sea anemone thrives among the broad, flat, strap-like blades of a turtlegrass bed. Turtlegrass is the most common species of seagrass in the Caribbean. **(b)** The cylindrical blades of manateegrass give this seagrass bed the appearance of a field of spaghetti. **(c)** Paddlegrasses have oval blades.

One disadvantage of the aerenchyme is the potential for invasion by pathogenic fungi and labyrinthulids. As a chemical defense, many seagrasses produce antimicrobials called **tannins**, which are commonly found in plants living in anoxic sediments and which give the waters of cypress swamps their brown tint.

Seagrasses also face the danger of seawater entering the aerenchyme when the plant is damaged and interrupting the flow of oxygen to parts that need it, thereby drowning the plant. The danger is real, for many animals dig up and through marine sediments, a process called **bioturbation**. The feeding activities of horseshoe crabs, stingrays, migratory geese, manatees, and green turtles break stems and leaves. How is it that local damage from these activities does not result in death of extensive areas of surrounding seagrass that are connected by rhizomes to the damaged plants?

Although the lacunae are spacious, the aerenchyme is reduced to microscopic pores at the nodes and where plant parts join **(Figure 7-17a)**. The pores measure 0.5 to 1.0 micrometers (about 2 to 4 one hundred thousandths of an inch). Oxygen can pass through the pores but the high surface tension prevents water from passing, thus sparing the plant from suffocation by flooding with seawater if one plant part is damaged.

About a quarter of a century ago, three seagrass biologists asked two related questions: Is the gas space of the lacunae really continuous throughout the seagrass? Will seawater also flow through that space? To answer these questions, D. G. Roberts, A. J. McComb, and J. Kuo of the University of Western Australia developed a simple approach. (In science, the simplest approach to observation is often the most elegant and the most instructive.) They connected a segment of the paddlegrass *Halophila ovalis* to a piece of plastic tubing at the cut end of the rhizome (see **Figure 7-17b**). Except for the single cut, the plant was otherwise intact. A syringe filled with air was inserted into the tubing, and the entire experimental setup was submerged in seawater. When the plunger of the syringe was depressed, they observed no effect. On cutting a root, a blade, or a distant internode, the investigators observed bubbles of gas rising from the cut ends, thus demonstrating the continuity of the gas spaces in the aerenchyme. When the syringe was filled with water containing a soluble red dye, the biologists could depress the plunger only a short distance before gas bubbles stopped appearing; no red dye escaped the cut ends until a cut was made to the same internode to which the tubing was connected. Fluids could not, therefore, pass from one plant part to the next.

Tannins are compounds in plants that reduce herbivory and microbial infection.

Bioturbation is the mixing up of sediment by the activities of seafloor animals.

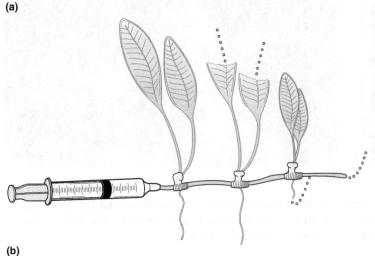

FIGURE 7-17 FUNCTIONS OF SEAGRASS AERENCHYME. The upper diagram **(a)** is a three-dimensional interpretation of the structure of two internodes and the intervening node. Microscopic spaces (pores) between the small cells of a node connect the lacunae of one internode with those of the adjacent internode, allowing gas to diffuse but preventing water passage. The lower diagram **(b)** illustrates the observational procedure of Roberts, McComb, and Kuo in demonstrating the continuity of gas space within the body of a seagrass. Bubbles emerge from the cut ends of plant parts when a submerged plant is charged with air from a syringe.

Vascular bundle

Pore

Pore

Lacuna

(a)

(b)

© Cengage Learning

Reproduction in Seagrasses

Seagrasses vary widely in flowering and setting of seed. Flowers are unknown in some species, and colonization of new sites must occur only by fragmentation, drifting, and rooting of plants. Those species that do flower lack the showy petals of their relatives, the lilies; and the flowers are small and inconspicuous (see **Figure 7-15**), sometimes remaining below the sediment. In most species, flowers are either male or female and are borne on separate plants. Sperm-bearing grains of **pollen** (actually a male gametophyte) of most seagrasses are greatly elongate in shape, or many spherical grains are embedded in long sticky filaments. In both cases, the pollen is carried by water currents to female flowers. It has been shown in a few species that a two-part, water-insoluble adhesive binds pollen to a **stigma**, the female pollen receptor. This process is called **hydrophilous** ("water-loving") **pollination** and is common among submerged aquatic plants. In the unusual seagrass *Enhalus acoroides*, the whole male flower is released and drifts at the surface until reaching a female flower to fertilize. In surfgrass, rafts of pollen grains either at the surface or barely submerged are painted on the female flowers as the tide and wavelets rise and fall. Seed production is extensive in some seagrasses—so much so that migratory ducks sift the sediment for food in the accumulated seed banks. A few species of seagrass produce seedlings on the mother plant, a process called **viviparity**.

Pollen is the male gametophyte of a seed-bearing plant.

The **stigma** is the female part of a flower that receives the pollen.

Hydrophilous pollination is the mechanism of dispersal of a pollen grain by water currents from the male flower to the female flower.

Viviparity is reproduction by the initial retention and nourishment of offspring on or in the parent.

Ecological Roles of Seagrasses

Seagrasses play several key ecological roles in the marine environment. As primary producers, they are food for herbivores, long lists of which may be found in the scientific literature. Seagrasses are important in depositing and stabilizing coastal sediments, and seagrass beds provide habitats for many marine species.

Role of Seagrasses as Primary Producers Seagrasses live beneath less than 2% of the sunlit surface of the ocean and contribute little more than 1% of oceanic primary production. Where they occur, however, they are highly productive. Although the rates of production in seagrass meadows rival those of many terrestrial ecosystems, the direct contribution by seagrasses to the nutrition of animals is much less than for seaweeds for two reasons. First, seagrasses cover less of the seafloor (about one tenth) than seaweeds. Except for the paddlegrasses (*Halophila*), seagrasses do not thrive in areas of low light intensity. The maximal depth of occurrence is 90 meters (297 feet). Most live at depths of less than 20 meters (66 feet), and dense beds of high biomass occur at much shallower depths. Second, because of its poor digestibility, few animals can use seagrass tissue as a primary component of their diet.

Birds and humans may consume the starchy rhizomes and seeds of seagrasses. Major herbivores of the leaves are the manatees and dugong **(Figure 7-18)**, green sea turtle, and some species of fish and sea urchin. Sometimes it is difficult to distinguish the relative contributions of tough seagrass blades and the more easily digested epiphytes to the diets of these consumers. It is believed that direct consumption of seagrasses by manatees and dugongs was far higher between 50 million and 2 million years ago, when these marine mammals were more diverse and abundant. Their grazing activities—for example, using tusks for digging up rhizomes—are possibly responsible for many features of growth and reproduction that seagrasses have today. These features are similar to those of plants in grasslands that have evolved under grazing pressure of terrestrial herds: growth zones and rhizomes that are beneath the sediment, production of roots and stems at nodes, growth of blades at the base where they meet the sheaths, the short life cycle of leaves, and rapid recovery from loss of leaves through grazing.

Seagrasses also contribute to marine food webs through fragmentation and loss of leaves. This material is an important source of detritus in microbial food chains in which bacteria and fungi convert the cellulose, hemicellulose, and lignin of the leaves into microbial biomass. The microbes and detritus are consumed by suspension and deposit feeders of the seagrass meadow and nearby habitats. Detritus is also exported to offshore communities.

(a)

Marine Ecology Progress Series, 124, pg. 204–205./Anthony Preen

(b)

Impact of Dugong Grazing Seagrasses

FIGURE 7-18 IMPACT OF DUGONG GRAZING ON SEAGRASSES. Dugongs can almost denude an area of seagrass blades. **(a)** This dense bed of *Halophila ovalis*, a seagrass preferred by dugongs in Australia, was cropped heavily **(b)** in 3 hours of grazing by a herd of 70 dugongs. Rapid growth from underground rhizomes renews this resource for the herd, which might return someday to graze the patch again.

Role of Seagrasses in Depositing and Stabilizing Sediments Seagrasses are important in the deposition and stabilization of sediments. The blades standing in the water column serve as baffles that reduce water velocity. Because slow currents can hold a smaller burden of particles than swift currents, inorganic and organic particles settle in the meadow. Decay of seagrass leaves, stems, and roots adds more organic matter, and the sediment builds up around the slowly growing vertical stems. The rhizomes and roots of seagrasses help stabilize the bottom, preventing currents and waves from stirring up sediments (resuspension). In these ways, seagrass beds reduce the **turbidity** (cloudiness) of the overlying water. If the water in a seagrass bed becomes too turbid, reducing the available light, it can destroy the seagrasses and their dependent organisms. A massive die-off of seagrasses took place in Maryland's Chesapeake Bay estuary during the 1960s, largely due to diminished light associated with excessive sediment runoff from the land surrounding the bay. Other estuaries around North America have experienced gradual decline of seagrass beds due to persistent turbidity in recent decades. The loss of seagrasses makes the turbid conditions worse and can lead to erosion if sediments are resuspended.

Role of Seagrasses as Habitat The considerable acreage and single-species dominance of seagrasses make them a significant habitat for other marine primary producers, consumers, and decomposers. In the presence of seagrasses, an otherwise two-dimensional surface of sand or mud is converted into a three-dimensional space with a greatly increased area on which other organisms can settle, hide, graze, and crawl. Complexity within the sediment increases physically and chemically with the presence of the **rhizosphere**, the system of roots and rhizomes along with the surrounding sediment. Seagrass beds are home to epiphytes such as foraminiferans, diatoms, fungi, delicate seaweeds, and many kinds of invertebrate animals. Microbes, several groups of worms, crustaceans, and molluscs live in the rhizosphere—a greater number and variety than in surrounding bare flats. The water among the blades holds abundant fishes and crustaceans. Species diversity is high, and the young of many commercial species of shellfish and finfish thrive in this refuge. It is no wonder that coastal states seek ways to restore seagrass communities that have declined from wasting disease and from the impact of local human populations.

Human Uses of Seagrasses

Seagrasses have great economic value to the varied fisheries that depend on the presence of coastal meadows. In addition to indirect uses of seagrasses, human populations have put various parts of these plants to good use over the centuries. Blades are by far the part most often used, primarily harvested as wrack blown ashore by storms. Some cultures have used seeds, rhizomes, and sheath fibers, as well as peat accumulated over hundreds or thousands of years.

The most common use of blades seems to be as filler for insulation, stuffing, and packaging. In the first half of the 1900s, the Samuel Cabot Company manufactured a quilted blanket of dried eelgrass sandwiched between layers of paper for insulation of walls. The product, called Cabot's Quilt, was favored for its thermal and soundproofing qualities, failure to support combustion, and resistance to rot. These properties are partly attributable to aerenchyme, silica content, lignin, and tannins. Eelgrass wrack also was used in coastal America as a temporary layer around the foundations of homes to insulate them against the winter. In addition to wall construction, many cultures stuff dried blades into pillows and furniture and apply it as a thatch to roofs. Blades are used wet as a moist packing for shipment of live shellfish and dry as a lightweight packing for shipment of other goods. Other products incorporating seagrass blades include cordage, fishing line, footwear, dolls, wet-weather gear, mats, fodder, weed-free mulch, and caulk for boat hulls. Some of these applications specifically use the fiber (from vascular tissue) of blades, although the source of fiber is sometimes other parts of the plant. One source of such fiber is *Posidonia* balls, tennis ball–sized accumulations of the dead vascular bundles of the Mediterranean species, *P. oceanica*. *Posidonia* balls form by constant agitation of decayed plants over the seafloor before they are cast ashore. This process may be duplicated in a washing machine by laundering disaggregated fibers of seagrass through three 10-minute cycles. *Posidonia* balls have been known for centuries, but they seem mostly to serve as curios for beachcombers.

Material extracted from seagrasses has been used for food, medicine, and industrial applications. Some native cultures eat seagrass seeds and grind rhizomes into flour for their starch content. In addition to the harvest of seagrass peat as fuel, burning of peat allows extraction of minerals from seagrass. Seagrasses have been used experimentally or in the short term for production of methane, nitrocellulose, and fertilizer. Medicinal uses include herbal remedies for rheumatism and skin ailments.

SALT MARSH PLANTS

Plants of the salt marsh are much less adapted to marine life than are seagrasses, because they must be exposed to air by the ebbing tide to flourish. This requirement restricts salt marshes to the intertidal zone, where they are a biological filter for terrestrial runoff on its way to the sea. Although these plants grow in waterlogged, anoxic sediments, as do seagrasses, they must tolerate much higher sediment salinities created by evaporation of surface water at low tide. Exposure makes the plants available to terrestrial herbivores, but they otherwise play the same ecological roles as seagrasses.

Classification and Distribution of Salt Marsh Plants

Salt marshes are well developed along the low slopes of river deltas and shores of lagoons and bays in temperate regions of the world. Their distribution is limited at high latitudes by ice scouring and hard bottoms and at low latitudes perhaps by high temperature and weak tides. Plants of the salt marsh **(Figures 7-19 and 7-20)** include true grasses (cordgrasses of the genus *Spartina*), needlerushes (*Juncus*), and many kinds of shrubs and herbs such as saltwort (*Batis*) and glassworts (*Salicornia, Sarcocornia*). Grasses and rushes form the main structure of the marsh community, but the shrubs and herbs generally grow in the middle to upper intertidal zones, where they are more protected from waves, tidal currents, and prolonged submersion.

Structure of Salt Marsh Plants

Smooth cordgrass, *Spartina alterniflora*, is the plant that initiates the formation of salt marshes and dominates the lower marsh along temperate Atlantic coasts of North and South America. It grows in tufts of vertical stems connected by rhizomes. Each vertical stem (**culm**) produces additional stems (**tillers**) at its base, giving a tufted or clumped

Turbidity is cloudiness of water from suspended particles.

The **rhizosphere** is the area below ground that is physically and chemically influenced by the complex of roots of a vascular plant.

A **culm** is the primary vertical stem of a marsh grass.

Tillers are secondary stems surrounding a culm.

FIGURE 7-19 SALT MARSH. Cordgrass (*Spartina*) dominates this Maryland salt marsh.

Connie Toops

A **facultative halophyte** is a plant that thrives in the presence or absence of salt.

Salt glands are epidermal cells that release salt solutions to control mineral balance in the plant.

A **succulent** plant has enlarged water-filled cells in its tissues.

appearance **(Figure 7-21)**. Roots are concentrated at the culms. Vegetative growth by rhizomes allows smooth cordgrass to command extensive acreage of the marsh. Long, smooth, flat blades arise from nodes on the culms and tillers, which produce flower heads in warm months. Plants in the well-drained, nutrient-rich soils at the seaward edge of the marsh stand as high as 3 meters (10 feet), whereas those in the anoxic, waterlogged, nitrogen-poor soils of most of the lower marsh may grow only one-tenth

as high. Aerenchyme allows diffusion of oxygen from the blades to the subterranean rhizomes and roots. Flowers are pollinated by the wind, and seeds drop to the sediment or are carried to distant shores by water currents. Other grasses and rushes of the marsh inhabit higher ground. They vary anatomically in the organization of their culms, tillers, and rhizomes and physiologically in their tolerance to salinity, nutrients, and anoxic sediments. Their differences in growth and physiology result in zonation of the marsh.

Adaptations of Salt Marsh Plants to a Saline Environment

Plants of salt marshes are considered **facultative halophytes**, meaning that they can tolerate salty as well as freshwater conditions. Most seagrasses, in contrast, are intolerant of freshwater. Because they live in sediments with a high salt content, marsh plants tend to lose water to their environment by osmosis and display adaptations similar to those of desert plants to help them retain water. Cordgrasses and rushes have leaves covered by a thick cuticle to help retard water loss, and the stems and leaves contain well-developed vascular tissues for the efficient transport of water within the plant. The leaves of *Spartina alterniflora* have **salt glands** that secrete salt, whereas vascular tissues maintain a high osmotic pressure not with salts but with organic solutes. This replacement prevents salt toxicity. Shrubs and herbs of the marsh often have thick, water-retaining, **succulent** parts similar to those of plants that live in hot, arid, terrestrial habitats. In many, such as saltwort, the leaves are succulent; in glassworts, which have no leaves, the stems are succulent.

Ecological Roles of Salt Marsh Plants

Salt marsh plants, particularly the grasses and rushes, contribute heavily to detrital food chains, accumulate and stabilize sediments,

(a)

Courtesy of Richard Turner

(b)

Courtesy of Richard Turner

(c)

Courtesy of Richard Turner

(d)

Courtesy of Richard Turner

FIGURE 7-20 PLANTS OF THE SALT MARSH. (a) Needlerush, *Juncus,* is the dominant plant in this high salt marsh. Its name comes from the very sharp tips of its cylindrical blades (inset), shown here with the egg sac of a spider adapted to marsh life. **(b)** Succulence is one adaptation of marsh plants such as saltwort (*Batis*), with its water-storing leaves (inset). Note the runners, an adaptation for extending territory on the marsh. **(c)** Annual glasswort (*Salicornia*) and **(d)** perennial glasswort (*Sarcocornia*) have succulent stems and nearly invisible scalelike leaves.

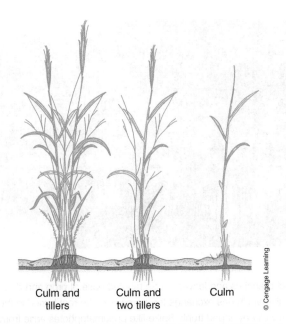

FIGURE 7-21 GROWTH FORM OF SMOOTH CORDGRASS. Upright stems (culms) of *Spartina alterniflora* arise from rhizomes and eventually produce additional culms, called tillers, at their bases. This growth pattern gives cordgrass a clumped appearance on a short scale, but its spread by rhizomes allows the grass to form extensive uniform meadows over the marsh.

and serve as a refuge, feeding ground, and nursery for other marine organisms.

The detritus from salt marsh plants supplies the nutrient needs of a variety of fish and shellfish. One study suggests that as many as 95% of sport and commercial fish species, many of which spend part or all of their lives in coastal and estuarine waters, are supported by the productivity of detrital food chains in salt marshes.

Like seagrasses, salt marsh plants usually have shallow roots and rhizomes, an arrangement that helps to stabilize coastal sediments and prevent shoreline erosion. Cordgrass rhizomes also play an important role in recycling the nutrient phosphorus by transferring phosphates deposited in the bottom sediments into the leaves and stems of the plant. Marsh plants filter runoff from coastal areas, removing some of the potentially toxic organic pollutants of terrestrial origin and preventing them from entering the sea. They may also help maintain water quality in shallow-water areas by removing excess nutrients entering from terrestrial sources that might contribute to algal blooms.

Unlike seagrasses, marsh plants do not support a heavy community of epiphytes because of their periodic tidal exposure to air. For the same reason, they are more prone to herbivory by terrestrial animals that are well adapted to consuming plants. A significant portion of primary production is eaten by grasshoppers, beetles, and other terrestrial consumers of leaves, flowers, and seeds. Leafhoppers and aphids suck nutritive fluid from plant vascular tissue. Marsh plants form at least part of the diet of some crabs, and geese feed seasonally on rhizomes.

Human Uses of Salt Marsh Plants

Salt marshes and their associated creeks have been used for hunting and gathering by coastal communities of temperate latitudes for thousands of years. The plants, however, provide few direct uses. American colonists hayed marsh grass as bedding, fodder, and mulch, and they used the high marsh as grazing pasture for livestock. An advantage of using marsh grass for mulching is the absence of weed seeds. Some of the succulent herbs of the marsh make a salty addition to salads. Seeds, flowers, roots, and other parts of various shrubs and herbs of the high marsh, such as the mallow, may be used as food or herbal medicines.

MANGROVES

Mangroves are in some ways the tropical equivalents of temperate marsh plants, but they more often are trees and shrubs than grasses and herbs. Very little of the mangrove is submerged by the tide. Both similarities and differences in anatomy, physiology, and ecological roles of mangroves and marsh plants are instructive of how terrestrial plants have adapted in reinvading the seas.

Classification and Distribution of Mangroves

The word **mangrove** comes from the Portuguese *mangue*, meaning "tree," and the English word "grove." It is applied to a group of plants that are taxonomically diverse. Defining "mangrove" is subjective, but a conservative account of what constitutes the mangroves of the world lists 54 species of trees, shrubs, palms, and ferns in 16 families. Half of these plants belong to two families: red mangroves, including *Rhizophora mangle* **(Figure 7-22a)**, and black mangroves, including *Avicennia germinans* **(Figure 7-22b)**. Both red and black mangroves are well known for specialized roots that descend to or rise from the sediment. In addition to the red and black mangrove, three other large mangroves of the tropical Americas are white mangrove (*Laguncularia racemosa*), buttonwood (*Conocarpus erectus*), and *Pelliciera rhizophoreae*, of very limited distribution.

Mangroves thrive along protected tropical shores with limited wave action, a low slope, high rates of sedimentation, and soils that are waterlogged, anoxic, and high in salts. Mangroves occur at low latitudes of the Caribbean Sea, Atlantic Ocean, Indian Ocean, and western and eastern Pacific Ocean. They are frequently associated with saline lagoons and are commonly found in tropical estuaries, as well as in full ocean water on the leeward sides of islands and atolls. Mangroves form the dominant vegetation in communities called mangrove swamps, or **mangals**. The Indo-West Pacific is the hotspot of mangrove biodiversity, as it is also of seagrasses.

Structure of Mangroves

The representative mangroves featured here are trees with simple leaves and complex root systems. These plant parts must help the tree conserve water in a warm saline environment, supply oxygen to roots in poorly drained, anoxic sediments, and remain stable in shallow, soft mud. In addition, the youngest stages in the life cycle of the plants must be adapted to deal with dispersal by tides, navigation among the tangle of above-ground roots, and establishment in their respective zones in the mangal.

Roots Mangrove roots are adapted to grow in loose, shallow, anoxic, saline sediments in regions often visited by hurricanes and typhoons. Root parts are diverse, and different species have their own combinations of parts **(Figure 7-23)**. Tap roots and deep roots generally do not occur in

A **mangrove** is an intertidal salt-tolerant tree or shrub of tropical coastlines.

A **mangal** is a forest of mangroves.

(a)

(b)

FIGURE 7-22 MANGROVES. (a) This forest of red mangrove consists of a canopy of branches and leaves over a tangled base of prop and drop roots. These above-ground roots not only help to support the large plant in loose mud but also aid in gas exchange for the parts of the root buried in the anaerobic sediments. **(b)** The black mangrove looks more like a typical tree, having a well-developed trunk. Spike-like pneumatophores arise from underground cable roots radiating from the base of the trunk. Pneumatophores are the gas-exchange organs of the black mangrove.

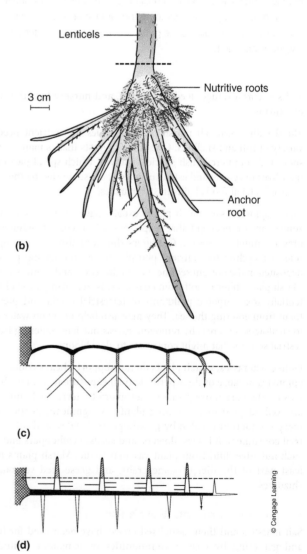

(a)

(b)

FIGURE 7-23 MANGROVE ROOT SYSTEMS. (a) Aerial roots of mangroves have scarlike lenticels, which allow air into the aerenchyme of the roots. Lenticels appear as small pale dots on these prop roots. **(b)** Within the sediment, mangroves produce anchor roots for stability and nutriive roots for absorption of water and minerals. **(c)** A red mangrove's looping arches (black) of a prop root give general support, and the part that descends to the substratum (cross-hatch) allows for gas exchange and the production of anchor roots (white) and nutritive roots (black lines). **(d)** A black mangrove's main structure of the root system is the cable root (black), from which anchor roots (white) descend and pneumatophores arise. The pneumatophores consist of an anchoring part (white), which produces nutritive roots (black lines), and an aerial part (cross-hatch), which serves for gas exchange.

(c)

(d)

mangroves, and root hairs are lacking. Many are **aerial** (above ground) **roots** and have aerenchyme. The aerial **stilt roots** of the red mangrove arise high on the trunk (**prop roots**) or from the underside of branches (**drop roots**). As the growing tip of the stilt root grows into the mud, aerenchyme is rapidly produced and may constitute as much as 40% of the root volume. The aerenchyme communicates with the atmosphere by way of scar-like **lenticels** on the surface of the stilt root. Lenticels supply oxygen to the subterranean root system. After penetration of the mud, the root begins to branch, producing a cluster of **anchor roots**, which themselves bear yet smaller branches below ground called **nutritive roots**. As sediment accumulates in the mangal, anchor roots keep producing more nutritive roots, which absorb mineral nutrients from the rich surface layers of mud. Stilt roots branch above ground and extend the root system far out from the trunk in a series of looping arches and columns that give the tree great stability in an otherwise unstable habitat.

The root system of the black mangrove arises below ground with **cable roots** spreading from the base of the trunk. Anchor roots penetrate below the cable root, and aerial roots called **pneumatophores** grow from the upper side out of the sediment and into the water or air above (see **Figure 7-23d**). Lenticels and aerenchyme of the pneumatophores ventilate the root system of the black mangrove. Nutritive roots branch from the underground parts of the pneumatophores, which keep up with the accumulation of rich sediment with continued upward growth.

Other root types and parts of mangrove trees are plank roots and knees, and some species have fluted bases to the trunk. Mangroves that grow on higher ground of the mangal usually lack specialized parts because they live in well-drained, firmer, deeper soil.

The mangrove's primary adaptation for life in saline waters is the ability of its roots to prevent salts from entering the system osmotically from the sediment. Because much of the sediment of the mangal is bare, in contrast to the heavily vegetated terrain of the salt marsh, evaporation of water at low tide may cause the salinity of water in the poorly drained sediment to be more than twice the salinity of ocean water.

Leaves Mangrove leaves are simple, oval, leathery, and thick, displaying the same succulence of marsh plants (**Figures 7-24a** and **7-24b**). The leaves are never submerged by the flood tide. The epidermis of the leaf is usually covered with a thick cuticle, and the **stomata**, openings in the leaves for gas exchange and water loss, are sunken and usually confined to the undersurface. These adaptations help retard the loss of water by evaporation from the leaves. Old leaves accumulate salt and fall from the tree, ridding it of some of its salt load. Mangroves are evergreen, and the loss of senescent leaves is continuous, peaking in summer. The black mangrove has salt glands on the surface of the leaf that secrete a concentrated salt solution, allowing the plant to eliminate some of the salt that does cross the barrier at the roots. Hairs on the undersurface of the leaf hold the secretion away from the epidermis while the droplets dry. Grains of salt may accumulate so heavily that the leaves glisten, but periodic rains wash the salt from the leaves to the soil below.

Although all mangroves prevent much salt in the sediment from entering the tree at the roots (a process called salt exclusion), few secrete salts at the leaves. In addition to these two mechanisms and the fall of salt-laden succulent leaves, some mangroves deposit excess salts in the bark, and all tolerate high concentrations of salt in the sap. Salt concentrations in the sap of mangroves are 10 to 100 times greater than those in

terrestrial plants. Such high concentrations of salt might be toxic to the mangroves except that the sap is confined to the nonliving xylem tissue. The living cells must balance the osmotic pressure of the sap by using small organic molecules in place of salts, as is the case with other halophytes and many marine animals.

Reproduction in Mangroves

Mangroves have simple flowers that are pollinated by wind or bees. Mangroves that occupy higher elevations in the mangal usually produce small, buoyant seeds that drift to distant shores and may be stranded by the highest tides into a zone for which the species is adapted (**Figures 7-24c–7-24d**). Mangroves of the seaward fringe and middle elevation of the mangal have viviparity, which prepares the new plant for more rapid establishment under tidal disturbance. Rather than entering a period of dormancy, the embryo grows on the parent plant and is called a **propagule** rather than a seed. Propagules of red mangroves (see **Figure 7-24a**) grow by absorption of water, minerals, and some organic nutrients through its attachment to the parent plant. It eventually breaks through the fruit wall and grows an elongated cigar-shaped stem (**hypocotyl**, meaning "below the leaf") hanging pendulously up to 25 centimeters (10 inches) long from the parent branch. The propagule eventually separates from the fruit and drifts with currents by its buoyant hypocotyl for as long as 100 days. The length of the hypocotyl prevents most propagules from penetrating the tangle of stilt roots and pneumatophores. Once stranded near the seaward edge of the mangal, the propagule takes root within 15 days. The propagule of the black mangrove (see **Figure 7-24b**) is the entire fruit, but the embryo absorbs water quickly and bursts from the thin fruit wall shortly after separation from the parent. The embryonic leaves are the floating organs, and they buoy the young plant for only 2 weeks. The smaller size (3 centimeters, or 1.2 inches) of black mangrove propagules allows them to drift out of and into the mangal to become stranded in the middle zone, where their shorter rooting time of 7 days and their loss of buoyancy help them take root under less frequent tidal disturbance.

An **aerial root** is a root that occurs above ground.

Stilt roots are aerial roots that hold up a mangrove.

A **prop root** is a stilt root that arises from the trunk.

A **drop root** is a stilt root that arises from a branch.

Lenticels are scar-like openings on the surface of roots that allow gas to enter the aerenchyme.

Anchor roots are short branches from the main root that hold the tree in the sediment.

Nutritive roots are the finest divisions of roots for absorption of minerals.

A **cable root** is a subterranean horizontal part of a root system that extends from the trunk.

Pneumatophores are aerial roots that grow out of the sediment from a cable root and provide air to the root.

Stomata are depressions in a leaf for passage of water and gases.

A **propagule** is the dispersal stage of a mangrove.

The **hypocotyl** is the initial stem of a young plant.

Ecological Roles of Mangroves

Mangroves play ecological roles similar to those of seagrasses and salt marsh plants. Their root systems stabilize the sediments, and aerial roots aid deposition of particles derived from external sources, as well as from

(a)

Richard Turner

(b)

Laurie Petrone, Florida Institute of Technology

(c)

Courtesy of Richard Turner

(d)

Richard Turner

FIGURE 7-24 MANGROVE LEAVES AND PROPAGULES. (a) Red mangroves demonstrate the typical oval leaves of many mangrove species. A cigar-like young tree, or propagule, sprouts from the fruit while attached to the parent plant. Precocious growth on the parent plant is called viviparity, a condition typical of mangroves that inhabit the seaward fringe of the mangal. **(b)** The name black mangrove comes from the dark color of the bark. The embryonic tree looks like a pointed lima bean and will break from the thin wall of the fruit after it falls into the water. Because this propagule was not dormant and is ready to sprout on release, this is another example of viviparity among mangroves. **(c)** These cone-like fruits of the buttonwood have the size and general appearance of blackberries or raspberries. They produce small seeds in clusters. Mangroves that live at the landward fringe of the mangal are not viviparous. **(d)** The fruit of the white mangrove look like shelled almonds in size and shape. They, too, contain seeds and are not viviparous. As with seeds of the buttonwood, such small buoyant propagules easily leave and enter mangals to reach high ground, where they become stranded and germinate.

IMPACT BUBBLE

HUMAN ACTIVITIES AFFECT seaweeds and marine plants in a variety of ways. The introduction of non-native species, nutrient pollution, and the destruction of coastal habitats all have led to altered ecosystems.

Humans have directly or indirectly caused the introduction of numerous species to locales outside their native range. Attempts to eliminate the introduced species fail more often than they succeed. Species of the toxic alga *Caulerpa* that have been introduced into the Mediterranean Sea and to the southern Florida and California coasts have become nuisance species because there is a lack of herbivores to control their populations. Some biologists look toward ascoglossan molluscs as a possible means to control nuisance populations of *Caulerpa*.

Nutrient runoff from human communities into coastal waters fertilizes seaweeds and phytoplankton more so than seagrasses because of the different ways in which these two groups of primary producers absorb nutrients. The resulting algal blooms and shading may cause seagrasses to decline. Although harmful algal blooms probably are natural occurrences, eutrophication near human communities increases the frequency and intensity of some blooms in coastal systems, causing significant reduction in the acreage and depth of seagrass beds.

Unlike coral reefs, mangals have not attracted a following in the media or among the public. Consequently, destruction of mangals for the production of marine organisms (mariculture), agriculture, salt production, or urban development does not raise the ire of human interest groups. Ivan Valiela and coauthors estimated in 2001 that about 35% of forested mangrove area had been lost over the preceding 20 years. The majority of the lost area has been from conversion of mangals into ponds surrounded by dikes for mariculture, especially the culture of shrimp. Despite attempts to compare the economic value of conversion with the ecological value of the ecosystem, it is clear that the ecological services mangals provided to other coastal communities and to former inhabitants of the region are lost, perhaps forever.

---

the fall and decay of their own leaves and twigs. Epiphytes such as seaweeds, oysters, snails, barnacles, and crabs grow or crawl on the aerial roots. The canopy is host to lichens, many insects, and roosting and nesting birds. The mangal and surrounding seagrass beds serve as a nursery and refuge to many animals. A variety of insects and crabs and a few terrestrial mammals consume leaves, fruit, and propagules. Mangroves are less subject to direct herbivory than are the grasses of salt marshes, probably in part because of their high salt content. As is the case with seagrasses and salt marsh plants, the greatest contribution of mangroves to the ecology of the seas is through detrital food chains. Because mangals are such open systems with a large proportion of bare sediment, far less mangrove detritus is retained in this habitat than in seagrass beds and salt marshes.

Human Uses of Mangroves

Mangals, like seagrass beds and salt marshes, are used for hunting and gathering. Direct uses of the trees mostly derive from their wood: Trunks and branches are harvested for production of charcoal, as firewood, and for many kinds of construction. Leaves and propagules are used as fodder for livestock, as well as for human consumption. Extracts from various parts of the trees serve as fish poison, dyes, industrial preservatives (tanning), drinks (alcohol, tea), cooking ingredients, paper, glue, cosmetics, and herbal remedies.

IN SUMMARY

- The plants living in marine environments are vascular plants that produce seeds.
- Seagrasses are mainly found in shallow waters of temperate and tropical oceans. (Have You Wondered? #3)
- Temperate salt marsh plants generally grow in the middle to upper intertidal zones, where they are protected from wave action. (Have You Wondered? #3)
- Mangroves are tropical trees and shrubs that grow in sandy and muddy coastal areas that are protected from wave action and have high sedimentation rates. (Have You Wondered? #3)
- Marine plants survive in ocean and brackish waters because of adaptations that allow for control of salt concentrations, use of water currents for dispersal of life-history stages, and provision of oxygen to submerged parts of the plant. (Have You Wondered? #4)
- Marine plants act as sediment traps, help stabilize sediments, and filter runoff from the land. (Have You Wondered? #2)
- Marine plants are primary producers, and they contribute to marine food webs by way of detrital pathways both locally and offshore.
- They also provide habitats for many organisms, especially nursery grounds for many species in the commercial and sport-fishing industries. (Have You Wondered? #2)
- The plants that form these communities have been used directly by human societies for thousands of years in agriculture, construction, household goods, industry, and medicine. (Have You Wondered? #5)

IN PERSPECTIVE

Phylum	Representative Organisms	Form and Function	Reproduction	Biochemistry	Ecological Roles
Chlorophyta	Green algae: sea lettuce	Entire thallus photosynthetic. Highly variable shape. Often coenocytic.	Usually annual. Vegetative growth and sexual reproduction with flagellated spores and gametes, two multicellular stages.	Chlorophylls *a, b*, carotenoids. Cell walls of cellulose, sometimes mineralized with calcium carbonate.	Seasonal food for herbivores, grazers. Some form turfs and reef sediments.
Rhodophyta	Red algae: laver, dulse, corallines, Irish moss	Entire thallus photosynthetic. Highly variable shape, but often filamentous.	Usually annual. Sexual reproduction with three multicellular generations. No flagellated stages.	Chlorophylls *a, d*, phycoerythrin, phycocyanin. Cell walls of cellulose, phycocolloids, mucilage, sometimes mineralized with calcium carbonate.	Seasonal food for herbivores, grazers. Some form turfs and reef sediments or stabilize reef sediments. Often epiphytic.
Phaeophyta	Brown algae: rockweed, knotted wrack, kelp, sargassum weed	Entire thallus photosynthetic. Often differentiated into blade, stipe, holdfast.	Often perennial. Strongly differentiated sporophyte and gametophyte generations, producing flagellated spores and sperm that are heterokont.	Chlorophylls *a, c*, fucoxanthin. Cell walls of cellulose, alginates, mucilage.	Long-term food for herbivores, grazers. Extensive intertidal, subtidal, pelagic stands.
Anthophyta	Flowering plants: seagrasses, cordgrass, mangroves	Photosynthesis confined to leaves and young stems. Vascular tissue. Highly differentiated above- and below-ground parts. Salt glands for mineral balance.	Vegetative growth. Sexual reproduction with flowers, seeds, fruit. Female generation parasitic on sporophyte.	Chlorophylls *a, b*, carotenoids. Cell walls of cellulose, often with hemicellulose and lignin.	Major organizing components in single-species stands. Detrital food chains and some direct herbivory or grazing. Sediment deposition and stabilization. Nursery, refuge, and epiphytism.

Photos: **Chlorophyta:** Courtesy of Richard Turner; **Rhodophyta:** Mike Guiry/AlgaeBase; **Phaeophya:** Courtesy of Richard Turner; **Anthophyta:** Laurie Petrone/Florida Institute of Technology.

KEY CONCEPTS

1. Multicellular marine macroalgae, or seaweeds, are mostly benthic organisms that are divided into three major groups according to their photosynthetic pigments. (7-1)

2. The distribution of seaweeds depends not only on the quantity and quality of light but also on a complex of other ecological factors. (7-1)

3. Marine algae supply food and shelter for many marine organisms. (7-1)

4. Flowering plants that have invaded the sea exhibit adaptations for survival in saltwater habitats. (7-2)

5. Seagrasses are important primary producers and sources of detritus, and they provide habitat for many animal species. (7-2)

6. Salt marsh plants and mangroves stabilize bottom sediments, filter runoff from the land, provide detritus, and provide habitat for animals. (7-2)

1. Green, red, and brown algae are distinguished mainly by the pigments that they use to capture light for photosynthesis and to protect their primary photosynthetic pigments from harmful rays. In addition, they differ in several aspects of their life cycles and in the composition of cell walls.

2. Seaweeds and marine plants are used as food directly or indirectly by many organisms, but they play other ecological roles as well. Their structure provides physically complex habitats in which organisms live or find refuge. By stabilizing and accumulating sediments, they may build up the land and improve water quality.

3. Although one green and one red alga hold records for living hundreds of meters in the sea, most seaweeds and marine plants inhabit the inner continental shelf. Some kinds are strictly intertidal. Seaweeds and seagrasses as groups have wide geographic ranges, but mangroves are tropical and marsh plants temperate. Most green and red algae are annuals and may be highly seasonal in occurrence. Most brown algae and marine vascular plants are perennials and can form persistent marine communities.

4. Marine plants are tolerant of high salts in their environment, but they also can exclude salts, secrete salts, or store salts. Aerenchyme allows the distribution of air or oxygen to parts of the plant that are submerged in anoxic sediments. To various degrees, water currents assume the role of air currents and animals in dispersal of pollen, seeds, and propagules of marine plants.

5. Seaweeds and marine plants are used by coastal farming communities as fodder, mulch, and fertilizer. Household uses include food, insulation, stuffing, and construction material. There are extensive applications of seaweeds in the pharmaceutical, cosmetic, and food industries.

QUESTIONS FOR REVIEW

Multiple Choice

1. Seaweeds and marine plants are confined to shallow water because they are limited mainly by
   a. light
   b. herbivores
   c. pressure
   d. temperature
   e. minerals

2. The body of a multicellular alga is called a
   a. rhizome
   b. holdfast
   c. thallus
   d. stipe
   e. blade

3. The stage in the algal life cycle that produces gametes is called the
   a. sporophyte stage
   b. adult stage
   c. hydrophyte stage
   d. gametophyte stage
   e. vascular stage

4. A type of brown alga that grows quite large and forms undersea forests is
   a. sargassum weed
   b. Irish moss
   c. kelp
   d. rockweed
   e. sea lettuce

5. A coenocytic thallus consists of a single cell or several large cells that contain more than one
   a. cell wall
   b. chloroplast
   c. thallus
   d. holdfast
   e. nucleus

6. Plants that live submerged beneath the water are called
   a. halophytes
   b. hydrophytes
   c. anthophytes
   d. vascular plants
   e. algae

7. Seaweeds and marine plants share all of the following except
   a. chlorophyll *a*
   b. cellulose
   c. vascular tissue
   d. alternation of generations
   e. a role as primary producers

8. Each of the following is an important ecological role of marine plants except
   a. improving water clarity
   b. trapping nutrients
   c. stabilizing bottom sediments
   d. fixing nitrogen
   e. providing a habitat

9. Plant adaptations for life in the marine environment include
   a. gas-transporting tissue
   b. reduced dependence on pollination by animals
   c. specialized systems of roots and stems
   d. water storage in stems and leaves
   e. all of the above

Matching

Match the following seaweeds and vascular plants with their distinctive features:

a. brown algae

b. green algae

c. red algae

d. red mangroves

e. seagrasses

f. *Spartina alterniflora*

1. Form culms and tillers.

2. Have 3 multicellular stages in life cycles.

3. Occur mainly in freshwater habitats.

4. Produce prop and drop roots.

5. Related to diatoms.

6. Use hydrophilous pollination.

Match the following chemicals with the appropriate description:

a. agar

b. alginate

c. calcium carbonate

d. chlorophyll *a*

e. mannitol

f. tannins

1. A mineral in cell walls of some green and red algae used to deter herbivores.

2. A phycocolloid of brown algae.

3. A phycocolloid used as a microbial culture medium.

4. An alcohol used as food in brown algae.

5. An antimicrobial chemical found in various vascular plants.

6. The primary photosynthetic pigment found in all seaweeds and plants covered in this chapter.

Short Answer

1. What factors affect the distribution of algae in the marine environment?

2. List some of the important ecological roles of marine flowering plants.

3. Describe the alternation of generations that occurs in the life cycle of an alga.

4. Describe the adaptations that have evolved in algae to protect against wave shock.

5. Describe the adaptations that have evolved in salt marsh plants or mangroves to help them survive in areas where salt content is high.

Thinking Critically

1. Salt marshes play an important role as nurseries for many commercially important fishes and shellfish. What characteristics make them such ideal nurseries?

2. Predict the effects of coastal zone development on seagrass and mangrove communities.

3. One bay experiences runoff from surrounding communities with well-kept lawns. Another has a community with failing septic systems that discharge their contents through groundwater. How might these two bays differ in their growth of seaweeds and seagrasses?

4. In Chapter 2 you learned that overgrazing of kelp by sea urchins led to a decline in several other animal species associated with kelp forests. On the basis of what you have learned in this chapter, why would a decrease in the amount of kelp affect animals that do not feed on kelp?

SUGGESTIONS FOR FURTHER READING

Allen, J. A., and K. W. Krauss. 2006. Influence of Propagule Flotation Longevity and Light Availability on Establishment of Introduced Mangrove Species in Hawai'i, *Pacific Science* 60(3): 367–376.

Daerr, E. G. 2001. Splendor Is a Grass: Johnson's Seagrass, Found in Florida, Was the First Marine Plant to Be Listed as Endangered, *National Parks* 75(11–12): 42.

Green, E. P., and F. T. Short. 2003. *World Atlas of Seagrasses*. Berkeley: University of California Press.

Hemminga, M. A., and C. M. Duarte. 2000. *Seagrass Ecology*. Cambridge: Cambridge University Press.

Hogarth, P. J. 2007. *The Biology of Mangroves and Seagrasses*. New York: Oxford University Press.

Koehl, M. A. R., W. K. Silk, H. Liang, and L. Mahadevan. 2008. How Kelp Produce Blade Shapes Suited to Different Flow Regimes: A New Wrinkle, *Integrative and Comparative Biology* 48(6): 834–851.

Larkum, A. W. D., R. J. Orth, and C. M. Duarte. 2006. *Seagrasses: Biology, Ecology and Conservation*. Dordrecht, The Netherlands: Springer.

Ludlam, J. P., D. H. Shull, and R. Buchsbaum. 2002. Effects of Haying on Salt-Marsh Surface Invertebrates, *The Biological Bulletin* 203(2): 250–251.

Molyneaux, P. 2001. Seaweed Scene: Tag Along with the Dulse Pickers of New Brunswick's Grand Manan Island, *National Fisherman* 82(8): 22–24.

Orth, R. J., et al. 2006. A Global Crisis for Seagrass Ecosystems, *BioScience* 56(12): 987–996.

Schneider, P. 2000. In the Middle of a Marsh, *Audubon* 102(4): 100.

Short, F., T. Carruthers, W. Dennison, and M. Waycott. 2007. Global Seagrass Distribution and Diversity: A Bioregional Model, *Journal of Experimental Marine Biology and Ecology* 350: 3–20.

Thomas, D. 2002. *Seaweeds.* Washington, D.C.: Smithsonian Institution Press.

Thornber, C. S. 2006. Functional Properties of the Isomorphic Biphasic Algal Life Cycle, *Integrative and Comparative Biology* 46(5): 605–614.

Valiela, I., J. L. Bowen, and J. K. York. 2001. Mangrove Forests: One of the World's Threatened Major Tropical Environments, *BioScience* 51(10): 807–815.

Vroom, P. S., K. N. Page, J. C. Kenyon, and R. E. Brainard. 2006. Algae-dominated Reefs, *American Scientist* 94(5): 430–437.

Vroom, P. S., and C. M. Smith. 2001. The Challenge of Siphonous Green Algae, *American Scientist* 89(6): 524–531.

Waycott, M., et al. 2009. Accelerating Loss of Seagrasses Across the Globe Threatens Coastal Ecosystems, *Proceedings of the National Academy of Sciences USA* 106(30): 12377–12381.

Have You Wondered?

1. How you can tell that a sponge is an animal? (8-2)
2. If anything eats sponges? (8-2)
3. If there are male and female jellyfish? (8-3)
4. If jellyfish stings are lethal? (8-3)
5. If all gelatinous animals are jellyfishes? (8-4)

Lower Invertebrates

THE SEA IS inhabited by a large number of animal species, ranging from microscopic worms to giant squids. Regardless of their size, each fills a particular niche in the overall marine ecosystem. In this chapter and the next, we will examine more closely the biology and ecological roles of the myriad animals without backbones, the invertebrates, that are part of the fabric of life in the sea.

8-1 What Are Animals?

It may seem easy to distinguish animals from other forms of life, but this is not always the case. The animals known as sponges, for instance, were generally thought to be plants until 1765, when a British naturalist named Ellis observed that sponges could expel water from their bodies. This observation prompted Ellis to suggest that sponges were animals. Their place as animals, however, was not fully accepted until 1825, when R. E. Grant of Edinburgh University, using improved microscope technology, observed that sponge cells lacked a cell wall and demonstrated that sponges could take tiny colored particles from the water, circulate them through their bodies, and then expel them.

The earth is populated by a wide variety of animals, and it is difficult to arrive at a definition that applies to them all. Despite their great diversity, they all share four characteristics:

1. Animals are multicellular. This distinguishes them from bacteria, archaeons, and most protists, which are unicellular.

2. Animal cells are eukaryotic, and they lack cell walls. These characteristics distinguish animals from bacteria and archaeons, whose cells are prokaryotic and have cell walls, and from fungi, multicellular algae, and plants, all of which have cells with rigid cell walls.

3. Animals cannot produce their own food, so they depend on other organisms for nutrients. In other words, they are heterotrophs.

4. Animals, with the exception of some sessile adult forms, can actively move.

Invertebrates are animals that lack a vertebral column.

Vertebrates are animals that have a vertebral column.

Animals that lack a vertebral column (backbone) are **invertebrates**, whereas animals that have a vertebral column are **vertebrates**. The majority of animals in the sea are invertebrates (**Figure 8-1**).

8-2 Sponges

Sponges (phylum Porifera) are simple, asymmetric, **sessile** animals (meaning they are permanently attached to a solid surface, such as rock or shell). Sponges exhibit a wide variety of forms, their size and shape frequently being determined by the nature of the material on which they are growing and by the water currents flowing over them. Although many living sponges are drab, some species are brightly colored. Red, yellow, green, orange, and purple specimens are common.

Sessile animals are permanently attached to a solid surface.

Ostia are holes in the body of sponges through which water enters.

A **spongocoel** is a cavity in the body of a sponge.

SPONGE STRUCTURE AND FUNCTION

The structure of a sponge's body is unique in that it is built around a system of water canals. This arrangement is associated with the sponge's sessile lifestyle. A sponge's body is full of tiny holes, or pores, called **ostia** (singular, ostium; **Figure 8-2**), through which large amounts of water circulate. The water is a source of nutrients and oxygen, and it carries away the animal's wastes. Water enters a sponge's body through the ostia and eventually flows into a spacious cavity called the **spongocoel** (SPUN-joh-seel). Water then exits the spongocoel through a large opening called the **osculum**. Many species have several spongocoels and oscula.

Sponges lack **tissues**, groups of specialized cells that perform a specific function, and organs. Instead, they have several special cell types that

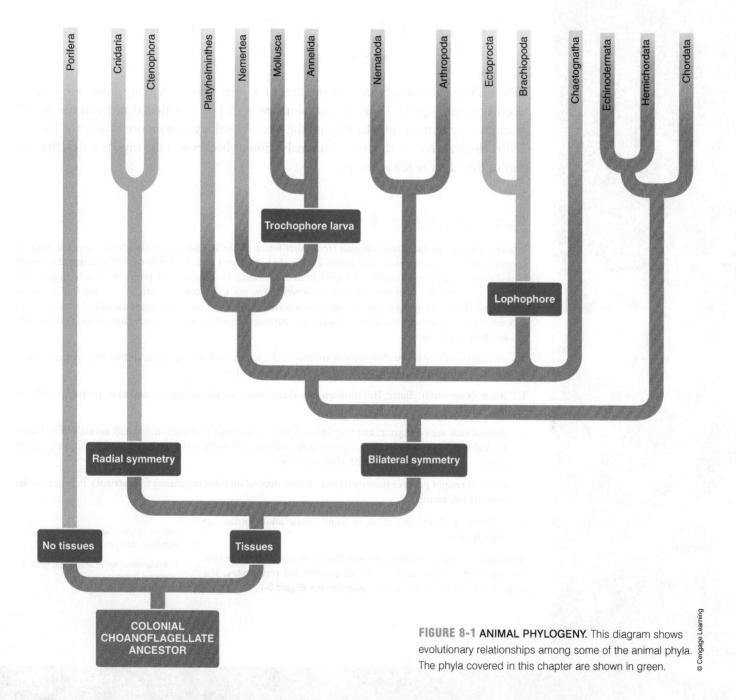

FIGURE 8-1 ANIMAL PHYLOGENY. This diagram shows evolutionary relationships among some of the animal phyla. The phyla covered in this chapter are shown in green.

© Cengage Learning

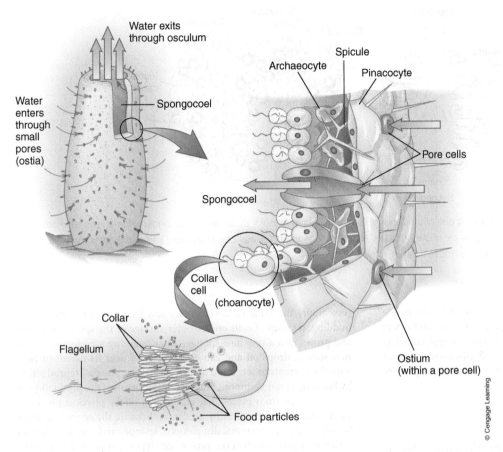

Water exits through osculum

Water enters through small pores (ostia)

Spongocoel

Spongocoel

Collar cell (choanocyte)

Collar

Flagellum

Food particles

Archaeocyte

Spicule

Pinacocyte

Pore cells

Ostium (within a pore cell)

© Cengage Learning

An **osculum** is an opening through which a sponge expels water.

Tissues are groups of specialized cells that perform a specific function.

Collar cells (choanocytes) are flagellated cells in a sponge's body that circulate water and trap food.

Pinacocytes are cells that make up the outer covering of a sponge and line internal chambers not lined by choanocytes.

Archaeocytes are cells that form any of the cell types in the sponge body.

Spicules are structures that support a sponge's body.

Spongin is a protein that makes up flexible spicules.

Asconoid is a type of sponge body that lacks invaginations in the body wall.

Syconoid is a type of sponge body with a single spongocoel containing many invaginations.

Leuconoid is a type of sponge body with multiple spongocoels and chambers leading to them.

FIGURE 8-2 ANATOMY OF A SPONGE. A sponge's body has many specialized cells, but no tissues. The beating of the collar cells' flagella circulates water through the sponge's body, bringing in food and oxygen and removing waste.

perform specific functions within the animal **(Figure 8-2)**. **Choanocytes**, or **collar cells**, have a flagellum. The movement of the choanocytes' flagella provides the force that moves water through the sponge's body. A layer of cells called **pinacocytes** provides an outer covering for the sponge. They also line internal chambers that are not lined by choanocytes. **Archaeocytes** are cells that resemble amoebas, and like an amoeba they can move through the sponge's body. Archaeocytes can form any of the cell types in the sponge body and play an important role in repair and regeneration. They also transport food and other materials within a sponge's body. Other specialized cell types produce **spicules**, skeletal elements that support a sponge's body. Spicules may be composed of calcium carbonate, silica, or a protein called **spongin**. Spongin forms flexible fibers, allowing the dried sponge skeleton to be flexible, as in commercial sponge species (class Demospongia). Sponges known as glass sponges (class Hexactinellida) have spicules made of silica that resemble glass fibers. These are usually fused to form elaborate networks, giving these sponges a delicate appearance. The Sclerospongia is a small group of sponges that live in shaded or dark habitats. They secrete massive skeletons of calcium carbonate, and it is thought that they were the original reef-forming organisms, having evolved before the corals.

SPONGE SIZE AND BODY FORM

The size of a sponge is limited by its ability to circulate water through its body, and this is determined by the body form **(Figure 8-3)**. The **asconoid**

form is the simplest. It is tubular and always small. Sponges with this type of body are usually not found alone but in clusters. As a sponge with an asconoid shape grows, the volume of the spongocoel increases; this is not, however, accompanied by a proportional increase in the surface area available for choanocytes. As a result, there are not enough choanocytes to move sufficient amounts of water through the sponge to provide enough food or to meet its other physiological needs. This is the reason asconoid sponges are always small.

During the evolution of sponges, the problem of water flow and surface area was overcome by folding the body wall and, in many species, reducing the size of the spongocoel. The folding increases the surface area of the choanocyte layer, and the reduction of the spongocoel that occurs because of the folding decreases the volume of water flow through the sponge. The more efficient movement of water allows the choanocytes to capture more food, and the sponge grows larger. Sponges display various stages in these changes. In sponges that exhibit the first stages of body-wall folding, called **syconoid** sponges, the body wall folds to form internal pockets that are lined with choanocytes. The highest degree of folding takes place in **leuconoid** sponges. In these sponges there are many chambers lined with choanocytes, and the spongocoel is frequently reduced to a series of water canals leading to an osculum. In some leuconoid sponges the number of chambers lined by choanocytes may be enormous. For example, the sponge *Microciona prolifera* contains 10,000 chambers per cubic millimeter. Because the leuconoid body plan is the most efficient arrangement, most sponge species conform to it.

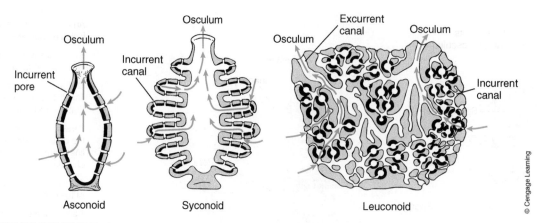

FIGURE 8-3 SPONGE BODY FORMS. Sponges exhibit one of three possible body plans: asconoid, syconoid, or leuconoid. Blue arrows represent the direction of water flow. Areas occupied by choanocytes are shown in black. Notice that as the body wall becomes more infolded there is more surface area available for choanocytes and the sponge can then move a larger volume of water.

The efficiency of the leuconoid body plan can be seen in *Leuconia*, one of the smaller leuconoid sponges. This sponge is 10 centimeters (4 inches) tall and 1 centimeter (⅓ inch) wide, contains approximately 2,250,000 chambers, and pumps about 22.5 liters (5 gallons) of water per day through its body. It is no wonder that all of the largest sponges have a leuconoid body type.

NUTRITION AND DIGESTION

Because sponges feed on material that is suspended in seawater, they are called **suspension feeders**. They are also referred to as **filter feeders** because they filter food from the water that passes through their body. Each choanocyte that lines the chambers within a sponge has a single flagellum and a collar of finger-like cellular projections (see **Figure 8-2**). The beating of the flagellum creates a current that not only moves water through the sponge's body but also across the choanocyte's collar. Approximately 80% of a sponge's food is trapped by filtration across the collar or on the choanocyte cell surface and consists of smaller particles (0.1 to 1.0 micrometers, or 0.000004 to 0.00004 inch), such as viruses and organic matter (see Figure 8-2). Large food particles (1 to 50 micrometers, or 0.00004 to 0.002 inch), such as bacteria and detritus, are ingested by phagocytosis by pinacocytes and archaeocytes along the sponge's system of canals that carry water from the ostia to spongocoels. Very few animals can capture food of this size, and the ability of sponges to utilize this largely untapped food source accounts for the success of these sessile animals. Although both choanocytes and archaeocytes digest food particles, most often they are passed to archaeocytes for digestion, and these cells then transfer the nutrients to other cells. Archaeocytes also function in the storage of food. Undigested material and nitrogenous wastes leave the sponge with the exiting water currents.

A **suspension feeder** is an organism that feeds on food suspended in the water.

A **filter feeder** is an organism that filters its food from the water.

REPRODUCTION IN SPONGES

Sponges can reproduce both sexually and asexually **(Figure 8-4)**. Asexual reproduction can involve either **budding** or fragmentation (essentially the same process that we saw in algae in Chapter 7). In budding, a group of cells on the outer surface of the sponge develops and grows into a tiny new sponge. After attaining a certain size, this new sponge drops off and can establish itself near its parent or float with the currents to settle and mature elsewhere. Asexual reproduction by budding is not common. Fragmentation involves the production of a new sponge from pieces that are broken off by physical processes such as waves or storms or by predators. Because the sponge is rather loosely organized anatomically, pieces of sponge can form new sponges if they contain enough of the various cell types. For years, sponge fishers have taken advantage of this characteristic to replenish sponge beds. Large commercial sponges are cut into pieces, tied to concrete blocks or other material that will anchor them, and then thrown back into the water. The pieces of sponge will then eventually grow into new adult sponges. Sponges exhibit marvelous powers of regeneration, and when a piece is broken off or eaten, the sponge can readily replace the missing part.

Most sponges are **hermaphrodites**, meaning they can produce both male and female gametes (sex cells) although usually not both at the same time. Sperm cells are formed from modified choanocytes. Eggs usually develop from archaeocytes but may also be formed from choanocytes. The stimulus to produce gametes is usually a change in the water temperature or **photoperiod** (the relative amount of light and darkness in a 24-hour period). When a sperm cell enters another sponge, it is engulfed by a choanocyte. Both cells lose their flagella and the choanocyte then transports the sperm to an egg. In most sponges development of the new sponge to the larval stage occurs in the adult's body, although some species release fertilized eggs to develop in the water column. The larval stage of most sponges is an **amphiblastula** that spends time in the water column as plankton before settling and forming a new adult sponge.

ECOLOGICAL ROLES OF SPONGES

Sponges interact with other marine organisms in several ways. They compete aggressively for space with other sessile organisms. They are links in some marine food chains. Sponges form many symbiotic relationships, and they provide habitat for other organisms. Sponges also play an important role in the recycling of calcium.

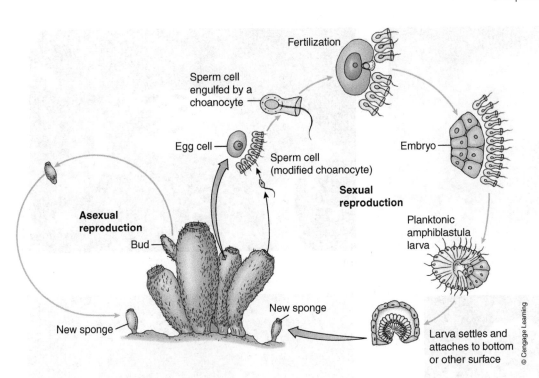

FIGURE 8-4 SPONGE REPRODUCTION. Sponges can reproduce asexually (by budding) or sexually.

Budding is a type of asexual reproduction in which a group of cells on the surface of the parent develop into a new individual.

Hermaphrodites are animals that can produce both male and female gametes.

Photoperiod is the relative amount of light and darkness in a 24-hour period.

An **amphiblastula** is the planktonic larval stage of a sponge.

Competition

The biggest problem that sponges face is finding enough suitable hard substrate for attachment. Consequently, their primary competitors for space are corals and bryozoans. Some sponge species produce chemicals that either kill corals or inhibit their growth. Other species such as the boring sponges (family Clionidae) create their own habitat by boring into corals and dead shells. Some species of crab known as sponge crabs (family Dromiidae) attach pieces of sponge to themselves for camouflage and protection and in the process provide a solid surface for the transplanted sponge species **(Figure 8-5a)**.

Predator–Prey Relationships

Very few animals feed on sponges, possibly because ingesting the spicules would be like eating a mouthful of needles. Many sponges produce chemicals that prevent organisms from settling on their surface or that deter grazing. A study of sponges found that 9 of 16 Antarctic sponges and 27 of 36 Caribbean species were toxic to fish. There are a few species of bony fishes and molluscs that will eat sponges, and the hawksbill sea turtle (*Eretmochelys imbricata*) feeds almost exclusively on them **(Figure 8-5b)**. As much as 95% of the turtle's feces contains glass spicules, indicating the large role of sponges in the turtle's diet. The tough lining of the turtle's mouth and digestive tract prevents the spicules from injuring the animal.

Symbiotic Relationships

Many sponges are hosts to other organisms. Some of these relationships are mutually beneficial for both the sponge and its symbionts (mutualism), whereas others seem to favor the symbionts without harming the sponge (commensalism). Some species of sponge host large numbers of mutualistic bacteria that provide food to the sponge while gaining nutrients and protection within the sponge's system of water canals. In fact, in some species, the volume of bacterial cells is twice that of sponge cells. Cyanobacteria make up 33% of *Verongia*, a sponge that lives in shallow, well-lit habitats. Many species of sponge that live on coral reefs contain symbiotic cyanobacteria in their bodies. A sponge provides protection and a sunlit habitat for the cyanobacteria, and the cyanobacteria provide the sponge with nutrients and oxygen. Studies of sponges on the Great Barrier Reef of Australia have found that reef sponges get between 48% and 80% of their food from their symbiotic cyanobacteria. There are also a few species of sponge that contain symbiotic dinoflagellates, a single-cell, photosynthetic organism.

The spongocoel and internal canals of a sponge can be home to a variety of organisms such as shrimp and fish that take advantage of the protection and continuous flow of water. For instance, one specimen of loggerhead sponge (*Spheciospongia vesparia*) from Florida was found to contain 16,000 snapping shrimp (*Synalpheus brooksi*). The glass sponge known as Venus's flower basket (*Euplectella*; **Figure 8-5c**), found in the deep waters of the tropical Pacific Ocean, exhibits an interesting symbiotic relationship with certain species of shrimp (*Spongicola*). A male and a female shrimp enter the sponge's spongocoel when they are young. They feed on plankton in the water that circulates through the sponge's body. Eventually the shrimp grow so large that they cannot escape through the network of spicules that covers the osculum. Their entire life is spent in the sponge. A specimen of this sponge along with the shrimp once was given in Japanese wedding ceremonies to symbolize lifelong fidelity.

Sponges and Nutrient Cycling

Members of the family Clionidae (class Demospongia), known as boring sponges **(Figure 8-5d)**, are important in recycling calcium in the marine

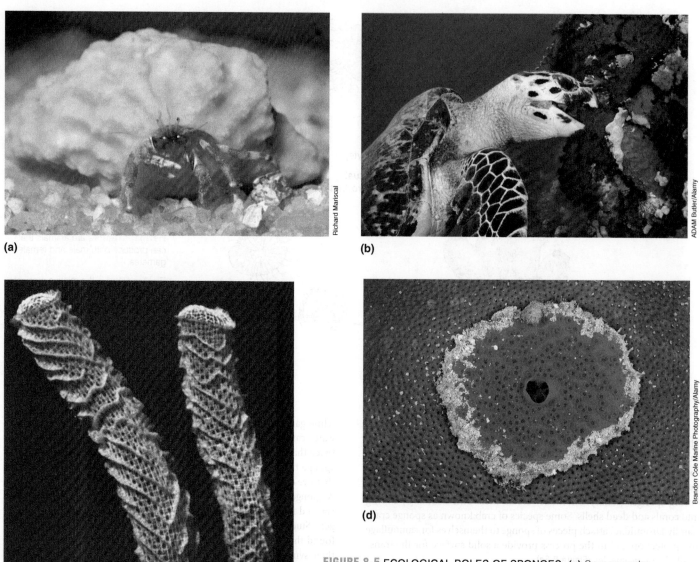

(a)

Richard Mariscal

(b)

ADAM Butler/Alamy

(c)

American Museum of Natural History

(d)

Brandon Cole Marine Photography/Alamy

FIGURE 8-5 ECOLOGICAL ROLES OF SPONGES. **(a)** Sponge crabs camouflage themselves by attaching sponges to their bodies. In this symbiotic relationship the crab gains protection from predators and the sponge gains mobility and a surface for attachment. **(b)** This hawksbill sea turtle is one of the few animals that will feed on sponges. **(c)** Venus's flower basket is a deepwater glass sponge from the Philippines. A pair of shrimp live within the sponge, gaining protection and feeding on the food circulated through the sponge's body. **(d)** Boring sponges burrowed into coral. Boring sponges help recycle calcium back into seawater.

environment. These small sponges burrow into coral and mollusc shells. In the process they convert the calcium into a soluble form that is returned to the seawater for use by other organisms.

COMMERCIAL USES OF SPONGES

Large species of sponge of the class Demospongia are harvested for commercial use **(Figure 8-6)**. Their dried spongin skeletons can hold as much as 35 times their weight in liquid. It takes approximately 5 years for a sponge in the wild to reach a marketable size (12.5 centimeters, or 5

inches), and it will retail for about $10.00. Although this is expensive when compared with the price of synthetic sponges, many people are willing to pay the extra money for a sponge that is superior to any synthetic product. For some applications, such as polishing metal, there is no good substitute for a natural sponge.

Sponges may also prove to be sources of novel medications for fighting diseases. A chemical called cytosine arabinoside blocks DNA synthesis in tumors and is used in the treatment of cancer. This substance is isolated from a Caribbean sponge, *Cryptotethya crypta*. Sponges produce antibacterial chemicals and are thus generally free from bacterial

of these animals close to shore or washed up on a beach after a storm. Jellyfish belong to a group of animals that also includes hydroids, corals, and sea anemones. They are called **cnidarians** (ny-DEHR-ee-uhns; phylum Cnidaria) because of the stinging cell, or **cnidocyte** (NYD-uh-syt), that they all possess. This unique cell is used not only in the capture of prey but also for protection.

ORGANIZATION OF THE CNIDARIAN BODY

Cnidarians have bodies that exhibit **radial symmetry (Figure 8-7a)**. This means that many planes can be drawn through the central axis that will divide the animal into two equivalent halves. The body parts of these animals are arranged in a circular pattern around a central axis, much like the spokes of a wheel. Radial symmetry is particularly beneficial to sessile organisms and organisms that are not active swimmers because it allows them to meet and respond to their environment equally well from all sides.

Cnidarians often exhibit two different body plans in their life cycles (**Figures 8-7b** and **8-7c**). The **polyp** (PAHL-uhp) is generally a benthic form characterized by a cylindrical body that has an opening at one end, the mouth, which is usually surrounded by a ring of tentacles. The **medusa** is a free-floating stage that is commonly known as a jellyfish. Many cnidarians exhibit both body plans during their life cycles, although some, such as corals and sea anemones, exist only in the polyp stage. Both the polyp and medusa have an outer layer of cells called the **epidermis**. Inside the body is a large cavity (**gastrovascular cavity**) that is lined by a layer of cells called the **gastrodermis**. Between these two layers is a gelatinous material called **mesoglea**. Most of the animal's body is the jelly-like mesoglea, which accounts for the common name, jellyfish, given to many cnidarian species.

STINGING CELLS

One of the most important features of cnidarians is their cnidocytes. The cnidocyte contains a stinging organelle called the **cnida** (NY-duh). There are more than two dozen types of cnidae (NY-dee, plural form of cnida) found in cnidarians. Some function in locomotion, whereas others function in the capture of prey and in defense. Most of the cnidae are of a spearing type known as **nematocysts (Figure 8-8)** that are located within a capsule in the cell. The capsule has a lid that opens when

A **cnidarian** is an animal that belongs to the phylum Cnidaria.

A **cnidocyte** is a stinging cell found in all cnidarians.

Radial symmetry is the organization of body parts around a central axis.

A **polyp** is the generally benthic form of cnidarian characterized by a cylindrical body with an opening at one end, usually surrounded by tentacles.

A **medusa** is the free-floating form of a cnidarian that resembles an umbrella.

The **epidermis** is an outer layer of cells.

A **gastrovascular cavity** is a large cavity found within the body of some animals.

The **gastrodermis** is the layer of cells that lines the gastrovascular cavity.

Mesoglea is the gelatinous material found between the epidermis and gastrodermis of cnidarians.

A **cnida** is the stinging organelle of a cnidocyte.

A **nematocyst** is a spearing type of cnida.

FIGURE 8-6 COMMERCIAL SPONGE FISHING. A sponge fisherman at Tarpon Springs, Florida, is drying some commercial sponges.

infections. These chemicals are being studied to see if they might have uses in the treatment of bacterial diseases of humans and livestock. Other potentially useful drugs from sponges are also being studied.

IN SUMMARY

- Sponges depend on their ability to filter large amounts of water through their bodies to survive. (Have You Wondered? #1)
- Sponge bodies are asymmetrical and contain several cell types that perform specific functions.
- Sponges provide habitats for many organisms and play a role in recycling calcium.
- Sponges have few predators with the exception of the hawksbill sea turtle and a few molluscs and fishes. (Have You Wondered? #2)
- Sponges are partners in many symbiotic relationships.

8-3 Cnidarians: Animals with Stinging Cells

Even a casual visitor to the seashore can usually identify a jellyfish and knows to avoid contact with one. Many of these creatures can cause injury with their toxic stings. It is not uncommon to find large numbers

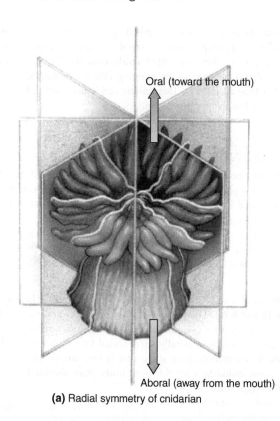

Oral (toward the mouth)

Aboral (away from the mouth)

(a) Radial symmetry of cnidarian

FIGURE 8-7 CNIDARIAN BODY PLANS. (a) Cnidarians exhibit radial symmetry. Any plane that passes longitudinally through the central axis of the animal will divide it into two equivalent halves. Cnidarians can exhibit two different body forms in their life cycle. **(b)** The polyp body form is cylindrical with a ring of tentacles around the mouth. **(c)** The medusa body form is shaped like an umbrella.

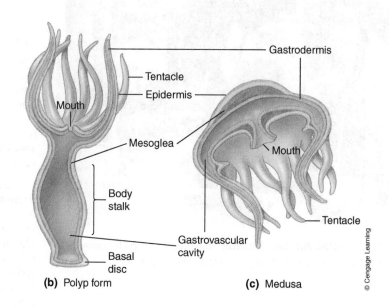

Gastrodermis

Tentacle

Epidermis

Mouth

Mesoglea

Mouth

Body
stalk

Tentacle

Gastrovascular
cavity

Basal
disc

(b) Polyp form

(c) Medusa

© Cengage Learning

the nematocyst is discharged. A short bristle-like structure called the **cnidocil** extends from one end of the cnidocyte and acts as a trigger. When the cnidocil comes into contact with prey or some other object, it causes the lid to open and the nematocyst to discharge. The cnidocyte can also be triggered through the action of certain chemical substances released by prey organisms. Some nematocysts have a thread attached that entangles stung prey. Once the cnidocyte has fired, it is reabsorbed by the animal's body, and a new one forms to take its place. Cnidocytes are most common on the tentacles of cnidarians, but they can also be found on the outer body wall and, with the exception of hydrozoans, in the lining of the gastrovascular cavity.

A **cnidocil** is a short bristle-like structure that acts as a trigger for a cnidocyte.

CNIDARIAN STINGS

Many species of cnidarian, such as the Portuguese man-of-war (*Physalia physalis*; **Figure 8-9a**), can cause painful stings. Even tentacles that have broken off from this animal or that are dead on the beach can sting. The symptoms include an immediate, intense, burning pain with redness and swelling in the regions of contact. Some individuals may also experience weakness, nausea, headaches, pain, spasms of abdominal

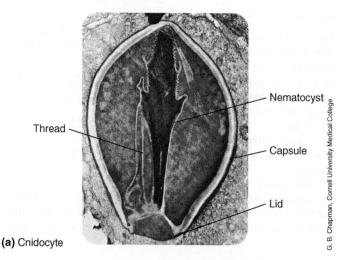

Thread

Nematocyst

Capsule

Lid

(a) Cnidocyte

G. B. Chapman, Cornell University Medical College

FIGURE 8-8 CNIDARIAN STINGING CELLS. (a) An electron micrograph of a stinging cell (cnidocyte) shows a nematocyst that has not been discharged. **(b)** The structure of a stinging cell and nematocyst; notice the position of the cnidocil, or trigger.

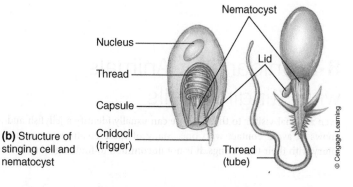

Nematocyst

Nucleus

Thread

Lid

Capsule

Cnidocil
(trigger)

(b) Structure of
stinging cell and
nematocyst

Thread
(tube)

© Cengage Learning

and back muscles, dizziness, and increased secretions from the nose and eyes. The sting is generally nonfatal to humans unless they are allergic to the toxin; however, the stings of the box jellyfish *Chironex fleckeri* (see Figure 8-14) can kill a person within minutes or leave scars that last for life. Death from *Chironex* can happen 3 to 20 minutes after stinging. Nonlethal stings can be severe and slow to heal **(Figure 8-9b)**. In Australia this animal is commonly known as a stinger and is seasonally abundant around parts of the country, causing the closing of some swimming beaches.

Another box jellyfish from Australia, the Irukandji jellyfish (*Carukia barnesi*) **(Figure 8-9c)**, is thought to be the most venomous animal in the world. The Irukandji jellyfish is tiny, only 2.5 centimeters (1 inch) in diameter including the bell and tentacles. The sting of the jellyfish causes a condition known as Irukandji syndrome, symptoms of which include vomiting, profuse sweating, headache, agitation, rapid heart rate, and high blood pressure. Symptoms begin to occur minutes to hours

following the stings (which themselves are typically mild) and usually last for hours to several days, requiring victims to be hospitalized. Death is generally the result of high blood pressure, irregular heartbeat, or pulmonary edema. Every summer in Australia over 60 people are hospitalized for Irukandji syndrome. Although generally thought to be restricted to the waters off northern Australia, the Irukandji jellyfish has recently been reported in waters off Japan, Britain, and the Florida coast.

Although it is more common to think of jellyfish stings as being the most harmful, some sea anemone stings can be toxic to humans as well, including the berried sea anemone (*Alicia mirabilis*) from Europe and the stinging anemone (*Lebrunia danae*) from the Caribbean. The west Australian armed anemone (*Dofleina armata*) is believed to be the most toxic sea anemone.

Not all animals are bothered by the toxic stings of cnidarians. For example, the leatherback sea turtle (*Dermochelys coriacea*) regularly feeds on

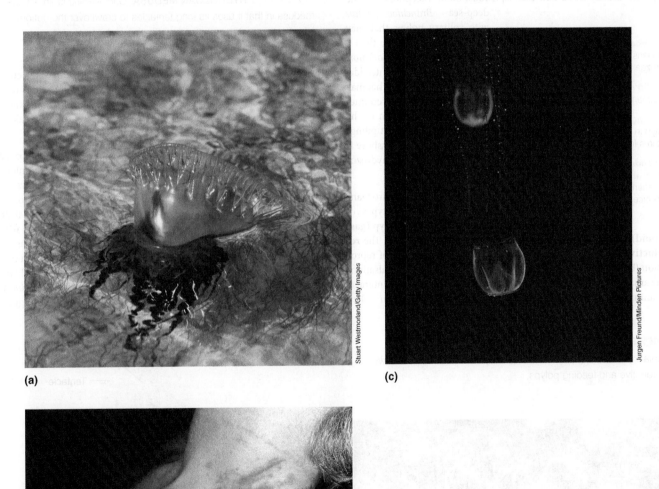

(a)

(b)

(c)

Stuart Westmorland/Getty Images

Custom Medical Stock

Jurgen Freund/Minden Pictures

FIGURE 8-9 DANGEROUS CNIDARIANS. (a) The hydrozoan known as the Portuguese man-of-war (*Physalia*) is a colony of many specialized individuals and can cause painful stings. **(b)** This victim's neck and chest show injuries produced by cnidarian stings. **(c)** The Irukandji jellyfish (*Carukia barnesi*) is thought to be the most venomous animal in the world. The stings of this cubozoan can be fatal.

jellyfish. Some nudibranchs feed on the Portuguese man-of-war and somehow are able to incorporate the nematocysts into finger-like projections called cerata for their own protection. Some flatworms can also store nematocysts from their prey and use them in their own defense.

TYPES OF CNIDARIANS

Representative cnidarians include hydrozoans, jellyfish, box jellyfish, sea anemones, and corals. These marine predators include colonial and solitary benthic organisms, as well as some of the largest members of the marine plankton.

Hydrozoans

Hydrozoans, or **hydroids** (class Hydrozoa), are mostly colonial organisms. A colony is composed of individual members that are physically connected and adapted to share resources such as food. Most hydrozoan colonies are 5 to 15 centimeters (2 to 6 inches) high, and individual polyps are usually small and inconspicuous. Some exceptions are the deep-sea *Branchiocerianthus* with solitary polyps that may reach a length of more than 2 meters (7 feet) and *Corymorpha palmata* from the intertidal mudflats in California that may reach a height of 14 centimeters (5.8 inches). Much of the marine growth seen on pilings and rocks along the seashore is composed of colonial hydrozoans (**Figure 8-10a**).

Colonial forms of hydrozoans usually contain two types of polyp. The **feeding polyp (gastrozooid)** functions in capturing food and feeding the colony. The **reproductive polyp (gonangium)** is specialized for the process of reproduction (**Figure 8-10b**). Hydrozoan medusae are usually small. Pulsations of the animal's bell-shaped body drive it upward. This action counteracts the animal's tendency to sink. Although they often turn when

> **Hydrozoans** are mostly colonial cnidarians that belong to the class Hydrozoa.
>
> **Hydroid** is another name for a hydrozoan.
>
> A **feeding polyp (gastrozooid)** is a polyp in a hydrozoan colony that captures food for the colony.
>
> A **reproductive polyp (gonangium)** is a polyp in a hydrozoan colony that asexually reproduces hydrozoan medusae.

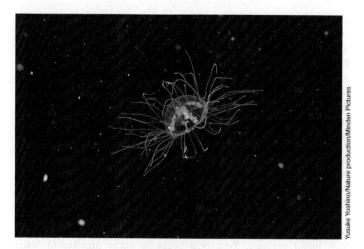

Yusuke Yoshino/Nature production/Minden Pictures

FIGURE 8-11 HYDROZOAN MEDUSA. *Gonionemus* is an unusual medusa in that it uses its long tentacles to crawl over the bottom, clinging to vegetation rather than floating in the water column.

swimming, horizontal movement mainly depends on water currents. The medusa of the hydrozoan *Gonionemus* (**Figure 8-11**) crawls over the bottom, attaching to vegetation with its tentacles. Hydrozoans known as hydrocorals (*Millepora* and *Stylaster*; **Figure 8-12**) secrete a calcareous skeleton around the polyps and resemble hard corals.

Not all hydrozoans form sessile colonies. Members of the hydrozoan orders Siphonophora and Chondrophora produce floating colonies. Perhaps the best known of these is the Portuguese man-of-war (*Physalia*; see **Figure 8-9a**). The colony is suspended from a large gas sac (up to 30 centimeters, or 12 inches) that acts as a float. It is believed that this structure develops from the original larval polyp, and specialized polyps with gas glands are responsible for keeping the gas sac inflated with a gas similar in composition to air but with a higher concentration of carbon monoxide. Like sessile colonies, these floating colonies contain several types of specialized polyps. Feeding polyps have a single long tentacle and function in digestion. Fishing polyps have tentacles strewn with stinging cells

FIGURE 8-10 HYDROZOANS. **(a)** *Tubularia* is an example of a colonial hydrozoan. **(b)** This drawing of hydrozoan anatomy shows reproductive and feeding polyps.

Zig Leszczynski/Animals Animals

(a)

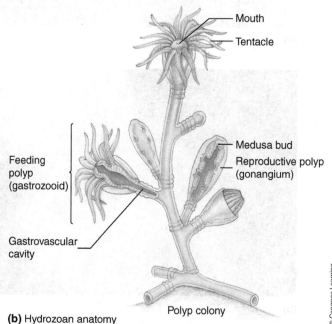

Mouth

Tentacle

Medusa bud

Reproductive polyp (gonangium)

Feeding polyp (gastrozooid)

Gastrovascular cavity

Polyp colony

(b) Hydrozoan anatomy

© Cengage Learning

FIGURE 8-12 HYDROCORALS. The fire coral *Millepora* is not a true coral but a colonial hydrozoan that produces a calcareous skeleton.

and are responsible for capturing prey, usually surface fishes such as mackerel or flying fish. Other parts of the colony are modified medusae that contain ovaries or testes and function in reproduction.

Jellyfish and Box Jellyfish

Scyphozoans, or true jellyfish (class Scyphozoa; **Figure 8-13**), generally can exist in both a polyp stage and medusa stage. The predominant stage in the life cycle of these animals is the medusa. Scyphozoan medusae are larger than hydrozoan medusae and are better swimmers. They can swim both vertically and horizontally. Although jellyfish are capable of swimming by pulsating their bodies, many of them are not strong swimmers, and most float with the currents and are thus part of the plankton. Most medusae are small (less than 10 centimeters, or 3 inches), but some, such as the medusae of the jellyfish *Cyanea*, have a bell diameter of 2 to 3 meters (7 to 10 feet) and tentacles that are 60 to 70 meters (200 to 230 feet) long. Jellyfish have sense organs called **photoreceptors** that allow them to determine if it is dark or light. Many species avoid bright sunlight and come to the surface on cloudy days and at twilight. During bright sunlight and at night they tend to move deeper in the water.

Closely related to true jellyfish are **cubozoans or** box jellyfish (class Cubozoa; **Figure 8-14**). Cubozoans are tropical animals that have box-shaped bells and are all strong swimmers. They are voracious predators,

feeding primarily on fish. Cubozoans have complex eyes and can probably fix an image so they can track their prey.

Anthozoans

Sea anemones, corals, gorgonians, and their relatives are collectively known as **anthozoans** (class Anthozoa). Anthozoans are benthic animals, and all of the adults are sessile (corals) or **sedentary** (sea anemones). Many species, such as the corals, form large, intricately shaped colonies. These animals exhibit only the polyp stage in their life cycle, and many of them bear a resemblance to brightly colored flowers; thus the name anthozoans, which means "flower animal."

Sea Anemones Sea anemones are polyps that are larger, heavier, and more complex than hydrozoan polyps (**Figure 8-15a**). Unlike hydrozoan polyps, the gastrovascular cavity of sea anemones is divided into compartments that radiate out from the central cavity. Most range from 1.5 to 10 centimeters ($^3/_4$ to 3 inches) in diameter, but the sand rose anemone (*Tealia columbiana*) on the North Pacific coast of the United States and carpet anemones (*Stichodactyla*) on the Great Barrier Reef may grow to a diameter of 1 meter (3.3 feet) or more. Sea anemones inhabit both deepwater and shallow coastal water worldwide and are most diverse in the tropics. Most species are found attached to a hard surface such as rocks, shells, or submerged wood, but some burrow in sand or mud. Sea anemones are capable of expanding, contracting, and

Scyphozoans are cnidarians known as jellyfish belonging to the class Scyphozoa.

Photoreceptors are light-sensing organs.

Cubozoans are cnidarians known as box jellyfish belonging to the class Cubozoa.

Anthozoans are benthic cnidarians belonging to the class Anthozoa.

Sedentary animals can move but spend the majority of time staying in one place.

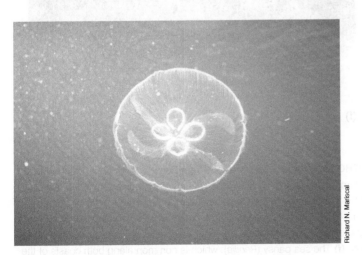

FIGURE 8-13 SCYPHOZOANS. A common jellyfish in coastal waters is the moon jelly (*Aurelia*). It is not uncommon to see large numbers of these jellyfish on the beach or close to shore after a storm.

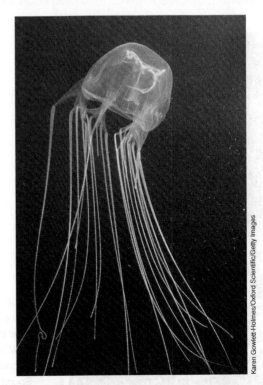

FIGURE 8-14 CUBOZOAN. The stinger (*Chironex*) from Australia is an example of a box jellyfish. These animals can produce painful stings and even cause death.

FIGURE 8-15 **ANTHOZOANS. (a)** A sea anemone from Puget Sound, Washington; note the young anemones temporarily attached around the column of the parent anemone. **(b)** Live coral polyps from the Caribbean with their tentacles extended for feeding. **(c)** Mushroom coral (*Fungia*) is an example of a large solitary coral polyp. **(d)** Sea fans are common colonial gorgonians found in tropical waters. **(e)** An orange sea pen (*Ptilosarcus*) from Puget Sound, Washington. **(f)** The sea pansy (*Renilla*), which is common along both coasts of the United States, has a body composed of a large primary polyp that gives rise to several small secondary polyps.

reaching with tentacles to capture their prey. When disturbed, sea anemones will withdraw their tentacles into their oral openings and contract their bodies. Many species of sea anemone can change location by gliding on their base, crawling on the side of their body, or walking on their tentacles. A few species can detach and swim briefly by rapid contractions of their body stalk or tentacles. This behavior is used to escape predators such as sea stars. Some species exhibit aggressive behavior toward other species of sea anemone or clones of other individuals of the same species. Cnidocytes on specialized tentacles known as **acrorhagi**, fire on contact with foreign anemones. Both animals then move away from each other. This behavior may provide for spatial separation to avoid crowding.

Coral Animals Coral animals **(Figure 8-15b)** are polyps that, unlike sea anemones, secrete a skeleton, which may be hard or soft, around their bodies. The hard or stony corals (**scleractinian corals**) produce a skeleton of calcium carbonate, and the majority of species form large colonies of small polyps. All members of the colony are connected by a horizontal sheet of tissue. Coral reefs are the products of reef-building hard corals, coralline red algae, and calcified green algae. Some hard corals such as mushroom (*Fungia*; **Figure 8-15c**) and feather corals (*Polyphyllia*) have solitary polyps as large as 25 centimeters (5 inches) in diameter that are similar in structure to sea anemones.

Soft Corals Soft corals form colonies that look more like plants than animals. Sea fans **(Figure 8-15d)** and sea pens **(Figure 8-15e)** are just two examples of this type of coral. Soft corals known as **octocorals** have 8 tentacles instead of the 12 or more that are typical of other coral species, and the tentacles are feathery. The most common octocorals are **gorgonians** (order Gorgonacea), such as whip corals, sea feathers, sea fans, and pipe organ coral. Octocorals are common members of reef fauna, especially in the Caribbean. Sea pens and sea pansies (order Pennatulacea) are inhabitants of soft bottoms. They have a large primary polyp with a stem-like base for attachment. The body is fleshy and its upper part gives rise to small secondary polyps. The sea pansy (*Renilla*; **Figure 8-15f**) is common along the Atlantic, Gulf, and southern California coasts. They live near shore, where their flattened horizontal surface reduces resistance to turbulent water.

NUTRITION AND DIGESTION

Cnidarians digest their prey in a central gastrovascular cavity, which functions in both digestion and the movement of materials within the animal. Waste products and indigestible material are forced back out through the mouth after digestion is completed.

Many hydrozoans are suspension feeders, feeding on plankton and organic material floating in the water. Gastrozoids in the hydrozoan colony capture and ingest prey and provide food for the colony. Most colonies feed on any zooplankton small enough to be handled by their gastrozoids. However, the hydrozoan *Rhizophysa eysenhardti*, from the order Siphonophora reaches a length of 1 meter (3.3 feet) and has up to 28 gastrozoids. *Rhizophysa eysenhardti* feeds diurnally, consuming approximately nine fish larvae (5 to 15 millimeters long) per day. Jellyfish (class Scyphozoa) and box jellyfish (class Cubozoa) are carnivorous, feeding mostly on fish and larger invertebrates. The prey is paralyzed by a toxin on the nematocyst, and in some cases the nematocyst has a long fiber that entangles the prey. The prey is drawn into the mouth and forced into the gastrovascular cavity, where it is digested. Some jellyfish, however, such as the upside-down jellyfish (*Cassiopeia*; **Figure 8-16**), feed on plankton that are trapped in mucus produced by modified tentacles. These animals lack a true mouth. The food is passed along fused tentacles by the action of cilia to multiple openings of the gastrovascular cavity, where it is digested. They also receive nutrients from symbiotic algae living within their tissues. Unlike typical jellyfish, the upside-down jellyfish spends little time swimming. Instead, it usually lies upside down in shallow lagoons—an ideal place for sunlight to reach the symbiotic algal cells that provide it with nutrition and oxygen. The margin of its bell pulsates regularly, drawing a current of water containing plankton across its tentacles. The contractions of the bell also help to hold the animal in place so it is not swept away by strong tidal currents.

Sea anemones generally feed on invertebrates, although some large species can capture fish, and nutrition from symbiotic algae is important to

Acrorhagi are specialized tentacles found in some anemones that are used to prevent other anemones from getting too close.

Scleractinian corals are corals with hard skeletons of calcium carbonate.

Octocorals are soft corals whose polyps have 8 tentacles.

Gorgonians are soft corals that belong to the order Gorgonacea.

Therisa Stack/Tom Stack & Associates

FIGURE 8-16 UPSIDE-DOWN JELLYFISH. The upside-down jellyfish (*Cassiopeia*) spends most of its time upside down in shallow lagoons. It feeds on plankton that it traps with its tentacles, and also receives nutrients from symbiotic algae that grow in its tentacles.

shallow-water species. Many intertidal species feed on crabs and bivalves that are washed down by waves. The prey is captured by the tentacles, paralyzed by nematocysts, and carried to the mouth. Some large species of sea anemone with short tentacles feed on fine organic particles and plankton. Sea anemones contain a groove within their body that is lined by cells with cilia on their free surface. The beating of the cilia creates a water current through the animal's gastrovascular cavity that brings in oxygen and removes waste. Corals feed like sea anemones, and, depending on the size of the polyp, the prey ranges from zooplankton to small fish. In addition to capturing zooplankton, many corals also collect fine particles in mucous films or strands. Cilia move the coral mucus and the trapped food to the animal's mouth. Stony corals derive the majority of their nutrition from symbiotic dinoflagellates known as zooxanthellae.

REPRODUCTION

Cnidarians can reproduce both asexually and sexually, and they exhibit a variety of reproductive strategies in their life cycles. Asexual reproduction generally occurs in the polyp stage and results in the formation of more polyps or tiny medusae. Sexual reproduction usually occurs in the medusa stage.

Hydrozoans

Hydrozoans generally exhibit both an asexual polyp stage and a sexual medusa stage in their life cycles **(Figure 8-17)**. Reproductive polyps in hydrozoan colonies asexually reproduce by forming tiny medusa-like buds. When mature, the buds are released into the water column, where they grow into adult medusae. Medusae can be hermaphrodites, or the sexes can be separate. Adults release their gametes into the water column, where fertilization and early development occur. The fertilized egg develops into a planktonic **planula larva** that grows in the water column. The planktonic larva allows these sessile animals to disperse their young to other areas. Eventually the planula settles on a solid surface and begins to form a new colonial hydroid by asexual reproduction.

Scyphozoans

In adult jellyfish the sexes are generally separate. The medusae are the sexual stage in the life cycle, and they release gametes into the water column, where fertilization and early development take place **(Figure 8-18)**. The fertilized eggs develop into planula larvae that spend time in the water column before settling on a solid surface to form the polyp stage of the life cycle. The polyps reproduce medusa-like buds by asexual reproduction. The immature buds are released into the water column to grow into mature medusae.

Anthozoans

Asexual reproduction is common in anthozoans. Sea anemones can reproduce asexually in several ways. **Pedal laceration** involves leaving parts of the base, also known as a **pedal disk**, behind to grow into new individuals. Some species can pinch off parts that will develop into new anemones. In **fission**, the anemone splits into two, and each half grows into a new individual. Most sea anemones are hermaphroditic but produce only one type of gamete each reproductive period. Fertilization can occur either in the

A **planula larva** is the planktonic larval stage of a cnidarian.

Pedal laceration is a type of asexual reproduction that occurs in sea anemones in which a portion of the animal's base is broken off and forms a new individual.

A **pedal disk** is the base of a sea anemone.

Fission is a type of asexual reproduction that occurs in sea anemones in which the anemone splits into two and each half develops into a new individual.

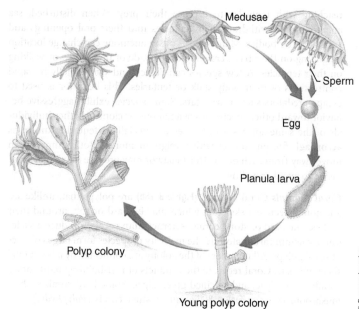

FIGURE 8-17 **LIFE CYCLE OF THE HYDROZOAN OBELIA.** The sexual stage of the life cycle is the medusa. Fertilized eggs develop into planula larvae that spend time in the water column before settling and growing into polyps. The polyp reproduces asexually to form a new colony.

gastrovascular cavity or in the water column. The larval stage is a planula larva that is a member of the plankton before settling on a solid surface to develop into a new adult.

Large colonies of hard corals are composed of identical individuals that have been produced asexually by budding. Coral species usually have separate male and female forms, although some species are hermaphrodites. Gametes are shed into the water column, where fertilization and early development occur. The larval stage is the planula larva. After feeding and growing in the water column, the planula settles on a hard surface and begins to form a new coral colony.

ECOLOGICAL ROLES OF CNIDARIANS

Cnidarians play several key ecological roles in the marine environment. They are predators that feed on a variety of prey, both large and small. Sessile cnidarians such as corals provide habitats for numerous marine organisms. Many cnidarians host symbionts that aid in their nutrition and help them grow. Some cnidarians are symbionts of other marine organisms, trading protection for mobility.

Predator–Prey Relationships

Cnidarians are predators, and because they are well protected by their stinging cells, few predators prey on them. Sea turtles, some fish, and molluscs will feed on hydrozoans and jellyfish. The crown-of-thorns sea star (*Acanthaster*) is one of the chief coral predators on Pacific reefs.

Habitat Formation

Coral polyps are some of the most ecologically important animals in the sea. They form complex three-dimensional structures that provide habitat for thousands of other organisms. Coral reefs provide a solid surface for sessile marine animals and places for pelagic animals to rest and hide. They act as a buffer to protect coastal organisms from the effects of wave action and storms.

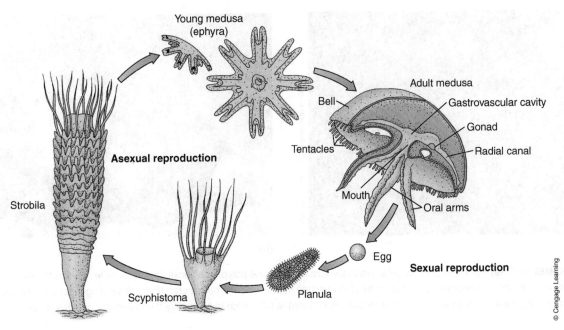

FIGURE 8-18 **A SCYPHOZOAN LIFE CYCLE.** The life cycle of the moon jellyfish (*Aurelia*) is thought to be representative of many scyphozoan life cycles. The medusa is the sexual stage in the life cycle. Fertilized eggs develop into planula larvae that eventually settle out of the water column and form a polyp known as a scyphistoma. The scyphistoma grows and becomes a strobila asexually reproducing more medusae. The immature medusae known as ephyras are then released back into the water column.

Symbiotic Relationships

There are many interesting examples of symbiotic relationships involving cnidarians. One is the Portuguese man-of-war (*Physalia*) and the man-of-war fish (*Nomeus*). The man-of-war fish swims among the tentacles of the Portuguese man-of-war where, unharmed by the hydrozoan, it is safe from predators. The fish may serve as bait to lure other fishes into the tentacles of the man-of-war.

Reef-forming corals harbor zooxanthellae in their tissues **(Figure 8-19a)**. These symbionts not only provide food to the corals but also are an important food source for other reef animals. Parrotfishes (family Scaridae), for instance, consume large numbers of polyps to feed on their symbiotic zooxanthellae.

Symbionts of sea anemones include clownfishes, cleaning shrimp, snapping shrimp, arrow crabs, other small crustaceans, and brittle stars. Certain clownfishes (family Pomacentridae) form symbiotic relationships with some of the large species of sea anemone that occur in the tropical Pacific Ocean. These sea anemones secrete fish attractants that differ for each species of anemone and attract only a particular species of clownfish. After a brief period of acclimation, the fish can swim unharmed among the anemone's tentacles, feeding on scraps from the prey captured by the anemone, which the fish itself may have lured.

One of the more unusual symbiotic relationships involving anemones is between the hermit crab, *Eupagurus prideauxi*, and the anemone, *Adamsia palliata* **(Figure 8-19b)**. A young sea anemone will attach itself to the mollusc shell occupied by the young hermit crab. The sea anemone is then carried around by the crab. The anemone's mouth is located just above and behind the crab's, and it feeds on the scraps the crab leaves. Eventually the anemone overgrows the mollusc shell and absorbs it, replacing it with its own body. At this point, the sea anemone and crab both grow at the same rate so that the crab remains covered. Not only

does the crab benefit by being camouflaged and protected but it also does not have to find new shells as it grows, as do other species of hermit crab. The sea anemone benefits by being carried from place to place and sharing the crab's food. If the sea anemone is removed from the crab, it will soon die, and the undisguised crab is easy prey. The hydrozoan *Hydractinia echinata* also lives on shells inhabited by hermit crabs. The crabs benefit from the camouflage and the protection of the hydrozoan's venomous stings. The hydrozoan benefits by being carried from one feeding ground to another, and it does not have to compete for space with other hydrozoan colonies.

IN SUMMARY

- Cnidarians exhibit radial symmetry.
- Cnidarians can be distinguished from other radially symmetrical animals by their highly specialized stinging cell, the cnidocyte, which they use for capturing prey and for defense.
- Cnidarians exhibit two body plans in their life cycles: the polyp and the medusa. (Have You Wondered? #3)
- Although painful, jellyfish stings generally are nonfatal, although the stings of some box jellyfish can be lethal. (Have You Wondered? #4)
- Cnidarians are carnivorous and feed on a variety of prey.
- Corals form complex habitats that are home to thousands of other species.
- Many cnidarians are involved in symbiotic relationships.

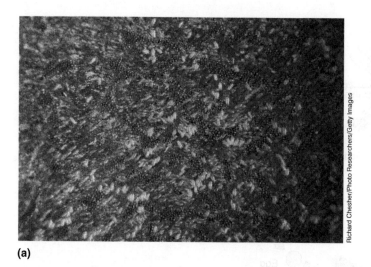

(a)

(b)

FIGURE 8-19 SYMBIOSIS. (a) Symbiotic zooxanthellae within the tissues of a coral polyp. The zooxanthellae provide nutrition for the coral polyp, as well as other reef animals. **(b)** Sea anemones (*Calliactus tricolor*) attached to the shell of a hermit crab (*Petrochirus diogenes*) from the Gulf of Mexico are another example of symbiosis. Note also the porcelain crab (*Porcellana*) at left, which is commonly found with this species of hermit crab.

IMPACT BUBBLE

SOME JELLYFISH POPULATIONS have exploded in recent years, leading to problems on swimming beaches and interfering with commercial operations such as fish farms and power plants. In some instances, jellyfish have replaced species of plankton-feeding (planktivorous) fishes in marine food chains. The population increases are for the most part the result of climate change, pollution, and overfishing.

The rise in global temperature due to human-induced climate change is partly to blame for the increase in the size of jellyfish populations. As the water warms, young jellyfish emerge earlier, grow quicker, and remain in the water column longer. The increased levels of carbon dioxide in the atmosphere contribute to ocean warming, and as excess carbon dioxide dissolves in seawater, the water becomes more acidic. The increased acidity makes it harder for some invertebrates to form shells or hard external coverings. Some of these animals compete with jellyfish for plankton, so as their numbers decrease more food becomes available to support larger jellyfish populations.

Excess amounts of nutrients entering coastal waters also contribute to the increased numbers of jellyfish. These nutrients promote algal blooms, which in turn rob the water of oxygen and decrease water clarity. Water conditions such as these have an adverse effect on planktivorous fish populations. Since jellyfish are more tolerant of water with low levels of oxygen and do not need to see their prey, they tend to thrive in these polluted areas and take advantage of the decreased competition for food.

Overfishing of planktivorous fish and bivalves that feed on plankton also leaves more plankton for jellyfish to consume. Overfished commercial species of planktivorous fish such as herring and sardines have been replaced in several locations by jellyfish species. This was a bit puzzling for biologists, because fish can move quickly and can generally see their prey whereas jellyfish move slowly and have to make contact with their prey to know they are present. Based on these observations, fish should be more efficient at prey capture than jellyfish. However, the relatively large body size of jellyfish and the action of pulsing their bodies to draw plankton-laden water past their tentacles increases their chances of capturing prey and competing successfully with planktivorous fish. Although jellyfish do not swim as quickly as fish, they are very efficient in the energy consumed for swimming, and they can feed regardless of light conditions. These factors allow jellyfish to compete closely with fish, and when fish numbers decline, jellyfish can become the dominant species. The increase in the numbers of jellyfish could change the nature of marine ecosystems in these areas.

In areas where fish populations are declining due to overfishing or pollution, jellyfish are taking their place. These areas include coastal waters off Japan, the Northeastern United States, the Black Sea, and the Mediterranean. In some of these areas the large numbers of jellyfish in the water and washing ashore cause problems for human beachgoers. In some regions, the jellyfish are causing problems for commercial fish farms. In November 2007, a mammoth swarm of baby mauve stinger jellyfish (*Pelagia noctiluca*) 26 square kilometers (20 square miles) in area and 10 meters (33 feet) deep in the Irish Sea washed into a salmon farm, killing all 100,000 salmon. In Japan, Scotland, and Israel, nuclear power plants drawing water from the sea to cool their reactors have had to periodically shut down their pumps due to large numbers of jellyfish clogging the water intake filters. Since human activities are to blame for the jellyfish increases, it will take a concerted effort by the human population to make the necessary changes to avert further damage to ecosystems.

8-4 Ctenophores

Although jellyfish are the most familiar of the gelatinous animals that float in the ocean's water, they are not the only ones. Comb jellies are radial animals that closely resemble jellyfish. Their transparent bodies make them difficult to see in the water, and unlike jellyfish, comb jellies lack stinging cells.

CTENOPHORE STRUCTURE

Ctenophores (TEEN-uh-forz), or **comb jellies** (phylum Ctenophora), are planktonic, nearly transparent marine animals named for the eight rows of **comb plates** (**ctenes**) they use for locomotion **(Figure 8-20a)**. The comb plates are made of very large cilia, and when the cilia beat, the animal is able to move. Ctenophores are weak swimmers and are mostly found in surface waters. They are not powerful enough swimmers to make much forward progress, but they can move up and down in the water column by beating their cilia. At the apex of each animal there is a small, transparent, bubble-like structure called a **statocyst** within which are four hairs, each with a granule of calcium carbonate balanced on its tip. If the ctenophore tilts in one direction or another, the granule on that side will press harder on the hair. This causes the comb row on that side to beat harder, thus righting the animal. Like cnidarians, ctenophores exhibit radial symmetry, but they lack the stinging cells that are the hallmark of cnidarians. The delicate bodies of ctenophores are iridescent during the day. At night, almost all ctenophores give off flashes of luminescence, possibly to attract mates or prey or frighten potential predators. Along with other bioluminescent plankton, they are responsible for the luminescence of many seas.

DIGESTION AND NUTRITION

Ctenophores are carnivorous, feeding on zooplankton, larval fish, and fish eggs. Some ctenophores, such as *Pleurobrachia* **(Figure 8-20b)**, capture prey with branched tentacles that form a large net when fully expanded. Prey, especially small crustaceans, are caught on specialized adhesive cells called **colloblasts** that are found on the tentacles and then pulled into the mouth, where they are digested in the gastrovascular cavity. Other species such as *Mnemiopsis* and *Leucothea* use both tentacles and their mucous-covered oral surfaces to capture prey, especially small crustaceans. The ctenophore *Beroe* is cylindrical, has no tentacles, and feeds on other ctenophores. When its mouth comes in contact with another ctenophore, the prey is sucked up whole. The ctenophore *Euchlora rubra* feeds on jellyfish and moves the nematocysts to its tentacles, using them instead of colloblasts to capture prey and for defense.

REPRODUCTION

Almost all species of ctenophore are hermaphroditic. Most species appear to shed their eggs and sperm directly into the water column, where fertilization takes place. A few species, though, brood the eggs in their bodies. A fertilized egg develops into a free-swimming larva called a **cydippid larva** that resembles the adult ctenophore.

ECOLOGICAL ROLES OF CTENOPHORES

Ctenophores are important predators of zooplankton and play a significant role in managing the size of zooplankton populations. They also prey on fish eggs and larvae and thus play a role in the regulation of fish populations. Ctenophores channel nutrients to larger plankton feeders as they are consumed for food.

Ctenophores (comb jellies) are gelatinous zooplankton belonging to the phylum Ctenophora.

Comb plates (ctenes) are rows of cilia used by ctenophores for locomotion.

A **statocyst** is an organ found in some animals that helps them maintain equilibrium.

Colloblasts are specialized adhesive cells found on the tentacles of some ctenophores and used to capture prey.

A **cydippid larva** is the planktonic larva of a ctenophore.

IN SUMMARY

- Ctenophores are gelatinous, planktonic animals that exhibit radial symmetry. (Have You Wondered? #5)
- Ctenophores lack the stinging cell of cnidarians and move by rows of cilia called comb plates.
- Ctenophores are predators feeding on planktonic organisms, as well as fish larvae and fish eggs.

(a)

Richard Mariscal

(b)

Bill Curtsinger/Getty Images

FIGURE 8-20 CTENOPHORES. (a) Ctenophores, also known as comb jellies, like this species of *Beroe* propel themselves through the water with specialized structures called comb plates. The comb plates appear as lines on this animal's body surface. **(b)** Ctenophores such as *Pleurobrachia* use their tentacles to capture food.

ECOLOGY & THE MARINE ENVIRONMENT

Attack of the Killer Ctenophores

In 1999, the Caspian Sea was invaded by one of the most feared invasive species, the comb jelly *Mnemiopsis leidyi* (**Figure 8-A**). Within a short period of time, 75% of the zooplankton in the southern portion of the Caspian Sea were wiped out, disrupting food chains and especially affecting fishes such as kilka (*Clupeonella*) (**Figure 8-B**) and sturgeon (family Acipenseridae). As a result of the invasion, some species of zooplankton have vanished from Caspian waters due to *Mnemiopsis'* ability to consume up to 15 times its body weight per day and reproduce quickly. Within 2 weeks after hatching, the hermaphroditic species reaches sexual maturity and can produce thousands of eggs each day. Although *Mnemiopsis* mainly preys on zooplankton, it will also eat fish eggs and larvae.

Researchers assume that *Mnemiopsis* may have entered the Caspian Sea in ballast water that was taken on in the Black Sea or the Sea of Azov by ships that later entered the Caspian Sea. Once introduced, *Mnemiopsis* quickly spread, with populations reaching densities of more than 2,000 individuals per square meter. Their numbers remain high for 6 months of the year, peaking in August. As a result of zooplankton predation by *Mnemiopsis*, the zooplankton biomass in the Caspian Sea has been reduced to about one-tenth of its normal amount. This has caused the Iranian kilka fishery to plummet from 85,000 tons in 1999 to 15,000 tons in 2004, causing an economic loss in excess of $125 million. Other countries that border the Caspian Sea, such as Russia and Azerbaijan, have reported similar declines. There is also concern that the Caspian seal (*Pusa caspica*), which feeds on kilka, may also become a casualty of the invasion. The seal population is already under pressure from pollution, hunting, and an outbreak of canine distemper virus, which has reduced the seal population by 83% over the last 50 years. Adding to the environmental nightmare are phytoplankton blooms that result from reduced grazing pressure due to the decline in zooplankton.

The situation is especially dire in the Caspian Sea because degradation of the environment and overfishing already have placed excessive pressure on the kilka and sturgeon populations.

Mnemiopsis is native to the waters off the East Coast of the United States. It was introduced to the Black Sea in 1989 in ship ballast water. The bloom that resulted that summer was overwhelming. As many as 800 million tons of *Mnemiopsis* filled the Black Sea, which represented a biomass 800 times that year's total fish catch. Local fisheries were decimated by the event. The problem continued until 1997, when another ctenophore, *Beroe ovata*, arrived in the Black Sea, probably from ballast water again. *Beroe* appears to prey exclusively on *Mnemiopsis*, and by 2001 the numbers of *Mnemiopsis* in the Black Sea had declined so greatly that researchers had trouble finding specimens for analysis.

In 2001, researchers organized a program to determine if *Beroe* could be introduced into the Caspian Sea. The water in the Caspian Sea is not as saline as that in the Black Sea, and in early experiments the *Beroe* survived but did not breed well. Since it is difficult to recreate the conditions of the Caspian Sea in the laboratory, the only way to find out if *Beroe* will survive well enough and reproduce in the Caspian Sea is to introduce it. Some environmentalists are concerned that once introduced, the *Beroe* will prey on zooplankton other than *Mnemiopsis*. Laboratory experiments indicate that *Beroe* does not feed on other zooplankton at all. *Beroe* was also tested for its ability to introduce other pests into the Caspian Sea. Parasites associated with *Beroe* die off, probably because of the less saline water, and the Black Sea and Caspian Sea have similar bacteria fauna, so it is unlikely that *Beroe* would introduce new bacteria. The strategy of introducing *Beroe* to control *Mnemiopsis* could be used in the Caspian Sea, but the Caspian is bordered by five nations and the politics of the region have interfered with the deployment of this plan.

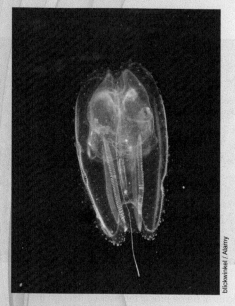

FIGURE 8-A KILLER CTENOPHORE. The ctenophore *Mnemiopsis leidyi* has consumed almost 90% of the zooplankton in the Caspian Sea.

FIGURE 8-B KILKA. The kilka (*Clupeonella engrauliformis*), an important commercial fish from the Caspian Sea, feeds on zooplankton. Their populations are in decline due to the introduction of *Mnemiopsis leidyi*, a ctenophore that has decimated zooplankton in the Caspian Sea.

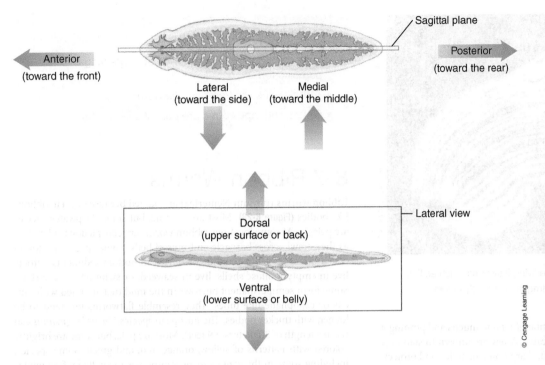

© Cengage Learning

FIGURE 8-21 BILATERAL SYMMETRY. Most animals, like this flatworm, exhibit bilateral symmetry. That is, only one plane (the midsagittal plane) passing through the midline of the long axis will divide the animal into two equivalent halves. Animals that exhibit bilateral symmetry have their sense organs concentrated in one region called the head.

Bilateral symmetry is a type of body organization in which the body parts are arranged in such a way that only one plane through the midline of the central axis divides the animal into similar halves.

A **midsagittal plane** is a plane through the midline of the central axis of an animal.

Cephalization is the evolutionary process whereby sense organs became concentrated in the head of an animal.

Turbellarians are nonparasitic flatworms.

Meiofauna are tiny invertebrates that live in the spaces between sediment particles.

Flukes are parasitic flatworms with complex life cycles.

Tapeworms are parasitic flatworms that live in the intestines of animals.

Chemoreceptors are sense organs that can detect chemicals in the environment.

8-5 The Evolution of Bilateral Symmetry

The majority of marine animals exhibit bilateral symmetry **(Figure 8-21)**. In **bilateral symmetry**, the body parts are arranged in such a way that only one plane through the midline of the central axis (**midsagittal plane**) will divide the animal into similar right and left halves. With the evolution of muscle tissue, animals became more mobile. Bilateral symmetry allowed for a more streamlined body shape and was more advantageous for animals with active lifestyles than radial symmetry. The streamlined body created less drag when moving through the watery environment, and thus less energy was required for movement.

The active lifestyle of animals with bilateral symmetry favored the concentration of sense organs at one end of the animal, the end that would first meet the environment. This evolutionary trend toward concentration of sense organs in the head region of an animal is known as **cephalization** (sef-uh-luh-ZAY-shun).

8-6 Flatworms

Flatworms (phylum Platyhelminthes) have flattened bodies and exhibit bilateral symmetry with a definite head and posterior end. Some flatworms such as the turbellarian flatworms (class Turbellaria) are free-living organisms, whereas others such as flukes (class Trematoda) and tapeworms (class Cestoda) are parasites.

TYPES OF FLATWORM

Turbellarians are free-living (that is, not parasitic) flatworms. They range in length from a few millimeters to 50 centimeters (from less than ⅛ inch to 20 inches) long but are only a few centimeters wide and 1 millimeter thick **(Figure 8-22)**. A few species are pelagic, but most are bottom dwellers that live in sand or mud, under stones and shells, or on seaweed. Turbellarians are also common members of the **meiofauna** (MY-oh-fawn-uh), the tiny invertebrates adapted to living in the spaces between sediment particles. A turbellarian's body is covered with a layer of cells called the epidermis. The ventral surface of the epidermis is frequently ciliated and contains gland cells that produce mucus. Very small turbellarians can swim or crawl by using their cilia. Larger turbellarians secrete slime trails of mucus over which they glide, using their cilia to propel them. This form of locomotion is not efficient for larger forms, which also use muscle contractions to propel themselves. Turbellarians have sensory receptors in their head region that can detect light, chemicals, and movement and that function in maintaining balance.

Flukes and **tapeworms** are parasitic flatworms. Flukes usually have complex life cycles, involving stages of both sexual and asexual reproduction and frequently one or more species as an intermediate host. Tapeworms are parasites that live in the digestive tract of their host; lacking a digestive tract of their own, they gain nutrients from their host. The largest tapeworms are those that infest whales. Individuals of the species *Polygonoporous giganteus* found in the intestines of sperm whales have measured as long as 30 meters (99 feet).

NUTRITION AND DIGESTION

Most turbellarians are carnivorous, feeding on small invertebrates that they locate with chemical-detecting organs called **chemoreceptors**. Small crustaceans, snails, and annelid worms are common prey. They can also feed on detritus or the bodies of dead animals that sink to the bottom. The oyster leech *Stylochus frontalis* feeds on live oysters, and *Stylochustriparitus* feeds on barnacles. The commensal flatworm *Bdelloura* lives on the book gill of horseshoe crabs and shares the food of its host. Some species feed on algae, especially diatoms. The turbellarian *Convoluta roscoffensis* relies on zooxanthellae for nutrition and does not ingest food as long as it has the symbiotic algae.

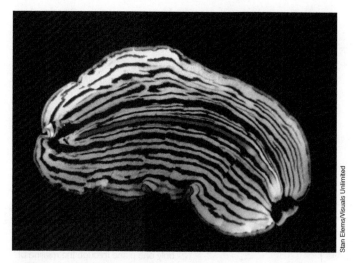

FIGURE 8-22 **TURBELLARIAN.** Free-living flatworms, such as this species from the Pacific Ocean, are frequently brightly colored.

Some species subdue their prey by entangling it in mucus and pinning it against a solid surface until it suffocates. A few are known to stab prey with a penis that terminates in a hard, sharp point, or stylet, and projects from the mouth. Once the prey is captured, the animal extends a muscular tube called the **pharynx** out of its mouth, pumps enzymes onto its prey, and sucks out body fluids or sucks off pieces of food. Turbellarians without a pharynx swallow their prey whole and digest it in the gastrovascular cavity. Some species have gastrovascular cavities with lateral branches. This increases the surface area for digestion and absorption and compensates for the lack of an internal transport system.

REPRODUCTION

Flatworms can reproduce asexually and can regenerate missing body parts. Sexually, flatworms are hermaphrodites. When two animals copulate, they usually fertilize each other, a process known as reciprocal copulation. Some species exhibit a behavior known as penis fencing in which two animals attempt to pierce the skin of each other with their dagger-like penis. The one successful in penetrating the other's body wall is able to deposit sperm and inseminate the other, a process known as hypodermic insemination. Fertilization is internal and generally development is direct; that is, there is no larval stage. Some species have a free-swimming planktonic larva. Because of their small size, turbellarians produce few eggs. By comparison, parasitic flatworm egg production is 10,000 to 100,000 times greater than that of the free-living turbellarian worms.

A **pharynx** is a muscular tube that forms part of an animal's digestive system.

Ribbon worms are animals belonging to the phylum Nemertea.

A **proboscis** is a tube-like structure used by ribbon worms to capture prey.

ECOLOGICAL ROLES OF FLATWORMS

Turbellarians that live in the spaces between sediments feed on food items that are too small for larger organisms, thus funneling nutrients to higher trophic levels when they become prey. Some turbellarians are important predators and are in turn prey for higher-level consumers. Other turbellarians are important mutualistic and commensal symbionts. Parasitic flatworms play a significant role in regulating population size by lowering the fitness of their hosts and sometimes causing death.

IN SUMMARY

- Flatworms are represented by both free-living and parasitic forms.
- Turbellarians are free-living flatworms.
- Flukes and tapeworms are parasitic flatworms.

8-7 Ribbon Worms

Ribbon worms (phylum Nemertea) are named because of their ribbon-like bodies **(Figure 8-23)**. Most are benthic, but some deepwater species are pelagic. At some locations ribbon worms are conspicuous at low tide as they crawl across barnacle and mussel beds. Some species are found coiled under stones at low tide or buried in bottom sediments. Others live in empty mollusc shells, live in seaweed, or swim near the surface. Some form semipermanent burrows in the mud that are lined with mucus or cellophane-like tubes. They resemble flatworms but tend to be longer, with thicker bodies. The European species *Lineus longissimus* can reach a length of 30 meters (99 feet). Most are pale, but some are brightly colored with patterns of yellow, orange, red, and green. Some species, including some in the genus *Lineus*, reproduce asexually by fragmentation. The sexes are separate in most species, and fertilization is external.

Ribbon worms are carnivorous and feed primarily on annelids and crustaceans. A tube-like structure, the **proboscis**, coils around the prey as sticky toxic secretions from the anterior end help to hold and immobilize it. Species that have a sharp stylet at the tip of the proboscis are said to be armed. After the prey is captured, it is either swallowed whole or its tissues are sucked in. The worm *Paranemertes peregrina* of the Pacific coast is an armed intertidal species that feeds on worms known as polychaetes. It leaves its burrow and follows the mucous trail of its prey. It must touch the polychaetes for the feeding response to be initiated. Once contact occurs, the proboscis extends and wraps around the prey, stabbing it repeatedly with the stylet. The prey is quickly paralyzed by the injected toxin and then swallowed as the retracting proboscis pulls the prey toward the mouth. After feeding, the worm locates its burrow by following its own mucous trail. Other armed ribbon worms feed on small crustaceans. They kill their prey by piercing the ventral body wall and then forcing their head through the opening. A portion of the digestive tract, the esophagus, is everted through the mouth, and the contents of the prey are

FIGURE 8-23 **RIBBON WORM.** A ribbon worm from the Pacific coast of Panama crawls over the surface of a gorgonian.

sucked out and digested. The large, unarmed ribbon worm *Cerebratulus lacteus* of the east coast of the United States enters the burrows of razor clams (*Ensis directus*) from beneath and swallows the anterior end of the clam. Some *Carcinonemerter* species are economically important because they feed on eggs brooded by commercially valuable female crabs.

ECOLOGICAL ROLES OF RIBBON WORMS

Ribbon worms are primarily predators of annelids and crustaceans and are prey organisms for higher consumers. The burrowing of benthic species helps move nutrients from deeper sediments to the surface and aerate bottom sediments. The burrows of ribbon worms can also serve as habitat for other organisms after they have been abandoned.

IN SUMMARY

- Ribbon worms resemble flatworms and are predators preying mainly on annelids and crustaceans.

- Ribbon worms use an eversible proboscis to capture and kill their prey.

8-8 Lophophorates

Lophophorates (lohf-uh-FOHR-ayts) are sessile animals that lack a distinct head and possess a feeding device called a **lophophore** (LOHF-uh-fohr). The lophophore is an arrangement of ciliated tentacles that surround the mouth and functions in feeding and gas exchange, and it can be withdrawn when the animal is disturbed. There are three phyla of lophophorate animals: phoronids, bryozoans, and brachiopods.

PHORONIDS

Phoronids (phylum Phoronida) are small worm-like animals **(Figure 8-24a)** that range from a few millimeters to 30 centimeters (12 inches) long. They secrete a tube of leathery protein or chitin that can be attached to rocks, shells, or pilings or buried in bottom sediments. Phoronids feed on plankton and detritus that they catch in the mucus on their tentacles. In addition to reproducing sexually, phoronids can reproduce asexually either by budding or by transverse fission. Phoronids have a planktonic larval stage.

BRYOZOANS

Bryozoans (phylum Ectoprocta) are extremely abundant small animals (averaging 0.5 millimeters, or 0.02 inches, high) **(Figure 8-24b)**. The majority live in shallow water, forming colonies on a wide variety of solid surfaces such as rock, shell, algae, mangroves, and ship bottoms. A few species live and move about in the interstitial spaces of marine sand, some bore into coral and other calcareous material, and a few live on soft bottoms. Along with hydroids, bryozoans rank among the most abundant marine epiphytic animals. Large brown algae are colonized by many species of bryozoan, which display distinct preferences for certain types of algae. Evidence indicates that at the time of settling, the larvae of epiphytic species are attracted to the appropriate algal surface, perhaps by some substance produced by the algae.

Bryozoan colonies may appear as white encrustations or as fuzzy growths, or they may resemble hydrozoan colonies. Each colony is composed of thousands of tiny individuals called **zooids**, each of which inhabits a boxlike chamber that it secretes. When feeding, a zooid extends its lophophore; at other times it is withdrawn into the chamber. Most bryozoans are hermaphrodites with only one sex organ functioning at a time. The vast majority of bryozoans brood their eggs, though a few species shed small eggs directly into the seawater. After a planktonic larval stage, the larvae settle to form new colonies.

BRACHIOPODS

Some species of **brachiopod** (phylum Brachiopoda), or lamp shells, have changed very little since they first evolved more than 400 million years ago **(Figure 8-24c)**. Most species are benthic and live in shallow water. They range in size from 5 millimeters to 8 centimeters (0.02 to 3 inches). They have shells with two valves that resemble clamshells. The

(a) (b) (c)

FIGURE 8-24 REPRESENTATIVE LOPHOPHORATES. (a) Phoronids are small, worm-like lophophorates. **(b)** A colony of *Bugula* (bryozoan) from Newport Harbor, California. **(c)** The brachiopod *Terebratulina* has a body protected by a shell composed of two asymmetrical valves and fastened to the bottom by a fleshy stalk (pedicle).

ECOLOGY & THE MARINE ENVIRONMENT

Induced Defenses

The bryozoan *Membranipora membranacea* forms encrusting colonies on the surface of kelp. Marine biologists have long known that some colonies of this bryozoan have large protective spines and others do not. They also observed that the colonies with spines were frequently being attacked by *Doridella steinberghae*, a nudibranch that specializes in feeding on *Membranipora* (**Figure 8-C**). These observations prompted Dr. C. Drew Harvell to hypothesize that the protective spines in *Membranipora membranacea* were formed only when the colony was threatened by predators such as *Doridella*. She reasoned that an induced defense such as the production of spines might be an effective strategy for surviving attacks by a predator such as *Doridella*, which feeds slowly and intermittently.

She found that colonies exposed to the predatory nudibranchs produced large spines within 2 days, whereas those in the control tank did not. It did not matter which predator was introduced: The bryozoan colonies responded to each in the same way, by forming protective spines.

Dr. Harvell observed that only colonies that suffered direct mechanical damage by nudibranchs produced spines. General mechanical damage alone, however, was not sufficient to induce spine production, because control colonies mechanically damaged during their removal from the kelp did not form spines. This observation suggested that chemical cues from the predator may also be involved in the response. To test this hypothesis, predatory nudibranchs were homogenized, and the homogenate was added to the water in the aquarium with the experimental colonies. The control colonies were maintained under identical conditions, but were given a homogenate made from a marine snail that is also common on kelp. The colonies in the experimental group formed protective spines within 2 days again, whereas none of the colonies in the control group formed spines. The results of this experiment suggest that a chemical substance from the nudibranchs is responsible for the spine-forming response.

In all of the experimental colonies, spines were induced on existing and developing zooids around the entire perimeter of the colony under attack, thus fortifying the edges. Induced spine formation always followed a consistent pattern, with regular bands of spined zooids forming completely around the colony. This pattern of response seems to indicate that information about a predator attack could be sent to a site removed from the actual attack, triggering a programmed response of spine development throughout the colony. The spines effectively control the pattern and rate of nudibranch feeding. In the field, nudibranch damage was restricted to the central unspined portions of the colony. The defended zooids along the margin of the colony can usually regenerate new individuals to replace the damaged zooids in the central area, thus maintaining the colony.

In each experiment, colonies stopped growing spines within a day after the nudibranchs were removed. Spines that were already formed were permanent, and in nature these indicate colonies that have had prior attacks by predatory nudibranchs.

There is an ecological cost associated with such protection. The spine response, which increases with the level of predator attack, is associated with reduced growth of colonies. Spined colonies also begin reproduction earlier, remain reproductive for shorter periods of time, and die sooner than unspined colonies. Taken together these results confirm that spined colonies in the absence of predators have reduced fitness, and they support the hypothesis that the benefits of inducible defenses in the presence of predators are partly balanced by fitness costs. This would explain why not all bryozoan colonies produce protective spines before being attacked by predators.

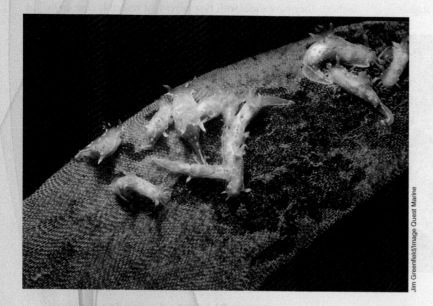

FIGURE 8-C INDUCED DEFENSE. Colonies of the bryozoan *Membranipora membranacea* being attacked by predatory nudibranchs. *Membranipora* will produce spines to protect the remainder of the colony within 2 days of an attack.

Jim Greenfield/Image Quest Marine

IN PERSPECTIVE

Phylum	Representative Organisms (Common Names)	Form and Function	Reproduction	Type of Feeding	Ecological Roles
Porifera	Sponges	Asymmetric or radial animals that lack tissues and organs	Asexual by budding or fragmentation. Sexually hermaphrodites. Larval stage is the amphiblastula.	Filter feeders and suspension feeders	Provide habitat for other organisms, symbionts, recycle calcium
Cnidaria	Hydrozoans, jellyfish, corals, sea anemones, gorgonians	Radially symmetric animals with specialized stinging cells called cnidocytes	Asexual by budding or fission. Sexual reproduction, sexes usually separate. Larval stage is a planula larva.	Mostly carnivorous	Important predators, prey for larger animals, symbionts, corals form habitats for other organisms
Ctenophora	Comb jellies	Radially symmetric animals but without stinging cells; move by means of comb plates	Sexually reproducing, most hermaphrodites. Larval stage is a cydippid larva.	Mostly carnivorous	Important predators of zooplankton, channel nutrients to larger organisms
Platyhelminthes	Turbellarians, flukes, tapeworms	Bilaterally symmetric flattened body	Most reproduce sexually and are hermaphrodites.	Turbellarians are carnivorous and scavengers; flukes and tapeworms are parasites	Predators of smaller organisms, channel nutrients to larger organisms, symbionts
Nemertea	Ribbon worms	Bilaterally symmetric; adults have an eversible proboscis for feeding	Usually sexual and sexes are separate.	Mainly carnivorous	Predators of annelids and crustaceans; burrowing helps recycle nutrients and burrows provide habitat for other organisms
Phoronida	Phoronids	Bilaterally symmetric, worm-like animals that secrete a leathery tube around their body	Reproduce asexually by budding or transverse fission. Can reproduce sexually; about half are hermaphrodites; in the other half, sexes are separate. They have a planktonic larval stage known as an actinotropha larva.	Use a lophophore to feed on plankton and detritus	Channel food to higher organisms
Bryozoa	Bryozoans (moss animals)	Form colonies of small individuals called zooids that appear as white encrustations on solid surfaces	Most are hermaphrodites and their cyphonautes larval stage is planktonic.	Use a lophophore to filter food from the water column	Channel nutrients from plankton to higher-order consumers
Brachiopoda	Lamp shells	Body covered by a shell composed of two asymmetrical valves; many species have a stalk or pedicle that anchors the animal to a solid surface	Sexes are separate and the planktonic larval stage resembles a tiny brachiopod.	Use a lophophore to feed on detritus and algae	Channel nutrients from plankton to higher-order consumers

Photos: **Porifera:** American Museum of Natural History; **Cnidaria:** Richard Mariscal; **Ctenophora:** Richard Mariscal; **Platyhelminthes:** Stan Elms/Visuals Unlimited; **Nemertea:** Kjell Sandved/Visuals Unlimited; **Phoronida:** Roger Klocek/Visuals Unlimited; **Bryozoa:** Kjell Sandved/Visuals Unlimited; **Brachiopoda:** Richard Herrmann/Visuals Unlimited/Getty Images.

A **pedicle** is a fleshy stalk that attaches a brachiopod to the substrate.

two valves of the brachiopod shell differ in size and shape and are dorsal and ventral. Clams have right and left valves that are generally similar in size and shape. Another difference between clams and brachiopods is the fleshy stalk called the **pedicle** found in many brachiopods. The pedicle attaches the brachiopod shell to a hard surface or is buried in soft sediments. The animal's body occupies only a portion of the shell, and an extension of the body wall, the mantle, lines the shell and is responsible for secreting it. Lying in the mantle cavity is a large, horseshoe-shaped lophophore that is used for feeding and gas exchange. Brachiopods feed on detritus and algae that are swept into a groove leading to the mouth by the ciliated lophophore. Sexes in brachiopods are generally separate, and eggs and sperm are generally shed into the seawater. A planktonic larval stage lasts for 24 to 30 hours, after which the larvae settle to develop into adults.

ECOLOGICAL ROLES OF LOPHOPHORATES

Lophophorates as a group are filter feeders. In turn they supply food for a number of large invertebrates, especially molluscs and crustaceans. From an economic standpoint marine bryozoans are one of the groups of organisms most responsible for fouling ship's bottoms. About 120 species, of which different species of *Bugula* are among the most abundant, have been removed from ship bottoms.

IN SUMMARY

- Lophophorates are sessile animals that possess a feeding device called a lophophore.

- Lophophorates include phoronids, bryozoans, and brachiopods.

KEY CONCEPTS

1. Sponges are asymmetric sessile animals that filter food from the water circulating through their bodies. (8-2)

2. Sponges provide habitats for other animals. (8-2)

3. Cnidarians and ctenophores exhibit radial symmetry. (8-3)

4. Cnidarians possess a highly specialized stinging cell used to capture prey and for protection. (8-3)

5. Ctenophores resemble jellyfish but lack the stinging cells of cnidarians. (8-4)

6. Most marine animals exhibit bilateral symmetry. (8-5)

7. Turbellarians are free-living flatworms; flukes and tapeworms are parasitic flatworms. (8-6)

8. Ribbon worms are marine predators that somewhat resemble flatworms. (8-7)

9. Phoronids, bryozoans, and brachiopods have a specialized feeding structure called a lophophore. (8-8)

ARE YOU STILL WONDERING?

1. Sponges are animals because they have cells lacking a cell wall and can actively pump water through their bodies.

2. Not many organisms feed on sponges. Some molluscs and fishes do, as does the hawksbill sea turtle.

3. Jellyfish have separate sexes when they reproduce sexually, usually in the medusa stage.

4. Jellyfish stings can be quite painful but are rarely fatal unless a person is allergic to the venom. The stings of box jellyfish are more dangerous and some, such as that of the Irukandji jellyfish, can be fatal to humans.

5. Ctenophores resemble jellyfish but represent a different group of animals. They lack the stinging cells of cnidarians and are able to move with comb plates.

QUESTIONS FOR REVIEW

Multiple Choice

1. Each of the following statements concerning sponges is true except one. Identify the exception.
   a. Sponges are living animals.
   b. Sponges have asymmetric bodies.
   c. Sponges reproduce only sexually.
   d. Spicules help to give a sponge support.
   e. The size of a sponge is related to the amount of water it can circulate through its body.

2. If a sponge lacked collar cells it would not be able to
   a. form spicules
   b. produce colored pigments
   c. feed
   d. exchange respiratory gases
   e. protect itself

3. Jellyfish use their nematocysts to
   a. capture prey
   b. digest food
   c. coordinate swimming movements
   d. circulate water through their bodies
   e. eliminate wastes

4. Ctenophores can be distinguished from jellyfish by their lack of
   a. radial symmetry
   b. statocysts
   c. comb plates
   d. cnidocytes
   e. a mouth

5. Which of the following animals exhibits radial symmetry as an adult?
   a. sponge
   b. flatworm
   c. bryozoan
   d. brachiopod
   e. jellyfish

6. An example of a parasitic flatworm would be a(n)
   a. phoronid
   b. nemertean
   c. turbellarian
   d. fluke
   e. annelid

7. Animals known as *brachiopods*
   a. possess an exoskeleton composed of chitin
   b. exhibit radial symmetry as adults
   c. produce a shell composed of two valves
   d. are all members of the plankton
   e. possess stinging cells known as cnidocytes

8. Which of the following cnidarians has no medusa in its life cycle?
   a. hydroid
   b. sea anemone
   c. box jellyfish
   d. moon jellyfish
   e. Portuguese man-of-war

Matching

Match each of the following animals with their mode of feeding:

   a. hydrozoan
   b. ribbon worm
   c. brachiopod
   d. ctenophore
   e. sponge

   1. Uses a tube-like proboscis to capture prey.
   2. Filters food particles from the water on the collar of a choanocye.
   3. Stings and paralyzes prey.
   4. Acquires food with a structure called a lophophore.
   5. Captures food with tentacles lacking stingers but having specialized adhesive cells to which the prey sticks.

Match each of the following terms with the appropriate definition or description:

   a. polyp
   b. medusa
   c. statocyst
   d. pedicle
   e. meiofauna

   6. An organ that helps animals maintain equilibrium.
   7. The fleshy stalk of a brachiopod.
   8. The umbrella-shaped, free floating stage of a cnidarian's life cycle.
   9. Tiny organisms that live in the spaces between sediment particles.
   10. The benthic form of cnidarians, characterized by a tubular body with a ring of tentacles around the mouth.

Short Answer

1. How does a sponge's body structure affect its size?
2. What role do boring sponges play in marine environments?
3. What are the advantages of bilateral symmetry?
4. Describe the ecological role of meiofauna.
5. Describe how sponges feed and reproduce.
6. Describe how the cnidarian stinging cell (cnidocyte) functions.
7. Why is radial symmetry advantageous to a sessile organism?
8. Distinguish between hydrozoans and scyphozoans.
9. What is a lophophore, and how does it function in feeding?
10. How do armed ribbon worms capture prey?

Thinking Critically

1. Development of beachfront property frequently increases the amount of sediment in coastal water. How would you expect this to affect sponges and hydrozoans that are living in the shallow water close to shore?

2. Based on the information in this chapter, construct a food web that includes meiofauna, marine worms, and larger predators such as fishes.

3. Explain with examples how symbiotic relationships can allow marine animals to live in habitats where they normally could not survive.

SUGGESTIONS FOR FURTHER READING

Estabrook, B. 2002. Jellies on a Roll: Jellyfish are "Blooming" around the World and for Many Marine Ecosystems, That Means Trouble, *Audubon* 104(3).

Grosberg, R. K. and R. R. Strathmann. 2007. The Evolution of Multicelluarity: A Minor Major Transition?, *Annual Review of Ecology, Evolution, and Systematics* 38: 621–654.

Mallon, T. 2007. Do Jellyfish Rule the World?, *Discover* September: 42–47.

Perkins, S. 2002. Model Predicts Chance of Encountering Jellyfish, *Science News* 162(14).

Raffaele, P. 2005. Killers in Paradise, *Smithsonian* 36(3): 80–88.

Ruppert, E. E., R. S. Fox, and R. D. Barnes. 2004. *Invertebrate Zoology*, 7th ed. Belmont, Calif.: Cengage Learning.

Have You Wondered?

1. If any snails are venomous? (9-1)
2. How octopus and squid change colors? (9-1)
3. If there is such a thing as a carnivorous worm? (9-2)
4. Why crabs and other crustaceans molt? (9-5)
5. If anything eats sea stars? (9-7)

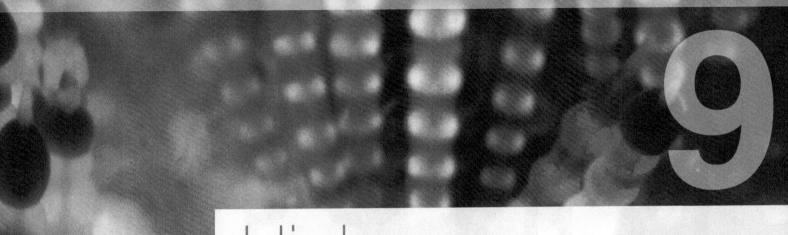

Higher Invertebrates

THE INVERTEBRATE ANIMALS that will be introduced in this chapter are among the most familiar creatures of sea and shore **(Figure 9-1)**. If you have ever picked up a seashell or sea star or watched a crab scurry across the beach, you have encountered some of these animals. Not only are these animals very familiar, many are commercially valuable. Scallops, oysters, mussels, crabs, lobsters, and shrimp are only some of the invertebrate animals that provide food for large numbers of people. Many of these animals are also threatened as a result of human activities. In this chapter, you will learn more about these animals and the roles they play in marine environments.

9-1 Molluscs

Molluscs (phylum Mollusca) are one of the largest and most successful groups of animals. The term *mollusca* (Latin for "soft") refers to the animals' soft bodies, which in most cases are covered by a shell made of calcium carbonate. Molluscs are represented by familiar animals such as chitons, snails, clams, octopods, and squid. They range in size from microscopic snails to the giant squid, *Architeuthis*. These animals can be found in most marine habitats and exhibit a variety of lifestyles. Some molluscs, such as oysters, scallops, and conchs, are food for humans, whereas others, such as shipworms, cause commercial damage.

MOLLUSCAN BODY

The generalized molluscan body plan consists of two major parts: the **head-foot** and the **visceral mass (Figure 9-2)**. As the name implies, the head-foot region contains the head, with its mouth and sensory organs, and the foot, which is the animal's organ of locomotion. The dorsal visceral mass contains the other organ systems, including the circulatory (heart and vessels), digestive (stomach, digestive gland, intestine, and anus), respiratory (gill), excretory (nephridium), and reproductive (gonad) systems.

The soft parts of a mollusc are covered by a protective tissue called the **mantle**. The mantle extends from the visceral mass and hangs down on each side of the body, forming a space between the mantle

Molluscs are animals that belong to the phylum Mollusca.

The **head-foot** is the part of the molluscan body that contains the animal's head and a muscular foot.

The **visceral mass** is the part of the molluscan body that contains all of the organs, with the exception of the animal's head and foot.

The **mantle** is a tissue found in molluscs that is responsible for forming the shell in animals that have one and that in cephalopods functions in locomotion.

The **mantle cavity** is the space between the mantle and the mollusc's body.

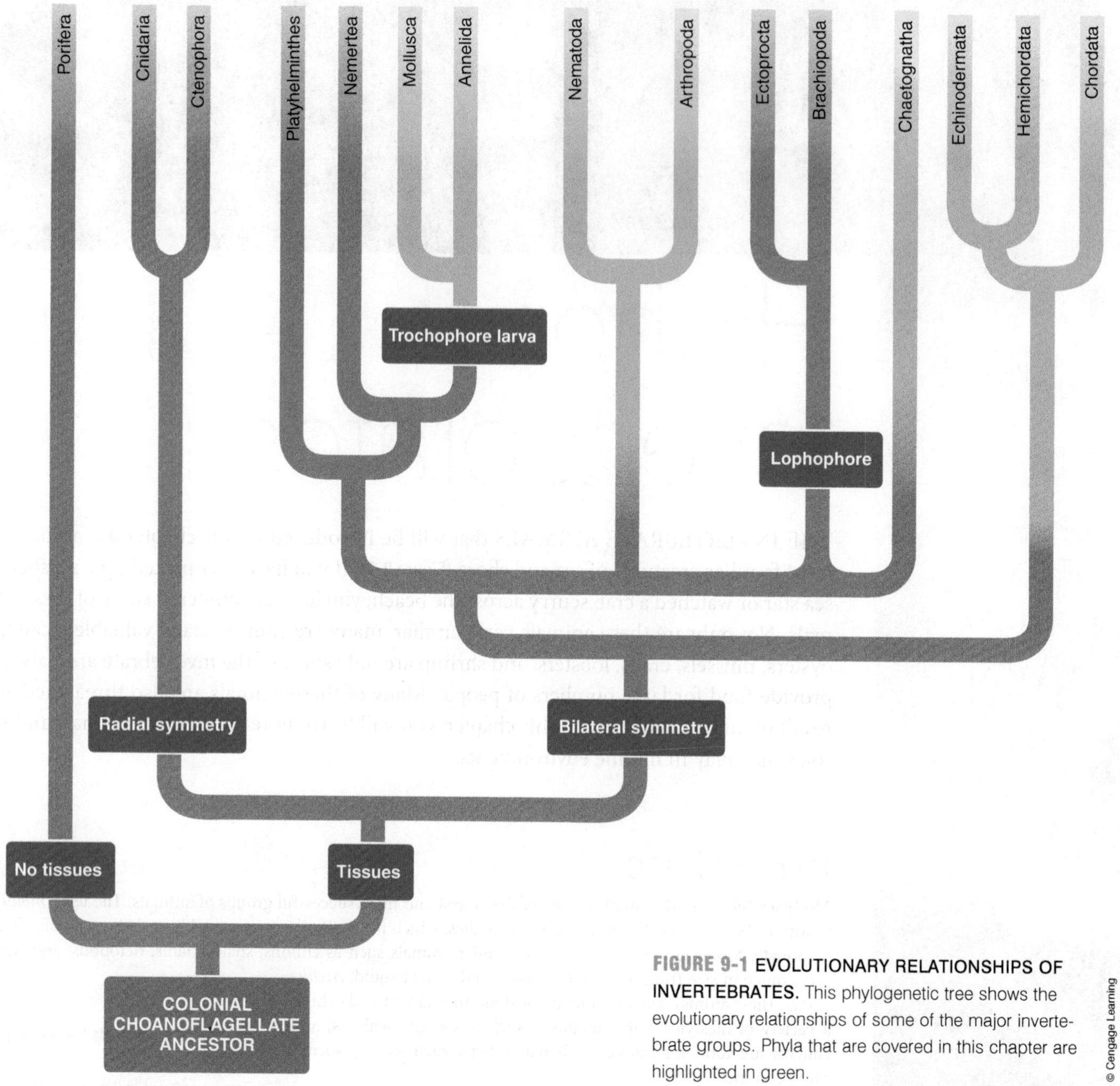

FIGURE 9-1 EVOLUTIONARY RELATIONSHIPS OF INVERTEBRATES. This phylogenetic tree shows the evolutionary relationships of some of the major invertebrate groups. Phyla that are covered in this chapter are highlighted in green.

© Cengage Learning

and the body known as the **mantle cavity**. The mantle is responsible for forming the animal's shell in those species that have one. Molluscs such as squid and octopods use the mantle in locomotion, and those molluscan species without gills also use the mantle for gas exchange.

A structure unique to molluscs is the **radula (Figure 9-3)**, a ribbon of tissue that contains teeth and is present in all molluscs except bivalves. Depending on the species, the radula is adapted for scraping, piercing, tearing, or cutting pieces of food, such as algae or flesh, which are then moved into the digestive tract. As the anterior portions of the radula wear out, new teeth are produced at the posterior end. The characteristics of the radula are key in the classification of molluscan species.

MOLLUSCAN SHELL

The molluscan shell is secreted by the mantle and normally comprises three layers (**Figure 9-4**). The outermost layer, the **periostracum**, is composed of a protein called **conchiolin** that protects the shell from dissolution and boring organisms. The middle layer, the **prismatic layer**, is composed of calcium carbonate and protein and makes up the bulk of the shell. The innermost **nacreous layer** is also composed of calcium carbonate, but its crystal structure is different from that found in the prismatic layer, and it forms in thin sheets.

A **radula** is a ribbon of tissue that contains teeth.

The **periostracum** is the outermost layer of a molluscan shell.

Conchiolin is a protein that makes up the periostracum.

The **prismatic layer** is the middle layer of the molluscan shell and is composed of calcium carbonate and protein.

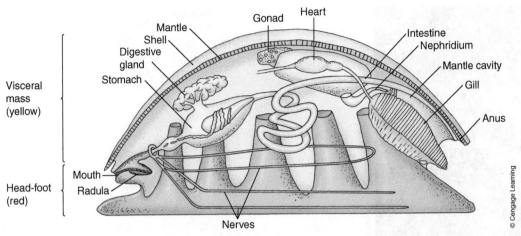

FIGURE 9-2 GENERALIZED MOLLUSCAN BODY PLAN. The generalized molluscan body consists of a head-foot composed of the animal's head and muscular foot and a visceral mass containing the animal's other organs.

The **nacreous layer** is the innermost layer of the molluscan shell and is composed of calcium carbonate, but its crystal structure is different from that found in the prismatic layer.

The **mother-of-pearl layer** is the nacreous layer of oysters.

Chitons are members of the molluscan class Polyplacophora and they have flattened bodies that are most often covered by eight shell plates.

Scaphopods are members of the molluscan class Scaphopoda commonly called tusk shells.

As the animal grows, new periostracum and prismatic layers form at the margin of the mantle. The nacreous layer is secreted continuously and accounts for the increased thickness of the shell in older animals. The iridescent nature of the nacreous layer is due to the arrangement of the calcium carbonate crystals. The thin sheets are a prism, refracting light and producing an iridescent effect. The nacreous layer of oysters is also known as the **mother-of-pearl layer**. Pearls are formed when the nacreous material is layered over sand grains and other irritating particles beneath the animal's mantle.

CHITONS

Chitons (KY-tuhnz; class Polyplacophora) have flattened bodies that are most often covered by eight shell plates **(Figure 9-5)**. These plates are held together by a tough girdle that is formed from the mantle, and in some animals, such as *Cryptochiton stelleri* from the Pacific Northwest, the plates are hidden beneath the mantle tissue. Chitons use their large flat foot to attach tightly to rocks, usually in the intertidal zone. When chitons are removed from a rock, they roll up into a ball for protection.

Most chitons feed on algae and other organisms that they scrape off the surface of rocks with their radulae. An examination of the gut contents of three species of chiton from the coast of Maine revealed the remains of 14 different species of algae and animals, and about 75% of the contents were sediments. The genus *Placiphorella* on the West Coast feeds on small crustaceans and other invertebrates.

SCAPHOPODS

Scaphopods (class Scaphopoda; *scapho* meaning "sheath" and *poda* meaning "foot") are commonly called tusk shells. They derive their name from the shape of their shell, which resembles an elephant's tusk

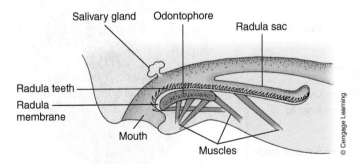

FIGURE 9-3 THE RADULA. The radula is a ribbon of teeth that is unique to molluscs. It is formed in the radula sac and supported by a mass of muscle and cartilage called the odontophore. Muscles attached to the odontophore move the radula into and out of the animal's mouth.

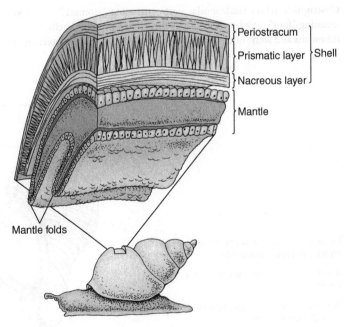

FIGURE 9-4 THE MOLLUSC SHELL. The molluscan shell is produced by the animal's mantle and is composed of three layers: (1) a layer of protective protein called the periostracum, (2) the prismatic layer that makes up the bulk of the shell, and (3) a smooth inner nacreous layer.

FIGURE 9-5 **CHITONS.** Chitons, such as *Tonicella lineata* from California, are usually found on intertidal rocks, where they feed on algae. Their shells are composed of eight separate plates held together by a leathery girdle.

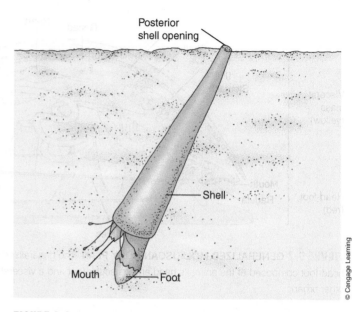

FIGURE 9-6 **SCAPHOPODS.** Scaphopods, or tusk shells, use their foot to burrow into bottom sediments. They feed primarily on foraminiferans.

(Figure 9-6). The tube-like shell is open at both ends, and the animal's foot, which is used for burrowing, protrudes from the larger end. Water enters and exits at the smaller end, bringing in oxygen and removing wastes. They bury themselves in bottom sediments ranging from the intertidal zone to several thousand meters deep, where they feed primarily on foraminiferans. Scaphopods use their foot or special tentacles that emerge from the head to capture their prey.

GASTROPODS

Gastropods (class Gastropoda; *gastro* meaning "stomach" and *poda* meaning "foot") exhibit tremendous diversity. Most have shells, but nudibranchs and some others have lost them. Some are microscopic; others

such as the marine snail *Syrinx aruanus* reach lengths approaching 1 meter (3.3 feet).

Gastropods live in a wide variety of habitats, from the littoral zone (shore) to the ocean bottom, and even pelagic (open ocean) species are known. When they possess a shell, it is always one piece (**univalve**) and may be coiled (as in snails; **Figure 9-7a**) or uncoiled (as in limpets; **Figure 9-7b**). As the animal grows, the **whorls**, or turns, of the shell increase in size around a central axis. Many snails can withdraw into the safety of their shells and close the opening, or **aperture**, with a cover called the **operculum**. The operculum can be horny (stiff protein) or calcareous in composition, and it is not found in all species of snail. Some gastropods such as limpets, abalones, and slipper snails are

Gastropods are members of the molluscan class Gastropoda.

A **univalve** is a shell composed of a single piece.

A **whorl** is a turn of a gastropod shell around a central axis.

The **aperture** is the opening to a gastropod shell.

An **operculum** is a structure that some gastropods have for closing their aperture.

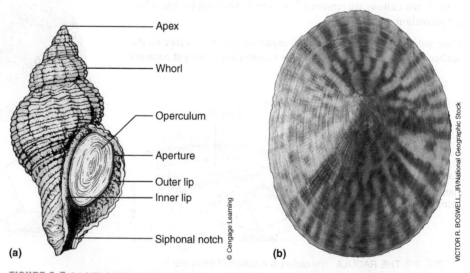

(a)

(b)

FIGURE 9-7 **GASTROPOD SHELLS.** Gastropod shells may be coiled **(a)** as in this oyster drill *Urosalpinx cinerea* or uncoiled **(b)** as in the limpet *Acmaea*. Each turn of a coiled shell is called a *whorl*, and the whorls wrap around a central axis. Many species have an operculum that is used to close the opening or aperture when the animal withdraws within its shell.

adapted for clinging to solid surfaces such as rocks and shells. All have low broad shells that can be pulled down tightly and offer less resistance to waves and currents. In these species the animal's large foot is an adhesive as well as a movement organ.

Feeding and Nutrition

Gastropods exhibit a wide variety of feeding styles. There are herbivores, carnivores, scavengers, deposit feeders, and suspension feeders.

Herbivores Most marine gastropods feed on fine algae that can be scraped from a rock or other solid surface with their radula **(Figure 9-8a)**. A few are adapted to feeding on large algae such as kelps.

Carnivores Carnivorous gastropods usually locate their prey by following chemical trails in the water. In response to the chemical cues, the animal extends its proboscis to search for the food source. The whelk *Buccinus undulatum* can locate a food source as far as 30 meters (99 feet) upstream, but it takes several days for the whelk to reach it.

Gastropods prey on a wide variety of animals, and they have evolved many interesting behaviors to trap and subdue their prey. The flamingo tongue (*Cyphoma*) and its relatives feed on colonial cnidarians, and large snails such as Triton's trumpet (*Charonia*) feed on echinoderms. Helmet snails (*Cassis*) feed on sea urchins. The helmet follows a chemical trail to its prey and pins the urchin down with its heavy shell. With the aid of sulfuric acid secretions, helmets can cut a hole in a sea urchin's shell within 10 minutes. They then insert their mouths into the hole and suck out the contents. Whelks (*Busycon*), tulips (*Fasciolaria*), and murex (family Muricidae) feed on clams. They grip the bivalve with their foot, pulling it open, and then keep it open by wedging the two valves apart with the

FIGURE 9-8 **GASTROPOD FEEDING STRATEGIES. (a)** Cowries are herbivores that graze on algae. They use their radula to scrape the algae off hard surfaces such as rock and coral. **(b)** A striated cone (*Conus striatus*) and other cone snails use their harpoon-like radula to inject a paralyzing toxin into their prey. After the prey is paralyzed it is slowly eaten by the snail. **(c)** Horn snails (*Cerithium*) are deposit feeders. At low tide, large numbers can be seen on sand or mud feeding on the organic material left behind by the tides. **(d)** Slipper limpets use their gills to filter food from the water.

(a)
NOAA

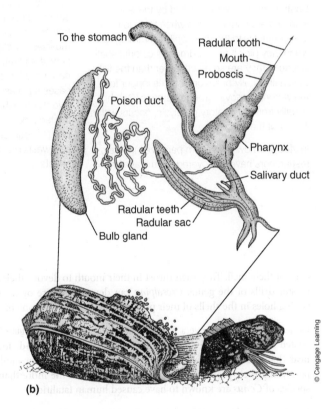
(b)
To the stomach — Radular tooth — Mouth — Proboscis — Poison duct — Pharynx — Salivary duct — Radular teeth — Radular sac — Bulb gland
© Cengage Learning

(c)
Brandon D. Cole/CORBIS

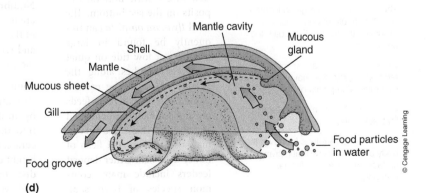
(d)
Mantle cavity — Mucous gland — Shell — Mantle — Mucous sheet — Gill — Food groove — Food particles in water
© Cengage Learning

MARINE BIOLOGY & THE HUMAN CONNECTION

Deadly Snails and Medicine

The marine snail *Conus geographus*, or geography cone **(Figure 9-A)**, is one of nature's most sophisticated predators. It uses its long, brightly colored proboscis to attract the attention of fish. A fish apparently mistakes the proboscis for an edible worm. When the fish moves in to bite, the snail fires its venomous radula into the fish's mouth. The radula penetrates the soft tissues in the mouth and paralyzes the fish. After its prey has succumbed to the venom, the snail begins the leisurely process of digesting its meal.

To date, at least 20 unwary swimmers in the South Pacific have been killed by this animal. Humans are usually stung when they come into contact with the snail while swimming, or when they pick it up to admire its colorful shell. Because they are so much larger than the fish the snail usually preys on, it takes longer for the venom to move through humans. Death, usually from respiratory paralysis, occurs within 1 hour of the sting.

In studying how the venom paralyzes its prey, researchers have discovered chemicals that can be used to probe the mysteries of the human nervous system, especially the relationship between nerves and muscles. Researchers have isolated three different toxins from the snail's venom, one of which is quite similar to the neurotoxin produced by cobras. Investigators have found that some components of the venom interfere with the action of nerve cells, and others interfere with the movement of stimuli from nerve cells to muscles. The toxins are highly specific and bind only to certain receptors on cell surfaces. Scientists can use these toxins to locate receptors in different cells and study the role receptor molecules play in normal and abnormal function of nerves and muscles. By studying the effects of the toxin on nerves and muscles, researchers hope to gain new insight into how some neuromuscular diseases such as myasthenia gravis (a progressive disease that causes muscle weakness in the face, neck, and chest) and muscular dystrophy (a group of diseases that cause progressive wasting of the muscles), produce their effects.

FIGURE 9-A **THE GEOGRAPHY CONE SNAIL.** The geography cone (*Conus geographus*) from the Indian and Pacific Oceans is one of the six species of cone snails known to cause human fatalities. The toxin produced by this animal causes death by paralyzing the victim's respiratory muscles.

---

edge of their shell. They then thrust in their mouth to devour their prey. Oyster drills of the genus *Urosalpinx* can devastate entire oyster beds, boring holes in the shells of their prey and sucking out the contents.

Cone snails (*Conus*) have a radula shaped like a harpoon **(Figure 9-8b)**, coated with a toxin produced in a poison gland (**bulb gland**) located near the mouth. These snails feed on worms, fish, and other molluscs, paralyzing their prey with a venomous sting. Six of the more than 600 species of *Conus* are known to have caused human fatalities.

Scavengers and Deposit Feeders Snails of the genus *Nassarius* are scavengers and **deposit feeders**. Deposit feeders feed on organic material that mixes with mineral deposits on the sea bottom. The snail *Ilyanasa obsoleta* can frequently be found in large numbers at low tide on quiet protected beaches along the Atlantic coast of the eastern United States, where it feeds on organic material deposited by the tide. This same species may also feed on the flesh of newly dead fish. Other deposit feeders include many common species of horn snail (*Cerithium*; **Figure 9-8c**). Conchs (*Strombus*) are deposit feeders, as well as grazers on fine algae. They sweep their large mobile proboscis across the bottom like a vacuum cleaner, sucking up food.

Filter Feeders Some gastropods are filter feeders. Slipper limpets (*Crepidula*) feed by ciliary action, using their gills to trap food particles that are then formed into a mucous ball and carried to the mouth for digestion **(Figure 9-8d)**. Planktonic gastropods, such as sea butterflies (*Gleba* and *Corolla*), form a mucous net that they use to catch the small organisms on which they feed.

Naked Gastropods

Nudibranchs are marine gastropods that lack a shell **(Figure 9-9)**. They are usually brightly colored and sometimes have bizarre shapes because of their gills and appendages. The name *nudibranch* means "naked gill" and refers to the projections from the animal's body called **cerata** that increase the surface area available for gas exchange.

Some nudibranchs feed on cnidarians. Instead of digesting their prey's stinging cells, the nudibranch leaves the cells intact and moves them by means of ciliated tracts in its digestive system. These tracts extend into the cerata. The stinging cells are deposited in the tips of the cerata, where they are used by the nudibranch for its own defense. The bright colors of many species warn that the animals are venomous or distasteful: Potential predators recognize the bright colors or patterns and avoid them.

The **bulb gland** is a gland located near the mouth of cone snails that produces the venom that coats the snail's radula.

Deposit feeders are animals that feed on organic material that mixes with mineral deposits on the seafloor.

Nudibranchs are marine gastropods that lack a shell.

Cerata are projections found on the body of nudibranchs that increase the surface area available for gas exchange.

FIGURE 9-9 NUDIBRANCHS. Nudibranchs are gastropods that lack a shell. The projections from the animal's body are called *cerata* and function in gas exchange.

Reproduction and Development

The sexes are separate in most gastropods, although some are hermaphroditic. Most species exhibit internal fertilization. Males of these species use a tentacle-like penis to deposit sperm in or near the female genital opening. The eggs of females that exhibit internal fertilization are usually surrounded by a jelly-like substance or covered by protective egg cases **(Figure 9-10a)**. In some species, sperm and eggs are released into the water column, and fertilization occurs externally.

In primitive molluscs that shed their eggs directly into the sea, a free-swimming larval stage, the **trochophore** (TROHK-eh-fohr) **larva (Figure 9-10b)**, hatches from the egg. More characteristic, however, of marine gastropods is a free-swimming stage called the **veliger larva** (see **Figure 9-10c**). Veligers can sometimes travel great distances on currents, allowing the otherwise slow-moving gastropods to disperse to other suitable habitats. In some species of marine snail there is no free-swimming larva; instead a juvenile that resembles the adult hatches from the egg.

A **trochophore larva** is a free-swimming larval stage associated with primitive gastropods that shed their eggs into the water.

A **veliger larva** is a free-swimming larval stage characteristic of many marine gastropods.

(a)

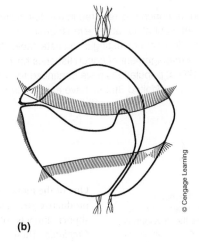

(b)

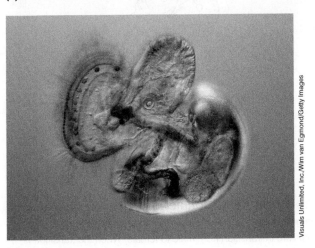

(c)

FIGURE 9-10 MOLLUSCAN LARVAE. (a) Many gastropods lay their eggs in protective cases such as this one of the whelk (*Busycon contrarium*). **(b)** A trochophore larva is characteristic of many primitive molluscs. **(c)** The larval stage of many marine molluscs is a veliger larva.

IMPACT BUBBLE

THE QUEEN OR PINK conch (*Strombus gigas*) is a large marine snail that provides a significant amount of the protein in the diet of many Caribbean peoples. It is also a popular food item in the United States and other countries. In 2001, the United States imported 250 metric tons of conch meat at a value of 6.6 million dollars. Not only is there a market for the meat of this animal, but its attractive pink shell is used in making jewelry and curios. Although not currently threatened with extinction, the population of queen conch has been steadily in decline throughout its geographical range. Queen conch are found in shallow-water grass beds and are an easy catch for conch fishers. Conch fishing is a major component of the economy of some Caribbean nations, and for some their major fishery.

Fishing of the queen conch in the Caribbean has been going on for centuries, but in the mid 1970s commercial conch fishing commenced. The increase in commercial fishing was the result of increased demand both in Caribbean countries and in foreign markets, mainly the United States, for conch meat, and an increase in tourism to Caribbean islands, where demand for the attractive shells and jewelry grew. The conch fishery continued to expand between 1992 and 2002, developing into a large-scale commercial fishery with almost industrial characteristics in some Caribbean countries. By the early 21st century, the conch fishery was one of the most important marine fisheries in the Caribbean region.

The high demand for conch has resulted in overharvesting of the species throughout most of its range, even though data indicate that maintaining such a level of harvesting is unsustainable. Although regulations for the management of conch fisheries have been in place since the 1980s, populations have continued to decline. Many countries, including the United States, have enacted legislation to protect conch populations and regulate the import of conch products, but the populations continue to be under pressure. In some areas the overharvesting has led to temporary closing of the fisheries.

Although some areas have enacted a total ban on harvesting conch (Bermuda in 1978 and the United States in 1986), there is little indication of significant recovery in these areas almost 30 years later. Part of the problem is that the queen conch is a slow-growing species requiring about 2.5 years to reach a market size of 845 g. At this size, the animal will yield about 100 g of meat. Another problem is that the diminishing population makes it harder for the conch to locate mates. Female queen conch require contact with a male to stimulate egg production, and in small populations fewer of these critical connections occur. A study conducted in the Bahamas in 2000 indicated that no mating behavior occurred if the population density was less than 56 animals per hectare (2.47 acres). An increase in reproductive behavior was observed at densities near 200 conch per hectare. Predation, other than human, may also be impacting populations that are already dangerously low.

For some Caribbean nations such as Belize, Cuba, Dominican Republic, and Turks and Caicos, conch is the major fishery, a primary source of food, and an important part of their economy. Even though these countries now regulate the conch fishery by limiting catches and closing conch fishing for part of the year, conch populations still decline due to illegal fishing, which is thought to be a widespread problem in the Caribbean. It will take significant cooperation among Caribbean nations to regulate the conch fishery and eliminate illegal fishing if the conch population is to recover and continue to provide food and economic benefit for these nations.

A **pheromone** is a hormone, released into the environment by an animal, that controls the development and behavior of other animals of the same species.

One of the more interesting reproductive strategies occurs in slipper limpets of the genus *Crepidula*. These mostly sessile animals are hermaphroditic and tend to congregate in groups in which several individuals are stacked on top of each other **(Figure 9-11)**. The right margins of the shells are next to each other, allowing the penis of the individual on top to reach the female opening of the individual below. The young of these species are always male. This initial male phase is followed by a transition period during which the male reproductive tract degenerates, and the animal develops into either a female or another male, depending on the sex ratio of the group. If males are in short supply, the animals develop into males. If females are limited, the animals develop into females. The sex of the individual is probably controlled by a **pheromone**, a hormone released into the environment by some members of a population to control the development and behavior of other members. Older males remain males longer if they remain attached to a female. If an older male is removed from the female or isolated, it develops into a female. Once an individual becomes a female, it cannot return to being a male.

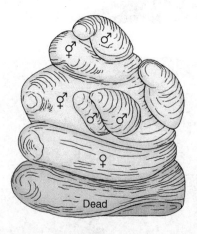

FIGURE 9-11 CREPIDULA FORNICATA. Slipper limpets are normally found in groups such as this. The male is on the top, and the next two individuals are in transition from male to female. When the transition is complete, all of the females will be fertilized by the male at the top.

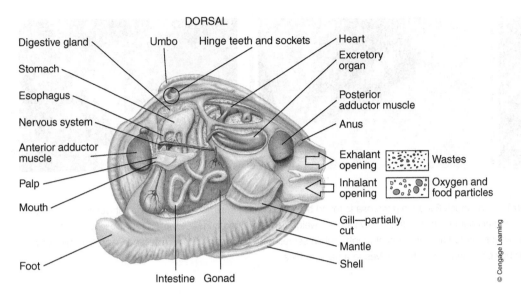

DORSAL

Digestive gland

Stomach

Esophagus

Nervous system

Anterior adductor muscle

Palp

Mouth

Foot

Umbo   Hinge teeth and sockets

Heart

Excretory organ

Posterior adductor muscle

Anus

Exhalant opening   ▢ Wastes

Inhalant opening   ▢ Oxygen and food particles

Gill—partially cut

Mantle

Shell

Intestine   Gonad

© Cengage Learning

FIGURE 9-12 BIVALVE ANATOMY. The internal anatomy of a typical marine clam, *Mercenaria mercenaria*. Notice that food and oxygen enter through the inhalant opening, while wastes are eliminated by way of the exhalant opening. The gills not only function in gas exchange but also filter food, mainly plankton, from the water.

Bivalves are molluscs in the class Bivalvia that have two hinged shells covering their body.

A **valve** is one of the two jointed halves of a bivalve shell.

The **umbo** is the area around the hinge and the oldest part of a bivalve shell.

Adductor muscles close the two valves of a bivalve shell.

The **inhalant opening** is an opening formed by the mantle that allows water to enter the mantle cavity.

The **exhalant opening** is an opening formed by the mantle that allows water to exit the mantle cavity.

Palps are a pair of structures, located near a bivalve's mouth, that form a food mass from the food filtered by the bivalve and move it to the animal's mouth.

Siphons are tubular structures formed from a bivalve's mantle that are fused around the inhalant and exhalant openings.

Byssal threads are tough threads composed of protein.

BIVALVES

Bivalves (class Bivalvia) are molluscs that have shells divided into two jointed halves called **valves**. This group includes clams, oysters, mussels, scallops, and shipworms. They range from tiny, 1-millimeter clams to the giant clam *Tridacna gigas*, which can grow as large as 1 meter (3.3 feet) and weigh 230 kilograms (507 pounds) or more.

Bivalve Anatomy

Bivalves have no head and no radula. Their bodies are laterally compressed, and the two halves of the shell that cover it are usually attached dorsally at a hinge by ligaments. The oldest part of the shell is the area around the hinge called the **umbo (Figure 9-12)**, and growth generally occurs in a concentric fashion around it at the edge of the shell. Large muscles called **adductor muscles** close the two valves. When the muscles relax, the elasticity of the ligaments and the weight of the valves cause them to open. The animal's foot is located ventrally and, in clams, functions in burrowing and locomotion.

For the most part bivalves are filter feeders. In many species, the mantle forms separate **inhalant** and **exhalant openings**. Cilia on the surface of the animal's gills move water through the animal's body, and the openings direct the water flow into and out of the mantle cavity. Water entering the inhalant opening carries food. The food particles, mainly plankton, are filtered from the water by the gills and formed into a mass by a pair of structures called **palps**. The palps move the food mass to the bivalve's mouth where it enters the digestive system.

Bivalve Adaptations to Different Habitats

The evolution of filter feeding has allowed bivalves to successfully colonize several different habitats. Many species burrow into soft bottom sediments and are conspicuous members of the infauna. Others spend their lives attached to solid surfaces as members of the epifauna. Still others are unattached epifauna. A few bivalves have even adapted to burrowing into wood and rock.

Soft-Bottom Burrowers Most species of bivalve are infauna, inhabiting soft bottoms to take advantage of the protection gained from burrowing in sand and mud. They feed on food suspended in the water they draw in from above. A major problem associated with burrowing in soft bottoms is that sediments are brought in with the incoming water currents. To deal with this problem, the mantle in many species is fused around the inhalant and exhalant openings to form tubular structures called **siphons** that can project above the surface of the sediments. With siphons the animals can be completely buried in the mud and only the siphon tips need to project above the bottom.

The geoduck (GOO-ee-duhk; *Panopea generosa*; **Figure 9-13a**) from the Pacific coast of the United States burrows as deep as 1 meter (3.3 feet) into the mud and can have a siphon more than 1 meter long. Their siphons are so large that they can no longer be retracted between the shell valves. The enormous siphon can account for as much as one half of this animal's 2.75-kilogram (6-pound) body.

Attached Surface Dwellers A number of bivalves are epifauna, attaching to firm surfaces, such as rock, coral, wood, shell, sea walls, or pilings. These bivalves attach either by fusing one of their valves to the hard surface or by **byssal threads** composed of a tough protein secreted by a gland in the bivalve's foot. Of the bivalves that attach with byssal threads, the mussels (family Mytilidae; **Figure 9-13b**) are the most familiar. Mussels can be found attached to wharf pilings, sea walls, and rocks or among oysters. Other bivalves that attach by byssal secretions include many of the heavy-bodied arks (family Arcidae), which are common on tropical corals; mangrove oysters (*Isognomon*), which hang in clusters from mangrove roots; and winged oysters (family Pteridae), which live attached to sea fans and other gorgonians.

(a) (b) (c)

FIGURE 9-13 BIVALVE ADAPTATIONS. (a) The geoduck *Panopea generosa* is found burrowed in soft sediments along the West Coast of the United States. Its siphon can be as long as 1 meter and account for half of this bivalve's body weight. **(b)** Attached surface-dwelling bivalves such as this mussel attach to solid surfaces by way of a byssus formed from threads of tough protein secreted by the animal's foot. **(c)** Surface-dwelling bivalves such as this spiny oyster (*Spondylus*) attach by cementing one of their valves to a hard surface.

Surface-dwelling bivalves that attach by fusion lie on one side and cement one of their valves to a hard surface. Which valve is cemented, the right or the left, depends on the species. Oysters are the most familiar bivalves that attach in this manner **(Figure 9-13c)**. In attached bivalves the foot is reduced to varying degrees, and it is completely absent in those bivalves, such as the oyster, that are attached by one valve.

Unattached Surface Dwellers Bivalves that live free on the bottom or other surface include scallops (family Pectinidae) and file clams (family Limidae). Although some members of both families live anchored by byssal threads, most are unattached and able to move. Scallops and their relatives are able to move by a type of jet propulsion. The animal moves by contracting its adductor muscle, causing the two valves to close forcefully and emit two streams of water backward and to each side. The animal then moves forward in a series of jerky movements. The swimming ability of scallops and file clams is used primarily to escape predators. Some scallops use the water jets to blow out a depression in the sand surface into which they settle to become less visible. File shells typically nest in crevices beneath stones and swim only when disturbed.

Shipworms are bivalves belonging to the genera *Teredo* and *Bankia*.

Cephalopods are members of the molluscan class Cephalopoda, which includes octopods and squid.

Nautiloids are cephalopods whose body is covered by a shell.

Coleoids are cephalopods that do not have an external shell.

Septa are partitions that separate the chambers of a nautilus shell.

A **siphuncle** is a cord of tissue that runs through the chambers of a nautilus shell and removes seawater from new chambers as they form.

Boring Bivalves Some bivalves are capable of burrowing into wood or stone and can cause a great deal of commercial damage. Members of the genera *Teredo* and *Bankia* are known as **shipworms** and can cause damage to wooden ships and wharves. These bivalves resemble tiny worms with a shell, and the valves bear microscopic teeth that help the animal burrow **(Figure 9-14)**. Pieces of wood that are removed by the burrowing action are swallowed and digested with the aid of enzymes produced by symbiotic bacteria. These bacteria live in a special organ located in the animal's digestive system. The bacteria also fix nitrogen, which helps their hosts compensate for their low-protein diet. Another group of boring clams belongs to the genus *Pholas*. These bivalves can burrow into soft rocks, such as limestone, and play a role in recycling calcium.

Reproduction in Bivalves

In the majority of bivalves the sexes are separate. During reproduction sperm and eggs are shed into the water column by way of the excurrent water. Fertilization takes place in the water column for most species. The fertilized eggs develop into a free-swimming trochophore larva, and this is followed by a second larval stage known as the veliger larva. The veliger larva develops a shell and settles on the bottom where the animal begins an adult existence. Shipworms and some species of cockle (*Cardium*), scallop, and oyster are hermaphroditic. The oyster *Ostrea edulis* does not release eggs but broods them in a cavity near the gills. In this species sperm are brought in with the incurrent water to fertilize the eggs.

CEPHALOPODS

Molluscs such as the octopus, squid, and cuttlefish are **cephalopods** (SEF-uh-loh-pahdz; class Cephalopoda). The term *cephalopod* means "head-footed" and refers to the animal's foot, which is modified into a head-like structure. A ring of tentacles projects from the anterior edge of the head, and they are used to capture prey, for defense, in reproduction, and in some species as an aid in locomotion. With the exception of the nautiloids, modern cephalopods either have small internal shells, such as some squid, or lack shells entirely, such as the octopods.

Types of Cephalopod

There are two major groups of living cephalopods. The **nautiloids** (subclass Nautiloidea) are represented by the genus *Nautilus*. These animals are covered by a shell. The **coleoids** (subclass Coleoidea) includes squid, octopods, and cuttlefish. These animals do not have an external shell.

Nautiloids Nautiloids produce large coiled shells composed of chambers separated from each other by partitions called **septa (Figure 9-15a)**. The chambers are filled with gas and aid in maintaining buoyancy for swimming. The animal inhabits only the last chamber but remains connected to the others by means of a cord of tissue called the **siphuncle** (SY-fuhn-kuhl; **Figure 9-15b**). The siphuncle removes the seawater from each new chamber as it forms. The living tissue of the siphuncle removes salts from the seawater in the new chamber by active transport. The water that remains is then more dilute than the animal's body fluids and readily moves by osmosis into the animal. The excess water is ultimately eliminated by the kidney. The gases that replace the water in the chambers are produced by the respiration of the siphuncle's living tissue.

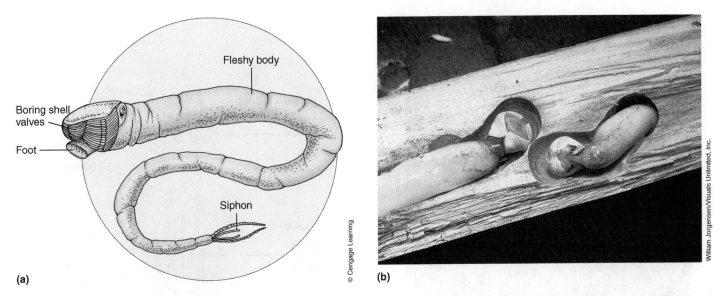

(a)

(b)

FIGURE 9-14 DESTRUCTIVE BIVALVES. (a) Shipworms are bivalves with long worm-like bodies and small shells that are used to bore through wood. **(b)** Shipworms can cause commercial damage to wooden vessels and docks.

The amount of gas in the chambers can be regulated, allowing the animal to rise or sink in the water column.

The animal's head has 60 to 90 tentacles coated with a sticky substance that helps them to adhere to objects. Many of these are chemosensory and tactile (sense of touch) in function but some around the mouth bring food to the mouth. Like other cephalopods, nautiloids move by means of jet propulsion. Water is drawn in through an incurrent siphon and forced out through an excurrent siphon, pushing the animal shell first through the water. A structure known as the funnel directs the jet of water, allowing the nautiloid to steer.

Nautiloids tend to come to the surface at night and dwell on the bottom during the day. Originally it was thought that they came to the surface at night to feed, but examination of stomach contents indicates that they prey on hermit crabs and scavenge other food on the bottom. After

ingesting their food, nautiloids store it temporarily in a large sac-like structure called the **crop** before sending it along to the stomach and intestine for complete digestion.

Coleoids Cuttlefish, squid, and octopods are coleoids. Cuttlefish (*Sepia*; **Figure 9-16a**) have a bulky body, fins, and 10 appendages; the 8 short, heavy appendages are called arms, and the 2 larger appendages are called tentacles. They have small internal shells, some of which have chambers similar to those of the nautilus shell. The shell of a cuttlefish, however, is embedded in its mantle.

Squid **(Figure 9-16b)** have large cylindrical bodies with a pair of fins derived from mantle tissue. Squid possess 10 appendages called arms that are arranged in 5 pairs around the head. One pair of arms, larger than the rest, are

> A **crop** is a sac-like structure that stores food.

(a)

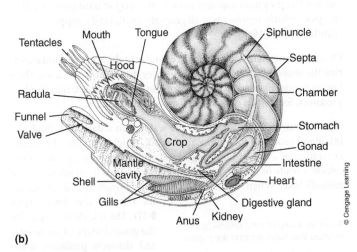

(b)

FIGURE 9-15 SHELLED CEPHALOPODS. (a) This chambered nautilus from the tropical Pacific Ocean has changed very little since it first evolved. **(b)** The nautilus occupies only the last chamber of its shell. The other chambers are filled with gases from the blood, such as carbon dioxide. The animal can regulate the gas content of the chambers with its siphuncle, allowing it to rise toward the surface, sink to the bottom, or maintain neutral buoyancy as necessary.

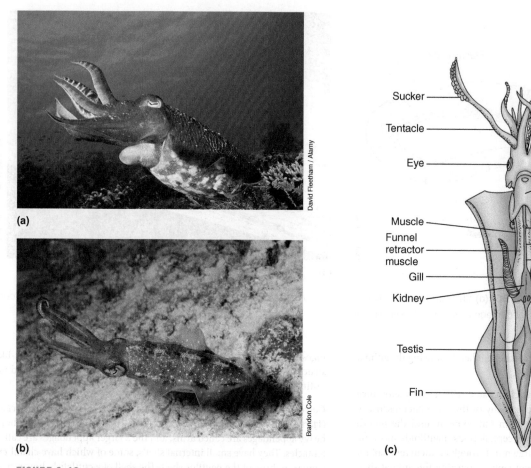

(a)

David Fleetham / Alamy

(b)

Brandon Cole

(c)

© Cengage Learning

FIGURE 9-16 **CEPHALOPODS. (a)** The cuttlefish resembles a squid and has a small internal shell. **(b)** Squids have streamlined bodies and are active swimmers. The part of the animal's body that looks like a head is actually formed partly from its foot. **(c)** The anatomy of a male squid.

called tentacles. The inner surface of each arm is flattened and covered with cup-shaped suckers that are attached by a short stalk. Toothed structures surround the rim of the sucker. Suckers are present only on the flattened ends of the highly mobile tentacles, which can be projected out with great speed to seize prey. The arms aid in holding the prey after capture. An internal strip of hard protein called the **pen**, which represents a degenerate shell, lends support to the mantle **(Figure 9-16c)**.

Octopods have eight arms, which are similar to the arms of squid except that the suckers lack stalks and teeth. They do not have tentacles. Their bodies are sac-like and lack fins. All of these animals are extremely adept predators, feeding on a variety of fish and invertebrates.

When disturbed, coleoids cloud the water with a dark inky fluid. An ink gland produces the fluid, called **sepia**, which contains a high concentration of a brown-black pigment called **melanin** (deepwater squid release a white or luminescent fluid). Sepia is stored in an ink sac in the animal's body. When the animal is disturbed or frightened, it can release the ink into the water **(Figure 9-17)**. The ink cloud resembles the general shape of the animal and distracts predators while the animal escapes.

Squid, octopods, and their relatives can swim, like nautiloids,

A **pen** is a strip of hard protein found in squid that helps support the mantle.

Sepia is a dark fluid produced by the ink gland of coleoids.

Melanin is a brown-black pigment.

by jet propulsion produced by forcing water through a ventrally located siphon. The funnel directs the flow of water and thus the direction of movement. Most slow swimming in squid species, however, is achieved by fin undulation. Squid and cuttlefish are very good swimmers, and

Jeff Rotman

FIGURE 9-17 **HOW CEPHALOPODS AVOID PREDATION.** Many cephalopods, such as this octopus, escape predators by releasing an inky substance into the water. The ink obscures the predator's view, allowing the cephalopod to escape.

their bodies are streamlined for efficient movement. Although octopods can swim, they are better adapted to crawling over the bottom.

Coleoids have the most advanced and complex nervous system of any invertebrate. Octopods can be trained to perform simple tasks and are used as model systems to study neural development associated with learning and memory. They have highly developed eyes and can recognize shapes and colors. Octopods seem to have keen tactile sense and can discriminate objects on the basis of touch.

Color and Shape in Cephalopods

Cephalopods can communicate with each other through movements of their arms and bodies and by color changes. Color changes involve special cells in the skin called **chromatophores** that contain pigment granules. When the granules are dispersed, the color of the animal's skin darkens. When the granules are concentrated, the color lightens because of the permanent light background color **(Figure 9-18)**. The animal not only changes general body color but also produces stripes and patterns that communicate information to other members of its species and a warning to other species. For instance, when agitated, the blue-ring octopus (*Hapalochlaena maculosa*) of the Pacific Ocean alternately

changes its background color from dark to light so that the blue rings on its skin become more obvious **(Figure 9-19a)**. This is a warning to potential antagonists to stay away. The bite of this octopus is extremely toxic and can be fatal to humans. The mimic octopus from Indonesia (*Thaumoctopus mimicus*) can change its body shape to mimic other animals, such as the venomous sea snake, allowing it to forage for food in broad daylight in a habitat that has few hiding places **(Figure 9-19b)**.

Feeding and Nutrition

Cephalopods are carnivores. They locate prey with their highly developed eyes and capture it with their tentacles or arms. A radula is present in cephalopods, but more important is the pair of powerful beak-like jaws in the oral cavity. The jaws bite and tear off large pieces of tissue, which are then pulled into the oral cavity by a tongue-like action of the radula and swallowed. The diet of a cephalopod depends on the habitat in which it lives. *Nautilus* is a scavenger. When feeding, it swims forward searching for food with extended tentacles. Pelagic squid such as *Loligo* and *Alloteuthis* feed on fish, crustaceans, and other squid. *Loligo* will dart into a school of young mackerel, seize a fish with its tentacles, and quickly bite off the head or a chunk behind the head. Cuttlefish swim

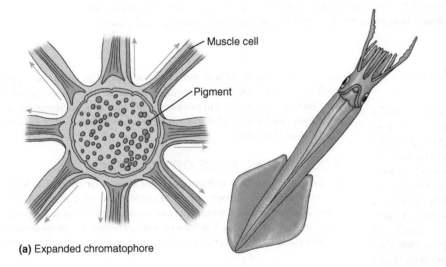

(a) Expanded chromatophore

Muscle cell

Pigment

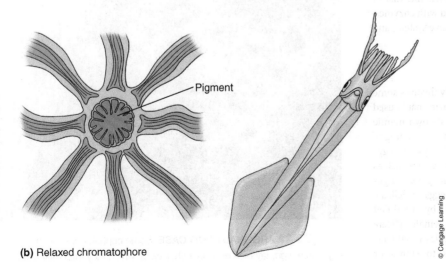

(b) Relaxed chromatophore

Pigment

FIGURE 9-18 HOW CEPHALOPODS CHANGE COLOR. (a) Tiny muscle cells pull at the edges of the chromatophore, causing the cell to expand and the pigment to disperse, producing a darker color. **(b)** When the muscle cells relax, the chromatophore returns to its original size, concentrating the pigment; the color of the animal is then lighter.

© Cengage Learning

Chromatophores are pigment containing cells that function in color changes.

FIGURE 9-19 FUNCTIONS OF COLOR AND SHAPE CHANGE IN OCTOPODS. (a) The blue-ring octopus, *Hapalochlaena maculosa,* from Australia, uses color to advertise its toxicity. By changing its background color from dark to light, the bright blue rings on its body become more evident. **(b)** The octopus *Thaumoctopus mimicus* protects itself and sneaks up on prey by changing its body shape to resemble other animals. Notice how the arms of this specimen mimic venomous sea snakes.

over the bottom and feed on invertebrates they find on the surface, especially shrimp and crabs.

Octopods live in dens located in crevices and holes. They either forage in search of food or lie in wait near the entrances of their dens. They mainly feed on clams, snails, and crustaceans, and the diet of a particular species may include as many as 55 different prey items, although a few items predominate. When feeding, an octopus will leap on its prey, such as a crab, enveloping it in its arms. Sometimes an octopus leaps on clumps of algae or other objects and then feels beneath the object for a possible catch. One of the octopod's salivary glands secretes venom, which enters the tissues of the prey through the wound inflicted by the jaws and aids in subduing the prey. To remove gastropods from their shells, octopods drill a hole through the shell with their radula and inject venom to kill the occupant. In contrast to cuttlefish and squid, which tear prey with jaws, the feeding habit of octopods is rather like that of spiders. The prey is injected with venom and then flooded with enzymes. The partially digested tissues are taken up into the digestive system, and the indigestible remains are discarded.

Reproduction in Cephalopods

Sexes are separate in cephalopods, and mating frequently involves some form of courtship display. Male squid have a modified arm that is used to take a package of sperm, the **spermatophore**, from its own mantle cavity and place it in the mantle cavity of the female near the opening of the **oviduct** (tube that carries eggs to the outside of the body). The eggs are fertilized as they are released from the oviduct. Some species such as members of the genus *Argonauta* lay their eggs in delicate shells secreted by modified tentacles of the females (**Figure 9-20**). Other species usually attach their eggs to stones or

A **spermatophore** is a package of sperm.

An **oviduct** is a tube that carries eggs to the outside of the body.

other objects. The females of some octopus species incubate their eggs until they hatch, constantly pumping a stream of water over them to keep them oxygenated and to prevent fungus and other microorganisms from infecting them. Cephalopods usually reproduce only once in their life cycle and then die.

ECOLOGICAL ROLE OF MOLLUSCS

Molluscs are important to humans as well as to other animals as food, and snail shells are a major source of calcium for some marine birds. According to one study, sperm whales alone consume as large a mass of squid as humans do of all species of fish combined. Other molluscan foods include clams, oysters, mussels, abalone, conch, and scallops. The

FIGURE 9-20 ARGONAUT EGG CASE. Some cephalopods, such as *Argonauta argo,* lay their eggs in a shell case, which they then cast afloat.

shells of many molluscs provide habitat for other species. Some snails are intermediate hosts to parasites. Shipworms cause extensive damage to wooden pilings and boat hulls and are becoming increasingly troublesome as improved water quality allows more of them to survive. On the other hand, because few marine organisms are able to break down wood, wood-boring bivalves help prevent the accumulation of driftwood in the marine environment. A small number of bivalves have evolved commensal relationships. Most commensals are attached by byssal threads, but some crawl on their foot like a snail. The hosts are usually burrowing echinoderms such as heart urchins, brittle stars, sea cucumbers and shrimp-like crustaceans. The only known parasitic bivalve is a species of *Entovolva* that lives in the gut of sea cucumbers.

IN SUMMARY

- The phylum Mollusca includes chitons, tusk shells, gastropods, bivalves, and cephalopods.

- The generalized molluscan body plan consists of two parts: a head-foot and a visceral mass.

- Molluscs have soft bodies that, in most cases, are covered by a shell of calcium carbonate that is formed by the mantle.

- In cephalopods, the mantle functions in locomotion.

- Cepahlopods are able to change color with the aid of cells called chromatophores. (Have You Wondered? #2)

- Some molluscan species rely on the mantle for gas exchange, though most marine molluscs exchange gases by means of gills.

- The radula, or ribbon tooth, is a structure unique to molluscs that is used for feeding. (Have You Wondered? #1)

9-2 Annelids: The Segmented Worms

Annelids (phylum Annelida) are worms, the majority of which have bodies that are divided internally and externally into segments that allow them more mobility by enhancing leverage **(Figure 9-21)**. Most worms gain support for their body from fluid (usually water) that is contained in a body compartment. This condition is known as a **hydrostatic skeleton**.

The body wall of annelids contains both longitudinal and circular muscles, allowing them to crawl, swim, and burrow efficiently. The skin of many contains small bristles called **setae** (SEET-ee) that are used for locomotion, digging, anchorage, and protection. Fireworms (family Amphinomidae), which live in coral and beneath stones, have brittle, tubular, calcareous setae containing venom that are used for defense. After penetrating the skin, the setae break off, releasing the venom and causing searing pain. Fish avoid these worms.

POLYCHAETES

The most common annelids in marine environments are **polychaetes** (PAHL-eh-keets; class Polychaeta), but their secretive habits cause them to go unnoticed by most casual visitors to the seashore. Most species are less than 10 centimeters (4 inches) in length, although some larger species may grow to more than 3 meters (10 feet). Some are brightly colored red, pink, green, a combination of colors, or iridescent. They live burrowed in

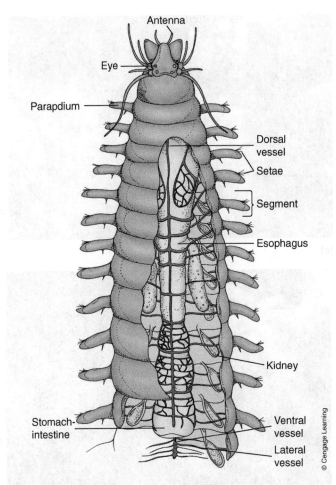

FIGURE 9-21 **POLYCHAETE WORM.** Polychaetes are segmented worms (annelids). Most body segments have a pair of appendages called parapodia that can function in locomotion, gas exchange, and feeding depending on the species.

sand and mud, under rocks and corals, in crevices, in shells of other marine organisms, and in tubes they produce themselves.

Traditionally, polychaetes are divided into two groups: the **errant** (free-moving) **polychaetes** that move actively and the **sedentary polychaetes** that are sessile forms **(Figure 9-22)**. Some errant polychaetes are strictly pelagic, some crawl beneath rocks and shells, some are active burrowers in sand and mud, but some also occupy stationary tubes. Sedentary species construct and live in stabilized burrows or tubes. Some species cannot leave the tube and can project only their head from the opening. Tube worms are sedentary polychaetes that construct their tubes from a variety of materials. Some species sort sand grains from the food particles that they trap and use the sand to enlarge their tubes. Some tubes are made of protein and have a consistency similar to stiff paper (parchment). Others

Annelids are worms belonging to the phylum Annelida.

A **hydrostatic skeleton** is a means of supporting an animal's body using fluid contained in a body compartment.

Setae are small bristles in the skin of some annelids.

Polychaetes are annelid worms belonging to the class Polychaeta.

Errant polychaetes are polychaetes that are active movers.

Sedentary polychaetes are sessile polychaetes.

(a)

(b)

FIGURE 9-22 ERRANT AND SEDENTARY POLYCHAETES. (a) This bearded fireworm (*Hermodice*) is an example of an errant (mobile) polychaete. The setae of this worm contain a chemical that can cause a severe burning sensation when touched, thus the name fireworm. **(b)** The horseshoe worm (*Pomatostegus*) is a filter feeder and an example of a sedentary polychaete.

are solid tubes composed of calcium carbonate, whereas still others are built from stones, coral, or sand granules that are held together by mucus. Feather duster, or fan worms (Sabellidae and Serpulidae), live in tubes that they form in or on the bottom sediments or other solid surface. From the opening of the tube emerges a ring of tentacles, each of which is very delicate and frequently brightly colored. The tentacles are ciliated and used for straining food from seawater, as well as providing a surface for gas exchange. They are equipped with light-sensitive organs, and when the worm is disturbed or when a shadow falls on it, it quickly withdraws its tentacles into its tube and out of sight.

Burrowing and tube-dwelling polychaetes commonly occur in enormous numbers on the ocean floor and compose a major part of the soft-bottom infauna. A study in Tampa Bay, Florida, reported the average density of polychaetes to be 13,425 individuals per square meter, representing 37 species living in the sediments beneath that area. On the upper continental slope and the deep ocean floor, polychaetes make up 40% to 80% of the infauna. In general, such large populations do not

appear to be limited by the resources available, at least not in shallow water. Predation and other pressures usually prevent infaunal populations from ever reaching the carrying capacity of the habitat. When areas in the York River estuary of the Chesapeake Bay were protected from predatory fish and crabs by wire cages, more than half of the species in the polychaete population increased to two or more times the population size in unprotected conditions. Studies such as these support the hypothesis that predation by fishes and crabs plays an important role in limiting the size of infaunal worm populations.

Feeding and Digestion

Some errant polychaetes have mouths equipped with jaws or teeth, and they are active predators **(Figure 9-23)**. They tend to feed most at night, when they come out of hiding to search for small invertebrates, including other polychaetes that they capture with their jaws. Tube dwellers may leave the tube partially or completely when feeding, depending on the species. *Diopatra* uses its hood-shaped tube as a lair. Chemoreceptors monitor passing water currents and, when approaching prey is detected, the worm partially emerges from the tube opening and seizes the victim. Species of *Diopatra* may also feed on dead animals, algae, organic debris, and small organisms such as foraminiferans that are in the vicinity of the tube or become attached to it. Species of *Glycera* live within a system of tubes constructed in muddy bottoms and containing numerous loops that open to the surface. Lying in wait at the bottom of a loop, the worm uses its four tiny antennae to detect the movement of prey, such as small crustaceans and other invertebrates. When prey are detected, the worm slowly moves to the burrow opening and seizes the prey with its pharynx. Not all errant polychaetes that possess jaws are carnivores. The jaws may, for example, be used to tear off pieces of algae. These polychaetes generally belong to the same families as the carnivores.

FIGURE 9-23 JAWS OF ERRANT POLYCHAETES. Some errant polychaetes are equipped with jaws that adapt them for a predatory lifestyle.

FIGURE 9-24 PARCHMENT WORM. The parchment worm (*Chaetopterus*) lives in a U-shaped tube composed of protein. The tube is on the left and the darker-colored worm is on the right.

Many sedentary burrowers and tube-dwelling polychaetes are filter feeders or suspension feeders. The head is usually equipped with special feeding structures that collect detritus and plankton from the surrounding water. The particles stick to the surface of the feeding structures and are carried to the mouth along ciliated tracts. The crown-like appendages of fan worms (families Serpulidae, Sabellidae, and Spirobidae) form a funnel of one or two spirals when expanded outside the end of the tube. The beating of cilia on the surface of the appendages produces a current of water that flows through the feather-like appendages and then flows upward and out. Particles are trapped on the appendages and are carried by the cilia into a groove running the length of each appendage. The polychaete *Chaetopterus* **(Figure 9-24)** lives in a U-shaped parchment tube and produces a mucous filter. The mucous film is continuously secreted and is shaped like a bag. Water passes through the mucous bag, which strains suspended detritus and plankton. When large objects are brought in by the water current, the mucous bag is pulled back to let them pass by. The food is continuously rolled into a ball, and when the ball reaches a certain size, the bag is cut loose and rolled up with the food. The ball is then carried to the mouth and the food consumed.

The digestive tract of polychaetes is usually a straight tube extending from the mouth at the anterior end of the worm to the anus located on the last segment. Digested food is absorbed in the animal's intestine. Polychaetes such as *Chaetopterus* that live in tubes with double openings use the water current moving through their tube to carry away fecal wastes. Many polychaetes turn around in their tube or burrow and thrust the last segment out of the opening during defecation. Some species produce fecal pellets or strings, which reduce the risks of fouling. Fan worms have a ciliated groove that carries fecal pellets from the anus out of the tube.

Small organic particles from decomposing algae, plants, and animal bodies, as well as planktonic organisms, are continuously settling on the bottom. Feces are also an important source of organic matter that settles to the bottom, where it mixes with mineral deposits on the sea bottom and becomes an important food source.

> **Nonselective deposit feeders** are animals that ingest both organic and mineral particles and then digest the organic material.
>
> **Fecal casts,** or **castings,** are masses of organic material and mineral particles that are defecated by deposit feeders.

Nonselective deposit feeders ingest both organic and mineral particles and then digest the organic material, especially the bacteria that grow on the surface of the mineral particles. The remaining organic material and minerals that are not digested are defecated and frequently form piles outside the burrow called **fecal casts,** or **castings.** The small burrowing polychaete *Euzonus mucronatus*, which is about 25 millimeters (1 inch) long and not more than 2 millimeters in diameter, inhabits the intertidal zone on the Pacific coast of the United States and forms colonies that occupy large areas of beaches that are protected from waves. In these colonies the number of worms averages 7,500 to 9,000 per square meter. Worms occupying a typical strip of beach 1.6 kilometers (1 mile) long, 3 meters (10 feet) wide, and 30 centimeters (1 foot) thick ingest approximately 14,600 tons of sand each year. Lugworms, members of the family Arenicolidae, are common deposit feeders **(Figure 9-25a).** They live in an L-shaped burrow whose vertical part opens to the surface. Sand that is rich in organic material is continually ingested. At regular intervals the worm backs up to

(a)

(b)

FIGURE 9-25 DEPOSIT FEEDERS. (a) The lugworm (*Arenicola*) is a direct deposit feeder. It feeds on the organic material that coats the sand particles it ingests as it digs its burrow. **(b)** The spaghetti worm (*Amphitrite*) is a selective deposit feeder, feeding on organic matter that sticks to its tentacles.

Selective deposit feeders are animals that separate organic material from minerals and ingest only the organic material.

Epitoky is a type of reproduction in some polychaetes that involves the production of a reproductive individual that is adapted for a free-swimming existence.

An **epitoke** is a free-swimming, reproductive individual.

Swarming is a behavior that brings sexually mature individuals together.

the surface to defecate castings containing mineral material. **Selective deposit feeders** separate organic material from minerals and ingest only the organic material. Specialized head structures extend over the surface or into the bottom sediments. Deposit material sticks to mucous secretions on the surface of these feeding structures and is then carried to the mouth along ciliated tracts or grooves. Polychaetes such as *Amphitrite* (**Figure 9-25b**) and *Terebella* have large clusters of contractile tentacles that stretch over the surface of the bottom sediments. The tentacles of some large tropical species such as *Eupolymnia crassicornis* reach a meter or more in length and move across the surface of the sand like living spaghetti. The deposit-feeding cone worm *Pectinaria* lives buried head down in the sand. It selects clumps of organic material and mineral particles coated with organic matter and, by selecting the organic material specifically, concentrates it from about 32% in the sediment to 42% in its gut. Of the ingested organic matter, only about 30% is used by the worm; the remainder passes through its gut and ends up in fecal casts.

Reproduction in Polychaetes

Asexual reproduction is known in some polychaetes. It takes place by budding or division of the body into two parts or into a number of fragments. Polychaetes have relatively great powers of regeneration. Tentacles and even heads ripped off by predators are soon replaced. Such replacement is a common occurrence in burrowers and tube dwellers. Some polychaetes, such as *Chaetopterus* and *Dodecaceria*, can regenerate the entire body from a single segment. Cells for regeneration are supplied by the remains of whatever tissues have been lost. Experimental studies indicate that the nervous system is crucial in the regeneration process.

Most polychaetes reproduce only sexually, and the majority of species have separate sexes. Gametes are often shed into the body cavity, where they mature. In a mature worm, the body cavity is packed with eggs or sperm, and in species in which the body wall is thin or not heavily pigmented, the presence of the gametes is apparent. The blue eggs of the red, white, and blue worm, *Proceraea fasciata* (**Figure 9-26**), show through its red-banded whitish body, thus the common name of this species.

Gametes are released in many different ways. A few species have separate, specialized ducts for discharging gametes into the water column. In many species gametes leave through the worm's excretory pores. In some species, such as *Nereis*, gametes are released when the body wall ruptures, although this is probably a specialized adaptation in a few species. Rupturing is more common among polychaetes that become pelagic at sexual maturity. After rupture of the body wall, the adults die.

Some errant polychaetes exhibit an interesting reproductive phenomenon known as **epitoky** (EP-i-toh-kee), the formation of a pelagic reproductive individual, or **epitoke**, that differs from the nonreproducing form of the worm. The epitoke is usually most modified in the posterior segments that contain reproductive structures (**Figure 9-27**). The posterior segments increase in number in the epitoke and are filled with either eggs or sperm. The males and females come to the surface in large numbers at night to shed their sperm and eggs. This behavior, known as **swarming**, brings sexually mature individuals together, increasing the likelihood of fertilization and the species' chances for survival.

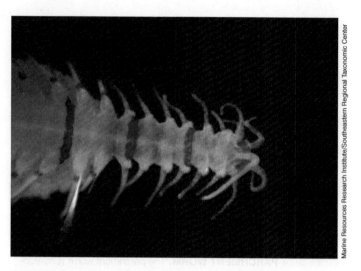

FIGURE 9-26 RED, WHITE, AND BLUE WORM. The blue eggs within this worm's body show through the thin body wall.

Swarming occurs only at specific times of the year and appears to be related to the lunar cycle and tides. The polychaete *Odontosyllis enopla*, which is found in the Caribbean and Bermuda, swarms in the summer about 1 hour after sunset following a full moon. The palolo worm, *Palolo viridis*, an errant polychaete found in the waters of the South Pacific around the islands of Samoa, reproduces in October or November during the last-quarter moon. The palolo worms luminesce when they reach the surface, and when males and females swim around each other releasing gametes, they create small circles of light. The female epitoke releases a

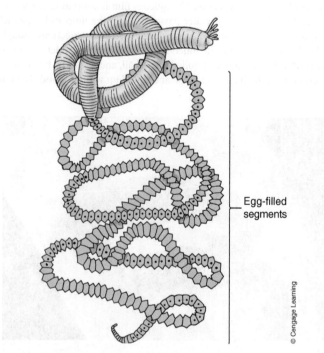

Egg-filled segments

FIGURE 9-27 EPITOKY. The epitoke (pelagic reproductive form) of the Samoan palolo worm is shown here. In this species, the anterior part of the worm remains unchanged, while the posterior portion grows to form a chain of egg-filled segments.

FIGURE 9-28 PEANUT WORM. Sipunculids are commonly called peanut worms because they can contract their bodies into a small mass that resembles a peanut.

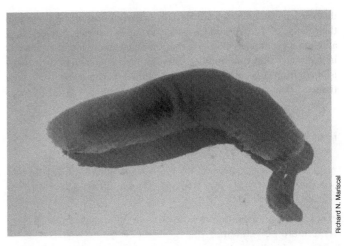

FIGURE 9-29 ECHIURAN. This echiuran worm uses its flat proboscis for feeding.

substance called **fertilizin** that stimulates the male to shed his sperm. The sperm, in turn, stimulate the females to release their eggs, and fertilization occurs externally. The epitokes provide food for many fish, birds, and some indigenous peoples of tropical islands. For the West Indian palolo worm, *Eunice schemacephala*, swarming takes place in July near the last-quarter moon. The worm backs out of its burrow in coral, and the posterior epitoke breaks free and swims to the surface. By dawn the ocean surface is covered with them. As the sun rises, the epitokes burst. Fertilization immediately follows rupture, and acres of eggs may cover the sea. The fertilized eggs develop into a ciliated larval stage by the next day, and in 3 days the larvae sink to the bottom to develop into adult worms.

SIPUNCULIDS

Sipunculids (class Sipuncula **Figure 9-28**) are solitary nonsegmented annelids that live in burrows in mud or sand, in empty mollusc shells, or in coral crevices. When disturbed, these animals contract their bodies so that they resemble a peanut kernel, which gives rise to the common name for some of these species, peanut worms. All are benthic, and the majority live in shallow water. Some such as *Sipunculus* live in sand and mud and burrow actively, others inhabit mucus-lined burrows, and still others live in coral crevices, empty mollusc shells, tubes of other annelids, or other protected areas. Several species bore into coralline rock, and at least one bores into wood. Borers direct the anterior end of their body toward the opening to feed. Populations of sipunculids as high as 700 per square meter of coralline rock have been reported from Hawaii. In parts of the tropical Indo-Pacific they are consumed for food.

Sipunculids are either suspension or deposit feeders. They have a proboscis with a ring of tentacles that extends from the mouth. Organic material is trapped in the mucus that coats the tentacles, and the food is then drawn into the mouth by ciliary action. Burrowing species such as *Sipunculus* are direct deposit feeders, ingesting the sand and silt through which they burrow. Some rock borers such as *Phascolosoma antillarum* spread their tentacles around the opening while suspension feeding. Other hard-bottom species spread tentacles over rock surfaces and ingest deposited material.

The sexes are separate in sipunculids, and they spawn into seawater. Fertilization is external, and some species develop directly into small worms, whereas others go through a larval stage.

ECHIURANS

Echiurans, or **spoonworms** (class Echiura), are sausage-shaped annelids that resemble sipunculid worms **(Figure 9-29)**. Some occupy burrows in sand and mud. Others live in rock or coral crevices. *Lissomyema mellita* from along the southeast coast of the United States inhabits the remains of dead sand dollars and grooves in discarded mollusc shells. The majority of spoonworms live in shallow water, but some deepwater species are known. They range from 1 centimeter (⅜ inch; *Lissomyema*) to 50 centimeters (20 inches; *Urechis*). Most echiurans are deposit feeders, but at least one is a filter feeder: *Urechis caupo*, from the California coast, forms a mucous net on its burrow wall and then pumps water through at the rate of 1 liter (1.06 quart) per hour. Virtually all particles, including typical plankton, are trapped on the net as the water passes through. When loaded with food, the net is detached and swallowed. *Urechis* is known as an innkeeper because of the large number of symbiotic organisms that share its burrow. Typical deposit feeders such as *Thalassemi* and *Bonnellia* extend their flat ribbon-like proboscis onto the surface of sediments around their burrows. The proboscis curls, forming a gutter along which deposited particles are carried to the mouth. In echiurans the sexes are separate. When reproducing, they shed their gametes into the water column and fertilization is external. The developing young go through a planktonic larval stage before settling on the bottom and taking up an adult existence.

Fertilizin is a chemical released by some female epitokes that stimulates males to shed sperm.

Sipunculids are solitary nonsegmented annelid worms.

Echiurans (spoonworms) are sausage-shaped annelids that belong to the class Echiura.

Pogonophorans (beardworms) are annelids that belong to the class Pogonophora

POGONOPHORANS

Pogonophorans, or **beardworms** (class Pogonophora), live in tubes buried in the ocean bottom at depths of 200 meters (656 feet) or more. The body of these animals is cylindrical, with a ring of tentacles around the anterior end. Beardworms are unusual in that they lack a mouth and a digestive tract. Evidence suggests that they derive nutrition from dissolved nutrients in the seawater by actively transporting them into their tentacles. Large beardworms (*Riftia*) up to 3 meters (10 feet) long and 5 to 8 centimeters (2 to 3 inches) in diameter are found exclusively in deepwater vent communities. These animals obtain all of their nourishment from large numbers of chemosynthetic bacteria that live in their tissues.

IN SUMMARY

- Annelids are worms the majority of which have a body that is divided internally and externally into segments, allowing them more mobility.

- The most abundant annelid worms in the marine environment are polychaetes.

- Polychaetes may be either active (errant polychaetes) or sessile (sedentary polychaetes).

- Some polychaetes are carnivores, whereas others are filter feeders, suspension feeders, or deposit feeders. (Have You Wondered? #3)

- Polychaetes are an important source of food for other organisms, and burrowing polychaetes play an important role in recycling nutrients.

- Sipunculids (peanut worms), echiurans (spoonworms), and pogonophorans (beardworms) are other types of marine annelid.

9-3 Nematodes

Nematodes (phylum Nematoda), also known as roundworms, are the most numerous animals on earth. Although the number of species described is only about 12,000, the number of individuals representing those species is staggering, and nematodes can be found in virtually every conceivable habitat. Their bodies are round, slender, elongated, and tapered at both ends **(Figure 9-30)**. Most nematodes are less than 5 centimeters (2 inches) long, although some parasitic species more than 1 meter (3.3 feet) long have been reported. Nematodes play a critical role as scavengers, and many are important parasites of plants and animals. Nonparasitic nematodes are benthic, living in the interstitial spaces of algal mats and sediments. One square meter of bottom mud off the Dutch coast has been reported to contain as many as 4.4 million nematodes. The great numbers of nematodes that are found in estuaries are sometimes the result of excess detritus accumulation. This is especially true in small harbors where human waste is abundant. Many free-living nematodes are carnivorous, feeding on small animals, including each other. Others feed on diatoms, algae, fungi, and bacteria. Most are hermaphroditic, but some species have separate sexes.

> **Nematodes** are worms that belong to the phylum Nematoda.

IN SUMMARY

- Nematodes are important members of the meiofauna.

- Nematodes are the most numerous animals on earth, with dense benthic populations, especially in areas with large amounts of organic material.

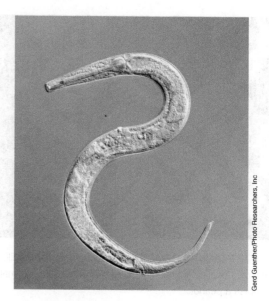

FIGURE 9-30 NEMATODES. Nematodes are prominent members of the meiofauna. This specimen was taken from bottom sediments in the North Sea.

9-4 Ecological Roles of Marine Worms

The many different types of marine worm serve in a variety of ecological roles. The burrowing activity of worms aids in nutrient cycling. Many species of marine worm feed on material too small to be gathered by other marine organisms and channel these nutrients to larger animals in marine food chains. Worm burrows provide habitat for several other species, many of which are commensal symbionts of the worms.

NUTRIENT CYCLING

Many marine worms are burrowing animals and, like other burrowing organisms, play an important role in the cycling of nutrients. Most oceanic production occurs above the bottom sediments in the photic zone, yet most decomposition occurs on or in the benthos. Over time, many of the nutrients released by the process of decomposition become buried in the bottom sediments. As burrowing organisms tunnel through sediments, many of these nutrients are brought back to the surface, where currents and other natural processes can move them to areas where they are used by producers to make new organic molecules.

PREDATOR–PREY RELATIONSHIPS

Marine worms are important links in marine food chains. Many consume organic matter that is unavailable to larger consumers. They in turn provide food for larger marine organisms, channeling nutrients up the food chain.

Marine nematodes, the most abundant members of the meiofauna, feed on microorganisms and detritus that are too small for larger animals. The nematodes in turn provide food for larger heterotrophs such as fish, birds, and large invertebrates. In food chains, free-living nematodes are important in channeling nutrients from microscopic producers and consumers to larger consumers.

Echiurans may constitute a significant portion of the diet of some fish. In one dietary study of leopard sharks caught along the California coast, *Urechis* was found to be the largest of any single food. Apparently the sharks use suction to remove the worms from their burrows. *Urechis* and other echiurans provide shelter for many commensal species in their burrows.

Polychaetes are a major source of food for other marine organisms, both invertebrate and vertebrate. Some errant and many sedentary polychaetes are deposit feeders. They feed on the bottom sediments, which include deposit material (organic material that settles to the bottom). The organic material is digested, and the mineral portion is eliminated as castings. These organic nutrients are then channeled to larger organisms when polychaetes are consumed by predators.

Symbiotic Relationships

Noncarnivorous tube-dwelling and burrowing polychaetes provide a protected and ventilated retreat for many commensal organisms. These commensals include scale worms, bivalves, and crustaceans, especially small crabs. The 80-centimeter-long (32-inch) Maitre d' worm (*Notomastus lobatus*) lives in clay sediments along the southeast coast of the United States. It occupies a permanent burrow that it shares with at least eight commensals, including one scale worm, two other polychaetes, three clams, one crab, and one amphipod. Commensal polychaetes are found in many families, but the scale worms (Polynoidae) contain the largest number. They live in tubes and burrows of other polychaetes and crustaceans, with hermit crabs, on echiuran worms, in the grooves in the arms of sea stars, on sea urchins and sea cucumbers, and with other animals. Many display colors similar to those of the host, making them difficult to see.

IN SUMMARY

- Marine worms are important in recycling nutrients and as links in marine food chains.
- Many burrowing worms have evolved symbiotic relationships with other organisms.

9-5 Arthropods: Animals with Jointed Appendages

Arthropods (phylum Arthropoda) represent the most successful group of animals in the animal kingdom: Almost 75% of all identified animal species (most of them insects) belong to this phylum. Several factors contributed to the enormous success of marine arthropods, including the evolution of a hard exterior, jointed appendages, and sophisticated sense organs.

The hard, protective, exterior skeleton, or **exoskeleton**, of arthropods is composed of protein and a tough polysaccharide called **chitin** (KY-tin). In many marine species, the exoskeleton is impregnated with calcium salts to give it extra strength. The exoskeleton is flexible enough in the region of the joints to allow movement, and it provides points of attachment for muscles, allowing more efficient movement. The exoskeleton does have its drawbacks, however. Because the exoskeleton is not living, it does not grow with the animal. As the animal grows, it must shed its old exoskeleton and produce a new larger one, a process called **molting**. As the exoskeleton increases in size, so does its weight, making it more difficult for the animal to carry around. It is not surprising that the largest arthropods are found in aquatic habitats, where the buoyancy of the water helps to counteract the weight of their heavy exoskeletons.

The arthropod body is divided into segments. Generally, each segment has a pair of jointed appendages, many of which function in locomotion, allowing the animals to move quickly and efficiently. Other appendages have been modified into mouthparts for more efficient feeding, sensory structures for monitoring the environment, and body ornamentation that helps to attract a mate or to camouflage the animal.

Arthropods have highly developed nervous systems. Their sophisticated sense organs allow them to respond quickly to changes in their environment. The high degree of development of the arthropod nervous system has given rise to a number of behavior patterns that play an important role in their daily activities. Experiments have shown that some species are capable of learning.

There are two major groups of marine arthropods: chelicerates and mandibulates. Chelicerates have a pair of oral appendages called chelicerae and lack mouthparts for chewing food. As a result, most predigest their food and suck it up in a semiliquid form. Mandibulates have appendages called mandibles that can be used for chewing food.

CHELICERATES

Chelicerates (keh-LI-suh-rehts; subphylum Chelicerata) are a primitive group of arthropods that includes spiders, ticks, scorpions, horseshoe crabs, and sea spiders. These animals have six pairs of appendages. One pair, the **chelicerae** (ke-hLI-suh-ree), is modified for the purpose of feeding and takes the place of mouthparts.

Horseshoe Crabs

Horseshoe crabs (*Xiphosura*; **Figure 9-31**) are not true crabs but chelicerates that live in shallow coastal waters. They have not changed much since they first evolved more than 230 million years ago. The horseshoe-shaped body of the animal comprises three regions: the **cephalothorax** (sef-uh-loh-THOR-aks) (a fusion of the head and thorax), the **abdomen**, and the **telson**. The cephalothorax, the largest region, contains the more obvious appendages. The abdomen is smaller than the cephalothorax and contains the gills. The telson is a long spike that is used for steering and defense. The entire body is covered by a hard outer covering, the carapace.

Horseshoe crabs move by walking, or they swim by flexing the abdomen. They are scavengers that feed primarily at night on worms, molluscs, and other organisms, including algae. They pick up food with their chelicerae and pass it to the walking legs. The walking legs have structures at their bases that crush the food before passing it to the mouth.

Males are smaller than females, and during the mating season,

Arthropods are animals with jointed appendages belonging to the phylum Arthropoda.

An **exoskeleton** is a hard exterior skeleton.

Chitin is a tough polysaccharide found in the exoskeletons of arthropods.

Molting is the process whereby an arthropod sheds its old exoskeleton and forms a new one.

Chelicerates are arthropods that have a pair of oral appendages called chelicerae that function in feeding.

Chelicerae are appendages found in chelicerates that are modified for feeding.

A **cephalothorax** is a body region composed of a fused head and thorax.

An **abdomen** is the body region of an animal that corresponds to the belly.

A **telson** is a long spike used by horseshoe crabs for steering and defense.

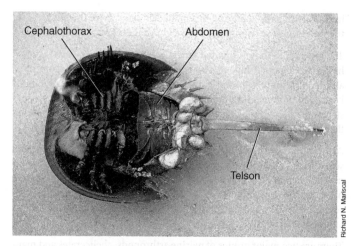

FIGURE 9-31 HORSESHOE CRABS. Horseshoe crabs are not really crabs but are more closely related to spiders. The large anterior portion with the walking legs is the *cephalothorax*. The abdomen contains the gills, and the long tail-like structure is the *telson*.

FIGURE 9-32 SEA SPIDER. Sea spiders are marine chelicerates that use their proboscis to suck fluids from their prey, much like terrestrial spiders.

a male or several males will attach to the carapace of a female. The animals come to shore during high tide to mate. The female digs up the sand with the front of her carapace, depositing eggs in the depression. The males riding on her back shed their sperm onto the eggs before they are covered by the female. The eggs are then incubated by the sun and hatch into larvae that return to the sea during another high tide to grow into adults.

Sea Spiders

Sea spiders (class Pycnogonida; **Figure 9-32**) are chelicerates that can be found from intertidal waters to depths of more than 6,000 meters (19,700 feet) in all oceans, but especially the polar seas. They have small thin bodies and usually four pairs of walking legs, although some species may have more. Males in several species have an extra pair of appendages that are used to carry the developing eggs. These are the only marine invertebrates known where the male carries the eggs. Other appendages include chelicerae for capturing prey and sensory structures called **palps**. Sea spiders feed on the juices of cnidarians and other soft-bodied invertebrates. They extract the juices with a long, sucking proboscis.

MANDIBULATES

Mandibulates have pairs of appendages on the head called **mandibles** that are modified for feeding. The mandibulates found in the marine environment are mostly crustaceans (subphylum Crustacea), a group represented by a variety of animals that range in size from microscopic zooplankton to large lobsters.

Palps are sensory structures.

Mandibulates are arthropods that have appendages called mandibles for chewing their food.

Mandibles are appendages found in mandibulate arthropods that are modified for feeding.

Swimmerets are arthropod appendages modified for swimming.

Chelipeds are claws.

Decapods are arthropods that belong to the class Decapoda.

Crustacean Anatomy

Crustaceans have three main body regions: head, thorax, and abdomen **(Figure 9-33)**. In some species the head and thorax are fused to form a cephalothorax. Each body segment usually bears a pair of appendages, although in some species abdominal appendages are lacking. Crustacean appendages include sensory antennae (they are the only arthropods that have two pair) and mandibles and maxillae that are used for feeding. Depending on the species, other appendages may include walking legs and legs modified for swimming (**swimmerets**), reproduction, and defense (**chelipeds**).

Small crustaceans exchange gases through their body surface, especially through areas at the base of their legs, where the exoskeleton is thin. Larger crustaceans have gills for gas exchange. These feathery structures are usually located beneath the carapace or at the base of specialized appendages.

Molting

Molting is a very important part of the crustacean life cycle. Stages are diagrammed in **Figure 9-34**. While the animal is molting, it is very vulnerable and usually seeks a hiding place until a new exoskeleton has hardened. Young crustaceans molt frequently, and the time between molts is relatively short. As the animal ages, molting is less frequent, and some species eventually reach an age where they no longer molt. Molting is directly controlled by specific hormones produced by glands in the crustacean's head. The process appears to be initiated by changes in environmental conditions such as temperature and photoperiod, which alter the level of the hormones.

Decapods

Crabs, lobsters, and true shrimp are called **decapods** (order Decapoda; *deca*, meaning "ten," and *poda*, meaning "feet") because they have five pairs of walking legs. The first pair of walking legs is modified to form chelipeds, pincers used for capturing prey and for defense. Although most decapods are relatively small, some can get quite large. The largest of all crustaceans is the giant spider crab (*Macrocheira kaempferi*) from Japan **(Figure 9-35a)**. Specimens are known to have exceeded 4 meters (13 feet) in width and weighed more than 18 kilograms (40 pounds). The North Atlantic lobster (*Homarus americanus*) can also get quite large. Two record specimens were captured off the coast of Virginia in 1934. One measured more than 1 meter (3.3 feet) long; one weighed 19 kilograms (42 pounds), and the other 17 kilograms (37 pounds).

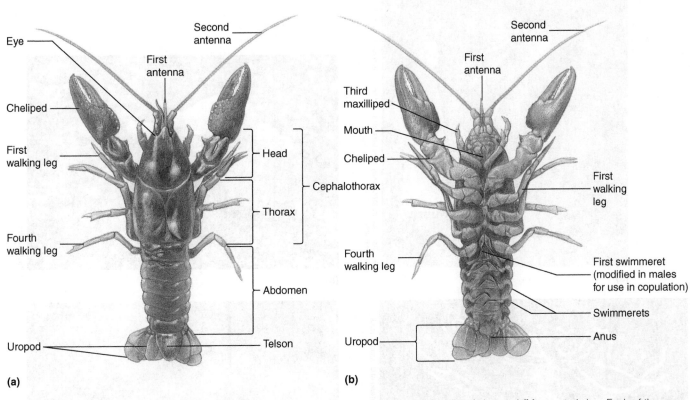

(a)

(b)

© Cengage Learning

FIGURE 9-33 **CRUSTACEAN ANATOMY.** The external anatomy of a lobster as seen from **(a)** a dorsal view and **(b)** a ventral view. Each of the animal's body segments has a pair of appendages. Some of the appendages are sensory, some are used for feeding, and others are used for locomotion and copulation.

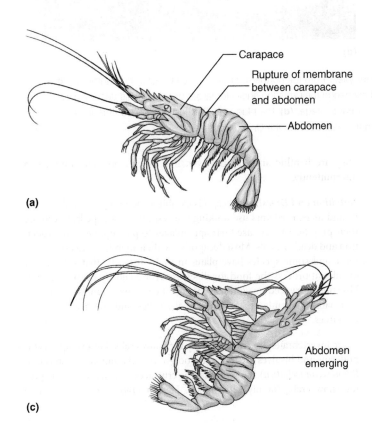

(a)

(b)

(c)

© Cengage Learning

FIGURE 9-34 **MOLTING.** The nonliving exoskeleton of crustaceans must be shed periodically in a process called molting. **(a)** Old membranes rupture along a joint line. **(b)** The animal pulls itself free of the anterior portion of the exoskeleton and then **(c)** crawls out of the posterior part. Once the old exoskeleton has been loosened, it usually takes about 15 minutes for the animal to shed it. After molting, the crustacean's body is soft, and the animal is particularly vulnerable until, in several days, the new exoskeleton is formed.

FIGURE 9-35 DECAPODS. **(a)** The Japanese spider crab, *Macrocheira kaempferi,* from the northern Pacific, is the largest living arthropod. **(b)** Hermit crabs inhabit the empty shells of molluscs. As they grow they must exchange their existing shells for larger ones. **(c)** This decorator crab from the Gulf of Mexico attaches a variety of sessile organisms to its shell for camouflage. **(d)** The blue crab *Callinectes sapidus* (the name means "beautiful swimmer that tastes good") is fished commercially for food along the east coast of the United States.

Specialized Behaviors Many decapods display interesting adaptations and behaviors that help them to survive in their habitat. Hermit crabs (family Paguridae) have a soft abdomen. They inhabit empty gastropod shells or other suitable enclosures **(Figure 9-35b)**, and as they grow they not only molt but also must find a larger shell to accommodate their ever-increasing bodies. When disturbed, these animals withdraw into their shells and close the opening with a large cheliped. Decorator crabs (family Majidae) attach bits of sponge, anemones, and hydrozoans to their carapace for camouflage **(Figure 9-35c)**.

Most crabs cannot swim, but members of the family Portunidae, which includes the common edible blue crab (*Callinectes sapidus*; **Figure 9-35d**) of the Atlantic coast, are the most powerful and agile swimmers of all crustaceans. The last pair of legs in members of this family terminate in broad flattened paddles that during swimming move in a figure-eight pattern. The action is essentially like that of a propeller. Portunids can swim sideways, backward, and sometimes forward with great rapidity.

They are benthic animals, however, like other crabs, and only swim intermittently.

Nutrition and Digestion Decapods exhibit a wide range of feeding habits and diets. Predators use walking legs, especially chelipeds, to capture their prey. Scavengers use their appendages to pick up pieces of vegetation and dead animals. Most decapods use their mandibles to crush their food, and some species have plates in their stomachs that will further grind and process the food for digestion. The majority of species combine predatory feeding with scavenging. The relative importance of these two habits depends on the individual species and the available food resources.

Large invertebrates are common prey for many crabs. For example, echinoderms and bivalves are the main food of the Alaskan king crab (*Paralithodes camtschatica*), and polychaetes, crustaceans, and bivalves are prey for snow crabs (family Lithodidae). Hermit crabs and many shrimp

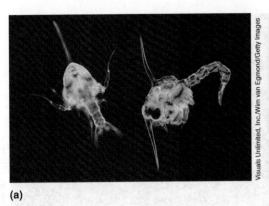

(a)   **(b)**

FIGURE 9-36 CRUSTACEAN LARVAE. (a) The zoea larva is a common larval stage in crabs and some shrimp. The spines may help deter predators such as fish. **(b)** The nauplius larva is the larval stage of shrimp, copepods, and barnacles.

Copulatory pleopods are the anterior two pair of abdominal appendages in male decapods which are adapted for delivering spermatophores to a female.

A **zoea** is the planktonic larval stage of some crustaceans such as crabs and mantis shrimp.

A **nauplius** is the planktonic larval stage of some crustaceans such as shrimp and barnacles.

Krill are shrimp-like crustaceans belonging to the order Euphausiacea.

Photophores are bioluminescent organs found in krill.

Swarms are large masses of krill.

species are scavengers or feed on detritus. Fiddler crabs (*Uca*) are deposit feeders. They scoop up mud and detritus with their small chelipeds, and the food is filtered and processed in the oral cavity. After organic material has been removed, the mineral residue is spit out in the form of round pellets, which may eventually surround the crab's burrow or cover the surface of the beach. Filter feeders include mole crabs, many burrowing shrimp, some commensal pea crabs, and most porcelain crabs.

Reproduction In decapods the sexes are usually separate. Males transmit sperm in packages known as spermatophores, which in most species are delivered to the female by the male's anterior two pairs of abdominal appendages called **copulatory pleopods**. Fertilization is generally internal.

Most decapods brood their eggs and have brood chambers or modified appendages for this purpose. Shrimp shed their eggs into the water column. When most crabs hatch they are in the **zoea** larval stage **(Figure 9-36a)**, which is easily recognized by the very long rostral spine and sometimes lateral spines. The spines appear to reduce predation by small fish. In shrimp species the **nauplius** (NAW-plee-uhs) larva **(Figure 9-35b)** hatches from the egg. During subsequent molts the decapod larval stages gradually mature into the adult forms.

Mantis Shrimp

Mantis shrimp (order Stomatopoda) are highly specialized predators of fish, crabs, shrimp, and molluscs, and many of their distinctive features are related to their predatory behavior. Most mantis shrimp are tropical, but *Squilla empusa* is common along the North American Atlantic coast. Most mantis shrimp live in rock or coral crevices or in burrows excavated in the bottom sediments.

In mantis shrimp the second pair of thoracic appendages is enlarged and has a movable finger that can be extended rapidly to capture prey and for defense **(Figure 9-37)**. Mantis shrimp can either spear (spearers) or smash (smashers) their prey with these thoracic appendages. Spearers feed on soft-bodied invertebrates such as shrimp and fish. Smashers, which live in holes and crevices on rocky or coralline bottoms, stalk their prey, mostly snails, clams, and crabs. Their hard-bodied victims are smashed with the heavy heel of the unfolded appendage. The blows of a mantis shrimp's thoracic appendages are so powerful that captured specimens have cracked the glass walls of aquaria.

Some mantis shrimp pair for life, sharing the same burrow. Others come together only at the time of mating. The female produces an egg mass

that can be as large as a walnut and contain as many as 50,000 eggs. A zoea larva hatches from the egg and may stay in the planktonic form for 3 months before settling and taking up an adult existence.

Krill

Krill (order Euphausiacea) are pelagic shrimp-like creatures that are about 3 to 6 centimeters (1 to 2 inches) long **(Figure 9-38)**. They are filter feeders that feed primarily on zooplankton. Most species of krill are bioluminescent, producing light in a specialized organ called a **photophore**. It is thought that the luminescence is a signal to attract individuals into large masses called **swarms**. It may also function in reproduction to attract mates. Many Antarctic species, such as *Euphausia superba*, live in large swarms that can cover an area of several hundred meters and be as thick as 5 meters (16.5 feet). The density of krill in these swarms can be as great as 60,000 individuals per cubic meter of water. Krill such as *Euphausia superba* constitute the main diet of some whales, seals, penguins, and many fishes. A single blue whale (*Balaenoptera musculus*) can consume a ton of krill in one feeding and may feed four times a day. *Euphausia superba* can molt so quickly that, when alarmed, individuals literally jump out of their skins. The shed molt may then function as a decoy.

FIGURE 9-37 MANTIS SHRIMP. The second pair of thoracic appendages in mantis shrimp are modified into structures that can be used as weapons to capture and kill prey.

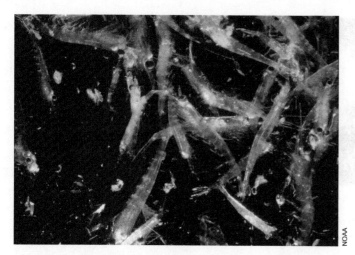

FIGURE 9-38 **KRILL.** Krill are important members of the zooplankton and the favorite food of many marine mammals and penguins. Most species of krill, such as this one, are bioluminescent.

FIGURE 9-39 **AMPHIPOD.** The beach flea (*Talitrus*) is a common amphipod along sandy shores.

Amphipods

Amphipods (order Amphipoda) have bodies that tend to resemble shrimp. The posterior three pairs of appendages are directed backward and are used for jumping, burrowing, or swimming, depending on the species. Most amphipods are between 5 and 15 millimeters (0.2 and 0.6 inches) long, although *Alicella gigantea* may reach 14 centimeters (6 inches).

Many amphipods are burrowers, and some live in tubes that they build. Several tube dwellers such as species of *Siphonoectes* and *Cerapus* build tubes of shell fragments and sand grains that they carry with them. A species of *Pseudamphithoides* builds its tube from a species of brown alga. The amphipod uses foul-tasting chemicals produced by the alga to protect itself from reef fish.

The beach flea *Talitrus* **(Figure 9-39)** is known to use its eyes to obtain astronomical clues for locating the high tide zone where they normally live. If moved above or below the high tide line, they migrate accurately to their normal zone. They use the angle of the sun as a compass, along with a map sense of the east–west orientation of the beach they inhabit. Beach fleas have retained gills although they live out of water; as a result, they are restricted to living in moist sand beneath drift or in damp forest leaf litter away from the sea. They feed at night when there is less danger of desiccation.

Amphipods are crustaceans belonging to the order Amphipoda.

Gnathopods are special appendages, found in amphipods, that are used in feeding.

Copepods are small planktonic crustaceans belonging to the class Copepoda.

Some amphipods are herbivores, but most are detritus feeders or scavengers. Animal and plant remains are picked up with special appendages called **gnathopods**, or detritus is raked from the bottom with antennae. Some species use their appendages to filter food from the water.

Amphipod males are attracted to females by pheromones. In some species of *Gammarus*, once the male has found a mate, he will carry her beneath him for days. Eggs are fertilized in the female's brood chamber, and when the young hatch they resemble the adults.

Copepods

Copepods (class Copepoda) are the largest group of small crustaceans. Marine copepods exist in enormous numbers and are usually the most abundant members of the marine zooplankton. Copepods in the genus *Calanus* **(Figure 9-40)**, for instance, constitute a major portion of the zooplankton and are a major food source for several species of commercially valuable fishes, as well as some whales, sharks, and birds. Although most planktonic copepods live in the upper 50 meters (165 feet) of the sea, many species live at greater depths, even in the deep sea. Vertical movement is oriented by light, and many species exhibit daily vertical migrations.

Planktonic copepods are chiefly suspension feeders, feeding on phytoplankton, although some rely heavily on detritus as well. By using radioactive diatom cultures, *Calanus finmarchicus*, a copepod about 5 millimeters

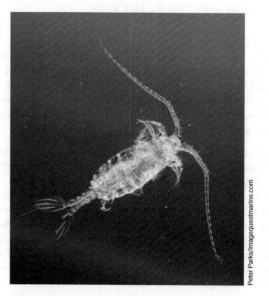

FIGURE 9-40 **COPEPOD.** Copepods in the genus *Calanus* constitute a major portion of the zooplankton that channels energy from phytoplankton to larger consumers.

(less than ¼ inch) long, was found to collect and digest from 11,000 to 373,000 diatoms every day, depending on their size. A few species of copepod are omnivorous or are strictly predators. Species of *Anomalocera* and *Pareuchaeta*, for example, even capture young fishes. Species of *Tisbe* are known to swarm over a small fish and eat its fins, immobilizing it. They then devour the body as it sinks to the bottom. Carnivorous species swim continually as they seek prey.

Male copepods are commonly smaller than, and usually outnumbered by, females. The males are among the few small crustaceans that form spermatophores. The spermatophores are transferred to the female by the thoracic appendages, and they adhere to the female by means of a special cement. Most copepods shed their eggs into the water column, where the young hatch.

Barnacles

Barnacles (class Cirripedia), the only sessile crustaceans, can be found attached to rocks, shells, coral, floating timber, and other solid objects. Some species of barnacle attach to ship hulls in large numbers, increasing the ship's drag and slowing it. Other barnacles attach to plants or to animals such as whales and large fish. The bodies of most barnacles are covered by a shell of calcium carbonate. The shell can be attached directly to a hard surface, as in acorn barnacles (*Balanus*; **Figure 9-41a**), or be attached by a stalk, as in the goose barnacles (*Lepas*; **Figure 9-41b**). When the shell is open, feathery appendages known as **cirripeds** extend into the water to filter food such as microorganisms and detritus from the water.

Barnacles are hermaphroditic and usually cross-fertilize using a long extensible penis (the largest per body length of any animal) to copulate. A nauplius larva hatches from the brooded egg and develops into a **cyprid larva** that has compound eyes and a carapace composed of two shell plates. When the cyprid larva finds a suitable surface, it attaches by its antennae, which contain adhesive glands, and metamorphoses into an adult (**Figure 9-41c**).

ECOLOGICAL ROLES OF ARTHROPODS

Arthropods are essential links in food chains and a food source for animals both marine and terrestrial. They are common symbionts of marine organisms in almost every phylum. Some arthropods play an important role in nutrient recycling. Others are fouling organisms that cause commercial damage.

Arthropods as Food

Many crustaceans are food sources for other marine animals, and some, such as crabs, lobsters, and shrimp, are important food sources for humans. Blue crabs, shrimp, and lobster are fished commercially along many parts of the eastern United States and the Gulf of Mexico. Other species, including the Alaskan king crab and snow crab, are fished commercially along the Pacific coast of the United States. The diet of many marine animals relies heavily on copepods. Some planktonic crustaceans feed on phytoplankton, and they are the principal link between phytoplankton and higher trophic levels in many marine food chains. Without them, animal populations in aquatic habitats would collapse. By reproducing efficiently these small crustaceans manage to maintain high populations despite heavy predation.

Krill is harvested for human consumption in the waters around Antarctica. As many as 12 tons have been netted in a single hour, but so far the total annual catch by humans is minuscule compared with consumption by other predators such as whales. It is assumed that the decline in whale populations, due largely to commercial whaling, has left a surplus of krill available for fishing.

Arthropods as Symbionts

A number of unrelated groups of shrimp, called cleaning shrimp, remove external parasites (ectoparasites) and other unwanted materials from the bodies of certain reef fishes. Some copepods are ectoparasites on fishes and attach to the gill filaments, the fins, or the skin in general. Other copepods are commensals or endoparasitic within polychaete worms, the intestines of echinoderms, in tunicates, and in bivalves. Cnidarians, especially anthozoans, are hosts for many species of copepod. Commensal crabs, called pea crabs (family Pinnotheridae) because of their small size, live in the tubes and burrows of polychaetes. Species also live in the mantle cavities of bivalves and snails, on sand dollars, in tunicates, and in other animals. An Antarctic species of amphipod, *Hyperiella dilatata*, carries a sea butterfly (*Clione antarctica*) attached to its back for protection.

> **Barnacles** are sessile crustaceans belonging to the class Cirripedia.
>
> **Cirripeds** are feathery appendages found in barnacles that function in feeding.
>
> A **cyprid larva** is a planktonic larval stage that develops from a nauplius larva in the life cycle of a barnacle.

Barnacles are commensal with a wide range of hosts: sponges, hydrozoans, octocorals, scleractinian corals, crabs, sea snakes, sea turtles, manatees, porpoises, and whales. There are also parasitic species of barnacle.

Role of Arthropods in Recycling and Fouling

The grass shrimp (*Palaemonetes pugi*) is found in large numbers in tidal marshes along the East Coast of the United States. It feeds on small bits of cellulose material it picks from large fragments of detritus and thereby plays an important role in the breakdown of algae and grasses of tidal marsh ecosystems.

Barnacles are among the most serious fouling problems on ship bottoms, buoys, and pilings. The speed of a badly fouled ship may be reduced by 30%. Much effort and money has been expended toward the development of special paints and other antifouling measures. Many barnacle species have been transported all over the world via shipping.

IN SUMMARY

- Arthropods have an exoskeleton, jointed appendages, and sophisticated sense organs.

- The arthropods known as chelicerates lack mouthparts, having instead appendages called chelicerae that are modified for feeding.

- Mandibulate arthropods have a pair of appendages called mandibles that are used for feeding.

- Arthropods must molt because their exoskeleton does not grow as they grow. (Have You Wondered? #4)

- In the marine environment, crustaceans are the most abundant mandibulates; examples include lobsters, crabs, shrimp, barnacles, copepods, amphipods, and numerous other small animals widely distributed in the marine environment.

- Crustaceans are important members of marine food webs.

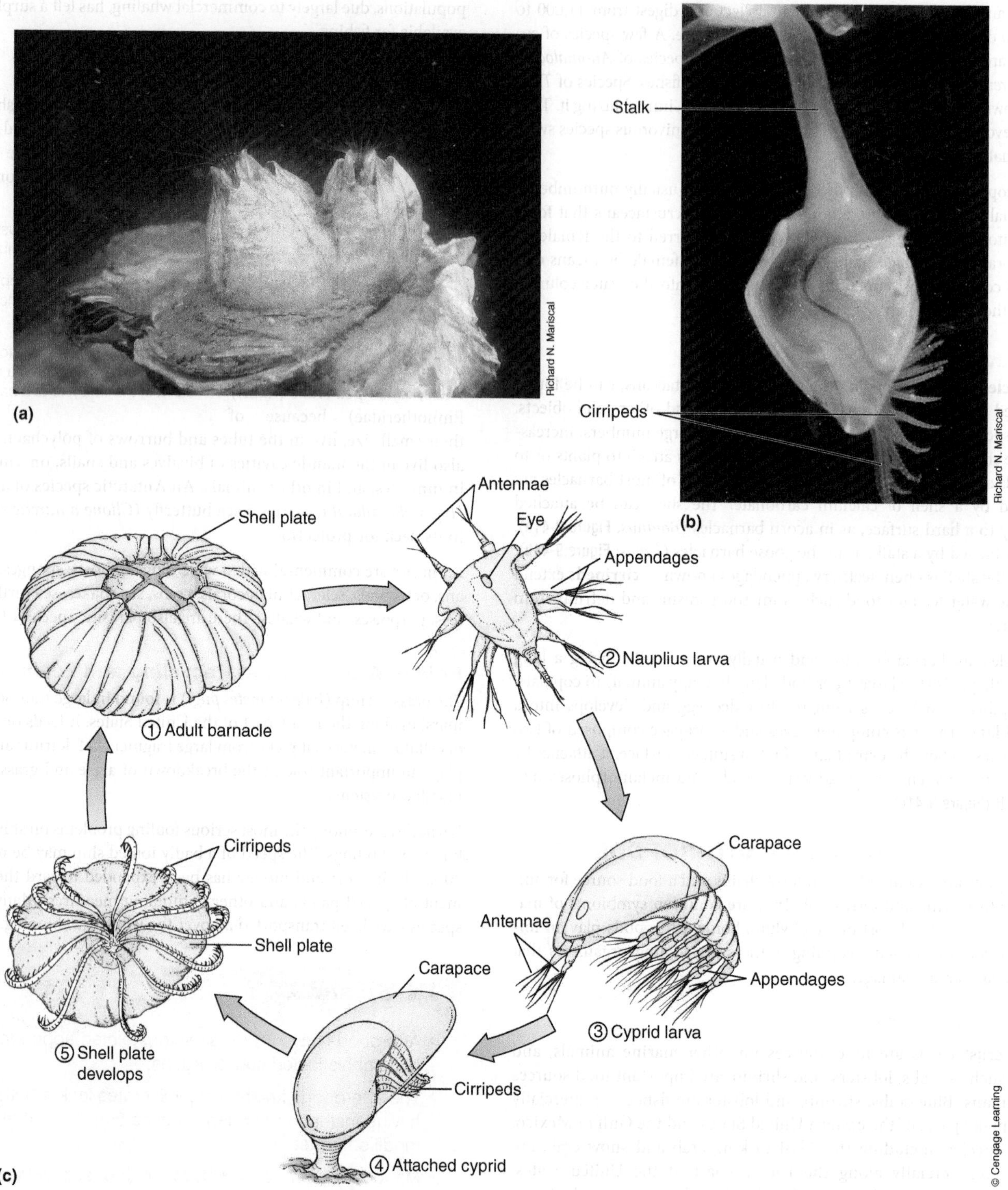

(a)

Richard N. Mariscal

Stalk

Cirripeds

(b)

Richard N. Mariscal

Shell plate

① Adult barnacle

Antennae

Eye

Appendages

② Nauplius larva

Carapace

Antennae

Appendages

③ Cyprid larva

Cirripeds

Shell plate

⑤ Shell plate develops

Carapace

Cirripeds

④ Attached cyprid

(c)

© Cengage Learning

FIGURE 9-41 BARNACLES. **(a)** Acorn barnacles cement their shell to solid surfaces. **(b)** Goose barnacles attach to solid surfaces by means of a long, fleshy stalk. **(c)** A generalized life cycle of an acorn barnacle. (1) After fertilization, eggs are brooded in the adult's shell until they develop into nauplius larvae. A single adult may release as many as 13,000 larvae. (2) Like other crustaceans, the nauplius has an eye, antennae, jointed appendages, and an exoskeleton. (3) After several molts, the nauplius develops into a cyprid larva. The cyprid has larger antennae and more body segments and appendages than the nauplius and a thin, but tough, carapace that protects its body. (4) Shortly after becoming a cyprid, the larva settles to the bottom and attaches to a solid surface, using specialized cement glands located on the antennae. Once attached, the cyprid molts and rotates its body so that its appendages, now called *cirripeds*, face upward. (5) The cyprid's carapace acts as a form around which the animal's shell develops. The feathery cirripeds filter food from the surrounding water.

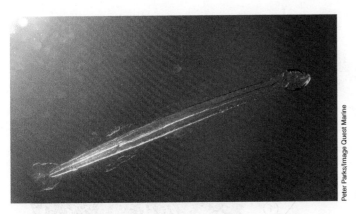

FIGURE 9-42 **ARROWWORM.** The arrowworm is a planktonic carnivore that feeds on copepods, other zooplankton, and small fish.

Peter Parks/Image Quest Marine

9-6 Arrowworms

Arrowworms (phylum Chaetognatha) are common animals found in marine plankton. Most are found in tropical waters in the upper 900 meters (2,970 feet). The body of an arrowworm is torpedo shaped **(Figure 9-42)** and ranges from 2 to 120 millimeters (0.08 to 2.4 inches) long. Hanging down from each side of the head and flanking the vestibule (the large chamber that leads to the mouth) are 4 to 14 large, curved hooks, called **grasping spines**, that are used to seize prey.

Arrowworms are all carnivorous and feed on other planktonic animals, particularly copepods, which they detect from the vibrations the copepods make. The arrowworm *Sagitta* consumes young fish and other arrowworms, including those of its own species, as large as itself. When capturing prey, arrowworms dart forward and spread their grasping spines. The prey is seized with the grasping spines and its body may be pierced with the arrowworm's teeth. The arrowworm then injects a toxin called tetrodotoxin, which is derived from bacteria, to immobilize its prey. Arrowworms are voracious feeders. *Sagitta nagae*, for example, consumes 37% of its own weight each day in prey. Arrowworms are important ecologically as a trophic link between the primary and higher consumers in marine food chains.

IN SUMMARY

- Arrowworms are predatory members of the zooplankton that feed on a variety of pelagic animals, including small fish.

9-7 Echinoderms: Animals with Spiny Skins

Echinoderms (phylum Echinodermata) are represented by well-known animals such as sea stars, sea urchins, and sea cucumbers. The name *echinoderm* means "spiny skinned" and refers to the spiny projections found on many species. It is thought that echinoderms evolved from a bilateral ancestor, because all of the larval forms still exhibit bilateral symmetry. Most adults, however, exhibit a modified form of radial symmetry. The move toward radial symmetry may have been an adaptation to a sessile lifestyle similar to that exhibited by some of the living echinoderms, such as sea lilies. As mentioned in Chapter 8, radial symmetry benefits a sessile animal because it allows the animal to meet its environment equally well on all sides.

Echinoderms are mostly benthic organisms and can be found at virtually all depths. In fact, sea cucumbers and brittle stars are usually the most common form of animal life in deep-sea dredging samples.

ECHINODERM STRUCTURE

The spiny covering of echinoderms is the result of the **endoskeleton** (internal skeleton) that lies just beneath the epidermis. The endoskeleton is composed of plates of calcium carbonate (**ossicles**) held together by connective tissue, and the spines and tubercles that produce the spiny surface of the echinoderm project outward from these plates. Around the bases of the spines are tiny pincer-like structures called **pedicellariae** (ped-uh-suh-LEHR-ee-eh; **Figure 9-43**). These structures are found only in some echinoderms. They keep the surface of the body clean and free of parasites and larvae of various fouling species. In some species, they may also aid in obtaining food.

The **water vascular system**, a hydraulic system that functions in locomotion, feeding, gas exchange, and excretion, is unique to echinoderms. Water enters the water vascular system by way of the **madreporite** and passes through a system of canals that runs throughout the animal's body. Attached to some of these canals are tube feet (**podia**). Each tube foot is hollow and has a sac-like structure, the ampulla, that lies within the body, and a sucker at the end that protrudes from an opening along the arm called the **ambulacral groove**. In some species, however, the terminal suckers are lacking. Later in this section we will examine how different species use their tube feet for locomotion and feeding.

SEA STARS

A typical sea star (class Asteroidea) is composed of a central disc and five arms, or rays. The mouth is located on the underside, and radiating from the mouth along each ray is an ambulacral groove with tiny tube feet (see **Figure 9-43**). The **aboral surface** (the side opposite the mouth) is frequently rough or spiny.

Sea stars move when water is pumped into the tube feet from the ampullae, causing them to project from the ambulacral groove. The suckers on the end hold firmly to solid surfaces, and muscles in the tube feet contract, forcing water back into the ampullae and causing the tube feet to shorten. As a result, the animal is pulled over the surface in the desired direction. This type of locomotion is best suited for hard surfaces such as rocks. When the animal is moving across sand, the tube feet move in a walking type of motion. Sea star locomotion is generally slow, moving the animal along at the rate of a few centimeters per minute.

Feeding in Sea Stars

The majority of sea stars are either carnivores or scavengers. They feed on all sorts of invertebrates and even fish. Most sea stars locate their prey by sensing

Arrowworms are planktonic organisms belonging to the phylum Chaetognatha.

Grasping spines are structures on the heads of arrowworms used to catch prey.

Echinoderms are animals with spiny skins belonging to the phylyum Echinodermata.

An **endoskeleton** is an internal skeleton.

Ossicles are plates of calcium carbonate that make up the echinoderm endoskeleton.

Pedicellariae are pincer-like structures found on the skin of echinoderms.

A **water vascular system** is a system of tubes found in the body of echinoderms that circulates water through the body and functions in locomotion, feeding, gas exchange, and excretion.

The **madreporite** is the site at which water enters the water vascular system of echinoderms.

Podia are tube feet.

The **ambulacral groove** is a groove in which the podia are located.

The **aboral surface** is the side opposite the mouth in an echinoderm.

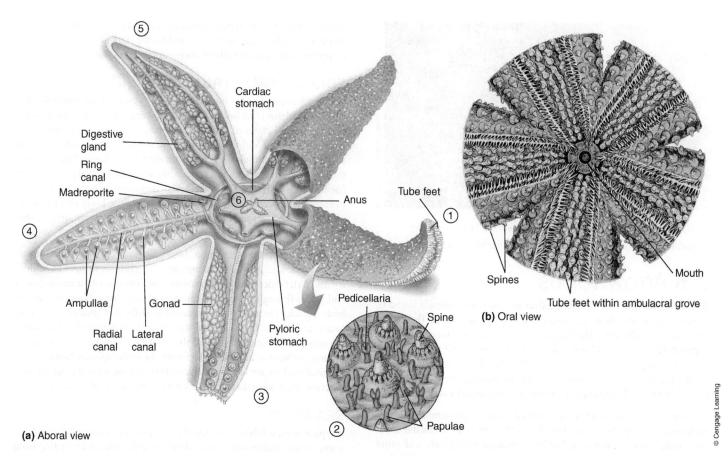

(a) Aboral view

FIGURE 9-43 **SEA STAR. (a)** An aboral view showing the general internal and external anatomy of a sea star, a representative echinoderm. (1) The aboral surface of a sea star ray with the end turned up shows the tube feet on the oral surface. (2) A close-up view of the aboral surface of a sea star ray shows that the spines are outward projections of the bony plates (ossicles) beneath the skin. The pincer-like pedicellariae keep the surface free of parasites and larvae. The papulae function in gas exchange and excretion. (3) Paired gonads are located in each ray. (4) Radiating from the ring canal, like spokes on a wheel, are the radial canals. The radial canals are attached by short side branches (lateral canals) to the ampullae of the tube feet. (5) Each ray contains digestive glands. (6) The central disc contains the stomachs, anus, madreporite, and ring canal. **(b)** An enlarged oral view of a sea star. The mouth is centrally located. Notice the rows of tube feet within the ambulaccral grooves of each ray.

substances the prey releases into the water. Some sea stars that inhabit soft-bottom habitats, including species of *Luidia* and *Astropecten*, can even locate prey that is buried in sediments. Sea stars use their tube feet in feeding. When feeding on large bivalves such as mussels the sea star wraps around its prey and pries the valves apart **(Figure 9-44)**. The sea star everts a portion of its stomach out of its mouth, and the stomach is inserted into the bivalve, where it digests the prey. Digestive enzymes are supplied by pairs of digestive glands located in each of the sea star's rays. When the sea star has finished feeding, it withdraws its stomach back into its mouth and moves away from its prey.

Reproduction and Regeneration

Sea stars have great powers of regeneration. They can produce new rays if one or more are removed, and some species are able to cast off injured rays and then grow new ones. Several species are capable of regenerating entire individuals from a single ray as long as a portion of the central disc is present. A number of sea stars normally reproduce asexually. Commonly, this involves a division of the central disc so that the animal breaks into two parts. Each half then regenerates the missing portion of the disc and arms, although extra arms are commonly produced. Species of *Linckia*, a genus of common sea stars in the Pacific and other parts of the world, are remarkable in being able to cast off arms near the base of

FIGURE 9-44 **SEA STAR FEEDING.** Some sea stars use their tube feet to pry open the shells of bivalves. When the valves are opened, the sea star everts a portion of its stomach and secretes enzymes to break down the flesh of its prey.

the disc. Unlike those of other sea stars, the severed arm regenerates a new disc and rays. Regeneration is a slow process, usually requiring 1 year before the organism is completely reformed.

With few exceptions the sexes in sea stars are separate. In the majority of sea stars the eggs and sperm are shed freely into the seawater, where fertilization takes place. There is usually only one breeding season per year, and a single female may shed as many as 2.5 million eggs. In most sea stars the liberated eggs and larval stages are planktonic, although some Arctic and Antarctic species brood their eggs beneath the central disc.

OPHIUROIDS

The class Ophiuroidea contains the greatest number of echinoderm species. Brittle stars, basket stars, and serpent stars are all examples of echinoderms known as **ophiuroids**. Ophiuroids are benthic organisms that can be found from shallow water to the ocean's depths. Like sea stars, they have five arms, but the arms are very slender and distinct from the central disc and frequently are covered with many spines (**Figure 9-45a**). Ophiuroids lack pedicellariae, and their ambulacral grooves are closed. The tube feet play a role in feeding and locomotion but do not have suckers. Ophiuroids tend to avoid light, coming out at night to feed and hiding under rocks and in crevices during the day. Many species are burrowers, burying themselves in bottom sediments with only their arms trailing over the seafloor to trap food. Ophiuroids known as brittle stars get their name because they will detach one or more arms when disturbed, and some species may even shed a portion of their central disc. Once an arm is shed, it undulates wildly, no longer controlled by the central nervous system. The undulating arm distracts potential predators while the brittle star moves away to safety. It later regenerates the missing part. Because the arms of many ophiuroids tend to writhe in a serpentine fashion, they are also called serpent stars.

Feeding in Ophiuroids

Ophiuroids can be carnivores, scavengers, deposit feeders, suspension feeders, or filter feeders. Brittle stars are mostly suspension feeders, feeding on food particles suspended in the water, or deposit feeders, feeding on organic material that they find on the bottom. When suspension feeding, they lift their arms from the bottom and wave them through the water. Strands of mucus strung between the spines on adjacent arms form a net to trap plankton and organic material. The food is then either swept to the mouth by the action of cilia or collected from the spines by tube feet and passed to the mouth. Deposit feeders use their tube feet in feeding to collect organic particles from bottom sediments, compact them into food balls, and move them toward the mouth. Brittle stars that are predominantly carnivores feed largely on polychaetes, molluscs, and small crustaceans.

Basket stars (**Figure 9-45b**) are suspension feeders that can capture relatively large zooplankton (10 to 30 millimeters, or 0.4 to 1.2 inches). They climb up on corals or rocks at night and fan their arms toward the prevailing water current. They capture their prey by coiling the ends of their arm branches around it. Minute hooks on the arm surface prevent the prey from escaping. The basket star removes the collected plankton from the arms by passing them through their comb-like oral papillae.

Reproduction and Regeneration in Ophiuroids

Many ophiuroids can cast off, or **autotomize**, one or more arms if disturbed or seized by a predator. A break can occur at any point beyond the disc and the lost portion is then regenerated. In some ophiuroids, notably six-armed species of *Ophiactis*, asexual reproduction takes place by division of the disc into two pieces, each with three arms.

Although sexes are separate in the majority of ophiuroids, hermaphroditic species are not uncommon. Many species shed their gametes into the water column, where fertilization and development take place. Some species, however, brood their eggs in their ovaries or body cavity. The larval stages in all species are planktonic, and metamorphosis occurs in the water column before settling.

SEA URCHINS AND THEIR RELATIVES

Sea urchins, heart urchins, sand dollars, and their relatives are known as **echinoids** (class Echinoidea, meaning "like a hedgehog"). Echinoids have a body that is enclosed by a hard endoskeleton, or **test**. These

> **Ophiuroids** are echinoderms belonging to the class Ophiuroidea.
>
> **Autotomize** is to cast off a body part.
>
> **Echinoids** are echinoderms belonging to the class Echinoidea.
>
> A **test** is the hard endoskeleton of an echinoid echinoderm.

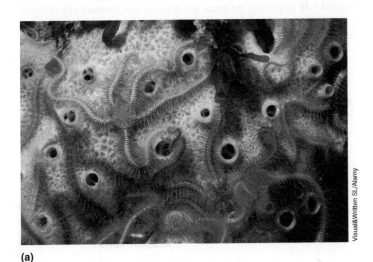

(a)

(b)

FIGURE 9-45 BRITTLE STARS. (a) Several serpent stars (*Ophiothrix*) on the surface of a sponge. The snake-like movement of their rays (arms) is the source of their common name. **(b)** A basket star *Gorgonocephalus arcticus* from Friday Harbor, Washington. Basket stars use their highly branched rays to capture plankton.

(a)

(b)

FIGURE 9-46 ECHINOIDS. (a) This purple sea urchin *Strongylocentrus purpuratus* lives off the West Coast of the United States. During the day it wedges itself into rock crevices. At night it grazes on the algae that grow on the surface of the rocks. (b) Irregular echinoids such as the heart urchin are adapted to burrowing and have much smaller spines than sea urchins.

animals are benthic organisms that can be found from shallow water to the ocean's depths. Sea urchins tend to occupy solid surfaces such as rocks, whereas sand dollars and heart urchins prefer to bury themselves in sandy bottoms.

Regular (radial) **echinoids** are known as sea urchins. A sea urchin's spheroid body is armed with relatively long, movable spines **(Figure 9-46a)**. Most are 6 to 12 centimeters (2.5 to 5 inches) in diameter, but some Indo-Pacific species may reach nearly 36 cm (14.5 inches).

The bilateral or **irregular echinoids** include heart urchins and sand dollars. These animals are adapted for burrowing in sand. Unlike sea urchins, the test in these species is covered with many small spines, which are used in locomotion and keep sediment off the body surface. Heart urchins (family Spatangidae) are ovoid **(Figure 9-46b)**. The typical sand dollar (family Mellitidae) has a flattened circular body. A few species such as sea biscuits (*Clypeaster*) are shaped somewhat like heart urchins.

Echinoid Structure

The tube feet of echinoids project from five pairs of ambulacral areas that are derived from the same embryonic structures as the arms of sea stars, and spines project from the test **(Figure 9-47)**. Sea urchins have relatively long spines, and they move mostly by means of their tube feet, with some help from their spines. The spines function in protection and, in some species (*Diadema*), contain a venom. When a person comes into contact with one of these stinging urchins, the spine tip breaks off in the skin and injects the venom, which causes a severe burning sensation. Some spines have barbed spinelets along their length that make them difficult to remove. The spines allow some species to live along the shore in regions pounded by surf. They help dissipate the energy of the waves and protect the animal from wave shock. Sand dollars and heart urchins have very short spines that give them a fuzzy appearance, an adaptation for a burrowing lifestyle.

The sexes are separate in all echinoids. Regular echinoids have five gonads, but most irregular echinoids have only four. Sperm and eggs are shed into the seawater, where fertilization takes place. The planktonic larvae swim and feed for as long as several months before gradually sinking to the bottom.

Feeding in Echinoids

Most regular echinoids are herbivores that feed on algae and some marine plants. Irregular echinoids are deposit feeders or suspension feeders.

Feeding in Regular Echinoids The majority of sea urchins are grazers, scraping the surfaces on which they live with their teeth. A sea urchin's mouth contains five teeth that form a chewing structure called **Aristotle's lantern** (see **Figure 9-47**) because of its resemblance to ancient Greek lanterns and in honor of that early student of echinoderm biology. Although algae is usually the most important food, most sea urchins are generalists and feed on a wide range of plant and animal material. The sea urchin *Lytechinus variegatus* inhabits turtle-grass beds and is a herbivore, consuming about 1 gram of grass per day. The ecological role of grazing by some sea urchins was dramatically revealed

Regular echinoids are echinoids with spherical bodies, commonly referred to as sea urchins.

Irregular echinoids are echinoids whose bodies are not spherical.

Aristotle's lantern is a feeding structure formed from the five teeth of a sea urchin.

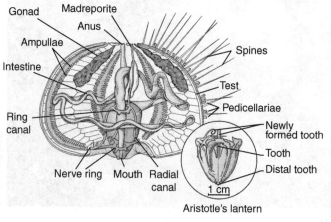

FIGURE 9-47 **SEA URCHIN.** The anatomy of a sea urchin showing details of Aristotle's lantern. Compare this with the sea star anatomy in Figure 9-43.

following the 1983 crash in Caribbean populations of the long-spined urchin *Diadema antillarum*. In some areas where the urchins died out, the thickness of the algal mat increased from 1 to 2 millimeters to 20 to 30 millimeters. Boring sea urchins feed on algae that grow on the walls of their burrows, as well as fragments of algae and organic matter that are washed into the burrow.

Feeding in Irregular Urchins Irregular urchins are selective deposit feeders. Most species feed on organic material in the sand in which they burrow. Heart urchins use modified tube feet on their oral surface to collect food. Sand dollars use their tube feet to pick up particles beneath their oral surface. On the West Coast of the United States, the sand dollar *Dendraster excentricus* feeds on suspended particles. This sand dollar lives in quiet water with the posterior half of its body projecting above the sand. Food includes not only particles collected between the spines but also diatoms and algal fragments collected by the tube feet and small crustaceans caught by the pedicellariae.

SEA CUCUMBERS

Sea cucumbers (class Holothuroidea; **Figure 9-48a**) have elongated bodies, and, in most species, the body wall is leathery. They range from 3 centimeters (1.2 inches) long to 1 meter (3.3 feet) long and 24 centimeters (10 inches) in diameter (*Stichopus* from the Philippines). Because of their body shape, sea cucumbers usually lie on one side. They move slowly, using their ventral tube feet and the muscular contractions of their body wall. Gas exchange in most sea cucumbers is accomplished by a system of tubules called **respiratory trees** located in the body cavity on the right and left sides of the digestive tract (**Figure 9-48b**). The sexes are generally separate. In species that do not brood, the eggs are incubated on the dorsal body surface. Some species brood their eggs in their body cavity, and the larvae leave the body of the mother through a rupture in the anal region. After hatching, the larvae of all sea cucumbers continue development in the water column as members of the plankton.

Feeding in Sea Cucumbers

Sea cucumbers are mainly deposit or suspension feeders. Located around the mouth of a sea cucumber are 10 to 30 modified tube feet called **oral tentacles** (see **Figure 9-48b**). Most sea cucumbers feed on small particles of food that they trap with these tentacles. The tentacles are coated with a sticky mucus, and any small organism that comes into contact with them becomes stuck. Sea cucumbers retract their tentacles into their mouths to remove the food and then extend them to capture more. Other species use their tentacles to shovel sand and bottom sediments into their mouths. They then digest the organic material, leaving conspicuous mounds of sand or fecal-mud castings behind. Many sedentary species that live on hard surfaces or beneath stones are suspension feeders.

Defensive Behavior

When disturbed, some species of sea cucumber release tubules (**Cuvierian tubules**) from their anus (**Figure 9-49**). The tangle of tubules resembles spaghetti, and, on contact with seawater, they become sticky. The tubules will stick tenaciously to a predator, which is distracted by the need to clean itself, and the sea cucumber escapes. The tubules are strong enough to ensnare and immobilize crustaceans, and they are distasteful to fishes. This behavior helps slow-moving sea cucumbers deter potential predators.

Other species will **eviscerate**, that is, release some of their internal organs through either the anus or the mouth. In each instance, the animal ultimately regenerates the lost body parts. In some species, evisceration appears to be seasonal and may represent the release of a digestive tract that is full of wastes following a long feeding period. These waste-laden intestines serve as an important food source for detritus feeders.

> A **respiratory tree** is a system of tubules in a sea cucumber that functions in gas exchange.
>
> **Oral tentacles** are modified tube feet located around the mouth of a sea cucumber that function in feeding.
>
> **Cuvierian tubules** are sticky tubules ejected from the anus of some sea cucumbers that function in defense.
>
> **Eviscerate** is to release internal organs through the anus or the mouth.
>
> **Crinoids** are echinoderms belonging to the class Crinoidea.

CRINOIDS

Crinoids (class Crinoidea) are commonly called sea lilies and feather stars, and their bodies look very much like flowers. They are the most primitive of the echinoderms and have a long fossil history. Attached stalked crinoids called sea lilies flourished during the Paleozoic era, and some 80 species still exist today (**Figure 9-50a**). The stalk of sessile sea lilies may reach a length of 1 meter (3.3 feet). Modern sea lilies, however, live

(a)

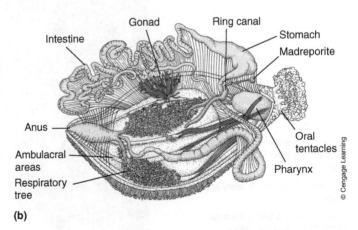

Intestine · Gonad · Ring canal · Stomach · Madreporite · Oral tentacles · Pharynx · Anus · Ambulacral areas · Respiratory tree

© Cengage Learning

(b)

FIGURE 9-48 SEA CUCUMBER. (a) Although it resembles its namesake, this sea cucumber is an animal. Like other echinoderms it has five rows of tube feet that it uses for locomotion. Sea cucumbers use their oral tentacles to obtain food. **(b)** The anatomy of a sea cucumber.

Richard N. Mariscal

FIGURE 9-49 SEA CUCUMBER DEFENSE. This sea cucumber from the Red Sea is releasing Cuvierian tubules. The sticky tubules distract potential predators, giving the sea cucumber time to escape.

at depths of 100 meters (330 feet) or more and are not commonly encountered. The majority of living crinoids are feather stars (**Figure 9-50b**). These are free-living crinoids whose habitat ranges from the intertidal zone to great depths. Some occur in large numbers on coral reefs. Feather stars are free moving. They swim and crawl only for short distances, and swimming is largely an escape response. They cling to the bottom for long periods by means of the grasping **cirri**. Many of the shallow-water species are nocturnal. For instance, three Red Sea species that inhabit coral reefs hide during the day in crevices and deep within branching corals, keeping their arms tightly rolled. Stimulated by the lowered light intensities at sunset, they crawl out of their hiding places to exposed positions and extend their arms to feed.

Crinoids are suspension feeders. They feed on small organisms that they capture with their tube feet and with mucous nets. There is still some uncertainty about the precise nature of crinoid food. Gut contents of Red Sea crinoids are largely zooplankton, but *Antedon bifida* from the Indo-Pacific appears to feed largely on detritus.

Crinoids possess considerable powers of regeneration and in this respect are similar to sea stars. Part or all of an arm can be cast off if seized or if the animal is subjected to unfavorable environmental conditions. The lost arm is then regenerated. Such regeneration may be important in surviving fish predation. Sexes are separate in all crinoids, and eggs and sperm are shed into the seawater. After a free-swimming stage, the larva settles to the bottom and attaches. This is followed by a metamorphosis resulting in the formation of a minute crinoid.

ECOLOGICAL ROLES OF ECHINODERMS

Because of their spiny skins, echinoderms are prey for few animals. Some molluscs and sea otters eat sea urchins and sea stars. Many spider crabs feed on echinoderms, which they tear apart or crush with their

FIGURE 9-50 CRINOIDS. (a) Crinoids such as this sea lily have changed very little since they first evolved. Most modern species live at depths of greater than 100 meters. **(b)** The majority of living crinoids are feather stars, such as this one from the Pacific Ocean.

Cirri are hook-like structures on feather stars used for clinging to hard substrates.

Roe are an organism's ovary and eggs.

Holothurin is a toxic substance produced by sea cucumbers.

(a)

(b)

chelipeds. Humans eat sea urchin gonads and sea cucumbers. Some fish have mouth-parts modified for feeding on echinoderms.

On the other hand, many echinoderms are predators of molluscs, other echinoderms, cnidarians, and crustaceans. The crown-of-thorns sea star (*Acanthaster planci*) is a major predator of corals. Sea urchins have destroyed several valuable kelp forests and beds in many parts of the world, including the West Coast of the United States. In the Atlantic, excessive populations of sea urchins have become a nuisance by robbing lobster traps. Many former lobster fishers are now sea urchin fishers. They sell the **roe** (ovary with eggs) to Japan, where it is used in sushi, for $100 to $150 a pound. Throughout the Caribbean the populations of the black sea urchin (*Diadema antillarum*) exploded because of depletion of predatory fishes. The population of *Diadema* then suddenly collapsed because of an infectious disease, leaving the coral reefs without any effective control of algae. The resulting overgrowth of algae contributed to a 90% reduction in coral reef area since 1980. Only recently have the populations of *Diadema* started to rebound. The complexity and unpredictability of these population fluctuations and their effects remind us of how little we understand the role of echinoderms in marine ecology.

In the Pacific near China, sea cucumber beds have been depleted and fishers have begun to threaten beds off South America and the Galápagos Islands. In the future sea cucumbers may be collected as a source of medicine. Pacific Islanders have long known that cut-up sea cucumbers can be used to poison fish in tide pools. The poison, **holothurin**, has various effects on nerves and muscles and also suppresses the growth of certain tumors.

IN SUMMARY

- Echinoderms exhibit radial symmetry as adults, although their larvae exhibit bilateral symmetry, suggesting that they evolved from bilateral ancestors.

- Echinoderms have an internal skeleton and a unique water vascular system that functions in locomotion and food gathering.

- Echinoderms are represented by sea stars, brittle stars, sea urchins, sea cucumbers, and feather stars and sea lilies (crinoids).

- Sea stars are mainly predators.

- Echinoderm predators include molluscs, crabs, sea otters, and other echinoderms. (Have You Wondered? #5)

- Some echinoderms are herbivores, scavengers, suspension feeders, or deposit feeders.

9-8 Hemichordates

Hemichordates, or acorn worms (phylum Hemichordata), are sessile bottom dwellers **(Figure 9-51)**. They live burrowed in the sediments of intertidal mud or sand flats or under stones. An acorn worm uses its large proboscis to collect food from its surroundings. The food is trapped in mucus and then transported by cilia, located in grooves, to the collar and then into the mouth, where it is swallowed. Some species use their proboscis to dig burrows, which usually have two openings. The head extends from one opening for feeding, and the anus deposits a characteristic mound of fecal material around the second opening. At one time hemichordates were classified as chordates

DARLYNE A. MURAWSKI/National Geographic Stock

FIGURE 9-51 HEMICHORDATE. Acorn worms bury into sediments and use their proboscis to trap food from their surroundings.

because they possess the distinctive structures found in chordates (discussed below). However, it was determined that their notochord-like structure was different from the notochord of chordates, so they were placed into a separate phylum. It is now believed that they may share a common ancestor with chordates.

9-9 Invertebrate Chordates

Chordates (phylum Chordata) are animals that sometime in their life cycle have four key anatomical characteristics. These hallmarks of the chordate line are (1) a **notochord**, a rod-shaped structure composed of cells and covered by a sheath that forms the axial skeleton of chordate embryos and adult cephalochordates; (2) **pharyngeal gill slits**, slit-like openings in the neck area that are associated with gills in aquatic chordates; (3) a **postanal tail**, a tail that extends past the digestive tract; and (4) a **dorsal, hollow nerve tube**. It is generally believed that chordates evolved from a sessile filter-feeding ancestor that may have been similar to modern-day urochordates. Although the most familiar chordates are the vertebrate animals, there are some invertebrate members of the phylum as well.

Hemichordates are worm-like animals belonging to the phylum Hemichordata.

Chordates are animals that belong to the phylum Chordata and, at some time in their life cycle, possess a notochord, pharyngeal gill slits, a postanal tail, and a dorsal, hollow, nerve tube.

A **notochord** is a rod-shaped structure that forms the axial skeleton of chordates some time in their life cycle.

Pharyngeal gill slits are slit-like openings found in the neck of a chordate some time in its life cycle.

A **postanal tail** is a tail that extends past the digestive tract.

A **dorsal, hollow nerve tube** is a neural structure found in chordates some time in their life cycle.

Tunicates are chordate animals belonging to the subphylum Urochordata.

A **tunic** is the body covering of urochordates.

TUNICATES

Tunicates (subphylum Urochordata) are mostly sessile animals that are widely distributed in all seas. These animals are named for their body covering, called the **tunic**, which is largely composed of a substance similar to cellulose.

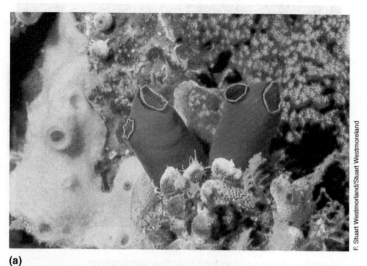

(a)

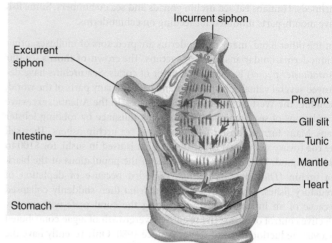

(b)

Incurrent siphon
Excurrent siphon
Pharynx
Intestine
Gill slit
Tunic
Mantle
Stomach
Heart

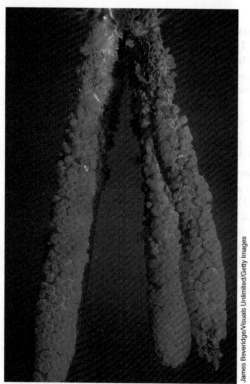

(c)

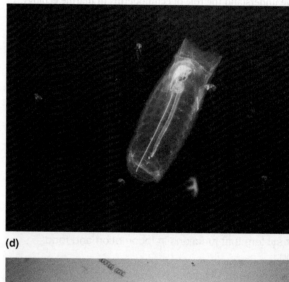

(d)

(e)

FIGURE 9-52 UROCHORDATES. (a) Two solitary blue and gold sea squirts on a reef in the Philippine Islands. **(b)** The blue arrows in this drawing of sea squirt anatomy represent the pattern of water flow through the animal's body, and the red arrows represent the movement of food. **(c)** A colonial mangrove tunicate on the root of a mangrove. **(d)** Thaliaceans, also known as salps are planktonic tunicates.
(e) Larvaceans are also planktonic tunicatres.

Sea Squirts

Tunicates known as sea squirts (class Ascidiacea; **Figure 9-52a**) get their name because many species, when irritated, will forcefully expel a stream of water from their excurrent siphon. The bodies of sea squirts are generally round or cylindrical and have two projecting tubes: an incurrent siphon that brings in water and food, and an excurrent siphon

that eliminates water and wastes. Food is trapped on a mucous net that is formed in the animal's pharynx **(Figure 9-52b)**. The pharynx is also used in gas exchange.

Sea squirts can be solitary, colonial, or compound organisms, which are composed of several individuals called zooids that share a common tunic. A great diversity of species inhabit shallow tropical seas, and many minute colonial forms inhabit crevices in old coral heads and the underside of coralline rock. Others form large conspicuous clusters on gorgonian corals and mangrove roots **(Figure 9-52c)** or massive rubbery lobes on rocks and pilings. Most of the larger sea squirts, such as *Ascidia* and *Molgula*, are solitary.

Sea squirts are filter feeders and remove plankton from the water passing through their pharynx. Some members of the colonial family Didemnidae contain symbiotic algae within the tunic. One of these symbionts, prokaryotes in the genus *Prochloron*, is found only in ascidians. These same ascidians may also have symbiotic cyanobacteria in their tunic. The excess food produced by these photosynthetic symbionts is thought to be used by the tunicate as an auxiliary food source. At least one *Didemnum* species may migrate over the bottom as it grows to optimize the light intensity for photosynthesis by its *Prochloron* symbionts.

All sea squirts can regenerate damaged body parts, but asexual reproduction is characteristic only of colonial tunicates. In these animals, asexual reproduction takes place by means of budding. With few exceptions tunicates are hermaphrodites that release their gametes into the water column, where fertilization takes place. The larval stage resembles a tadpole, and after a free-swimming period of 36 hours or less the larva settles and metamorphoses to the sessile stage.

Salps and Larvaceans

Salps (class Thaliacea; **Figure 9-52d**) are free-swimming tunicates that have incurrent and excurrent siphons located on opposite ends of their barrel-shaped bodies. As they swim, they pump water through their bodies, extracting food and eliminating wastes. When food supplies are abundant, the salp populations increase dramatically. Some species are bioluminescent.

Larvaceans (class Larvacea), another group of free-swimming tunicates **(Figure 9-52e)**, produce delicate enclosures made of mucus that are used in feeding. When these mucous networks are clogged (approximately every 4 hours) they shed them and produce another in a matter of minutes. Like salps, larvacean populations explode when food supplies are abundant, Scuba diving at these times among the shed mucous networks has been compared to swimming through a snowstorm.

CEPHALOCHORDATES

Cephalochordates (subphylum Cephalochordata) are fish-like chordates that are collectively known as **lancelets**. Lancelets are slender, laterally compressed, and eel-like in appearance and behavior **(Figure 9-53)**. Adult lancelets range in size from 4 to 8 centimeters (0.4 to 3.2 inches). Lancelets are benthic animals that burrow in coarse, shelly, current-swept sands, usually in shallow near-shore areas. Their heads project above the sand into the water, from which they filter suspended food particles. Lancelets ingest large quantities of particles from which they extract organic material while eliminating water through gill slits and mineral particles through the anus. The sexes are separate in cephalochordates, and fertilization is external. Their life cycles are complex, including a benthic adult and planktonic swimming larva.

Cephalochordates are an important food in parts of Asia. Despite their small size, lancelets are clean muscular animals that lack bones. Fresh animals may be fried for immediate consumption or dried for later use. One lancelet fishery in southern China recorded an annual catch of 35 tons of lancelets (*Brachiostoma belcheri*)—approximately 1 billion animals—from a fishing ground 1 mile wide and 6 miles long. Five thousand individuals of *Brachiostoma caribbaeum* per square meter have

FIGURE 9-53 CEPHALOCHORDATES. The lancelet *Amphioxus* has a body shaped like an eel. It lives in coastal sediments where it feeds on organic material it removes from particles suspended in the water.

OSF/G.I. Bernard/Animals Animals

been reported from Discovery Bay in Jamaica. In parts of Brazil, chickens are herded onto beaches to feed on lancelets.

ECOLOGICAL ROLES OF HEMICHORDATES AND INVERTEBRATE CHORDATES

Acorn worms are either deposit feeders or suspension feeders and are important in channeling nutrients to other organisms in food chains. The burrowing activity of some acorn worms helps to move nutrients to the surface of sediments, where they can mix back into seawater. The burrows they form may also provide habitat for other organisms.

Urochordates are filter feeders that channel nutrients to other organisms in benthic and planktonic food chains. Cephalochordates are filter feeders and provide food for other organisms, including humans.

Salps are free-swimming urochordates that belong to the class Thaliacea.

Larvaceans are free-swimming urochordates that belong to the class Larvacea.

Cephalochordates are fish-like chordates that belong to the subphylum Cephalochordata.

Lancelet is a common name for a cephalochordate.

IN SUMMARY

- Acorn worms live in bottom sediments, where they use their proboscis to obtain food.

- Chordates are animals that have a notochord, pharyngeal gill slits, a postanal tail, and a dorsal, hollow nerve tube sometime in their life cycle.

- Tunicates have bodies that are covered with a tunic composed of molecules similar to cellulose.

- Sea squirts are sedentary tunicates that filter their food from the water.

- Salps and larvaceans are planktonic tunicates.

- Cephalochordates, also known as lancelets, are small animals that resemble eels.

- Lancelets are found in the bottom sediments along coastal areas, where they filter food from the water.

IN PERSPECTIVE

Phylum or Subphylum	Representative Organisms	Form and Function
Mollusca	Chitons, snails, nudibranchs, bivalves, tusk shells, nautilus, octopus, squid, cuttlefish	Most are covered with a shell; notable exceptions are nudibranchs and cephalopods. With the exception of bivalves, all have a unique tooth structure called a radula. Gas exchange with gills or mantle cavity. Cephalopods have a highly developed nervous system.
Annelida	Polychaetes, spoon-worms, beardworms	Body generally segmented internally and externally. Polychaetes have pairs of appendages called paddle feet on most segments. The paddle feet are used for gas exchange and locomotion.
Nematoda	Nematodes	Generally small, round, nonsegmented worms
Arthropoda	Horseshoe crab, sea spider, crab, lobster, shrimp, krill, copepod, amphipod, barnacle	A hard or tough exoskeleton protects the body. Paired, jointed appendages aid in locomotion and feeding. Gas exchange is by gills or through body surface. Growth is by molting.
Chaetognatha	Arrowworm	Torpedo-shaped body with grasping spines around the mouth
Echinodermata	Sea stars, brittle stars, basket stars, sea urchins, sand dollars, sea cucumbers, and crinoids	Adults exhibit secondary radial symmetry. All have an endoskeleton and a spiny body surface. A unique water vascular system is used for locomotion, feeding, and circulating internal fluids.
Hemichordata	Acorn worms	Worm-like animals that have a large proboscis.
Subphylum Urochordata	Tunicates, sea squirts, salps, and larvaceans	Body is covered by a tunic composed of a polysaccharide similar to cellulose. Salps and larvaceans are planktonic. Sea squirts are benthic and can be solitary or colonial.
Subphylum Cephalochordata	Lancelets	Body resembles a small eel; adults are benthic.

Reproduction and Development	Type of Feeding	Ecological Roles
Mainly sexual reproduction; separate sexes or hermaphrodites depending on species; larval stages include trochophore and veliger	All types of feeding: herbivores, carnivores, filter feeders, suspension feeders, scavengers, and deposit feeders	Important links in food chains, food organisms for humans; many higher-level predators; shells of dead molluscs provide habitat for other species, shipworms cause commercial damage, several commensal species
Some reproduce asexually, the majority sexually; sexes are generally separate	All types: carnivores, herbivores, filter feeders, suspension feeders, deposit feeders, scavengers	Important predators, links in food chains, burrows provide habitat, recycle nutrients, symbionts
Sexes are usually separate	Carnivores, herbivores, scavengers, parasites	Important members of the meiofauna; channel nutrients to larger consumers
Mainly sexual reproduction; sexes separate or hermaphrodites depending on the species; larval stages include nauplius, zooea, and cyprid, depending on the class	All types: herbivores, carnivores, filter feeders, suspension feeders, scavengers, and deposit feeders	Planktonic crustaceans are main links between phytoplankton and higher-order consumers; many symbiotic species; some species involved in recycling nutrients; barnacles are important fouling organisms; food for human populations and other higher-order consumers
All are hermaphrodites; young resemble adults when they hatch	Predators that prey on zooplankton, especially copepods and small fishes	Control populations of zooplankton; important links in food chains
Can reproduce asexually and sexually; the bilateral larvae are planktonic; the larval stage of sea stars is the bipinnaria; ophiuroids and echinoids have a pluteus larva and the larval stage of sea cucumbers and crinoids is the pentacula or vitellaria larva	All types: herbivores, carnivores, filter feeders, suspension feeders, deposit feeders, and scavengers	Not preyed on by many animals because of their spiny exteriors; major predators of molluscs (especially bivalves), cnidarians, crustaceans, and other echinoderms as well as other invertebrates and sometimes fishes. Sea urchins can cause extensive damage to kelp. Their grazing in kelp forests and on coral reefs helps keep algal populations under control.
Most reproduce sexually and sexes are separate; larval stages have characteristics very similar to chordates	Mainly deposit feeders	Nutrient cycling and links in food chains.
Colonial forms can reproduce asexually; sexual forms are hermaphrodites; larvae are planktonic and referred to as tadpole larvae	Filter feeders	Channel nutrients to higher-level consumers; some harbor symbiotic photosynthetic bacteria
Sexes are separate and larval stage is planktonic	Feed on organic material extracted from particles filtered from the water	Channel nutrients to higher-level consumers

KEY CONCEPTS

1. Molluscs have soft bodies that are usually covered by a shell. (9-1)

2. Molluscs are important herbivores and carnivores in the marine environment. (9-1)

3. Polychaete diversity stems from the evolution of a segmented body that allows increased motility. (9-2)

4. In addition to being important consumer organisms, polychaetes are the primary prey of many marine animals and play an important role in recycling nutrients. (9-2)

5. Nematodes are abundant and important members of the meiofauna. (9-3)

6. Marine worms are important in recycling nutrients and as links in marine food chains. (9-4)

7. Arthropods have external skeletons, jointed appendages, and sophisticated sense organs. (9-5)

8. Crustaceans make up a majority of the zooplankton that are a major link between phytoplankton and higher-order consumers in oceanic food webs. (9-5)

9. Arrowworms are carnivorous zooplankton. (9-6)

10. Echinoderms exhibit radial symmetry as adults. (9-7)

11. Echinoderms have internal skeletons and a unique water vascular system that functions in locomotion, food gathering, and circulation. (9-7)

12. Hemichordates (acorn worms) are benthic suspension feeders and deposit feeders. (9-8)

13. Invertebrate chordates include tunicates, salps, larvaceans, and lancelets. (9-9)

ARE YOU STILL WONDERING?

1. Snails known as cone snails (family Conidae) produce a toxin that coats their harpoon-like radulae. When the radula pierces a prey organism, it causes death by paralysis. Several species of large cone snail are known to have caused death in humans.

2. Cepahlopods have specialized cells called chromatophores that they can use to lighten or darken the color of their bodies and also produce different patterns. Some octopods can also change the shape of their bodies to match their surroundings.

3. Some species of polychaete worm are carnivorous, feeding on bivalves and other worms.

4. Arthropods must molt because their exoskeleton is nonliving and does not grow as the animal grows. When an arthropod begins to grow larger than its exoskeleton, it must shed the old one (molt) and form a new one.

5. Sea star predators include molluscs, crabs, some fishes, sea otters, and other echinoderms.

QUESTIONS FOR REVIEW

Multiple Choice

1. Molluscs that have shells composed of eight plates held together by a fleshy girdle are
   a. snails   b. bivalves   c. scaphopods
   d. cephalopods   e. chitons

2. A type of bivalve that can damage wood is the
   a. scallop   b. shipworm   c. nudibranch
   d. oyster   e. surf clam

3. Molluscs that have tentacles and a highly developed nervous system are
   a. gastropods   b. bivalves   c. scaphopods
   d. cephalopods   e. chitons

4. Worms with segmented bodies are called
   a. flatworms   b. nematodes   c. annelids
   d. acorn worms   e. ribbon worms

5. Most sedentary polychaetes are
   a. herbivores   b. carnivores   c. filter feeders
   d. parasites   e. symbionts

6. Nematodes play an important role in
   a. recycling calcium in marine environments
   b. removing carbon dioxide from seawater
   c. producing oxygen
   d. recycling organic matter
   e. controlling marine parasites

7. Arthropod characteristics include
   a. a water vascular system
   b. a hard external shell composed of calcium carbonate
   c. jointed appendages
   d. soft segmented bodies with hydrostatic skeletons
   e. a tunic composed of cellulose

8. During molting, arthropods
   a. grow extra appendages
   b. shed their old exoskeleton
   c. are well protected from predators
   d. shed excess appendages
   e. reproduce

9. While on a field trip to the seashore, you discover an animal with a spiny skin and a water vascular system. This animal is probably a(n)
   a. mollusc   b. echinoderm   c. arthropod
   d. lophophorate   e. tunicate

10. Which of the following adult animals is likely to be a member of the zooplankton?
    a. crab   b. squid   c. sea star
    d. larvacean   e. sea cucumber

Matching

Match the following structures with the appropriate definition or description:

a. mantle

b. radula

c. cheliped

d. epitoke

e. Cuvierian tubules

1. The claw of some crustaceans such as crabs and lobsters.

2. Sticky tubules ejected from the anus of some sea cucumbers for defense.

3. A ribbon of teeth unique to molluscs.

4. A free-swimming, reproductive errant polychaete.

5. The organ in a mollusc that secretes the shell.

Match the following larvae with the appropriate description:

a. veliger

b. nauplius

c. zoea

d. cyprid

e. trochophore

6. The larval stage in a barnacle life cycle that develops from a nauplius larva.

7. A free-swimming larval stage characteristic of marine gastropods and other molluscs.

8. The planktonic larva of a crab.

9. A free-swimming larval stage associated with primitive gastropods and annelids.

10. The planktonic larval stage of shrimp.

Short Answer

1. What adaptations allow squid to be successful predators?

2. Name four commercially important crustaceans.

3. Explain how the radula is modified in gastropods for different types of feeding.

4. Describe how a sea star uses its water vascular system to move.

5. Explain how slow-moving animals such as sea cucumbers avoid predation.

6. Describe how sea squirts feed.

7. Why are arthropods such a successful group of animals?

8. Distinguish between chelicerates and mandibulates.

9. How are regular and irregular echinoids particularly adapted to their individual lifestyles?

10. Why do bivalves that burrow in soft sediments need siphons?

11. What important ecological contribution do burrowing organisms make to the environment?

12. Distinguish between selective and nonselective deposit feeders.

13. Why are nematodes important marine organisms?

14. Why were acorn worms at one time considered to be members of the phylum Chordata?

Thinking Critically

1. What is the advantage of being able to alter the sex ratio in populations of the slipper limpet, *Crepidula fornicata*?

2. On a field trip to the ocean you discover a gastropod that you have never seen before. Because you are very interested in gastropod feeding, you want to know whether this animal is a herbivore or a carnivore. How could you determine with a fair amount of certainty which type of feeder this animal is?

3. What type of shell characteristics would you expect to observe in a gastropod that spends most of its life burrowing through soft sediments?

4. A sample of a deep-sea dredging contains an animal that is small, flat, and round. It has a ventrally located mouth, a stomach, but no intestines. To which of the groups of invertebrates introduced in this chapter does this animal probably belong? What other characteristics would the animal need to have to confirm your identification?

5. Some invertebrates covered in this chapter brood their eggs, and others do not. What are the advantages in each of these reproductive strategies?

SUGGESTIONS FOR FURTHER READING

Ballarini, R. and A. H. Heuer, 2007. Secrets of the Shell, *American Scientist* 95(5): 422–429.

Holland, J. S. 2007. Hawaii's Unearthly Worms, *National Geographic* 211(2):118–131.

Holland, J.S. 2008. Living Color, *National Geographic* 213 (6): 92–105.

Leonard, G. H., M. D. Bertness, and P. O. Yund. 1999. Crab Predation, Waterborne Cues, and Inducible Defenses in the Blue Mussel, *Mytilus edulis, Ecology* 80(1).

Mather, A. 2007. Eight Arms, With Attitude, *Natural History* 116 (1): 30–36.

Ruppert, E. E., R. S. Fox, and R. D. Barnes. 2004. *Invertebrate Zoology*, 7th ed. Belmont, Calif.: Cengage Learning.

Vecchione, M., R. E. Young, A. Guerra, D. J. Lindsay, D. A. Clague, J. M. Bernhard, W. W. Sager, A. F. Gonzalez, F. J. Rocha, and M. Segonzac. 2001. Worldwide Observations of Remarkable Deep-Sea Squids, *Science* 294(5551).

Ward, P. 2008. Chambers of Secrets, *New Scientist* 198 (2650): 5–11.

Have You Wondered?

1. What are the major groups of marine fishes? (10-1)

2. How jawless fishes like hagfish and lampreys feed? (10-2)

3. How to distinguish a skate from a ray? (10-3)

4. Which group of marine fishes is most closely related to land vertebrates? (10-4)

5. What factors distinguish ray-finned fishes from other fish groups? (10-5)

6. What adaptations marine fish use to avoid predators? (10-6)

Marine Fishes

FISHES ARE A DIVERSE group of organisms that evolved more than 530 million years ago from an invertebrate chordate ancestor. This ancestor also gave rise to the tunicates and lancelets, discussed in the previous chapter. Fishes, as well as land vertebrates (amphibians, reptiles, birds, and mammals), are classified as Chordates, organisms characterized, at least some time in their development, by the presence of pharyngeal gill slits, a dorsal hollow nerve cord, a notochord (a slender skeletal rod), and a tail lying posterior to the anus. Most chordates possess a cranium, a cartilaginous or bony covering that encloses the brain. These craniates have become the dominant organisms in both the sea and on land.

Fishes display an amazing array of adaptations that allow them to exploit virtually every ecological niche in the ocean. Their habitats vary from the intertidal to the deepest ocean trenches. There are more species of fish than all other chordates combined. Fish are commercially important for human nutrition, fertilizer, recreational fishing, and for numerous other reasons. Demand is so great that commercial fishing has greatly depleted the stocks of many important species. Destruction of estuaries and pollution of coastal waters has also had a detrimental impact on many fish populations. Some experts have suggested that more than 90% of the world's commercial fisheries will be depleted by the year 2050.

10-1 Fishes and Other Vertebrates

All craniates except hagfish are distinguished by the presence of **vertebrae**, a series of bones or cartilages that surround the spinal cord and help support the body.

Vertebrae provide attachment sites for muscles, thereby increasing and improving the animal's mobility. All fishes except hagfishes, as well as land vertebrates (tetrapods), are termed vertebrates.

The earliest vertebrates were fishes that lacked both jaws and paired appendages. They probably spent their time scavenging food in the bottom sediments of early seas, a niche they filled for millions of years. About 425 million years ago, some fishes appeared that possessed jaws and paired fins. Fishes with these adaptations could more efficiently obtain food and ultimately replaced all but a few of the early jawless forms. Modern fishes now make up the majority of the highest-level consumers in the sea.

Although the classification of the animals we call fishes is still a work in progress, we can distinguish six groups (**Figure 10-1**), five with marine representatives. Two basal groups (agnathans: hagfishes and lampreys) lack jaws and paired appendages. Among vertebrates with jaws (gnathostomes), the ray-finned fishes (class Actinopterygii) exceed all others in numbers of individuals and species. The cartilaginous fishes, sharks, rays, and their relatives, are the sea's top predators, contributing to the regulation of both coastal and pelagic food webs. The lobefins (coelacanths) and freshwater lungfishes have so many similarities with tetrapods that many scientists now include all of them in a single class, the Sarcopterygii.

Vertebrae are the series of bones or cartilages that surround the spinal cord and constitute the spinal column of vertebrate organisms.

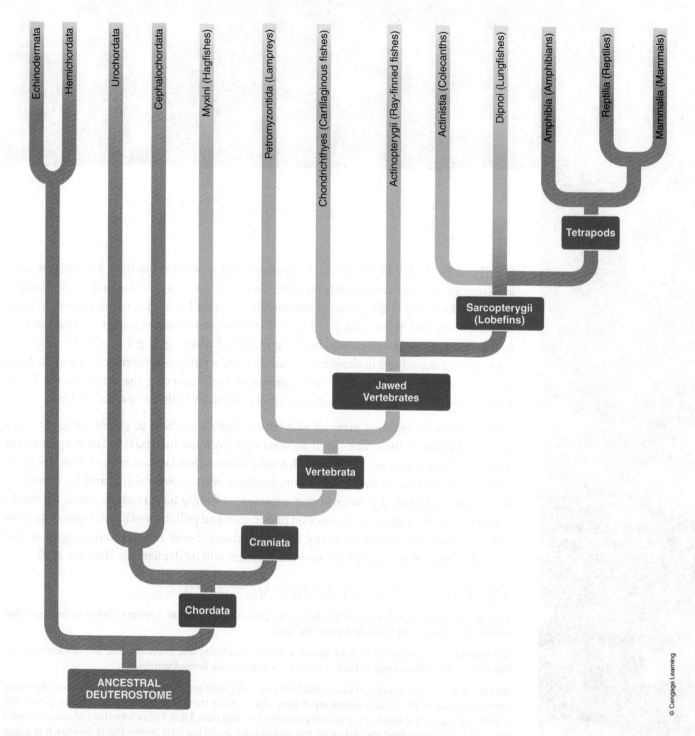

FIGURE 10-1 CLASSIFICATION OF FISHES. Phyogenetic relationships of the six major groups of fishes (shown in green) to other members of the deuterostome clade.

IN SUMMARY

- All craniates except hagfish possess vertebrae, a series of bones or cartilages that surround the spinal cord.

- The earliest vertebrates were fishes that lacked both jaws and paired appendages.

- The first fishes with jaws and paired fins appeared about 425 million years ago.

- The oceans are inhabited by five major groups of fishes, hagfishes, lampreys, cartilaginous fishes, ray-finned fishes, and lobefins. (Have You Wondered? #1)

10-2 Jawless Fishes

Hagfishes and lampreys are jawless fishes that lack both paired appendages and scales. Their skeletons are composed entirely of cartilage. Although superficially similar, they are as morphologically distinct from one another as they are from jawed vertebrates. Because hagfish lack vertebrae, many scientists now classify them in their own subphylum, the Myxine. The lampreys are included in the subphylum Vertebrata along with all other vertebrates. Recent comparisons of DNA sequences among chordates indicate that hagfish are most closely related to the lampreys.

HAGFISH

Hagfish **(Figure 10-2)**, sometimes called "slime eels," are bottom-dwelling fish found in ocean waters throughout the world, with the possible exception of the Arctic and Antarctic. Although they are seldom found at depths shallower than 600 meters (1,980 feet) in the tropics, some species enter the intertidal zone in the cold coastal waters off South Africa, Chile, and New Zealand. In some parts of the world, hagfish support a thriving commercial fishery, their tanned hide being used for leather goods. The demand for hagfish skin has risen to the point that many populations are depleted.

When feeding, hagfish extend an apparatus composed of two structures, termed *dental plates*, containing horny cusps. They use these dental plates to grasp their prey, usually small soft-bodied invertebrates. As the feeding apparatus is withdrawn into the mouth, the dental plates close tightly, and the cusps tear away the flesh of the prey. A fang above the dental plates keeps live prey from wriggling away between bites. In addition to feeding on live prey, hagfish act as scavengers on dead or dying whales and other large vertebrates. Because hagfish cannot penetrate the skin of whales or fish, they enter the body cavity through the mouth, anus, or openings made by other scavengers. They literally eat the carcass from inside out, leaving only the bones and skin.

Hagfish possess slime glands positioned along their body that can produce large amounts of a milky gelatinous fluid when the animal is disturbed. Although some slime is produced during feeding, its most important function seems to be physical protection. When seized, they produce so much slime that it coats the gills of predatory fish and either suffocates them or encourages them to release their prey. Hagfish remove the slime from their bodies by tying their tails in an overhand knot and sliding the knot forward and over the head **(Figure 10-3)**.

Little is known about reproduction in the approximately 77 species of hagfish. Ovaries and testes often occur in the same individual, but only one gonad is functional. We do not know where or when hagfish lay their eggs, and relatively few fertilized eggs or juveniles have ever been found. Why females outnumber males by 100 to 1 in some species is also a mystery.

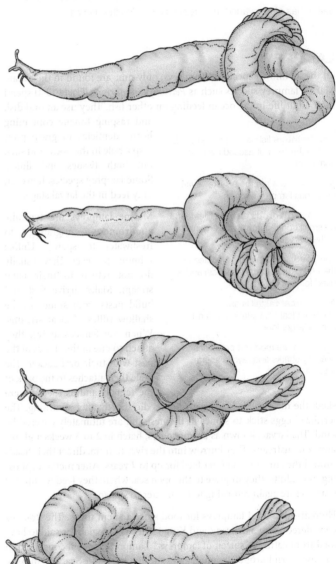

FIGURE 10-3 REMOVING SLIME. Hagfish have flexible bodies that they can tie into overhand knots. They tie the knot at their tail, then slide it forward and over their head. This behavior is used to remove excess slime from their body, to escape predators, and to gain leverage when tearing flesh from their prey.

FIGURE 10-2 HAGFISH. Hagfish, also known as slime eels, are jawless, bottom-dwelling fishes found in ocean waters throughout the world.

Zig Leszczynski/Animals Animals

FIGURE 10-4 **LAMPREY.** The sea lamprey (Petromyzon marinus) is a parasitic species found in the North Atlantic Ocean and Mediterranean Sea. It has also invaded the Great Lakes of North America.

LAMPREYS

Of the 43 known species of lampreys, only nine are found in the ocean. Marine lamprey species such as *Petromyzon marinus* **(Figure 10-4)** spend their adult life in the ocean feeding on other fish. They use an oral disk and rasping tongue containing horny denticles to grasp prey, rasp a hole in the body, and suck out both tissues and fluids. Some lamprey species, however, only feed in the larval stage.

Anadromous fishes spend their adult lives in the sea but ascend rivers for spawning.

Ammocoetes are the eel-like larvae of various lamprey species.

Zooplanktivorous animals filter-feed on small nonphotosynthetic organisms in the plankton.

The **caudal fin** (tail fin) is located at the end of the caudal peduncle and is used for propulsion.

A **heterocercal** tail is an asymmetrical fin; vertebrae extend into its larger lobe.

Claspers are modified pelvic fins used by sharks and rays in sperm transfer.

Marine lampreys are typically **anadromous**, migrating to freshwater to spawn. Unlike salmon, however, they usually do not return to their natal stream. Males arrive first and build nests from stones in the shallow riffles of clear streams. When the females arrive, they attach to one of the stones of the nest with their oral sucker. The male then attaches to the back of the female, and as the eggs are shed, the male sheds his sperm. Adults die shortly after spawning. The fertilized eggs stick to stones in the nest and are ultimately covered by sand. The larvae, known as **ammocoetes**, hatch in 2 to 3 weeks and migrate downstream. They burrow into the river bottom, direct their heads toward the current, and filter feed for up to 7 years. After metamorphosing into adults, they migrate to the open sea, where they feed for up to 2 years before again returning to freshwater to spawn.

Humans have used lampreys for food since ancient times. The Romans considered them a delicacy, and King Henry I of England is said to have died from eating a "surfeit of lampreys." Lampreys are still consumed in southwestern Europe, in countries bordering the Baltic Sea, and in Asia. Overfishing has severely reduced lamprey stocks in these areas.

IN SUMMARY

- Hagfish and lampreys are jawless fishes that lack both paired appendages and scales.

- Hagfish use *dental plates*, containing horny cusps, to grasp and tear away the flesh of their prey, usually small soft-bodied invertebrates. Hagfish also act as scavengers on dead or dying whales and other large vertebrates. Marine lampreys use horny teeth on their oral disk and tongue to rasp a hole in the body of their prey (other fish) and suck out both tissues and fluids. (Have You Wondered? #2)

- Most lamprey species are confined to freshwater, but some species spend their adult life in the sea. Marine lampreys are anadromous, returning to freshwater to spawn and die.

10-3 Cartilaginous Fishes

Sharks, skates, rays, and chimaeras are the modern representatives of the cartilaginous fishes, or class Chondrichthyes. Their skeletons are composed entirely of cartilage, although it is often strengthened by the deposition of calcium salts. They possess jaws and paired fins, and their skin is covered with sandpaper-like **placoid scales (Figure 10-5)**. Placoid scales can take on several additional forms, such as teeth in the jaws or the spines and large denticles on the backs of skates and rays. Cartilaginous fishes can be divided into two major groups, the holocephalans (chimaeras, or ratfish) and the elasmobranchs. The elasmobranchs have evolved into two general body forms, the typically streamlined bodies of sharks and the dorsoventrally flattened bodies of skates and rays. There are over 900 species of cartilaginous fish including, with the exception of whales, the largest living vertebrate animals.

SHARKS

Sharks are usually the top predators of the ocean's food webs. Although often characterized as "primitive" fish, modern sharks are actually highly specialized organisms. All are carnivorous, most taking relatively large prey such as other fish, marine mammals, seabirds, and various invertebrates. A few large species, such as the whale shark (*Rhincodon typus*; **Figure 10-6**), megamouth shark, and basking shark, are **zooplanktivorous**, filtering small planktonic animals from the water column. Sharks vary in size from the 16- to 20-centimeter (6- to 8-inch) dwarf dogshark (*Etmopterus perryi*) to the massive plankton-feeding whale shark, which may exceed 14 meters (46 feet) in length and weigh over 12,000 kilograms (5,500 pounds).

Most sharks have streamlined bodies and are excellent swimmers. Using their massive trunk muscles, sharks swim with powerful, sideways sweeps of the tail, or **caudal fin**. The caudal fin of many sharks has a dorsal lobe that is longer than the ventral, a condition that is termed **heterocercal** (see **Figure 10-5**). Additional fins include one or two dorsal fins as well as paired pectoral and pelvic fins. A sharp spine may be associated with the dorsal fins, as in the spiny dogfish (*Squalus acanthias*). The pelvic fins of male sharks are partly modified to form **claspers (Figure 10-7)**, which transfer sperm from the male to the female during reproduction.

Most sharks have a ventral mouth that is filled with multiple rows of teeth (see **Figure 10-5**). The teeth of all sharks are constantly replaced, either individually or in entire rows, depending on the species. Nonfunctional teeth in inner rows move forward to displace and replace functional teeth. It is estimated that a typical shark may produce as many as 30,000 teeth during its lifetime. A single tooth may last only a

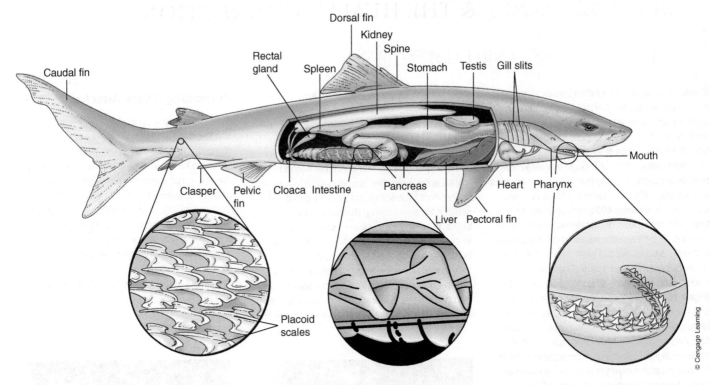

FIGURE 10-5 **SHARK ANATOMY.** The external and internal anatomy of a typical shark. Sharks have streamlined bodies and are well adapted for rapid swimming. The caudal fin of this shark is heterocercal and the skin is covered with placoid scales, which are similar to the teeth of other vertebrates. Modified placoid scales are found on the jaws and serve as teeth.

few days. In many sharks the teeth in the upper jaw are flat bladed with serrated edges, whereas the lower ones are spike-like. When a shark bites, it may shake its head and rotate the body, using the upper jaw teeth to slice through the prey's flesh and tear away large chunks. Food chunks are swallowed whole because sharks cannot move their jaws back and forth to chew.

Sharks are found in all oceans, the greatest number of species inhabiting temperate and tropical waters of less than 2,000 meters (6,600 feet) in depth. More than half of all species occur over the continental shelves

and slopes at depths of less than 200 meters (660 feet). Only a few species inhabit ocean depths as great as 3,500 meters (about 11,550 feet; e.g., *Centroscymnus coeolepis*). None reach the depths, up to 8,000 meters (26,247 feet) inhabited by ray-finned fishes.

Humans exploit shark populations for their fins, meat, oil, leather, cartilage, and sport. In Asia, the tremendous demand for the delicacy shark fin soup has endangered and threatened many species with

FIGURE 10-6 **WHALE SHARK.** A whale shark cruises a reef off Western Australia.

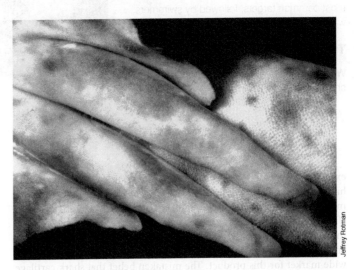

FIGURE 10-7 **CLASPERS.** Male sharks have modified pelvic fins termed *claspers* that are used to transfer sperm from the male to the female during sexual reproduction.

MARINE BIOLOGY & THE HUMAN CONNECTION

Shark Attacks on Humans

Sharks have acquired a rather unsavory reputation due to accounts of attacks on humans in the popular press, as well as in various movies and novels. In truth, the annual risk of death from lightning strikes is 47 times greater than that from shark attack. Although any large shark may be a potential risk to humans, most species are actually rather timid and cautious animals. Of the approximately 400 shark species, less than 10% have been documented in attacks on humans. The three species most often involved in deadly attacks are the white shark (*Carcharodon carcharias*), tiger shark (*Galeocerdo cuvier*), and bull shark (*Carcharhinus leucas*) **(Figure 10 A)**. All of these sharks are large and cosmopolitan in distribution, and they feed on large prey such as marine mammals, sea turtles, and large fish. Other shark species that have been implicated in human attacks include the mako (*Isurus oxyrhynchus*), great hammerhead (*Sphyrna mokarran*), oceanic whitetip (*Carcharhinus longimanus*), Galápagos (*Carcharhinus galapagensis*), and various "reef" sharks.

Since 1958, the International Shark Attack File, administered by the American Elasmobranch Society and the Florida Museum of Natural History at the University of Florida, has compiled statistics on shark attacks throughout the world. According to these data, about 40% of attacks reported worldwide have occurred in North American waters, but only about 8% of those were fatal. Australia had the second highest number of attacks (40% fatal), followed by South Africa (25% fatal). Surfers are the most common targets, followed by swimmers, waders, and divers.

Types of Shark Attacks

Worldwide, there are usually less than 100 shark attacks and fewer than 15 fatalities reported annually. Three types of attacks are often described: hit-and-run, bump-and-bite, and sneak attacks. Hit-and-run attacks occur most often in shallow water under conditions of low visibility on swimmers or surfers who are splashing the water. The shark bites and then releases the swimmer, usually causing relatively minor lacerations on the limbs that are seldom life threatening. Bump-and-bite attacks, in contrast, occur on swimmers or divers in deeper water. The shark bumps the victim before attacking. Sneak attacks differ in that they occur without warning. Both sneak and bump-and-bite attacks may occur repeatedly and cause deep lacerations that are often severe enough to result in the death of the victim.

Preventing Shark Attacks

Preventing shark attacks involves such commonsense practices as never swimming alone; avoiding areas where people are fishing or where blood and human wastes may be in the water, not swimming at dusk or at night or where the water is murky or turbid; refraining from splashing; and not wearing shiny jewelry because it may be mistaken for the scales of prey fish. Sightings of porpoises do not indicate the absence of sharks as both often eat the same foods. If sharks are sighted, swimmers should leave the water quickly and calmly.

(a)

Mike Parry/Minden Pictures

(b)

JAMES D WATT/Stephen Frink Collection/Alamy

(c)

Visual & Written/SuperStock

FIGURE 10-A DANGEROUS SHARKS. The three shark species most often involved in shark attacks are the **(a)** great white shark (*Carcharodon carcharias*), **(b)** tiger shark (*Galeocerdo cuvier*), and **(c)** bull shark (*Carcharhinus leucas*).

extinction. The scalloped hammerhead (*Sphyrna lewini*), for example, has seen a population decline of 98% in some areas. To meet the demand for shark fins, tens of millions of sharks are caught each year, "finned," and thrown back into the sea to die a horrible death. Although both the United States and European Union now limit the amount of shark fins landed, there continues to be a flourishing worldwide market for this product. The mistaken belief that shark cartilage has medicinal properties has further increased the exploitation of these fishes. The major problem with the current level of exploitation is that, as top predators, most shark populations are probably not abundant enough to withstand even a 5% annual removal of the existing population. The history of commercial shark fishing is therefore one of initial boom followed by collapse. Currently, North American shark populations are in such a state of decline that a moratorium on exploitation is desirable.

SKATES AND RAYS

Skates and rays differ from sharks by having flattened bodies, greatly enlarged pectoral fins that attach to the head, reduced dorsal and caudal fins, eyes and **spiracles** on top of the head, and gill slits on their ventral

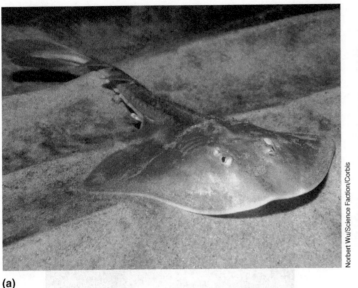

(a)

(b)

FIGURE 10-8 SKATES AND RAYS. Both skates and rays have enlarged pectoral fins. **(a)** Skates lack a stinging spine associated with the tail, whereas **(b)** rays, such as this stingray, have a spine that can inflict a painful injury. Rays are ovoviviparous, whereas skates are oviparous.

side **(Figure 10-8a and b)**. The location of the spiracles on the dorsal surface and gill slits on the ventral side is an adaptation for a bottom existence. Water is drawn in through the spiracles and passed out over the gills. This arrangement helps prevent the delicate gill filaments from becoming clogged with sand or debris. Other characteristics include the lack of an anal fin and the presence of specialized pavement-like teeth that are used for crushing prey, usually invertebrates inhabiting sandy or muddy bottoms. In contrast to other rays, the manta ray (*Manta birostris*, **Figure 10-9a**) feeds on plankton. Skates and rays, with the exception of sawfishes (*Pristis*, **Figure 10-9b**) and guitarfishes (*Rhinobatos*), swim with their pectoral fins. Most of the approximately 530 species are adapted to a bottom existence, but a few species such as the eagle rays (family Myliobatidae) and manta rays live in open water.

Skates and rays have evolved a variety of defenses to protect themselves from predators. Electric rays (*Torpedo* and *Narcine*) have a pair of electric organs in their head that can deliver up to 220 volts. In addition to defense, electric rays use their electric charge to navigate and to stun prey. Stingrays (*Dasyatis*; **Figure 10-8b**) and their relatives produce venom that can cause severe pain or even death (see discussion below). Sawfishes and guitarfishes are very atypical-looking rays. Sawfish have a series of barbs along their pointed snout **(Figure 10-9b)**. When disturbed or when feeding, they shake their heads from side to side, using the sharp points of the "saw" to inflict injury.

> **Spiracles** are small openings located behind the eyes of sharks and rays that serve as an opening for water entering the gill chamber when the animal is at rest.

(a)

(b)

FIGURE 10-9 MANTA RAY AND SAWFISH. (a) The manta ray, a plankton feeder, is the largest of all rays. **(b)** Sawfishes are related to skates and rays. Both are tropical and subtropical in distribution.

Some sawfish can reach a length of 7.6 meters (24.7 feet) and weigh more than 600 kilograms (1,300 pounds).

Like sharks, skates and rays are fished commercially, primarily for food. Skate "wings" are a popular dish, and sometimes their flesh is sold as "scallops." Several species (white skate, *Raja alba*, and common skate, *Dipturus batis*) are already overfished to the point that they are considered endangered. Because there is little regulation of the taking of rays and skates, many species are considered threatened.

Differences between Skates and Rays

Skates and rays differ in several ways. Rays swim by moving their fins up and down, much like a bird moves its wings while flying. Skates, on the other hand, swim by creating a wave that begins at the forward edge of the pectoral fins and sweeps down the edge to the back of the fins, allowing the animal to glide easily along the bottom as its fins ripple. The tails of skates are fleshier than those of rays and have small fins. In contrast, the tails of rays are streamlined and contain serrated, hollow spines (barbs) connected to venom glands. When disturbed, stingrays whip their tails around. If the barb punctures skin, it injects venom that causes smooth muscle to contract severely, excruciating pain, swelling, and cramping. Flesh wounds are usually treatable, but if the venom enters the abdominal cavity or heart it can be fatal, as in the sad case of "Crocodile Hunter" Steve Irwin. Irwin died in September 2006 while filming a television program off Australia's north coast after a stingray's barb pierced his chest. A few weeks after that incident, a stingray jumped into a boat in the Florida Keys and stabbed 81-year-old James Bertakis in the chest. Miraculously, the life of Mr. Bertakis was saved through quick medical intervention. Treatment for flesh wounds usually involves submerging the injured area in hot water to denature the toxin. Wounds from stingrays heal very slowly and are prone to bacterial infection.

Most skates release their eggs in a leathery rectangular egg case called a **mermaid's purse (Figure 10-10)**. In contrast, rays give birth to live young. At birth, the young rays' tail barbs are flexible and covered with a sheath to prevent injury to the mother during the birth of the pup. Rays grow much larger than skates. The manta ray, for example, can reach a width of 7 meters (22 feet) and weigh more than 1,360 kilograms (3,000 pounds). In contrast, the largest skate (*Raja binoculata*) reaches a width of about 2.4 meters (8 feet).

CHIMAERAS

Chimaeras (subclass Holocephali; **Figure 10-11**) are given such common names as ratfish, rabbitfish, and spookfish because of their large pointed heads and long, slender tails. Unlike other cartilaginous fish, their upper jaws are immovably attached to the braincase, adults lack a spiracle, and the gills are covered with an **operculum**, a flap covering the gills, as in ray-finned fishes. Unusually, water is taken in through the nostrils rather than the mouth. Instead of teeth, chimaeras have flat plates that they use to crush their prey. They feed on a wide variety of foods, including crustaceans, molluscs, echinoderms, and fish. A few denticles are the only remnants of scales. There is no **cloaca**; instead, there are separate anal and urogenital openings. In addition to the typical claspers on the pelvic fins of males, there is also a clasper on the head for use in reproduction. Chimaeras produce large eggs in a leathery

A **mermaid's purse** (or devil's purse) is the common name given to the egg case or capsule of some sharks, skates, and chimaeras.

An **operculum** is a stiff flap of tissue that covers the gills of ray-finned fishes and chimaeras.

A **cloaca** is a common chamber for the products of the intestinal and urogenital systems.

(a)

FLPA/Alamy

(b)

Richard N. Mariscal

FIGURE 10-10 SKATE EGG CASE. **(a)** Skates release their eggs in a leathery egg case called a *mermaid's purse*. **(b)** This one contains four developing young, each attached to a large yolk sac.

case. Generally bottom dwellers, they inhabit depths ranging from the shallows to 2,545 meters (8,400 feet) or deeper. Commercially, the 33 known species are of little value. Some are marketed as food in parts of China and New Zealand, and their oils make a fine lubricant.

IN SUMMARY

- Sharks, skates, rays, and chimaeras are the modern representatives of the cartilaginous fishes (class Chondrichthyes).

- Sharks have streamlined bodies and are efficient swimmers. Most are predators on other fish, marine mammals, seabirds, and various invertebrates, but a few species filter zooplankton from the water column.

FIGURE 10-11 **CHIMAERAS.** Relatives of the sharks, chimaeras such as this male spotted ratfish (*Hydrolagus colliei*) from British Columbia, Canada, are bottom dwellers that feed on a variety of fishes and invertebrates.

FIGURE 10-12 **LOBEFINS.** Coelacanths, such as *this Latimeria chalumnae* from the Comoros Islands, are classified as lobefins due to the muscular bundles at the base of their paired fins.

- Skates and rays are similar to sharks but have flattened bodies, greatly enlarged pectoral fins that attach to the head, reduced dorsal and caudal fins, and gill slits on their ventral side.

- Most skates and rays are bottom dwellers that feed on molluscs and crustaceans. Skates and rays differ in the way they swim, the fleshiness of the tail, fin structure, and their reproductive mode. (Have You Wondered? #3)

- Chimaeras, or ratfish, have large pointed heads and long, slender tails. Unlike other cartilaginous fish, the gills are covered with an operculum and water is taken in through the nostrils.

- Chimaeras feed on crustaceans, molluscs, echinoderms, and fish, using flat plates instead of teeth to crush their prey.

10-4 Lobefins

Coelacanths (**Figure 10-12**) are classified as lobefins due to the presence of rod-shaped bones surrounded by thick muscle in their pelvic and pectoral fins. As these characteristics are shared with the tetrapods, many scientists now consider all to be members of a **monophyletic group**, the class Sarcopterygii. The two extant species of coelacanths (*Latimeria*) are the only modern representatives of this ancient group of fishes. Coelacanths were known only as fossils until the curator of a small museum in South Africa, Marjorie Courtenay-Latimer, discovered a living specimen in 1938. This specimen, later named *Latimeria chalumnae* by the famous chemist turned ichthyologist J. L. B. Smith, created quite a stir in the scientific community. Despite a sizeable reward, a second specimen was not found until 1952, this time in the deep waters of the Indian Ocean around the Comoros Islands. Since that time, several other specimens have been taken there, in the Mozambique Channel, and as far south as Sodwana Bay, South Africa. In 1998, American and Indonesian scientists discovered a second species of coelacanth, *Latimeria*

menadoensis, off Sulawesi, Indonesia, some 10,000 kilometers (6,000 miles) east of where they were originally known.

Most coelacanths live at depths of 150 to 250 meters (495 to 825 feet) in rocky areas of the Indian Ocean with steep lava slopes that contain caves. One study suggested that caves might be a limiting factor for the distribution of coelacanths. They range in size from 42 to 183 centimeters (16 to 72 inches), and their weight averages about 80 kilograms (176 pounds). Their skeletons are made of both bone and cartilage, but the vertebral column is essentially cartilage. They feed on squid, sharks, and ray-finned fishes found in their deepwater habitats. They possess a rostral organ on the head that detects weak electrical currents and may aid in finding prey. Their life span may be up to 60 years, and it may take 20 years or more to reach full sexual maturity. They produce about 5 to 26 live young after a gestation period of about 13 months. With an estimated population size of 500 or fewer individuals, they are considered to be in danger of extinction. An international treaty on trade of endangered species currently forbids commercial harvesting of these fishes.

Monophyletic groups (clades) are groups of organisms consisting of a common ancestor and all of its descendants.

IN SUMMARY

- Lobefins are the group of marine fishes most closely related to land vertebrates. (Have You Wondered? #4)

- The two species of coelacanth are relatively large fish that occur in depths of 150 to 250 meters (495 to 825 feet) in rocky areas of the Indian Ocean with steep subsurface gradients.

- Like tetrapods, lobefins have rod-shaped bones surrounded by thick muscle in their pectoral and pelvic appendages.

10-5 Ray-Finned Fishes

With more than 26,000 named species, 56% or more of them inhabiting seawater, the ray-finned fishes are by far the most numerous and dominant group of vertebrates in the ocean. Ray-finned fishes constitute about half of all vertebrates and more than 95% of all the organisms we call fishes. They occupy virtually every available ecological niche and habitat in the sea. As a group, they are so diverse that no single characteristic separates them from other vertebrates. Most forms, however, can be characterized by the presence of a swim bladder, fin rays, bony skeleton, bony scales embedded in the skin, a terminal mouth, and an operculum that covers and protects the gills.

Ganoid scales are thick bony plates composed of an inner layer of bone and an outer layer of enamel.

Homocercal extend beyond the end of the vertebral column and are symmetrical.

Cycloid scales are thin plates of dermal bone with smooth margins found in the skin of some ray-finned fishes.

Ctenoid scales are thin plates of dermal bone with tiny teeth on their posterior edge found in the skin of some ray-finned fishes.

We can divide ray-finned fishes (class Actinopterygii) into two major groups. The first (the Chondrostei) contains forms such as marine sturgeons **(Figure 10-13)** that possess heterocercal tails similar to those of sharks, a skeleton made primarily of cartilage, and **ganoid scales (Figure 10-14a)**. Ganoid scales are very thick and heavy, giving the fish an armored appearance. Members of the second group (the Neopterygii) typically have **homocercal** tails, **cycloid** or **ctenoid scales (Figure 10-14bc)**, and more maneuverable fins. Homocercal tails extend beyond the end of the vertebral column, and the dorsal and ventral lobes are nearly equal in size. Cycloid and ctenoid scales are thinner and more flexible than ganoid scales and are less cumbersome for active swimmers.

blickwinkel/Alamy

FIGURE 10-13 MARINE STURGEONS. The scales of this sturgeon (*Acipenser gueldenstaedti*) are modified into bony plates that give the fish an armored appearance.

Ray-finned fish have median fins and paired fins **(Figure 10-15)**. The median fins consist of one or more dorsal fins, a caudal fin, and usually one anal fin. Median fins help fishes maintain stability while swimming. The paired fins consist of pectoral and pelvic fins, both of which are used in steering. Pectoral fins also help to stabilize the fish. Teeth are set in the jaw and not regularly replaced, as in cartilaginous fish.

The world's great fisheries are associated with the taking of ray-finned fishes. The populations of many species are overexploited and in danger of collapse. Climatic and other environmental changes have also detrimentally affected fish stocks in certain areas. Better governmental regulation is necessary to allow stocks to recover and provide sustainable yields.

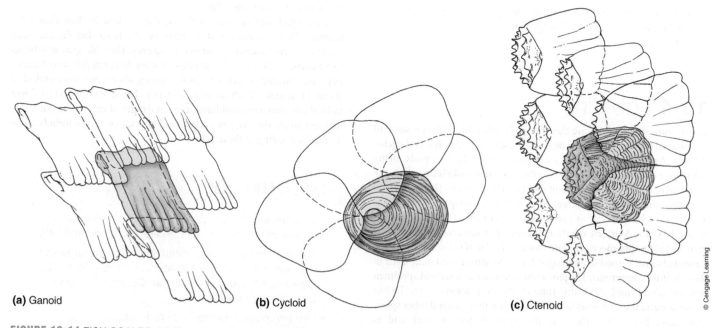

(a) Ganoid   **(b) Cycloid**   **(c) Ctenoid**

© Cengage Learning

FIGURE 10-14 FISH SCALES. (a) Thick, heavy, ganoid scales are characteristic of primitive ray-finned fishes such as sturgeons. Modern ray-finned fishes have lighter, more flexible scales such as **(b)** cycloid scales and **(c)** ctenoid scales.

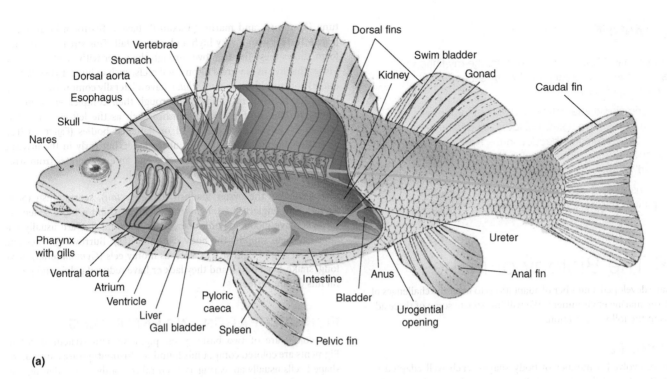

(a)

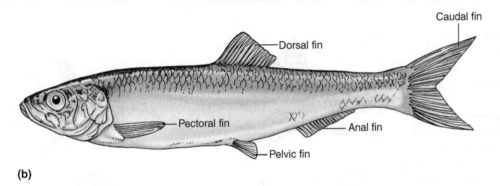

(b)

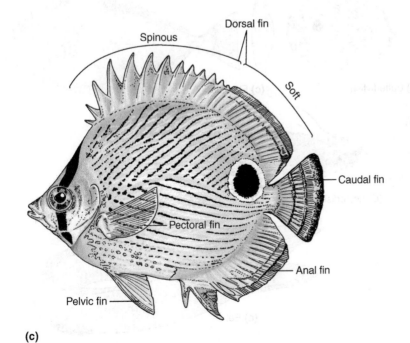

(c)

FIGURE 10-15 ANATOMY OF RAY-FINNED FISHES. **(a)** The general internal and external anatomy of a ray-finned fish. **(b)** In basal species such as the herring (Clupeidae) the pectoral fins are lower on the body and the pelvic fins are more posterior. **(c)** In more derived species such as butterflyfishes (*Chaetodon*) the pelvic fins are closer to the throat and the pectoral fins are higher on the body and more vertical.

© Cengage Learning

IN SUMMARY

- With more than 26,000 named species, the ray-finned fishes (class Actinopterygii) are the largest and most diverse group of fishes.

- Although no single characteristic defines them, most ray-finned fishes have a swim bladder, fin rays, a bony skeleton, bony scales embedded in the skin, a terminal mouth, and an operculum covering the gills. (Have You Wondered? #5)

- Populations of many species of ray-finned fishes are overexploited and in danger of collapse.

10-6 The Biology of Fishes

Fishes have developed a number of adaptations to meet the challenges of living in the marine environment. We will investigate some of these adaptations in the following sections.

BODY SHAPE

Fishes have evolved a number of body shapes, each well adapted to the characteristics of their habitat. Open water fishes, such as sharks, tuna (*Thunnus*), and marlin (*Makaira*), have a **fusiform** body shape (**Figure 10-16a**) with a very high and narrow tail. This streamlined body form allows these fish to move through the water with great efficiency. Fishes that live in sea-grass or on coral reefs, such as butterflyfish (*Chaetodon*) and angelfish (*Pomacanthus*), have a laterally compressed or deep body that helps them to navigate through their complex environment (**Figure 10-16b**). Bottom-dwelling fishes such as the left-eye flounders (family Bothidae) have depressed or flattened bodies (**Figure 10-16c**). Flounders begin life looking like normal fish, but early in the juvenile stage, they begin to swim on their side and an eye migrates from what will become the bottom side to the upper side.

Fishes such as the oyster toadfish (*Opsanus tau*), scorpionfish (*Scorpaena*), and anglerfish (*Antennarius*), which exhibit a more sedentary lifestyle, have globular bodies, and their pectoral fins are usually enlarged to help support the body (**Figure 10-16d**). Burrowing fishes and fishes that live in tight spaces, such as moray eels (*Gymnothorax*), have long, snake-like bodies, and they lack or have reduced pelvic and pectoral fins (**Figure 10-16e**).

FISH COLORATION AND PATTERNING

Fish colors are of two basic types, pigments and structural colors. Pigments are colored compounds found in **chromatophores**, irregularly shaped cells usually appearing as a central cell body with radiating processes. Fish are able to alter their color by moving pigments between the

FIGURE 10-16 FISH SHAPES. (a) Fish that are active swimmers such as this marlin have a fusiform body. **(b)** Reef fish, such as this butterflyfish, that swim among the corals have laterally compressed or deep bodies. **(c)** Bottom dwellers such as this flounder have horizontally compressed or depressed bodies. **(d)** Sedentary fish such as this anglerfish have globular bodies. **(e)** Burrowing fish and fish that live in tight crevices such as this moray eel have snake-like bodies.

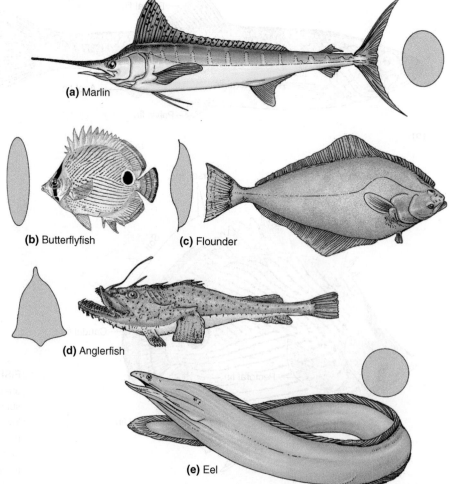

(a) Marlin

(b) Butterflyfish

(c) Flounder

(d) Anglerfish

(e) Eel

© Cengage Learning

Fusiform refers to a body shape characterized by tapering at both the head and the tail.

Chromatophores are the pigment-containing and light-reflecting cells that generate skin and eye color.

FIGURE 10-17 OBLITERATIVE COUNTERSHADING. This shark demonstrates the principle of obliterative countershading. When viewed from above, the dark back blends in with the surrounding dark water; when viewed from below, the white belly blends in with the brightly lit surface.

central core and these processes. The flounder, for example, is well known for its ability to alter body color and pattern to match its immediate environment. The control of pigment movement is complex but appears to be under both hormonal and nervous influence. The most common pigments found in fishes are the melanins and carotenoids.

Structural colors are produced by light reflecting from crystals located in specialized chromatophores called **iridophores**. Unlike pigments, these crystals are colorless and relatively immobile within the cells. Depending on their orientation, they can produce the mirror-like silver color of many pelagic fish or the iridescent colors seen in many reef fish.

Many fish that live in the open ocean such as sharks, tuna, marlin, and swordfish (*Xiphius gladius*) display a type of coloration known as **obliterative countershading**. A fish with obliterative countershading has a dark-colored back (dorsum) that graduates along the sides to a pure white belly **(Figure 10-17)**. When viewed from above, the dark back blends in with the surrounding dark water. When viewed from below, the white belly blends in with the brightly lit surface. This effectively camouflages the animal even though it is in open water. This same type of coloring is also found in many marine mammals and in birds such as penguins.

Many species of coral reef fish exhibit disruptive coloration in which vertical lines interrupt the background color of the body. This helps to break up the pattern and make it more difficult for predators to see the fish. Often, one of the lines passes through the eye, making it more difficult to be seen, and a dark dot, or eyespot, is present in the area of the tail **(Figure 10-18)**. Many aquatic predators use the eye to determine which end of their prey is the head. The eyespot on the tail and the line through the eye draw a predator's attention to the wrong end of the fish, making it more likely that the prey will survive an attack.

Fish associated with coral reefs may also exhibit bright and showy **poster colors** in patterns that may advertise territorial ownership, aid foraging individuals to keep in contact, or be important sexual displays. Lionfish (*Pterois volitans*, **Figure 10-19**) and some other species seem to use bright colors as **aposematic**, or **warning coloration**, to advertise to predators that they are too venomous or spiny to be worth eating.

Iridophores are specialized skin pigment cells that use crystalline plates made from guanine to reflect light.

Obliterative countershading is the use of different coloration on upper and lower body surfaces as a means of camouflage.

Poster colors are the bright, showy, color patterns often seen in reef fish that advertise territorial ownership, aid in maintaining schools, or are used in sexual displays.

Aposematic (warning) coloration is the bright coloration used by organisms to warn potential predators that they are distasteful or poisonous.

FIGURE 10-18 DISRUPTIVE COLORATION. The vertical lines on this butterflyfish break up the background color, making the fish more difficult to see. Notice the band that runs through the eye and the eyespot on the opposite end of the body, which make it more difficult for a predator to identify the fish's head.

FIGURE 10-19 APOSEMATIC (WARNING) COLORATION. Some species such as this lionfish use bright colors to advertise to predators that they are too venomous or spiny to be worth eating.

(a)

(b)

FIGURE 10-20 CAMOUFLAGE. **(a)** This pipefish from the Solomon Islands in the Pacific Ocean resembles a piece of seaweed. **(b)** A scorpionfish from the Caribbean mimics its environment.

Some fishes use cryptic coloration to blend with their environment, camouflaging themselves to avoid predators or ambush prey. Several pipefish species (*Syngnathus*; **Figure 10-20a**), for example, avoid predation by mimicking seaweed, both in body pattern and behavior. Scorpionfish (family Scorpaenidae; **Figure 10-20b**) often use their irregularly shaped bodies and coloration to blend almost perfectly with the environment. Smaller fishes become an easy meal when they swim too close to a hidden scorpionfish.

LOCOMOTION

Fishes move about by drifting with the current, burrowing, crawling on the bottom, gliding, and swimming. In swimming, the trunk muscles propel the fish through the water. These muscles are arranged as a series of muscle bands, each band looking like a letter W lying on its side **(Figure 10-21a)**. Movement results when the bands of muscles contract alternately from one side of the body to the other. The muscle contractions originate at the anterior end of the fish and move toward the tail, flexing the body and pushing against the water.

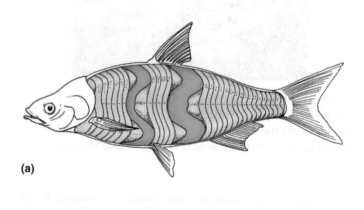

(a)

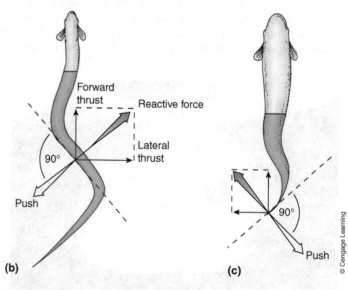

(b)

(c)

FIGURE 10-21 FISH LOCOMOTION. **(a)** The trunk muscles of fishes are arranged as a series of W-shaped bands. These muscles contract in sequence from anterior to posterior and alternately from one side of the body to the other. By pushing its body against the water, the fish is propelled forward. **(b)** Eels undulate the body one full wavelength in swimming. **(c)** Jacks and other swift swimmers throw the body into a shallow wave of less than one-half wavelength.

The type of swimming characteristic of elongate fish such as eels involves undulating the entire body **(Figure 10-21b)**. In contrast, swift swimmers such as jacks (family Carangidae), snappers (family Lutjanidae), tuna and mackerels (family Scombridae), and drums (family Sciaenidae) swim by flexing only the posterior portion of the body **(Figure 10-21c)**. Fish such as cods (family Gadidae) flex their bodies, making a movement somewhere between the full-body undulation of the eel and the posterior-body flex of the jacks. Trunkfish (family Ostraciidae, **Figure 10-22a**) are encased in a dermal skeleton, so only the area before the caudal fin can be flexed. Movement is relatively slow, but these fish have toxins and spines to protect themselves from predators.

Many species swim by using their fins alone without body flexure. Triggerfish (family Balistidae, **Figure 10-22b**) can move by undulating only their dorsal and anal fins. Wrasses (family Labridae) sometimes move by using the pectoral fins as oars. Flying fish (family Exocoetidae) use their expanded pectoral fins when they glide through the air.

RESPIRATION AND OSMOREGULATION

Fishes use their gills to extract oxygen (O_2) from the water, to eliminate carbon dioxide (CO_2), and as an aid in maintaining proper salt balance within the body. Gills **(Figure 10-23a)** are composed of thin, highly vascularized, rod-like structures called **gill filaments**. In these structures, blood flows in the opposite direction from the incoming water **(Figure 10-23b)**, creating a **countercurrent exchange system**. The countercurrent flow maintains a stable gradient that favors the diffusion of oxygen into and carbon dioxide out of the body. Water passing over the gills, therefore, constantly meets blood coming from the body with a lower oxygen and higher carbon dioxide concentration. This mechanism is very efficient; studies have demonstrated that up to 80% of the oxygen in the incoming water can be extracted through this arrangement versus less than 10% when the flow is concurrent (in the same direction).

Water must continuously move past the gills to keep the blood properly oxygenated. Most ray-finned fishes ventilate their gills by pumping water across them. Water enters the open and expanded mouth cavity and is then pushed across the gills by contracting the mouth cavity and expanding the chamber surrounding the gills. As the gill chamber contracts water is released from the opercula. Some shark species, tuna, swordfish, and other very active fish ventilate their gills by ram ventilation, meaning they continuously swim forward at high velocity with their mouth open. Less active sharks and rays ventilate their gills by using a **gill pump**; water enters the mouth or spiracle (or both) due to expansion of the mouth cavity and is pushed across the gills with the contraction of the pharyngeal muscles and gill slits.

Osmoregulation refers to the process whereby an organism maintains the proper concentration of solutes and water in its body fluids. Because the salt concentration of their blood is about one third the concentration of salts in seawater, marine fishes tend to lose water to their environment by osmosis. Coelacanths, sharks, skates, and rays solve this problem by retaining enough urea and trimethylamine oxide (TMAO) in their blood and body fluids to either balance the solute concentration of seawater **(Figure 10-24a)** or become slightly hypertonic to it. Unlike other vertebrates, their tissues are tolerant to high concentrations of these nitrogenous waste products. Species such as the bull shark (*Carcharhinus leucas*) have the ability to enter freshwater by reducing the levels of these nitrogenous wastes in their body fluids. The gills and **rectal gland**, a large structure that empties into the intestine, are involved in the excretion of excess sodium chloride, while the kidney excretes other salts in dilute urine.

Marine ray-finned fishes compensate for water loss by drinking seawater, removing the excess salt, and retaining the water **(Figure 10-24b)**. Specialized chloride cells on the gills eliminate most of this excess salt. The kidneys and digestive tract remove salts not excreted by this mechanism. Ray-finned fishes excrete negligible amounts of urine because they need to retain as much water as possible.

CARDIOVASCULAR SYSTEM

The cardiovascular system of fishes consists of a heart, arteries, veins, and capillaries. Deoxygenated blood, initially collected from the veins by a thin-walled chamber (the sinus venosus), is passed to a second chamber

Gill filaments are the thin specialized tissues of gills that act as the respiratory surface for gas exchange.

A **countercurrent exchange system** is a system in which two flows of fluids, like water or blood, move in opposite directions, thereby establishing a stable concentration gradient for some property, usually heat or dissolved substances, between them.

A **gill pump** is a series of muscles found in cartilaginous fishes that are used to suck in water and push it past the gills.

Osmoregulation is the process whereby an organism maintains the proper fluid and electrolyte balance within its cells and internal body fluids.

The **rectal gland** of sharks is an osmoregulatory organ located in the hindgut that concentrates and secretes excess salt.

(a)

(b)

FIGURE 10-22 ALTERNATIVE METHODS OF LOCOMOTION.
(a) Trunkfish are enclosed in a "box" made of dermal bone so only the caudal peduncle can be flexed. The entire muscle mass on each side is used to flex the short caudal peduncle, producing a sculling motion. **(b)** Triggerfishes swim by undulating only their dorsal and anal fins.

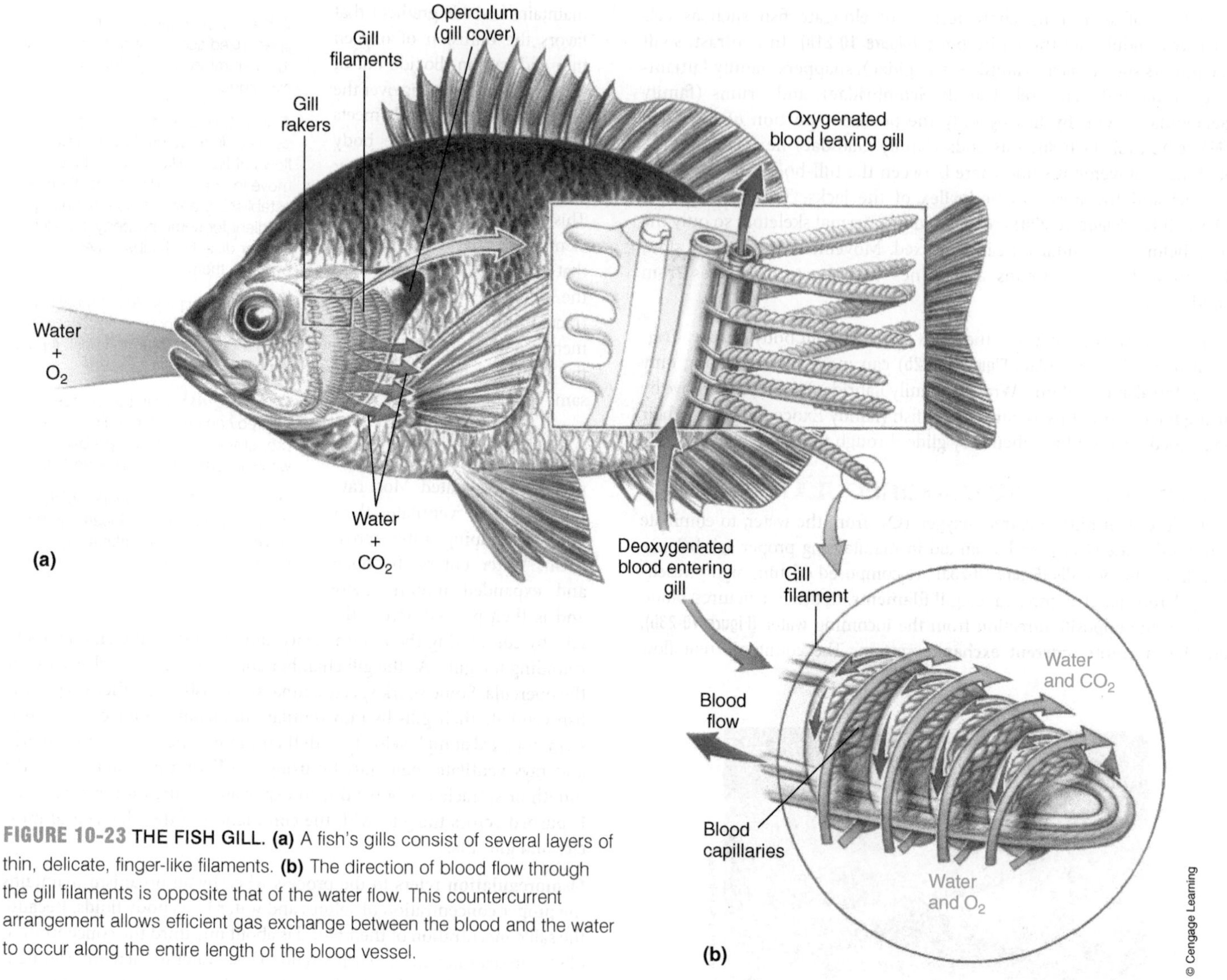

Operculum
(gill cover)

Gill
filaments

Gill
rakers

Oxygenated
blood leaving gill

Water
+
O_2

Water
+
CO_2

(a)

Deoxygenated
blood entering
gill

Gill
filament

Water
and CO_2

Blood
flow

Blood
capillaries

Water
and O_2

(b)

FIGURE 10-23 THE FISH GILL. **(a)** A fish's gills consist of several layers of thin, delicate, finger-like filaments. **(b)** The direction of blood flow through the gill filaments is opposite that of the water flow. This countercurrent arrangement allows efficient gas exchange between the blood and the water to occur along the entire length of the blood vessel.

© Cengage Learning

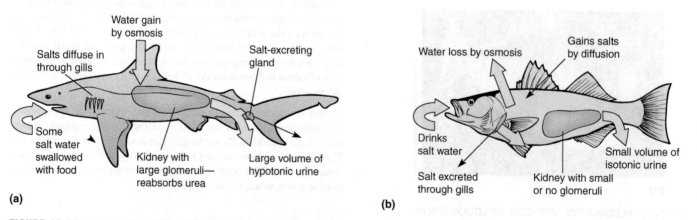

Water gain
by osmosis

Salts diffuse in
through gills

Salt-excreting
gland

Some
salt water
swallowed
with food

Kidney with
large glomeruli—
reabsorbs urea

Large volume of
hypotonic urine

(a)

Water loss by osmosis

Gains salts
by diffusion

Drinks
salt water

Small volume of
isotonic urine

Salt excreted
through gills

Kidney with small
or no glomeruli

(b)

© Cengage Learning

FIGURE 10-24 OSMOREGULATION IN MARINE FISH. **(a)** Cartilaginous fish such as sharks and rays maintain high levels of urea and TMAO in their tissues and become slightly hypertonic to the sea. As a result, water will enter the shark by osmosis. These fish excrete dilute urine to maintain salt balance. **(b)** Marine ray-finned fish lose water to the sea by osmosis. They compensate for this loss by drinking seawater and excreting the salts through the gills. Very little water is lost in the urine.

(the atrium) and then to a muscular ventricle (see **Figure 10-15a**). The ventricle propels the blood forward to the gill capillaries, where it is oxygenated. From the gills, the blood is collected by the dorsal aorta and passed to the rest of the body through arteries and capillaries. Blood pressure is lower in fishes than other vertebrates because blood does not return to the heart after oxygenation. Veins collect blood from the capillaries to complete the circuit.

Many active swimmers such as tuna have a countercurrent exchange arrangement of their blood vessels (**Figure 10-25**) to maintain body-core temperature at 2°C to 10°C above that of the surrounding water, thereby increasing the efficiency of their swimming muscles. The veins containing relatively cool blood from the body's surface pass close to and in the opposite direction from the arteries containing warm blood coming from the body's core. Heat is transferred from the blood in the arteries to blood in the veins. In this way warmed venous blood flows back into the core of the body, helping to maintain a higher internal temperature.

BUOYANCY REGULATION

Sharks sink if they stop swimming because their bodies are denser than seawater. They compensate for this problem by maintaining large quantities of an oily material called **squalene** in their livers. In some species the liver may account for 20% of a shark's weight. Squalene has a density less than seawater's (density of squalene is 0.8 g/cm³; the density of seawater is 1.020 to 1.029 g/cm³), and this helps to offset the shark's density. The large, appropriately directed pectoral fins and broad head of many species provide additional lift. Still, many sharks have to swim constantly to maintain buoyancy. In a similar fashion, coelacanths use a fat-filled swim bladder along with a reduced skeleton to maintain neutral buoyancy.

Most ray-finned fishes, with the exception of some pelagic species, bottom dwellers, and deep-sea fishes, use a gas-filled **swim bladder** (see **Figure 10-15a**) to offset the density of their bodies and regulate buoyancy. By adjusting the amount of gas in the swim bladder, a fish can remain indefinitely at a given depth without any muscular movement and with minimal expenditure of energy. When the fish descends, more gas must be added to the swim bladder, or else the bladder will compress and the fish will become denser and sink. On the other hand, as the fish ascends, it must remove gas from the swim bladder or else it will expand, become less dense, and rise too rapidly. Two mechanisms have evolved to allow adjustments in the gas volume of the swim bladder. Some fishes such as herrings and eels adjust the gas volume of their swim bladders by gulping air from the surface or "spitting it out" as needed. Others use a specialized **gas gland** to fill the swim bladder from gases dissolved in the blood. In these fishes, the swim bladder is deflated by diffusion of gases directly into the bloodstream.

Active pelagic fishes such as mackerels (*Scomber*) and skipjacks (*Katsuwonus pelamis*) lack swim bladders. These animals must keep swimming or they sink. Bottom dwellers such as scorpionfishes lack a swim bladder because they do not need to maintain buoyancy in the water column. Many fishes that live in the deep ocean also lack a swim bladder.

NERVOUS SYSTEM AND SENSES

Like other vertebrates, the nervous system of fishes consists of a brain, spinal cord, associated peripheral nerves, and various sensory receptors. The brain can be divided into several regions, each involved in coordinating an important body function. Areas associated with the senses of smell (olfaction) and vision are particularly well developed.

Olfaction and Taste

The olfactory receptors of sharks are very well developed and located in sacs or pits usually located in front of the mouth. Almost two-thirds of the cells in a shark's brain are involved in processing olfactory information. Indeed, these animals can detect the presence of a drop of blood diluted in 1 million parts of water and accurately find the source. It is no wonder that some biologists refer to sharks as "swimming noses." The role that the sense of smell plays in locating prey may explain the shape of the hammerhead shark's head (**Figure 10-26**). This shark has a nostril at the tip of each end of the "hammer," and it moves its head from side to side as it swims. When the strength of a smell is equal in both nostrils, the shark senses that its prey is straight ahead.

The olfactory receptors of ray-finned fishes are located in olfactory pits, blind sacs that open to the external environment. Swimming, movement of cilia within the pit, or constriction of the nasal sacs allows water to move across the olfactory receptors. The olfactory sacs and their receptors may be greatly elongated in fishes such as eels that rely heavily on olfaction to find their prey. In contrast, some puffers (family Tetraodontidae) have greatly reduced olfactory organs, most likely a result of their primary reliance on sight for feeding.

> **Squalene** is a low-density lipid that is often stored in the tissues of cartilaginous fishes.
>
> A **swim bladder** is an internal gas-filled organ that allows a ray-finned fish to control its buoyancy.
>
> A **gas gland** is the highly vascular structure found in many ray-finned fishes that is capable of secreting gasses (primarily oxygen) into the swim bladder, increasing its internal pressure and allowing the animal to achieve neutral buoyancy at differing depths.

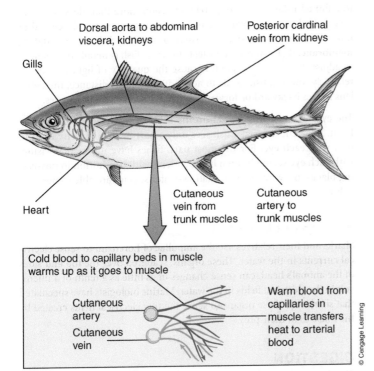

Dorsal aorta to abdominal viscera, kidneys

Posterior cardinal vein from kidneys

Gills

Heart

Cutaneous vein from trunk muscles

Cutaneous artery to trunk muscles

Cold blood to capillary beds in muscle warms up as it goes to muscle

Cutaneous artery

Cutaneous vein

Warm blood from capillaries in muscle transfers heat to arterial blood

© Cengage Learning

FIGURE 10-25 COUNTERCURRENT HEAT EXCHANGE IN TUNA.
Tuna are able to use a countercurrent heat exchange to keep the core of the body several degrees higher than their environment and thereby improve muscle efficiency.

FIGURE 10-26 HAMMERHEAD SHARK. The hammerhead shark has a nostril at the tip of each end of its "hammer." As it swims, it moves its head from side to side. When the strength of a smell is equal in both nostrils, the shark senses that its prey is straight ahead.

Taste receptors of ray-finned fishes may be located on the surface of the head, jaws, tongue, mouth, and **barbels**, whisker-like processes about the mouth. These specialized receptors are used to detect both food and noxious substances.

Lateral Line System and Hearing

Fish have a sensory **lateral line system** consisting of canals running along the length of the animal's body and over the head **(Figure 10-27)**. At regular intervals, the canals open to the outside and water moves freely in and out of them. Within the canals are sensory receptors called **neuromasts** that can detect vibrations in the fluid that fills the canals. Even the slightest movement in the water stimulates these lateral line sensory receptors. Fish use the lateral line system to locate prey and avoid potential predators.

Fish ears are internal structures that evolved from the lateral line system. They consist of fluid-filled semicircular canals and membranous sacs that contain receptors similar to the neuromasts of the lateral line system. All vertebrates except lampreys and hagfish have three semicircular canals that are used to regulate equilibrium and balance. Lampreys have two semicircular canals, and hagfish have one. The membranous sacs (utriculus, sacculus, and lagena) contain calcareous **otoliths** (ear stones) associated with sensory receptors termed **maculae**. Sound is detected from the vibration of the sensory maculae in relation to the otoliths overlying them. Fishes are capable of detecting sounds in the

Barbels are slender tactile organs containing taste buds.

The **lateral line** is a sensory organ found in aquatic organisms that is used to detect vibration and movement in the surrounding water.

Neuromasts are specialized receptor organs in the lateral lines of vertebrates that detect the direction of water movement.

Otoliths are calcareous structures found in the inner ear of fish that are used along with sensory membranes to detect sound, linear acceleration, and balance stimuli.

Maculae are sensory membranes within the inner ear containing neuromast-like sensory receptors that are used to detect fluid movement.

A **nictitating membrane** is a transparent third eyelid that protects the eye while maintaining visibility.

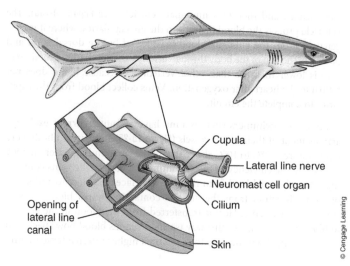

FIGURE 10-27 THE LATERAL LINE SYSTEM. The lateral line system consists of two canals, one on each side, that run the length of the animal's body and over the head. The fluid in the canals freely communicates with the water surrounding the animal. Tiny vibrations in the surrounding water cause the fluid in the canals to move the cupula of the neuromast cells, sending a signal to the animal's brain.

range of 200 to 13,000 hertz. In comparison, humans hear sounds ranging from 20 to 20,000 hertz.

Vision

Ray-finned fishes, as a group, rely on vision more than do sharks and rays to find prey and avoid predators. Eyelids are poorly developed in most fish, but some sharks possess a third eyelid, the **nictitating membrane**, that covers and protects the eyes. Fish generally do not need to adjust the size of the pupil because the quantity of light in water is relatively low. If a fish needs to adjust its vision for distance, the entire lens moves backward or forward, much like a camera lens focuses.

The eyes of most fish are set on the sides of the head instead of the front. It is believed that a fish sees only a narrow field directly in front of it with each eye. For the most part, fishes have monocular vision, with each eye seeing its own independent field. In addition to possessing black-and-white vision, shallow-water species are able to perceive color.

Ampullae of Lorenzini

Sharks and their relatives use the ampullae of Lorenzini to sense electrical currents in the water. These organs, scattered over the top and sides of the animal's head, can sense changes of as little as a tenth of a microvolt in the electrical fields in the water. Marine biologists have speculated that sharks use these organs to sense the tiny electrical fields created by the muscles of their prey.

DIGESTION

The digestive system of fishes is typical of vertebrates and consists of a mouth and pharynx, esophagus, stomach, and intestine (see **Figure 10-15a**). Many ray-finned fishes have blind sacs called pyloric caeca that secrete digestive enzymes, and a few lack a true stomach. The intestine may be modified in various ways depending on the diet. Whereas carnivorous fishes tend to have short intestines, herbivorous fish intestines are typically longer and coiled. The intestine of sharks and their relatives is relatively short but has a **spiral valve** (see **Figure 10-5**) to increase the surface area for nutrient absorption. The liver and pancreas are both important in digestion: the former for its production of bile for use in fat emulsification, the latter for its production of enzymes. The liver is well formed in fishes, but pancreatic tissue, depending on the species, is dispersed to different areas about the digestive system.

Feeding in Fishes

While all cartilaginous fishes are carnivores, the great diversity of ray-finned fishes is reflected by their ability to exploit virtually every food resource available in the marine environment. They can be carnivores, herbivores, detritivores, or omnivores.

Prey of carnivorous fish are usually seized and swallowed whole, because spending time chewing food would block the flow of water past their gills. Many species such as pufferfish (*Sphoeroides*) and box-fish (*Ostracion*) crush their prey with powerful jaws. Some butterfly-fish (*Chaetodon*) use their tiny mouths to feed on individual coral polyps. Groupers (*Epinephelus*) have large mouths with small teeth. They lie in wait in their lairs until their prey comes along. When a mullet (*Mugil*), grunt (*Haemulon*), or large crustacean comes by, the grouper opens its huge mouth, creating suction that draws in the prey. If the prey is too large to fit completely into its mouth, the grouper holds the prey's tail with its jaw teeth while pharyngeal teeth in the throat crush, grind, and shear the victim. Flounders lie camouflaged on the bottom, motionless, until a meal in the form of a crustacean, worm, or small fish comes along. The flounder then springs up and grabs the unsuspecting victim.

Herbivorous fishes such as surgeonfishes (*Acanthurus*) feed on the algae that grow on rocks and coral. In most species, the teeth are broad and flat with a sharp edge (like a shovel), making them ideal tools for scraping food from these surfaces. Several species have a gizzard-like stomach to grind the vegetable matter. Parrotfish (*Scarus*; **Figure 10-28**) have teeth fused into a beak-like structure that scrapes algae from the hard surfaces of coral reefs. Some parrotfish species bite off and ingest pieces of coral-line algae or hard coral, along with the coral's inhabitants. As the material passes through the fish's digestive tract, it is pulverized, the algae are

extracted and digested, and fine white sand is passed as part of the indigestible wastes. Because they extract so little organic material from each mouthful of food, parrotfish feed almost constantly. A large parrotfish may weigh as much as 27 kilograms (59 pounds) and expel as much as 2 or 3 tons of sand per year. Parrotfish have contributed sand to many of the white sand beaches of the world. Obvious scars on the reef surface are an indication of active parrotfish feeding.

> A **spiral valve** is a modification of the lower portion of the intestine whereby internal surfaces are twisted and coiled to increase the surface area for nutrient absorption.
>
> **Gill rakers** are bony or cartilaginous processes, used in filter feeding, that project from the gill arches of ray-finned and cartilaginous fishes.

The basking shark, whale shark, megamouth shark, manta rays, and many pelagic fishes such as herring and anchovies (*Engraulis*) are filter feeders. Filter feeders typically use projections from the gill arches called **gill rakers** (see **Figure 10-23a**) to filter both phytoplankton and zooplankton from the water column. Species of mullet (*Mugil*) and some gobies (e.g., *Evorthodus lyricus*), among others, suck up bottom sediments and strain out organic matter with their gill rakers. They are termed *detritivores* because they feed on dead organic matter and associated live organisms found in the sediments.

ADAPTATIONS TO EXTREME COLD

The fact that the temperature of polar waters ranges from about +1.5°C to −1.8°C presents a unique challenge to the fish fauna in these areas. Marine invertebrates are able to survive here because their bodies contain the same concentration of salts and minerals as that of the surrounding seawater, as well as a variety of organic compounds that depress the freezing point of their body fluids. In contrast, the body fluids of marine fish are more dilute than seawater and freeze at approximately −0.8°C. The low temperature of polar seas, therefore, has a greater impact on the fish fauna than it does on invertebrates.

Today, about 240 species of fish inhabit Arctic seas, and about 322 species reside in the seas around Antarctica. In Antarctica, about 35% of the fish species are members of just six families of the ray-finned fish suborder Notothenioidei. Notothenioids (**Figure 10-29**) have been so successful that they now make up approximately 90% of the fish biomass in Antarctic seas. The dominance of these Notothenioids is probably due to the development of a series of adaptations that allowed survival in these icy waters as well as a lack of competition from species that could not adapt. These adaptations and lack of competition allowed this group to experience extensive radiation over the past 24 million years.

The first adaptive innovation seen in Notothenioids, and possibly the most important, was the development of antifreeze glycoproteins. These glycoproteins (molecules composed of sugar and protein) protect cells by lowering the temperature at which ice crystals enlarge. At these temperatures, if an ice crystal were even to brush a fish's skin, it would quickly propagate and pierce the skin like a spear. The development of antifreeze glycoproteins is an interesting story in molecular evolution. They seem to have been derived by chance modifications of the preexisting gene sequence that encoded messenger RNA (mRNA) for pancreatic trypsinogen, an enzyme involved in digestion. It has been postulated that an early version of this gene may have first functioned in preventing freezing of intestinal fluid. Although Arctic fishes are less well studied than their Antarctic counterparts, it appears that they likewise express an antifreeze glycoprotein of similar structure in their blood. Recent studies indicate that the pancreas is the primary source of this antifreeze.

Richard N. Mariscal

FIGURE 10-28 PARROTFISH. Some parrotfish feed on the symbiotic algae of corals. Their beak-like mouthparts allow them to crush the coral to extract the algae.

Dr. Julian Gutt

Chris Gilbert/British Antarctic Survey

FIGURE 10-29 NOTOTHENIOIDS. All Notothenioids have a number of modifications that allow them to inhabit Antarctica's frigid waters. **(a)** This "white blooded" icefish (*Pagetopsis macropterus*, family Channichthyidae) lacks the oxygen-carrying protein hemoglobin and red blood cells. The oxygen that its cells need is carried dissolved in the relatively large volume of blood that circulates through its body. **(b)** This "red-blooded" black rockcod (*Notothenia coriiceps*, family Nototheniidae) has a reduced red blood cell volume but carries oxygen to its tissues using hemoglobin.

Most vertebrates transport oxygen to their tissues bound to molecules of hemoglobin located in red blood cells. The packed cell volume (hematocrit) of fishes typically ranges from about 18% to 53%, depending upon the species, the waters inhabited, and the activity level. In humans, the hematocrit is 45% of the blood's volume. The low temperature of polar waters makes blood with such high cell volumes far too viscous to flow efficiently. To compensate for the low temperature blood "sludge" factor,

Notothenioids and other polar fishes have a reduced number of red blood cells and a lower concentration of hemoglobin molecules in comparison to their temperate and tropical relatives. Low-temperature metabolic rates and relatively lethargic lifestyles allow polar fishes to survive with this reduced ability to transport oxygen.

One family of Notothenioids, the crocodile icefish (family Channichthyidae, **Figure 10-29a**), has lost red blood cells and hemoglobin entirely. Indeed, most of the gene sequence that encodes hemoglobin has been deleted from their genome. Notothenioids (including the crocodile icefish) also have lost myoglobin (a molecule that stores oxygen) from their skeletal muscles but, with the exception of six icefish species, most retain it in the ventricle of the heart. The 16 species of icefish are able to survive without red blood cells and hemoglobin because the water they inhabit is maximally saturated with oxygen, and they have such cardiovascular adaptations as an enlarged heart, enlarged and more numerous blood vessels, and a fourfold increase in blood volume. Their tissues also have an increased mitochondrial density. Oxygen is transported to the tissues of icefish dissolved in the blood plasma alone.

The icefish's loss of red blood cells and hemoglobin does not seem to confer any improved chance of survival. Studies on red-blooded Notothenioids (**Figure 10-29b**) have shown that they are able to survive without using hemoglobin to transport oxygen. The random loss of the ability to produce hemoglobin, therefore, probably has no particular advantage or disadvantage in this unique environment. But, in an age of global climate change, icefish may be an evolutionary dead end. The complete loss of hemoglobin coding sequences means that there are no genes to modify and restore this function. As temperatures continue to rise in Antarctica we may see the extinction of these unique creatures.

ADAPTATIONS TO AVOID PREDATION

Just as fishes have evolved a variety of feeding styles, they have also evolved many clever strategies to avoid being eaten. Although many marine fishes exhibit elaborate camouflage that obscures their presence, others have evolved more direct methods of avoiding predation. The pufferfish and porcupinefish (*Diodon hystrix*), for example, can swallow large amounts of air or water and inflate their bodies to a size that deters potential predators (**Figure 10-30a**). The rapid change in size frightens some

Dave B. Fleetham/Tom Stack and Associates

© Therisa Stack/Tom Stack & Associates

FIGURE 10-30 AVOIDING PREDATION. **(a)** A porcupinefish inflates itself to become a large prickly mouthful. **(b)** The large pectoral fins of the flying fish help it to avoid predators by allowing it to leave the water and glide through the air.

IMPACT BUBBLE

MORE THAN 50,000 humans are sickened annually from eating toxic fish flesh. Most cases of fish poisoning are due to eating reef fish containing various forms of a toxin called *ciguatoxin*, the result being termed *ciguatera poisoning*. Spanish explorers in Cuba first reported symptoms of this disorder more than 500 years ago, after eating snails they called cigua. Most people do not die from ciguatera poisoning, but symptoms may persist for weeks, months, and, in some cases, years. Symptoms include joint and muscle pain, weakness, intense itching, numbness in the extremities, and reversal of cold and hot sensations. Abdominal cramps, diarrhea, vomiting, and heart arrhythmia may accompany these symptoms. There is currently no effective medical treatment or antidote for ciguatera poisoning.

Gambierdiscus toxicus grows on dead coral surfaces colonized by filamentous and calcareous algae. Relatively few cases of ciguatera are reported from reefs with living coral. Large groupers, snappers, amberjack, barracuda, and Spanish mackerel are most commonly associated with ciguatera poisoning, but more than 400 species are known to carry the toxin. The fish acquire toxic levels of the ciguatoxin through food-chain amplification. Grazing fish ingest the toxin and are eaten by carnivorous fish that are, in turn, eaten by humans. Cooking has no effect on the toxin, and there are no safe reliable methods to detect it in fish flesh. Cases are reported from Australia, the tropical Pacific, and the Caribbean. In the United States, most cases occur in Hawaii and Florida.

Approximately 100 people die every year, primarily in Japan and the Philippines, from consuming fugu, or Japanese puffer fish (*Takifugu rubripes*). The liver and ovaries, particularly, contain a powerful toxin, tetrodotoxin, which blocks nerve-impulse transmission. It is 1,200 times more lethal than cyanide. The toxin itself is produced by several species of symbiotic bacteria. There is no antidote; medical treatment mainly involves supporting the respiratory and circulatory systems until the effects wear off.

Fugu is considered a delicacy; a full-course meal may cost up to $450. Licensed chefs must prepare the fish; they must serve a 2- to 3-year apprenticeship before taking the licensure examination, and only about 30% pass. The meal may be prepared with just enough toxin to cause numbness to the tongue, and there is usually a ritual associated with consuming it. Although many Westerners find fugu rather bland and tasteless, connoisseurs rave about it. Most deaths occur outside the main Japanese cities, where individuals may not be as well trained in proper fugu preparation. Recently, farm-raised fugu has become available that is nontoxic, accomplished by severely restricting the fish's diet, but is generally considered inferior by the public.

---

potential predators, whereas others now find the fish too large to fit in their mouths. In the case of the porcupinefish, or spiny boxfish, not only does the fish enlarge itself but also in the process extends spines that normally lie flat against the body, adding an extra measure of protection.

Flying fishes (*Cypselurus*) avoid predation by using enlarged pectoral fins to glide through the air **(Figure 10-30b)**. When frightened or disturbed, these fishes swim forward very quickly and leap out of the water, spreading their large pectoral fins at the same time. This behavior allows the flying fish to glide out of the range of many predators. Pearlfish (*Carapus*) avoid being eaten by slipping into the bodies of such animals as sea cucumbers or bivalve molluscs. At night, some species of parrotfish secrete a mucous cocoon about them, perhaps to discourage nocturnal predators. The surgeonfish is so named because of a pair of razor-sharp spines located on the sides of its tail. When this herbivorous fish is disturbed, the spines snap out like a switchblade knife and cut or stab the intruder. Clingfish (family Gobiesocidae) possess a powerful sucker (formed from modified pelvic fins) that is used to secure the body to rocks in tide pools. Predators have a difficult time dislodging clingfish and therefore tend to leave them alone. Many species of triggerfish (*Balistes*) have dorsal fins with the first three spines modified; these spines can be pulled flat against the body or projected up. When a predator tries to swallow a triggerfish, it projects the spines and is rejected as a possible meal. The spines also help the triggerfish to wedge itself into the tight nooks and crannies of coral reefs, making it nearly impossible for a predator to dislodge.

To deter predators, many fish species produce toxic substances. Others accumulate toxic substances from their environment in the course of their normal feeding activities. Scorpionfishes (family Scorpaenidae), for example, use the neurotoxins produced in specialized venom glands associated with grooved spines to discourage potential predators. The venom of a close relative of the scorpionfishes, the Indo-Pacific stonefishes (*Synanceia*), is extremely toxic. There are many recorded instances of people accidentally stepping on them and being severely injured or killed. Parrotfish (family Scaridae), surgeonfish (family Acanthuridae), and trunkfish (family Ostraciidae) produce toxic secretions from their skin. Puffers (family Tetraodontidae, **Figure 10-31**) store toxic substances produced by symbiotic bacteria in their internal organs. Reef fish may accumulate toxins while feeding in areas associated with blooms of dinoflagellates such as *Gambierdiscus toxicus*.

REPRODUCTION

The gonads—testes in males and ovaries in females—are paired structures suspended by membranes from the dorsal part of the body cavity. The manner in which sperm and eggs exit the body differs among fish groups. In jawless fishes and salmon, the gametes are shed directly into the body cavity and pass to the outside through abdominal pores. In cartilaginous fishes, sperm and eggs pass through special ducts, those in males shared with the kidney, to the cloaca. From this chamber, feces, urine, and gametes pass to the outside. In contrast, ray-finned fishes typically have separate openings that lie behind the anus for their urinary and reproductive systems. Unlike in sharks and most land vertebrates, there is typically no association of the reproductive and excretory systems.

Development of eggs and sperm is usually seasonal (except in some tropical species), the timing of the reproductive period being influenced by temperature and photoperiod. As in other vertebrates, variation in

Jeremy Sutton-Hibbert

FIGURE 10-31 JAPANESE PUFFERFISH. Puffers like this Japanese pufferfish, *Takifugu rubripes*, store toxic substances produced by symbiotic bacteria in their internal organs. This and similar species of puffers (termed fugu) are considered a delicacy in Japan. Licensed chefs remove the highly toxic liver and prepare the flesh, fins, and testis for human consumption.

the level of pituitary and gonadal hormones controls the reproductive process. Three reproductive modes are seen in fishes: oviparity, ovoviviparity, and viviparity.

Oviparity

In oviparous reproduction, eggs are shed into the water and embryos develop outside the mother's body. This mode is the most common in ray-finned fishes. Both eggs and sperm are shed directly into the water column or on the bottom, and fertilization occurs there. Whale sharks, bullhead sharks, skates, and a few other species of cartilaginous fishes also practice this mode, but fertilization, as in all cartilaginous fishes, is internal. Males transfer sperm to females through a groove in their claspers, the modified pelvic fins described earlier. The male holds the pectoral fin of the female in his mouth as he uses his claspers to introduce sperm into the female's cloaca (**Figure 10-32**). After fertilization, the shell gland secretes a horny or membranous egg case. Egg cases with fertilized eggs are shed on the seafloor, where they attach to hard surfaces. Shark or skate pups produced in this manner tend to be smaller than those produced by the other reproductive modes because of the limited availability of nutrients within the egg case.

Ovoviviparity

In ovoviviparity, fertilization is internal and eggs hatch within the mother's uterus, where they are nourished by yolk stored in the egg. This is the most common mode observed in sharks. In the sand tiger (*Carcharias*) and a few other species, the yolk is quickly used and the embryos then feed on unfertilized eggs or other embryos in the uterus. In these species, only a single pup may ultimately be born from each uterus. Sharks exhibiting ovoviviparity include basking sharks, thresher sharks, and saw sharks. Only a few ray-finned fishes, such as some rockfishes (family Scorpaenidae), practice this reproductive mode.

Viviparity

In viviparity, the most recent mechanism to evolve, either the young directly attach to the mother's uterine wall or the mother's uterus produces "uterine milk" that is absorbed by the embryo. Viviparous sharks

Nick Caloyianis/National Geographic Stock

FIGURE 10-32 SHARK REPRODUCTION. When nurse sharks copulate, the male holds the pectoral fin of the female in his mouth as he uses his claspers to introduce sperm into the female's genital opening.

include requiem and hammerhead sharks. Among ray-finned fishes, surfperches (family Embiotocidae) of the Pacific coasts of North America and Asia exhibit viviparity. Their young develop in a uterus-like structure in the ovary and obtain nutrition from the mother through enlarged, highly vascularized fins.

REPRODUCTIVE STRATEGIES

Fishes exhibit an amazing variety of reproductive strategies. Spawning fish may simply release their eggs into the water column or deposit them on the bottom without further care. Some species hide their eggs, whereas others guard them until they hatch. Still others brood their eggs in the body until they hatch. We can divide the reproductive strategies of fishes into the following categories.

Pelagic Spawners

Pelagic spawners include such commercially important species as cod, tuna, and sardines and coral reef species such as parrotfish and wrasses. These species release vast quantities of eggs into the water, the males fertilize them, and the fertilized eggs drift with the currents. There is no parental care. An advantage of this strategy is that the offspring are widely dispersed, but its disadvantage is that the mortality rate is very high. These species compensate by producing large numbers of offspring over a lengthy spawning period.

Benthic Spawners

Benthic spawners such as smelt (family Osmeridae) live closer to shore and produce eggs that are generally larger than those of pelagic spawners and have a large quantity of yolk. These eggs are usually nonbuoyant and are spread over surfaces such as vegetation or rocks. Large numbers of eggs are produced, and there is no parental care. When hatched, the embryos and larvae may be either pelagic or benthic.

Brood Hiders

The grunion (*Leuresthes tenuis*) is classified as a brood hider, a species that hides its eggs in some way but exhibits no parental care. Grunions swim ashore at high tide during a full moon, burrow partially into the sand, and deposit their eggs. Males curl around the partially buried female and fertilize the eggs as they are laid. The eggs remain in the sand until the next high tide, when they hatch and are washed out to sea.

Guarders

Guarders care for their offspring until they hatch and, frequently, through their larval stages. Many species of damselfish, blennies, and gobies exhibit this behavior. Some species of damselfish defend territory, both as spawning sites and sites to cultivate algae. Spawning sites may be prepared nests or simply a bare surface. Females lay eggs at these spawning sites, but only the male guards the offspring. The amount of time spent guarding the eggs varies from a few days to more than 4 months, as in the Antarctic plunderfish (*Harpagifer bispinis*).

Bearers

Jawfish (*Ophistognathus macrognathus*) and seahorses (*Hippocampus*) are examples of bearers **(Figure 10-33)**. The female jawfish lays her eggs in the mouth of the male, who incubates them in his mouth until they hatch. When the male must feed, he temporarily deposits the eggs in his protected burrow. The female seahorse lays eggs in a special pouch on the male's abdomen. After the eggs are deposited, the male fertilizes them and then carries and incubates them until the young hatch.

Hermaphroditism

Hermaphroditism, in which individuals have both testes and ovaries at some time in their lives, is known in at least 14 families of ray-finned fishes. Hermaphrodites may be synchronous, possessing functional gonads of both sexes at one time, or sequential, changing from one sex to another. Sequential hermaphrodites may change from females to males (**protogyny**) or from male to female (**protandry**). An evolutionary advantage of synchronous hermaphroditism is that the animal can spawn with any individual, which is particularly important when population densities are low, as in the deep sea. An advantage of sequential hermaphroditism is that large males are usually more successful in mating with females than are smaller ones and therefore able to produce more offspring.

> **Protogyny** is a type of sequential hermaphroditism in which an organism that is born female changes sex to become a male.
>
> **Protandry** is a type of sequential hermaphroditism in which an organism that is born male changes sex to become a female.

Hamlets (*Hypoplectrus*; **Figure 10-34a**), a small group of tropical fish in the sea bass family (family Serranidae), are a good example of a synchronous hermaphrodite. A partner in a monogamous pair alternately has its eggs fertilized and fertilizes its partner's eggs. Sexual roles may change up to four times during a single mating encounter.

At least seven fish families, including the wrasses (family Labridae), parrotfish (family Scaridae), and sea basses (family Serranidae) exhibit protogynous sequential hermaphroditism. Typically, a large dominant male of a species such as the striped cleaner wrasse (*Labroides dimidiatus*; see **Figure 10-34b**) controls a harem of females. If this male dies, the largest female changes to a male and replaces him. Males, therefore, may be primary, beginning as a male, or secondary, resulting from a female.

FIGURE 10-33 MALES INCUBATING EGGS. In some fish species it is the job of the male to incubate and care for the eggs. **(a)** The male jawfish incubates the eggs in its mouth until they hatch. **(b)** Male seahorses have a special abdominal pouch in which the eggs are incubated.

(a)   (b)

(a)

(b)

FIGURE 10-34 HERMAPHRODITISM.
(a) Hamlets are synchronous hermaphrodites. Members of monogamous pairs alternately offer eggs to be fertilized and, in turn, fertilize their partner's eggs. **(b)** The striped cleaner wrasse (*Labroides dimidiatus*) is a sequential hermaphrodite practicing protogyny. If a male controlling a harem of females dies, the largest female changes to a male and replaces him.

Smaller primary males that look like females may use this concealment to occasionally "sneak" into group spawning aggregations and contribute some of their gametes to the next generation.

Anemone fish (*Amphiprion*, family Pomacentridae) exhibit protandrous sequential hermaphroditism. They inhabit large sea anemones where they are protected from predators by the host's stinging cells. Each inhabited anemone contains a single large female, a smaller adult male, and many juvenile males. If the female dies or disappears, the adult male changes to a female and is replaced by the largest of the juvenile males. As anemone fish seldom leave the protection of their anemone, this large female–smaller male combination can produce a larger number of offspring than the reverse and thus has an evolutionary advantage.

LARVAL DEVELOPMENT

Larval fish are nourished by a yolk sac attached to their abdomen (**Figure 10-35**). As the larva develops, a mouth and digestive tract form, and ultimately the yolk sac is absorbed. In many species, the larvae initially become members of the plankton community, where they are heavily preyed upon. Those that survive ultimately transform into juveniles, leave the plankton community, and feed as adults. Unlike birds and mammals that cease to grow or grow very little after achieving maturity, fish grow for as long as they live.

SCHOOLING

A **school** of fish is a grouping of individuals that operates in a highly polarized and synchronized fashion. It is one of a continuum of behaviors known as **shoaling**, an association of fish for various social reasons (**Figure 10-36**). About 25% of all fish form these aggregations at some time in their lives. Schooling fish include such commercially important species as herring, mullets, and mackerel. Some species may school (or shoal) throughout their lives, whereas others come together specifically for spawning, migration, or other reasons.

A **school** is a group of fish swimming in the same direction in a coordinated manner.

Shoaling refers to the activity of any group of fish that stay together for social reasons.

Many reasons have been proposed for schooling. It has been suggested that having more eyes increases feeding efficiency and the detection of predators. The fact that it is difficult for a predator to focus on one individual reduces the probability that the individual will be eaten. This factor is sometimes called the "dilution and confusion" effect. One would also expect that a tightly formed school of fish would have greater hydrodynamic efficiency and reduced energy expenditure, but that has not been demonstrated experimentally. Finally, schooling can be useful in reproduction to bring eggs and sperm into close proximity.

Vision seems to be the most important sense required for schooling as schools (but not necessarily aggregations) tend to disperse somewhat in the dark. Jacks feed on schooling herring at twilight when they are less coordinated and it is harder to see the predator. Other senses, such as the lateral line, sound, and pheromones, may also be important in schooling.

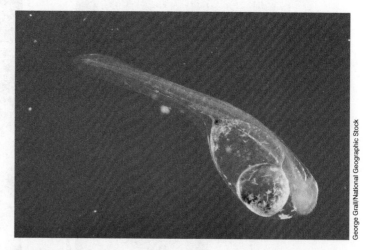

FIGURE 10-35 FISH LARVA. After hatching, many fishes still have a yolk sac containing nutrients attached to their abdomen. The size of the yolk sac impairs the larva's ability to swim, and it remains a member of the plankton until the mouth and digestive tract are fully formed and the yolk sac is absorbed.

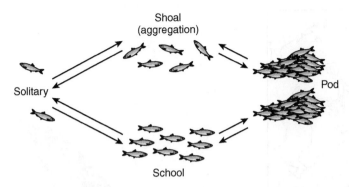

FIGURE 10-36 SCHOOLING. Schools are highly polarized and synchronized groups of fish. They are one end of a continuum of such behaviors known as shoaling. Fish may range from a solitary existence, to loose aggregations, to schools, to clumps of individuals known as pods. (Adapted from Breder, Jr., C. M. 1959. Studies On Social Grouping in Fishes, *Bulletin of the American Museum of Natural History*, Volume 117, Article 6.)

FISH MIGRATIONS

Migratory movements of marine fish are common and may occur daily or seasonally. Daily migrations are usually associated with feeding and predator avoidance. Grunts (family Haemulidae), for example, hover over reefs during the day but at night move to surrounding seagrass beds to feed. Lanternfish (family Myctophidae) are known to perform rapid vertical migrations of 200 meters (660 feet) or more as they follow their planktonic prey. Seasonal migrations of marine fish are usually associated with spawning, changing temperatures, or feeding. In some species, migrations occur entirely within saltwater. In other species, migrations occur between freshwater and saltwater. The North Pacific population of albacore tuna (*Thunnus alalunga*), for example, winters in midocean but travels to the California-Oregon coast or the coast of Japan for the summer months. Many other species such as the herring (*Clupea harengus*) annually travel north and south, following sea temperature variation.

Some fish species move between freshwater and saltwater for a purpose other than reproduction. Young mullets (*Mugil cephalus*), for example, spend part of their time in freshwater or estuaries. But as adults, they live most of their life in the ocean and spawn there. Fishes that move from freshwater to seawater to spawn are said to be **catadromous**, whereas those that move from seawater to freshwater to spawn are anadromous.

Freshwater Eels

The freshwater eels of North America (*Anguilla rostrata*) and Europe (*A. anguilla*) are probably the best-studied catadromous fishes. Adult eels of both species migrate down coastal rivers to the sea during the fall, changing from a dull olive color to silver and acquiring larger eyes. The American species takes about 2 months to reach the deepwater spawning grounds in the Sargasso Sea, which lies southeast of Bermuda **(Figure 10-37)**. The European species takes approximately 1 year to make the journey. Here the eels spawn at depths of 300 meters (990 feet) or more, and the adults die after reproducing.

Hatchlings develop into leaf-like **leptocephalus** larvae and begin their migration back to the rivers of Europe and North America. It takes approximately 3 years for larvae to reach Europe, but only 1 year for the North American species to complete the trip. When they arrive at the coastal rivers they undergo a metamorphosis, becoming juvenile eels, called "**elvers**," that migrate into streams and estuaries. Males usually remain in freshwater for 4 to 8 years, whereas females require up to 12

years or more to reach sexual maturity. During the fall season following sexual maturity, the eels migrate downstream and return to the breeding grounds of the Sargasso Sea.

The Japanese freshwater eel (*Anguilla japonica*) has a similar pattern, spawning near the Mariana Islands and drifting back to the Orient via prevailing currents. Unlike the American and European eels, some populations of the Japanese eel never enter freshwater but remain in the sea or estuaries. These populations, along with those inhabiting freshwater, all seem to contribute individuals to the populations that return to the waters near the Mariana Islands to spawn.

Salmon

Atlantic and Pacific salmon **(Figure 10-38)** are examples of anadromous fishes. The six species of Pacific salmon (*Onchorhyncus*) die after returning to their freshwater spawning grounds. In contrast, the single Atlantic species of salmon (*Salmo salar*) may return multiple times to the stream where it was hatched. Salmon lay their eggs in a shallow depression in the gravel termed a **redd**. Hatchlings develop first into **alevins**, then **parrs**, and finally into **smolts** as they prepare to return to the sea. Some salmon return to the sea almost immediately, but others may remain in freshwater for up to 5 years. Some populations even become permanently landlocked. Young salmon mature at sea and, after several years, return to the headwaters of their native stream to spawn and complete their life cycle.

A **catadromous** fish lives its adult life in freshwater but returns to the sea to reproduce.

A **leptocephalus** is the flat transparent larva of tarpon, bonefish, and various types of eels.

Elvers are young eels, especially those migrating into freshwater from the sea.

Redds are depressions dug in the gravel of streams by female salmon for the deposition of eggs during spawning.

Alevins are newly hatched salmon that still have a yolk sac attached. They are known as fry after they lose the yolk sac and begin feeding.

Parrs are 1- to 5-year-old salmon that inhabit freshwater.

Smolts are salmon that develop from parrs, acquire silvery scales, and migrate to the sea.

Many experiments have shown that salmon are guided to their native (natal) stream for spawning by olfaction. One hypothesis suggests that salmon recognize the characteristic odors originating from chemicals in the soil and plants along their native stream's edge. Salmon are thought to be "imprinted" with this odor knowledge during development. Another hypothesis suggests that population-specific pheromones are released by smolts returning to the sea. Adult salmon near the coastline from that same population are attracted by these pheromones to the streams occupied by the smolts. Currently, there is evidence that supports both hypotheses.

How salmon locate the mouth of their natal stream from the open sea is also controversial. Some suggest that the fish use cues such as the sun and the earth's magnetic field to find the portion of the coast near their home stream. Others suggest that currents, temperature gradients, and food supplies ultimately bring a salmon to the mouth of its native stream.

Populations of salmon have been greatly reduced because of the damming of rivers, pollution, and the transformation for human use of land bordering streams. The Chinook salmon (*O. tshawytscha*), which once ascended the Columbia River for more than 1,600 kilometers (1,000 miles), has been particularly affected by these activities. The 2008 collapse of the Chinook population off the coasts of California and Oregon has necessitated the closure of that $150-million fishery. Hatcheries have had limited success in maintaining stocks, but there is concern about the loss of the genetic diversity of the wild stock.

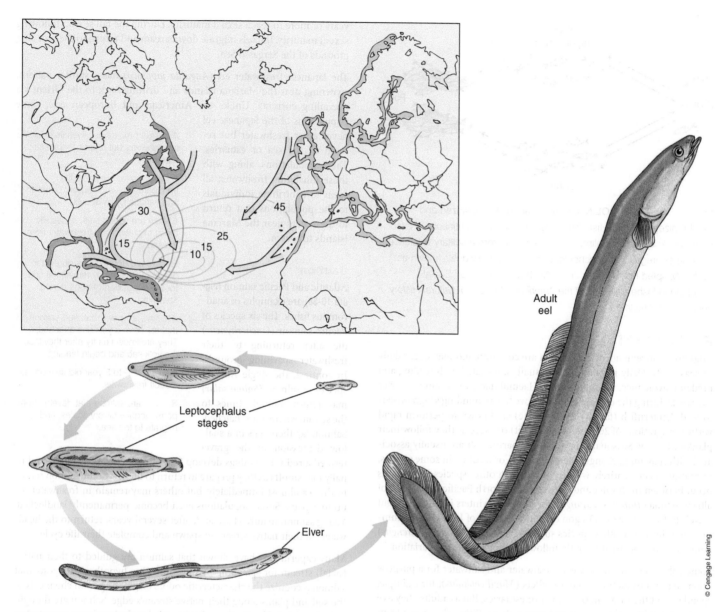

Adult eel

Leptocephalus stages

Elver

© Cengage Learning

FIGURE 10-37 **LIFE CYCLE OF THE AMERICAN AND EUROPEAN EELS.** Both the American eel (*Anguilla rostrata*) and the European eel (*Anguilla anguilla*) breed in the Sargasso Sea. Yellow arrows indicate the migratory paths the eels follow to the Sargasso Sea. Curved lines and numbers in the ocean indicate larval distribution and size of the larva in millimeters. Colored coastal areas show where the elvers enter rivers.

Paul Nicklen/National Geographic/Getty Images

FIGURE 10-38 **SALMON.** Salmon are anadromous fishes. They reproduce and mature in freshwater but spend their adult lives in the marine environment. These Pacific salmon have just returned to their native river to spawn.

IN PERSPECTIVE

Fish Group	Representative Organisms (Common Names)	Form and Function	Reproduction and Development	Type of Feeding	Ecological Roles
Class Myxini	Hagfish	Lack vertebrae, jaws, paired fins, dorsal fin, oral disk, and lateral line system	Sexual, with sexes normally separate No larval stage present Reproductive behavior is poorly known	Carnivorous, generally feeding on soft-bodied invertebrates	Important as scavengers on carcasses of large vertebrates
Class Petromyzontida	Lampreys	Possess two dorsal fins, cartilaginous vertebrae, oral disk, and lateral line system Lack jaws and paired fins	Sexual Reproduce in fresh water and have a larval stage	Larvae are filter feeders Adults are non-feeding or parasitic	Predators on large fish
Class Chondrichthyes: subclass Elasmobranchii	Sharks, rays, skates	Possess jaws, paired and median fins, cartilaginous skeletons, placoid scales, and lateral line system	Sexual Fertilization internal with claspers Oviparous, ovoviviparous, and viviparous reproductive modes present	Most are carnivorous but a few are zooplanktivorous	Dominant predator in the sea
Class Chondrichthyes: subclass Holocephali	Chimaeras, ratfishes, etc.	In addition to the entries for Elasmobranchii, possesses an operculum, crushing plates instead of teeth, and males have head claspers	Sexual Fertilization internal with claspers Reproductive mode ovoviviparous	Carnivorous, crush their prey with flat plates	Predators in deep water
Class Actinistia	Coelacanths	Possess jaws, paired and median fins, skeleton composed of cartilage and bone, an operculum, lateral line system, and fins attached with fleshy lobes	Sexual reproductive mode ovoviviparous, but males lack an obvious insertion organ	Carnivorous, feeding on large bottom fishes	Predator on other fish
Class Actinopterygii: subclass Chondrostei	Sturgeons	Differs from the above in skeleton being primarily cartilaginous, scales ganoid, and fins attached with fin rays	Sexual reproductive mode oviparous Fertilization external	Bottom feeders on benthic invertebrates	Predators and scavengers
Class Actinopterygii: subclass Neopterygii	Modern ray-finned fishes	Differs from the above in skeleton being primarily bony, scales cycloid, ctenoid, or absent	Sexual reproductive mode generally oviparous but other modes exist Fertilization generally external	Include herbivores, carnivores, omnivores, and detritivores	Exploit virtually every feeding niche in the sea

Photos: **Class Myxini:** Tom Stack and Associates; **Class Petromyzontida:** Zig Leszczynski/Animals Animals; **Class Chondrichthyes/Subclass Elasmobranchii:** Jeffrey Rotman; **Class Chondrichthyes/Subclass Holocephali:** Wolfgang Pölzer/Alamy; **Class Actinistia:** Peter Scoones/Photo Researchers, Inc.; **Class Actinopterygii: Subclass Chondrosetei:** blickwinkel/Alamy; **Class Actinopterygii: Subclass Neopterygii:** Richard N. Mariscal.

IN SUMMARY

- Fishes exhibit a variety of body shapes, each suited for their particular habitat.

- Fishes use coloration and patterning for camouflage, signaling to other species that they are dangerous, or to attract potential mates.

- Fishes move about by drifting with the current, burrowing, crawling on the bottom, gliding, and swimming.

- Ray-finned fishes osmoregulate by using their gills to remove excess salt from ingested seawater. Sharks and their relatives do not drink but maintain their internal concentration of solutes at greater than or equal to that of seawater by retaining relatively high levels of urea and TMAO. Shark kidneys and rectal glands excrete salts that are not removed by other mechanisms.

- Most ray-finned fishes, with the exception of some pelagic species, bottom dwellers, and deep-sea fishes, use a gas-filled swim bladder to help them offset the density of their bodies and maintain neutral buoyancy. Cartilaginous fish and coelacanths maintain buoyancy by retaining large amounts of lipids in the liver or swim bladder, respectively, and having appropriately shaped fins.

- All cartilaginous fishes are carnivores, but included among ray-finned fishes are detritivores, herbivores, carnivores, and omnivores.

- Fishes residing in frigid polar seas have developed antifreeze glycoproteins to prevent the formation of ice crystals in their tissues and reduced hematocrits to compensate for the increased viscosity of blood at low temperature.

- To avoid predation, marine fishes have evolved such adaptations as camouflage patterning, sharp, often poisonous, spines, the ability to rapidly change their size, specialized structures to secure themselves in crevices, production of toxic substances, and specialized fins for gliding outside the water. (Have You Wondered? #6)

- Fishes exhibit three modes of reproduction: oviparity, ovoviviparity, and viviparity. Most ray-finned fishes are oviparous, and most cartilaginous fishes are ovoviviparous.

- Schooling may be important in avoiding predators, improving hydrodynamic efficiency, and improving reproductive success.

- Migratory movements of marine fishes are common and may occur daily or seasonally. Daily migrations are usually associated with feeding and predator avoidance, whereas seasonal migrations are usually associated with spawning, changing temperatures, or feeding.

KEY CONCEPTS

1. The oceans are inhabited by five major groups of fishes: hagfishes, lampreys, cartilaginous fishes, ray-finned fishes, and lobefins. (10-1)

2. Hagfish and lampreys are jawless fishes that lack both paired appendages and scales. (10-2)

3. Sharks, skates, and rays have skeletons composed entirely of cartilage. Their streamlined bodies and highly developed senses help them to be efficient predators. (10-3)

4. Lobefins are the most closely related group of fishes to land vertebrates. (10-4)

5. Most ray-finned fishes have skeletons composed primarily of bone. With over 26,000 described species, they are the dominant group of vertebrates in the ocean. (10-5)

6. Fishes have evolved a variety of adaptations in anatomy, physiology and behavior to meet the challenges of living in the ocean. (10-6)

ARE YOU STILL WONDERING?

1. Hagfishes, lampreys, cartilaginous fishes, ray-finned fishes, and lobefins are the five major groups of marine fish.

2. Hagfish use horny cusps (*dental plates*) to grasp and tear away the flesh of their prey (soft-bodied invertebrates) or to scavenge deceased marine mammals. Marine lampreys use horny teeth on their tongue to rasp a hole in their host's body and suck out both tissues and fluids.

3. Skates and rays differ in the way they swim, the fleshiness of the tail, fin structure, and reproductive mode. Rays swim by moving their fins up and down, but skates swim by creating a wave that begins at the forward edge of the pectoral fins and sweeps down the edge to the back of the fins. The tails of skates are fleshier than those of rays and have small fins. The tails of rays are streamlined and contain serrated, hollow spines (barbs) connected to venom glands. Most skates release their eggs in a leathery case (mermaid's purse) but rays give birth to live young.

4. Lobefins are the group of marine fishes most closely related to land vertebrates.

5. No single characteristic defines all ray-finned fishes, but most have a swim bladder, fin rays, a bony skeleton, bony scales embedded in the skin, a terminal mouth, and an operculum covering the gills.

6. Adaptations used by marine fish to avoid predators include the use of camouflage, ability to rapidly change size, the possession of sharp, often poisonous, spines, specialized structures to secure themselves in crevices, production of toxic substances, and specialized fins for gliding outside the water.

QUESTIONS FOR REVIEW

Multiple Choice

1. Lampreys and hagfish lack
   a. paired fins
   b. tails
   c. jaws
   d. mouths
   e. both a and c

2. The skeletons of sharks and rays are composed of
   a. bone
   b. cartilage
   c. soft tissue
   d. fluid
   e. cellulose

3. Shark's teeth are actually modified
   a. cartilage
   b. fins
   c. ctenoid scales
   d. placoid scales
   e. gill supports

4. A sense organ that allows fishes to detect vibrations in the water is the
   a. statocyst
   b. lateral line
   c. ampullae of Lorenzini
   d. olfactory organ
   e. gill arch

5. The presence of _____ in the intestine of the shark slows the passage of food through this organ and increases the surface area for absorption.
   a. stones
   b. chambers
   c. spiral valves
   d. gill rakers
   e. squalene

6. What type of cartilaginous fish has a head clasper and an operculum?
   a. bull shark
   b. sawfish
   c. whale shark
   d. chimaera
   e. manta ray

7. Bony fish can maintain neutral buoyancy by regulating the gas content of their
   a. blood
   b. gills
   c. intestines
   d. swim bladder
   e. liver

8. Many pelagic fish have a dark dorsum and a light-colored belly. This type of coloration is known as
   a. cryptic
   b. disruptive
   c. warning
   d. obliterative countershading
   e. mimicry

9. Most fish that live in seagrass beds or on coral reefs typically have bodies that are
   a. fusiform
   b. globular
   c. snake-like
   d. flattened
   e. laterally compressed

10. An anadromous fish spawns in _____ and lives its adult life in _____.
    a. freshwater; saltwater
    b. saltwater; saltwater
    c. freshwater; brackish water
    d. saltwater; freshwater
    e. freshwater; freshwater

11. The organ primarily responsible for osmoregulation in ray-finned fish is the
    a. rectal gland
    b. gills
    c. kidneys
    d. intestines
    e. skin

12. Fish that exhibit a sedentary lifestyle (e.g., toadfish) are most likely to have this body shape.
    a. laterally compressed
    b. fusiform
    c. globular
    d. depressed
    e. snake-like

Matching

Match the following fish with the description that fits them:

a. clingfish

b. sharks

c. flying fish

d. lionfish

e. hamlet

f. parrotfish

g. tuna

h. flounder

i. anemone fish

j. puffers

k. pearlfish

l. pipefish

m. crocodile icefish

n. trunkfish

o. butterflyfish

1. Has a fusiform body shape, lacks a swim bladder, and regulates swimming muscle temperature with a counter-current exchange system.

2. Lives on the bottom, has a flattened body shape, lacks a swim bladder, and is known to change color rapidly to match its surroundings.

3. Uses cryptic coloration to mimic seaweed.

4. Displays bright colors to warn potential predators that it is toxic.

5. Has a body encased in a hard dermal skeleton and a skin that produces toxins.

6. Uses their expanded pectoral fins to glide through the air after leaping from the water.

7. Osmoregulates by retaining high urea and trimethylamine oxide (TMAO) concentrations in its blood and body fluids.

8. Uses a tiny mouth to feed on individual coral polyps.

9. Possesses a ventral sucker to hold onto rock surfaces.

10. Has teeth fused into a beaklike structure to scrape algae from hard surfaces.

11. Possesses specialized glycoproteins to prevent freezing.

12. Can swallow large amounts of air or water and inflate its body to a size that deters potential predators.

13. Hides in the bodies of sea cucumbers or bivalve molluscs to escape predators.

14. Is a good example of a synchronous hermaphrodite.

15. Is a good example of a sequential hermaphrodite.

Short Answer

1. What are two adaptations that help prevent fish from sinking because of their relatively high density?

2. Explain what is meant by "disruptive coloration" and give an example.

3. Describe how a fish uses its trunk muscles to swim.

4. What mechanism allows fish gills to extract up to 80% of the oxygen in the water passing over them?

5. How is the bull shark able to enter and live in freshwater?

6. Describe how reproduction in an oviparous fish differs from reproduction in an ovoviviparous fish.

7. How do the swimming modes of skates and rays differ?

8. Why do most carnivorous fishes swallow their prey whole?

9. What special adaptations do crocodile icefish possess to compensate for their lack of red blood cells?

10. Why is the blood pressure of fish less than that of other vertebrates?

Thinking Critically

1. The jawfish and seahorse depend on males to take care of their eggs. What is the advantage of this reproductive strategy?

2. What are some advantages of hermaphroditism as a reproductive strategy in fish?

3. What characteristics would you expect to observe in a fish that was adapted to a sedentary life hiding among rocks and coral on a coral reef?

4. Several species of marine fish display different colors as juveniles and as adults. What might be the benefit of such an arrangement?

5. Why are crocodile icefishes considered an "evolutionary dead end?"

SUGGESTIONS FOR FURTHER READING

Butler, Carolyn. 2011. Ancient Swimmers, *National Geographic* 219(3): 86–93.

Dupree, J. 2008. The Most Important Fish in the Sea, *National Wildlife* 46(2): 38–45.

Ellis, R. 2008. The Bluefin in Peril, *Scientific American* 298(3): 70–77.

Fields, R. D. 2007. The Shark's Electric Sense, *Scientific American* 297(2): 74–81.

Haedrich, R. L. 2007. Deep Trouble, *Natural History* 116(8): 28–33.

Hutchings, J. A., and J. D. Reynolds. 2004. Marine Fish Population Collapses: Consequences for Recovery and Extinction Risk, *BioScience* 54(4): 297–309.

Martin, R. A., and A. Martin. 2006. Sociable Killers, *Natural History* 115(8): 42–48.

Montaigne, F. 2007. Still Waters: The Global Fish Crisis, *National Geographic* 211(4): 42–51.

Moyer, Michael. 2010. The Deadliest Catch, *Scientific American*, 302(3): 14–17.

Skomal, Gregory B., Stephen I. Zeeman, John H. Chisholm, Erin L. Summers, Harvey J. Walsh, Kelton W. McMahon, and Simon R. Thorrold. 2009. Transequatorial Migrations by Basking Sharks in the Western Atlantic Ocean, *Current Biology* 19: 1019–1022.

Shadwick, R. E. 2005. How Tunas and Lamnid Sharks Swim: An Evolutionary Convergence, *American Scientist* 93(6): 524–531.

Tennesen, Michael. 2011. The White Shark Café, *National Wildlife*, 49(5): 20–21.

Tennesen, Michael. 2011. Natural Inquiries: The Bull Shark's Double Life, *National Wildlife* 50(1): 16–17.

Warne, K. 2007. Blue Haven, *National Geographic* 211(4): 70–81.

Wilson, S. G. 2006. The Biggest Fish, *Natural History* 115(3): 42–47.

Have You Wondered?

1. If there are any marine lizards and snakes? (11-1)
2. Why so many sea turtle species are endangered? (11-1)
3. Why pelicans have pouches beneath their bills? (11-2)
4. If there are any penguins in the Northern Hemisphere? (11-2)
5. Why emperor penguins incubate their eggs during the Antarctic winter? (11-2)

Marine Reptiles and Birds

REPTILES AND BIRDS generally occupy the higher trophic levels in oceanic food chains. Many are tertiary or higher-order consumers. Their large size allows them to feed on a wide variety of organisms and reduces the number of predators that can feed on them. They are well adapted to moving freely about in search of food and have highly developed senses to supply them with a constant flow of information about their environment. A relatively complex nervous system processes this information, and their proportionately larger brains give them a greater capacity for learning. The ancestors of these marine animals were terrestrial; their move back to the sea led to changes in body form, locomotion, insulation, osmotic balance, and feeding. In this chapter, we will examine how reptiles and birds have successfully met the demands of the marine environment.

11-1 Marine Reptiles

Reptiles have been successful in both terrestrial and marine environments. The same characteristics that allowed the ancestors of reptiles to conquer land were also useful in allowing their descendants to return to the sea. The ancestors of modern reptiles first began to appear about 100 million years ago. Modern-day reptiles include crocodilians, turtles, lizards, and snakes, all of which are represented in the marine environment (Figure 11-1).

AMNIOTIC EGG

A major reason for the success of reptiles was the evolution of an **amniotic egg** (Figure 11-2 is on page 5). An amniotic egg is covered by a protective shell and contains a liquid-filled sac called the **amnion** in which the embryo develops. The egg also contains a supply of food in the form of yolk, stored in a **yolk sac**, and an additional sac, the **allantois** (eh-LAN-toys), for the disposal of waste. A membrane called the **chorion** lines the inside of the shell and provides a surface for gas exchange during development.

The evolution of an amniotic egg about 340 million years ago gave reptiles a great advantage. It allowed development within the protective egg to continue longer, eliminating the need for a free-swimming larval stage that would be highly vulnerable to predation. It also allowed the eggs to be laid in a dry place out of the reach of aquatic predators.

An **amniotic egg** is an egg covered by a protective shell and containing a liquid-filled sac in which the embryo develops.

The **amnion** is a liquid-filled sac that contains the developing embryo of some vertebrate animals.

The **yolk sac** is a sac-like structure in amniotic eggs that contains a supply of food.

The **allantois** is an embryonic support membrane that functions in elimination of wastes and is found in some vertebrates.

The **chorion** is an embryonic support membrane that functions in gas exchange and is found in some vertebrates.

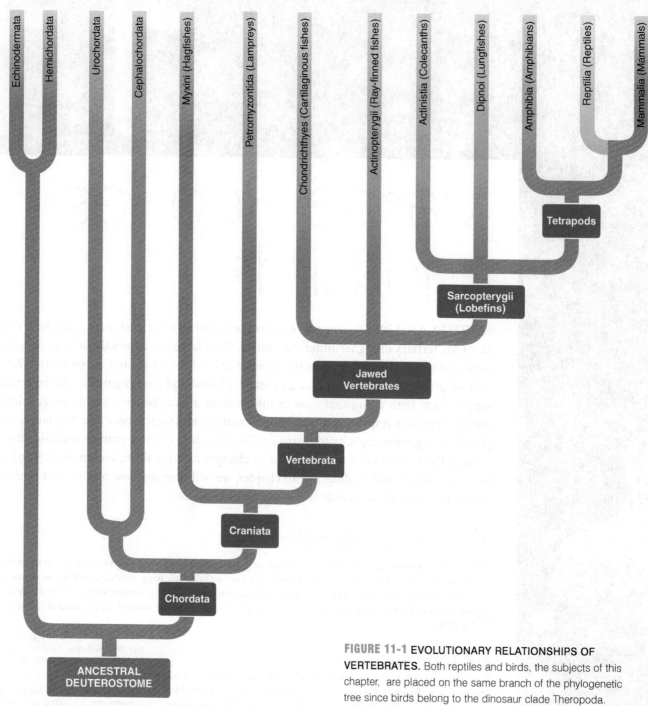

Echinodermata
Hemichordata
Urochordata
Cephalochordata
Myxini (Hagfishes)
Petromyzontida (Lampreys)
Chondrichthyes (Cartilaginous fishes)
Actinopterygii (Ray-finned fishes)
Actinistia (Colecanths)
Dipnoi (Lungfishes)
Amphibia (Amphibians)
Reptilia (Reptiles)
Mammalia (Mammals)

Tetrapods

Sarcopterygii (Lobefins)

Jawed Vertebrates

Vertebrata

Craniata

Chordata

ANCESTRAL DEUTEROSTOME

FIGURE 11-1 EVOLUTIONARY RELATIONSHIPS OF VERTEBRATES. Both reptiles and birds, the subjects of this chapter, are placed on the same branch of the phylogenetic tree since birds belong to the dinosaur clade Theropoda.

© Cengage Learning

To produce an embryo within a shelled egg required a means of fertilizing the ovum before the shell was added. The evolution of copulatory organs by reptiles increased the efficiency of internal fertilization before the ovum was encased by a shell and laid by the female.

PHYSIOLOGICAL ADAPTATIONS

Several other adaptations have helped reptiles survive both on land and in the ocean. The circulatory system of reptiles is more advanced than that of fishes. Circulation through the lungs is nearly completely separate from circulation through the rest of the body (in crocodilians the two circulatory paths are for the most part completely separate). This pattern

of circulation results in a more efficient method of supplying oxygen to the animals' tissues and helps to support their active lifestyles. Their kidneys are very efficient in the elimination of wastes and conservation of water, allowing them to inhabit dry regions and the salty environment of the ocean. Reptile skin is covered with scales and generally lacks glands. This adaptation decreases body water loss in marine environments.

MARINE CROCODILES

Several species of crocodile, including the American crocodile (*Crocodylus acutus*) and the Nile crocodile (*Crocodylus niloticus*), venture into the marine environment to feed. The one best adapted to the marine

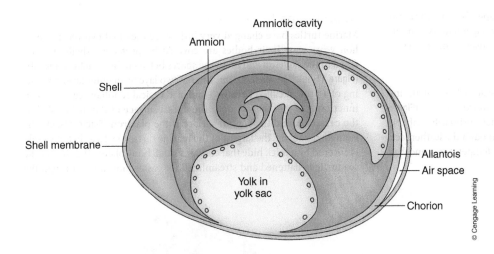

Amniotic cavity

Amnion

Shell

Shell membrane

Yolk in
yolk sac

Allantois

Air space

Chorion

© Cengage Learning

FIGURE 11-2 **AMNIOTIC EGG.** An amniotic egg is covered by a protective shell and contains a watery interior (amniotic fluid) in which the embryo develops. The yolk sac contains a supply of nutrients for the developing embryo. The allantois serves as a waste receptacle for products of metabolism. The membrane known as the chorion functions in gas exchange.

environment, however, is the Asian saltwater crocodile (*Crocodylus porosus;* **Figure 11-3**). Saltwater crocodiles have a high tolerance for salinity and can be found in rivers as well as estuaries and coastal seas from northern Australia and the islands of New Guinea and Indonesia to India and Southeast Asia. This animal is a strong swimmer and sometimes makes oceanic migrations of several hundred kilometers. One specimen is known to have traveled more than 1,100 kilometers (683 miles) from Malaysia to the Cocos Islands in the Indian Ocean.

Saltwater crocodiles are the largest living reptiles. Adult males can grow to lengths of 6 to 7 meters (20 to 23 feet) and weigh more than 227 kilograms (500 pounds). Females are smaller and average 2 to 2.5 meters (6.6 to 8 feet). They feed mainly on fishes, and some individuals have been known to attack and kill sharks close to their own size. They will also eat crabs, turtles, snakes, birds, wild boars, and monkeys. When saltwater crocodiles hunt for terrestrial animals, they usually hide in the water with only the nostrils, eyes, and a portion of their back exposed. When the prey approaches, the crocodile lunges and attacks, usually killing its prey with a single snap of the jaws. The crocodile then drags its prey under the water where it is more easily consumed. Saltwater crocodiles are very aggressive and will attack and kill humans. Like other marine reptiles, saltwater crocodiles drink saltwater, eliminating the excess salt through salt glands on their tongues.

Female saltwater crocodiles do not reach sexual maturity until they are between 10 and 12 years old. Males reach sexual maturity on average around 16 years. They breed in rivers during the wet season, which falls between the months of November and March. Males establish a territory and defend it vigorously against other males. After mating, females normally lay between 40 and 60 eggs, but as many as 90 have been recorded. The eggs are laid in nests made of mounds of mud and plant material, and then buried. Since the eggs are laid during the wet season, the nests must be elevated to prevent loss during flooding. The males leave after mating, but the females stay and protect the eggs from predators, including humans. Incubation averages 90 days, and the sex of the animal is determined by nest temperature, with males being produced at temperatures around 31.6°C. At temperatures slightly above or below this point, females will be produced. On hatching, the young produce a characteristic chirping sound. When the female hears this, she excavates the nest, allowing the young to get free. She then carries them in her mouth into the water and stays with them until they learn to swim. The young are raised in the freshwater rivers and feed on amphibians, crustaceans, small fishes, and other reptiles.

Recent studies indicate that saltwater crocodiles are gifted navigators and will return to their home estuaries if displaced. Researchers from the University of Queensland captured three large males on the Cape York Peninsula of Australia and fitted them with satellite transponders. They were then taken by helicopter 35, 60, and 90 miles, respectively, away from their capture sites. After staying in their new habitat for as long as 3 months, they eventually made their way back to their home estuaries. One specimen that had been airlifted across the peninsula swam 250 miles clear around the coast to return to its estuary, averaging 42 kilometers (19 miles) per day. The researchers speculate that like their close relatives, turtles and birds, saltwater crocodiles use clues from the sun and the Earth's magnetic field, as well as their senses of sight and smell, to find their way home.

Saltwater crocodile populations fluctuate, but they do not appear to be in immediate danger, as are some other marine reptiles. The greatest threats to the saltwater crocodile are habitat destruction due to coastal development, and hunting for their hide, which is considered quite valuable. In some regions, they are raised in farms for the crocodile leather market, a controversial practice because the animals are raised only to be skinned for leather products—the remains are discarded. The Australian management program is the world leader in conservation of the saltwater crocodile. The program has opened crocodile farms to maintain a breeding population, established national sanctuaries to ensure an undisturbed habitat, and developed educational programs for the general

Jon L. Hawker

FIGURE 11-3 **SALTWATER CROCODILE.** Saltwater crocodiles inhabit estuaries and rivers along the coast of Australia and New Guinea and north to Malaysia.

Keratin is a tough protein found in reptilian scales.

The **carapace** is the dorsal surface of a turtle's shell.

The **plastron** is the ventral surface of a turtle's shell.

public. Papua New Guinea also has a management system in place to conserve the species.

SEA TURTLES

Seven species of sea turtle inhabit the world's oceans **(Figure 11-4)**. They primarily inhabit shallow tropical and subtropical coastal waters, although the leatherback turtle (*Dermochelys coriacea*) can tolerate colder temperatures and has been recorded as far north as Canada and Alaska.

Adaptations to Life at Sea

Marine turtles have changed very little since they first evolved 150 million years ago. Their bodies are covered by protective shells that are fused to the skeleton and fill in the spaces between the vertebrae and ribs **(Figure 11-5a)**. The shell is composed of two layers: an outer layer consisting of a protein called **keratin**, similar to typical reptilian scales, and an inner layer composed of bone. The dorsal surface of the turtle's shell is the **carapace**, and the ventral surface is the **plastron (Figure 11-5b)**. The leatherback turtle, the largest of the marine turtles, lacks a shell. Its body is covered by a thick hide that contains small bony plates. The shell of the sea turtle is flattened and streamlined for easier movement through the

(a) Leatherback

(b) Hawksbill

(c) Kemp's Ridley

(d) Green

(e) Loggerhead

(f) Flatback

(g) Olive Ridley

© Cengage Learning

FIGURE 11-4 SEA TURTLE SPECIES. The seven species of sea turtle are the **(a)** leatherback sea turtle (*Dermochelys coriacea*), **(b)** hawksbill sea turtle (*Eretmochelys imbricatus*), **(c)** Kemp's Ridley sea turtle (*Lepidochelys kempii*), **(d)** green sea turtle (*Chelonia mydas*), **(e)** loggerhead sea turtle (*Caretta caretta*), **(f)** flatback sea turtle (*Natator depressus*), and **(g)** Olive Ridley sea turtle (*Lepidochelys olivacea*).

(a)   (b)

FIGURE 11-5 SEA TURTLE ANATOMY. **(a)** The turtle's shell is composed of an outer layer of hard protein and an inner layer of bone that is directly attached to the animal's backbone and ribs. **(b)** The limbs of sea turtles are modified to form flippers.

water. It is also reduced in both size and weight compared with that of its terrestrial cousins, an adaptation for buoyancy and swimming. Other adaptations for buoyancy include large fatty deposits beneath the skin and light spongy bones.

The front limbs of sea turtles are modified into large flippers, whereas their back limbs are paddle shaped and used for steering and digging nests. On land their movement is awkward, but in the ocean they move with great grace and speed.

Behavior

Sea turtles are generally solitary animals that remain submerged for most of the time they are at sea. They rarely interact with each other except for courtship and mating. Even when large numbers of turtles congregate on feeding grounds or during migrations, individual animals usually do not interact with each other. During the day sea turtles alternate their time between feeding and resting. Sea turtles sleep on the bottom, wedged under rocks or coral. In deep water, sea turtles can sleep floating at the surface. Turtles breathe air but can stay submerged for as long as 3 hours. Although they generally prefer the safety of open water, it is not uncommon to find some species close to reefs.

Feeding and Nutrition

Turtles lack teeth but have a beaklike structure that they use to secure food. Six of the seven sea turtle species are carnivorous **(Table 11-1)**. The single herbivorous species is the green sea turtle (*Chelonia mydas*), which is named for its large deposits of greenish fat. Leatherback sea turtles feed primarily on jellyfish. Their pharynx is lined with sharp spines to hold slippery prey, and their digestive system is adapted to withstand the stings of their prey. They may also consume floating plastic bags or plastic trash that they mistake for jellyfish. The turtles cannot digest the plastic, and if enough is consumed, it will block the animal's intestine, causing death from starvation.

Sea turtles consume large amounts of salt with their food and the water they drink. To eliminate excess salt, they have salt glands located above their eyes that secrete a concentrated salt solution. The tears produced by these salt glands can have a concentration of salt twice that of seawater.

Reproduction

Marine turtles generally mate at sea, and breeding occurs in cycles that vary from 1 to 5 years. Males remain in the water past the surf line, and females mate with them between trips to land to nest.

TABLE 11-1 FOOD PREFERENCES OF MARINE TURTLES

Turtle Species	Food Preference
Green sea turtle (*Chelonia mydas*)	Turtlegrass and manateegrass
Hawksbill sea turtle (*Eretmochelys imbricates*)	Sponges
Kemp's Ridley sea turtle (*Lepidochelys kempii*)	Crabs, shrimp, snails, clams, jellyfish, sea stars, and fish
Olive Ridley sea turtle (*Lepidochelys olivacea*)	Jellyfish, snails, shrimp, and crabs
Leatherback sea turtle (*Dermochelys coriacea*)	Jellyfish
Loggerhead sea turtle (*Caretta caretta*)	Conchs, clams, crabs, horseshoe crabs, shrimp, sea urchins, sponges, fish
Flatback sea turtle (*Natator depressus*)	Sea cucumbers, jellyfish, molluscs, prawns, bryozoans, and seaweed

© Cengage Learning

FIGURE 11-6 SEA TURTLE REPRODUCTION. Sea turtles, such as these green sea turtles, generally mate at sea. The female then comes ashore to lay her eggs in shallow nests in the sand.

Courtship During the mating season, males will court females by nuzzling their heads or nipping at their necks and flippers. If the female does not try to escape, the male will grasp the back of the female's shell with the claws on his front flippers. Once he has successfully mounted the female, he copulates with her **(Figure 11-6)**. Sometimes several males will compete for a single female, and they may even fight with each other for the chance to mate. Some females mate with and lay eggs fertilized by several different males. This behavior may increase the genetic diversity of the population. The sperm that the female receives during mating are stored in her body and will fertilize eggs that will be laid on a future trip 2 or 3 years later.

Nesting Nesting occurs most often at night. During nesting the female crawls to a dry area of the beach and digs a shallow pit by pushing away loose sand with her flippers and rotating her body. She then digs a hole in the pit for the eggs using her rear flippers. The female lays two to three eggs at a time **(Figure 11-7a)**. During egg laying, the female constantly secretes mucus on the eggs to keep them moist. The clutch size is between 80 and 150 eggs, depending on the species of turtle. When she is finished laying the eggs, the female uses her rear flippers to cover them with sand. She packs the sand down and uses her front flippers to cover the pit and disguise the nest **(Figure 11-7b)**. Once the nest is concealed, she crawls back to sea to rest before nesting again. A single female will generally lay several clutches of eggs at intervals of 2 to 3 weeks.

Development and Hatching Although the average incubation period is 60 days, the temperature of the nest determines how quickly the turtle embryos will develop. The temperature of the nest also influences the sex ratio. At temperatures above 29.9°C the embryos develop as females. Lower temperatures cause embryos to develop as males.

The eggs usually hatch at night, and digging out of the nest is a group effort. Free of the nest, the hatchlings orient themselves to the brightest horizon, the sea, and then move quickly toward its relative safety **(Figure 11-8)**. As they try to find their way to the water, turtle hatchlings are easy prey for predators such as birds and crabs. If they do not make it to water before daylight, they may die of dehydration when the sun rises. Once in the sea, the hatchlings swim several miles offshore until they are caught in currents that carry them farther out to sea. Even in the sea, hatchlings are not completely safe because they may become prey for large fishes and seabirds. Some will die as the result of eating tar balls or plastic garbage. Only about 0.1% of hatchlings survive to adulthood.

Young sea turtles spend most of their time feeding and growing in nearshore habitats. After reaching adulthood, they migrate to new feeding areas that will remain their primary feeding grounds for life, leaving only to migrate to their nesting beaches to reproduce, then returning.

FIGURE 11-7 SEA TURTLE NEST. (a) Sea turtles come to land to lay their eggs. After digging a depression in the sand, a pit is hollowed out with the hind flippers and 80 to 150 golf ball–sized eggs are laid in the nest. **(b)** Female sea turtles disguise their nests to prevent terrestrial predators from disturbing them. In many nesting areas, volunteers go out early in the morning to locate new nests and mark them to prevent accidental intrusions by humans.

(a)

(b)

FIGURE 11-8 TURTLE HATCHLINGS. Hatchlings must move quickly to the sea to avoid bird and terrestrial predators. They orient themselves to the ocean by looking for a bright reflection off the horizon.

TABLE 11-2 CONSERVATION STATUS TERMS

Status	Description
Extinct in the wild	Survives only in captivity
Critically endangered	Faces an extremely high risk of extinction in the wild
Endangered	Faces a very high risk of extinction in the wild
Threatened	Faces a high risk of extinction in the wild
Near threatened	Likely to qualify as threatened in the near future
Common	Widespread and abundant
Unknown	Need more data to be able to evaluate, or still needs to be evaluated

© Cengage Learning

Turtle Migrations

Sea turtles migrate hundreds and sometimes thousands of kilometers from their feeding grounds to their nesting beaches. Females return to the same beaches where they hatched to mate and lay eggs. Each breeding season, the females return again and again to the same beaches, sometimes to within a few hundred meters of the spot where they nested previously. Why sea turtles nest on some beaches and not others is not known. The Florida loggerheads (*Caretta caretta*), for instance, nest on the central east coast of Florida but not on beaches to the north that appear to be identical. Their preference for the central coastal area may reflect conditions that existed in the distant past, such as temperature, shape of the shore, and numbers of predators.

The migrations of green sea turtles have been studied extensively. Green sea turtles feed on the turtlegrass and manateegrass found in warm, shallow continental waters but breed on remote islands that can be thousands of kilometers away. One population of turtles feeds on manateegrass along the coast of Brazil, but migrates 2,330 kilometers (1,400 miles) to Ascension Island, a small island between South America and Africa, to breed. Tagging studies showed that not all of these turtles migrate every year; some breed on a 2- or 3-year cycle. The migration begins in December and ends in February to April. Food supplies are scarce around Ascension Island, so after the females have finished laying their clutches of eggs, they return to the Brazilian coast to feed.

There are many hypotheses as to how the green sea turtles, as well as other species, are able to navigate over such long distances. The senses of smell and taste may play an important role, as well as audible signals produced by other animals. Some researchers suggest that the turtles can sense both the angle and intensity of the earth's magnetic field or the forces produced by the earth spinning on its axis. Others believe the turtles may be able to navigate using the sun. Studies to answer these and other questions concerning the natural history of sea turtles are currently being conducted.

Sea Turtles in Danger

Most species of sea turtle are endangered **(Table 11-2)** as a result of several human activities. Beach erosion after commercial development and alteration of beaches for improved boat access has destroyed many nesting sites. In a current effort to save the endangered Kemp's Ridley sea turtle (*Lepidochelys kempii*), eggs are being removed from current nesting sites to remote Texas beaches. The hope is that when the turtles hatch they will

imprint on the new nesting area and later return there to breed. Artificial lighting near nesting beaches causes young turtles leaving the nest to crawl toward the lights and away from the water. Many coastal U.S. states, including Florida, and several international nesting sites have strict laws protecting sea turtle eggs and hatchlings. Some Florida coastal communities require turning off bright lights along the beaches during the seasons when the eggs are laid so that the females and hatchlings will not become confused and lost. Turtles are hunted by humans for their meat, eggs, leather, and shells. Humans, however, are not the only predators of turtle eggs; dogs, cats, and wild animals such as raccoons also dig them up to eat.

MARINE IGUANA

The only marine lizard is the marine iguana (*Amblyrhynchus cristatus*) of the Galápagos Islands, off the coast of Ecuador **(Figure 11-9)**. These large lizards are descended from the green vegetarian iguanas that still inhabit the tropical forests of the mainland. The marine iguana can grow to 1 meter (3.3 feet) long, and most are usually entirely black, although some are mottled red and black, showing a hint of green during the breeding season. It is thought that their dark coloration allows these lizards to absorb more heat energy to raise their body temperatures so they can swim and feed in the cold Pacific waters.

FIGURE 11-9 MARINE IGUANA. The marine iguana, the only marine lizard, is native to the Galápagos Islands. The scar on this animal's shoulder and side is the result of a shark bite.

MARINE BIOLOGY & THE HUMAN CONNECTION

Endangered Sea Turtles

Six of the seven species of marine turtle are endangered by human activities. Pollution, poaching, and destruction of nesting sites are all serious problems for sea turtles. In many areas of the world, one of the greatest threats to turtles is commercial fishing. The Pacific population of leatherback sea turtles, for instance, is in such serious decline that researchers estimate that by 2012 fewer than 50 females will return to the nesting site at Playa Grande, Costa Rica, once one of the most popular nesting sites for Pacific turtles. Marine biologists agree that 50 females is not enough to maintain the population and that, unless there is a change, the Pacific population will become extinct. During the 1988–1989 nesting season, 1,367 leatherbacks nested at Playa Grande. By the 2009-2010 nesting season, the number had fallen to 225 animals. Although the total number of breeding females in the Pacific population is disputed, there is no doubt that the numbers are declining. Records from other Pacific nesting sites in India, Sri Lanka, and Malaysia show declines similar to those in Costa Rica. Records for a major nesting beach in Mexico indicate that in the 1988–1989 nesting season, 70,000 turtles returned to nest, whereas in 1998–1999, fewer than 250 turtles were recorded nesting at the site. Studies cite fishing as the reason for the precipitous decline in the numbers of turtles at Playa Grande and other beaches. Turtles tangle in fishing nets and lines and drown, or they are hauled aboard fishing vessels with the nets and killed. Most fishers are paid by the boatload of fish. This discourages them from spending time saving hooked or tangled turtles.

Pacific leatherbacks are not the only sea turtle in danger. Five species of sea turtle, including the critically endangered Kemp's Ridley in the Gulf of Mexico and waters off the southeastern United States, are also threatened or endangered. Fishing has been cited as contributing to the decline of these turtle populations as well, but over the last 10 years the U.S. government and some shrimp fishers have been trying to change this. The drowning of sea turtles in

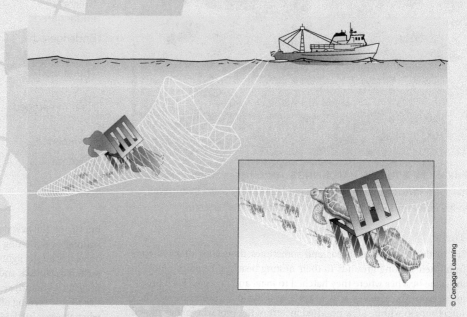

© Cengage Learning

FIGURE 11-A TURTLE EXCLUSION DEVICE. Turtle exclusion devices (TEDs) are used in shrimp trawls to save sea turtles from drowning. The TED fits into the neck of the trawl. A wall of bars is slanted toward the back of the net. While the shrimp pass through the bars, the turtles slide upward along the bars and out through a trap door.

shrimp trawls was identified as a serious problem in the 1970s. Shrimp nets stay in the water for hours, and if a turtle becomes trapped in the net, it drowns in 40 minutes. During the 1970s and early 1980s, it was estimated that as many as 400 turtles per day were dying in shrimp nets. During the 1980s, a device called the **turtle exclusion device (TED)**, was invented. This rather simple device, when incorporated into the shrimp net, reduces turtle mortality rates by as much as 95%. The TED is a grating placed in the neck of the net **(Figure 11-A)**. The net is cut just above and in front of the grate. When a turtle is caught in the net, it swims to the grate and is deflected up and out through the opening. The pressure of the water closes the net flap behind the turtle. The much smaller shrimp are pulled through the grate and into the catch pouch of the trawl net. Shrimp caught in TED nets are designated "turtle safe." In 1991, the federal government began

requiring all U.S. shrimpers to use TED nets, provoking a great deal of hostility from most shrimpers. They claimed that the TEDs would seriously interfere with their catches. Studies conducted after the introduction of TEDs did not confirm these fears. For instance, shrimp catches in the Gulf of Mexico in 1992, when the TED law was in effect, were higher than in the previous 3 years, when the law was not in effect. In the Atlantic Ocean off South Carolina the catch was the highest in 6 years. Claims for lost and damaged fishing gear declined in the years following enactment of the TED law. Since the introduction of TED nets, strandings and drownings of endangered turtles have been down, and turtle nesting activity at some key beaches is up. Although the U.S. government requires that shrimp supplied to U.S. markets be caught in TED nets, there is much resistance in the international community, and not all countries use TEDs, especially in the Pacific.

Marine iguanas have very few natural predators. On Santa Fe Island, one of the Galápagos group, the predators are hawks, short-eared owls, snakes, and crabs. These predators mainly feed on the iguana's eggs. With so few natural predators the marine iguana is very vulnerable to feral predators (introduced wild animals) such as rats, dogs, and cats. The feral animals can affect egg survival and adult mortality rates. Females are especially at risk of predation when going to open nesting areas.

Feeding and Nutrition

Like their cousins on the mainland, marine iguanas are herbivores. Instead of having a long snout like the forest iguana, however, the marine iguana has a short, heavy snout, similar to that of a bulldog, that is better suited for grazing on the dense mats of seaweed that hug the rocks. Larger animals generally dive at high tide to feed on algae that is deeper in the water, whereas smaller animals are restricted to intertidal feeding at low tide. Marine iguanas feed by biting with one side of the mouth first and then the other. As the animal tears at the tough algae, it clings to the slippery rocks with powerful clawed toes. To feed under water, the marine iguana swallows small stones to make its body less buoyant.

Whether feeding at low tide or under water, the animal consumes large amounts of saltwater. The excess salt from the water is extracted and excreted by specialized tear and nasal glands. Salt that is extracted by nasal salt glands is periodically expelled by nasal spraying, an unusual but effective behavior.

Behaviors

Marine iguanas are good swimmers, using lateral undulations of their bodies and tails to propel themselves through the water. They avoid heavy surf and rarely venture more than 10 meters (33 feet) from shore. When leaving the water, they tend to ride in with the swell, then swiftly crawl up on the rocks. If they do not find their territory immediately, they touch the rocks with their tongues, which, like a snake's, carry scent to a receptor in the roof of the mouth. When they locate their own scent, they follow it to their territory, where they rest on the rocks, lying almost motionless above the high tide.

Each male occupies a small territory on the rocks, usually in the company of one or two females. If an intruder wanders by, even by accident, he is immediately attacked and driven off. Most fights between male marine iguanas are the result of deliberate challenges by an intruder. Combat begins with a great deal of posturing, threatening, and bluffing. Each male rises high on its legs and exposes the bright red interior of its mouth. They may also spray each other with a stream of moisture from their nostrils, and as the battle escalates, the combatants butt heads and try to push each other away. In the end, the loser, often the challenger, lies down in a submissive posture, and the winner stands high and nods vigorously. Because such encounters rarely result in serious injury, the survival of the population remains unaffected.

A **turtle exclusion device (TED)** is a grate placed in shrimp trawls to guide sea turtles out of nets so that they won't drown.

Reproduction

The breeding season for marine iguanas is from January through April. Males establish mating territories in sandy areas that are 300 meters or more inland and defend them vigorously against other males. Females will lay one to six eggs in burrows that are 30 to 80 centimeters deep. The females guard the burrow for several days and then leave. The incubation period is approximately 95 days.

SEA SNAKES

Snakes are descendants of lizards that have lost their limbs as an adaptation to a burrowing lifestyle, a lifestyle that was later abandoned by many species. Although most species of snake are terrestrial or arboreal, about 65 species live in the marine environment. Most sea snakes remain close to shore, in the shallow waters of the western Pacific and the Indian Ocean, but the yellowbellied sea snake (*Pelamis platurus;* **Figure 11-10a**) has been sighted on several occasions hundreds of miles from land. This species has migrated east and west from the coast of Asia and can be found off the eastern coast of Africa and the western coast of tropical America. Most of the other species are found in warm coastal waters from the Persian Gulf to Japan and east to Samoa.

Snakes from five different families have become adapted to life in the marine environment. True sea snakes (family Hydrophiidae) evolved from Australian terrestrial snakes known as elapsids (a family of snakes that contains cobras) around 30 million years ago. They are the largest group of sea snakes, with 54 species **(Figure 11-10b)**. True sea snakes have fixed front fangs and the same venom as their terrestrial ancestors. They

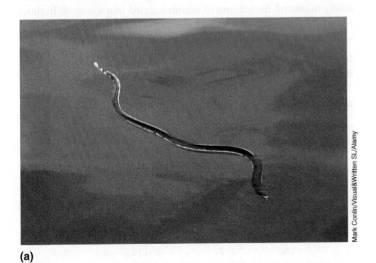

(a)

Mark Conlin/Visual&Written SL/Alamy

(b)

Jon L. Hawker

FIGURE 11-10 SEA SNAKE. (a) The yellow-bellied sea snake (*Pelamis platurus*) is known for making long ocean migrations. **(b)** True sea snakes, such as this olive sea snake (*Aipysurus laevis*), evolved from Australian terrestrial snakes.

Hemipenes are paired penises found in snakes and lizards.

are also viviparous and do not need to come to land to breed. The sea kraits (family Lauticau-didae) are represented by four marine species distinguished by their bold banding patterns and commonly seen in large numbers on the beaches of Southeast Asia and some Pacific Islands. Sea kraits also evolved from terrestrial elapsids and have their characteristic fixed fangs and a highly toxic venom. They are placid animals and unlikely to bite unless provoked. Sea kraits are the only oviparous sea snakes, returning to land to breed. File snakes (family Acrochordidae) are represented by a single marine species and two others that inhabit estuaries and freshwater habitats. File snakes are not venomous and are viviparous. Mangrove snakes (family Homalopsidae) are related to colubrids, a group of mostly nonvenomous snakes (although the highly venomous African boomslang is a colubrid) that are found almost entirely in estuaries. The three marine species are venomous but their fangs are in the rear of their mouths. Salt marsh snakes (family Natricidae) are represented by three marine species that are confined to temperate and subtropical salt marsh environments of North America.

Adaptations to Life in the Sea

Sea snakes have evolved several adaptations that help them survive in a marine environment. As an adaptation to streamline the body, scales on sea snakes are either absent or greatly reduced. The tail is laterally compressed and used as a paddle for swimming. The nostrils of sea snakes are higher on the head than in terrestrial snakes to aid in breathing while drifting. Specialized valves in the snake's nostrils prevent water from entering when they are submerged. Although all sea snakes breathe air, some species can remain submerged for several hours. The animal's single lung reaches almost to its tail, and its trachea (windpipe) has become modified to absorb oxygen, thus acting as an accessory lung. Sea snakes can also exchange gases through their skin when they are underwater. These adaptations allow the sea snake to absorb large amounts of oxygen in a very efficient manner. Sea snakes are able to lower their metabolic rate so that they consume less oxygen when submerged, which is another reason they can remain underwater for long periods.

Like other snakes, sea snakes periodically shed their skins. Sea snakes shed every 2 to 6 weeks, which is more frequent than land snakes and more often than needed for growth alone. Skin shedding allows sea snakes to rid themselves of fouling marine organisms such as algae, barnacles, and bryozoans that could interfere with their swimming efficiency and possibly contribute to disease. The process of shedding begins when the animal rubs its lips against coral or other hard surface to loosen the skin around the head. The snake then catches the skin against something to anchor it and crawls forward, leaving the skin turned inside out behind it.

Feeding and Nutrition

Sea snakes feed mainly on fish, fish eggs, and eels. As noted previously, most species have venomous fangs, but a few that have become specialized for feeding on fish eggs have lost them. Because sea snakes are slow swimmers they rely more on ambush to capture their prey. Some lie quietly at the surface, resembling a piece of drifting wood. Small fish that come near to seek shelter become an easy meal. Others probe crevices and holes on the seafloor for food, sometimes maneuvering their prey into a corner so that it will be within easy striking distance. Like their land-dwelling cousins, sea snakes can swallow prey more than twice their diameter. Sometimes two snakes try to feed on the same prey.

When this happens, the smaller of the two is usually seen disappearing into the larger one. Some fishes that sea snakes feed upon have spines, and instead of digesting or voiding the spines, the sea snakes push them out through their body wall. Eyesight is the most highly developed sense in sea snakes. Thus they feed during the day and spend the night on the ocean bottom, rising occasionally to the surface to breathe. Sea snakes eliminate excess salt that they acquire while feeding by way of a salt-excreting gland located posteriorly under the tongue.

Reproduction

Sea snakes are less tied to the land than other marine reptiles, with only about four oviparous species, such as the banded sea krait (*Laticauda colubrina*), coming to land to lay their eggs. The rest are viviparous, with the females retaining their eggs within their body until they hatch. Sea snakes sometimes congregate in enormous numbers, presumably to mate. In 1932, a group of sea snakes that was 3 meters (10 feet) wide and 97 kilometers (60 miles) long was reported near Indonesia. There are many other reports of smaller aggregations. Like other snakes and lizards, male sea snakes have two penises called **hemipenes**. Each hemipene is capable of functioning independently and only one is used during mating. Mating takes place for long periods of time during which the female controls breathing. As she swims to the surface to take a breath, the male is pulled along, attached by his hemipenis. Since males are unable to disengage from the female until mating is finished, he needs to quickly gulp air when they reach the surface or he will have to wait until the next time the female surfaces to take a breath. Gestation periods in sea snakes range from 4 to 11 months, depending on the species. The young emerge able to swim and feed immediately.

Sea Snakes and Humans

Although sea snake toxin is specifically adapted to killing fish, it is also highly toxic to other vertebrate animals, including humans. Sea snakes, however, are timid by nature and rarely bite humans. They do become aggressive during the mating season, which occurs in the winter. There are no accounts of sea snakes attacking swimmers or divers, although they are curious and seem to be fascinated by elongated objects, such as high-pressure hoses, possibly mistaking them for eels or another snake. This fascination may draw them close to scuba divers. Most sea snake bites, however, occur on trawlers when sea snakes are caught in fishing nets. Fishers tend to handle sea snakes with indifference because only a small proportion of bites are fatal. This may be the result of the snake's ability to control the amount of venom injected into a wound. In those recorded instances in which sea snakes caused human deaths, the snakes were inadvertently grabbed and were biting in self-defense.

In many parts of their range, sea snakes are eaten by humans. In Japan the consumption of sea snakes is so large that it supports a major fishery. The hunting of sea snakes for their skins has led to their near extinction in some areas.

IN SUMMARY

- The evolution of an amniotic egg allowed reptiles to completely sever all ties with their aquatic environment, giving them a great reproductive advantage.

- Adaptations such as more efficient respiratory, circulatory, and excretory systems, as well as a highly developed nervous system, helped reptiles survive on land and, later, in the marine environment.

- Saltwater crocodiles feed mainly on fish.

IN PERSPECTIVE

Classification	Common Name	Food	Reproduction	Distribution
Order Crocodilia	Saltwater crocodile	Fish, birds, and mammals	Oviparous; eggs laid in elevated nests on land along freshwater rivers	Australia, Papua New Guinea, Indonesia, India, Southeast Asia
Order Testudines	Sea turtles	All but one species carnivorous; green sea turtle is a herbivore	Oviparous; eggs laid in sand nests above the high-tide mark	Worldwide in tropics and subtropics
Suborder Sauria	Marine iguana	Seaweed and other algae	Oviparous; eggs laid in nest in sand or lava ash about 100 meters inland	Galápagos Islands
Suborder Serpentes	Sea snakes	Fish, fish eggs, and eels	Four species are oviparous and lay eggs on land; the remaining species are viviparous and give birth at sea	Tropical and subtropical Pacific and Indian Oceans

Photos: **Order Crocodilla**: Jon L. Hawker; **Order Testudines**: George Karleskint; **Suborder Sauria**: Jon L. Hawker; **Suborder Serpentes**: Jon L. Hawker.

- Turtle bodies are streamlined, and their appendages are modified into flippers.
- Most sea turtles are carnivores or omnivores with the exception of the green sea turtle, which is a herbivore.
- Sea turtles return to the same beaches where they hatched to lay their eggs and sometimes migrate long distances from their feeding grounds to their nesting beaches.
- Sea turtles are currently endangered because of many human endeavors. (Have You Wondered? #2)
- The only marine lizard is the marine iguana of the Galápagos Islands. (Have You Wondered? #1)
- Marine iguanas feed on seaweeds and exhibit several adaptations that allow it to feed in the cold marine waters of its habitat.
- Sea snakes are found in the Pacific and Indian Oceans, primarily in shallow water. (Have You Wondered? #1)
- Sea snakes feed on fish, fish eggs and eels.

11-2 Seabirds

Birds evolved from the same line of reptiles as dinosaurs and crocodiles, and many modern taxonomists consider them to be a type of reptile. Only about 250 of the approximately 8,500 bird species are adapted for life in and around the ocean. Although less varied than terrestrial and freshwater species, they are quite numerous. Seabirds feed in the sea and sometimes spend months out of sight of land, but they must return to land to breed.

ADAPTATIONS FOR FLIGHT

Birds are homeothermic; that is, they maintain a constant body temperature. Their bodies are covered with feathers, which help insulate the animal. In addition, the lightweight and strength of the feather make it the perfect body covering for an airborne animal. Birds have a high rate of metabolism to supply the large amounts of energy needed for active flight and to maintain a rapidly functioning nervous system for flight control. Adaptations such as strong muscles, quick responses, and superior coordination greatly aid birds in flight. An advanced respiratory system and a circulatory system that includes a four-chambered heart provide more oxygen to the active muscles to support the high metabolic rate that flying demands. Their senses, especially sight and hearing, are keener than those of their reptilian ancestors, and they have large brains relative to body weight that allow them to process sensory information effectively.

ADAPTING TO LIFE IN THE SEA

Like marine reptiles, seabirds consume large amounts of salt with their food and water. The seawater that these birds consume is about three times saltier than their body tissues. Their kidneys are not able to concentrate such high levels of salt in the urine; therefore, like marine turtles and marine iguanas, seabirds have salt glands (Figure 11-11). These are located one above each eye, and they are capable of producing a solution of sodium chloride that is twice the concentration in seawater. The salt solution produced by these glands is released by way of the internal or external nostrils, so that seabirds, like gulls and petrels, appear to have constantly runny noses.

© Cengage Learning

FIGURE 11-11 SALT GLANDS. **(a)** Salt glands located above the eyes of seabirds help excrete the salt the birds ingest when drinking seawater and feeding on marine animals. The glands drain into the bird's nasal passageways, which explains why many marine birds appear to have a constant runny nose. **(b)** A petrel blowing salt droplets from the nasal openings to eliminate the excess salt from its body.

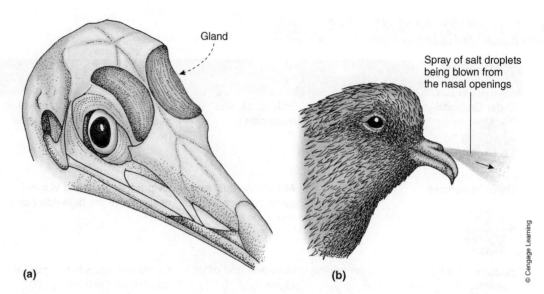

Gland

Spray of salt droplets being blown from the nasal openings

(a)

(b)

SHOREBIRDS

Shorebirds, or waders, feed on the abundance of marine life in the intertidal zone. They range in size from as small as a sparrow to larger than a chicken, and they exhibit countershading, with dark dorsums and white ventral surfaces. The group includes oystercatchers, tattlers, curlews, godwits, turnstones, sandpipers, jacanas, surfbirds, phalaropes, herons, and even inland species such as snipes and woodcocks.

Oystercatchers

Oystercatchers (Family Haematopodidae) have long, blunt, orange, vertically flattened bills **(Figure 11-12)**. They use their bills to slice through the adductor muscles of partially opened clams, mussels, and oysters so they can feed on the soft flesh. They can also use their bills to pry limpets off rocks, crush crabs, and probe mud and sand for worms and crustaceans.

Plovers and Turnstones

Plovers (Family Charadriidae) are shorebirds with worldwide distribution. They have short plump bodies, bills that resemble a pigeon's, and are

FIGURE 11-12 OYSTERCATCHER. The oystercatcher uses its sharp bill to slice through the adductor muscle of bivalves.

Jon L. Hawker

considerably shorter than other waders. These birds can be found on beaches, mudflats, and grassy fields. It is common to see flocks of hundreds of plovers following the receding waves and plucking small animals from the damp sand **(Figure 11-13a)**. One of the most intriguing plovers is the wrybill (*Anarhynchus frontalis*) of New Zealand **(Figure 11-13b)**, the only bird in the world with a beak that is bent to the right. The adaptive significance, if any, of this feature is not known.

The nests of plovers are characteristic of waders in general. Most nests are built on the ground in depressions or hollows. The females lay four pear-shaped eggs that fit together like the pieces of an orange, making it easier for a single bird to incubate the clutch. The chicks hatch in about 3 weeks and are covered with a camouflage down. They are active as soon as they hatch and leave the nest within hours to follow their parents and search for their own food.

Turnstones (*Arenaria*) are heavyset birds with short necks and slightly upturned bills, which they use like crowbars for turning over stones, sticks, and beach debris in search of food.

Sandpipers and Curlews

Sandpipers (Family Scolopacidae) **(Figure 11-14)** are relatives of the plovers and oystercatchers. As the surf and tide retreat, these birds scurry across the sand to feed on small crustaceans and molluscs. Sandpipers are sociable birds that share the beaches with many other species of shorebird. They are found in most parts of the world and make annual migrations that take some species as far as 10,000 kilometers (6,000 miles). The long-billed curlew (*Numenius americanus*) uses its 20-centimeter (8-inch) bill like a forceps to extract shellfish from their burrows.

Avocets and Stilts

Avocets (*Recurvirostra*) and stilts (*Himantopus*) have very long legs, elongated necks, and slender, graceful bodies **(Figure 11-15)**. Avocets feed by wading through shallow water and moving their partially opened beak from side to side through the water. Stilts have legs that are longer in proportion to their body size than any other bird except flamingos. They use their straight bills to probe the mud for insects, crustaceans, and other small animals. Both stilts and avocets live in warm climates and breed colonially in marshes or at the edge of lagoons.

(a)

(b)

FIGURE 11-13 PLOVERS. **(a)** Plovers are common shorebirds with worldwide distribution. **(b)** The wrybill is a plover from New Zealand. Notice the unique shape of this bird's bill.

The American avocet (*Recurvirostra americana*) is anything but friendly during the breeding season. When another bird comes too close to their nesting sites, the birds angrily charge the intruder. When there are no eggs or young to defend, these birds are quite indifferent to intruders, including humans. This characteristic, combined with destruction of their habitats, has made them rare in the eastern United States.

Herons

Herons (family Ardeidae), which include egrets and bitterns, are one of the most widespread families of wading bird and are represented on every major continent. Although they may appear rather ungainly, their skinny legs and long necks are aids for hunting. The great blue heron (*Ardea herodias*), for instance, is a stalker that uses its bill as pincers to capture small fish and crustaceans **(Figure 11-16)**. Its appetite for fish is enormous, and it catches all sizes, even those that appear too large for it to swallow.

Most herons feed by standing still and waiting for their prey, a small fish or crustacean, to come into range. An exception is the little egret (*Egretta garzetta*), an impatient bird that stalks its prey along mudflats. Sighting

a fish, it will move cautiously until within range, and then, with a lightning jab of the beak, seize its prey. It will also vibrate its feet as it moves through shallow water to scare bottom-dwelling prey up to the surface. The snowy egret (*Leucophoyx thula*) also feeds by stirring the bottom with its feet and then rushing about grabbing the small fishes and crabs that it has frightened into motion.

GULLS AND THEIR RELATIVES

Gulls (family Laridae) are probably the best-known seabirds because they are found mostly along shores and seaports, with some species venturing far inland. They have a worldwide distribution and are found everywhere the land and ocean meet, including polar regions. Although gulls have webbed feet and oil glands to waterproof their feathers, most species are not true ocean-going birds and so do not stray far from land. They have enormous appetites and are not very selective eaters. Many species will land on the water and feed on floating debris. Relatives that resemble gulls in many respects are the terns. These birds can be found in large numbers along tropical and subtropical shores. Also related to gulls are the hawk-like skuas and jaeger birds, skimmers, and alcids.

FIGURE 11-14 SANDPIPERS. Sandpipers are a common sight along beaches, where they search for food in the sand.

FIGURE 11-15 AVOCET. Avocets have very long legs and an elongated neck, adaptations for feeding in the shallow water along the shoreline.

FIGURE 11-16 GREAT BLUE HERON. Herons are wading birds that feed on fishes.

Gulls

Herring gulls (*Larus argentatus*) are the most widespread and best known of the gull family **(Figure 11-17)**. These vocal gray and white birds can be found along the shores of both North America and Europe, covering the waterfronts of large seaports and small fishing villages and ranging inland as well. They always travel in large groups. Hordes of these birds can be seen following fishing boats and garbage scows or migrating in groups of tens of thousands of individuals.

Feeding Gulls are noisy, aggressive birds that are efficient predators and scavengers. They are aggressively carnivorous, eating the eggs and young of other birds, stealing prey from other birds, and occasionally, when food is scarce, even eating the young of other gulls. In some areas they are so highly regarded for cleaning garbage from beaches and coastal waterways that they are protected by law.

Gulls exhibit interesting feeding behaviors. Some have been observed to pick up clams, mussels, oysters, or sea urchins from shallow water and then fly over rocks, dropping their food repeatedly until the shell breaks.

FIGURE 11-17 SEA GULLS. These lesser black-backed gulls can be found along the Atlantic coasts of North America and Europe. Like other gulls, they nest close to shore and feed on fish, shellfish, and whatever they can scavenge.

FIGURE 11-18 TERNS. Terns nest in large colonies of thousands of birds. Although similar in color to sea gulls, terns have head feathers that project from the back of the head that help distinguish them from gulls.

They have also been observed using parking lots as shell-dropping areas. Gulls are so successful at getting food and surviving that in some areas they have reached nuisance proportions, menacing other birds, colliding with airplanes, and fouling buildings and sidewalks with excrement.

Nesting Gulls are highly gregarious and gather in colonies of hundreds of thousands of individuals. They are not picky about nesting sites or nesting materials, using any available material. The female lays two or three eggs, and both sexes help in the incubation. The eggs hatch in 3 or 4 weeks, and the female has the capacity to lay another clutch of eggs immediately if the original clutch is lost. This survival mechanism can be repeated as many as three or four times a year. The newly hatched chicks remain in the nest until they are almost fully grown, camouflaged by their speckled down. Predators, and even other gulls, will feed on the hatchlings, and it is not uncommon for only one out of every five to survive the 8 weeks after hatching until they are able to fly.

Terns

Related to gulls are the terns (family Laridae), small, graceful birds with brightly colored and delicately sculpted bills **(Figure 11-18)**. Because they have forked tails, they are sometimes referred to as sea swallows. Although a few species are freshwater birds, most terns live along the shoreline, where they hunt for small fish and invertebrates by plunging into the water but rarely landing on it. They are generally not the avid hunters gulls are, but they do steal food from other birds. Terns are gregarious nesters; often millions of birds swarm together. Not all terns, however, are so gregarious. The arctic tern (*Sterna paradisaea*) makes solo flights from north of the Arctic Circle to Antarctica. These seasonal migrations allow the bird to spend 10 months of the year in continuous daylight, more than any other living creature.

Skuas and Jaegers

Skuas and jaegers (subfamily Stercorariinae) are very aggressive relatives of gulls. Skuas (*Catharacta*; **Figure 11-19**) are found in both hemispheres, particularly around the poles, where they breed. One might consider these birds to be the hawks, falcons, and even vultures of the sea. They are omnivorous and keen predators, feeding on garbage, berries, other birds, eggs (especially penguin eggs), lemmings and other small mammals, insects, carrion, and even unprotected chicks of their own species. Jaegers (*Stercorarius*) are predators that steal fishes from

FIGURE 11-19 **SKUAS.** Skuas are related to gulls and fill the same niche in the marine environment that hawks and falcons do on land.

FIGURE 11-20 **SKIMMER.** Skimmers fly along the water with their lower mandible just beneath the surface. When the mandible strikes a fish or shrimp, it snaps shut, capturing the prey.

other birds. They can frequently be seen pursuing terns and other marine birds and robbing them of their prey.

Skimmers

Skimmers, or scissorbills (subfamily Rynchopinae; **Figure 11-20**), are perhaps the most interesting of the gull's relatives. They are small birds that grow to a length of 50 centimeters (20 inches) and have two unusual features: pupils that are vertical slits similar to those of cats and a flexible lower jaw that protrudes much farther than the upper bill. These birds fish along coastal rivers and streams, flying along with their lower bill just beneath the surface of the water and creating a thin ripple that attracts small fish to the surface. The skimmer then reverses direction of flight and follows the same line that it previously made in the water. When the mandible strikes a small fish or shrimp, it immediately snaps shut. The birds feed at dusk, at dawn, and during the night. They race upstream and then turn to race downstream, occasionally pausing on the bank to rest and eat their catch.

Alcids

Although auks, puffins, and murres (family Alcidae) look more like penguins, they are actually related to gulls. Alcids are countershaded with black dorsums and white ventral surfaces. Like penguins, these birds are awkward on land. In the water they are remarkably agile, using their stout swimming wings to fly through the water as they pursue their prey of fish, squid, and shrimp. Alcids and penguins are the result of **convergent evolution**, in which similar selective pressures bring about similar adaptations in unrelated groups of animals (**Figure 11-21**). They are also **ecological equivalents,** different groups of animals that have

> **Convergent evolution** is an evolutionary process whereby similar selective pressures bring about similar adaptations in unrelated groups of animals.
>
> **Ecological equivalents** are different groups of animals that have evolved independently along the same lines in similar habitats and therefore display similar adaptations.

FIGURE 11-21 **CONVERGENT EVOLUTION.** The bodies of both puffins and penguins have been shaped by convergent evolution, a process in which similar selective pressures brought about similar adaptations in these two unrelated groups of birds.

Horned puffin

Colorful head and beak
(for attracting a mate)

Salt glands
(for removing extra salt after drinking ocean water)

Stiff, broad beak
(to catch and hold fish)

Torpedo-shaped body
(aerodynamic design to "fly" through water)

Black and white color
(counter-shading camouflage in ocean)

Flipper-like wings
(to help swim)

Dense overlapping feathers
(for insulation and waterproofing)

Short legs at rear of body
(for steering underwater)

Webbed feet
(for swimming)

Dense bones
(for diving)

Gentoo penguin

© Cengage Learning

evolved independently along the same lines in similar habitats and therefore display similar adaptations.

A major difference between all of the modern alcids and penguins is that alcids can fly, although the largest flightless bird in the Northern Hemisphere was an extinct alcid, the giant auk (*Pinguinus impennis*; **Figure 11-22**). Standing 750 centimeters (30 inches) high, it was an easy mark for hunters and feather collectors and was hunted to extinction by the middle of the 19th century.

Nesting and Reproduction Alcids spend their winters in offshore waters and then gather in dense, noisy colonies in the cliffs along the northern Atlantic and Pacific oceans in early spring. Auks (*Alcae*) and murres (*Uria*) nest on ledges and among boulders, whereas puffins (*Fratercula*; **Figure 11-23**) prefer crevices and burrows on higher bluffs. The females lay a single egg that is cared for by both parents. Because these nesting sites are rather precarious, eggs of the ledge dwellers are pear shaped, an adaptation that makes them less likely to roll over the edge. As one might expect, both egg and hatchling mortality rates are high. Adults jostling with each other on the ledges cause many eggs and chicks to fall, and others are preyed on by the ever-present gulls.

A **gular pouch** is a sac of skin that hangs between the flexible bones of a pelican's lower mandible.

Parental Care of the Young Young murres that survive leave the ledges in July and literally plunge into the ocean because they have not yet molted into their flight feathers. Each youngster's parents encourage it to the brink of the ledge where, after a moment of hesitation, it leaps into the ocean. The youngster is quickly joined by one or both parents, and they swim out to sea, where they will spend the winter months.

Like other young alcids, chicks of the common puffin (*Fratercula arctica*) have enormous appetites. Both parents spend most of their time gathering food for their hungry chicks, sometimes carrying as many as 18 fish in a single load. The parents provide the chick with enough fish,

FIGURE 11-22 GIANT AUK. The giant auk is now extinct, the victim of hunters interested in its meat and feathers.

Jon L. Hawker

FIGURE 11-23 PUFFIN. Although this puffin looks like a relative of penguins, it is actually more closely related to the gulls.

mussels, and sea urchins per day to equal its body weight. After 6 weeks of constant care, the adult puffins abruptly leave their chicks and head to the open ocean. The young puffin spends about another week in the burrow or crevice by itself before venturing out to the water. It learns to swim, fly, and dive without the guidance of any adult birds.

PELICANS AND THEIR RELATIVES

Pelicans (order Pelecaniformes) are members of a group of birds that includes gannets, boobies, cormorants, darters, frigatebirds, and tropicbirds. Members of this group have webs between all four toes, and pelicans, cormorants, and frigate-birds have a hooked upper mandible. Many species are brightly colored. Some, such as the darters (family Anhingidae), have conspicuous head adornments, whereas others, such as the boobies (family Sulidae), have brightly colored faces and feet. Members of this group have worldwide distribution but most live in the tropics and warm temperate regions. Although some species can be seen in the midocean, most are found in coastal areas and some (pelicans, cormorants, and darters) in inland waters.

Pelicans

Pelicans (family Pelecanidae) prefer warm latitudes and estuary, coastal, and inland waters. They are large animals; some species weigh as much as 11 kilograms (24 pounds). Because they are so large and nest in sizable colonies, the waters where they live must support a large fish population to feed them.

Sometimes pelicans have to fly some distance from their nesting sites to find enough food. After breeding, the birds disperse or migrate, sometimes crossing deserts. Their numbers have declined in Europe, Africa, and Asia because of their dependence on inland waters that have been drained or are polluted.

Pelicans feed just under the surface of the water, using their large **gular** (throat) **pouches** as nets. The gular pouch is a sac of skin that hangs between the flexible bones of the bird's lower mandible and is characteristic of pelicans. The brown pelican (*Pelecanus occidentalis*) of the Americas is the most fully marine of the pelicans (**Figure 11-24**). It patrols about 9 meters (30 feet) above the water, and when it sights a school of small fishes, it folds its wings back and crashes into the water with its gular pouch gaping. The impact stuns the fish, allowing the pelican to scoop up large numbers of them, as well as 8 to 12 liters

FIGURE 11-24 PELICAN. A brown pelican with a fish in its gular pouch.

(2 to 3 gallons) of water, with its pouch. The pelican then returns to the surface with its catch, buoyed by hundreds of air sacs located just beneath the skin (subcutaneous air sacs). It must then let the water drain from its pouch before it can become airborne again. An adult brown pelican will consume 4.5 to 6.5 kilograms (10 to 14 pounds) of fish per day.

The six other species of pelican hunt in groups in shallow water by forming semicircles and dipping their bills to flush and trap fish. Occasionally they will beat their wings on the water to drive the fish together, making them easier to catch.

Boobies

Pelicans are not the only birds in this order that exhibit interesting fishing techniques. Boobies and gannets dive into the sea from heights of 18 to 30 meters (59 to 98 feet), and they can catch flying fishes just before they reenter the water.

The red-footed booby (*Sula sula*; **Figure 11-25a**), the smallest of the boobies, nests in trees of the Galápagos Islands. The female usually lays only one egg. Blue-footed boobies (*Sula nebouxii* **Figure 11-25b**) are widely distributed in the Pacific. They nest close to shore on the ground, and the female of this species lays two or three eggs. Masked boobies (*Sula dactylatra*; **Figure 11-25c**), the biggest and heaviest species of booby, live close to cliffs on many of the Galápagos Islands, and the females generally produce two eggs in a clutch. In each of these species, both parents take turns incubating the eggs and feeding the young. Even with such good care, usually only one nestling of the blue-footed and masked boobies will survive to become a fledgling.

The difference in the number of eggs laid by each species is thought to be due to the reliability of their food supply. Because red-footed boobies forage farthest from land and have the most consistent food supply, they lay one egg. Blue-footed boobies feed close to the coastline, which is easier, but the food supply is very unpredictable. The masked booby feeds offshore in relatively deep water in an area that is intermediate between the feeding grounds of the red- and blue-footed boobies. The food resource is generally stable but not as predictable as it is for red-footed boobies, and masked boobies spend much more time foraging.

Cormorants

Cormorants (family Phalacrocoracidae) are very adept avian fishers. Swimming along the surface, they scan the water for fish and then plunge to spectacular depths in pursuit of their prey. These birds have been found tangled in fishing nets at depths of 21 to 30 meters (69 to 98 feet). Cormorants swallow their prey headfirst so that the fish scales and spines will not catch in their throats. Cormorants lack oil glands to waterproof their feathers, an adaptive advantage because it reduces positive

IMPACT BUBBLE

ONE OF THE GREATEST threats to seabirds and their habitats comes from oil and other chemicals produced during the drilling process. Seabirds are particularly vulnerable to the toxic effects of oil because it floats on the surface of the water. If feeding seabirds dive into water coated with oil, their feathers become oil covered. The birds then ingest oil while preening their oil-soaked feathers or while feeding on oil-contaminated prey. The seabirds most impacted by oil spills include pelicans, gulls, terns, and a variety of shorebirds.

Seabirds are protected from the cold temperature of the seawater by outer feathers that trap warm air next to their skin. The soft downy feathers close to the skin prevent water from coming into contact with the skin. If the bird comes into contact with oil, the oil mats its feathers, destroying their insulating property. Oil-covered feathers are no longer able to maintain a waterproof barrier. Water passes the outer feathers and saturates the downy feathers, causing the animal to lose insulation and making it more susceptible to hypothermia. Heat is lost much faster in water than in air, so oiled feathers are particularly problematic for birds that find food in the water, such as pelicans and cormorants. In addition, if too much of the body surface becomes

covered with oil, birds such as pelicans lose buoyancy, sink, and ultimately drown. Birds with an excessive covering of oil are unable to fly and. frequently become stranded on shore. Unable to feed, they become dehydrated and malnourished. In a warm climate such as the Gulf coast, birds forced to sit in the hot sun are in danger of overheating. Depending on whether they are in the water or on land, hyperthermia or hypothermia is usually the main cause of death.

For a good example of just how damaging oil can be to seabirds, we need look no further than the Deepwater Horizon disaster in the Gulf of Mexico off the coast of Louisiana that occurred during the spring and summer of 2010. As of August 2010, 1,643 live but contaminated birds were found in the Gulf, as were 3,271 carcasses. Since most carcasses sink and are not washed ashore, it is estimated that the actual number of seabird fatalities may have exceeded 23,400. Brown pelicans were particularly impacted because their breeding season had just begun and many pairs were already incubating eggs. Oil spills such as this should help us realize how important it is for government agencies to carefully regulate offshore drilling and develop plans to deal with disasters quickly and effectively should they occur.

(a)

Jon L. Hawker

(b)

Jon L. Hawker

(c)

Flip Nicklin/Minden Pictures

FIGURE 11-25 BOOBIES. Boobies include the **(a)** red-footed booby, **(b)** blue-footed booby, and **(c)** masked booby with two chicks.

buoyancy, allowing them to expend less energy while diving. Without oil glands, though, the feathers become saturated with water, and so the birds must periodically dry their wings on land before they are light enough to fly again. In Japan and some other Asian countries, cormorants are used by fishers to help catch fish **(Figure 11-26)**. The birds are released with a rope or a steel ring around their throats to prevent them from swallowing the fish that they catch.

The guano cormorant (*Phalacrocorax bougainvillei*) of the coast of Peru is one of the world's most valuable birds because it produces **guano**, or bird manure, that is particularly phosphate-rich and used in making fertilizer. The high phosphate content of this material is the result of the birds' anchovy-rich diet. Commercial guano harvesters construct large wooden platforms in sealed-off peninsulas where the birds can roost safely, isolated from predators.

Frigatebirds

Frigatebirds (family Fregatidae; **Figure 11-27**) have lightweight bodies and a large wingspan approaching 2 meters (7 feet) that allow them to soar for hours on the slightest of breezes before returning to land to rest. Frigatebirds are another species with no oil glands in their skin to waterproof their feathers. If they are forced to settle on the ocean's surface, there is a good chance that they will drown. Because of this, frigatebirds feed without getting much more than their bills wet. They skim along just above the water, picking up jellyfish, squid, fish, young sea turtles, and bits of carrion. They are particularly fond of flying fish and can catch them in midflight.

FIGURE 11-26 CORMORANT. In China, cormorants are used for fishing. This fisher is prodding the cormorant to release the fish it has just caught.

Frigatebirds are pirates and will frequently steal fish from other birds. They have been observed attacking boobies even after the booby has swallowed its catch. The frigatebird beats and jostles its victim in flight until the booby regurgitates the fish. The frigatebird then catches the meal before it drops back into the water. A frigatebird will sometimes perch on the upper bill of a feeding pelican and steal fish while the pelican is emptying water from its gular pouch.

Tubenoses

Petrels, albatrosses, and shearwaters (order Procellariiformes) are referred to as tubenoses because of the obvious tubular nostrils on their beaks. The tubes join with large nasal cavities within the head. The significance of these tubes is not known, and they vary in form and length from one species to the next. It has been suggested that these birds have a much better-developed sense of smell than most birds, but this does not explain the peculiar nostrils. These birds fly low over the ocean, and the tube-shaped nostrils may help prevent salt spray from entering the nasal cavity. All tubenoses have large nasal glands that secrete a concentrated salt solution, and the tubes possibly keep the solution away from the eyes. Others hypothesize that the nostrils aid in discerning the strength of air currents.

FIGURE 11-27 FRIGATEBIRD. A male frigatebird inflates its red throat to attract mates.

In addition to the tubular nostrils, members of this order can be distinguished from other **Guano** is bird manure.

birds by the structure of their stomachs, which contain a large gland that produces an oil composed of liquefied fat and vitamin A. This yellow oil is used to feed newly hatched young, and because it has a strong unpleasant odor, adult birds use it for defense by regurgitating it through the mouth and nostrils when they are disturbed.

Albatrosses

Albatrosses (family Diomedeidae) are superb gliders with wings nearly 3.5 meters (11 feet) long, the largest of any living bird. The wandering albatross (*Diomedea exulans*) is the largest of the albatrosses and master of the skies **(Figure 11-28a)**. It glides over ocean waves for hours at a time with barely the flick of a wing. With their very large wings, albatrosses are able to take advantage of light sea winds to keep them aloft. If the wind dies, the birds are forced into a labored, flapping flight; if they land on the water in calm conditions they cannot leave it. For this reason, most species of albatross are restricted in their range to the Southern Hemisphere, the only part of the ocean where winds circle the earth without encountering land. These birds are not likely to move northward because the doldrums, areas of rising air at the equator, form an impassable barrier for soaring birds, especially in the Atlantic Ocean. In the

(a)

(b)

FIGURE 11-28 ALBATROSS. (a) The long wings of the albatross allow the bird to glide on the air currents with minimal effort. **(b)** A male and female albatross clap bills during a mating ritual.

FIGURE 11-29 STORM PETREL. These Wilson's storm petrels exhibit a characteristic flapping type of flight. Early mariners thought the storm petrel could walk on water and named this bird after St. Peter.

FIGURE 11-30 DIVING PETREL. Diving petrels tend to resemble the auks of the Northern Hemisphere, but these birds are tubenoses and are found exclusively in the Southern Hemisphere.

Pacific Ocean, however, 3 of the 13 species of albatross have moved northward. Interestingly, these birds still migrate to the Southern Hemisphere to mate and breed at the same time that the southern species do, even though it is the middle of winter in the Northern Hemisphere.

As a rule, albatrosses travel to land only to breed, and some young birds may spend the first 2 years of their lives at sea before they return to nest. During the mating season, males court females with elaborate displays that include gyrations, wing flapping, and bill clapping (Figure 11-28b). After mating, the female normally lays a single egg on a nest that is shaped like a volcano, and both parents take turns at incubating the egg. The young hatch 60 to 80 days later, usually in March, and are fed first on stomach oil and then regurgitated fishes until they are able to feed themselves. Because it takes so long for the parents to raise a single chick, most species mate every other year rather than annually.

Petrels

Storm petrels (family Hydrobatidae) are small birds with long legs. They exhibit a characteristic erratic type of flight that looks more like a fluttering moth than a flying bird. Wilson's storm petrels (*Oceanites oceanicus*) are the most abundant seabirds. They feed with their legs extended and their feet paddling rapidly just below the surface of the water, appearing to walk on water (Figure 11-29). For such small birds, storm petrels have an extremely long life span (20 years or more). They usually form long-term pair bonds and return to the same nesting sites every year. They do not breed until they are 4 or 5 years old, which is late for a bird of this size. They breed seasonally and, like other species in this group, lay only a single egg.

Although diving petrels (*Pelecanoides*) have the nostrils, beak, and stomach glands that are characteristic of tubenoses, they look more like the auks of the Northern Hemisphere (Figure 11-30). Diving petrels are found exclusively in the Southern Hemisphere. They require cold water throughout the year and prefer gray skies, rough seas, and high precipitation, and generally they do not stray far from their colonies. Diving petrels can spot small crustaceans and fish from the air. Sighting its prey, the petrel performs a headlong dive and continues the chase by "flying" underwater. Grasping its meal in its beak, the bird emerges from the water and reenters the air without missing a wing beat.

Penguins

The birds that are most highly adapted to life in the sea are penguins (order Sphenisciformes). Although penguins are usually associated with

Antarctica, only 2 of the 17 penguin species, the emperor penguin (*Aptenodytes forsteri*) and the Adelie penguin (*Pygoscelis adeliae*), live there. Fourteen other species of penguin live on barren, rocky islands in the Antarctic Sea and along the shores of South America and South Africa. One species, the Galápagos penguin (*Spheniscus mendiculus*), lives as far north as the equator. It is able to survive in this atypical region because of the cold, food-rich waters of the Humboldt Current that bathe the western coast of South America.

Penguins are well adapted for swimming in cold ocean water. Like other birds their body is covered by an outer layer of feathers that streamlines the body and helps to waterproof it. Penguin feathers are more densely packed than those of any other bird, with about 80 feathers per square inch. The packing of feathers is so dense that they may seem like a layer of fur. Beneath the outer layer of feathers is a layer of down feathers that aid in insulation. Like other birds, penguins have precise control over their feathers: By ruffling them, they can trap more or less air in the down layer, thus adjusting the amount of insulation. There also is a layer of fat beneath the skin that helps insulate the bird.

On land, penguins are somewhat awkward, moving by hopping or waddling on short legs or tobogganing on their bellies as they push on the ice with their flipper-like wings. In the sea, they are as swift and agile as many fishes, flying through the water by flapping their wings (Figure 11-31). They can maintain swimming speeds of 20 kilometers (9 miles) per hour and are capable of short bursts of speed of as much as 44 kilometers (20 miles) per hour while pursuing prey. Their torpedo-shaped bodies are streamlined to offer less resistance to the water, and their flat, webbed feet, stretched out behind them, are used for steering. While swimming, penguins break the surface to breathe. They arc through the air like a dolphin and disappear into the water again with scarcely a ripple.

Penguins feed on the fishes, squid, and krill that abound in the cold southern seas. In turn, penguins are preyed on by leopard seals, killer whales, sharks, and skuas. On land, adult penguins have virtually no enemies, but gull-like skuas (discussed previously in this chapter) that live on the outskirts of penguin colonies take a heavy toll on eggs and chicks.

Adelie and emperor penguins breed on the Antarctic continent. Adelies lay their eggs during the summer when the temperatures are above freezing, but emperor penguins lay their eggs in the middle of the Antarctic winter, when temperatures drop to −63°C (−80°F) and blizzard

Tsuneo Nakamura/Volvox Inc./Alamy

FIGURE 11-31 PENGUIN FEEDING. Penguins pursue their prey underwater, swimming with powerful strokes of their wings.

Pete Oxford/Minden Pictures

FIGURE 11-32 EMPEROR PENGUIN. This male emperor penguin is incubating an egg. The egg is covered by a flap of abdominal skin that has a rich blood supply and that will act as a blanket to keep the egg warm during the Antarctic winter.

winds blow at 130 to 170 kilometers (81 to 106 miles) per hour. This strategy may help ensure that the young will be mature enough by summer for life in the sea.

The female emperor penguin lays one egg and then leaves for her feeding grounds, a trip that may take her over vast stretches of ice. While she is gone, the egg is incubated by the male **(Figure 11-32)**. For 2 months he stands with the egg on his feet, a fold of skin on his lower abdomen covering the egg to keep the developing embryo warm. During the incubation period, the male fasts, living off his reserves of fat. If the egg hatches before the mother returns, the chick is fed a secretion from the father's **crop**, a digestive organ that stores food before it is processed. When the mother returns, her crop is filled with fish, and she feeds the chick. The male is now free to travel to open water to feed. When the chick is about 6 weeks old, it requires the services of both parents to provide it with enough food, but by the time summer arrives, the young penguin is capable of feeding itself. The young birds do not have to travel as far to reach the sea as the parents did earlier because of the seasonal breakup of the ice.

> A **crop** is a digestive organ that stores food before it is processed.

As a result of human activities, 10 of the 17 species of penguin are either endangered or vulnerable. Global warming is causing the Antarctic ice pack to melt, reducing the available breeding habitat for some penguin species. Overfishing in the Southern Ocean has decreased the amount of fish and krill that make up the penguin's diet. In some areas of the world, breeding habitat is being lost due to human destruction. For instance, in the process of harvesting guano for fertilizer (mentioned earlier), large amounts of the Humboldt penguin's (*Spheniscus humboldti*) breeding grounds have been destroyed. In the Galápagos, South Africa, and Argentina, oil spills have killed or injured large numbers of Magellanic (*Spheniscus magellanicus*), African (*Spheniscus demersus*), and Galápagos penguins.

IN SUMMARY

- Instead of a body covering of scales, birds have feathers that facilitate flight and provide good insulation.
- Shorebirds, or waders, feed on the abundant food in the intertidal zone.
- Gulls are efficient predators and scavengers.
- Terns are found along the shoreline, where they hunt for small fishes and invertebrates.
- Skuas are aggressively carnivorous and will even eat the young of their own species.
- Skimmers are small birds that fly along the surface of the water, scooping up food with their enlarged mandible.
- Penguins and puffins are examples of ecological equivalents, and their similar body form is the result of convergent evolution.
- Pelicans are large-bodied birds that feed by scooping up large amounts of water with their gular pouch. (Have You Wondered? #3)
- Boobies feed on fishes that they catch in coastal waters.

- Frigatebirds feed on a variety of prey and will also steal food from other birds.
- Cormorants swim underwater in pursuit of prey, sometimes diving to great depths.
- Birds known as tubenoses are represented by albatrosses and petrels.
- As a group, tubenoses feed primarily on fish.
- Penguins are the birds most adapted to life at sea.
- Penguins have streamlined bodies and flippers instead of wings and are excellent swimmers.
- Most penguin species are found in the Antarctic, where they feed on fish, squid, and krill. (Have You Wondered? #4)
- Emperor penguins incubate their eggs during the Antarctic winter so that the young will be mature enough by summer to feed. (Have You Wondered? #5)
- Penguins are preyed on by leopard seals, killer whales, sharks, and skuas.

IN PERSPECTIVE

Classification	Common Name	Food	Reproduction	Distribution
Order Ciconiiformes	Herons	Fish and invertebrates	Eggs laid in nest away from shore	Worldwide temperate and subtropical shores and shallow shoreline water
Order Charadriiformes	Oystercatchers, plovers, turnstones, sandpipers, curlews, avocets, stilts, gulls, terns, skuas, jaeger birds, skimmers, alcids	Shorebirds feed on infauna, fish, and epifaunal invertebrates; gulls eat fish and shellfish, as well as garbage and human food; skuas and jaeger birds eat fish, as well as the eggs and young of other birds; skimmers and alcids feed on fish	Shorebirds lay eggs in nests away from shore; gulls lay eggs in nests close to shore; alcids nest on the ledges of cliffs	Shorebirds worldwide sandy and rocky shores; gulls worldwide tropical and temperate coastal and inland areas; Alcids colder regions of Northern Hemisphere
Order Pelecaniformes	Pelicans, boobies, cormorants, frigatebirds	Fish; frigatebirds will also eat squid, jellyfish, and carrion	Eggs laid in nests near shore	Worldwide tropical coastal areas
Order Procellariiformes	Albatrosses, petrels, shearwaters	Fish	Eggs laid in protected areas on land	Mostly Southern Hemisphere
Order Sphenisciformes	Penguins	Krill and fish	Eggs laid on land or snowpack	Galápagos Islands, Australia, New Zealand, tip of South America and South Africa, Subantarctic Islands, and Antarctica

Photos: **Order Ciconiiformes**: Arco Images GmbH/Alamy; **Order Charadriiformes**: John L. Hawker; **Order Pelecaniformes**: Tom Bledsoe/Photo Researchers, Inc.; **Order Procellariiformes**: Jon L. Hawker; **Order Sphenisciformes**: Pete Oxford/Minden Pictures

KEY CONCEPTS

1. The evolution of the amniotic egg gave reptiles a great reproductive advantage. (11-1)

2. The Asian saltwater crocodile lives in estuaries and is adapted to life in the marine environment. (11-1)

3. Sea turtles have streamlined bodies and appendages modified into flippers. (11-1)

4. Sea turtles mate at sea and lay eggs on the same beaches where the females hatched. (11-1)

5. Sea turtle populations are declining, due to hunting, habitat destruction, and shrimping. (11-1)

6. The marine iguana of the Galápagos Islands is the only marine lizard feeding on seaweed and algae. (11-1)

7. Sea snakes are mostly found in the shallow coastal waters of the Western Pacific and Indian oceans, where they feed on fish and fish eggs. (11-1)

8. Shorebirds are adapted for finding food in shallow water and sand. (11-2)

9. A variety of bird species, including gulls, pelicans, and tubenoses, are adapted to feeding on marine organisms. (11-2)

10. Penguins are the birds most adapted to life in the sea. (11-2)

11. Many marine reptiles and birds are endangered by human activities. (11-2)

ARE YOU STILL WONDERING?

1. There is only one marine lizard, the marine iguana of the Galapagos islands. There are 65 species of snake that have adapted to life in the sea.

2. Sea turtles are endangered as the result of many human activities. Their eggs and meat are harvested for food. Since females return to the same beaches where they hatched to lay their eggs, commercial development of nesting sites prevents turtles from coming ashore to lay their eggs. Also, many turtles are killed in the process of shrimp fishing.

3. Pelicans have pouches beneath their bills that act as nets to catch the fish on which they feed.

4. There are no penguins in the Northern Hemisphere. Penguins are found in the Southern Hemisphere, where the cold ocean water supports large amounts of the krill and fish on which they feed.

5. Emperor penguins breed and lay their eggs in the middle of the Antarctic winter so that by summer the chicks will be mature enough to make their way to the ocean to feed.

QUESTIONS FOR REVIEW

Multiple Choice

1. The key to the evolutionary success of reptiles is
   a. a skin covered with scales
   b. a four-chambered heart
   c. an amniotic egg
   d. an efficient excretory system
   e. a large brain

2. Most marine reptiles must still return to land to
   a. sleep
   b. reproduce
   c. feed
   d. molt
   e. die

3. Marine iguanas feed on
   a. grass
   b. crustaceans
   c. fishes
   d. seaweed
   e. molluscs

4. Marine crocodiles drink saltwater and eliminate the excess salt by way of
   a. their urine
   b. their feces
   c. skin glands
   d. their gills
   e. specialized glands on the tongue

5. Birds that exhibit soaring flight have
   a. small bodies
   b. few feathers
   c. long wings
   d. large heads
   e. small feet

6. _____ use their gular pouch as a net to catch fish.
   a. Petrels
   b. Pelicans
   c. Albatrosses
   d. Gulls
   e. Penguins

7. The seabird that plays an important role in keeping beaches clean is the
   a. albatross
   b. pelican
   c. petrel
   d. tern
   e. gull

8. Two types of seabird that are ecological equivalents are
   a. gulls and skuas
   b. petrels and albatrosses
   c. penguins and puffins
   d. herons and sandpipers
   e. boobies and frigatebirds

9. Skuas
   a. are related to pelicans
   b. feed on krill
   c. are most frequently found along tropical shores
   d. are predators of penguin chicks
   e. nest on ledges along the Arctic shore

10. Which of the following birds would not be considered a shorebird?
    a. sandpiper
    b. turnstone
    c. heron
    d. murre
    e. oystercatcher

Matching

Match each of the following extra-embryonic membranes with the appropriate function:

   a. amnion   1. Functions in waste elimination.
   b. chorion   2. Provides a watery environment for embryonic development.
   c. allantois   3. Contains a supply of food.
   d. yolk sac   4. Functions in gas exchange.

Short Answer

1. Describe how marine turtles are adapted to life in the sea.

2. Describe the nesting behavior of sea turtles.

3. Compare marine iguanas with their more terrestrial cousins.

4. Explain how sea snakes that do not lay their eggs on land reproduce.

5. Explain how birds are well adapted for flight.

6. Describe how seabirds maintain their osmotic balance despite drinking seawater and eating marine animals.

7. Why are most soaring birds such as albatrosses primarily confined to the Southern Hemisphere?

8. Why do the birds known as alcids resemble penguins more than the gulls and terns they are more closely related to?

9. What human activities are contributing to the decline in sea turtle populations?

Thinking Critically

1. Why do seabirds that nest in trees or other protected areas tend to lay fewer eggs than seabirds that nest on the ground?

2. Why is it important for marine reptiles and birds to efficiently eliminate salt from their bodies?

3. Which of the following would you expect to suffer more from predation: albatross adults or chicks? Why?

SUGGESTIONS FOR FURTHER READING

Appenzeller, T., 2009. Ancient Mariner, *National Geographic* 215(5): 122–141.

Bouchard, S. S., and K. A. Bjornadal. 2000. Sea Turtles as Biological Transporters of Nutrients and Energy from Marine to Terrestrial Ecosystems, *Ecology* 81(8).

Cohn, J. P. 1999. Tracking Wildlife: High-Tech Devices Help Biologists Trace the Movements of Animals through Sky and Sea, *Bioscience* 48(1).

DeRoy, T, and M. Jones. 2008. Around Their Necks, *Natural History* 117(3): 36–41.

Ellis, R. 2003. Terrible Lizards of the Sea, *Natural History* 112(7): 36–41.

Hochscheid, S., F. Bentivegna, and J. R. Speakman. 2002. Regional Blood Flow in Sea Turtles: Implications for Heat Exchange in an Aquatic Ectotherm, *Physiological and Biochemical Zoology* 75(1).

Holmes, B. 2008. Flight of the Navigators, *New Scientist* 199 (2666): 36–39.

Safina, C. 2007. On the Wings of the Albatross, *National Geographic* 212(6): 86–113.

Shetty, S., and R. Shine. 2002. Philopatry and Homing Behavior of Sea Snakes (*Laticauda colubrine*) from Two Adjacent Islands in Fiji, *Conservation Biology* 16(5): 1422–1426.

Warne, K. 2007. The Amazing Albatross, *Smithsonian* 38 (6): 46–54.

White, M. 2006. Ungainly Grace: The American White Pelican, *National Geographic* 209(6): 84–97.

Have You Wondered?

1. Why polar bears are considered marine mammals? (12-3)

2. How seals differ from sea lions? (12-4)

3. How marine mammals can dive so deep and stay under water so long? (12-4)

4. How dolphins use echolocation to find their prey? (12-6)

5. What impact humans have had on populations of marine mammals? (12-6)

Marine Mammals

THE MAJORITY OF marine mammals live mostly or entirely in the water, depend completely on food taken from the sea, and display anatomical features, such as fins, that adapt them for their aquatic lifestyle. Most marine mammals are quite intelligent compared with other marine animals. This trait, combined with their generally friendly nature, makes them very popular with the public. Unfortunately, marine mammals share another common characteristic: The bodies of many contain materials that are commercially valuable to humans. As a result, they have been hunted in large numbers over the centuries and now many species are endangered. During the last 50 years, international conservation measures have helped to reverse the decline in many populations, and the numbers of some species are now gradually increasing.

12-1 Characteristics of Marine Mammals

Most mammals (class Mammalia, **Figure 12-1**), whether terrestrial or aquatic, have an insulating body covering of hair, and all maintain a constant, warm body temperature (**homeothermic**). Mothers feed their young with milk, a secretion produced by special glands in the female called **mammary glands**. It is this characteristic that gives the class Mammalia its name.

Marine mammals are **placental mammals**. Placental mammals retain their young inside their body until they are ready to be born. They are sustained by the mother's systems through a remarkable organ, the **placenta**, which is present only during pregnancy. Although mammals produce fewer offspring than many other animals, more of the young survive to adulthood because they receive a great deal of parental care.

The same characteristics that allowed mammals to adapt so well to terrestrial environments also allowed them to function well in the sea. The hair (fur) of some and the layers of blubber in others effectively decrease the loss of body heat to the surrounding water. Like other homeothermic animals, mammals expend about 10 times as much energy as similarly sized reptiles, and they need more food than a fish of comparable size to support their high metabolic rate. Being homeothermic allows marine mammals to be active feeders night and day and to adapt to a wide range of habitats. Some such as baleen whales feed closer to the base of the food chain, whereas others such as sea otters and toothed whales are second-order or higher consumers.

Homeothermic refers to the ability of an animal to maintain a constant internal temperature.

Mammary glands are the milk-producing glands of mammals.

Placental mammals are mammals that retain their young inside their body until they are ready to be born.

The **placenta** is an organ found in pregnant placental mammals that sustains the young until they are born.

FIGURE 12-1 **PHYLOGENY OF CHORDATES.** This phylogenetic tree shows the relationship of mammals (green) to the other groups of chordates.

IN SUMMARY

- Mammals are homeothermic vertebrate animals that nourish their young with a secretion called milk that is produced by special glands in the mother called mammary glands.

- Most mammals have a body covering of hair.

12-2 Sea Otters

Sea otters (*Enhydra lutris;* order Carnivora) are found along the coast of California and as far north as the Aleutian Islands of Alaska. They have short erect ears, five-fingered forelimbs that they use with great dexterity, and well-defined hind limbs with fin-like feet. Instead of having a thick layer of blubber like other marine mammals, their skin is covered by a thick fur with an underlying air layer that protects the animal from the cold. The fur of a sea otter can be as dense as 350,000 hairs per square inch; by comparison, the human head is covered by about 100,000 hairs. A patch of skin that measures 10 centimeters by 10 centimeters (5 inches by 5 inches) square contains about 5.6 million hairs—about the same number of hairs as on the heads of 56 people. This dense underfur traps an insulating layer of air while the coarser, heavier outer guard hairs cover and protect the underfur. The coating of hairs is so dense that the animal's skin stays dry even when underwater, thus decreasing heat loss. Since an otter's feet cool faster than other parts of its body, when the animal is resting at the surface it keeps its feet out of water to decrease excess heat loss to the water.

Sea otters seldom venture more than a mile from shore. They favor the areas around coastal reefs and kelp beds, where they spend most of their time floating lazily on their backs **(Figure 12-2)**. They do not come ashore very often except during storms, although some individuals appear to prefer to sleep on shore, possibly to avoid their primary predators, sharks and killer whales. Females normally give birth on shoreline rocks to one pup, and the offspring soon follows its mother into the sea.

Sea otters have a large appetite and consume nearly 25% of their body weight in food per day. Their diet consists almost entirely of sea urchins, molluscs (especially abalone), crustaceans, and a few species of fish. They carry their prey from the seabed to the surface, where, floating on their backs, they use their chests as eating surfaces. They have been observed bringing a stone up along with a mollusc and using it as a tool to smash open the shell. They also have extraordinarily powerful jaw muscles and strong cheek teeth (molars) to help them break through the hard shells of their prey.

Sea otters are diurnal and gregarious animals and can often be observed playing. They are also quite vocal. Pups cry when hungry, and females coo affectionately to their offspring and mates. In distress, both sexes scream loudly. Their companionship call is a high-pitched squeal that, from a distance, sounds like a whistle.

Because of the value of their fur, sea otters were hunted nearly to extinction. At one time sea otters ranged from Baja California, all along the western coast of North America, and west along the Aleutian Islands to Japan and parts of Russia. After 170 years of unrestricted killing, the sea otter was almost wiped out. In 1911, only about 1,000 animals were left in the world. Since that time, international treaties have protected the animal, and currently 130,000 animals now occupy about 20% of their original range. Their survival today is one of the successes of the marine mammal conservation movement.

IN SUMMARY

- Sea otters inhabit the northern Pacific Ocean.
- Instead of thick layers of blubber, sea otters have a thick coat of fur to keep them warm.
- Sea otters usually stay close to shore, favoring areas around coastal reefs and kelp beds.
- The diet of sea otters consists mainly of sea urchins, crustaceans, and molluscs (especially abalone).
- Hunted nearly to extinction because of their valuable fur, sea otters are now making a comeback after international protective measures were enacted.

12-3 Polar Bears

The polar bear (*Ursus maritimus*; order Carnivora) is a top predator in marine food chains in the Arctic region **(Figure 12-3)**. Polar bears are large animals. Adult males may reach lengths of 3 meters (9.9 feet) and weigh as much as 725 kilograms (1,595 pounds). They live on the shifting ice sheets and ice floes in the Arctic, where the air temperature can drop below –50°F (–46°C) and swim in the near-freezing Arctic Ocean. The polar bear is well adapted for survival in this cold environment. Its large body helps to minimize heat loss (small surface area to volume). A 4-inch layer of fat lies beneath the bear's black skin for insulation. The black skin absorbs radiant energy, and this helps warm the bear's body. The polar bear's coat, like the sea otter's, consists of a dense layer of underfur under long guard hairs. The underfur traps an insulating layer of air, while the guard hairs get matted when wet and help keep the skin dry, thus decreasing heat loss to the water. The polar bear is an excellent swimmer, using its large, rounded forepaws to move through the water. They can swim for up to 60 miles without resting.

FIGURE 12-2 SEA OTTER. Sea otters bring their food to the surface and eat it while floating on their backs. This otter is preparing to feed on a sea urchin.

Wayne Lynch/All Canada Photos/Alamy

FIGURE 12-3 POLAR BEAR. Polar bears are well adapted to the Arctic cold, with a thick layer of fat and a heavy coat of fur. They are top-level consumers in Arctic food chains, primarily preying on seals.

IMPACT BUBBLE

CLIMATE CHANGE DUE to human activities is having a significant effect on species all over the globe. In the Arctic the climate is changing faster than in other regions, and the change is having a negative effect on the organisms that live in this area, especially marine mammals. The most critical change in the Arctic is the loss of sea ice. One recent study estimates that Arctic ice cover and ice thickness have been declining at a rate of 11.3% per decade since 1979. Sea ice is declining at rates faster than expected, and the rate of sea ice loss is projected to accelerate. It is possible that the Arctic Ocean will be ice-free during the summer by the end of the 21st century. Among the species most vulnerable to the loss of sea ice are polar bear, walrus, bearded seal, and ringed seal.

Polar bears are vulnerable to global warming primarily because they depend on sea ice as a means of getting access to their main prey, ringed seals and bearded seals. Polar bears also need sea ice for mating and finding dens to give birth. As temperatures rise, the areas of open water and the rate at which ice floes drift increases. For polar bears, traveling in an environment that is constantly changing would be more expensive in terms of energy because they would have to walk or swim longer distances to find food, mates, and suitable habitats. Also, the seals on which they prey would be spread over a larger area as they follow the ice floes. Scientists predict that the combined effects of decreasing food availability and increasing energy demands will result in decreasing polar bear health and fitness.

Pregnant females give birth in maternity dens and for a period of 4 to 8 months they do not feed. In order to have enough energy to meet the demands of gestation, survival, and lactation, females need to feed heavily and accumulate sufficient energy stores before denning to give birth. The lightest female on record to produce viable offspring weighed 189 kg (416 pounds) at the time she entered the maternity den. With more bears having a difficult time finding prey, the number of pregnant females below this reproduction threshold will likely increase, and fewer cubs will be born. After the mother and her cubs leave the den, the cubs are still nursed for about 2.5 years, but stress caused by the lack of food could very well decrease the mother's milk production. This situation would negatively affect cub growth and survival rates.

Since polar bears are adapted to survive extended periods of time without feeding, the survival rate for adult bears would probably not be as greatly affected unless conditions became more severe. Juvenile bear mortality, however, may increase sooner than adult bear mortality because young bears are not as adept at finding food and thus are more vulnerable to adverse conditions. With such negative changes occurring in reproduction and survival, especially of young bears, population growth rates would likely decrease and the size of polar bear populations decline.

There is current evidence that some of these changes are occurring now. Polar bears in the Western Hudson Bay population have shown declines in individual body condition, reproductive success, survival, and population size. These changes are thought to be the result of stress resulting from decreased food availability and prolonged open-water fasting periods. Unless changes are made to limit the amount of greenhouse gases and stem the effects of global warming and climate change, the polar bear might join the list of recently extinct species.

Although polar bears will eat birds and walrus pups, their primary food is seal. Polar bears are constantly on the move in search of food, swimming from ice floe to ice floe as the ice advances and retreats with the seasons. Because polar bears are less agile in the water than seals, they rely on surprise to catch their prey on the ice. During the winter, polar bears will wait at breathing holes in the ice until a seal breaks the surface for air. Lunging forward, the bear will club the seal with its paw or deliver a crushing bite to the head. The polar bear then drags the carcass onto the ice to feed. In summer the polar bear stalks seals while they are resting on the ice out of the water. The polar bear cautiously draws near from the land, cutting off the seal's escape to the water. When close enough to strike, it rushes over the final few feet of ice and grabs its prey. An alternative strategy is for the bear to swim to the edge of the ice where seals are resting and quickly exit the water for a swift and deadly attack. Seals are clumsy on the ice and cannot outrun or outmaneuver the bear. The seal breeding season provides a yearly feast of young seal pups. The polar bear uses its keen sense of smell to locate the pups in their birthing dens beneath the snow. A polar bear can eat 10% of its bodyweight in 30 minutes, and a big male's stomach can hold up to 70 kilograms (154 pounds) of food.

Polar bears mate in the spring. Females with cubs do not mate, and cubs will stay with their mothers for almost 3 years. As a result, the number of available breeding females is usually limited. Males compete aggressively for the available females. After mating, a female feeds heavily to build her fat reserves, and she may nearly double her weight when pregnant. Cubs are born in snow dens between November and early January. Females usually give birth to two cubs weighing about 1 pound each. With the exception of breeding pairs and mothers with young, polar bears are solitary.

In the 1960s, the polar bear population was estimated at 10,000 animals. In 1981, the United States, Canada, Norway, Russia, and Denmark signed an agreement that limited hunting and recognized the need to protect the entire Arctic ecosystem. Today, the polar bear population is estimated to be around 40,000, but it is beginning to decline again. Polar bears are currently considered endangered.

IN SUMMARY

- Polar bears are found on Arctic ice sheets and ice floes.
- Polar bears are one of the top predators in oceanic food chains in the Arctic, preying primarily on seals. (Have You Wondered? #1)
- Although currently protected by international treaty, they are vulnerable to the effects of climate change.

External ear

Rear flippers
can rotate
forward

Front flippers
can rotate
backward

Patti Murray/Animals Animals

(a)

Front and back
flippers cannot
rotate

No external
ears

Marty Synderman/Visuals Unlimited

(b)

FIGURE 12-4 PINNIPEDS. Pinnipeds are well adapted to the marine environment, with torpedo-shaped bodies and limbs modified to form flippers. **(a)** Eared seals, such as this sea lion, have visible external ears and long necks. They also have front flippers that rotate backward and rear flippers that can be moved forward to better support the animals' body weight as they move on land. **(b)** True seals, or phocids, lack external ears, have short necks, and have front and rear flippers that cannot be rotated and thus make movement on land awkward.

12-4 Pinnipeds: Seals, Sea Lions, and Walruses

The term **pinniped**, meaning "featherfooted," is the name applied to the group of marine mammals (order Carnivora; suborder Pinnipedia) that includes seals, elephant seals, sea lions, and walruses. Although pinnipeds are more at home in the water than on land, they still retain the four limbs that are characteristic of terrestrial mammals. All pinnipeds come ashore to give birth and to molt. Most species also mate on shore, and some species prefer to sleep on land or ice floes, where they are safe from sharks and killer whales. Pinnipeds can be found in all oceans, although most species prefer the colder waters of the Northern and Southern Hemispheres. They feed on fish and larger invertebrates and a few, notably the leopard seal (*Hydrurga leptonyx*), feed on homeothermic animals such as penguins and other seals. The natural predators of pinnipeds are sharks, killer whales, and humans. Although hunted by humans for more than 200 years, the world's pinniped populations are estimated to be 30 million, though three species, the Guadalupe fur seal (*Arctocephalus townsendi*), Hawaiian monk seal (*Monachus schauinslandi*), and Steller's sea lion (*Eumetopias jubatus*), are still endangered.

PINNIPED CHARACTERISTICS

Pinnipeds are divided into three families within the order Carnivora: eared seals (family Otariidae), true seals (family Phocidae), and walruses (family Odobenidae). As their name suggests, eared seals, which include sea lions and fur seals, have visible but small external ears **(Figure 12-4a)**. True seals, or **phocids**, and walruses lack external ears and are thus more streamlined for swimming under water **(Figure 12-4b)**. Eared seals and phocids also differ in their manner of swimming. The main propulsive force for swimming in eared seals is produced by the forelimbs, and they appear to be flying underwater. Their hind limbs remain nearly motionless during swimming and are used for steering. Phocids, on the other hand, propel themselves through the water with a sculling movement of their hind flippers. Walruses use a combination of the two methods, relying primarily on their forelimbs to move their bulky bodies.

The body of pinnipeds is spindle shaped, and many species have several thick layers of fat beneath the skin. The head is round and is carried on a distinct neck so that it can move independently of the rest of the body. They have large brains and well-developed senses. The two sets of pinniped limbs are modified into flippers. The arm bones are shorter than those of whales, so that movement occurs only at the shoulder. Their hind limbs, which are set well back, overlap the stumpy vestigial tail. In phocids, the hind limbs serve only for swimming and are dragged about on land. Eared seals have hind limbs that can be rotated at right angles to the body and act as legs on land.

SWIMMING AND DIVING

Pinnipeds are fast swimmers and can achieve speeds of 25 to 30 kilometers per hour (15 to 18 miles per hour) for short distances **(Table 12-1)**. By way of comparison, a human Olympic swimmer at top speed averages 4 to 5 miles per hour. Pinnipeds are also expert divers, with some species remaining under water for as long as 80 minutes or more without coming up for a breath **(Table 12-2)**. Pinnipeds exhale before diving to decrease the amount of air in their lungs and thus make themselves less buoyant. One recent study, however, found that Antarctic fur seals (*Arctocephalus gazella*) consistently dive with full lungs and exhale during the latter part of their ascent. While underwater the animal's metabolism slows by 20%, and its heart rate decreases to conserve

A **pinniped** is a marine mammal with flipper-like appendages.

Phocids are seals that lack external ears and are known as true seals.

TABLE 12-1 SWIMMING SPEEDS (UNDERWATER) OF SOME MARINE MAMMALS AS COMPARED WITH HUMANS

Olympic rowing record	14 mph
Human Olympic swimmer	5 mph
California sea lion	25 mph
Pacific white-side dolphin	25 mph
Blue whale	23 mph
Leopard seal	23 mph
Beluga whale	14 mph
Harbor seal	11 mph
Sea otter	6 mph

© Cengage Learning

TABLE 12-2 DEPTHS AND DURATIONS OF PINNIPED DIVES

Species	Maximal Depth (meters)	Maximal Duration of Breath-hold (minutes)
Northern elephant seal (male)	1530	77
Southern elephant seal (female)	1430	120
Weddell seal	626	82
Crabeater seal	528	10.8
Harbor seal	508	7
Harp seal	370	16
California sea lion	482	15
Steller's sea lion	424	6
Northern fur seal	207	8
Antarctic fur seal	181	10
Guadalupe fur seal	82	18
Walrus	300	12

Source: Berta, A., et al. 2008. *Marine Mammals Evolutionary Biology*, 2nd edition, p. 238. Burlington, Mass.: Academic Press. Reprinted with permission.

oxygen. During the dive, blood is redistributed so that vital organs such as the brain and heart receive sufficient amounts of oxygen to maintain proper levels of function. Blood flow to the skin, flippers, and kidneys is reduced. In fact, the blood pressure in a seal's flipper can reach 0 during a dive. Pinnipeds store 10 to 30 times more oxygen in their muscles than humans or other land animals. This allows their muscles to function longer without needing to replenish the oxygen. They also have up to twice as much blood per pound of body weight compared with humans. During a dive, this large amount of blood acts as an oxygen storehouse. These physiological adaptations allow pinnipeds to feed on animals in deeper water as well as at the surface. Pinnipeds and other marine mammals increase swimming efficiency by diving below the surface, where there is less resistance and to save energy by reducing metabolism.

REPRODUCTION IN PINNIPEDS

Most pinnipeds leave the water during breeding season and congregate on well-established breeding beaches. The bulls arrive first to establish territories and to await the arrival of females. Some species such as elephant seals (*Mirounga*) and fur seals (*Callorhinus*) are **polygynous** (pah-LIJ-eh-nehs; *poly*, meaning "many," and *gyn*, meaning "female"), with bulls establishing harems of 15 or more females **(Figure 12-5)**. The females arrive ready to give birth to pups conceived the previous year. They then mate again almost immediately after giving birth. Although some species mate every 2 years, most mate annually.

The time frame between copulation and birth for most species of pinniped is about 1 year, even though an entire year is not required for the fetus to complete development. A process known as **seasonal delayed implantation** allows the mother to adjust the **gestation period** (time of pregnancy) of less than 1 year into an annual time frame. In seasonal delayed implantation, the zygote (fertilized egg) undergoes several cell divisions, forming a hollow ball of a few hundred cells called a **blastocyst**. The blastocyst can remain in the female's uterus for periods of a few weeks to several months depending on the species. Following this delay, the blastocyst implants in

the uterus and normal embryonic growth and development continue throughout the remainder of the gestation period. Delayed seasonal implantation allows mating to take place while adults are congregating at the rookeries (breeding beaches) rather than when they are dispersed at sea. This reproductive strategy allows for birth and mating to occur within a brief period of time and allows the young to be born when conditions are optimal for their survival. The physiological mechanisms that control this phenomenon are not well understood. Typically one or two pups are born at the end of the gestation period.

The length of time that the pups nurse (the **lactation period**) depends on the species and the habitat. Pinnipeds that breed in the coldest

Polygynous refers to the situation in which a male mates with more than one female.

Seasonal delayed implantation is a process in which the implantation of an embryo in the mother's uterus is delayed in order to allow for birth to occur at the most opportune time.

Gestation period is the period of pregnancy.

A **blastocyst** is an early developmental stage in mammals characterized by a hollow mass of cells.

FIGURE 12-5 NORTHERN FUR SEALS. Northern fur seals (*Callorhinus ursinus*) breed on the Pribillof Islands of Alaska. Males arrive first and establish territories on the beach. When the females arrive, males will collect a harem of as many as 15 to 20 cows with which they will mate.

© Tim Thompson/CORBIS

habitats, mainly phocids, have the shortest lactation periods, from a few days to 2 to 3 weeks, whereas the sea lions of temperate waters will generally nurse their young for as long as 6 months. One reason for the short lactation period of many phocids is that females fast or feed very little during lactation, placing a severe physiological stress on their body. There is also a limit to the amount of weight that the mother can lose before she will be too weak to survive. Another reason that pinnipeds in cold habitats have short lactation periods has to do with the nature of the habitat. Many species that live in cold environments breed on pack ice. The ice is unstable in many cases, and as it breaks up, there is the possibility that shifting ice will crush and injure or kill the pup. The shorter the lactation period, the faster a pup develops insulation, which it needs to enter the water and survive the cold climate.

EARED SEALS

There are two groups of eared seals: sea lions and fur seals. Sea lions have a coarse coat of nothing but hairs, whereas fur seals have thick dense underfur beneath the stiff outer guard hairs.

Sea Lions

The best known of the eared seals is the California sea lion (*Zalophus californianus;* **Figure 12-6a**). They are the trained seals of zoos and circuses, intelligent animals that learn quickly how to perform tricks. These sleek pinnipeds are naturally playful and chase each other in the water while

(a)

(b)

FIGURE 12-6 EARED SEALS. (a) Sea lions, such as the California sea lion (*Zalophus californianus*), are eared seals that are insulated with a blubber layer. **(b)** This southern fur seal (*Arctocephalus*) has a thick coat of fur for insulation.

vocalizing with a variety of honking and barking sounds. Before becoming popular attractions at animal parks, sea lions were hunted for their blubber.

> The **lactation period** is the period of time that a female mammal is producing milk and nursing her young.

Sea lions are highly social animals and usually congregate in groups when they come ashore. During the mating season, breeding beaches become battlegrounds as bulls compete with each other for territory and females. Harems consist of as many as 12 females protected by a dominant bull that can be 2.5 meters (8 feet) long and weigh as much as 1 ton. The bull aggressively guards his territory and harem.

Fur Seals

Fur seals **(Figure 12-6b)** can be distinguished from the shorthaired sea lions by their thick woolly undercoats. This thick fur is prized by hunters and is quite valuable on the fur market, a fact that has resulted in massive fur seal slaughters in the past. They are smaller than sea lions, with bulls averaging 2 meters (7 feet) and weighing about 270 kilograms (600 pounds). There are eight species of fur seal in the Southern Hemisphere, but only one in the Northern Hemisphere. Both northern and southern species exhibit similar habits, although the northern species has been studied more extensively.

Of the nine species, the northern fur seal (*Callorhinus ursinus*) is the most abundant. At one time it was hunted relentlessly for its valuable fur, and by 1914 only 200,000 animals remained. This animal is now protected by international treaties, and only a specified number of young males are allowed to be killed each year. As a result of these protective measures, the number of northern fur seals is estimated to be 1.5 million animals. Similar conservation measures have been taken to protect the several species of southern fur seal (*Arctocephalus*) such as those that breed on the islands off the coast of South America.

PHOCIDS, OR TRUE SEALS

The forelimbs of true seals, or phocids, are set closer to the head and are smaller than the hind limbs. As a result, they are less well adapted to life on land than other pinnipeds. On ice floes and land, they move by dragging their bodies and take every opportunity to slide or roll instead of crawl. Most phocids congregate during the breeding season, and males establish their territory on the land. Instead of forming a harem, however, a male mates with a single female, and the couple remains together for the entire breeding season.

The most abundant pinniped is probably the crabeater seal (*Lobodon carcinophagus;* **Figure 12-7a**), with a population estimated at more than 15 million animals. When these animals were first described in 1842, biologists erroneously thought they fed on crabs, thus the name. The crabeater seal actually feeds on plankton, primarily krill, which it strains from the water with its teeth.

Some of the most familiar species of seal are the harbor seals (*Phoca*) **(Figure 12-7b)** found along cooler coasts of the Northern Hemisphere. Although most species are marine, a few inhabit freshwater lakes in Canada and Finland. All harbor seals spend some time on land, inching along with a caterpillar-like locomotion. In water they use alternating strokes of their flippers to propel themselves. Harbor seals are generally solitary feeders that eat almost any type of fish, crustacean, or mollusc.

A close relative of the harbor seal is the harp seal (*Phoca groenlandica;* **Figure 12-7c**) found in the cold waters of the Atlantic and Arctic oceans. Newborn harp seals have long, silky, yellowish-white fur that turns pure white a few days after birth. The white fur of the pups, known as

ECOLOGY & THE MARINE ENVIRONMENT

Where Have the Steller's Sea Lions Gone?

Steller's sea lions (*Eumetopias jubatus*) **(Figure 12-A)** were once abundant from the North Pacific coast of California to Japan, with an estimated 70% of them living in Alaskan waters. Since the 1970s, the population has dropped to fewer than 70,000 animals, and during the same period fur seal and sea otter populations have also mysteriously declined in the region. Marine scientists have long debated the cause, some suggesting overfishing of the marine mammals' prey, others suggesting that climate change, predators, disease, or poaching may be the culprit.

In 2000, biologists working for the federal government concluded that commercial fishers who fished for bottom fishes such as pollock were the main threat to sea lion populations because they removed too much of the animals' food source. The proponents of the overfishing theory suggested that commercial-catch restrictions would ensure that the sea lions have sufficient prey. As a result, major catch restrictions were placed on Alaska's $1-billion fishery. Because the Alaskan waters are one of the world's most valuable fisheries, this action gained the attention of members of Congress, who asked a committee of marine scientists to look into the problem. In the committee's preliminary report, issued in December 2002, Dr. Robert Paine of the University of Washington stated that fishing is probably not the major reason for the population decline, but this is not to say that commercial fishing is not a significant problem. To investigate the problem further, Alaskan legislators and other members of Congress ordered a study to be done by the National Academies of Science to determine if overfishing was really the problem. In response, the National Academies of Science recommended that the U.S. government embark on a major ecological experiment by running a 10-year test in Alaskan waters to see if commercial fishing is indeed a threat to the Steller's sea lion populations. The experiment's design calls for setting up four experimental zones in Alaskan waters, each containing a breeding colony of Steller's sea lions. Fishing would be banned for up to 50 nautical miles around two of the colonies but permitted near the other two. Biologists would then monitor and compare trends in the sea lion populations.

If, as Dr. Paine suggests, overfishing is not the major villain in this drama, then what is? Several major studies of sea lions and their habitats indicate that the animals have plenty of food. This observation prompted Dr. Alan Springer of the University of Alaska in Fairbanks to ask if excess predation might be the cause of the population decline. Using data that were collected by the International Whaling Commission, Springer and his associates hypothesized that whalers had effectively eliminated fin, sei, and sperm whales from Alaskan waters by 1970. Interestingly, this coincides with the beginning of the population declines of the smaller marine mammals, including Steller's sea lions. Springer and his associates think that a shift in the diet of killer whales may be to blame not only for the decline in sea lion populations but also the populations of fur seals and sea otters.

Killer whales feed on a wide variety of prey, including the great whales. Although biologists debate how often large whales are eaten by killer whales, they do agree that killer whales are the most significant natural predators of great whales, with the exception of humans. According to data collected over the years, the Alaskan harbor seal population crashed in the 1970s, and populations of fur seals and Steller's sea lions began their precipitous decline in the 1980s. Sea otter populations crashed during the 1990s. Springer thinks these trends indicate predation by killer whales. He suggests that a decline in great whales forced the killer whales to start feeding on the fat harbor seals. As that prey became scarce, they shifted to eating the smaller fur seals and the more aggressive Steller's sea lions. As these populations dwindled, the orcas started feeding on the much smaller sea otters. The decline in the sea otter population has led to an increase in the number of sea urchins, one of the sea otter's favorite foods. This in turn has placed more pressure on the kelp forest habitats, because sea urchins graze on young kelps.

Springer and his colleagues estimated how many seals, sea lions, and sea otters have died in the last 30 years, and by calculating how many small marine mammals it would take to feed a killer whale, they were able to show that the population crashes could have been caused by a shift in the killer whale diet of less than 1%.

Dr. Andrew Trites of the University of British Columbia in Vancouver, however, questions these results. He points out that computer models of marine ecosystems show no significant impact on pinniped populations when great whales are removed. Dr. Springer and other researchers in the field admit that solving the mystery of the declining populations will not be easy. Perhaps the new initiative by the National Academies of Science will shed light on the problem and suggest solutions.

Thomas & Pat Leeson/Photo Researchers, Inc.

FIGURE 12-A STELLER'S SEA LIONS. Populations of Steller's sea lions have been declining in recent years, but marine scientists are still not sure of the cause.

(a)

(b)

(c)

(d)

(e)

FIGURE 12-7 TRUE SEALS. Examples of true seals are the
(a) crabeater seal, **(b)** harbor seal, **(c)** harp seal, **(d)** leopard seal, and
(e) male elephant seal with trunk inflated.

"white coat" to fur merchants, is highly prized and expensive. During the mid-1960s, conservationists drew worldwide attention to the cruelty with which harp seal pups were killed and skinned in the major hunting grounds of Canada. Deluged with complaints, the Canadian government implemented stricter hunting laws and stronger control over the killing and skinning of these animals.

The leopard seal (*Hydrurga leptonyx*; **Figure 12-7d**) is the only phocid that eats homeothermic prey. Although leopard seals feed on fish and krill

like other phocids, penguins, sea birds, and other seals make up the bulk of their diet. They are powerful swimmers and lurk patiently along the edges of ice floes in the Antarctic Sea, waiting for their unwary prey to enter the water. They are less dangerous on land because they must laboriously drag their bodies across the ice floes and cannot move quickly enough to catch their prey. Their primary predators are killer whales.

The giants among the pinnipeds are elephant seals (*Mirounga*; **Figure 12-7e**). The common name for these animals comes from not only their

Sirenians are marine mammals belonging to the order Sirenia.

Tail flukes are the lobes of a sirenian or cetacean tail.

large size but also the unique proboscis on the males, which is actually an enlarged and inflatable nasal cavity. It is usually limp and deflated, but during the mating season it can be inflated to form a trunk 50 centimeters (20 inches) long. The inflated trunk acts as a resonating chamber that amplifies the bull's roar, warning rival bulls to stay away from his territory. The inflated proboscis also plays a role in attracting mates.

There are two species of elephant seal. The largest of the two, the southern elephant seal (*Mirounga leonina*), grows to a length of 6 meters (20 feet) and weighs as much as 4 tons. It is found on islands at the tip of South America and in the Antarctic. The northern elephant seal (*Mirounga angustirostris*) rarely exceeds 5 meters (16 feet) and is estimated to weigh as much as 3 tons. This closely related species can be found on the Pacific coast of North America, from Vancouver Island in Canada to Baja California. Northern elephant seals breed from December to mid-March. This allows the young pups to first enter the sea at an optimal time, during the spring upwelling. During summer and fall the adults and pups remain at sea, where they feed primarily on squid.

WALRUSES

Walruses (*Odobenus rosmarus*) lack external ears and have a distinct neck, as well as hind limbs that can be used for walking on land **(Figure 12-8)**. They are 3 to 5 meters (10 to 16 feet) in length and weigh up to 1,364 kilograms (3,000 pounds). The canine teeth of their upper jaw have developed into tusks in the males, which are used for fighting with other males. Most bulls carry the scars of battles during the mating season. The tusks also help the animal to hoist its massive body onto ice floes. It plants its tusks in the ice like a pick ax and uses them as an anchor. Tusks first appear at 5 years of age and continue to grow for the rest of the animal's life.

The typical family group consists of a large dominant bull that presides over a harem of as many as three females and half a dozen calves of various ages. One or two calves are born at the end of an 11-month gestation period, and they will stay with their mother until they are between 4 and 5 years old. Initially, young walruses have yellow-brown fur. As they age, the fur turns pale and eventually disappears. Although females are devoted mothers, old bulls are known to kill the young, and the remains of infant walruses have been found in the stomachs of adult males.

FIGURE 12-8 WALRUS. A large male on an ice floe. Males use their tusks as weapons during the mating season as they contest other males for mates.

Walruses are found in the Arctic region of the Northern Hemisphere, where they feed on fishes, crustaceans, molluscs, and echinoderms. At one time walruses were killed by hunters for their ivory tusks and by Eskimos and other northern peoples for their succulent flesh. As a result, walrus populations declined to levels so low as to make the species endangered. Protective laws have virtually eliminated walrus hunting for sport and for ivory. However, native peoples who subsist on walrus meat are exempt from the restrictions and continue to take the animals for food.

IN SUMMARY

- Pinnipeds have four limbs that are modified into flippers.
- Although pinnipeds are more at home in the water, they can and do come onto land to mate, give birth, and molt.
- Eared seals, which include fur seals and sea lions, have visible external ears. True seals and walruses lack external ears. (Have You Wondered? #2)
- Eared seals use their front limbs primarily to propel themselves through the water. True seals use their hind limbs, and walruses use a combination of both. (Have You Wondered? #2)
- The hind limbs of eared seals can be rotated at right angles to the body axis and act as legs on land. (Have You Wondered? #2)
- Pinnipeds have evolved a variety of adaptations that allow them to make deep dives and hold their breaths for long periods of time. (Have You Wondered? #3)
- Walruses are restricted to the Arctic seas, where they feed on fishes, crustaceans, echinoderms, and molluscs.

12-5 Sirens: Manatees and Dugongs

Manatees and dugongs are collectively referred to as **sirenians** (order Sirenia). They get their name from the mythical sirens of Homer's *Odyssey*. These sweet-voiced creatures, who lured Ulysses and his men with temptations, were probably the forerunners of the mythical mermaids. Mermaids have been a part of the mythology of the sea for more than 2,500 years. Pliny the Elder, a respected Roman naturalist, gave detailed descriptions of them, and Christopher Columbus wrote in his log, "Today I saw three mermaids, but they were not as beautiful as they are painted." No doubt what Columbus and other seafarers saw and mistook for mermaids were manatees and their relatives, although it is difficult to imagine how anyone could mistake these large, gray animals with wrinkled skin and whiskers for the alluring maidens of legend.

At one time sirenians enjoyed a wide distribution. Now they are confined to coastal areas and estuaries of tropical seas. They share many similarities with whales. Both have streamlined, practically hairless bodies, forelimbs that form flippers, a vestigial pelvis with no hind limbs, and **tail flukes**. Sirenians are completely aquatic animals and are helpless on land, not even able to crawl, as do the pinnipeds. Sirenians are by nature gentle animals that become quite tame and trusting in captivity and in their natural habitat. There are several preserves in Florida where divers and snorkelers can swim with manatees and even touch them and rub their bellies. Only in those areas of the world such as India where they have been hunted nearly to extinction have sirenians become shy and elusive. There are two families of sirenians, one represented by the manatees (family Trichechidae) of the Atlantic Ocean and Caribbean Sea and the other by the dugongs (family Dugongidae) of the Indian Ocean.

DUGONGS

The primary differences between manatees and dugongs are in anatomy and habitat. Manatees inhabit both the sea and inland rivers and lakes. Dugongs are strictly marine mammals living in coastal areas, where they feed on shallow-water grasses. The head of the dugong is larger and the flippers are shorter than those of the manatee. The dugong's tail is notched, whereas the manatee's tail is rounded. There is only a single species of dugong (*Dugong dugon*; **Figure 12-9a**). Large numbers of this animal were once found along coastal areas of the Indian Ocean, Red Sea, and South China Sea. It has been hunted extensively for its tasty flesh, and today the species is endangered and found only in isolated areas of its once extensive range.

MANATEES

There are three species of manatee. The northern manatee (*Trichechus manatus;* **Figure 12-9b**) is found from the southeastern United States to northern South America. The Brazilian manatee (*Trichechus inunguis*) is a freshwater species endemic to the Amazon and Orinoco rivers of South America. The African manatee (*Trichechus senegalensis*) lives in coastal habitats, rivers, and lakes of western Africa. All three species of manatee have been extensively hunted, and the northern species is now protected by law in Florida. Manatees mate and give birth underwater, and the male remains with his mate even after the breeding season. The female typically gives birth to a single calf after an 11-month gestation period. Manatees are strict vegetarians and consume large amounts of shallow-water plants. A single manatee will consume at least 27 kilograms (60 pounds) of aquatic plants per day. When manatees eat, they guide the water plants to their mouths with their flippers—this behavior may account for the association with mermaids. Because of their voracious appetites, they provide a service by keeping coastal waterways free of nonnative plant species, such as the water hyacinth, that tend to take over their new habitats. In Guyana, manatees have been introduced to clear weeds from the canals that transport irrigation water to sugarcane fields.

The greatest threat to northern manatees are the propellers of motorboats. Many of these relatively slow-moving animals are mauled or killed by boats. Although there are strict laws governing the speed and use of motorboats in areas where the manatees congregate, there are few regulations governing the canals and waterways that connect these fresh water bodies to the sea. As the manatees migrate from the freshwater areas to the ocean, they must travel these waterways, and it is then that they are in the greatest danger.

STELLER'S SEA COW

Neither the manatee nor the dugong is found in Arctic waters. The only sirenian from this area, Steller's sea cow (*Rhytina gigas*), is extinct. During 1741–1742, Georg Wilhelm Steller, a German, was the physician and naturalist for Russia's expedition to the northern Pacific Ocean led by Captain Vitus Bering. Steller discovered the enormous mammal when his ship was wrecked off one of the Commander Islands. He made many notes on the physical characteristics and behavior of this animal and correctly identified it as a previously unknown species of Sirenia. Desperate for food, Steller and his companions killed and ate several of the creatures. When their ship was repaired, they returned to Russia with the furs of sea otters and Arctic foxes that they had also killed for food. Word spread quickly that a fortune in furs was available for the taking on the Commander Islands, and hundreds of hunters descended on them. The hunters found the meat of Steller's sea cow tasty and easy to obtain, and 27 years after its initial discovery, the species was extinct. Fossil evidence suggests that Steller's sea cow had a much wider range than just the Commander Islands and that they were most likely eliminated from most of the range before Steller "discovered" the species.

(a)

(b)

FIGURE 12-9 SIRENIANS. (a) Dugongs are found in coastal estuaries of India and Asia. **(b)** Northern manatees can be found along the Florida coast, where they feed on seagrass and other vegetation.

IN SUMMARY

- Sirenians are represented by manatees and dugongs.

- Although at one time sirenians enjoyed a wide distribution, they are currently confined to coastal areas and estuaries of the tropics.

- Sirenians are herbivores feeding on shallow-water grasses and a variety of water plants.

12-6 Cetaceans: Whales and Their Relatives

Of all marine mammals, **cetaceans** (seh-TAY-shenz; whales, dolphins, and porpoises) are the ones most extensively adapted to a marine environment. Humans have been fascinated and awed by these animals for centuries. Ancient Greeks and Phoenicians worshiped the dolphin or "sacred fish," as did many Polynesian cultures. In Australia more than 5,000 years ago, aborigines painted on stones

Cetaceans are marine mammals belonging to the order Cetacea.

recognizable pictures of dolphins. The biblical story of Jonah being swallowed by a whale, the appearance of marine mammals in the art of the Middle Ages, and the Herman Melville novel *Moby Dick* all attest to our fascination with these marine mammals.

GENERAL CHARACTERISTICS OF CETACEANS

The bodies of cetaceans closely resemble those of fishes, and early naturalists included them both in the same group of animals. Their exact ancestry is not known, but it is generally agreed that they evolved from an ancient group of land-dwelling carnivorous mammals. Recent discoveries in Pakistan, Egypt, and the Arabian desert of fossil whales with forelimbs, hind limbs, and teeth similar to those of carnivores support this hypothesis. Other evidence for the terrestrial origins of cetaceans can be seen in developing fetuses. Whale fetuses are remarkably similar to those of land mammals. Early in development they have four limbs. The rear appendages disappear externally before birth, although there are a vestigial pelvis and leg bones in some species **(Figure 12-10)**. The front appendages persist and develop into flippers (pectoral fins) with the bone structure of a five-fingered hand. Initially, the animal's nostrils are located at the end of its snout, but before birth they migrate to the top of the head to form the

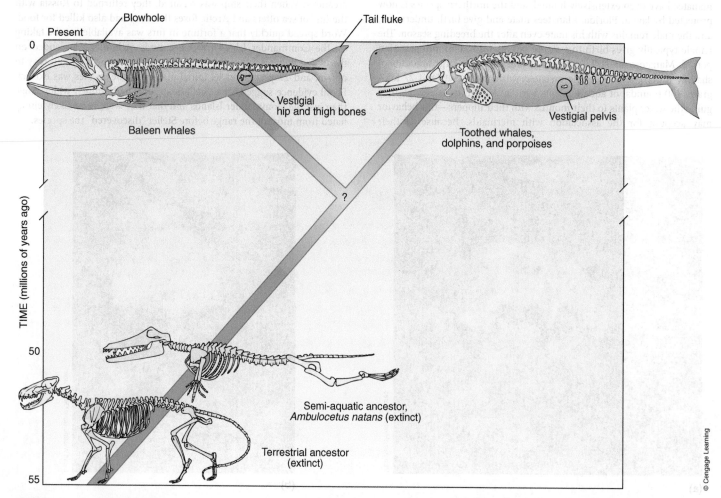

FIGURE 12-10 CETACEAN EVOLUTION. Whales evolved from four-legged terrestrial carnivores. Their ancestors may have moved about on land in addition to swimming in shallow seas, much as eared seals do today. Modern cetaceans are well adapted to life in the sea, with streamlined bodies, appendages modified to form flippers, and a large tail flipper, or fluke.

characteristic **blowhole**. The location of the blowhole allows these animals to surface and breathe with minimum effort.

The cetacean body is streamlined, and beneath the skin is a uniformly thick layer of fat called **blubber**. Water robs the body of heat 50 to 100 times faster than air, so the animal must be well insulated to survive in its environment. The fat provides insulation to conserve heat and is an energy reserve, as well as a source of water when the fat is metabolized. The thickness of the blubber layer varies from species to species and, in an individual, from one part of the body to another. For instance, the blubber layer of a bowhead whale (*Balaena mysticetus*) averages 50 centimeters (20 inches) thick, whereas the blubber layer of a sperm whale (*Physeter macrocephalus*) averages 12 to 18 centimeters (5 to 7 inches) thick. The average bowhead whale has 15 tons of blubber, one reason that it was a favorite catch of whalers. A 1-ton beluga whale (*Delphinapterus leucus*) has a 10-centimeter (4-inch) layer of blubber that can contribute 364 kilograms (800 pounds) to the animal's total weight. The thickness of the blubber layer also varies from season to season. Animals are thinner after periods of fasting, during migration, and during breeding season.

In becoming streamlined, the neck disappeared and the head became continuous with the rest of the body. The seven cervical (neck) vertebrae of cetaceans are very compressed, and in some species they are fused into a single structure. The result of these modifications is that many cetaceans cannot move their head relative to their body. Cetaceans have no external ears or projecting nostrils to hamper their movements through the water. The whale ear consists of a small opening on the side of the head that leads to an eardrum. The opening is plugged with wax to prevent water from entering and damaging the eardrum. The inner ear is contained in a bone that floats freely in the soft tissue surrounding the skull.

A **blowhole** is an opening of the top of a cetacean's head that contains the nostrils.

Blubber is a thick uniform layer of fat under the skin of some marine mammals.

The cetacean body is essentially devoid of hair, with the exception of a few hairs on the head. The skin of cetaceans lacks the sweat glands that are characteristic of other mammals. Because sweat glands work via evaporative cooling, they do not work underwater. The absence of these glands also helps cetaceans to conserve water in their salty environment. When whales need to lose heat, they shunt blood to their blubber layer.

The forelimbs of cetaceans have become modified into flippers that can move only up and down and twist a bit. They function as stabilizers. The cetacean tail consists of flat flukes composed of dense connective tissue and acts as the main propulsion organ. The tail flukes not only supply the force of propulsion but also regulate vertical movement. Most cetaceans also have a dorsal fin that apparently helps them to control roll.

The flippers, as well as the tail flukes, serve an important role in retaining body heat (**Figure 12-11**). The arteries that carry blood to the flippers are

FIGURE 12-11 COUNTERCURRENT BLOOD FLOW IN CETACEAN FLIPPERS AND TAILS. The blood vessels in a cetacean's **(a)** flippers and **(b)** tail fluke are arranged so that **(c)** some of the heat carried by arterial blood is transferred to the cooler venous blood returning to the body core, thus conserving body heat.

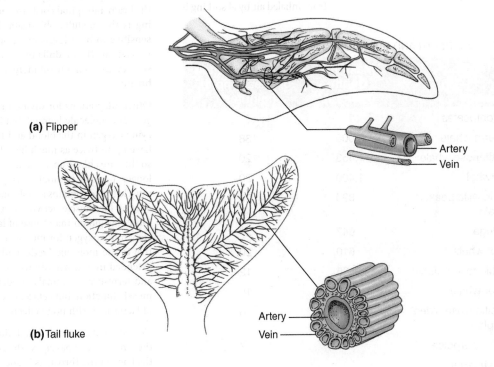

(a) Flipper

Artery
Vein

(b) Tail fluke

Artery
Vein

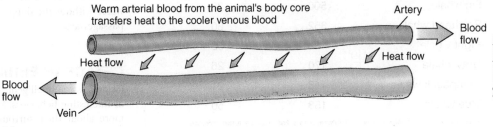

Warm arterial blood from the animal's body core transfers heat to the cooler venous blood

Artery

Blood flow

Heat flow

Heat flow

Blood flow

Vein

(c)

© Cengage Learning

surrounded by veins that carry blood back to the heart, an example of countercurrent flow. As warm blood in the arteries moves out into the flipper, it transfers heat to the cold venous blood that is returning to the core of the body. Most of the heat is returned to the body, and cold blood is carried to the uninsulated flippers and flukes.

ADAPTATIONS FOR DIVING

Perhaps the most striking adaptations of cetaceans involve those that allow them to dive to great depths and remain submerged for relatively long times (Table 12-3). As with the pinnipeds, the ability to dive to great depths allows cetaceans to feed on animals in deeper water and increase swimming efficiency. When a cetacean surfaces, the top of its head emerges first, thus exposing the blowhole to air. From this opening the animal emits a spout that consists of vapor condensing in the cool air, as well as mucous droplets. The mucous droplets may play a role in eliminating nitrogen gas from inhaled air by absorbing it

The bends is a condition that occurs when nitrogen gas dissolved in the blood comes out of solution and forms gas bubbles as the pressure decreases during an ascent from deep water, causing disruptions in circulation.

Alveoli are the tiny spaces, or gas sacs, within the lungs where gas exchange occurs.

Myoglobin is a protein in mammalian muscle that stores oxygen.

Lactic acid is a waste product produced when cells metabolize anaerobically.

before it reaches the lungs, thus preventing a condition known as **the bends**. The bends occur when nitrogen gas that is dissolved in the blood comes out of solution and forms gas bubbles as the pressure decreases during an ascent. These bubbles can interfere with the circulation of blood to body tissues and can cause tissue damage or even death.

Most whales have lungs that are smaller in proportion to their body size than the lungs of humans and other land animals. A major difference is that humans exchange 15% to 25% of the oxygen in their lungs with their blood, whereas whales exchange 80% to 90% of the oxygen (Figure 12-12). Whales typically dive with full lungs, although they may make necessary adjustments to achieve neutral buoyancy during a dive. The lungs and rib cage of whales are structured so that they collapse easily during descent, preventing the ribs from being broken due to pressure at great depth. As a cetacean descends, the increased pressure of the water causes the **alveoli**, or gas sacs, in the lungs to collapse. Any residual air remaining in the lungs is forced into airways, where gas exchange with the blood cannot occur. As a result, the lungs contain little air during a dive. This adaptation prevents nitrogen gas from entering the blood, thus helping the animal to avoid the bends and the problems of compression and decompression that can plague human divers while diving and surfacing.

While diving, the animal's metabolism decreases. The heart rate slows to as low as 10% of normal (from 100 beats per minute to 10 beats per minute). This way the heart expends less energy and uses less oxygen. Blood is preferentially shunted to vital organs and tissues such as the brain and spinal cord. The portion of the brain that controls breathing is the medulla oblongata. In most mammals this region is very sensitive to high levels of carbon dioxide in the blood, which trigger inhalation. The medulla of cetaceans is less sensitive to carbon dioxide, so they can remain submerged longer without feeling the urgency to breathe.

Other adaptations for diving involve the transport and storage of oxygen. The molecule hemoglobin in red blood cells is responsible for carrying oxygen in the blood and for giving blood its red color. Cetaceans have up to twice as much blood per pound of body weight as humans, so they are able to absorb and transport more oxygen, allowing them go longer periods without having to take another breath. Mammalian muscle tissue contains a molecule called **myoglobin** that is a reservoir of oxygen for muscle activity. Cetacean muscles contain 10 to 30 times more myoglobin than those of land animals and thus store proportionately more oxygen for muscle activity. Their muscles are also less sensitive to the molecule **lactic acid**, a waste produced during vigorous or extended muscle activity when there is not enough oxygen. In humans and terrestrial mammals, moderate amounts of lactic acid can impair muscle function, but cetaceans can tolerate much higher accumulations of lactic acid with no ill effects.

Cetaceans also exhibit adaptations for preventing water from entering their respiratory passages. The larynx, or voice box, does not open into the back of the throat, as it does in some mammals, but rather into the nasal chambers. This allows cetaceans to open their mouths under water without the danger of food or water entering their respiratory passageways.

CETACEAN BEHAVIORS

Cetaceans are intelligent, inquisitive animals that exhibit a wide range of interesting behaviors. Some of these behaviors help cetaceans learn more about their surroundings, whereas others may play a role in communication.

TABLE 12-3 DEPTH AND DURATION OF CETACEAN DIVES

Species	Maximal Depth (meters)	Maximal Duration of Breath-hold (minutes)
Odontocetes		
Sperm whale	3,000	138
Bottlenose whale	1,453	120
Narwhal	1,400	20
Blainsville beaked whale	890	23
Beluga	647	20
Pilot whale	610	20
Bottlenose dolphin	535	12
Killer whale	260	15
Pacific white-sided dolphin	214	6
Dall's porpoise	180	7
Mysticetes		
Fin whale	500	30
Bowhead whale	352	80
Right whale	184	50
Gray whale	170	26
Humpback whale	148	21
Blue whale	153	50

Source: Berta, A., et al. 2008. *Marine Mammals Evolutionary Biology*, 2nd edition, p.238. Burlington, Mass.: Academic Press. Reprinted with permission.

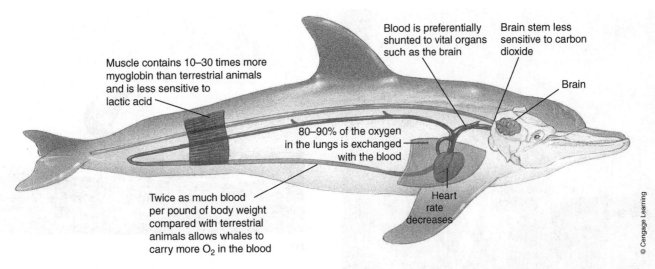

Muscle contains 10–30 times more myoglobin than terrestrial animals and is less sensitive to lactic acid

Blood is preferentially shunted to vital organs such as the brain

Brain stem less sensitive to carbon dioxide

Brain

80–90% of the oxygen in the lungs is exchanged with the blood

Twice as much blood per pound of body weight compared with terrestrial animals allows whales to carry more O_2 in the blood

Heart rate decreases

© Cengage Learning

FIGURE 12-12 DIVING PHYSIOLOGY. Physiological changes in the respiratory system, nervous system, circulatory system, and muscles allow whales to dive to great depths and remain submerged for over 1 hour.

Spy Hopping

Whales sometimes stick their heads straight up out of the water and survey their surroundings, a behavior known as **spy hopping (Figure 12-13a)**. During spy hopping the whale uses its strong flukes to push itself partially out of the water. The whale does not appear to maintain its position by swimming but rather relies on buoyancy control and positioning with its pectoral fins and tail flukes. Generally, the whale's eyes will be slightly above or just below the surface of the water, allowing it to see whatever is on the surface nearby. This behavior can last for several minutes, depending on what the whale is viewing. Spy hopping often occurs when the whale is interested in a boat or some other object rather than in other whales that may be nearby. This behavior may also help whales establish their bearings while in coastal waters.

Breaching

Breaching refers to a behavior in which whales completely or almost completely leave the water **(Figure 12-13b)**. They begin this behavior under water by swimming quickly, accelerating as they go. Whales are strong swimmers, and some species such as the adult humpback whale (*Megaptera novaeangliae*) can accelerate to full speed with as few as three or four pumps of their flukes. They hit the surface at maximum velocity in a nearly vertical position and exit the water as they pump their flukes for the last time. At some point the whale usually begins twisting its body in a corkscrew movement. This action may start in the approach and continue as the whale shoots from the water into the air, causing the pectoral fins to fling wide. Some whales can rise out of the water the length of their body, as much as 14 meters (45 feet) for humpback whales. The animal generally assumes a horizontal position as it falls back to the water, usually hitting the water on its back. Although many whale watchers think that breaching is a sign of playfulness, it is probably not. It may be a way for a whale to establish dominance. It may also communicate with special emphasis that an individual is arriving or leaving.

Sometimes a whale breaches several times in a row, a behavior called serial breaching. Serial breaching is usually observed in whales that have started to become active after resting, have just left a social group, or appear excited or irritated. Usually the first breach of the series is the largest and the rest become progressively smaller, with less of the body exiting the water.

A related behavior is the head lunge, in which a whale breaks the surface and falls forward instead of backward. This behavior includes a quick horizontal burst through the water with open mouth and is most often associated with feeding.

Slapping

Whales like to lift their huge tails high above the water and slap them down on the surface with enough force to make a huge splash and produce a noise that is loud enough to be heard for great distances. This behavior, called **tail slapping** or **tail lobbing (Figure 12-13c)**, may be associated with marking position and is sometimes interpreted as being an aggressive behavior. Tail cocking is another aggressive behavior in which a whale can cock its tail up in the air and bring it down on an opponent. The part of the whale's body closest to the tail fluke is called the **peduncle**. In the peduncle slap the whale swings the rear portion of its body, sometimes as far forward as its dorsal fin, out of the water and then drops it down sideways on the water or another whale. Tail slash and tail swish involve moving the tail from side to side across the surface of the water to create turbulence. These are also thought to be aggressive displays.

After arching their body, some whales may bring their flukes above the surface. If the tail is brought straight up so that the ventral (bottom) surface is visible, the behavior is called **fluke up (Figure 12-13d)**. If the fluke clears the water but remains turned down, the behavior is called **fluke down**.

Spy hopping is a cetacean behavior in which the animal raises its head out of the water.

Breaching is a cetacean behavior in which the animal completely or almost completely leaves the water.

Tail slapping (tail lobbing) is a cetacean behavior in which the animal lifts its tail above the water and slaps it down hard on the surface.

The **peduncle** is the part of the whale's body closest to the tail fluke.

Fluke up is a cetacean behavior in which the animal brings its fluke above the water so that the ventral surface is visible.

Fluke down is a cetacean behavior in which the animal brings its fluke above the water so that the dorsal surface is visible.

FIGURE 12-13 CETACEAN BEHAVIORS. Some cetacean behaviors are **(a)** spy hopping, **(b)** breaching, **(c)** tail slapping, **(d)** fluke up, and **(e)** flipper flapping.

In **flipper flapping** a whale rolls over onto its back and flaps its flippers in the air **(Figure 12-13e)**. Some whales lie on the surface and lift one or both pectoral fins out of the water and wave them about. In pec slapping, a whale rolls on its side and slaps the water with its pectoral fin to create a forceful splash and loud sound. No one knows why they do these behaviors; but some researchers speculate that they are a form of communication.

Pectoral fins can be used to stroke the body of another whale, a behavior known as **pectoral stroking**. Mothers and calves will stroke each other as a bonding behavior, and pectoral stroking in adults is associated with courtship and mating.

REPRODUCTION AND DEVELOPMENT

Since cetaceans mate and give birth in the water, our understanding of these events is quite limited. Most of what is known comes from observations of small whales in captivity. Baleen whales usually mate and give birth in the same locality at nearly the same time of the year. In toothed whales, breeding occurs throughout the year with only minimal seasonality. The gestation period of baleen whales ranges from 10 to 13 months, and for toothed whales the range is 7 to 17 months. Normally cetaceans bear only one offspring at a time, rarely two. The young are fed extremely rich milk, containing 40% to 50% fat and 10% to 12% protein. Cow's milk contains 2% to 4% fat and 1% to 3% protein. The high caloric content of whale's milk allows the infant whale to grow at an exceedingly fast rate and produce enough body heat to resist the cold until it develops its thick layer of blubber. Baby blue whales, for instance, double their birth weight in the first week and then gain some 91 kilograms (200 pounds) a day thereafter.

Many species of toothed whales live in structured social groups called schools or **pods** depending on the species. With few exceptions, notably cooperative feeding by humpback whales, baleen whales do not group together, school, or form pods.

TYPES OF WHALES

Whales are divided into two suborders **(Figure 12-14)**. Baleen whales (suborder Mysticeti) lack teeth. Instead they use structures called **baleen** to filter their food from the water. The largest whales are baleen whales. Toothed whales (suborder Odontoceti) have teeth and feed on larger prey. The familiar dolphins and killer whales are examples of toothed whales.

Baleen Whales

Baleen whales have enormous mouths to accommodate plates of baleen, which take the place of teeth **(Figure 12-15a)**. Each plate of baleen has an elongated triangular shape and is anchored at its base to the gum of the upper jaw. The baleen plate is composed of **keratin** (a tough protein in mammalian hair and nails) fibers that are fused, except at the inner edge, where they form a fringe **(Figure 12-15b)**. Hundreds of these plates, each one 1 to 4 meters (3.3 to 13 feet) long, form a tight mesh on the sides of the mouth. The diet of baleen whales consists mainly of plankton, especially krill, or fish. A baleen whale feeds by opening its mouth and swimming into dense groups of krill or schools of fish. When it has filled its mouth, it closes it and expels the water through the baleen plates, straining out the prey in the process. The retained food is then moved by the tongue to the back of the oral cavity and swallowed. The process is repeated until, at the end of a feeding, a large whale has as much as 10 tons of krill or fish in its stomach. When not actively feeding, the whale covers the baleen with the underlip, which projects above the lower jaw. Humpback whales sometimes capture their food by blowing a ring of bubbles called a **bubble net** that traps fish near the surface **(Figure 12-16)**. Baleen whales include right whales, bowhead whales, rorquals, and gray whales.

Right and Bowhead Whales
Right whales and bowhead whales (family Balaenidae) can be distinguished from other cetaceans by the lack of a dorsal fin and the absence of grooves on the throat and chest. They have two separate blowholes and produce a characteristic V-shaped spout. Little is known about most species of balaenids, and they are considered rare. Right whales received their name from early whalers. When lookouts sighted a whale, if it was the "correct" whale for hunting, they would yell, "It's the right whale!"; thus the origin of the name. Right whales and bowhead whales were the favorite catch of early whalers because they were relatively slow and easily harpooned, floated when killed, and could be sliced up with relative ease. As the result of overhunting, two of the three species of balaenids were nearly extinct by the end of the 19th century, and the bowhead whale is the rarest of all whales.

Rorquals Rorquals (family Balaenopteridae) have the dorsal fin and ventral grooves that the balaenids lack. The ventral grooves, or pleats, under the mouth allow the throat to expand while the animal is feeding. Rorquals are generally slender and streamlined and are capable of relatively fast swimming. Some of the better-known rorqual species are the blue whale, fin whale, and humpback whale. The blue whale (*Balaenoptera musculus*) has the distinction of being the largest of the whales and may be the largest animal that has ever lived **(Figure 12-17)**. Blue whales range in size from 24 to 30 meters (80 to 100 feet) and weigh more than 100 tons (as much as 1,600 humans or 25 elephants!). Because of excessive hunting, however, large individuals are rarely sighted. Whaling records indicate that one blue whale consisted of 50 tons of muscle; 8 tons of blood; 1 ton of lungs; and 60 tons of skin, bones, and internal organs (other than the lungs). The heart alone weighed 591 kilograms (1,300 pounds) and measured about 1 meter (3.3 feet) in each direction.

The fin whale (*Balaenoptera physalus*) is the second largest whale species, growing to between 19 and 22 meters (63 and 73 feet) long and weighing between 45 and 75 tons. It gets its name from the large dorsal fin that often slopes backward. The animal is dark gray to brownish black with a white ventral surface. Fin whales are found in every ocean in the world but rarely come close to shore. Like other whale species, fin whales migrate to polar waters in the summer, where they feed primarily on krill, and to warmer waters in breeding season.

The humpback whale can be distinguished by the low hump on its back; the large bumps, called **bosses**, on its snout; and the very long pectoral fins (flippers) that may be one-third the animal's length, or 5 meters

Flipper flapping is a cetacean behavior in which the animal rolls over onto its back and flaps its flippers in the air.

Pectoral stroking is a cetacean behavior in which one animal strokes the body of another with its pectoral fin.

A **pod** is a structured social group of toothed whales.

Baleen is a structure made of protein that is used by baleen whales to filter food from the water.

Keratin is a tough protein that makes up baleen.

A **bubble net** is a ring of bubbles produced by humpback whales that is used to concentrate fish near the surface for easier feeding.

Rorquals are baleen whales in the family Balaenopteridae that have a dorsal fin and ventral grooves on their throats.

Bosses are bumps on the snouts of humpback whales.

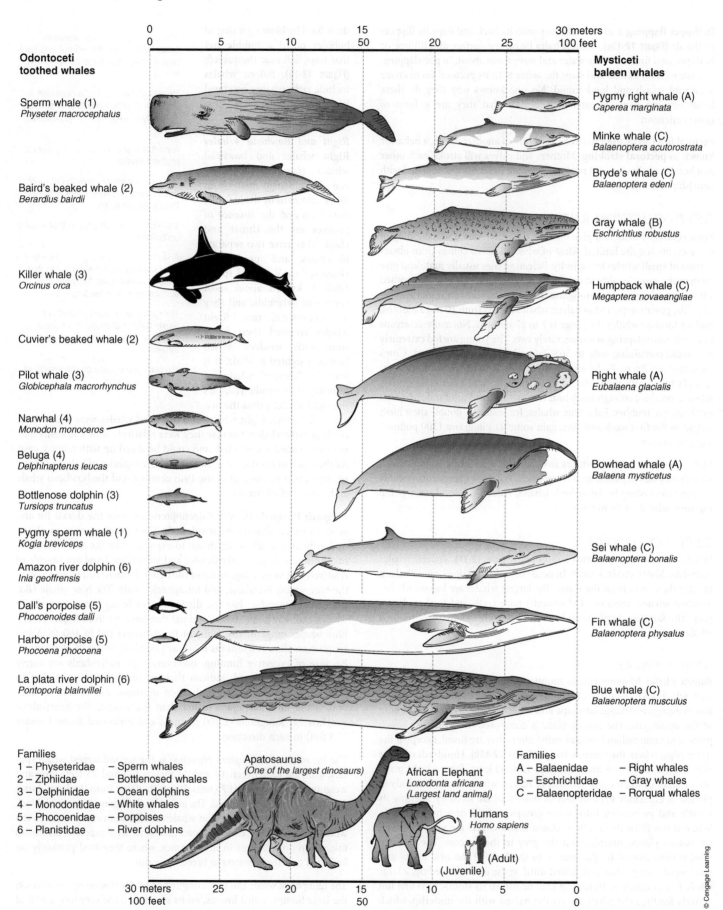

Odontoceti toothed whales

Sperm whale (1)
Physeter macrocephalus

Baird's beaked whale (2)
Berardius bairdii

Killer whale (3)
Orcinus orca

Cuvier's beaked whale (2)

Pilot whale (3)
Globicephala macrorhynchus

Narwhal (4)
Monodon monoceros

Beluga (4)
Delphinapterus leucas

Bottlenose dolphin (3)
Tursiops truncatus

Pygmy sperm whale (1)
Kogia breviceps

Amazon river dolphin (6)
Inia geoffrensis

Dall's porpoise (5)
Phocoenoides dalli

Harbor porpoise (5)
Phocoena phocoena

La plata river dolphin (6)
Pontoporia blainvillei

Mysticeti baleen whales

Pygmy right whale (A)
Caperea marginata

Minke whale (C)
Balaenoptera acutorostrata

Bryde's whale (C)
Balaenoptera edeni

Gray whale (B)
Eschrichtius robustus

Humpback whale (C)
Megaptera novaeangliae

Right whale (A)
Eubalaena glacialis

Bowhead whale (A)
Balaena mysticetus

Sei whale (C)
Balaenoptera bonalis

Fin whale (C)
Balaenoptera physalus

Blue whale (C)
Balaenoptera musculus

Families
1 – Physeteridae – Sperm whales
2 – Ziphiidae – Bottlenosed whales
3 – Delphinidae – Ocean dolphins
4 – Monodontidae – White whales
5 – Phocoenidae – Porpoises
6 – Planistidae – River dolphins

Apatosaurus
(One of the largest dinosaurs)

African Elephant
Loxodonta africana
(Largest land animal)

Humans
Homo sapiens
(Adult)
(Juvenile)

Families
A – Balaenidae – Right whales
B – Eschrichtidae – Gray whales
C – Balaenopteridae – Rorqual whales

© Cengage Learning

FIGURE 12-14 WHALES. A size comparison chart of some common baleen and toothed whales.

(a)

(b)

FIGURE 12-15 **BALEEN.** **(a)** The large mouth of this right whale contains numerous large plates of baleen. **(b)** This piece of baleen is from a humpback whale. Notice the triangular shape and the fringe of fibers on the posterior surface that function to strain plankton from the water column.

(16 feet). Humpbacks are slow-moving heavyset creatures that can reach up to 15 meters (50 feet) long and 12 meters (40 feet) around. It is one of the slowest whales, swimming 3 to 8 kilometers (2 to 5 miles) per hour. Humpbacks are mainly coastal inhabitants and will frequently enter harbors and even venture up the mouths of large rivers. These characteristics have made the humpback a very vulnerable species. Although it has been protected by the International Whaling Commission since 1966, it is still considered to be an endangered species. There are three distinct populations of humpback whale: in the North Atlantic, the North Pacific, and the Southern Hemisphere. Of these three, only the North Atlantic population is showing signs of recovery from the impact of whaling.

Humpbacks spend most of their time in polar seas feeding on krill and other plankton that are abundant there, but they come to warmer waters to mate and bear their young. They are a popular sight for tourists during March in the waters off Maui, Hawaii, where the North Pacific population comes to breed. Because humpbacks are coastal species, they are more accessible than other whale species and have been the subjects of many studies. Much research has focused on their vocalizations, or "songs," that may play a role in mating and intraspecies communication.

Until modern whaling techniques were introduced, rorquals, like the blue whale, were relatively safe from human hunters. They were too large to approach in small boats and fast enough to get out of range of handhurled harpoons. On the few occasions when a rorqual was killed, it usually frustrated the whalers by sinking or swamping the whaleboats. All of this changed in the last century with the introduction of harpoon

Herring

Bubbles

Whale ascends in a spiral pattern, blowing bubbles from its blow hole

FIGURE 12-16 **BUBBLE NET.** When feeding, humpback whales will frequently blow rings of bubbles as they rise to the surface. Small fish hesitate to cross the bubble barrier and become easy prey for the whales.

FIGURE 12-17 **BLUE WHALE.** The blue whale, which may be the largest animal that has ever lived, feeds on plankton, especially krill.

FIGURE 12-18 WHALING SHIP. The development of modern whaling vessels allowed whalers to kill more whales, pushing many species to the brink of extinction.

guns, motorized boats, factory ships, and techniques for injecting air into the whale carcass so that it would float and could be towed for great distances **(Figure 12-18)**. Because the average blue whale could furnish 120 barrels of high-grade oil, blue whales became the primary targets of the whaling industry, and their numbers plummeted. In the 1930–1931 whaling season, the worldwide catch of blue whales was 29,649. By the 1964–1965 season, whalers could find only 372 blue whales to kill. In 1966, the International Whaling Commission gave the blue whale worldwide protection, and their numbers have started to increase again. **Table 12-4** lists the current conservation status of the blue whale and other cetacean species.

Gray Whale At one time, three different populations of gray whales (*Eschrictius gibbosus*) existed in the western and eastern Pacific Ocean and the North Atlantic Ocean. Today, only the eastern Pacific population survives. The other two populations were hunted to extinction by the whaling industry. The eastern Pacific population, which initially contained an estimated 24,000 individuals, was on the brink of extinction in 1946. The number of individuals in the population at that time was

TABLE 12-4 MARINE MAMMALS WITH ENDANGERED STATUS

Species	Population Size
Pinnipeds	
Hawaiian monk seal	1,200–1,800
Siamaa seal	200–250
Steller's sea lion	87,000
Guadalupe fur seal	Not available
Cetaceans	
North Pacific right whale	200
Bowhead whale	Less than 600
Blue whale	Less than 9,000
Humpback whale	10,000
Sei whale	25,000
Fin whale	119,000
Western Pacific population of gray whales	Less than 100
Sperm whale	1.8 million
Vaquita	Less than 500

© Cengage Learning

estimated to be 250. In that year, strict laws were enacted to protect the remaining gray whales, and today the population consists of more than 22,000 individuals, allowing them to be removed from the list of endangered species in 1994.

Gray whales migrate from their summer feeding grounds in the Bering Sea to the waters off Baja California to mate and give birth. Females aggressively defend their young, a trait that prompted whalers to refer to them as "devilfish." Gray whales carry large accumulations of barnacles on the skin, more than any other cetacean, and for this reason are sometimes referred to as mossback whales **(Figure 12-19)**. Although the gray whale has been recovering from the impact of whaling, its popularity with tourists is now causing problems. Large numbers of whale watchers

(a)

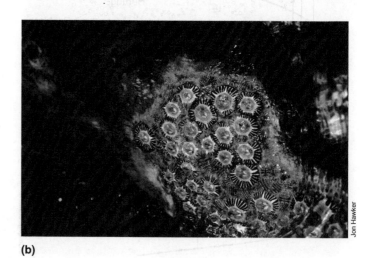

(b)

FIGURE 12-19 GRAY WHALE. Gray whales are sometimes called mossback whales because of the large number of barnacles attached to the animals. The white patches on this whale's head are barnacles. **(b)** A close view of the barnacles on a Pacific gray whale. This species of barnacle is found only on whales.

FIGURE 12-20 **SPERM WHALE.** **(a)** A sperm whale nears the surface of the water in the North Atlantic Ocean. **(b)** Details of the head of a sperm whale. The spermaceti organ contains an oily substance called spermaceti that solidifies on contact with air. The spermaceti is believed to function in echolocation and was prized by whalers as a high-quality wax.

(a)

Doug Perrine/JeffRotman.com

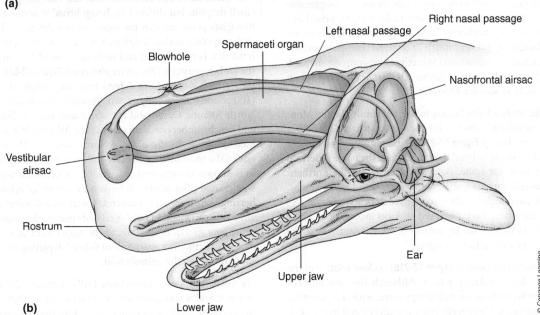

Blowhole · Spermaceti organ · Left nasal passage · Right nasal passage · Nasofrontal airsac · Vestibular airsac · Rostrum · Upper jaw · Ear · Lower jaw

(b)

© Cengage Learning

and their motorboats disturb the whales while they attempt to mate and give birth, resulting in fewer offspring.

Toothed Whales

Toothed whales include sperm whales, dolphins, porpoises, killer whales, and narwhals. As with other aquatic mammals, the teeth of these whales tend to be simplified. Dolphins and porpoises have between 100 and 200 identical teeth that are fused to the jaws. The teeth grasp the fish on which these animals feed. Sperm whales have functional teeth only in the lower jaw; these teeth fit into sockets in the upper jaw. It is believed that this is an adaptation for feeding on the slimy cephalopods, which are their primary food. Beaked whales have only a few teeth, sometimes only two in each jaw, and the narwhal is peculiar in that one of its teeth forms a tusk that projects straight out from its head.

Sperm Whales The giant of the toothed whales is the sperm whale, or cachalot (family Physeteridae), the third largest animal on earth behind the blue whale and fin whale. Sperm whales **(Figure 12-20a)** are sometimes confused with right whales and rorquals, but they have teeth instead of baleen and a massive blunt snout that projects forward beyond the mouth. Sperm whales have no real dorsal fin but instead have a series of humps on the rear third of their body. Unlike most other species of whale, sperm whales are aggressive, attacking squid, fishes, and, on

occasion, whalers in small boats. They are polygynous, and males are always accompanied by several females. Although mainly found in tropical and temperate waters, they do range occasionally into the polar seas.

At one time, the sperm whale was abundant in most seas, and for almost a century and a half it was the favorite catch of American whalers. Sperm whales were challenging targets for whalers of the 18th and 19th centuries because of their large size (up to 18 meters, or 60 feet), toothed jaw, and powerful tail. Whalers judged the risks well worth taking because the animal's commercial value was so great. During the 20th century, modern whaling techniques decreased the risks, whereas profits remained high. As a result, the once large populations have decreased significantly, and although protected by international treaty, the sperm whale is still considered to be a rare and endangered species.

The sperm whale received its name from an oily wax-like substance in its head. The snout contains a cavity that is developed from one of the nasal chambers. This cavity contains a thin, colorless, transparent oil that forms a waxy material when it comes into contact with air. Early whalers erroneously thought it to be an enormous reserve of semen and named it **spermaceti**

Spermaceti is a waxy substance found within the rostrum of sperm whales.

(spur-meh-SEH-tee; **Figure 12-20b**). At the height of the whaling industry, spermaceti was one of the most prized products of the sperm whale catch because it was found to be a high-grade wax. It was used in the manufacture of ointments, face creams, luxury candles, and lubricants. Now synthetics have taken its place. It is generally believed that the spermaceti plays a role similar to that of the melon in dolphins and thus is part of the whale's echolocation system (see the later discussion of echolocation).

Ambergris is a secretion of the sperm whale's digestive tract.

The sperm whales' fondness for squid and cuttlefish generates a digestive by-product called **ambergris**, which in the past was even more valuable than spermaceti. Ambergris is a secretion of the whale's digestive tract and may function in protecting the enormous digestive system from undigested squid beaks and cuttlefish cuttlebone. When this secretion is exposed to air, it forms a solid, gray, waxy material with a disagreeable odor, but over time it acquires properties that make it highly prized as a base for the most expensive perfumes. Even though synthetic substitutes are available, ambergris is still very much in demand. Lumps of ambergris that are evacuated by the animals sometimes wash up on shore. A 418-kilogram (920-pound) lump of ambergris that washed up on an Australian beach in 1953 sold for $120,000.

White Whales The white whales belong to a small family (family Monodontidae) that contains two genera with one species in each. The beluga whale (*Delphinapterus leucas;* **Figure 12-21a**) is unique among whales for its white color and its ability to bend its neck. In fact, the name is derived from the Russian word for *white*. Beluga whales live in the northern polar seas. They usually forage in small family groups and feed on crabs, cuttlefish, flounder, and halibut. Their main predators are killer whales and polar bears. They have been observed to form groups of as many as 10,000 animals as they travel to warmer shallower water near the mouths of rivers to give birth to their calves.

The narwhal (*Monodon monoceris;* **Figure 12-21b**), a close relative of the beluga whale, also lives in Arctic waters. Although the fetal animals have tooth buds, by birth these have all disappeared, with the exception of the two upper incisors. These teeth remain undeveloped in the skull of the females, but in males a tusk as long as 3 meters (10 feet) develops from one of the two teeth, almost always the left. In some individuals, the right tooth also develops a tusk, giving an already odd-looking creature an even more bizarre appearance.

During the Middle Ages, the narwhal tusk was passed off as the magical horn of a unicorn. Because the tusk was worth literally more than its weight in gold, the narwhal was hunted relentlessly. In later times, the narwhal was killed incidentally by whalers in the Arctic Ocean who were searching for bowhead whales. Since the demise of the bowhead fishery, the narwhal population has recovered, and the animals survive relatively undisturbed. Their only remaining predator appears to be Eskimos, who kill the animal for its nutritious skin.

Porpoises Porpoises (family Phocoenidae) are related to the familiar dolphins, and both are placed in the same superfamily, Delphinoidea. The word *porpoise* comes from a contraction of the Latin word *porcopiscis* (meaning "pigfish"), apparently referring to the animal's stocky body. The Greek naturalist Aristotle observed that the porpoise resembled a small dolphin but differed by being broader across the back. Modern biologists point out that the most obvious difference between dolphins and porpoises is that dolphins have a beak, whereas porpoises have a head that is rounded off and no beak. One of the smallest cetaceans is the harbor porpoise (*Phocoena phocoena;* **Figure 12-22a**), which measures only 1 to 2 meters (3 to 7 feet) long and weighs nearly 45 kilograms (100 pounds) when fully grown. This animal is widely distributed in the North Atlantic Ocean and elsewhere and is a familiar sight to boaters and beachcombers on both sides of the Atlantic. It is also widely distributed along the Pacific coast of North America. The harbor porpoise is one of the most intelligent cetaceans, and its capacity for learning, as far as is known, is surpassed only by that of the bottlenose dolphins, killer whales, and pilot whales. It feeds on a variety of schooling fish such as herring, sardines, and mackerel, sometimes feeding on fish that have already been caught in nets. A related species, the vaquita, or "little cow" (*Phocoena sinus*), is endemic to the upper Gulf of California and is in danger of extinction as a result of fishers' depleting the stocks of fish and squid on which the animals feed.

For years, Europeans considered Dall's porpoise (*Phocoenoides dalli*) a delicacy, and it was quite rare. The rarity was probably more a factor of the porpoise eluding its human predators than few numbers. This animal is perhaps the first to be protected by law. In Normandy as far back as 1098, laws were passed regulating the size and number of the Dall's

(a)

(b)

FIGURE 12-21 WHITE WHALE. (a) The beluga whale, *Delphinapterus leucas*, is the only whale species that is totally white. **(b)** The upper incisor (usually the left) of male narwhals develops into a large tusk that projects straight out from the head.

(a)

(b)

FIGURE 12-22 COMPARISON OF DOLPHIN AND PORPOISE. Dolphins and porpoises are both delphinids. **(a)** The head of a porpoise is rounded and the animal is broader across the back, whereas **(b)** dolphins can be distinguished by their beaks.

porpoise harvest. Dall's porpoises feed on fish, preferring herring, whiting, and sole. These animals have been observed swimming up rivers as far as 58 kilometers (36 miles), and one specimen was observed in the Maas River in the Netherlands 323 kilometers (200 miles) from the sea.

Dolphins Because most people think of whales as being giant animals, few are aware that the playful, acrobatic dolphins are also toothed whales. In fact dolphins are the most numerous cetaceans. Dolphins belong to the family Delphinidae and are collectively referred to as **delphinids**. The name is derived from ancient Greek mythology. According to legend, Apollo rose from the sea in the shape of a dolphin to lead settlers to Delphi, home of the Delphic oracle.

The common dolphin (*Delphinus delphis*) has a definite beak that is separated from the snout by a groove. This animal is small compared with other cetaceans; adults grow to no more than 3 meters (10 feet). They can be found in all temperate oceans and are among the swiftest cetacean swimmers. They feed on schooling fish such as herrings and sardines that they devour in enormous quantities. The common dolphin is frequently seen escorting ships. When common dolphins find a ship, they encircle it. Some of the dolphins will lead at the bow, while others follow along at the sides. They will leap completely out of the water in a graceful arc that covers several meters and will repeat this behavior for hours and hundreds of kilometers without ever seeming to tire.

The most abundant dolphin species along the eastern coast of the United States and in the Caribbean is the bottlenose dolphin (*Tursiops truncatus;* **Figure 12-22b**). These dolphins also inhabit the Mediterranean, Baltic, and Caspian seas, and a subspecies lives in the temperate

waters of the Pacific Ocean. These are the dolphins that are most frequently used for scientific studies of cetacean intelligence. Bottlenose dolphins

> **Delphinids** is a collective term for members of the cetacean family Delphinidae, specifically dolphins.

are often the species used as performers at aquariums or stars of television and movies. These intelligent animals are quite playful and appear to like the company of others of their species. They have been observed to aid disabled members of their species, pushing them to the surface so they can breathe or helping them to shallow water. Occasionally, dolphins will socialize with other species, including humans. In a coastal region of Brazil, female bottlenose dolphins and their young have assisted fishers for almost 150 years. The fishers cast hand-thrown nets from the shore and wait for the dolphins to chase fish, usually mullet, into them. The dolphins feed on the fish that escape the nets. There are also many stories, some authenticated, about bottlenose dolphins coming to the aid of injured or drowning humans.

Bottlenose dolphins live their lives in carefully structured, complex social groups. Studies of dolphins near Sarasota, Florida, indicate that females associate with other females that are in a similar reproductive state. For instance, mothers with calves swim with other mothers and their calves. Calves remain with their mothers for 3 to 6 years before joining juvenile groups. During this time, a calf learns to identify the other dolphins in the group, its relationship to the other members, how to find food and capture it, and where there is danger and where safety.

Orca is another name for a killer whale.

Echolocation is a process that allows toothed whales to use sound to distinguish objects from a distance.

Orientation clicks are low-frequency clicks used in echolocation that give an animal a general idea of its surroundings.

The largest of the dolphins is the killer whale, or **orca** (*Orcinus orca*; **Figure 12-23**). Males sometimes reach lengths of 10 meters (33 feet), whereas females are only half as large. These animals received their common name because they are the only cetaceans that feed on homeothermic animals.

They can be found in all seas, but are most common in the cold waters of the North Pacific, Arctic and Antarctic Oceans. Killer whales off the Pacific coast of North America can be classified into two distinct types called residents and transients. These two types are sometimes referred to as races, and some biologists think they may be separate species. The two types differ in morphology, genetics, behavior, and diet. Residents feed almost entirely on fish while transients feed almost entirely on marine mammals.

Killer whales have a high dorsal fin (as much as 1.8 meters, or 6 feet) and broad rounded flippers. Their coloration is particularly distinctive. They are primarily black with a white patch over the eye and a white ventral stripe. These powerful animals are quite agile and are the fastest cetacean swimmers, reaching speeds of 48 kilometers per hour (30 miles per hour). Both the upper and lower jaws of killer whales contain large conical teeth. Although they tend to eat more fish than anything else, some will also prey on seals, sea lions, baby walruses, penguins and other sea birds, and other cetaceans. The stomach of one captured specimen contained the remains of 13 porpoises and 14 seals. Pods of killer whales have also been known to attack larger whales.

There are no authenticated reports of killer whales killing or injuring humans in the wild. There have been a few incidents in which killer whales have rammed boats, usually after the inhabitants of the boat antagonized the whale. There are also reports of killer whales ramming ice floes on which humans are standing, but this appears to be a case of mistaken identity. In February 2010, a captive killer whale at Sea World in Orlando, Florida killed its trainer, but such attacks are extremely rare. In captivity, killer whales are usually quite docile, and their high level of

intelligence and ability to learn make them popular performers in aquarium and marine park shows. A relative, the false killer whale (*Pseudorca crassidens*), has a similar appearance but is much smaller and feeds on squid.

Pilot whales (*Globicephala*) are dolphins related to the killer whales. These animals have a globular head, a projecting forehead, and a muzzle that forms a small beak. Pilot whales are inoffensive animals that live in pods of several hundred individuals that appear to follow a single leader. They feed mainly on cuttlefish and, when attacked, group closely together. Pilot whales exhibit the odd behavior of following individual members that stray from the pod. Whalers exploit this behavior by harpooning one or two pilot whales, towing them to shallow water, and then waiting for the rest of the pod to follow to mass slaughter. These animals are not protected by any international agreement and are still hunted for dog food and oil. Large numbers of the North Atlantic species are caught each year.

Pilot whales will frequently beach themselves, sometimes in large numbers. There are several theories on these mass strandings. One theory suggests that the cause of the beachings may be a parasite that attacks the inner ear and interferes with the animal's ability to navigate.

Echolocation

Toothed whales use sound to navigate and identify objects. In the late 1940s, Arthur F. McBride, the first curator of Marine Studios in Florida, speculated that dolphins used sound to avoid obstacles and to navigate. Every time he tried to capture dolphins for an exhibit by driving them toward a net, the animals would stop short of the net and swim away. It didn't matter whether he tried during day or night, and even in murky water where they could not see the net they would still avoid it. McBride hypothesized that the animals might somehow use sound to sense the presence of the nets in the water.

Some observers thought the animals might be using sonar, a process in which sound is bounced off an object to determine characteristics such as distance and shape, to locate their prey and "see" their surroundings. The idea was certainly plausible because light does not penetrate very far in the water, while sound can travel great distances. Later in the 1950s, Winthrop N. Kellogg of Florida State University suggested that dolphins use a series of clicking sounds in the same way that humans use sonar. The final proof came in 1960, when Kenneth S. Norris and his colleagues used rubber suction cups to blindfold a captive Atlantic bottlenose dolphin at Marineland in California and found that the dolphin's ability to move about within its enclosure was not affected. Norris discovered that dolphins not only use sound to determine distance and direction but they could also identify the size, shape, and texture of an object.

Even in clear water, toothed whales cannot see much farther than 30 meters (100 feet). To compensate for this, they have ears that are modified to receive a wide range of underwater vibrations. This adaptation not only improves the animal's hearing under water but refines it for **echolocation**, which, like sonar, allows the animal to distinguish and home in on objects from distances of several hundred meters. Dolphins can emit a wide spectrum of sounds, known as pulsed sounds, lasting from a fraction of a millisecond to as long as 25 milliseconds. To obtain information about their environment, dolphins emit sounds with frequencies that range from less than 2,000 to greater than 100,000 hertz. Only a portion of these sounds can be perceived by the human ear, and these we hear as clicks. The low-frequency clicks are **orientation clicks** that give the animal a general idea of its surroundings. The high-frequency

NOAA

FIGURE 12-23 KILLER WHALE. Killer whales are the only cetaceans known to feed on homeothermic prey. This whale is exhibiting spy hopping behavior.

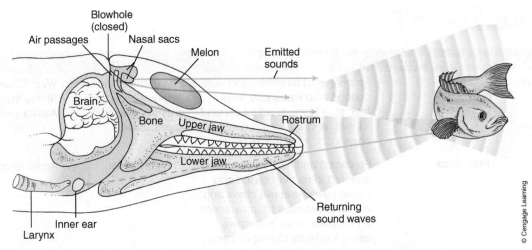

FIGURE 12-24 ECHOLOCATION. Toothed whales use sound to locate objects, determine shapes and distance, and navigate. Sounds produced by the animal's larynx (voicebox) are focused by a structure in the head called the melon. When the sounds strike a target, they are reflected as echoes (returning sound waves). The echoes are picked up by the animal's lower jaw and transmitted to the inner ear. The vibrations received by the inner ear are then converted into signals that can be carried by nerves to the brain for interpretation.

clicks are **discrimination clicks** that give the animal a precise picture of a particular object. The clicks can be produced as a single sound or as a series of sounds strung together.

The dolphin's larynx does not have vocal cords but instead has a ring of muscles that acts as a valve, enabling the animal to control the air flow through the larynx. Air under pressure circulates through the animal's nasal passages, producing the clicks and other characteristic sounds **(Figure 12-24)**. The sounds are directed by being focused in the **melon**, an oval mass of fatty waxy material that is located between the blowhole and the end of the head. The sounds are directed toward objects in front of the animal but not on objects below a line level with its jaws. This may explain why some toothed whales cannot sense gradually sloping bottoms and sometimes run aground. The clicking sounds travel through the water and bounce off anything solid. The reflected sounds (echoes) are picked up by sensitive areas of the lower jaw that transmit the vibrations to the inner ear. The animal's brain then processes the information from the inner ear to produce a mental image of the target object.

When the sounds bounce off a target, they produce echoes that provide at least four types of information: the direction from which the echo is coming, the change in frequency, the amplitude, and the time elapsed before the emitted sound returns. With this information, dolphins, as well as other toothed whales, can determine not only the range and bearing of an object, but also its size, shape, texture, and density. In tests, blindfolded dolphins have been able to tell the difference between a sheet of copper and a sheet of aluminum and discriminate between two very small objects about 5 centimeters (2 inches) in diameter that have different shapes at a distance of 3 meters (10 feet). Recent studies have found that blindfolded dolphins can discriminate between metal cylinders that differ by just .0076 millimeters (1/3000 inch) in diameter.

When a dolphin travels, it usually moves its head from side to side and up and down, scanning a broad path ahead of it. The scanning motions become fast and jerky when the dolphin becomes interested in a small target such as a fish. As the dolphin scans, it can determine the direction from which echoes are returning and thus determine the bearing of the

target. Changes in the frequency of the returning sound give the dolphin information about the size and shape of the object. The amplitude of the echo and time elapsed before it returns help the dolphin determine distance.

Dolphin Communication

Although dolphins probably communicate with each other in a number of ways, the most important are auditory signals. Bottlenose dolphins produce two types of vocalizations: whistles and pulsed sounds (which include echolocation clicks). In the 1960s, it was discovered that each dolphin produces its own unique **signature whistle**. The signature whistle identifies the dolphin that is vocalizing, gives its location, and may relay other information as well. Current research indicates that perhaps as many as 90% of the whistles made by captive dolphins are signature whistles. It appears that a female's whistle is quite different from her mother's, whereas a male's is very similar to his mother's. Because females generally stay with their mothers when they mature, it may be important for them to have distinct signature whistles to avoid confusion. Males usually leave their mothers' group and thus do not have to differ as much in their signature whistle. Calves develop signature whistles between ages 2 months and 1 year. The whistles remain unchanged for at least 12 years and possibly for the animal's life. Although scientists are beginning to learn the meaning of signature whistles, little is known of the pulsed sounds the dolphins make. Pulsed sounds other than the echolocation clicks may also be used for communication.

Many animals, including humans, communicate by body language. Dolphins are no exception. For instance, a loud opening and closing of the jaws, known as a jaw clap, or the slapping of tail flukes on the water surface can indicate threat or displeasure.

Discrimination clicks are high-frequency clicks used in echolocation that give an animal a precise picture of a particular object.

A **melon** is an oval mass of fatty, waxy material in a dolphin's head that functions in directing and focusing the sounds used in echolocation.

A **signature whistle** is a unique sound produced by a dolphin to identify an individual.

IN PERSPECTIVE

Classification	Common Names	Characteristics	Diet	Distribution
Family Mustelidae	Sea otter	Five-fingered forelimbs and well-defined hind limbs with fin-like feet; able to move about on land as well as in the water; dense fur covers and insulates the body	Sea urchins, molluscs (especially abalone), crustaceans, and some fishes	West Coast of the United States from California to Alaska and Japan
Family Ursidae	Polar bear	Forelimbs and hind limbs adapted to moving on ice and snow; paws have five claws and act as snowshoes for movement on ice pack; agile swimmers using forelimbs to propel themselves through the water; thick layer of fat and dense fur protect against the Arctic cold	Primarily seals	Arctic region
Family Otariidae	Eared seals, including sea lions and fur seals	Torpedo-shaped body; external ear flaps; limbs modified to form flippers can be rotated to support body on land; swim using their front flippers and steer with their hind flippers	Fish	Sea lions along western coasts of the Americas; fur seals in Arctic and Antarctic
Family Phocidae	True seals	Torpedo-shaped body; no external ear flaps; limbs modified to form flippers, but cannot be rotated to support the body on land; swim using hind limbs and steer with their forelimbs	Fish, krill, squid, except for leopard seal, which also feeds on penguins and other sea birds and other pinnipeds	Worldwide mainly in coastal waters; more species in temperate and polar seas than in the tropics
Family Odobenidae	Walrus	Bulky bodies; no external ears; a distinct neck; stiff bristles on muzzle, limbs modified to form flippers; can move awkwardly on land; males have tusks	Primarily bivalves	Arctic regions
Order Sirenia	Dugongs and manatees	Large bodies; forelimbs modified to form flippers; no hind limbs; tail flukes; totally aquatic	Aquatic plants	All are tropical; dugongs are found along coastal areas of India and Malaysia; manatees are found in the Caribbean Sea and along the coast of western Africa
Suborder Mysteceti	Baleen whales	Streamlined bodies that resemble fish; nostrils located in a blow-hole, forelimbs modified to form flippers; no hind limbs; tail flukes; plates of baleen hang down from the upper jaw and take the place of teeth; totally aquatic	Plankton, especially krill, and small fish	Worldwide; generally feed in polar seas and migrate to warmer tropical waters to mate and give birth
Suborder Odontoceti	Toothed whales	Streamlined bodies that resemble a fish; nostrils located in a blow-hole, forelimbs modified to form flipper; no hind limbs; tail flukes; conical teeth sometimes located in only one jaw; totally aquatic	Fish and squid; killer whales also feed on other marine mammals	Worldwide distribution in all seas

Photos: **Family Mustelidae:** Jon L. Hawker; **Family Ursidae:** Wayne Lynch/All Canada Photos/Alamy; **Family Otariidae:** Marty Synderman/Visuals Unlimited; **Family Phocidae:** Marty Snyderman/Visuals Unlimited; **Family Odobenidae:** NOAA; **Order Sirenia:** Mike Parry/Minden Pictures; **Suborder Mysteceti:** NOAA; **Suborder Odontoceti:** Doug Perrine/Jeff Rotman Photography.

Dolphin Behavior

Dolphin feeding behaviors exhibit signs of learned behaviors. Some dolphins in southwestern Florida beat their prey to death with their tail flukes. This appears to be a local behavior that the young learn from their mothers. In the Gulf of Mexico, dolphins have learned to follow shrimp boats for a free meal. This also appears to be a learned behavior that is passed from a mother to her offspring. As they mature, females frequently return to their mother's group, but males travel alone or with one or two other adult males with whom they have grown up. Bottlenose dolphins appear to be quite clever, sometimes devising ingenious solutions to problems. One example was observed in two captive dolphins trying to extract a moray eel from a rock crevice in their tank. One dolphin captured a scorpionfish and, with the fish in its mouth, nudged the moray eel from behind with the spine of the fish. As the eel fled from its crevice, the other dolphin caught it.

Studies conducted in Hawaii by researcher Lou Herman provided a basis for our knowledge of dolphin learning abilities. A dolphin named Ake (short for Akeakamai) and her mate, Phoenix, learned that they would receive rewards when they mimicked sounds generated by a computer. They also were able to make associations. When taught the hand signal to go through the gate in their tank, they also responded correctly to a similar hand signal that meant go through a hoop. They were able to learn that *ball* meant not only the small red ball that they were first trained with but any other ball, regardless of size, color, or its ability to sink or float.

Though originally taught to respond to hand signals given by their trainers, Ake and Phoenix could also recognize and respond to the same signals when they saw them on a television screen. In a test, dolphins outscored new trainers in interpreting televised gestural commands.

IN SUMMARY

- Cetaceans are the mammals best suited to life in the sea.
- Some of the most striking cetacean adaptations allow them to dive to great depths and remain submerged for more than 1 hour.
- Cetaceans are intelligent inquisitive animals that display a variety of interesting behaviors.
- Cetaceans known as baleen whales have enormous mouths that contain plates of baleen instead of teeth.
- Many species of baleen whale are endangered because of the impact of whaling. (Have You Wondered? #5)
- Toothed whales include sperm whales, dolphins, porpoises, killer whales, and narwhals.
- Dolphins are strongly social animals, exhibit problem-solving skills, have long periods to mature with many learning experiences, and are capable of intraspecies and interspecies cooperation.
- The ears of toothed whales are modified to receive a wide range of water vibrations which aids in the process of echolocation. (Have You Wondered? #4)
- Echolocation is similar to sonar and allows toothed whales to distinguish and home in on objects from distances of several hundred meters. (Have You Wondered? #4)
- Dolphins use sound to communicate with each other.

KEY CONCEPTS

1. Mammals have a body covering of hair, maintain a constant warm body temperature, and nourish their young with milk produced by the mammary glands of the mother. (12-1)

2. Sea otters have thick coats of fur and feed on marine invertebrates near shore. (12-2)

3. Polar bears feed mainly on seals and are top predators in Arctic food chains. (12-3)

4. Pinnipeds have limbs modified to form flippers and are better adapted to life at sea than to life on land. (12-4)

5. Sirenians are totally aquatic mammals that feed on a variety of aquatic vegetation. (12-5)

6. Cetaceans have a fish-like body shape and are the mammals best suited to life in the sea. (12-6)

7. Special physiological adaptations allow some marine mammals to dive to great depths and remain submerged for long periods. (12-4 and 12-6)

8. Cetaceans are intelligent animals that display a range of behaviors for communication and investigating their environment. (12-6)

9. Some cetaceans use echolocation to navigate, find prey, and avoid predators. (12-6)

10. Baleen whales have plates of baleen instead of teeth and feed primarily on plankton, such as krill. (12-6)

11. Toothed whales have teeth allowing them to feed on larger prey, primarily fish and squid, although killer whales will eat marine birds and mammals. (12-6)

12. Dolphins are intelligent animals that are capable of learning and sophisticated intraspecies communication. (12-6)

ARE YOU STILL WONDERING?

1. Polar bears are considered marine mammals because they feed on seals and are top predators in Arctic marine food chains.

2. Seals can be distinguished from sea lions by their lack of external ears, their inability to rotate their limbs under their body to support their weight on land, and the way they swim.

3. Many marine mammals have evolved a variety of physiological mechanism that allow them to dive to great depths and stay submerged for long periods of time. These include shunting blood during a dive from less vital organs to the brain and heart, muscles that store more oxygen and are less sensitive to lactic acid, a brain stem that is less sensitive to carbon dioxide, and a more red blood cells than terrestrial animals.

4. Dolphins locate their prey using echolocation. They produce a series of sounds that are directed by the melon in their head into the water ahead of them; when these sounds strike prey or another object, they bounce back as echoes to the dolphin, which senses the echoes with its lower jaw. Characteristics such as the amplitude and frequency of the returning sounds give the dolphin a mental image of what is in front of it.

5. Human activities such as hunting, whaling, and destruction of habitat have pushed some marine mammals to the brink of extinction and severely lowered the populations of others.

QUESTIONS FOR REVIEW

Multiple Choice

1. Mammals feed their young on secretions produced by the mother's
   a. digestive system
   b. oil glands
   c. mammary glands
   d. placenta
   e. salivary glands

2. Phocid seals lack
   a. tails
   b. hind limbs
   c. forelimbs
   d. external ears
   e. nostrils

3. Instead of teeth, whales in the suborder Mysticeti have
   a. tusks
   b. bony plates
   c. baleen
   d. strainer nets
   e. suckers

4. The spermaceti in the head of a sperm whale is thought to play a role in
   a. digestion
   b. reproduction
   c. excretion of salt
   d. echolocation
   e. swimming

5. The cetacean most often seen in captivity is the
   a. California sea lion
   b. common dolphin
   c. bottlenose dolphin
   d. beluga whale
   e. walrus

6. The only pinniped to feed on homeothermic animals is the
   a. California sea lion
   b. killer whale
   c. elephant seal
   d. walrus
   e. leopard seal

7. Sea otters are protected from the cold by a
   a. thick skin
   b. thick fur
   c. layer of blubber
   d. countercurrent exchange mechanism
   e. high metabolism

Matching

Match the following animals with the appropriate description:

a. whale
b. otter
c. walrus
d. dugong
e. gray whale

1. Use sensory bristles on their muzzles to locate bivalves in sediments.
2. Have baleen instead of teeth.
3. Feed on a variety of marine vegetation.
4. Have teeth only in the lower jaw and feed on large squid.
5. Feed on abalone and sea urchins.

Short Answer

1. What factor contributed to the near extinction of the sea otter?

2. How do polar bears withstand the Arctic cold?

3. List three characteristics that would help you to distinguish a fur seal from a true seal.

4. Explain the function of the large proboscis in the male elephant seal.

5. Explain how the arrangement of blood vessels in the flippers and tail flukes of cetaceans helps them to retain body heat.

6. Describe the changes that occur in the bodies of diving mammals when they dive.

7. What is spy hopping?

8. Explain how toothed whales use echolocation to navigate.

9. Describe how baleen whales feed.

10. How could you distinguish between a manatee and a dugong?

11. Why do mammals produce fewer offspring than most other animal groups?

Thinking Critically

1. Some marine mammals are polygynous. What is the advantage of this lifestyle?

2. From an ecological standpoint, why aren't more of the large marine mammals predators of birds and other mammals?

3. Toothed whales use echolocation to find their prey, but baleen whales do not. Why?

SUGGESTIONS FOR FURTHER READING

Berta, A., J. L. Sumich, and K. M. Kovacs. 2006. *Marine Mammals Evolutionary Biology*, 2nd ed. Burlington, Mass.: Academic Press.

Brower, K. 2009. Blue Whales, *National Geographic* 215(3): 134–152.

Casey, S. 2008. Elephant Seal Sojourn, *National Geographic* 214(5): 124–134.

Chadwick, D. H. 2007. What Are They Doing Down There?, *National Geographic* 211(1):72–93.

Chadwick, D. H. 2008. Right Whale Watch, *National Geographic* 214(4): 100–121.

Eliot, J. L. 2005. Refuge in White, *National Geographic* 208(6):46–57.

McGrath, S. 2011. Not Too Late for Polar Bears, *National Geographic* 220(1): 64–75.

Molnar, P. K., A. E. Derocher, G. W. Thiemann, and M. A. Lewis. 2010. Predicting Survival, Reproduction, and Abundance of Polar Bears Under Climate Change, *Biological Conservation* 143:1612–1622.

Mueller, T. 2010. Valley of the Whales, *National Geographic* 218(2): 118–137.

Nicklen, P. 2007. Hunting Narwhals, *National Geographic* 212(2): 110–129.

Norris, S. 2002. Creatures of Culture? Making the Case for Cultural Systems in Whales and Dolphins, *Bioscience* 52(1).

Righthand, J. 2011. Picky Eaters, *Smithsonian* 42(5): 90–93.

Wagner, E. 2011. Call of the Leviathan, *Smithsonian* 42(8): 68–76.

Have You Wondered?

1. **What factors affect the survival of organisms on rocky shores? (13-1)**

2. **What adaptations rocky intertidal organisms use to tolerate desiccation and high temperatures? (13-1)**

3. **What organisms are characteristic of each zone of the rocky intertidal? (13-1)**

4. **What factors influence the distribution of organisms on sandy shores? (13-2)**

5. **How organisms inhabiting sandy shores survive dry periods and the intense heat of the sun? (13-2)**

13

Intertidal Communities

INTERTIDAL COMMUNITIES are found within the area of the shoreline reached by the waters of the highest high tide and left uncovered at the lowest low tide. These communities are almost entirely marine; terrestrial organisms only occasionally invade their upper reaches. The interaction of wind, waves, sunlight, and other physical factors creates a complex and stressful environment. Inhabitants must not only endure the crushing force of waves but also desiccation and potentially searing heat or freezing cold while exposed to air. Despite these conditions, this environment is home to a diversity of organisms that equals or exceeds that found in completely submerged habitats. Organisms found here must be particularly hardy and have special adaptations that allow them to make a living in such a hostile environment.

Although intertidal communities make up only a relatively small portion of the marine ecosystem, they are by far the best studied. Their easy accessibility allows scientists to carefully monitor organisms in their daily endeavors without the necessity of specialized equipment. Studies on intertidal communities have produced many insights and unifying principles that can be applied to other marine systems, as well as ecology in general. Intertidal communities can be found on sandy beaches, rocky shores, sand flats, mud flats, salt marshes, and mangrove swamps. This chapter focuses on those inhabiting rocky shores and sandy beaches.

13-1 Rocky Shores

Rocky shores, coastal regions that are composed of hard materials, are more densely inhabited and have a much greater diversity of organisms than those composed of soft sediments such as sand or mud. They are found throughout the world, wherever hard substrate exists. In North America, hard substrate is found from California to Alaska on the West Coast and from Cape Cod northward on the East Coast. The Pacific coast is geologically younger than the Atlantic coast and lies on an **active continental margin** where geological processes have produced a coastline dominated by rock. In contrast, the Atlantic coast is geologically older, lies on a **passive continental margin** and is, therefore, dominated by sediments. The rocky shore north of Cape Cod was formed when glaciers scraped away the overlying sediments. Rocky coasts can also be formed by lava flows from active volcanoes, such as those in Hawaii. Many rocky coasts in the Caribbean and Mediterranean seas were formed from ancient coral

Active continental margins are regions where the leading edge of a continent collides with an oceanic plate.

Passive continental margins are regions within a tectonic plate where the dominant geological processes are weathering, erosion, and the accumulation of sediments.

reefs or the shells of dead marine organisms. Rocky coasts make up about 75% of the world's shorelines.

ADAPTATIONS TO LIFE ON ROCKY SHORES

Organisms inhabiting rocky shores are faced with conditions not encountered by those in the deeper ocean. They are alternately submerged by the incoming tides and exposed to air, heat, cold, and drying conditions when the tides are out. Rains can change the salinity of waters in which organisms live, particularly in tide pools, and the force of waves pounding the shore, **wave shock**, may either crush or carry them away. The inhabitants of sandy beaches can burrow to avoid these stresses, but those inhabiting rocky shores need special adaptations to survive in this harsh environment. The ability to withstand overheating, desiccation, and cold while exposed to air at low tide and wave shock determines the locations where organisms can survive on rocky shores.

Avoiding Overheating and Desiccation

The duration of an organism's exposure to air is determined by both its location on the shoreline and the pattern of the tides. We learned in Chapter 4 that tides are generated by the gravitational forces of the moon and sun and vary in height daily and throughout their annual cycle. Tides reach their highest and lowest levels twice a month during the so-called

Wave shock is the force of waves as they crash against the coastline and the organisms living on it.

spring tides, when the sun and moon are in alignment. The lowest tidal range occurs during the neap tides, when the sun and moon are at right angles. The geographical location and structure of the basin can also affect the height of the tides. The Bay of Fundy in eastern Canada, for example, has a tidal range of about 17 meters (55 feet), but the Mediterranean Sea has practically no tidal action, shoreline fluctuation being primarily due to winds. Variation in tide height can result in exposure to air up to half of the time. The ability to tolerate this exposure, therefore, is a major determinant of which organisms can survive in the intertidal and the relative vertical position they can occupy—the higher the position on the rocks, the longer the exposure to the effects of temperature and desiccation.

Organisms attached to open rock surfaces on high intertidal locations face the greatest challenge to maintaining body temperatures below critical levels and preventing excessive water loss; those inhabiting crevices or living under seaweeds are somewhat buffered from these effects. Animals inhabiting the highest part of the intertidal have acquired in the course of their evolution a number of adaptations that allow them to survive. Knobby periwinkles (*Cenchritis muricatus*, **Figure 13-1a**) of Caribbean seashores, for example, are particularly well adapted to the harsh conditions experienced at the highest part of the intertidal. Their large globular body exposes less surface area to heating by the sun and hot rocks relative to their volume. They also reduce body contact with hot rock by attaching to it with a mucous thread when wave action is not severe. Their light-colored shell reduces heat gain or loss, and the "beads" on

(a)

(b)

(c)

FIGURE 13-1 TROPICAL INTERTIDAL MOLLUSCS. Tropical intertidal molluscs have special adaptations to live in the harsh environment of the rocky intertidal. **(a)** The globular body of the knobby periwinkle (*Cenchritis muricatus*) exposes less surface area to heating relative to volume, and the light-colored shell reduces heat gain or loss. They attach to the rocks with a mucous thread, reducing body contact with the hot surface, and the "beads" on their shells act like radiator fins for cooling. **(b)** Nerites, like these ornate nerites (*Nerita scabricosta*), have a globular sculptured body with the ability to hold water extraviscerally, an important adaptation for evaporative cooling. They often aggregate into large clumps to further reduce heat stress and water loss. **(c)** Tropical periwinkles such as these *Echinolittorina riisei* orient their conical shells toward the sun in such a manner that the amount of exposed surface area is reduced.

their shells act like radiator fins for cooling. They can maintain a body temperature higher than their surroundings at temperatures below 29°C and less than their surroundings at 33°C. They also have a remarkable ability to withstand desiccation using special adaptations in structure and function of the kidney. Likewise, nerites (*Nerita* species), an exclusively tropical group, have a globular sculptured shell and the ability to hold a relatively large volume of water in their shell for evaporative cooling. Nerites may also aggregate into large clumps to further reduce heat stress and water loss (**Figure 13-1b**). Larger individuals of many species tend to live higher in the intertidal, as a larger body heats more slowly than a smaller one. Periwinkles of the genus *Echinolittorina* (**Figure 13-1c**) orient their conical shells toward the sun in such a manner that the surface area exposed is reduced. They attach themselves to the rock with a mucous thread to reduce tissue contact with the hot surface and tightly close their shells with an operculum to reduce water loss.

To avoid overheating and desiccation, crabs and other mobile animals move downward as the tide retreats into crevices and sheltered areas where water is available, or under a moist algal cover. Some anemones (*Anthopleura*) cover themselves with shell fragments when exposed to air. Mussels and barnacles avoid water loss by closing their shells at low tide. Barnacles close the opening of their shells with calcareous plates, trapping enough water inside to support the animal until covered again by water at high tide. If the temperature becomes too warm, the barnacle opens its shell just a little, allowing some trapped water to evaporate and cool the animal. Some limpets (e.g., *Patella*, **Figure 13-2**) return to a **home scar**, an area that fits their body exactly, where by clamping down they can reduce the exposure of their tissues to water loss. Although lacking a home scar, chitons also clamp down tightly on rocks to reduce water loss. Despite these many adaptations, intertidal organisms still are exposed to significant desiccation and heating during air exposure. While some organisms can tolerate significant levels of desiccation and heating at the upper limits of the rocky intertidal, others are limited to lower rock surfaces by their tolerance for the factors. Some species of algae, such as rockweed (*Fucus*, **Figure 13-3**) and

FIGURE 13-3 ROCKWEEDS. Rockweeds such as *Fucus* are common along rocky coasts of the North Atlantic and North Pacific oceans. To prevent desiccation, rockweeds produce a gelatinous covering that retards water loss.

Porphyra, can tolerate up to 90% water loss. Rockweeds also produce a gelatinous covering that retards water loss, allowing them to survive longer exposure to air than other algal species. Many limpet and chiton species can withstand up to 75% water loss and temperatures considerably higher than air temperature.

Coping with Cold
Intertidal animals, particularly those at high latitudes, may be exposed to ice and freezing temperatures during the winter. In contrast to the pattern of lighter colored, globular and sculptured shells seen in tropical molluscs, temperate species such as the common periwinkle (*Littorina littorea*) tend to have darker, more compact shells that are better adapted to absorbing and holding heat. Higher latitude species may also produce antifreeze compounds in their tissues to prevent severe cell damage during freezing. The mechanical abrasion by ice, however, may explain the relative deficit of long-lived intertidal species in far northern and far southern shores as compared with those at similar latitudes less affected by ice.

Avoiding Wave Shock
The force of waves hitting intertidal rocks varies depending on the level of exposure. Headlands receive the most wave force because a wave tends to refract, becoming more parallel to shore. Headlands dissipate much of the wave's energy and thereby protect bays. Not only do rock-dwelling organisms have to deal with the crushing force of a wave as it strikes the rocks, but they also must deal with the drag that is created as the water moves back out to sea. Animals that live in areas of severe wave shock tend to exhibit compressed or dorsally flattened bodies or shells that dissipate the force of waves (**Figure 13-4a**). These animals have also acquired mechanisms for adhering tightly to the rocks to reduce the chance of being dislodged and washed away by the waves. Barnacles and oysters cement themselves to the rock's surface. Animals such as limpets, chitons (**Figure 13-4b**), and intertidal snails found in areas of high wave stress have an enlarged foot for attachment. Sea stars and their relatives use tube feet to attach to rocks. Rock urchins (*Arbacia*, *Echinometra*) hollow out cavities to protect themselves from wave force. During low tide they wedge themselves firmly in place with

> A **home scar** is an area of rock that has been eroded by a limpet to fit its body exactly.

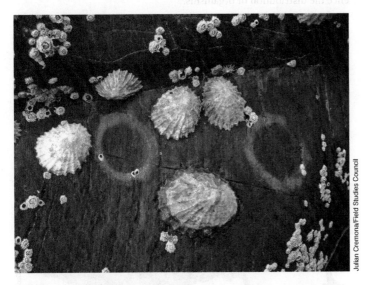

FIGURE 13-2 LIMPETS. To avoid overheating during air exposure, limpets, such as these *Patella vulgate*, clamp down tightly on rocks with their foot to reduce the exposure of their tissues to water loss. Some species reduce exposure even further by returning to a "home scar," an area that fits their body almost exactly. Two home scars are seen in this photograph, as well as limpets and barnacles.

(a)

(b)

(c)

(d)

FIGURE 13-4 COPING WITH WAVE SHOCK. (a) The shingle urchin (*Colobocentrotus atratus*) lives on intertidal rocks where it is constantly pounded by waves. It holds tightly to the rock with tube feet and its dorsally flattened body dissipates the wave's force. **(b)** The fuzzy chiton (*Acanthopleura granulata*) presents a low body profile to oncoming waves while holding tight to the rock with a powerful, muscular foot. **(c)** To avoid wave shock, pale rock-boring urchins (*Echinometra mathaei*) hollow out cavities in the rock. **(d)** Mussels (*Mytilus*) temporarily attach themselves to the rocks with tough byssal threads. These byssal threads can later be broken to allow limited movement.

their spines (**Figure 13-4c**). This behavior, combined with their low profile, helps them to survive wave shock. Mussels (*Mytilus*) temporarily attach themselves to the substrate with byssal threads (**Figure 13-4d**), which can later be broken to allow limited movement. Intertidal algae cope with wave shock by being flexible, bending with the incoming wave and reducing their exposed surface area and drag (**Figure 13-5**). Algae strongly attach to the substrate with their holdfasts. That many species of snails, mussels, and algae congregate in large groups also lessens the wave's impact on individuals. Animals such as crabs with no protection against being crushed or carried away survive by avoiding waves or seeking shelter under rocks or in crevices.

ROCKY SHORE ZONATION

As the tide retreats and the shoreline covered at high tide emerges from water, prominent horizontal bands defined by color or the distribution of organisms appear. This separation of organisms into definite bands is called **vertical zonation (Figure 13-6)**. Unlike shores of sand or mud, rocks provide a relatively stable substrate to which organisms can attach, as well as a variety of hiding places. As the tide retreats, organisms in the upper regions are exposed to air, changing temperatures, solar radiation, and desiccation for prolonged periods. The lower regions, on the other hand, are exposed for only a short time before the tide returns to cover them. The fact that zonation patterns in intertidal zones are similar worldwide led Alan and Anne Stephenson to propose a universal system to classify this phenomenon, replacing various schemes proposed by earlier authors. Zones were established primarily on the basis of the distributional limits of certain common organisms, rather than on the tides. The width of these zones varies depending on the amount of exposure to wave action (**Figure 13-7a**), slope of the shore (**Figure 13-7b**), and the tidal range of different areas.

Vertical zonation is a feature of intertidal communities whereby different species occur in distinct vertical bands along the slope of the shore.

If the slope of the rock is gradual, the zones will likely be broad; on steep rock faces they will be narrow. Zones may also expand due to greater wave action or tidal height, as upper areas remain wet for longer periods of time. The particular organisms present may also differ depending on climate, length of exposure, amount of light, shape of the shore, size and shape of the ocean basin, and type of rock. The presence of crevices, overhangs, and caves that retain moisture may also influence the distribution of organisms.

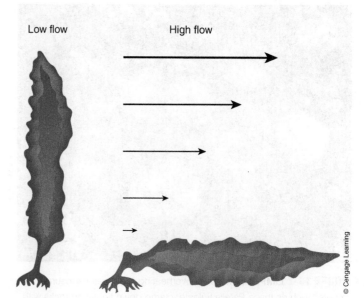

Low flow

High flow

FIGURE 13-5 INTERTIDAL ALGAE AND WAVE SHOCK. Intertidal algae cope with wave shock by being flexible, bending with the incoming wave and reducing their exposed surface area and drag. By getting closer to the substrate, they experience less wave force.

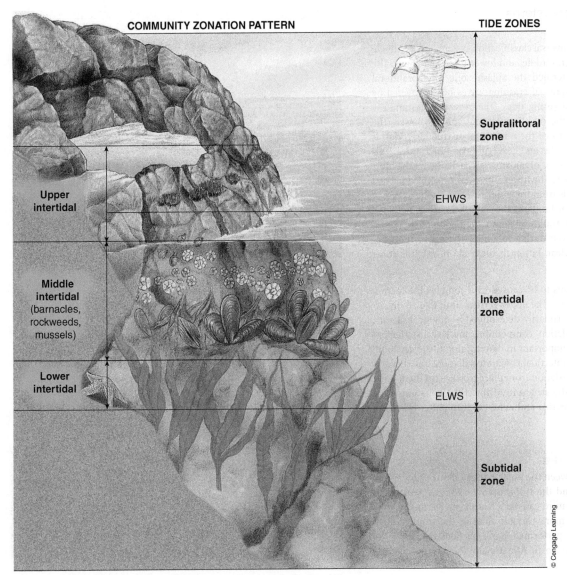

COMMUNITY ZONATION PATTERN

TIDE ZONES

Supralittoral zone

Upper intertidal

EHWS

Middle intertidal
(barnacles, rockweeds, mussels)

Intertidal zone

Lower intertidal

ELWS

Subtidal zone

© Cengage Learning

FIGURE 13-6 ROCKY SHORE ZONATION. Since rock provides a relatively stable surface on which organisms can attach, rocky shores exhibit definite bands or zones, each inhabited by organisms adapted to the special conditions of the environment. EHWS: extreme high water of spring tides; ELWS: extreme low water of spring tides.

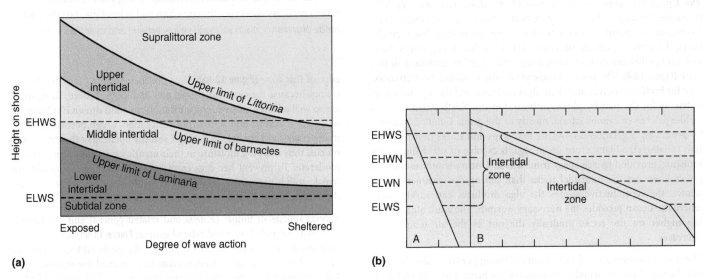

(a)

Supralittoral zone

Upper intertidal

Upper limit of *Littorina*

Height on shore

EHWS

Middle intertidal

Upper limit of barnacles

Upper limit of *Laminaria*

Lower intertidal

ELWS

Subtidal zone

Exposed — Degree of wave action — Sheltered

(b)

EHWS
EHWN
Intertidal zone
ELWN
Intertidal zone
ELWS

A   B

FIGURE 13-7 VERTICAL INTERTIDAL ZONES. (a) The vertical width of intertidal zones may vary depending on the amount of exposure to wave action. **(b)** The slope of the shore can also determine the width of vertical intertidal zones. A very steep shore (A) will have narrow intertidal zones, whereas broad bands will be found on shallow sloping shores (B). EHWS: extreme high water of spring tides; EHWN: extreme high water of neap tides; ELWN: extreme low water of neap tides; ELWS: extreme low water of spring tides. (Source: **(a)** Adapted from Lewis, J. R. 1964. *The Ecology of Rocky Shores.* Copyright © 1964 by Hodder Education. Reprinted with permission of Hodder Education; **(b)** Courtesy of Dr. R.C. Newell.)

Following the Stephensons' universal classification system, we can divide the rocky intertidal into upper, middle, and lower intertidal zones. The upper intertidal zone (also termed the splash zone or supralittoral fringe) is dampened by the spray of crashing waves but is covered by water only during the highest spring tides. The terrestrial community forms its upper border, and the middle intertidal zone its lower. The middle intertidal zone (also termed the midlittoral or true intertidal) is regularly exposed during low tides and covered during high tides. This region usually contains a variety of organisms, barnacles being the most prominent in temperate regions. The lower intertidal zone (also termed the infralittoral fringe) extends from the point reached by the lowest of low tides to the upper vertical limits of the large kelps (laminarians). Below it lies the subtidal (or infralittoral zone), the region of shore that is always covered by water, even during low tide. The distinctions between zones are not always clear, but each rocky shore exhibits some variation of this plan.

Rocky intertidal zonation seems to be caused by a complex interaction of biological and physical factors. The upper vertical limit that an organism can inhabit seems to be primarily determined by physical factors; such biological factors as predation, competition, and larval settlement patterns seem to be the most important in determining the lower limit. Although the basic pattern or intertidal zonation is demonstrated everywhere, there is considerable variation from place to place. In the following sections we will discuss the structure of intertidal zonation along temperate coasts of North America and compare it to patterns seen in the tropics.

Temperate Rocky Shores

The temperate region lies between the polar circles (about 66.5 degrees north and south latitudes) and the tropics (about 23.5 degrees north and south latitudes). Temperature can vary from very warm to below freezing, and the weather is more variable than in the tropics. Most intertidal studies have been performed here, particularly along the coast of Great Britain and the North Atlantic and Pacific coasts of the United States.

The Upper Intertidal As this region of the shore receives very little moisture, it is exposed to the drying heat of the sun in the summer and to extreme low temperatures in winter. As a result of these harsh conditions, the upper intertidal supports only a few hardy organisms. Gray and orange lichens composed of fungi and algae are common in this zone **(Figure 13-8)**. The fungi in these symbiotic relationships trap moisture for both themselves and their algal partners, and the alga produces nutrients for the pair by photosynthesis. On the North Atlantic coast, tarlike patches of cyanobacteria, mostly of the genus *Calothrix*, survive by producing a gelatinous covering that traps and stores moisture. Sea hair (*Ulothrix*), a filamentous green alga, is capable of surviving on the moisture provided by sea spray from waves. During winter months, it grows lower on the intertidal rocks than during the summer. In the winter, the reproductive spores the alga produces can survive only where the ocean provides the necessary warmth. The adult algae growing higher on the rocks gradually die out as the air temperature decreases.

The most common animal inhabitants of the upper intertidal throughout the world are periwinkles, molluscs of the genus *Littorina* and associated genera. The term *littorina zone* is often applied to the upper intertidal, although these molluscs also range into the middle intertidal. On the Atlantic coasts of North America and Europe, rough periwinkles (*Littorina saxatilis*) graze on various types of algae growing at the lower

FIGURE 13-8 **THE UPPER INTERTIDAL.** The upper intertidal, or splash zone, receives very little moisture and is inhabited by only a few hardy organisms, such as these yellow lichens and cyanobacteria.

edge of this zone **(Figure 13-9a)**. They use their mantle cavity for gas exchange because they lack functional gills and lungs. Indeed, these snails are so well adapted to breathing air that they would drown if submerged in water for several hours. To avoid desiccation, rough periwinkles hide in the cracks and crevices of rocks or seal the opening of their shell with mucus, thus trapping moisture in their mantle cavity until temperatures moderate. To prevent her eggs from being damaged by exposure, the female rough periwinkle retains them inside her mantle cavity, where they can be kept moist and oxygenated until they hatch.

Several species of limpet (*Patella* and related genera) and crustaceans known as isopods (*Ligia* and related genera; **Figure 13-9b**) are also common inhabitants of the upper intertidal. Like periwinkles, limpets are grazers, but isopods are scavengers that feed on available organic material. Limpets retain their gills (or have structures called pseudogills that can act as gills), but intertidal isopods are adapted to exchanging gas with the air and drown if submerged under water. Female isopods have thoracic pouches in which they carry their eggs to prevent them from drying out.

(a)   (b)

FIGURE 13-9 **ANIMALS OF THE UPPER INTERTIDAL. (a)** Rough periwinkles (*Littorina saxatilis*) graze on the algae that cover the rocks. **(b)** Isopods (*Ligia*) are scavengers that feed on available organic material.

While their location protects inhabitants from predatory marine organisms such as crabs or predatory snails, terrestrial organisms, such as birds, raccoons, or rats, often venture here to feed.

The Middle Intertidal In temperate regions, typical organisms found in the upper portion of the middle intertidal are acorn barnacles (*Chthamalus, Balanus*) and rock barnacles (*Semibalanus*), which form a line at and below the high tide mark. On some rocky shores, the barnacle population may be as dense as 9,000 individuals per square meter. Barnacles permanently cement themselves to solid surfaces. They are particularly common on shores pounded by heavy surf where other, less well-adapted organisms would not be able to survive.

In the middle and lower portion of the middle intertidal, bivalves (oysters, *Ostrea*, and mussels, *Mytilus*), limpets (*Patella* and *Acmaea*), and various periwinkles dominate. When the tide is in, bivalves open their shells to filter feed on plankton in the incoming water. At low tide, the bivalves close their shells, trapping enough water inside to provide for their needs until high tide. Limpets and chitons also occur in this zone and graze on algae at high tide.

The common periwinkle (*Littorina littorea*) is found on North Atlantic shores of North America and Europe. Unlike its relative the rough periwinkle, which inhabits the upper intertidal, this snail cannot breathe air and must remain moist to exchange gases. During low tide it buries itself in masses of seaweed that trap enough moisture for the snail to survive until the tide returns. At high tide, the common periwinkle feeds on algae that grow on the rocks. Rock-boring urchins (*Arbacia*) may also occur here. They feed on both the algae covering the rocks and drifting bits of seaweeds that they capture among their spines and tube feet. They actively feed during high tide but at low tide return to the shelter of the depressions that they hollow out of the rock.

The most characteristic seaweeds of the middle intertidal on the northern Atlantic and northern Pacific coasts, especially in colder temperate areas, are brown algae such as rockweeds (*Fucus*). Rockweeds are usually less than 30 centimeters (1 foot) long and grow on rocks that do not have full exposure to the sea as they do not tolerate large waves. In protected areas, such as bays, rockweeds may be as long as 1.8 meters (6 feet) and almost completely cover the rocks. Some species of rockweed use gas-filled chambers called air bladders to buoy them during high tide. As the tide goes out, seaweeds form large mats that trap water and provide a haven for many species of animal, such as juvenile sea stars, brittle stars, sea urchins, and bryozoans. The shelter provided by large algae such as rockweeds may significantly reduce the stress of increased temperature and desiccation on these associated animals during low tide.

The Lower Intertidal The lower intertidal is a transitional area between the middle and subtidal zones **(Figure 13-10)**. Because it is usually submerged (except at spring tides), there is no distinct boundary between this region and the subtidal. The lower intertidal typically has a rich flora and fauna of organisms that tolerate very limited air exposure. In some areas the rocks are covered with a dense turf of seaweeds, including red, green, and brown algae. Along coasts where the water is cold, large kelps such as *Laminaria* **(Figure 13-11)** may form a dense cover that protects smaller algal species and animals living among them. Coralline algae **(Figure 13-12)** are also abundant in many areas. In these cooler waters a variety of molluscs are typically found, as well as sea stars and brittle stars. Attached to many of the algae are lacey colonies of bryozoans. Hydrozoans attach to both rocks and broad blades of algae. On the West Coast of the United States, anemones cover the rocks, sea urchins graze on algae and bits of decaying material that they find, and spider crabs (*Libinia*) and Jonah crabs (*Cancer borealis*) scavenge for food.

Tropical Rocky Shores

The tropics are generally defined as the area between 23.5 degrees N and S latitudes, although tropical conditions can extend to higher latitudes as a consequence of warm ocean currents or other conditions. Tropical conditions exist in Bermuda (32 degrees N latitude), for example, as a result of the warm water carried by the Gulf Stream. The average incident solar radiation on the sea surface is approximately 30% greater in the tropics than in the temperate zone, and temperatures do not fall much below 18°C (64°F) for more than a month at a time. The temperature variation is also much less than in the temperate zone, and rainfall is seasonal. These conditions make tropical intertidal systems more stressful in some ways (e.g., higher temperatures) but less stressful in others (e.g., less variation in temperature and fewer storms).

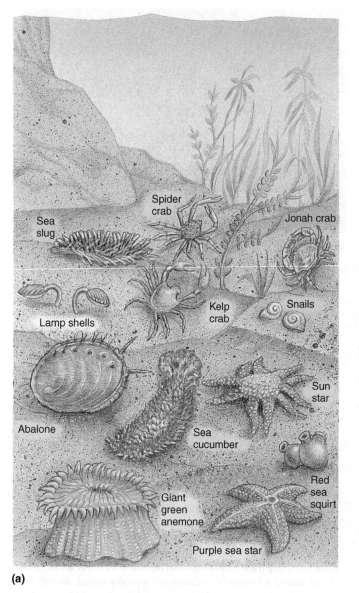

(a)

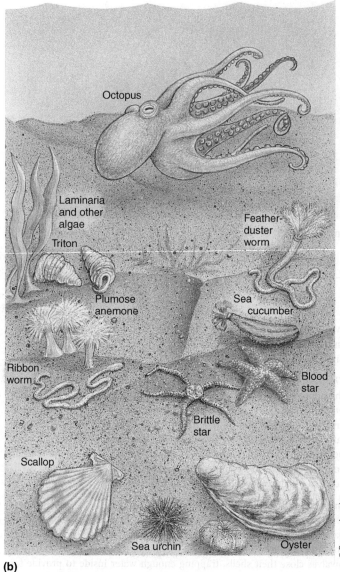

(b)

FIGURE 13-10 THE SUBTIDAL ZONE. The subtidal zone of rocky shores offers a more stable environment than the intertidal zone and thus exhibits greater diversity. This diagram shows common organisms inhabiting the subtidal zone of the **(a)** West Coast and **(b)** East Coast of the United States.

Zonation patterns in the tropics have many similarities to those seen in the temperate zone but also some significant differences. Tropical rocky intertidal systems have been best studied on the western shore of Panama and in the Caribbean. We will briefly examine zonation patterns in the latter.

The Tropical Upper Intertidal In the Caribbean, the upper intertidal can be divided into the white, gray, and black zones. The color of each zone is related to the dominant organisms that inhabit it. The white zone is the true border between the land and the sea. No fully marine organisms occur here, but hermit crabs, isopods (*Ligia*), and the knobby periwinkle (*Cenchritis muricatus*) are common. Knobby periwinkles also occur in the gray zone along with other periwinkle species (*Echinolittorina*) and nerites (*Nerita*), the exclusively tropical group that tends to replace limpets as the dominant form in higher tropical intertidal zones. Their globose shape and ability to hold water in their shell for evaporative cooling make them better adapted to high-temperature stress and desiccation than the limpets. The gray zone is also the farthest zone from the

low tide line where macroscopic marine algae (*Bostrychia*) grow. The black zone is immersed only at the highest spring tides and lacks the knobby periwinkle. Several species of algae (*Bostrychia, Polysiphonia*) and cyanobacteria dominate here. Animal inhabitants include smaller periwinkle snails (*Echinolittorina* species), nerites (*Nerita*), and the fuzzy chiton (*Acanthopleura granulata*). There are many more species of periwinkles in the tropics than in the temperate zone, and they tend to be smaller in size.

The Tropical Middle Intertidal The middle intertidal of the tropics is usually divided into yellow and pink zones. The yellow zone is typically yellow or green because of the microscopic boring algae covering its surface **(Figure 13-13a)**. Small populations of barnacles may occur here, as well as limpets, the fuzzy chiton, and predatory rock snails (*Thais*). Another inhabitant is the irregular worm snail (*Petaloconchus irregularis*), which forms calcareous tubes on the rock surface. They feed by trapping plankton on mucous threads. In some areas there is a pink zone **(Figure 13-13b)** underlying the yellow that is distinguished by the

FIGURE 13-11 GIANT KELPS. Along coasts where the water is cold, large kelps such as Laminaria may form a dense cover that protects smaller algal species and animals living among them, particularly at low tide.

FIGURE 13-12 CORALLINE ALGAE. Coralline algae are red algae characterized by having calcium carbonate deposited in their cell walls. Many species commonly found in the intertidal (e.g., *Mesophyllum lichenoides*) encrust over the hard surfaces and appear boulder-like. They range in color from grey to bright pink.

(a)

(b)

FIGURE 13-13 YELLOW AND PINK ZONES OF TROPICAL SHORES. (a) The yellow zone gets its name from the large number of yellow and yellow-green boring algae that occur there. The pink zone is seen below the yellow zone in this photo from Barbados. (b) The pink zone is named for the pinkish color of the coralline algae that can dominate this area. This photo is from Hawaii.

widespread encrustation of coralline algae. Characteristic animals include the irregular worm snail, mats of anemones, keyhole limpets, and various gastropods. In other areas the zone may be overgrown by short reddish-brown mats of algae. Barnacles, scorched mussels, several chitons, rock snails, and other forms may be found here.

The Lower Intertidal The lower intertidal includes the edge of the lower rocky platform and parts of fringing reefs. *Sargassum* species and turf algae may cover the rock surface. Boring urchins (*Echinometra*), anemones, sponges, bryozoans, sea cucumbers, and keyhole limpets are common animal inhabitants.

Unlike the species-rich subtidal of the temperate zone, the tropical subtidal is relatively barren, with most algae being confined to the lower and middle intertidal. Relatively large brown algae dominate in the temperate zone, whereas small, turf-forming red algae dominate in the tropics. A number of explanations have been proposed for these differences, including the effect of seawater temperature, the need for a high surface-to-volume ratio due to low nutrient availability in tropical water, and the relatively constant light intensity that has inhibited the evolution of large algae. The explanation most favored by researchers suggests that the high level of grazing in the tropics prevents large, fleshy algae from colonizing the subtidal zone.

Important tropical herbivores include surgeonfish, parrotfish, damselfish, sea urchins, and many species of gastropods. In areas where grazers were experimentally excluded, erect algae were able to establish themselves where none were previously found. The ubiquitous presence of such defenses as thallus calcification and noxious chemicals among tropical algae also supports this hypothesis. The abundance of "turf" algae on shallow wave-swept platforms of the intertidal zone is probably a result of grazer exclusion throughout much of the tidal cycle. Turf algae are also better able to resist wave shock on these higher platforms than their larger and fleshier relatives.

High predation levels would also explain the greatly reduced abundance of barnacles and mussels in tropical intertidal zones. In some parts of the Caribbean, barnacles and mussels are very rare. The fact that slow growing but well protected crustose algae typically dominate the middle and lower intertidal zones can also be attributed to the effects of grazing gastropods and other herbivores. The activity of herbivorous molluscs may also inhibit settling by sessile invertebrates.

Comparison of Temperate and Tropical Rocky Intertidal Systems

High temperatures, less temperature variation, and high predation make the tropical intertidal environment rather different from the temperate. Mobile invertebrates are very abundant, but sessile species may be rare and their competition for space less severe, particularly in upper intertidal zones. Holes and crevices seem to be more important as refuges for these animals than in temperate latitudes. There are more periwinkle species, and they are more abundant in the tropics than in temperate zones. Exclusively tropical groups such as the nerites are among the fauna but barnacles and mussels, so abundant in the temperate zone, may be scarce or lacking in the rocky intertidal of the tropics. In the temperate zone, large body size or residence in higher regions of the intertidal zone are important means of escaping predators, but these characteristics are less significant in the tropics. Because of their greater numbers and diversity of types, consumers have more influence on the availability of refuges in the tropics. In contrast to their role in the temperate zone, macroalgae have significantly less impact on community structure. Although there is some annual variation in the physical environment of tropical systems, these communities change little in space and time.

TIDE POOLS

Not all areas of the rocky intertidal zone are exposed when the tide retreats. Depressions in the rocks retain water, forming areas called **tide pools (Figure 13-14)**. Tide pools prevent organisms within them from being exposed to air, but they present their own unique set of difficult environmental conditions. As the small pool of water heats in the sun, it loses oxygen. This can lead to the suffocation of some inhabitants. During heavy rains, the salinity of tide pools decreases drastically as rainwater accumulates and dilutes the seawater. On the other hand, on a hot sunny day, so much water can evaporate that the salinity increases to dangerous levels. If the tide pool contains a large amount of algae, the oxygen content of the pool will be high during the day while sunlight powers photosynthesis. At night, however, the level of oxygen declines and the level of carbon dioxide rises, resulting in a lower pH. Large tide pools near the low-tide mark usually show the least amount of fluctuation, whereas smaller ones and those near the high-tide line exhibit the widest range of fluctuation. Regardless of their position on the shore, most tide pools return abruptly to marine conditions as the tide rises and seawater floods the pool. This return to the ocean produces almost instantaneous changes in salinity, temperature, and pH. Many intertidal organisms have the ability to tolerate these changes or move from pool to pool, but others, such as sea stars and urchins, may die from the osmotic stress.

Common tide pool inhabitants include various species of alga, sea star, anemone, tubeworm, hermit crab, and a variety of mollusc species. Many are filter feeders on phytoplankton and zooplankton. Anemones

Tide pools are depressions on a rocky shoreline filled with seawater.

FIGURE 13-14 A TIDE POOL. Organisms that inhabit tide pools must be able to adjust to rapid changes in temperature, pH, salinity, and the oxygen content of the water.

IMPACT BUBBLE

SHORELINES RECEIVE AMAZING numbers of human visitors. One California study found that, annually, there are between 25,000 and 50,000 visitors per 100 meters (328 feet) of shoreline for popular sites. Even those with low visitation receive 2,000 to 10,000 visitors per 100 meters per year. The problem has become so acute that some locations, like Moss Beach in California, are now limiting visitors. Visitors crush and dislodge animals and algae from rocky shorelines. If organisms are not completely dislodged, their attachment to the rocks may be weakened, making them more susceptible to loss from wave activity. Trampling rocky shores has been shown to damage the byssal threads of mussels in California, increasing their eventual losses. Humans also collect organisms for food, souvenirs, aquaria, and fish bait, thereby reducing populations of organisms and disrupting the ecological balance of competitors, predators, and the food supply. Because collectors prefer large specimens, their actions can also change the size and age distribution of populations. Common species collected include mussels, limpets, snails, octopuses, sea hares, chitons, abalone, sea stars, sea urchins, crabs, and lobsters. Another human disturbance is the overturning of rocks, crushing organisms attached to the surface, preventing algae from getting sunlight, and exposing hidden animals to predation, wave action, and desiccation.

Shoreline development for recreational and commercial use not only destroys locally existing intertidal communities but may also have long-distance effects along the coastline. Jetties and sea walls, for example, accelerate erosion by altering the movement of sediment along the coast. Shoreline structures can also alter the local hydrodynamic regime and the degree of wave action experienced by the shore. Sewage and storm water runoff contains nutrients, bacteria and viruses, mercury, lead, chromium, copper, PCBs, hydrocarbons, and other toxic chemicals. Oil and other toxic chemicals released from ships and drilling rigs suffocate and kill marine organisms. Toxic chemicals not only endanger the health of the marine environment but human health as well. Excess nutrients overfeed algal communities, contribute to their overgrowth and increase the accumulation of sediments, changing the structure of both intertidal and subtidal communities. Potential pathogens in sewage may be accumulated by filter-feeding organisms and eventually enter the human food chain.

Humans have inadvertently released invasive species that adversely affect intertidal communities. Asian shore crabs (*Hemigrapsus sanguineus*), for example, consume native mussels, barnacles, oysters, snails, clams, polychaetes, and crustaceans. They have a very high reproductive rate, producing 3 to 4 clutches of 50,000 eggs per year. Introduced in 1988, this species is now well established from Maine to North Carolina. The common periwinkle (*Littorina littorea*), introduced from Europe in the 19th century, competes with native littorine snails and has fundamentally changed North Atlantic intertidal systems through its grazing activities. This mollusc has also invaded the Pacific coastline from Washington to California. A Japanese green alga, dead man's fingers (*Codium fragile*), attaches to mussels and oysters with holdfasts. As the algae grow, they became buoyant and drift away with the shellfish attached. The introduction of Norway rats (*Rattus norvegicus*) into the Aleutian Islands has caused massive changes in intertidal communities. The rats reduced marine bird densities, which allowed intertidal invertebrates to become more abundant. The increased grazing pressure from intertidal invertebrates reduced the fleshy algal cover, changing the community from an algal to an invertebrate-dominated system.

These are just a few examples of the many types of detrimental effects human activities have on intertidal communities. It is up to humans as stewards of this planet to be aware of the effects of our actions on the marine environment and preserve it for future generations to enjoy.

---

and tubeworms expose a large amount of surface area while feeding. If disturbed, they retract their tentacles into their bodies or tubes to protect these delicate appendages from damage. This behavior is also an efficient way to avoid desiccation.

LIVING ON THE ROCKY SHORE

Although physical factors such as tolerance to air exposure, high temperatures, desiccation, substrate structure, and wave force may determine where organisms can potentially live, their interplay with biological interactions ultimately determines which organisms actually occur at a given location within the intertidal zone. Biological factors include the effect of competition, grazing, and predation; variation in larval settlement (recruitment); and positive interactions among species.

Competition, Grazing, and Predation

Tides consistently supply plankton, detritus, and nutrients for photosynthesis to intertidal communities. Where rocks are kept wet enough by spray, algae grow abundantly. The base of the intertidal food web usually consists of plankton and benthic algae. Benthic algae vary from microscopic diatoms to fleshy seaweeds. Grazers, filter feeders, detritivores, and predators are all abundant in rocky intertidal food webs **(Figure 13-15)**.

Although some slow-moving herbivorous animals, such as chitons and limpets, may be limited by their inability to move over a large enough grazing area, for most rocky intertidal inhabitants food is abundant. The availability of space, however, is another matter. The competition for space among community inhabitants is often the dominant biological factor in the organization of intertidal communities.

In areas where wave action is heavy, barnacles dominate upper intertidal zones primarily because only they have the ability to survive there. Less-resistant species such as periwinkles are washed away by wave action, and most predators cannot withstand the high temperatures and drying conditions of the upper intertidal. In the upper portion of the intertidal zone in Scotland, there is a distinct zonation of two barnacle species, *Chthamalus stellatus* and *Semibalanus balanoides*. Larvae of the former have the ability to settle in both the upper and the middle intertidal but are ultimately excluded from the lower zone by the competitively superior *Semibalanus balanoides*. *Chthamalus stellatus* persists in the upper zone because *Semibalanus* is intolerant to the high temperature and drying conditions there. Likewise, *Balanus glandula* persists in the upper regions of the intertidal of the Pacific coast of North America because only it can tolerate conditions there. This species and its competitor, *Balanus cariosus*, are heavily preyed on by several gastropods of

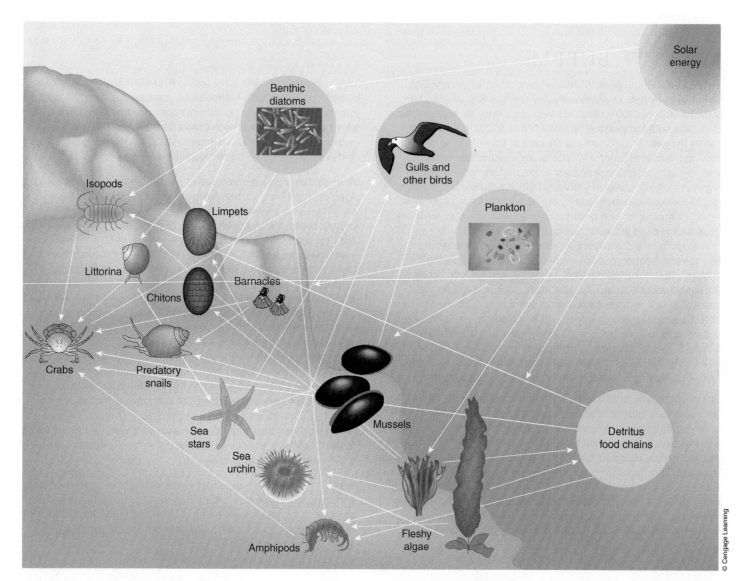

FIGURE 13-15 ROCKY INTERTIDAL FOOD WEB. The base of the intertidal food web consists of plankton and benthic algae. Benthic algae vary from microscopic diatoms to fleshy seaweeds. Grazers, filter feeders, detritivores, scavengers, and predators are all abundant in rocky intertidal food webs.

the genus *Nucella* (**Figure 13-16**). While *Balanus glandula* escapes its predators by its position in the intertidal, *Balanus cariosus* cannot do so. It escapes by growing to a size that is too large to be consumed. It is unclear how *Balanus cariosus* reaches this large size without being consumed, but they persist in random patches below the band of *Balanus glandula*.

Below the barnacles, the competitively superior mussels (*Mytilus*) are most abundant. Mussels displace barnacles by growing over them. Their dense clumps provide habitat for a number of other species, including anemones, crustaceans, and segmented worms. Although competitively superior, mussels do not completely exclude barnacles because of the activity of predators. On the Pacific coast of North America, ochre sea stars (*Pisaster ochraceus*) prey on *Mytilus californianus*, preventing them from completely overgrowing the barnacles. Sea stars almost completely eliminate *Mytilus* from the subtidal zone, but the mussels persist in the intertidal because sea stars can only feed when the tide is in. In many intertidal systems, however, dominant competitors may be controlled by a number of species rather than a single top carnivore. The effect of a keystone predator such as *Pisaster* may vary over its geographical range,

FIGURE 13-16 PREDATORY GASTROPODS. Predatory snails such as this dog whelk (*Nucella lapillus*) are found on rocky shores, breakwaters, and piers. Dog whelks feed primarily on barnacles and mussels.

and predation can vary with other environmental variables. In New England, the mussel *Mytilus edulis* can outcompete both the barnacle *Semibalanus balanoides* and the alga *Chondrus crispus* but is prevented from eliminating them by predatory sea stars and snails. Where these predators are eliminated by heavy wave conditions, *Mytilus edulis* dominates. In tropical intertidal systems, total predation is very strong and the control of competitively dominant species is spread over a number of consumers; hence there are no keystone predators in the usual sense.

In the lower portion of the temperate rocky intertidal habitat, particularly where wave action is less severe, algae compete with barnacles and mussels for space on the rocks. Opportunistic smaller species of algae can initially occupy bare space but are quickly overgrown by larger brown algae. Once established, the large blades of these algae move across the rocky surface in a sweeping action that is powered by waves. This action prevents the larvae of barnacles and other species from settling down and gaining a foothold, thus providing the algae with additional space for growth. Palm seaweeds (*Postelsia palmaeformis*) may become established if storms strip away mussels. Once established, spores slide down the frond and colonize any bare rock or the shells of mussels nearby. The attached seaweeds may weaken the hold of the mussels, and, when storms strip them off the rock, palm seaweeds' spores quickly colonize the vacant area and produce a clump of individuals **(Figure 13-17)**.

Doug Sokell/Visuals Unlimited, Inc.

FIGURE 13-17 PALM SEAWEEDS. When storms strip rocks of their inhabitants, palm seaweeds (*Postelsia palmaeformis*) may become established on bare areas. They produce multiple spores that slide down the frond and colonize any bare rock or the shells of nearby mussels. These attached seaweeds may weaken the hold of the mussels and, when storms strip them off the rock, palm seaweed spores quickly colonize the vacant area and produce a clump of individuals as shown here.

Grazing by limpets and other molluscs can reduce algal cover and allow barnacles and other organisms to colonize the rocks. To lessen grazing, some species of algae produce chemicals that deter herbivores from feeding on them. Studies have revealed that when herbivorous molluscs begin to graze on rockweed (*Fucus*), this alga produces more of a toxic chemical in the area where the molluscs are feeding, deterring further grazing. Indeed, these chemical defenses are so effective that noxious algae primarily enter the food web as detritus. Grazing-resistant algae, such as *Fucus* and *Ascophyllum*, tend to dominate these shores, forming a canopy that protects many types of consumers from desiccation. These consumers in turn may limit the populations of palatable algae, as well as barnacles and mussels, providing open space for algal colonization.

In New England tide pool communities, two algal species dominate depending on the level of grazing **(Figure 13-18)**. *Enteromorpha*, a green alga, outcompetes Irish moss (*Chondrus crispus*, a red alga) in tide pools lacking the common periwinkle (*Littorina littorea*). Where this snail is present, Irish moss is the prevalent alga. *Enteromorpha* persists in certain pools because green crabs (*Carcinus maenas*) eat periwinkles and prevent them from removing this alga. Irish moss persists in other tide pools because gulls eat the crabs and allow the periwinkles to persist. *Enteromorpha* covers the crabs and protects them from gulls in their pools, whereas gulls can readily capture the crabs in Irish moss pools. Removal and transplantation experiments have demonstrated the importance of these ecological relationships in maintaining both types of pools.

Variation in Larval Settlement (Recruitment)

Most intertidal organisms produce planktonic larval stages during spring tides when currents move them offshore. Larvae can be carried hundreds of kilometers with the currents and spread over a wide area. Great numbers are lost to a variety of causes, including lack of food and heavy predation by a host of organisms. Oceanographic phenomena can cause much variability in the availability of larvae and thereby affect the structure of intertidal communities. El Niño nutrient depletion, for example, reduced the growth of kelp forests off the coast of Southern California and the associated populations of rockfish, a major predator on barnacle larvae. Reduction in rockfish numbers increased barnacle recruitment to West Coast intertidal systems. Another study on barnacle larvae found that in an upwelling year most larvae were transported offshore and perished, whereas in a non-upwelling year larvae were abundant for settlement. The importance of recruitment to a particular intertidal community is also variable. A study in New England found that larval settlement explained about 11% of the community structure, whereas in Panama it explained between 39% and 78%. In contrast, predation and competition explained between 50% and 78% of the variation in New England, but less than 10% in Panama.

As larvae grow in the offshore waters, they become less buoyant, sink to deeper waters, and are carried back to shore. Habitats with high water flow usually have more recruitment than those with low water movement because high flow ultimately delivers more larvae to a particular location. In a study in New South Wales (Australia), it was found that the species of barnacle dominant at a given location depended on the level of wave action. Where wave action was weak, grazing gastropods were the dominant organisms rather than barnacles. Settlement can also be affected by subtle differences in the substrate. It has been demonstrated that larvae have a definite preference for the type of substrate on which they settle. The structure of the substrate can also affect local patterns of water flow, causing different patterns of larval settlement.

(a)   **(b)**

FIGURE 13-18 BIOLOGICAL INTERACTIONS IN TIDE POOLS. **(a)** In New England tide pools containing common periwinkles (*Littorina littorea*), fleshy green algae such as *Enteromorpha* and *Ulva* are grazed out and the bare surfaces colonized by unpalatable Irish moss (*Chondrus crispus*) and algal crusts. Irish moss and algal crusts persist in these pools because gulls eat green crabs (*Carcinus maenas*) that feed on periwinkles. **(b)** In tide pools lacking common periwinkles, the green algae outcompete slower-growing Irish moss and algal crusts. The resulting algal canopy protects green crabs from gulls, and the crabs prevent periwinkles from removing the green algae. Both types of tide pool persist because of the activities of green crabs, gulls, and periwinkles.

Variation in larval settlement can have repercussions throughout the intertidal system. Heavy barnacle settlement in central California increased sea star predation and resulted in oscillating populations of barnacles. Lower levels of settlement, in contrast, did not increase sea star predation and resulted in a stable population of barnacles with a mixed age distribution. High levels of recruitment can increase competition and, where space is limited, result in the classic zonation patterns discussed here. If recruitment is low, however, zonation may break down and produce very different distributional patterns **(Figure 13-19)**. Low recruitment may also reduce positive interactions among species, preventing many of them from inhabiting certain parts of the intertidal zone.

Positive Interactions among Species

The effect of positive interactions among species on the structure of intertidal communities has not been studied to the same extent as the effects of such negative interactions as competition, predation, and grazing. Group aggregations, one type of positive interaction, can have a significant benefit to its members by buffering high temperatures and desiccation. Seaweeds, for example, can protect such organisms as snails, barnacles, anemones, and other seaweeds from drying and heat stress at low tide. Mussel beds provide refuge to many organisms and protect them from wave stress. Barnacles and many snails, such as nerites, often live in dense groups that serve to minimize individual water loss. Another type of positive association allows organisms to survive by being associated

> A **climax community** is a stable, self-sustaining community that is the end result of ecological succession.

with a noxious species. Many palatable algae are protected from grazing by their association with *Ascophyllum*, a noxious seaweed.

Ecological Succession in the Intertidal Zone

Succession occurs when natural or human induced disturbances strip intertidal rocks of their inhabitants. Mussels can grow on top of one another, increasing the chance that waves will tear them from the rocks. Loose rocks and drifting logs can batter the intertidal community. Ice can form during the winter and scour the rocks of their organisms. Sessile invertebrates such as barnacles may be smothered by oil spills. Whatever the cause, bare rocks offer a location for opportunistic species to settle. Although the pattern of succession differs from place to place, the first colonizers are usually diatoms or some sort of filamentous algae. These early colonizers may prepare the surface for colonization by perennial species of red algae or other organisms. Different species of red algae may replace early colonizers before animal colonizers, such as barnacles, arrive. Dominant competitors, such as mussels, may replace them in turn to form what is often termed a **climax community**, the final stage in ecological succession. In reality, these communities are only temporary; they are subject to replacement during the next cycle of environmental disturbance. The exact sequence of events and ultimate structure of a community will vary depending on what plant propagules and larvae are available in the water column at a given time, the particular physical characteristics of the environment, and the various biological interactions between species in the community. A recent study in the Gulf of Maine found that seaweed canopy communities and mussel communities generally replaced

© Cengage Learning

High recruitment

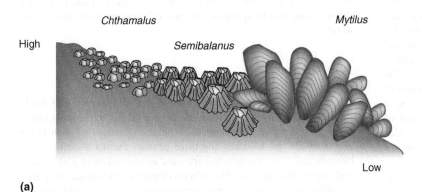

Chthamalus   *Mytilus*

High   *Semibalanus*

Low

(a)

Low recruitment

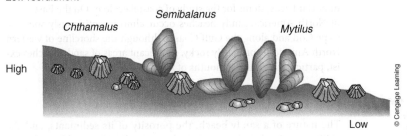

Semibalanus

Chthamalus   *Mytilus*

High

Low

© Cengage Learning

(b)

FIGURE 13-19 EFFECT OF LARVAL RECRUITMENT ON INTERTIDAL ZONATION. **(a)** With unlimited availability of larvae and the saturation of space, competition among species results in classic intertidal zonation. **(b)** If larval recruitment is limited, competition does not occur and zonation breaks down. (Source: Bertness, M.D. 2007. *Atlantic Shorelines.* © 2007 by Princeton University Press. Reprinted by permission of Princeton University Press.)

The **diversity** of a community is the total number of species inhabiting it.

The **competitive exclusion principle** states that when two species compete for a limited resource, the better competitor drives the other to extinction.

Patchiness is the condition in which organisms occur in small, isolated groups within a larger habitat.

themselves when disturbances removed them. Seaweed canopies persisted where there was low water flow, and mussel communities persisted where there was high flow. Both communities were under consumer control to a significant extent. Indeed, with crabs and snails present, neither community completely recovered after 3 years of monitoring.

Consumers and various physical factors (availability of crevices, substrate type, etc.) can create a moderate level of disturbance that often results in an increase in **diversity**, the number of species living in an area, because it interferes with the **competitive exclusion principle**. The rocky intertidal community, therefore, demonstrates **patchiness**, a condition in which organisms occur in isolated groups within a larger contiguous habitat. The greatest diversity occurs where disturbance prevents dominant competitors from excluding other species but is not severe enough to prevent less successful competitors from establishing populations.

INTERTIDAL FISHES

Fishes visiting the intertidal zone can be divided into two categories: residents and temporary inhabitants. Resident species include clingfish, blennies, gobies, sculpins (*Oligocottus*), and rock eels (*Xiphister*), among others. True residents are reported to make up between 20% and 67% of the fish fauna of tide pools. True residents typically have special adaptations for surviving the harsh conditions of life in the intertidal. Because wave conditions make large body size and lack of mobility disadvantageous, intertidal fish are rarely longer than 20 to 30 cm (8 to 12 inches). Scales are absent (blennies and clingfish), reduced, or very firmly attached (some gobies). Body shape may be compressed and elongated (blennies) or depressed (clingfish). The swim bladder is typically absent or reduced, and the body density greater than in pelagic forms.

Clingfish (**Figure 13-20**) and most gobies have pelvic fins modified into sucking disks that allow them to maintain position against surging water. The head may be enlarged and, in some clingfish, articulated (able to pivot)—an unusual adaptation in fish. The eyes may also be enlarged and set high on the head, an aid to spotting prey above the water level. Most species have a wide tolerance for changes in salinity and temperature, and some can withstand desiccation to the same extent as amphibians. A few species even leave the water to feed and to avoid the most extreme wave conditions (e.g., *Sicyases sanguineus*). Most fish that leave

Fred Bavendam/Minden Pictures

FIGURE 13-20 CLINGFISH. Clingfish, like this Tasmanian clingfish (*Aspasmogaster tasmaniensis*), have pelvic fins that are modified into a sucker. This adaptation allows them to attach firmly to solid surfaces, thus reducing the chances of being swept away by strong currents.

Swash is the water running up a beach after a wave breaks.

Backwash is the water flowing down a beach after a wave has broken.

Dissipative beaches are beaches on which large waves are dissipated in a wide surf zone.

Reflective beaches are beaches that lack an offshore surf zone.

the water exchange oxygen and carbon dioxide through the skin, but some forms (mudskippers) use a modified oral-pharyngeal cavity. It has also been suggested that gas exchange may take place across the intestinal wall in the Chilean clingfish. Excretion of nitrogenous wastes can be a problem when fish are out of the water. Some forms excrete urea rather than ammonia under these conditions. Intertidal fish can be herbivorous, omnivorous, or carnivorous.

Temporary inhabitants can be divided into tidal visitors, seasonal visitors, and accidental visitors. Tidal visitors move into the intertidal zone at high tide to feed, whereas seasonal visitors move into it for breeding. Accidental visitors are trapped by storms.

IN SUMMARY

- Rocky shores vary in their tidal range and the time they are exposed to air.

- During air exposure, organisms inhabiting the rocky shores must survive high temperatures, cold, desiccation, wave shock, and salinity changes. (Have You Wondered? #1)

- Structural and behavioral adaptations that have evolved to deal with environmental stresses include means of rock attachment, specialized body designs, and tolerance to temperature extremes and desiccation. (Have You Wondered? #2)

- Complex interactions of biological and physical factors cause vertical zonation on intertidal rocks.

- The upper intertidal zone receives only the little moisture delivered by sea spray. The dominant species found in this zone are periwinkles (*Littorina*). (Have You Wondered? #3)

- The middle intertidal zone, or true intertidal, is the area that is alternately exposed and submerged by tides. The characteristic species found here are barnacles. (Have You Wondered? #3)

- The lower intertidal extends from the lowest of low tides to the upper limits reached by the large kelps (laminarians). Large algae and kelps are abundant in the temperate region, but this zone is relatively barren in the tropics. (Have You Wondered? #3)

- The subtidal is covered by water even at low tide. Because this zone is usually more stable, it supports a greater diversity of organisms in the temperate zone but may be barren in the tropics due to the abundance of grazers and predators. (Have You Wondered? #3)

- Intertidal communities are well supplied with food by the incoming tide, but space is often at a premium. Competition, predation, grazing, and recruitment, affect both the vertical zonation and dynamics of the community. Positive interactions among species may also determine which species can exist in a given location.

Ecological succession occurs when natural or human induced disturbances strip intertidal rocks of their inhabitants. Tide pool inhabitants have evolved a variety of mechanisms that allow them to survive abrupt changes in temperature, salinity, pH, and oxygen content of the water.

13-2 Sandy Shores

About two-thirds of the world's ice-free coastlines consist of sandy shores as opposed to rock. Sand beaches are subject to the same physical factors as rocky shores but the relative importance of each factor differs in the two habitats. The three primary factors affecting sandy beaches and their communities are wave action, sediment particle size, and slope of the coastline. These factors interact to determine the physical structure of this dynamic environment and its community of organisms.

Sandy beaches are important recreational areas for humans as well as important ecosystems for the study of coastal ecology. On the East Coast of North America, sandy beaches occur almost continuously south of Cape Cod and along the Gulf Coast. Although the shoreline of western North America is primarily rocky, significant areas of sandy beaches exist, particularly near the mouths of rivers.

ROLE OF WAVES AND SEDIMENTS

The nature of a sandy beach, the porosity of its sediments, and the ability of animals to burrow into them are all influenced by sediment particle size. Waves play a significant role in determining the types of sediment on a sandy beach. Heavy wave action carries off much of the finer sediment, leaving behind only coarser material. Beaches with fine sand are therefore found along coasts with little wave action or in areas protected from heavy surf. The slope of a beach is determined by the interaction of waves, size of sediment particles, and the relationship of **swash** and **backwash** water. *Swash* is the water running up a beach after a wave breaks, and *backwash* is the water flowing down the beach. Swash brings particles with it that may accumulate on the beach, whereas backwash may remove particles, depending on their size.

A continuum of beach types results from variation in wave action, particle size, and slope **(Figure 13-21)**. At one extreme of this continuum is the **dissipative beach**, where wave energy is strong but is dissipated in a broad surf zone some distance from the beach face. Dissipative beaches have a gentle slope and fine sediments because they receive less wave action and have moderate swash. A **reflective beach** is at the opposite end of this continuum. On a reflective beach, heavy wave action sends large swashes up the beach face, producing a steep slope. The heavy waves directly impact the beach face, as there is no offshore surf zone. Coarse sediment is deposited as swash and backwash water collide. Fine sand remains in suspension and is carried away from the beach. Several intermediate beach types have been described between these extremes.

On all sandy beaches, a cushion of water separates the grains of sand below a certain depth. This is especially true of beaches with fine sand, where capillary action is the greatest. Fine sand beaches, therefore, have a greater abundance of organisms due to both greater water retention and the fact that this type of sediment is more suitable for burrowing. Coarse sand beaches drain well, dry out quickly, and therefore support relatively fewer organisms.

(a)

(b)

FIGURE 13-21 BEACH TYPES. A continuum of beach types results from variation in wave action, sediment particle size, and slope. **(a)** On dissipative beaches, wave energy is strong but is dissipated in a broad surf zone, as seen in the photo. **(b)** Reflective beaches, in contrast, lack a broad surf zone. Heavy wave action sends large swashes of water up the beach face, producing a steep slope. Several intermediate beach types have been described between these extremes.

LIVING IN THE SANDY INTERTIDAL

Most **macrofauna** found on sandy beaches are also called **infauna** because they burrow in the sand **(Figure 13-22)** to survive dry periods and the intense heat of the sun. Typically, infauna reside in permanent or semipermanent tubes or burrows, or they are able to quickly burrow into the sand. They obtain oxygen either through their skin or through gills that are sometimes bathed in water drawn in by elaborate siphons. The most important factor determining their distribution is wave action. On exposed beaches, few large organisms can inhabit the surface layer of sediment because it is in constant motion. Organisms that do live here either are mobile or burrow deeply into the sand. Temperature has much less effect on inhabitants of sandy

shores than those of rocky shores because of both the insulating properties of sand and the effect of water held in the spaces between sediment particles of its deeper layers. For these same reasons, desiccation and changes in salinity are easier to deal with than they are on rocky shores. Oxygen availability in the sediments, however, may be limited where respiration is high and exchange of water with the sea is reduced because of fine sediments. On sand flats and beaches with very fine sediments, exchange of water may be so low that anoxic conditions exist below the uppermost layers. Many animals possess siphons or pump water to oxygenate their burrows, whereas others use specialized respiratory pigments to more efficiently remove oxygen from the interstitial water. Anaerobic bacteria are common in such anoxic conditions and often produce foul-smelling **hydrogen sulfide** gas as a by-product of their breakdown of organic detritus in the sediments.

TIDES AND THE ACTIVITY OF ORGANISMS

Detritus is the main food source for most organisms in the sandy intertidal although, in limited areas, diatoms can sometimes make an important contribution. The activity of organisms on a sandy beach is keyed to the movement of the tides. As the tide moves in, the pace of life begins to quicken. During high tide, bivalves such as cockles (*Cardium*), tellins (family Tellinidae), and surf clams (*Spisula*) project their siphons from their burrows and begin to filter the water for food and bathe their gills with oxygen. Tubeworms and other polychaetes either trap their food by filter and deposit feeding or by searching the sand particles for detritus. Sand dollars use tiny spines and their tube feet to pick

FIGURE 13-22 ANIMAL TRAILS. This trail was produced by a gastropod mollusc known as an auger (*Terebra*). Tracks in the sand such as these indicate the presence of burrowing animals, frequently molluscs, crustaceans, or echinoderms.

Macrofauna are organisms large enough to be seen by the naked eye.

Infauna are organisms that live within the substrate of a marine system.

Hydrogen sulfide (H₂S) is a foul smelling gas produced by the bacterial breakdown of organic matter in the absence of oxygen.

(a)

(b)

FIGURE 13-23 LUGWORM. **(a)** A common burrowing organism in the sandy intertidal zone is the lugworm (*Arenicola marina*). **(b)** The worm ingests sand as it constructs its burrow. The sand passes through the worm's digestive system and is deposited during defecation outside the burrow entrance.

detritus out of the sand. Heart urchins (family Loveniidae) tunnel though the sediments, feeding on detritus that is mixed with the grains of sand. As they travel just beneath the surface of the sand, they angle their spines backward to avoid resistance, leaving a V-shaped trail as they move along the bottom, half buried in sand. Lugworms (*Arenicola marina*, **Figure 13-23a**) are deposit feeders, ingesting sand and removing the organic material that it contains. The sand passes through the worm's digestive system and is deposited during defecation at the

entrance to the worm's burrow. The coiled castings are prominent on sandy beaches at low tide **(Figure 13-23b)**.

Two filter-feeding organisms that move up and down the beach with the movement of the tide are mole crabs (*Emerita*) and coquinas (*Donax*). As the tide moves in, large numbers of mole crabs emerge from the sand **(Figure 13-24)**. Incoming waves carry these creatures up the sand, and as the waves dissipate, the animals force their way backward into the sand

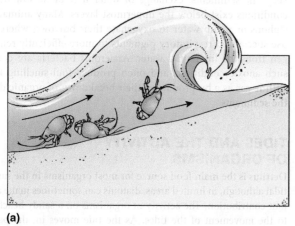

(a)

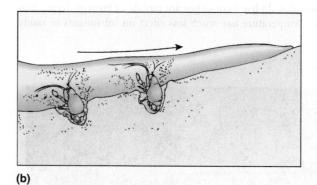

(b)

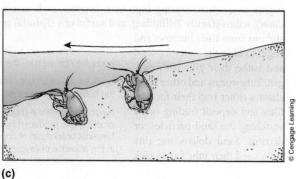

(c)

FIGURE 13-24 FEEDING BEHAVIOR OF MOLE CRABS. Mole crabs are filter feeders that move up and down the beach with the tides. **(a)** Incoming waves carry the mole crabs up the beach. **(b)** As the wave dissipates, the crabs burrow into the sand. **(c)** They then extend their antennae and filter food from the water as the wave retreats.

FIGURE 13-25 COQUINAS. Like mole crabs, coquinas emerge in advance of a wave so the wave will carry them up the beach. As the wave retreats, they burrow into the sand, straining the receding water for food.

and use their large antennae to strain food from the water as it returns to the sea. Mole crabs repeat the process in the opposite direction when the tide retreats. Coquinas move along with the tide in a similar fashion **(Figure 13-25)**. As the water covers the sand, thousands of these tiny bivalves come out of their burrows, and the waves carry them up the beach. Before the backwash can carry them back into the water, they burrow into the sand and put out their siphons to feed. They follow the retreating tide, always staying at the edge of the water.

The incoming tide brings not only food and oxygen but also predators. Carnivorous snails, called whelks (family Buccinidae), glide along the bottom in search of bivalves, which they locate by sensing the currents produced by their prey's siphons. In Florida and the Caribbean, sand sea stars (*Luidia*) feed mostly on small molluscs and crustaceans that they find burrowed in the sand. These sea stars swallow their prey whole and then regurgitate the shells and hard parts. Unlike the sea stars that inhabit rocky shores, these animals lack suckers at the ends of their tube feet, so they are not able to cling to rocks. They are, however, very efficient burrowers and can cover themselves with sand in a matter of seconds. Moon snails (*Polinices*, **Figure 13-26**) and olive snails (*Oliva*) are especially common in this habitat and leave hills of sand that are raised by the animal's foot as it crawls along the bottom. Moon snails feed on bivalves by using their radula to drill a neat hole into the bivalve shell. Once the shell is pierced, the moon snail inserts its proboscis through the opening and rasps out the flesh from inside the shell. The average moon snail consumes enough clams each week to equal one third of its body weight. Olive snails are also carnivorous. They feed on a variety of organisms but especially prefer small bivalves.

At high tide, blue crabs and green crabs (family Portunidae) move in to feed on smaller crustaceans and clams. Small fish feed on crustaceans, worms, and a variety of larval forms. Skates and rays cruise the bottom, preying on crustaceans and molluscs. Gulls and a variety of shorebirds replace these predators when the tide recedes **(Figure 13-27)**.

Ghost crabs (*Ocypode* species; **Figure 13-28**) and fiddler crabs (*Uca*) dominate the upper reaches of the sandy beach in the tropics and warmer parts of the temperate zone. The ghost crab gets its name from its light color and its habit of foraging at night for food. Ghost crabs feed on mole crabs, coquina clams, and a wide variety of other prey. During the day, a ghost crab spends most of its time in its burrow, avoiding the

heat and predatory birds. The ghost crab is not as aquatic as many of its relatives. Its legs are not adapted for swimming but are useful for scurrying sideways along the beach. When the crab enters the water to escape predators, it runs along the bottom. Female ghost crabs deposit their eggs in the ocean, and the young develop in the water. Young crabs rely on currents to carry them about and eventually to deposit them on a beach, where they can take up an adult existence. Once the crabs arrive on land, they return to the sea only occasionally to moisten their gills or to avoid predators. In either case, their visit is short, and they quickly return to the shore. Like its aquatic relatives, the ghost crab extracts oxygen from water through its gills. A special chamber surrounding the gills traps enough seawater to supply the crab's needs until it can enter the water and replenish its supply.

MEIOFAUNA

In addition to the macrofauna described above, many microscopic organisms inhabit the spaces between the sediment particles **(Figure 13-29)**. These organisms constitute the meiofauna and include those animals that pass through a 0.5-millimeter (0.02 inches) screen but are retained by a 62-micrometer (0.0024 inches) screen.

Factors Affecting the Distribution of Meiofauna

The meiofauna are entirely aquatic, so they require water within the interstitial spaces of the sand to survive. Grain size is, therefore, a very important determinant of both the size and types of organisms present there. Coarse-grain sediments have a greater interstitial volume that allows larger organisms to move between the particles, whereas fine-grain sediments have less space and exhibit more burrowing forms (e.g., kinorhynchs). Coarse-grain sediments also drain more rapidly, a factor that may affect both the presence and types of organisms. Other factors affecting the distribution and diversity of the meiofauna include water circulation, oxygen availability, temperature, salinity, and wave action. Fine-grain sediments can inhibit water flow and produce anoxic conditions. Anoxic conditions likewise exist at greater sand depth. The lack of oxygen and space severely limits the types of organisms present. Indeed, it has been found that the meiofauna disappears if the median grain diameter is less than 1 millimeter (0.04 inches). The upper layers of sand are more variable in temperature, but lower layers are well insulated. The fauna of upper layers, therefore, differs from that at lower levels and

FIGURE 13-26 MOON SNAIL. Moon snails, such as this Northern Moon Snail (*Polinices heros*) from Nova Scotia, Canada, feed primarily on bivalves.

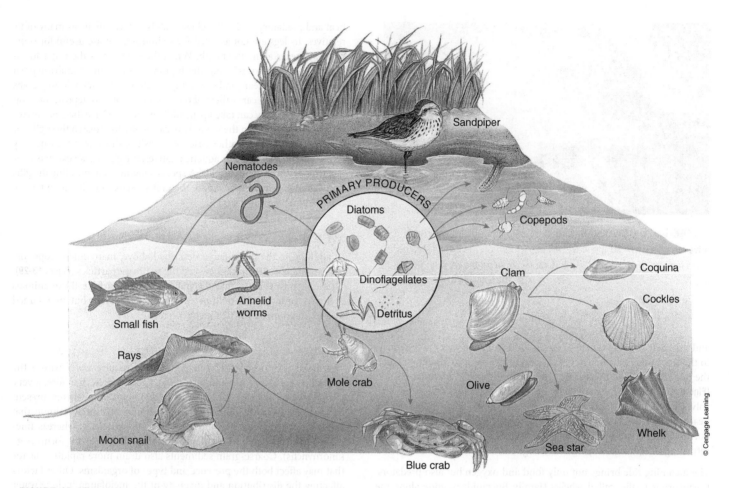

FIGURE 13-27 SANDY SHORE FOOD WEB. Detritus and single-celled algae provide food for numerous burrowing animals. These burrowing animals are a source of food for many larger animals, including crabs, fishes, and birds.

FIGURE 13-28 GHOST CRAB. Ghost crabs dominate the upper intertidal zone of sandy shores in temperate and tropical regions. They spend most of the daytime in their burrows, coming out at night to scavenge for food.

includes forms with wider tolerance to temperature change. Salinity may also affect the distribution and diversity of organisms, particularly in areas of freshwater runoff or if rain is heavy. Finally, wave action can suspend both the sediments and the organisms living there, making them more susceptible to predation.

Characteristics of the Meiofauna

Inhabitants of the interstitial spaces of the sand include invertebrates from many phyla. Ciliates, flatworms, and nematodes are particularly abundant in the meiofauna. Minor phyla represented include the gastrotrichs, kinorhynchs, rotifers, tardigrades, and priapulids, among others. Representatives of the phylum Annelida (both oligochaetes and polychaetes) and several small crustaceans (copepods and ostracods) may also be present. Meiofaunal organisms are generally elongated, with few lateral projections, such as setae or spines, which would interfere with their ability to move freely and quickly in the sand. Many are armored to protect themselves from being crushed by the moving grains of sand. The meiofauna include predators, herbivores, suspension feeders, and detritivores. Because of their small size, the number of offspring produced is very small. Most forms,

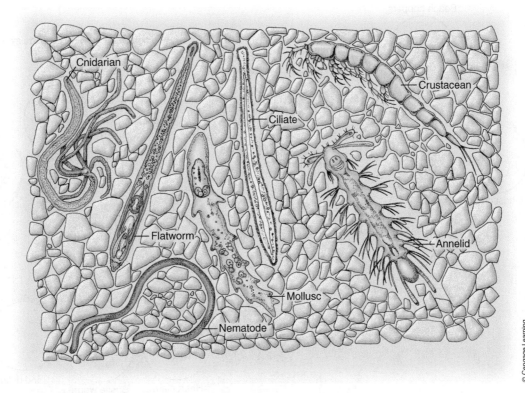

FIGURE 13-29 **MEIOFAUNA.** Meiofauna are tiny organisms that inhabit the spaces between sediment particles. Members of the meiofauna that inhabit sandy shores include tiny protozoans, nematodes, cnidarians, flatworms, annelids, crustaceans, and molluscs.

therefore, exhibit some sort of brood protection. Only a very few forms produce planktonic larvae, and they tend to have relatively larger brood sizes.

Factors Affecting the Size of Meiofaunal Populations

Meiofauna populations exhibit seasonal variations, reaching a peak during summer months. Beaches that are protected from wave action have the greatest abundance of meiofauna. Predation on the meiofauna by the macrofauna can be severe in the upper layers of the sediments. For example, the spot (*Leiostomus xanthurus*), a member of the drum family of fishes (family Sciaenidae), has been demonstrated to significantly reduce the meiofauna in the upper 2 centimeters (0.8 inch).

SANDY SHORE ZONATION

In sharp contrast to the rocky shore, sandy beaches lack a readily apparent pattern of macrofaunal zonation and superficially appear barren and devoid of life. Most studies, however, support the view that some level of macrofaunal zonation exists, but the nature of that zonation varies with beach type. We can typically divide a sandy beach into upper (supralittoral), middle (midlittoral), and lower (sublittoral) intertidal zones. Three physical boundaries are roughly associated with these zones. The upper intertidal zone lies above the high swash line (drift line). The upper boundary is typically where terrestrial vegetation begins but, in some areas, sand dunes **(Figure 13-30)** border the

uppermost extent of this zone. The middle intertidal zone lies between the drift line and the **water table** (also termed the effluent line). The lower intertidal zone lies below the water table and ends at the low-tide swash **(Figure 13-31)**. The width of these zones varies with the type of beach and daily

> The **water table** (or effluent line) is the level below which the ground remains saturated with water.

FIGURE 13-30 **SAND DUNES.** Sand dunes form where waves foster the accumulation of sand and onshore winds blow the sand inland.

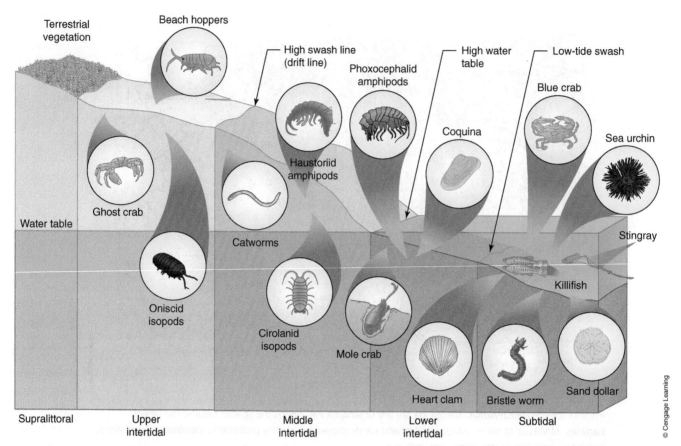

FIGURE 13-31 SANDY SHORE ZONATION. Although sandy shores do not exhibit the obvious pattern of zonation seen on rocky shores, most studies support the view that vertical zonation exists. An upper intertidal zone lies above the high swash line (drift line), and a middle intertidal zone lies between the drift line and the water table. A lower intertidal zone lies below the water table and ends at the low-tide swash. The kind of organisms buried at each level depends on the amount of water trapped, time of year, and type of beach.

FIGURE 13-32 BEACH HOPPERS. Beach hoppers, often called sand fleas, are common inhabitants of the upper intertidal on sandy beaches. These amphipod crustaceans (Talitridae) are mostly nocturnal, scavenging for food at night along the water's edge.

variations in swash level. It should be noted that zonation is only recognized at low tide because, at high tide, zones compress, some populations migrate, and others enter the water column.

The Upper Intertidal: Much of this zone is unsuitable for habitation because the sun bakes the surface sand, both raising its temperature and drying it. Just below this level is a zone of drying sand; moisture reaches this zone only during the highest tides and gradually evaporates over time. Insects, as well as some species of isopods (beach pillbugs) and amphipod crustaceans (beach hoppers, **Figure 13-32**) are the dominant macrofauna in temperate regions. Ghost crabs and fiddler crabs replace many of these species in the tropics and warmer parts of the temperate zone.

Middle Intertidal: The middle intertidal supports a rich fauna on all but reflective beaches with course-grained sediments. Course-grained sediments dry quickly and do not retain enough water to support life. Both an upper zone of retention that retains some moisture at low tide due to capillary action and a lower zone of resurgence that retains a greater amount of water are found here **(Figure 13-33)**. The fauna are true intertidal species not normally found in the lower intertidal. Common inhabitants include various species of isopods, amphipods, and polychaete worms.

Lower Intertidal: The lower intertidal is the upward extension of the surf zone. It is a true marine environment exposed during only the lowest spring tides. The sediments are mostly saturated with water. The variety and distribution of organisms in this zone is primarily influenced by the characteristics of the bottom sediments. Common organisms found include species of bivalves (e.g., coquinas), crabs (e.g., mole crabs), mysids (opossum shrimps), polychaetes, and amphipods. Along some coasts, fields of seagrass are found in the lower portion of the sublittoral zone. In addition to arthropods and molluscs, seagrass beds may host sea urchins, sea stars, brittle stars, sea cucumbers, and anemones. The addition of rocks and broken pieces of coral increases the complexity of the bottom and the diversity of life. Because the lower intertidal is rarely exposed, many species of fish are found here, primarily species that move into the middle intertidal at high tide to feed. The pace of life in the lower intertidal is relatively constant, and during only the lowest ebb tides do these organisms temporarily suspend their activities.

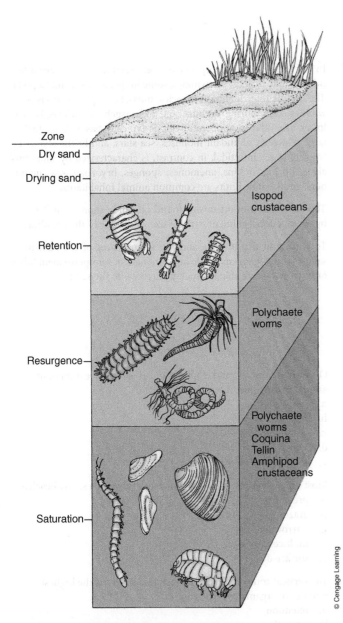

FIGURE 13-33 VERTICAL ZONATION OF THE SANDY INTERTIDAL ZONE. The sandy intertidal zone exhibits vertical zonation, with each zone characterized by its moisture content. The zones of retention, resurgence, and saturation are home to a variety of burrowing organisms.

IN SUMMARY

- The distribution of organisms found on sandy shores is influenced by wave action, sediment particle size, and slope of the coastline. (Have You Wondered? #4)

- Most macrofauna found on sandy beaches burrow in the sand to survive dry periods and the intense heat of the sun and for this reason, they are often termed infauna (Haver You Wondered? #5).

- Detritus, and to a much lesser extent diatoms, are the main food sources for most organisms in the sandy intertidal. Both filter and deposit feeding are common practices among sandy intertidal macrofauna. Their activity pattern is keyed to the movement of the tides.

- The meiofauna are microscopic organisms that live among the sand grains and include representatives from most invertebrate phyla. Their distribution is affected primarily by physical factors rather than biotic factors.

- Although sandy beaches superficially lack a well-defined pattern of macrofaunal zonation, most scientists believe relatively distinct upper, middle, and lower macrofaunal zones can be observed.

KEY CONCEPTS

1. To survive, intertidal organisms must be able to tolerate wave shock, desiccation, and radical changes in temperature and salinity (13-1).

2. Organisms on rocky shores tend to be found in prominent horizontal bands, or zones, on the rocks (13-1).

3. Tide pool organisms must be able to adjust to abrupt changes in temperature, salinity, pH, and oxygen levels (13-1).

4. Biotic factors rather than physical factors are most important in determining the distribution of organisms on rocky shores (13-1).

5. Ecological succession occurs on rocky shores after natural or human induced disturbances strip intertidal rocks of their inhabitants (13-1).

6. Most sandy shore organisms are characterized as infauna because they typically reside in permanent or semipermanent tubes or burrows, or are able to quickly burrow into the sand (13-2).

7. Physical factors rather than biotic factors are most important in determining the distribution of organisms on sandy shores (13-2).

8. Although not readily apparent, most investigators believe that some level of zonation exists on sandy shores (13-2).

ARE YOU STILL WONDERING?

1. Factors affecting the survival of rocky intertidal organisms include wave shock, desiccation during air exposure, overheating, cold, and rapid salinity changes.

2. To tolerate high temperatures and desiccation, rocky intertidal molluscs use such adaptations as large globular bodies that expose less surface area to heating, light-colored shells, beads on the shells that act like radiator fins for cooling, attachment to the surface by mucous threads rather than the foot, adaptations of the kidney, and active orientation of their bodies to minimize sun exposure. Mobile animals move closer to the water or hide in crevices as the tide retreats. Barnacles and mussels close their shells to trap water, allowing it to evaporate slowly for cooling. Rockweeds produce a gelatinous coat to lessen water loss through desiccation.

3. In the rocky intertidal, organisms characteristic of the upper intertidal include lichens, periwinkles, some limpet species, and isopods. The middle intertidal has barnacles, mussels, limpets, various periwinkles, rock boring urchins, and algae such as rockweeds. The lower intertidal in temperate regions typically has a dense turf of large seaweeds, abundant molluscs, sea stars, and brittle stars. The tropical lower intertidal, in contrast, is characterized by short turf algae. Boring urchins, anemones, sponges, bryozoan, sea cucumbers, and keyhole limits are common animal inhabitants.

4. The distribution of organisms found on sandy shores is influenced by wave action, sediment particle size, and slope of the coastline.

5. To withstand dry periods and the intense heat of the sun, most sandy shore organisms reside in permanent or semipermanent tubes or burrows, or are able to quickly burrow into the sand.

QUESTIONS FOR REVIEW

Multiple Choice

1. The width of intertidal zones on rocky shores is dependent on
   a. the amount of exposure to wave action
   b. the slope of the shore
   c. the variation in the tidal range in different areas
   d. all of the above factors determine the width of intertidal zones

2. Prominent herbivores that can be found grazing on algae at the lower edge of the upper intertidal of rocky shores are
   a. mussels
   b. barnacles
   c. periwinkles
   d. ochre sea stars
   e. rock crabs

3. A major competitor of barnacles for space on rocks are the
   a. mussels
   b. oysters
   c. periwinkles
   d. rock crabs
   e. rockweeds

4. Animals that live on intertidal rocks often exhibit compressed or dorsally flattened bodies. This is an adaptation that protects against
   a. sunlight
   b. high temperatures
   c. desiccation
   d. wave shock
   e. predation

5. The most important factor in determining the distribution of life on a sandy beach is
   a. temperature
   b. salinity
   c. pH
   d. wave action
   e. sediment characteristics

6. On warm temperate and tropical sandy beaches, the dominant animal in the upper intertidal zone is the
   a. periwinkle
   b. sea star
   c. sea urchin
   d. ghost crab
   e. mole crab

7. Most organisms that inhabit the intertidal zone of a sandy beach are
   a. predators
   b. grazers
   c. burrowers
   d. multicellular algae
   e. surface dwellers

8. The vertical zone of the sandy beach that supports the highest number of organisms is the zone of
   a. retention
   b. saturation
   c. resurgence
   d. dry sand
   e. drying sand

9. On intertidal rocks, mussels attach themselves with the use of
   a. an enlarged muscular foot
   b. byssal threads
   c. holdfasts
   d. tube feet
   e. flexible spines

10. Which of the following is **true** in reference to dissipative beaches?
    a. they have a gentle slope and fine sediments
    b. heavy wave action sends large swashes up the beach face, producing a steep slope
    c. heavy waves impact the beach face because there is no offshore surf zone
    d. fine sand remains in suspension and is carried away from the beach
    e. both c and d are correct

Matching

Match the following organisms with the adaptation used for living under the harsh conditions of the intertidal environment:

a.	knobby periwinkle	1.	Have light color shells and beads to reduce heat uptake and enhance cooling.
b.	chitons	2.	Cover themselves with shell fragments when exposed to air.
c.	nerites	3.	Clamp down hard on the rocks with a large muscular foot.
d.	rockweeds	4.	Attach to the substrate with a holdfast.
e.	anemones	5.	Have large globular bodies and aggregate in clumps.

a.	rock crabs	1.	Cement themselves to the rock's surface.
b.	limpets	2.	Burrow into the sand.
c.	rock urchins	3.	Hollow out cavities in the rock.
d.	coquinas	4.	Return to their home scar.
e.	barnacles	5.	Move into crevices and sheltered areas as the tide retreats.

Short Answer

1. Describe some adaptations that help organisms inhabiting rocky coasts survive wave shock.

2. Explain how knobby periwinkles avoid desiccation.

3. How does zonation of the rocky shore differ between the temperate zone and the tropics?

4. What environmental challenges do tide pool organisms encounter during the tidal cycle?

5. Compare the vertical zonation on a rocky shore to that of a sandy beach.

Thinking Critically

1. Do you think pollutants entering from the ocean would have a greater effect on rocky shores or on sandy shores? Explain your answer.

2. What abiotic factor is probably most important in terms of influencing the number of organisms that can inhabit a rocky shore?

3. What kinds of organisms that inhabit sandy shores would be least affected by the recreational use of beaches?

SUGGESTIONS FOR FURTHER READING

Aguilera, M. A. and S. A. Navarrete. 2011. Distribution and Activity Patterns in an Intertidal Grazer Assemblage: Influence of Temporal and Spatial Organization on Interspecific Associations, *Marine Ecology Progress Series* 431: 119–136.

Bertness, M. D. 2007. *Atlantic Seashores, Natural History and Ecology.* Princeton, N. J.: Princeton University Press.

Ellis, J. C, M. J. Shulman, M. Wood, J. D. Witman, and S. Lozyniak. 2007. Regulation of Intertidal Food Webs by Avian Predators on New England Rocky Shores, *Ecology* 88(4): 853–863.

Gosner, Kenneth. 1999. *A Field Guide to the Atlantic Seashore: From the Bay of Fundy to Cape Hatteras (Peterson Field Guide).* New York: Houghton Mifflin.

Helmuth, B., N. Mieszkowska, P. Moore, and S. J. Hawkins. 2006. Living on the Edge of Two Changing Worlds: Forecasting the Responses of Rocky Intertidal Ecosystems to Climate Change, *Annual Review of Ecology, Evolution, and Systematics* 37: 373–404.

Horn, M. H., K. L. M. Martin, and M. A. Chotkowski, editors. 1999. *Intertidal Fishes: Life in Two Worlds.* San Diego, Calif: Academic Press.

Kaplan, E. U. 1999. *A Field Guide to Southeastern and Caribbean Seashores.* Boston, Mass.: Houghton Mifflin.

Little, Colin, Gray A. Williams and Cynthia D. Trowbridge. 2009. *The Biology of Rocky Shores (Biology of Habitats).* New York: Oxford University Press.

McLachlan, Anton and Alec Brown. 2006. *The Ecology of Sandy Shores, Second Edition.* Burlington, MA: Academic Press.

Menge, B. A, M. M. Foley, J. Pamplin, G. Murphy, and C. Pennington. 2010. Supply-side Ecology, Barnacle Recruitment, and Rocky Intertidal Community Dynamics: Do Settlement Surface and Limpet Disturbance Matter?, *Journal of Experimental Marine Biology and Ecology* 392: 160–175.

Pilkey, O. H., W. J. Neal, J. T. Kelley, and J. A. G. Cooper. 2011. *The World's Beaches: A Global Guide to the Science of the Shoreline.* Berkeley and Los Angeles, CA: University of California Press.

Shumway, S. W. 2008. *The Naturalist's Guide to the Atlantic Seashore: Beach Ecology from the Gulf of Maine to Cape Hatteras.* Guilford, Conn.: Falcon.

Smith, J., P. Fong, and R. Ambrose. 2008. The Impacts of Human Visitation on Mussel Bed Communities along the California Coast: Are Regulatory Marine Reserves Effective in Protecting These Communities?, *Environmental Management* 41(4): 599–612.

Smith, T. B., J. Purcell, and J. F. Barimo. 2007. The Rocky Intertidal Biota of the Florida Keys: Fifty-two Years of Change after Stephenson and Stephenson (1950), *Bulletin of Marine Science* 80(1): 1–19.

Have You Wondered?

1. **What an estuary is? (14-1)**
2. **How estuaries are formed? (14-1)**
3. **What marine communities you would find in an estuary? (14-4)**
4. **What kinds of marine organisms live in estuaries? (14-3)**
5. **Why estuaries are so important to offshore fisheries? (14-3)**

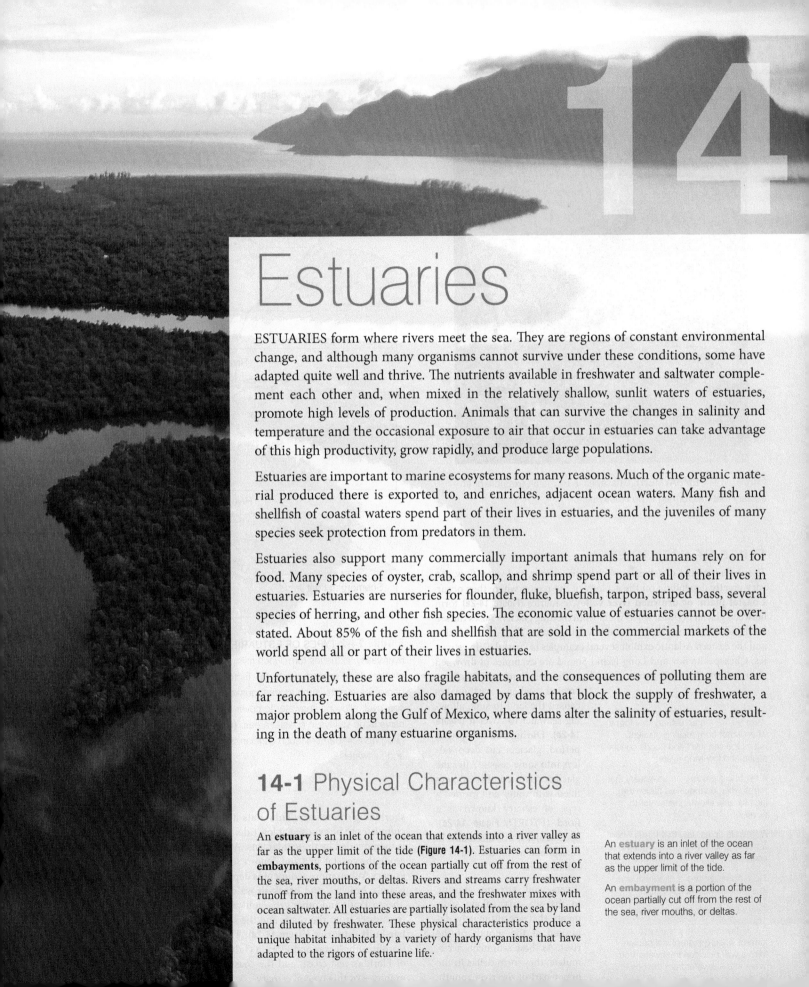

14

Estuaries

ESTUARIES form where rivers meet the sea. They are regions of constant environmental change, and although many organisms cannot survive under these conditions, some have adapted quite well and thrive. The nutrients available in freshwater and saltwater complement each other and, when mixed in the relatively shallow, sunlit waters of estuaries, promote high levels of production. Animals that can survive the changes in salinity and temperature and the occasional exposure to air that occur in estuaries can take advantage of this high productivity, grow rapidly, and produce large populations.

Estuaries are important to marine ecosystems for many reasons. Much of the organic material produced there is exported to, and enriches, adjacent ocean waters. Many fish and shellfish of coastal waters spend part of their lives in estuaries, and the juveniles of many species seek protection from predators in them.

Estuaries also support many commercially important animals that humans rely on for food. Many species of oyster, crab, scallop, and shrimp spend part or all of their lives in estuaries. Estuaries are nurseries for flounder, fluke, bluefish, tarpon, striped bass, several species of herring, and other fish species. The economic value of estuaries cannot be overstated. About 85% of the fish and shellfish that are sold in the commercial markets of the world spend all or part of their lives in estuaries.

Unfortunately, these are also fragile habitats, and the consequences of polluting them are far reaching. Estuaries are also damaged by dams that block the supply of freshwater, a major problem along the Gulf of Mexico, where dams alter the salinity of estuaries, resulting in the death of many estuarine organisms.

14-1 Physical Characteristics of Estuaries

An **estuary** is an inlet of the ocean that extends into a river valley as far as the upper limit of the tide **(Figure 14-1)**. Estuaries can form in **embayments**, portions of the ocean partially cut off from the rest of the sea, river mouths, or deltas. Rivers and streams carry freshwater runoff from the land into these areas, and the freshwater mixes with ocean saltwater. All estuaries are partially isolated from the sea by land and diluted by freshwater. These physical characteristics produce a unique habitat inhabited by a variety of hardy organisms that have adapted to the rigors of estuarine life.·

An **estuary** is an inlet of the ocean that extends into a river valley as far as the upper limit of the tide.

An **embayment** is a portion of the ocean partially cut off from the rest of the sea, river mouths, or deltas.

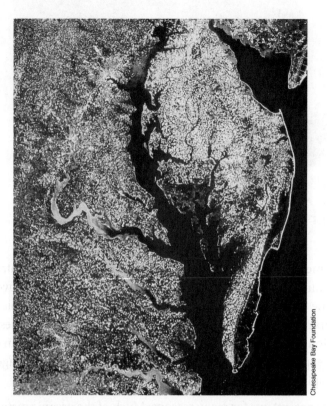

FIGURE 14-1 EMBAYMENT. Chesapeake Bay is an example of an estuary that has formed in an embayment.

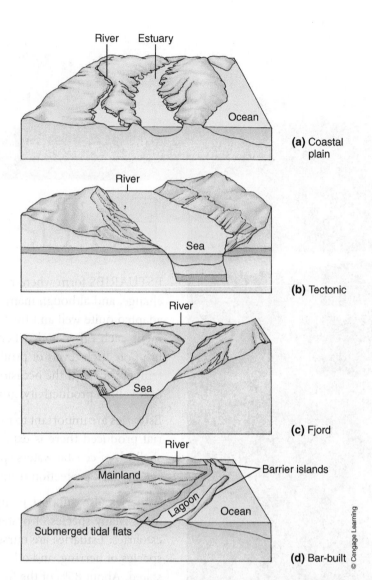

FIGURE 14-2 TYPES OF ESTUARIES. (a) Coastal plain, or drowned river valley, estuaries form when rising water levels from melting glaciers flood coastal plains. **(b)** Tectonic estuaries form when geologica events, such as earthquakes, cause the land to sink below sea level, allowing seawater to cover it. **(c)** Fjords form when glaciers carve large valleys in coastal areas. **(d)** Bar-built estuaries form when geographical barriers, such as islands, form walls between freshwater and saltwater.

TYPES OF ESTUARIES

The characteristics of estuaries, such as size, shape, and water flow, can vary greatly depending on the geology of the region where they occur. **Coastal plain**, or **drowned river valley**, estuaries **(Figure 14-2a)** form between glacial periods, when water from melting glaciers raises the sea level and floods coastal plains and low-lying rivers. The Gulf of Mexico and the eastern Atlantic exhibit several examples of coastal plain estuaries. Chesapeake Bay and Long Island Sound are examples of drowned river valleys. **Tectonic estuaries** such as the San Francisco Bay were created when earthquakes caused the land to sink, allowing seawater to cover it **(Figure 14-2b)**. During the last glacial period, glaciers cut deep valleys into some coasts. After the glaciers retreated, these valleys filled with water and formed a type of estuary known as a **fjord** (FYORD; **Figure 14-2c)**. Spectacular fjords are found in Alaska and along the coasts of Scandinavia.

Rivers and streams flowing to the sea carry along sediments that are ultimately deposited at the mouth of the river. As these sediments accumulate, they form deltas in the upper part of the river mouth,

Coastal plain estuary (drowned river valley) is an estuary that forms when water from melting glaciers raises the sea level and floods coastal plains and low-lying rivers.

A **tectonic estuary** is an estuary that forms when earthquakes cause the land to sink, allowing seawater to cover it.

A **fjord** is an estuary that forms when a glacier cuts a deep valley into the coastline that then fills with water.

Tidal flats are areas of an estuary that are exposed at low tide.

Brackish water is dilute saltwater.

A **bar-built estuary** is an estuary formed when geographical barriers form a wall between freshwater from rivers and saltwater from the ocean.

shortening the estuary. **Tidal flats** develop when enough sediment accumulates to be exposed at low tide. The original channel of the estuary is then divided by these tidal flats. At the same time, currents and tides erode the coastal area and deposit sediment on the seaward side of the estuary. When more sediment is deposited than is carried away, barrier islands, beaches, and **brackish water** (dilute saltwater) lagoons form. These islands and other geographical barriers form a wall between the freshwater from the rivers and the saltwater from the oceans, forming **bar-built estuaries (Figure 14-2d)**. The Cape Hatteras region of North Carolina, the Texas and Florida Gulf Coasts, the Indian River complex of Florida's east coast, and the coast of northern Europe are all good examples of this type of estuary.

SALINITY AND MIXING PATTERNS

As mentioned in Chapter 4, seawater has an average salinity of approximately 35‰. By comparison, the salinity of freshwater ranges from 0.065‰ to 0.30‰. The concentration of ions in river water also varies from one river drainage to the next and affects the chemistry and salinity of the water in estuaries. Although the quantity of dissolved salt in an estuary is about the same as that in seawater, it is distributed in a gradient that increases from the freshwater to the ocean. The salinity in estuaries varies both vertically and horizontally. The least salty waters are located near the mouth of the river where it joins the sea. Farther out to sea, salinity tends to increase. Salinity can be uniform, or it can be layered. Uniform salinity results when currents are strong enough to thoroughly mix the freshwater and saltwater from top to bottom. In some estuaries, vertical salinity is uniform during low tide, but at high tide the seawater at the surface moves upstream more quickly than the bottom water. The denser seawater at the surface tends to sink as the lighter freshwater beneath it rises, creating a mixing action from the surface to the bottom, a phenomenon called **tidal overmixing**. Strong winds can also mix freshwater and saltwater. Normally, though, freshwater flows seaward over the seawater moving upstream.

In most estuaries the influx of freshwater from the river more than replaces the amount of water lost to evaporation. As a result, the surface water is less dense and flows out to sea, whereas the denser saltwater from the ocean moves into the estuary along the bottom. An estuary with this pattern of circulation is called a **positive estuary**. Most estuaries are positive estuaries. Estuaries in hot, arid regions, however, can lose more water through evaporation than the river is able to replace. This type of estuary is called a **negative estuary**. A negative estuary has a flow opposite that of a positive estuary. The surface water flows toward the river, and the water along the bottom moves out to sea. This pattern of circulation occurs because evaporation increases the salinity of the surface water. Some negative estuaries may also experience seasonal or nighttime cooling that increases the density of the water at the surface. The denser surface water sinks and returns to sea along the bottom. The surface water lost to evaporation is replaced by water entering the estuary from the sea. This type of flow pattern traps few nutrients in the estuary, and because the surface water that is drawn in from the ocean is generally nutrient poor, negative estuaries are usually low in productivity. The Laguna Madre estuary in Texas is an example of a negative estuary.

The pattern of water circulation and vertical distribution of salinity are important characteristics of estuaries. On the basis of these mixing patterns, estuaries can be classified into the following types: salt-wedge estuaries, well-mixed estuaries, and partially mixed estuaries.

Salt-Wedge Estuary

Salt-wedge estuaries (Figure 14-3a) occur in the mouths of rivers that are flowing into saltwater. At the surface, freshwater flows rapidly out to sea, whereas at the bottom the denser saltwater flows upstream along the river bottom. The rapid flow of the river prevents saltwater from entering and produces a **salt wedge,** an angled boundary between the freshwater moving downstream and the seawater moving upstream. When the tide rises or the river flow decreases, the salt wedge moves upstream. When the tide falls or the river flow increases, the salt wedge moves downstream. The freshwater normally moves rapidly enough to create a sharp boundary between the freshwater and seawater. The moving freshwater takes saltwater from the face of the salt wedge and mixes it upward with the river water. This action increases the salinity of the surface water that is moving out to sea. Saltwater from the sea is constantly replacing the water removed from the salt wedge, so the salt wedge does not become smaller. Very little river water mixes downward with the saltwater, so the mixing of saltwater with freshwater is essentially a one-way process. In this type of estuary, the circulation and mixing of the water is controlled by the flow rate of the river, and tidal currents play only a small role. Some examples of salt-wedge estuaries are at the mouths of the Mississippi, Amazon, and Congo Rivers and at the mouth of the Sacramento River in San Francisco Bay.

Well-Mixed Estuary

In a **well-mixed estuary (Figure 14-3b)**, river flow is low and tidal currents play a major role in water circulation. The net result is a seaward flow of water and uniform salinity at all depths. The salinity of the water decreases as it approaches the river. Lines of constant salinity move toward land when the tide rises or when river flow decreases. On the other hand, when the tide falls or river flow increases, the lines of salinity move toward the sea. Delaware Bay is an example of a well-mixed estuary.

Partially Mixed Estuary

Partially mixed estuaries have a strong surface flow of freshwater and a strong influx of seawater **(Figure 14-3c)**. Tidal currents force the seawater upward, where it mixes with the surface water, producing a seaward flow of surface water. This system of circulation produces a rapid exchange of surface water between the estuary and the ocean. Salinity is increased by the influx of seawater. Chesapeake Bay, San Francisco Bay, and Puget Sound are examples of partially mixed estuaries.

Other Mixing Patterns

The mixing patterns of estuaries do not always fit neatly into one of the above categories. Some estuaries are intermediate between two of the types we have discussed. Others change from season to season as rainfall, tides, and winds alter the volume of water and strength of currents, as in Galveston Bay, Texas. In fjords, river water remains at the surface and moves seaward, mixing little with the saltwater beneath **(Figure 14-3d)**. Salinity increases slowly in the estuary with the slow influx of ocean water.

TEMPERATURE

Salinity is not the only environmental factor that varies in an estuary. Temperature is also important. Because estuaries are relatively shallow, the water temperature changes rapidly with changes in air temperature. Temperatures in estuaries can fluctuate dramatically seasonally and even daily. For instance, in northern temperate regions, water temperature

Tidal overmixing is a mixing action that occurs in some estuaries at high tide due to seawater at the surface moving upstream more quickly than bottom water from the river.

A **positive estuary** is an estuary in which surface water flows out to the ocean, and the bottom water moves toward the river.

A **negative estuary** is an estuary in which the surface water flows toward the river, and the bottom water moves toward the ocean.

A **salt-wedge estuary** is an estuary in which seawater moves in and out along an angled boundary.

A **salt wedge** is the angled boundary between saltwater and freshwater in a salt-wedge estuary.

A **well-mixed estuary** is an estuary in which there is a seaward flow of water with uniform salinity at all depths.

A **partially mixed estuary** is an estuary with a strong surface flow of freshwater and a strong influx of seawater.

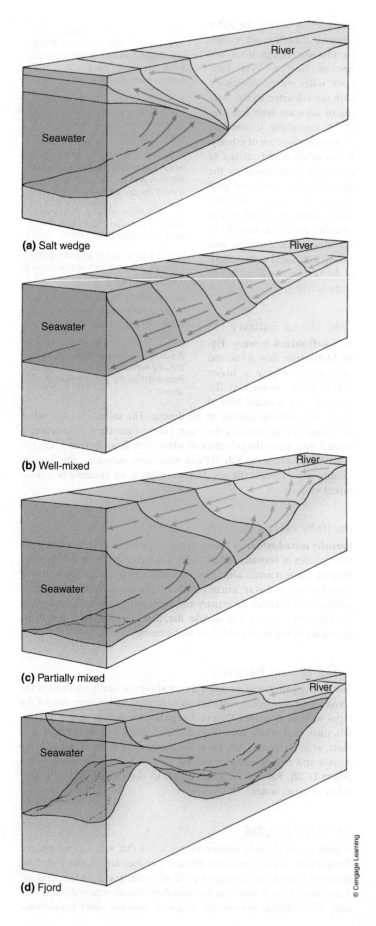

(a) Salt wedge

(b) Well-mixed

(c) Partially mixed

(d) Fjord

© Cengage Learning

FIGURE 14-3 MIXING PATTERNS IN ESTUARIES. Green represents low-salinity freshwater, and blue represents high-salinity saltwater. The darker the blue, the higher the salinity of the water. The arrows indicate the direction of water flow. **(a)** In salt-wedge estuaries, less-dense freshwater from a river flows rapidly to sea over the denser, slower-moving saltwater flowing upstream. This produces an angled boundary (salt wedge) between the two. **(b)** In well-mixed estuaries, river flow is low and tidal currents mix freshwater with saltwater, producing uniform salinity at all depths. **(c)** Partially mixed estuaries have a strong surface flow of freshwater and a strong influx of seawater. **(d)** In a fjord, the river water remains at the surface and mixes very little with the seawater beneath it.

may range from almost 22°C to 30°C, and portions of estuaries may freeze during the winter.

Like other bodies of water, estuaries are heated by the sun. Depending on the season, tidal currents warm or cool an estuary. In some estuaries, surface water is cooler in the winter and warmer in the summer than deeper water, as a result of the amount of solar energy reaching the surface. This situation produces a winter turnover in which cooler (denser) surface water sinks and is replaced by warmer (less dense) deeper water. Without vertical mixing, nutrients would be swept back out to sea with the next tide. The mixing that occurs because of turnover circulates nutrients vertically between the water and bottom sediments.

IN SUMMARY

- An estuary is an inlet of the ocean that extends into a river valley as far as the upper limit of the tide. (Have You Wondered? #1)
- The characteristics of estuaries vary with the geology of the regions in which they occur. (Have You Wondered? #2)
- Salt-wedge estuaries have sharp boundaries between freshwater and saltwater.
- In well-mixed estuaries, there is uniform salinity at all depths and a gradient of decreasing salinity as one proceeds toward the source of freshwater.
- Partially mixed estuaries occur where there is a strong surface flow of freshwater and a strong influx of seawater.
- Fjords have very little mixing between surface freshwater and deeper saltwater.
- Many estuaries exhibit a combination of mixing patterns or have seasonal mixing patterns.
- The salinity of estuaries varies both vertically and horizontally.
- Estuaries also exhibit fluctuations in temperature; in some, seasonal temperature change promotes a turnover that circulates nutrients.

14-2 Estuarine Productivity

Fishers and those who depend on the sea for food have long known how productive estuaries are. Freshwater runoff from the land carries detritus and nutrients, such as nitrogen, phosphorus, and silica, into the estuary. Ocean water often has less nitrogen and silica but more phosphorus and other nutrients, as well as detritus. When strong mixing currents in the shallow sunlit waters of an estuary combine the two types of water, the complementary nutrients support primary production that, together with the large amounts of detritus, can support productive marine communities.

In comparison with coastal seas, primary production in estuaries is low, with the exception of the extensive seagrass beds that form in the mouths of some estuaries. The basis of most estuarine food webs is detritus. The level of dry organic matter in an estuary can be as much as 100 times greater than that of the open ocean. The detritus in an estuarine mudflat can support a biomass of detritivores that is 10 times greater than offshore sediments. This large biomass of detritivores can support equally large numbers of predators such as fish and birds.

Another reason for the high productivity of estuaries is the ability of the sediments and some resident organisms to trap nutrients and organic matter and hold them in the estuary. Rivers dump loads of silt and clay as they meet the sea. These particles readily absorb any excess nutrients from the surrounding water and release them back to the water when nutrients are in short supply. In this way, the silt and clay are a sort of nutrient buffer, helping to maintain a more or less constant nutrient level in an estuary. The activities of some estuarine animals also contribute to keeping nutrients in an estuary. When there are large amounts of phytoplankton in an estuary, filter feeders such as bivalves remove more phytoplankton from the water than they are able to digest. The excess phytoplankton are eliminated in large, semisolid particles called **pseudofeces**. Pseudofeces contain substantial amounts of nutrients and are relatively large, so they can be easily manipulated by other organisms. Thus the normal feeding behavior of bivalves packages and stores nutrients for use by bottom feeders such as gastropod molluscs and polychaete worms.

IN SUMMARY

- Primary production in estuaries is generally low compared with coastal seas, and the basis of most estuarine food chains is detritus.
- The large amounts of detritus along with the primary production that does occur makes these areas some of the most productive in the marine environment.
- Nutrients and organic matter are trapped in estuaries by silt and clay that is brought in by rivers and streams.
- Pseudofeces produced by bivalves trap nutrients and make them available to animals that would not be able to filter them from the water.

14-3 Life in an Estuary

Because of constant fluctuations in the physical environment, such as changing salinity and temperature, estuaries contain fewer resident species than the nearby marine or freshwater ecosystems. Many estuarine

species tend to be generalists, feeding on a variety of foods depending on what is available. Species that can tolerate the salinity and temperature changes in estuaries can exploit the area's rich food supplies, grow rapidly, and multiply into enormous populations. At one time in the rich waters of the Chesapeake Bay, for instance, large populations of oysters (family Ostreidae) and blue crabs (*Callinectes sapidus*) literally carpeted the bottom **(Figure 14-4)**. Today, populations that large are seen infrequently because of pollution, disease, and overfishing. The Great South Bay along the coast of Long Island contains enough clams to support New York's largest single fishery. In both Chesapeake Bay and Great South Bay, the characteristics of the populations are typical of those in estuaries. Estuarine populations contain large numbers of individuals belonging to relatively

Pseudofeces are large semisolid particles containing phytoplankton and detritus that are produced by bivalves.

FIGURE 14-4 ESTUARINE POPULATIONS. Estuarine populations contain large numbers of individuals belonging to relatively few species, such as these blue crabs from the Chesapeake Bay.

few species, and the dominant animals are those with relatively broad tolerances and ecological requirements.

MAINTAINING OSMOTIC BALANCE

Many marine organisms have body fluids that contain about the same concentration of salts as seawater, and their body fluids are essentially isosmotic to the surrounding water. That is, the osmotic pressure of their body fluids is equal to the osmotic pressure of the seawater, and they neither gain nor lose water. Because the marine environment remains relatively constant, they have no problem maintaining water balance. Organisms that live in estuaries, however, must be able to tolerate dilution of their body fluids or have some physiological mechanism for dealing with the varying salinity. Organisms that survive in estuaries have tissues and cells that tolerate dilution (**osmoconformers**) or maintain an optimal salt concentration in their tissues, regardless of the salt content of their environment (**osmoregulators**).

Osmoconformers

Animals such as tunicates, jellyfish, many molluscs, and sea anemones are unable to actively adjust the amount of water in their tissues. When their environment becomes less saline, their body fluid gains water and loses ions until it is isosmotic to the surroundings. These organisms are examples of osmoconformers **(Figure 14-5a)**. Some algae can also survive variations in salinity. The cells of the green alga *Enteromorpha* can change the concentration of their ions and metabolites in response to changes in salinity in their environment. Estuarine *Enteromorpha* also have cell walls that are thinner and able to stretch, allowing them to take in more water when the salinity of their surroundings declines, without damage to the cells. The ability of osmoconformers to inhabit estuaries is limited by their tolerance to dilution of their body fluid.

Osmoregulators

In contrast to osmoconformers, osmoregulators employ a variety of strategies to maintain a constant salt concentration in their bodies. Osmoregulators that live in estuarine waters concentrate salts in their body fluids when the concentration of salts in the surrounding water decreases. For instance, some crabs and fishes regulate their salt content in less-saline water by actively absorbing salt ions through the gills to compensate for salt ions lost from their body **(Figure 14-5b)**. This helps them to maintain a relatively constant body fluid. Some animals can either concentrate salts when their environment is less saline or excrete salts when the environment is hypersaline. The latter are generally semi-terrestrial animals, such as some crustaceans, or live in areas such as salt marshes and mangrove swamps that occasionally receive large amounts of rain. Other animals such as the blue crab and polychaete worms in the genus *Nereis* are osmoregulators at lower environmental salinity and osmoconformers at higher environmental salinity. Many fish species are osmoregulators that can adjust to both high-salt and low-salt environments.

Some estuarine organisms wall themselves off from their external environment to decrease water and salt exchange with their surroundings. Many estuarine animals have a body surface with decreased permeability compared with purely marine forms, the result of increased amounts of calcium in the exoskeleton, as in arthropods, or increased numbers of mucous glands in the skin. Structural adaptations such as

Osmoconformers are organisms whose tissues and cells can tolerate dilution.

Osmoregulators are organisms that expend energy to maintain an optimal salt concentration in their tissues, regardless of the salt content of their environment.

(a)

(b)

FIGURE 14-5 OSMOCONFORMERS AND OSMOREGULATORS.
(a) These tunicates are osmoconformers, losing and gaining water and ions until their body fluid is isosmotic to the environment. **(b)** This hermit crab is an osmoregulator, actively adjusting the concentration of salts in its body fluid when the concentration in the surrounding water changes.

the operculum of a snail can be used to isolate the body surface from the environment when necessary to prevent salt and water loss or gain.

REMAINING STATIONARY IN A CHANGING ENVIRONMENT

In addition to changes in salinity, the problem of remaining stationary in a changing environment affects the distribution of organisms in estuaries. The more or less constant movement of water in an estuary makes it difficult for some organisms to remain stationary long enough to feed and carry on other vital functions. Because of this, natural selection favors benthic organisms. Marine plants and algae in estuaries have substantial root systems, rhizoids, stolons, or holdfasts to prevent moving water from pulling them up and carrying them out to sea. Animals live attached to the bottom, either in the available spaces around other sedentary animals and plants or buried in the small crevices between sediment particles.

Of the nonbenthic animals, crustaceans and fishes are the most dominant, especially their young. These animals usually spawn in the seawater offshore and then spend a portion of their development in the estuary, which is a more protected habitat than the open sea. These animals

Rob & Ann Simpson/Visuals Unlimited, Inc.

FIGURE 14-6 STRIPED BASS. Many commercial species, such as the striped bass (*Morone saxatilis*), spawn and complete the early stages of their development in estuaries where they can take advantage of the available food and relative protection from large predators.

maintain their position in the estuary by actively swimming or by moving back and forth with the movement of the tides.

ESTUARIES AS NURSERIES

Although estuaries are challenging habitats for many animals, they provide excellent nurseries for the juveniles and young of many species to grow and develop. The high level of nutrients and the lower number of predators allow juveniles and young to attain a size or stage that gives them a better chance of survival in the open sea.

The striped bass (*Morone saxatilis*; **Figure 14-6**), for instance, spawns at the border of freshwater and water of low salinity. As the larvae and young mature, they move downstream toward water with higher salinity. The shad (*Alosa*) is an anadromous species that spawns in freshwater but spends its adult life in the marine environment. Young shad spend their first summer in an estuary feeding and growing before moving out to the open ocean. Species such as the croaker (family Sciaenidae) spawn at the mouth of an estuary, and then their young move upstream to feed in plankton-rich, low-salinity water. In the eastern United States, young bluefish (*Pomatomus saltatrix*) come to estuaries to feed but spend the rest of their time in the ocean. Many species of shellfish, such as blue crabs and white shrimp (*Penaeus*), also spend a part of their life cycle developing in the relatively protected waters of estuaries.

IN SUMMARY

- Organisms who live in estuaries must be able to adapt to changing salinity. (Have You Wondered? #4)
- Osmoregulators use a variety of physiological mechanisms to maintain optimum salt concentrations in their tissues, regardless of the salinity of their surroundings.
- Osmoconformers have tissues and cells that can tolerate changes in salinity.
- The characteristics of estuaries tend to favor benthic organisms.

- Motile organisms must actively work to maintain position or move in and out with the tides.
- Because estuaries are highly productive and relatively protected from wave action, they make good nurseries for the juveniles and young of many species.
- Many important commercial fish and shellfish spend at least a portion of their life cycle in the protected waters of an estuary. (Have You Wondered? #5)

14-4 Estuarine Communities

Estuarine communities contain hardy organisms able to tolerate the changeable physical environment in which they live. Many of these organisms are **euryhaline** species. These are species that can tolerate a broad range of salinity such as that in an estuary. Some of the communities found in estuaries include oyster reefs, mud flats, and seagrass meadows.

OYSTER REEFS

A prominent mollusc in the intertidal zone of temperate estuaries is the oyster (family Ostreidae; **Figure 14-7a**). Like other bivalves, oysters have a free-swimming larval stage in their life cycle that allows these sessile animals to disperse to other areas **(Figure 14-7b)**. Oyster larvae attach themselves to any solid surface and form extensive oyster beds, or large reefs composed of oysters growing on the shells of previous generations. Oyster reefs are usually oriented at right angles to tidal currents and occur generally at the point of lower salinity. The currents bring food to the oysters and carry away their waste. Tidal currents also play a role in clearing sediment from the oysters. If the sediments were allowed to accumulate, they would suffocate the oysters. Oyster reefs provide habitat for a variety of other organisms, including algae, sponges, hydrozoans, bryozoans, polychaetes, molluscs, echinoderms, and barnacles. Many of these organisms depend on the oysters not only for protection and a surface for attachment but also for food. The oyster drill snail (*Urosalpinx*) preys on the sedentary oysters, drilling through their shells with its radula and feeding on the contents **(Figure 14-8)**. The veliger larvae of oyster drills are quite sensitive to changes in salinity and are more affected by freshwater than adults. Short periods of low salinity kill predatory oyster drills and disease-causing organisms in the oyster beds. However, rapid changes in salinity as the result of prolonged rainfall or hurricanes kill off large numbers of both oysters and oyster drills, as well as the organisms that grow with them.

MUD FLATS

Mud flats are found in bays and around the mouths of rivers wherever the land is protected from wave action. Mud flats contain rich deposits of organic material mixed with small inorganic sediment grains. Detritus from nearby communities and nutrients carried in from the sea by tides contribute to the rich food reserves. Bacteria and other microorganisms thrive in the mud and produce a variety of sulfur-containing gases that give mud flats a characteristic odor of rotten eggs. Mud provides good mechanical support for animals of the flats, many of which have very thin shells or soft bodies. Mud is also cohesive, permitting the construction of a permanent burrow. Sand is frequently mixed with the mud, making it softer and providing a much better bottom material for burrowing organisms.

Euryhaline species are species that can tolerate a broad range of salinity.

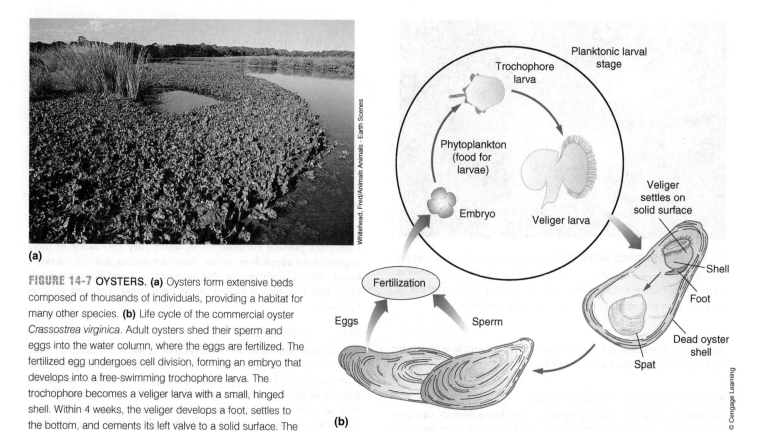

FIGURE 14-7 OYSTERS. (a) Oysters form extensive beds composed of thousands of individuals, providing a habitat for many other species. **(b)** Life cycle of the commercial oyster *Crassostrea virginica*. Adult oysters shed their sperm and eggs into the water column, where the eggs are fertilized. The fertilized egg undergoes cell division, forming an embryo that develops into a free-swimming trochophore larva. The trochophore becomes a veliger larva with a small, hinged shell. Within 4 weeks, the veliger develops a foot, settles to the bottom, and cements its left valve to a solid surface. The veliger then develops into an immature oyster called a *spat*. The spat grows and develops into an adult oyster.

Mud Flat Food Webs

The producers on mud flats consist of photosynthetic bacteria, chemosynthetic bacteria, phytoplankton such as diatoms and dinoflagellates, and, in some cases, large algae such as the green algae *Ulva* and *Enteromorpha* **(Figure 14-9)**. The main energy base, however, is detritus consisting of the decaying remains of local organisms and organic material brought in by river water or deposited during high tides. Much of this organic matter is channeled to other organisms as the result of bacterial decomposition. The action of bacteria is also important in recycling nutrients such as nitrogen

FIGURE 14-8 OYSTER DRILL. Oyster drill snails can cause extensive damage to oyster beds by feeding on the sessile oysters.

and phosphate back to the sea. Bacteria in mud flats are not just producers and decomposers but also primary consumers that serve as a food source for many higher-level consumers. Many deposit-feeding organisms such as nematodes, polychaetes, some gastropods, and arthropods ingest organic material to feed on the bacteria that it contains. A variety of small organisms, including filter-feeding animals and the larvae and juveniles of many animal species, also rely on these rich sources of food. The smaller organisms, in turn, are food for larger invertebrates, fishes, birds, and even some terrestrial carnivores such as raccoons.

Animals of the Mud Flats

Most animal inhabitants of mud flats are burrowing organisms **(Figure 14-10)**. They live just beneath the surface, where they avoid predators and exposure to the drying air during low tides. Unlike sandy beaches, mud flats are not very porous. The silt packs closely together and interferes with the circulation of water that is necessary for carrying oxygen through the sediments. As a result, burrowing animals that exchange gases through their skins generally cannot survive in this habitat unless they circulate water through their burrows or tubes or maintain a "snorkel" connection to the surface.

One common resident of the mud flats of North America is the soft-shelled clam (*Mya arenaria*). Like other bivalves, it is a filter feeder, removing oxygen and planktonic food from water drawn in through its incurrent siphon. Wastes are carried away by the water expelled from its excurrent siphon. When the tide retreats, the soft-shelled clam withdraws its siphon and metabolizes anaerobically. In laboratory experiments, soft-shelled clams have survived as long as eight days without oxygen.

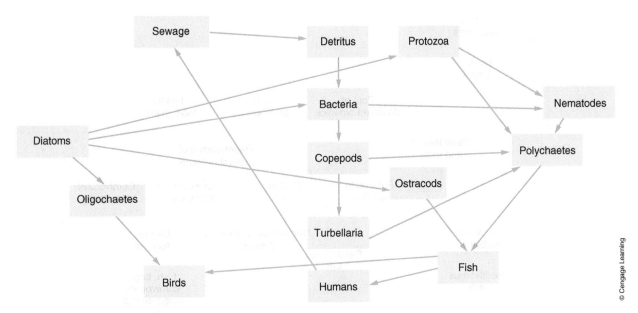

FIGURE 14-9 MUD FLAT FOOD WEBS. The energy base of mud flat food webs is detritus. A large amount of this organic matter is channeled to other organisms as the result of bacterial decomposition.

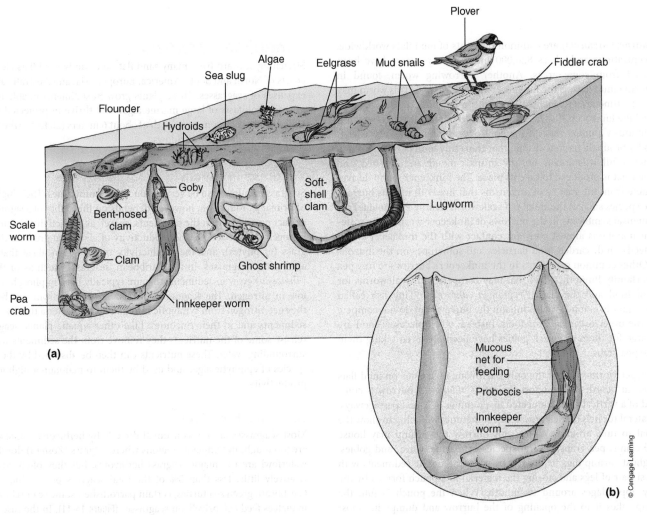

FIGURE 14-10 MUD FLATS. (a) Most inhabitants of mud flats are burrowing animals that feed on detritus or meiofauna. (b) Some, such as the innkeeper worm, live in burrows that provide protection for many other organisms.

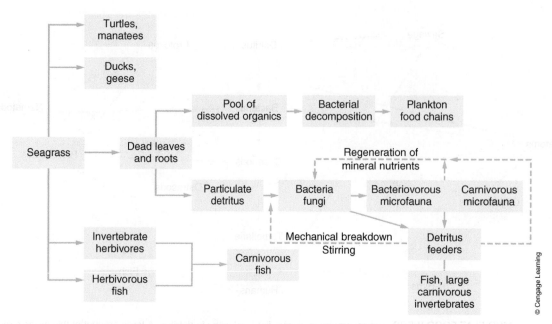

FIGURE 14-11 **SEAGRASS FOOD WEBS.** There are not many animals that feed on seagrass directly. Most of the photosynthetic output from seagrass meadows enters the food web as detritus.

Lugworms (*Arenicola*) are common residents of mud flats worldwide. Concentrations as high as 820,000 individuals per acre have been reported from some areas. Another burrowing worm, found in California mud flats, is the innkeeper worm (*Urechis*). This worm may be 30 centimeters (12 inches) long with a fat, pink, sausage-shaped body. Like lugworms, it lives in a U-shaped burrow through which it continuously pumps water to supply oxygen and remove waste. It feeds by producing a mucous net that traps tiny food particles. When the net is full, it is drawn into the animal's mouth and digested, and another net is produced to take its place. The innkeeper worm derives its name from the variety of organisms that live with it in its burrow. Some species such as the small red scaleworm (family Polynoidae) are commensal symbionts in the burrows of innkeeper worms. The scaleworm maintains almost constant contact with the innkeeper worm and feeds on discarded food particles and sometimes on the mucous net. Other common symbionts in the innkeeper's burrow are tiny pea crabs (family Pinnotheridae) that may compete with scaleworms for bits of food and the clam *Cryptomya californica*. Tiny fish called gobies (family Gobiidae) also inhabit the burrows but do not compete with the other organisms for food. Instead, when pieces of food are too large for them to ingest, gobies have been observed taking them to the pea crabs.

Innkeeper worms are not the only burrowing organisms on mud flats to take in boarders. The ghost shrimp (*Callianassa*) burrow is composed of a vertical shaft connected to a number of lateral passageways. The lateral tunnels branch repeatedly, sometimes widening to allow the animal to turn around. Within this burrow the shrimp may house small clams, pea crabs, several species of marine worm, and gobies. The ghost shrimp digs its tunnels by scooping up the sediments with its first pair of legs and storing the material in a pouch formed by the fleshy appendages around its mouth. When the pouch is full, the shrimp takes it to the opening of the burrow and dumps it. Ghost shrimp and mud shrimp play an important role in oxygenating the sediments.

SEAGRASS MEADOWS

Stretching seaward from many sand flats and sandy mud flats along the coasts of North and South America, Europe, Asia, and Australia are vast expanses of seagrasses. These plants grow in sediments unable to support the holdfasts of large marine algae and thrive in protected waters from the low-tide zone to a depth of about 6 meters (20 feet), where they form large underwater meadows.

Seagrass Productivity

Seagrass meadows can be very productive communities. Their high level of primary production, however, depends on the ability of seagrasses to extract nutrients from the sediments, as well as the activity of symbiotic nitrogen-fixing bacteria, the productivity of algae that grow on the seagrass (epiphytes), and the productivity of the green algae that grow among the seagrasses. In the Caribbean, seagrasses such as turtlegrass (*Thalassia*) grow in sediments that are typically high in phosphorus but low in nitrogen. The seagrasses flourish in this environment because they get nitrogen from symbiotic nitrogen-fixing bacteria that live in the sediments and in their rhizomes. Like other aquatic plants, seagrasses release some of the nutrients they remove from the sediments into the surrounding water. These nutrients can then be absorbed by the many species of epiphytic algae and used by them to maintain a high level of productivity.

Seagrass Food Webs

Most seagrasses are not consumed directly by herbivores because they are too tough. In temperate regions where eelgrass (*Zostera*) dominates, waterfowl are the major seagrass herbivores, but they often consume relatively little, less than 5% of the total seagrass production. In the Caribbean, green sea turtles, certain parrotfishes, some sea urchins, and manatees feed extensively on seagrasses (**Figure 14-11**). In the case of the green sea turtle and the bucktooth parrotfish (*Sparisoma radians*), about 90% of their diet consists of turtlegrass (**Figure 14-12**). Seagrasses respond

FIGURE 14-12 GREEN SEA TURTLE. The green sea turtle (*Chelonia mydas*) is one of the few marine animals to feed directly on seagrass. Its diet consists mainly of turtle grasses (*Thalassia*).

George Karleskint

to the grazing by producing more shoots to replace those that are eaten by herbivores. Although few animals feed directly on seagrasses, they provide a substantial food source for many organisms in the form of detritus. The dead leaves and other plant parts provide a food base for bacteria, crabs, brittle sea stars, sea cucumbers, and deposit-feeding worms. Some of the detritus is exported to nearby communities. Many species of snail—both herbivores such as conchs and carnivores such as whelks and tulip snails—thrive in Caribbean and temperate seagrass communities. As snails fall prey to carnivores, their shells become available for a variety of hermit crab species that are mainly scavengers. During high tide, predators such as eels, flounder, and other large fishes feed at the seagrass beds.

Seagrass Meadows as Habitat

The surfaces of seagrasses provide a place of attachment for many tiny organisms (epiphytes and epifauna; **Figure 14-13**). Older leaves may be completely covered by hydrozoans, tube worms, bryozoans, tunicates, and red algae. Juvenile scallops (*Argopecten*) often attach to seagrass leaves by byssal threads. Those attached well above the bottom sediments tend to avoid predation by benthic predators such as crabs. The epiphytes and epifauna on some turtlegrasses in the Caribbean may block enough radiant energy from reaching the plant to interfere with photosynthesis.

The sand among the blades of seagrass is home to a wide variety of filter feeders, such as adult scallops, jackknife clams (*Ensis*), and some species of sea cucumber **(Figure 14-14)**. Seagrass rhizoids and root complexes provide sites for more permanent attachments by tube worms and mussels. The thick growths of seagrass rhizomes make it difficult for small predators to dig out their prey. As a result, the number of species is greater in seagrass meadows than in areas where the sediments are exposed.

The high productivity of seagrass communities allows them to support a large and diverse group of organisms, including the larval and juvenile stages of many animal species. The combination of changing salinity, available hiding places, and shallower water allows small juveniles to be protected from larger predators. In the seagrass nurseries these animals feed and grow without the heavy predation pressure that will face them when they leave their protection and enter the sea.

ROBERT SISSON/National Geographic Stock

FIGURE 14-13 EPIPHYTES AND EPIFAUNA. Many organisms such as small epiphytic algae and epifauna such as bryozoans will attach to the surface of seagrasses like this turtle grass.

IN SUMMARY

- Estuaries support a variety of distinct communities that include oyster reefs, mud flats, and seagrass communities. (Have You Wondered? #3)

- Each community has unique characteristics and supports distinctive populations of producer and consumer organisms that have adapted to life in each habitat.

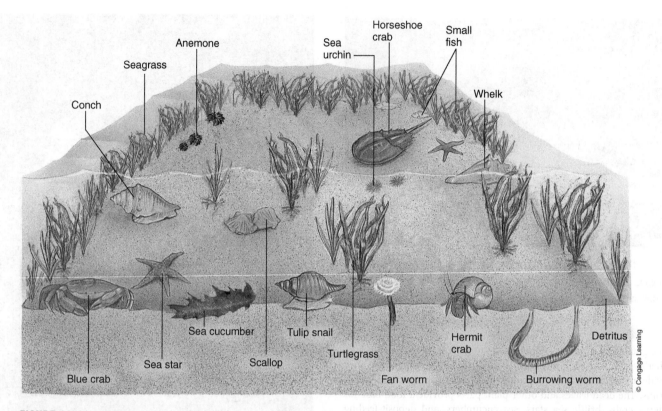

FIGURE 14-14 **SEAGRASS COMMUNITY.** The sandy mud in which seagrasses grow provides habitat to a wide variety of animals. Although the seagrass is too tough for most animals to eat, it is an important source of detritus.

14-5 Wetlands

Salt marshes and mangrove swamps are examples of **wetlands**, areas of land covered with water all or part of the year. Coastal wetlands are very productive ecosystems and supply large amounts of food in the form of detritus to both estuaries and offshore communities. They act as nurseries for many important commercial fish and shellfish. Evidence strongly suggests that coastal wetlands act as buffers in floods and can suppress the impact of tsunamis and hurricanes by absorbing wave energy and storing excess storm water, thus lessening the potential damage from these natural disasters.

SALT MARSH COMMUNITIES

Salt marsh communities are found on the shoreward side of mud flats in temperate and subarctic regions of the world. The dominant plant life in this community on many North American shores is cordgrass (*Spartina*) that has moved from land to the shallow intertidal area **(Figure 14-15)**.

Wetlands are land areas covered with water all or part of the year.

Many commercially important fish and shellfish spend a portion of their life cycle in the protection of salt marshes. The Wadden Sea, Europe's largest coastal salt marsh, is estimated to yield $140 million per year in North Sea fish. It is estimated that each acre of U.S. salt marsh has a commercial value of between $50,000 and $80,000 per year, and the value of fish associated with U.S. salt marshes is estimated to be several billion dollars.

FIGURE 14-15 **A SALT MARSH.** Wetlands known as salt marshes are found bordering estuaries in the temperate regions of the world. In U.S. salt marshes, the dominant plant is the cordgrass *Spartina*.

ECOLOGY & THE MARINE ENVIRONMENT

Predation Regulates Benthic Population Size

Robert Virnstein was interested in the population dynamics of benthic communities. He wanted to know whether physical environment, competition, predation, or some other factor most influenced the size of infaunal populations in estuaries. While doing fieldwork in the Chesapeake Bay estuary, Dr. Virnstein observed that populations of infauna living in shallow-water subtidal communities contained fewer animals in areas that lacked any vegetation for cover compared with similar areas nearby that were covered by vegetation. He reasoned that in the vegetated areas predators such as crabs would have difficulty digging through the rhizome mat of submerged grasses as they searched for food, so the animals in those habitats would be more protected. He also observed that the densities of infaunal populations were lowest in summer and fall, when predators that feed on bottom-dwelling prey such as crabs and fish have been feeding, and highest in winter and spring, when the predators were either absent or inactive.

On the basis of these observations, Dr. Virnstein hypothesized that large motile predators played a significant role in controlling the number of infaunal animals in this community. To test his hypothesis, Dr. Virnstein used wire-mesh cages to either exclude predators from or confine predators to a small area of estuary bottom. The specific predators that he studied were two fishes, the hogchoker (*Trinectes maculatus*) and the spot (*Leiostomus xanthurus*), and the blue crab (*Callinectes sapidus*). He selected the two fish species because they feed on the bottom and were abundant in the area he was studying. He selected the blue crab because it is known to feed heavily on infauna. For his study he selected a shallow, sandy, unvegetated area of bottom in the lower York River that he thought was representative of the Chesapeake Bay and its estuaries. He marked off 50 × 50–centimeter plots spaced 3 meters apart. Some plots were covered with wire cages

to keep predators out, whereas others were covered with wire cages containing one to four predators.

After 2½ weeks, he took samples from each plot and counted the number of infaunal animals and the number of different species in each, then compared his results to areas that had not been enclosed by a wire cage. He found that the number of species had not changed, but the number of animals in areas protected from predators was significantly greater than areas that were not enclosed by cages and significantly less in the caged areas that contained crabs and spot. The plots were sampled again at the end of 2 months and the same measurements , made. This time the differences were even more striking. As before, the cages that excluded predators showed large increases in the numbers of infauna compared with areas that were not enclosed by a cage. The caged areas containing crabs had the least infauna. The cages containing the hogchoker showed the same results as cages containing no predators, indicating that the hogchokers had little controlling influence on the infauna. The cages containing the spot had results that were intermediate between crabs and hogchokers. The infauna that lived near the surface were most affected by predation. Species that could bury deeply into the sediments and thus avoid predation showed few significant differences.

Based on the results of his experiments, Dr. Virnstein concluded that, although the physical stresses of the estuarine environment may be severe, they are not major factors limiting natural population densities of this community. In the York River community, densities of most species increased when protected from predators and no species decreased in density, suggesting a lack of competitive exclusion. He concluded that competitive pressures are not very important in the regulation of population densities in this community and that predation plays the major role.

Distribution of Salt Marsh Plants

Salt marshes can be divided into two regions: the **low marsh**, covered by tidal water much of the day and typically flushed twice each day by the tides, and the **high marsh**, covered briefly by saltwater each day and flushed only by the spring tides. In the marshlands of the Atlantic coast and the Gulf Coast of Florida, the low marsh is dominated by the tall cordgrass *Spartina alterniflora,* which can grow as high as 3 meters (10 feet), whereas the species *Spartina foliosa* is dominant along the coast of California. In the high marsh along both coasts is a thick carpet of short fine grasses that usually do not grow taller than 60 centimeters (2 feet). These may include salt meadow hay (*Spartina patens*), pickleweed (*Salicornia*), and spike grass (*Distichlis spicata*).

Along the Florida Gulf Coast, the most common species in the high marsh is *Juncus*. In areas where tall cordgrass grows thickly, and leaves, stalks, and debris are flushed by tidal currents, the water is very clear. Beds of short cordgrass are not flushed frequently by tidal currents, so dead leaves and debris accumulate on the marsh surface,

forming a moist mat that is an ideal habitat for many species. Other primary producers in salt marshes include green algae and benthic diatoms.

Salt Marsh Productivity

The changing tides bring fresh nutrients into salt marshes, where they are mixed and redistributed. This replenishing action helps to support a high level of production by marsh plants, green algae, and benthic diatoms. Experiments have shown that a Connecticut salt marsh can annually produce 3 to 7 tons of grass and algae per acre, and a typical southern salt marsh may produce as much as 10 tons of vegetation per acre. By comparison, a prime wheat field in the Midwest might

Low marsh is an area of salt marsh covered by tidal water much of the day and typically flushed twice each day by the tides.

High marsh is an area of salt marsh that is covered briefly by saltwater each day and only flushed by the spring tides.

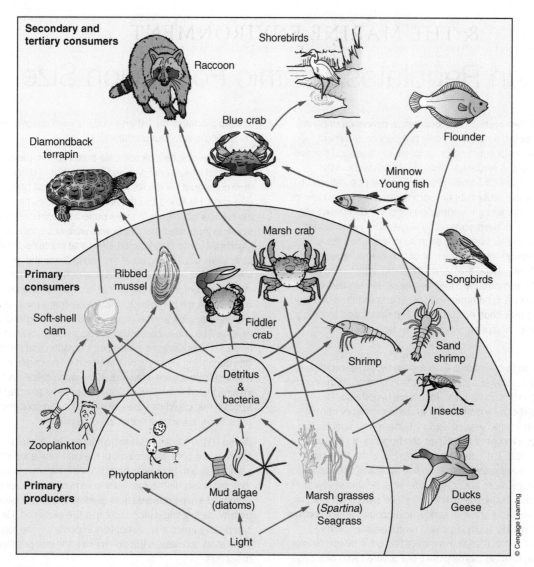

FIGURE 14-16 **SALT MARSH FOOD WEB.** As with other estuarine food webs, most of the primary production of salt marshes enters the marine environment as detritus. The detritus supports large numbers of bacteria, which in turn are food for numerous organisms that channel the nutrients to higher-order consumers.

produce 7 tons of vegetation annually. About 4% of the plant material in a salt marsh is consumed by insects, and another 1% to 2% is consumed by other herbivores **(Figure 14-16)**. Most of the primary production of salt marshes, however, supports detrital food chains. The dead and decaying plant material is a nutrient source for bacteria, which in turn are eaten by a host of deposit feeders. The bacteria convert some of the detritus into bacterial biomass, and the rest is decomposed to a mix of organic molecules (dissolved organic matter) and inorganic ions (inorganic nutrients). Some salt marshes export large amounts of nutrients in the form of detritus to nearby communities. The exported detritus primarily supports bacteria and deposit feeders in the neighboring waters. Along the east coast of the United States, as much as 40% of the primary production of a salt marsh is exported in the form of dead leaves and dissolved organic matter. Salt marshes farther north export

very little, and most of the primary production is consumed by resident organisms in the marsh.

Animals of the Salt Marsh

Marsh grasses form the basis of a complex and distinct community **(Figure 14-17)**. Some animals that live in this community are permanent residents, whereas others come to the salt marsh only at high tide or low tide to take advantage of the wealth of food.

Permanent Residents On the southern Atlantic coast, marsh periwinkle snails (*Littorina irrorata*) feed on algae that grow on the surface of marsh grass leaves **(Figure 14-18a)**. The marsh periwinkle, as with most periwinkles, survives out of water because its mantle cavity can function as a type of lung, allowing it to absorb oxygen from the air. At low

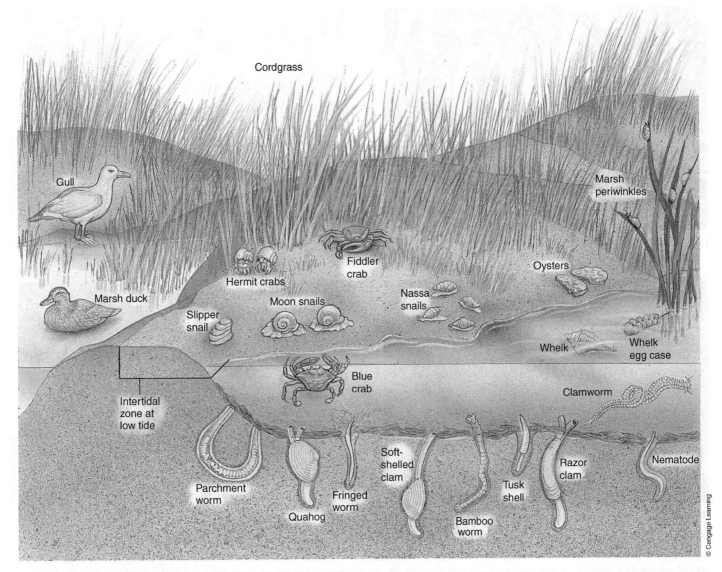

FIGURE 14-17 THE SALT MARSH COMMUNITY. Salt marshes provide habitat for a number of animals. Burrowing animals dominate the low marsh, whereas crabs, periwinkles, and marsh snails are common in the high marsh.

tide it moves to the underside of the leaves, where it is protected from the sun and predators. As the tide moves in, it crawls higher up the plant to avoid being submerged. When high spring tides submerge the entire plant and the snail, it can go without breathing for about an hour—long enough to survive until the tide begins to retreat. Marsh periwinkles exhibit an interesting behavior in that they appear to be able to sense an impending flood tide and crawl to the highest parts of the plant even before the water begins to rise. Periwinkles greatly restrict marsh grass productivity. In experiments in which the snails were removed from the marsh grass and prevented from returning, productivity increased as much as 25%.

The tidal marsh snail (*Melampus bidentatus*) is the dominant gastropod mollusc in many Atlantic coast marshes, frequently attaining densities of more than 1,000 per square meter **(Figure 14-18b)**. This snail is somewhat unusual in that, although it leads a largely terrestrial existence, it possesses an aquatic larval stage. Adults lay eggs in the water, and larvae hatch and begin to develop in the water column during spring tides when at least part of the high marsh is flooded. The larvae later settle back onto the marsh to take up a more terrestrial lifestyle.

Burrowed in the mud of the low marsh along the Atlantic coast is the ribbed mussel (*Geukensia demissa*). At low tide, the mussel closes its valves to trap moisture and prevent desiccation. During high tide, it opens its valves to filter food particles from the water. Particles that are not acceptable for food are trapped in mucous ribbons and ejected from the mussel as pseudofeces. Another prominent bivalve in some salt marshes along the Atlantic coast is the razor clam (*Solen*) **(Figure 14-18c)**. Like ribbed mussels, these clams burrow into the mud and filter feed at high tide. It is not unusual to find large clamworms (*Nereis*), sometimes as long as 19 centimeters (7.5 inches), feeding on the razor clams.

(a)

Jack Dermid/Visuals Unlimited, Inc.

(c)

Andrew J. Martinez/Photo Researchers, Inc.

(b)

Lightwave Photography, Inc./Animals Animals - Earth Scenes

(d)

Courtesy of David Campbell

FIGURE 14-18 ANIMALS OF THE SALT MARSH. **(a)** The marsh periwinkle (*Littorina irrorata*) is common on the cordgrass where it grazes on small algae. **(b)** The tidal marsh snail (*Melampus bidentatus*) is the dominant mollusc in Atlantic salt marshes. **(c)** Razor clams (*Solen*) are common inhabitants of salt marshes, where they filter feed on plankton at high tide. **(d)** Fiddler crabs (*Uca*) are also common inhabitants of salt marshes. The large claw (cheliped) of the male is used to attract mates.

At low tide in more southern salt marshes, purple marsh crabs (*Sesarma reticulatum*) come out of their burrows to feed on the marsh grass. The fiddler crab (*Uca*; **Figure 14-18d**) is a prominent inhabitant of salt marshes, where it digs its burrow at the base of tall cordgrass. As it excavates its burrow, it forms the mud being removed into neatly packed pellets that it carries to the opening and stacks around it. During high tide, the crab fashions a door from some of the mud pellets and seals the entrance, trapping air in the burrow. At low tide, hundreds to thousands of these animals emerge from their burrows and scurry across the mud in search of food. They are omnivores and feed on detritus, algae, and small animals.

The fiddler crab is named for the one oversized claw (cheliped) of the male, which is used for attracting a mate and defending his territory against other males. Fiddler crabs can exchange gas in both air and water and can survive several weeks without being immersed in water. The space beneath its carapace and just above the legs forms a lung cavity that traps air and has a rich blood supply. While the lining of the lung cavity remains moist, the crab can exchange sufficient quantities of gas across it to survive out of water.

Burrowing animals such as the fiddler crab play a vital role in the overall marsh ecology. Their burrowing activities constantly bring nutrient-rich mud from deeper down to the surface, where the nutrients have been depleted, and their burrowing allows oxygen to penetrate to deeper sediments. Amphipods called sandhoppers (*Orchestia*) are also abundant in some salt marshes. These small arthropods are common around the base of cordgrass and under debris. They feed on detritus and are an important source of food for several species of marsh bird. Another prominent arthropod is the grass shrimp (*Palaemonetes*), which is common in the water associated with both salt marshes and seagrass meadows. Seasonally, edible shrimp of the genus *Penaeus* are also very common. These animals feed on detritus and plankton and are important sources of food for fish and birds.

Tidal Visitors to the Salt Marsh Predators that feed in the salt marsh vary with the tides. The most important predators in salt marshes at low tide are birds. Marsh wrens (*Telmatodytes* and *Cistothorus*), clapper rails (*Rallus longirostris*), red-winged blackbirds (*Agelaius phoenicius*), and seaside sparrows (*Ammospiza maritima*) all nest in the tall marsh grass. Marsh hawks (*Circus cyaneus*) feed on the many rodents in this area.

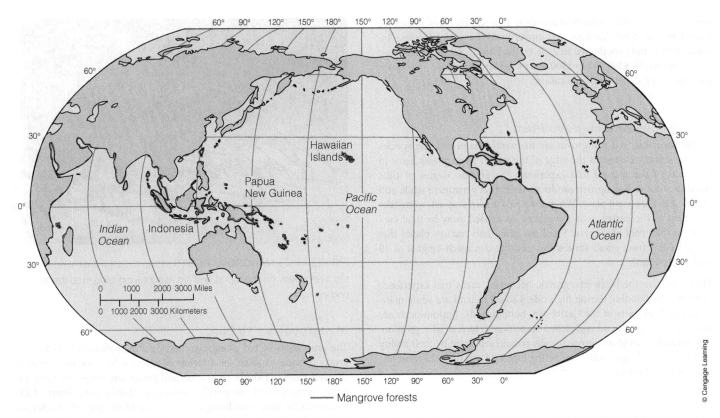

FIGURE 14-19 **MANGALS.** Mangrove forests, or mangals, are wetlands that are found in tropical regions of the world where there is little wave action and where sediments accumulate, and muddy sediments lack oxygen.

At low tide along the Atlantic coast, terrestrial predators such as diamondback terrapins (*Malaclemys terrapin*), raccoons (*Procyon*), otters (*Lutra*), and mink (*Mustela vison*) enter salt marshes in search of prey. Many insect and spider species are residents of salt marshes. These animals are not only predators but also food to many species of reptile, bird, and mammal. Herbivorous animals from the land such as swamp rabbits (*Sylvilagus aquaticus*), white-footed mice (*Peromyscus*), and sometimes deer (*Cervus*) browse on marsh plants at low tide. At high tide, fishes are the chief higher-order predators. One of the more important fish predators is the killifish (*Fundulus heteroclitis*). The killifish feeds at high tide on small fish, marsh snails, and various amphipods, restricting them to the high marsh where the water is too shallow for the killifish to enter. Blue crabs also come to the flooded marsh at high tide to feed on benthic invertebrates.

Succession in Salt Marshes

Salt marshes can be the first stage in a succession process that will eventually produce more land. The roots of marsh plants act as a sediment trap, holding sediments carried down by rivers and preventing their removal by tidal currents out to sea. Over time, the area becomes built up with sand and silt; this in turn becomes mud as it is enriched with decaying material from dead plants and animals. Eventually, small islands of mud appear, and as the cordgrass traps more sediment, the size of these islands increases and they begin to merge. As land masses build, high tide covers less and less of them. Tall cordgrass is ultimately replaced by short cordgrass, and that, in turn, is replaced by rushes. At this point, land plants begin to establish themselves, followed by an influx of terrestrial animals.

MANGROVE COMMUNITIES

Mangrove forests, or mangals, replace salt marshes in tropical regions, where they can cover as much as 75% of the coastline **(Figure 14-19)**. Mangrove forests appear where there is little wave action and where sediments accumulate and muddy sediments lack oxygen. The most highly developed mangals supporting the greatest number of mangrove species are along the coasts of Malaysia and Indonesia **(Figure 14-20)**. Mangals in this region may contain as many as 40 species or more and

FIGURE 14-20 **A PACIFIC MANGAL.** Mangrove forests such as this may contain as many as 40 different species of mangroves. This complex habitat is home to many species of marine and terrestrial life.

may exhibit a definite pattern of zonation. Like salt marshes, mangrove forests provide food, habitats, and nursery sites for many species. According to a study conducted in 2006 by the United Nations, mangrove forests are worth $200,000 to $900,000 per square kilometer (0.4 square mile) per year, depending on location.

Distribution of Mangrove Plants

In the Americas, red mangroves are frequently the pioneering species and are usually closest to the edge of the water. These plants grow in the parts of the mangal that experience the greatest degree of tidal flooding. The red mangrove produces seeds that germinate while still attached to the parent plant. When the seeds finally drop off, some take root next to the parent, whereas others are carried away by tidal currents to take root elsewhere. Seedlings grow into mature plants that range in size from small shrubs to trees that can reach heights of 30 meters (100 feet).

Shoreward are the black mangroves, occupying areas that experience only shallow flooding during high tide. Closest to land are white mangroves and, in parts of the Caribbean, buttonwoods. Buttonwoods are not true mangroves and represent a transition to terrestrial vegetation. The various types of mangrove usually remain separated by their ability to tolerate flooding by saltwater during high tide and their different tolerances to soil salinity.

Mangrove Root Systems

Because the mud in which mangroves grow is soft and oxygen poor, their roots are adapted to anchor the trees firmly and to supply sufficient oxygen to those parts of the plant buried in the mud. To help anchor them, mangroves have root systems that are shallow and widely spread. Red mangroves gain extra support from prop roots **(Figure 14-21)**, which grow from the trunk and branches. Prop roots also aid in gas exchange for buried roots. Black mangroves have many erect, aerial roots called pneumatophores that branch from horizontal roots beneath the mud. The pneumatophores not only help support the mangrove but also exchange gases for the buried roots. Without prop roots and pneumatophores, the buried roots of these plants would suffocate, and the plants would die.

Prop roots and pneumatophores form a tangle that slows the movement of water, causing suspended materials to sink to the bottom. The roots collect various sediments and organic material. This, in turn, slows the movement of tidal waters, allowing more sediment to build up. Eventually the area becomes a terrestrial habitat, as the colonizing mangroves continue their growth toward the sea.

Mangal Productivity

The primary producers in mangals are mangrove plants, algae, and diatoms. As is the case with salt marsh plants, the tough leaves of mangroves are hard for most organisms to digest. In tropical regions, burrowing and climbing crabs feed on mangrove leaves, but most of the leaves and other detritus are removed by tidal currents to the surrounding water and become the basis of a detritus food web **(Figure 14-22)**. This very productive food web supports a variety of important commercial fish and shellfish such as blue crab, shrimp, spiny lobster, mullet (*Mugil*), spotted sea trout (*Cynoscion nebulosus*), and red drum (*Sciaenops ocellatus*).

Richard N. Mariscal

FIGURE 14-21 MANGROVE ROOTS. Prop roots (shown here) give red mangroves extra support and aid in supplying oxygen to the roots underground.

Mangroves as Habitat

The prop roots of red mangroves and pneumatophores of black mangroves provide a habitat for a variety of animals. In mangrove forests of the west coast of Florida, prop roots and pneumatophores eventually become encrusted with the purple coon oyster (*Lopha frons*; **Figure 14-23**), which gets its name from being a favorite food of raccoons. Barnacles and mussels that filter feed when the tide is high compete with coon oysters for space on the roots of red and black mangroves. On the east coast of Florida, mussels and acorn barnacles are dominant. Periwinkle (*Littorina*) snails that are related to the marsh periwinkle graze on algae growing on the stems and prop roots of the mangroves and growing on the shells of the sedentary organisms attached to them. The king's crown conch (*Melongena corona*), a carnivorous snail, feeds on oysters by prying open their valves with its strong muscular foot and then digesting the contents. Mangrove crabs (*Aratus*) go through their larval stages in the water beneath the mangroves. When mature, they crawl up the mangroves and feed on the leaves. Mangrove roots provide a habitat for many of the same organisms found on mud flats and in salt marshes **(Figure 14-24)**. These include fiddler crabs, hermit crabs, marsh crabs, marsh snails, and ghost shrimp. Some animals that cannot tolerate the varying salinity of the salt marsh and mud flats can survive in the more stable environment of the mangrove forest. These include sea stars, brittle stars, and sea squirts.

In the mangals of Malaysia and Indonesia, fish known as mud skippers (*Periophthalmus chrysospilos*) live burrowed in the mud **(Figure 14-25)**. These fish come out of their burrows at low tide and scoot around on the surface of the moist mud, behaving more like amphibians than fish. The sheltered waters around mangrove roots are also an ideal nursery for crab larvae, shrimp, and fishes. On Florida's southern coasts, the role of mangroves as a nursery is equal to that of seagrass communities.

The rich food supplies of mangals along the Gulf Coast of the United States attract a variety of predators, including clapper rails, killifishes (family Cyprinodontidae), diamondback terrapins, water moccasins (*Agkistrodon*), and raccoons. Birds such as pelicans, herons, egrets, roseate spoonbills (*Ajaia ajaia*), and wood ibises (*Mycteria americana*) also find the upper branches of mangroves an ideal nesting site. Before much of this habitat was destroyed by drainage projects and pollution, mangals along the west coast of the Everglades provided major nesting sites or rookeries for many bird species.

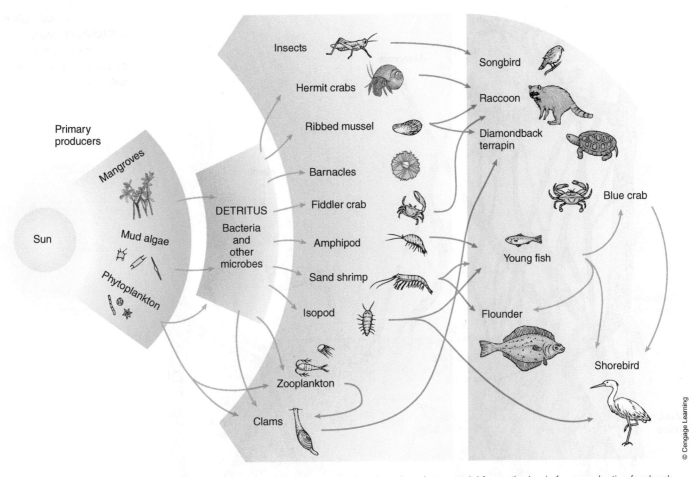

FIGURE 14-22 MANGROVE FOOD WEB. Detritus from falling leaves and other decaying plant material forms the basis for a productive food web in mangrove communities.

FIGURE 14-23 MANGROVE ORGANISMS. Many organisms, such as these oysters and barnacles, use mangrove roots for habitat.

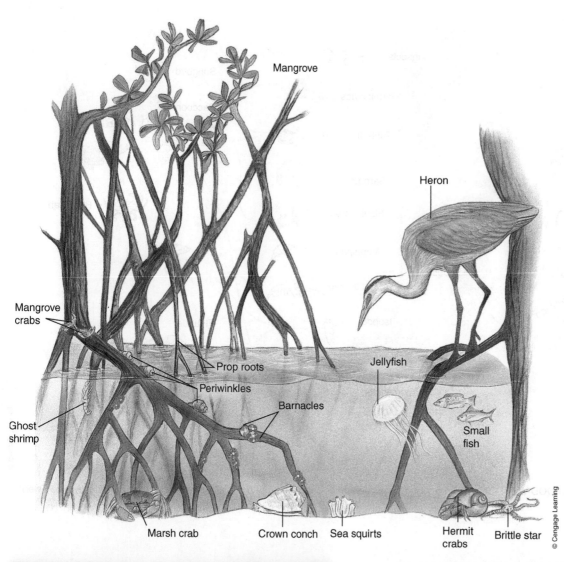

Mangrove

Heron

Mangrove crabs

Prop roots

Periwinkles

Jellyfish

Barnacles

Ghost shrimp

Small fish

Marsh crab

Crown conch

Sea squirts

Hermit crabs

Brittle star

© Cengage Learning

FIGURE 14-24 MANGROVE COMMUNITY. Many animals find shelter as well as food in the complex root system of mangroves.

JASON EDWARDS/National Geographic Stock

FIGURE 14-25 MUD SKIPPER. Mud skippers (*Periophthalmus chrysospilos*), residents of mangrove swamps in Asia, can spend time out of water by exchanging gases at their moist gills.

IN SUMMARY

- Wetlands are land areas covered with water all or part of the year.

- Salt marshes and mangrove swamps are examples of wetlands.

- Wetlands play important roles in producing food, providing habitat, and acting as nurseries for many species.

- Wetlands may also help protect terrestrial habitats from tsunami and hurricane damage.

14-6 Lagoons

Another common brackish water habitat that is similar to estuaries is the lagoon. A **lagoon** is a relatively small area of the ocean that has been partially isolated by the development of a barrier **(Figure 14-26)**. Lagoons can form when sand is deposited to build a bar that is not connected to land (similar to a bar-built estuary) or when sediments are deposited at the shore that then extend seaward, eventually

IMPACT BUBBLE

WETLANDS SHARE CHARACTERISTICS of both terrestrial and aquatic habitats. They serve many valuable ecological and economic functions, including maintaining water quality, protecting shorelines, providing habitat and nursery areas for animals, and providing recreational areas. Wetlands maintain water quality by filtering runoff from the land and removing sediments, nutrients, and pollutants. They reduce the severity of floods on the land side by acting as natural retention areas, and they reduce coastal flooding by absorbing energy from hurricane storm surges. This latter function is particularly important in coastal areas that are flat or low-lying. Coastal wetlands such as salt marshes and mangals serve as important nurseries for fish and shellfish. Approximately 75% of the nation's commercial fish and shellfish depend on wetlands or estuaries at some stage in their life cycle, and estuaries depend on adjacent wetlands to maintain water quality and provide the basis for detrital food chains. Coastal wetlands provide habitat for many species of animal, including migrating birds and many threatened or endangered species. They also provide for many recreational activities such as fishing, hunting, birding, boating, and wildlife photography.

Coastal wetlands are currently being stressed by several factors, including dredging and filling for coastal development, damage due to storms and tidal surges, and oil spills. Between 1998 and 2004, saltwater wetlands in the United States declined by nearly 45,000 acres (about 1.2%). The Gulf coast experienced a greater loss of coastal wetlands than other coastal areas due to development and the impact of hurricanes. Coastal wetlands and the organisms that live in them are quite vulnerable to destruction or alteration by oil spills. Oil covers the plants and can penetrate into the sediments, exposing the plant's roots to toxins in the oil. Plants and roots coated with oil die by

suffocation, and wetland loss occurs. Without plants to hold the soil in place and stabilize the shoreline, erosion occurs. The loss of vegetation allows the soil to become flooded with seawater, preventing many plants from recolonizing the area. As erosion increases, more coastal areas become flooded and more wetlands are lost, giving rise to a vicious cycle of destruction. Oil that reaches the wetlands also affects the animal population and is especially damaging to benthic organisms that reside in and on the sediments and are important components of wetland food chains.

Coastal wetlands are vital to the economy of many coastal states. In Louisiana, where wetland change and loss have been severe for decades and where wetlands are being lost at a more rapid rate than other areas, the impact of the Deepwater Horizon disaster has added even more stress to these fragile areas. The effects of the oil spilled onto the coast from the Deepwater Horizon well is still being assessed, and researchers are continuing to evaluate the environmental effects of the oil on this and other coastal ecosystems. Certain bacteria are capable of breaking down some of the compounds in oil, but this process removes oxygen from already low-oxygen environments, further stressing organisms in these communities. Studies currently underway to determine the long-term consequences of oil damage on large-scale ecosystems will be used to develop appropriate responses should there be future spills, not only in the Gulf of Mexico but in other ocean environments where oil and gas exploration occurs as well. The results of this research can assist various agencies in their work to remediate the consequences of oil damage in coastal and marine environments and in their attempts to restore these vital economic areas.

Bob Gibbons / Alamy

FIGURE 14-26 LAGOON. A brackish water lagoon is a relatively small area of the ocean that has been partially isolated by the development of a barrier and contains diluted saltwater. It is a habitat similar to an estuary.

enclosing a portion of the ocean. Lagoons are very shallow and exhibit a wide range of salinity. In the tropics, lagoons are often fringed by mangroves, whereas in temperate areas reed beds are the dominant plant life. Since lagoons are smaller, shallower, and typically more stable than estuaries they contain a more diverse plant community. The animal species that populate lagoons are generally the same as those found in neighboring waters, although some species, such as the lagoon cockle (*Cerastoderma glaucum*), are found only in lagoons.

> A **lagoon** is a relatively small area of the ocean that has been partially isolated by a barrier.

IN SUMMARY

- Lagoons are brackish water habitats that are similar to estuaries.

KEY CONCEPTS

1. Estuaries form where freshwater from rivers and streams mixes with seawater. (14-1)

2. The salinity of water in estuaries varies both vertically and horizontally. (14-1)

3. Mixing of nutrients from saltwater and freshwater, combined with plentiful sunlight and relatively shallow water, makes estuaries very productive ecosystems. (14-2)

4. Estuarine communities include oyster reefs, mud flats, and seagrass meadows. (14-4)

5. Wetlands such as salt marshes and mangrove forests (mangals) are frequently found bordering estuaries. (14-5)

6. The basis of many estuarine food chains is detritus. (14-2)

7. Animals and plants that live in estuaries must be able to adapt to changing salinity. (14-3)

8. The physical characteristics of estuaries tend to favor benthic organisms. (14-3)

9. Many commercially valuable fish and shellfish spend a portion of their life cycle in estuaries and wetlands. (14-3)

10. Lagoons are brackish water habitats that are similar to estuaries. (14-6)

ARE YOU STILL WONDERING?

1. Estuaries are portions of the ocean that extend into river valleys as far as the upper limit of the tide.

2. Estuaries can form when sea levels rise and flood coastal areas; when earthquakes cause portions of the coast to sink; when glaciers carve deep valleys into coasts that flood with seawater; or when a natural barrier cuts off a portion of the ocean.

3. Estuaries are homes to oyster reefs, mudflats, seagrass communities, salt marshes, and mangrove swamps.

4. The organisms that live in estuaries are typically species with broad tolerances and ecological requirements, including crabs, bivalves, snails, polychaetes, shrimp, and some fish species, to name a few.

5. Since estuaries contain large amounts of food and relatively few large predators, they are excellent nursery grounds for many commercial fish and shellfish.

QUESTIONS FOR REVIEW

Multiple Choice

1. In a well-mixed estuary, vertical salinity is
   a. stratified
   b. uniform

2. The environmental factor that affects the types of organisms in estuaries more than anything else is
   a. sunlight
   b. temperature
   c. salinity
   d. nutrient density
   e. type of sediment

3. Most organisms that inhabit mud flats are
   a. active swimmers
   b. terrestrial
   c. burrowers
   d. predators
   e. immobile

4. The mangrove species usually closest to the water in American mangrove forests is the
   a. white mangrove
   b. black mangrove
   c. red mangrove
   d. green mangrove
   e. buttonwood

5. An example of a deposit feeder found on mud flats would be the
   a. jackknife clam
   b. fiddler crab
   c. lugworm
   d. soft-shelled clam
   e. innkeeper worm

6. An animal that feeds directly on seagrasses is the
   a. queen conch
   b. green sea turtle
   c. sea cucumber
   d. fiddler crab
   e. mullet

7. If the concentration of salts in an animal's body tissues varies with the salinity of the environment, the animal would be an
   a. osmoregulator
   b. osmoconformer

8. Pseudofeces are formed by
   a. fish
   b. crabs
   c. bivalves
   d. snails
   e. anemones

Matching

Match each of the following terms with the appropriate definition:

a. embayment

1. A silt deposit at the mouth of a river.

b. estuary

2. An estuary that forms when a glacier cuts a deep valley in the coastline.

c. tidal flat

3. A part of the ocean partially cut off from the rest of the sea.

d. delta

4. Areas of an estuary that are exposed at low tide.

e. fjord

5. An inlet of the ocean that extends into a river valley as far as the upper tidal limit.

a. positive estuary

6. Estuary formed when a geographical barrier forms between freshwater from rivers and saltwater from the ocean.

b. tectonic estuary

7. Estuary in which the seawater moves in and out along an angled boundary.

c. salt-wedge estuary

8. An estuary in which surface water flows out to sea and the bottom water flows toward the river.

d. well-mixed estuary

9. An estuary that forms when earthquakes cause land to sink, allowing seawater to cover it.

e. bar-built estuary

10. An estuary in which there is a seaward flow of water with uniform salinity at all depths.

Short Answer

1. What are the distinguishing characteristics of an estuary?

2. What is a salt-wedge estuary, and how does it differ from other types of estuaries?

3. What factor(s) contributes to the productivity of estuaries?

4. What adaptations have evolved in mangroves that help them survive in their habitat?

5. Explain how osmoconformers survive in estuaries.

6. Explain how fiddler crabs are well adapted to life in the salt marsh.

7. Describe the process of succession in a salt marsh.

8. Compare a lagoon with an estuary.

9. Sketch a chart that traces energy flow in a mud flat.

10. What important roles do wetlands play?

Thinking Critically

1. Why is it difficult for burrowing animals that exchange gas through their skin to survive in mud flats?

2. Predict the effect that agricultural runoff would have on a neighboring estuary.

3. To control flooding, a series of dams is constructed along a river that feeds a large estuary. What effect do you think the dams will have on the estuary's productivity?

4. What effect would a hurricane likely have on the animal life in an estuary? Explain your answer.

SUGGESTIONS FOR FURTHER READING

Bertness, B., R. Sillman, and R. Jeffries. 2004. Salt Marshes Under Siege, *American Scientist* 92 (1): 54–61.

Bertness, M. D. 1999. *The Ecology of Atlantic Shorelines*. Sunderland, Mass.: Sinauer Publishing.

Buccheri, D. 2006. *Wetlands—The Coast's Great Protector, Oceans* 3: 6–13.

Cronk, J. K., and M. S. Fennessy. 2001. *Wetland Plants: Biology and Ecology*. Boca Raton, Fla.: Lewis Publishers.

Dybas, C. L. 2002. Florida's Indian River Lagoon: An Estuary in Transition, *Bioscience* 52(7).

Hart, J., and D. Sanger. 2003. *San Francisco Bay: Portrait of an Estuary*. Berkeley, Ca.: University of California Press.

Horton, T. 2005. Saving the Chesapeake, *National Geographic* 207 (6): 22–45.

Hughes, J. E., L. A. Deegan, B. J. Peterson, R. M. Holmes, and B. Fry. 2000. Nitrogen Flow through the Food Web in the Oligohaline Zone of a New England Estuary, *Ecology* 8(2).

Reddy, C. 2007. Still Toxic After All These Years, *Oceanus* 45 (3): 26–29.

Warne, K. 2007. Forests of the Tide, *National Geographic* 211 (2): 132–151.

Williams, T. 2006. The Last Line of Defense, *Audubon* 108 (3): 56–59.

Have You Wondered?

1. How corals obtain their food energy? (15-1)

2. Are there different types of coral reefs? (15-2)

3. Why coral reefs are seldom found below 60 meters (196 ft.)? (15-3)

4. How reefs in the Atlantic Ocean differ from those in the Pacific? (15-4)

5. What are the most important primary producers on coral reefs? (15-5)

6. Why coral reefs are important to humans? (15-8)

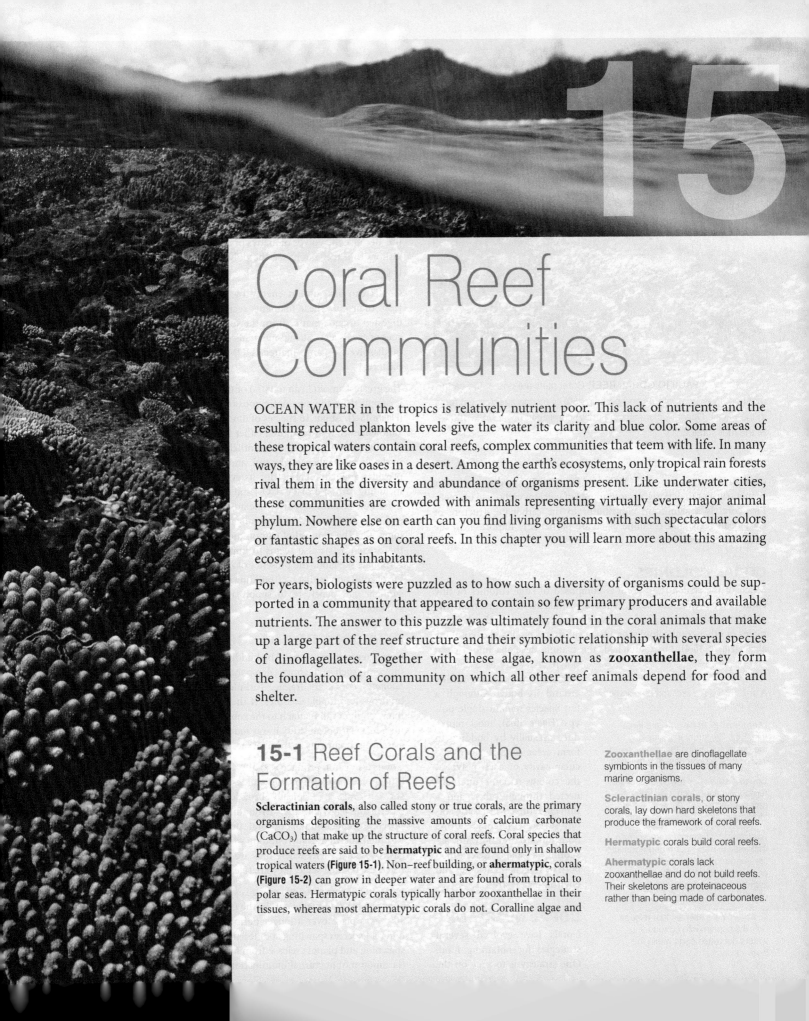

15

Coral Reef Communities

OCEAN WATER in the tropics is relatively nutrient poor. This lack of nutrients and the resulting reduced plankton levels give the water its clarity and blue color. Some areas of these tropical waters contain coral reefs, complex communities that teem with life. In many ways, they are like oases in a desert. Among the earth's ecosystems, only tropical rain forests rival them in the diversity and abundance of organisms present. Like underwater cities, these communities are crowded with animals representing virtually every major animal phylum. Nowhere else on earth can you find living organisms with such spectacular colors or fantastic shapes as on coral reefs. In this chapter you will learn more about this amazing ecosystem and its inhabitants.

For years, biologists were puzzled as to how such a diversity of organisms could be supported in a community that appeared to contain so few primary producers and available nutrients. The answer to this puzzle was ultimately found in the coral animals that make up a large part of the reef structure and their symbiotic relationship with several species of dinoflagellates. Together with these algae, known as **zooxanthellae**, they form the foundation of a community on which all other reef animals depend for food and shelter.

15-1 Reef Corals and the Formation of Reefs

Scleractinian corals, also called stony or true corals, are the primary organisms depositing the massive amounts of calcium carbonate ($CaCO_3$) that make up the structure of coral reefs. Coral species that produce reefs are said to be **hermatypic** and are found only in shallow tropical waters **(Figure 15-1)**. Non–reef building, or **ahermatypic**, corals **(Figure 15-2)** can grow in deeper water and are found from tropical to polar seas. Hermatypic corals typically harbor zooxanthellae in their tissues, whereas most ahermatypic corals do not. Coralline algae and

Zooxanthellae are dinoflagellate symbionts in the tissues of many marine organisms.

Scleractinian corals, or stony corals, lay down hard skeletons that produce the framework of coral reefs.

Hermatypic corals build coral reefs.

Ahermatypic corals lack zooxanthellae and do not build reefs. Their skeletons are proteinaceous rather than being made of carbonates.

Mark Conlin/Alamy

FIGURE 15-1 PACIFIC CORAL REEF. Coral reefs are one of the earth's most complex ecosystems and contain 25% of marine species. On reefs in the Pacific Ocean, Scleractinian (stony) corals can cover up to 100% of the reef surface.

organisms such as fire coral (*Millepora*; **Figure 15-3**), a hydrozoan rather than a true coral, deposit lesser amounts of calcium carbonate on reefs. Fire coral is a significant contributor to the carbonate structure of Caribbean reefs, and coralline algae are particularly important to the structure of Pacific reefs.

CORAL COLONIES

Much of the structure of coral is composed of large colonies of tiny coral polyps. Colonies begin when a planula larva settles from the plankton and attaches itself to a solid surface, frequently the dead remains of other coral colonies or a hard substrate composed of coralline algae **(Figure 15-4)**. Once attached, the planula develops into a coral polyp and secretes a cup of calcium carbonate, a **corallite**, around its body. Although a few species remain single polyps **(Figure 15-5)**, most reproduce asexually by budding to form colonies. The gastrovascular cavity of each polyp in the colony remains interconnected through tubes, and a thin epidermis overlies the colony surface. The colony grows both upward and outward, assuming a shape determined both by its genetic makeup and environmental stresses.

A **corallite** is the skeleton of a single coral polyp.

Mesenterial filaments are long, thin, tube like structures attached to the gut of cnidarians that are involved in digestion, absorption, and defense.

Fragmentation is a type of asexual reproduction in which a part of the parent breaks off and forms a new individual.

Broadcast spawners are animals that release their sperm and eggs into the surrounding seawater, where the eggs are fertilized and the resulting zygotes becomes part of the plankton.

Brooders are animals that release only their sperm into surrounding waters but retain eggs within the body cavity.

CORAL NUTRITION

Corals have evolved several strategies for obtaining food. One strategy is to prey on the tiny zooplankton or other small organisms that pass over the coral colony and venture too close to tentacles containing cnidocytes. Any small animal that brushes against these cells is paralyzed, captured by the tentacles, and moved to the mouth. Some coral polyps have short tentacles that produce mucus and are covered with cilia. During feeding, these cilia beat toward the mouth, creating a current that traps small organisms, as well as detritus, within the mucus and draws it into the mouth. When not being used for feeding, the cilia can reverse their motion so that silt and other debris are driven away from the polyp. When they are inactive or disturbed, polyps withdraw into their corallite. Because the polyps of many coral species are nocturnal feeders, day visitors to a reef rarely see the polyps extended outside the corallites **(Figure 15-6)**.

Corals can also obtain energy by extruding **mesenterial filaments** from their gut. These filaments secrete digestive enzymes and absorb digested organic matter from the sediments. Bacteria living in the tissues of corals also take up nutrients directly from the water in the form of dissolved organic matter (DOM), and corals can feed on these bacteria.

The greatest amount (up to 90%) of the nutritional needs of stony corals is supplied by the symbiotic zooxanthellae living within their tissues **(Figure 15-7)**. About 16 species of dinoflagellates are known to form zooxanthellae. They are acquired either from the environment or inherited directly from the parent. Zooxanthellae provide nutrients to the polyps in the form of glucose, glycerol, and amino acids, and the coral polyp provides a suitable habitat for its symbiont and a variety of nutrients, including carbon dioxide (CO_2) and nitrogen. The zooxanthellae absorb these nutrients directly from the animal's tissues so efficiently that several species of coral release virtually no nitrogen wastes to the surrounding water. The removal of carbon dioxide and production of oxygen by the zooxanthellae also seems to improve the growth rate of the coral's carbonate skeleton. The presence of zooxanthellae is so important to the life of coral polyps that reef-building corals cannot grow any deeper than sunlight can penetrate to supply the photosynthetic needs of their symbionts.

REPRODUCTION IN CORAL

Corals reproduce asexually by budding and by **fragmentation**. Branching corals such as staghorn (*Acropora cervicornis*) and elkhorn (*Acropora palmata*) corals are fragile. Turbulent water from storms can break branches and topple colonies. If the storms do not kill the fragments, many of them can reattach to the substrate and grow, forming new colonies. In a Caribbean study it was found that twice as many new colonies arise from fragments as from larvae.

Corals also can reproduce sexually. Many coral species are hermaphroditic, possessing both ovaries and testes, but other species have separate sexes. Corals are typically **broadcast spawners**, releasing gametes (sperm and eggs) into the surrounding seawater where the eggs are fertilized. A smaller number of coral species are **brooders**. Brooders broadcast their sperm into surrounding waters but retain their eggs within the gastrovascular cavity where they are fertilized. When the fertilized eggs reach the planula stage, they are released into the open water.

On many Pacific reefs, spawning is synchronous among species and occurs as a brief annual event. In the Caribbean, however, spawning usually takes place over a longer period of time and is nonsynchronous, different species spawning at different times. The exact time of broadcast spawning and planula release from brooders seems to be determined by the amount of nocturnal illumination.

FIGURE 15-2 AHERMATYPIC CORALS. Ahermatypic corals such as **(a)** gorgonians (sea fans) and **(b)** black corals have skeletons made of protein rather than calcium carbonate. Gorgonians are abundant on shallow Caribbean reefs, but black coral is found in deeper water. Black coral is becoming increasingly rare due to its use in making jewelry. **(c)** Delicate branching soft corals such as the ones shown here are more prevalent on Pacific reefs.

FIGURE 15-3 FIRE CORAL. Fire corals (*Millipora*) are not true corals but members of the class Hydrozoa. They are ubiquitous on Caribbean reefs and are important in the deposition of carbonate sediments. The name *fire coral* comes from the fact that contact produces a painful welt.

Matthew Landau

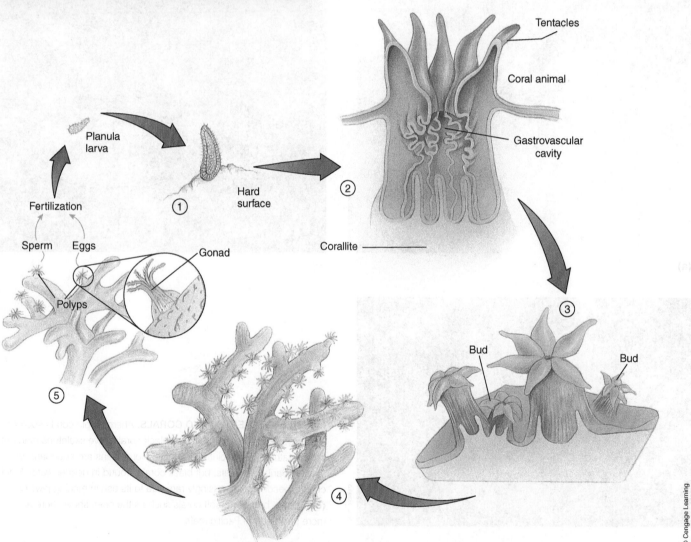

© Cengage Learning

FIGURE 15-4 FORMATION OF A CORAL COLONY. (1) A coral colony begins when a planula larva attaches to a hard surface. (2) The planula develops into a polyp, and the polyp secretes a cup of calcium carbonate. (3) The polyp reproduces asexually by budding, and each new polyp forms a calcareous cup around its body. (4) The colony continues to grow as the new polyps repeat the budding process. The shape of the colony is genetically determined. (5) As the colony continues to grow, some polyps develop gonads. The sperm and eggs produced by these polyps are typically shed into the surrounding water, where the eggs are fertilized. The fertilized eggs develop into new planula larvae that begin the process of colony formation again.

Richard N. Mariscal

FIGURE 15-5 SOLITARY STONY CORALS. Some stony corals, such as this mushroom coral, are composed of a single polyp and do not form colonies.

Richard N. Mariscal

FIGURE 15-6 CORAL POLYP FEEDING. At night, coral polyps extend their tentacles and feed on plankton.

REEF FORMATION

The formation of a coral reef involves both constructive and destructive phases. It is not a simple matter of new coral building on top of existing coral but rather a series of events involving many animal and algal species. Although stony corals form the basic material of the reef, many smaller species provide significant amounts of additional calcium carbonate to its structure. Large colonies of stony corals, coral species with solitary polyps, and many other skeleton-building and shell-building organisms all contribute to the structure of the reef.

The destructive phase of reef formation is called **bioerosion**. It is usually underway well before the death of a coral colony **(Figure 15-8)**. Boring clams or sponges regularly attack the exposed surfaces on the underside of large coral stands. These vulnerable areas are relatively bare and do not receive enough sunlight to support the growth of coral polyps. As

Peter Parks/Oxford Scientific/Getty Images

FIGURE 15-7 CORAL POLYP AND SYMBIONTS. Symbiotic dinoflagellates known as *zooxanthellae* live within the tissues of coral polyps and are responsible for the color of many live corals. The zooxanthellae provide the polyps with carbohydrates and remove carbon dioxide. In return, the polyps supply the zooxanthellae with nitrogen compounds and carbon dioxide.

the boring organisms riddle the base with tunnels, the coral stand weakens, and a storm or heavy ocean surge will eventually topple the coral, crushing and killing many living polyps.

> **Bioerosion** is the gradual wearing away of substrates such as coral reefs through the actions of living organisms.

As the debris from boring accumulates it eventually covers and smothers the boring organisms, and the boring stops. If this occurs quickly, the chunks of coral left over from the boring will be relatively large. If the process takes a longer time, the large corals will be broken up into much smaller pieces. The breakup of calcareous green algae such as *Halimeda* **(Figure 15-9)**, mollusc shells, foraminifera, and other calcareous organisms also add significantly to the calcium carbonate deposits on the reefs. This sediment covers the reef and settles into the cracks and crevices. Sponges and other organisms may hold this rubble together until encrusting coralline algae deposit more calcium carbonate over it, cementing it together. Planula larvae can then settle on this hard surface and begin to produce new stands of coral.

IN SUMMARY

- Scleractinian (or stony) corals produce the calcium carbonate deposits that make up much of the hard structure of a coral reef. Along with algae, they form the foundation of a community on which all the other reef animals depend for food and shelter.
- Reef-building, or hermatypic, corals are strictly tropical in distribution and harbor zooxanthellae but ahermatypic corals, those that do not deposit calcium carbonate, are more widespread and generally lack zooxanthellae.
- Corals obtain up to 90% of their energy from zooxanthellae, the remainder coming from feeding on plankton, detritus, and dissolved organic matter. (Have You Wondered? #1)
- Coral colonies can reproduce both asexually, by budding and fragmentation, and sexually.
- Coral reefs are constantly forming and breaking down, a process known as *bioerosion*.

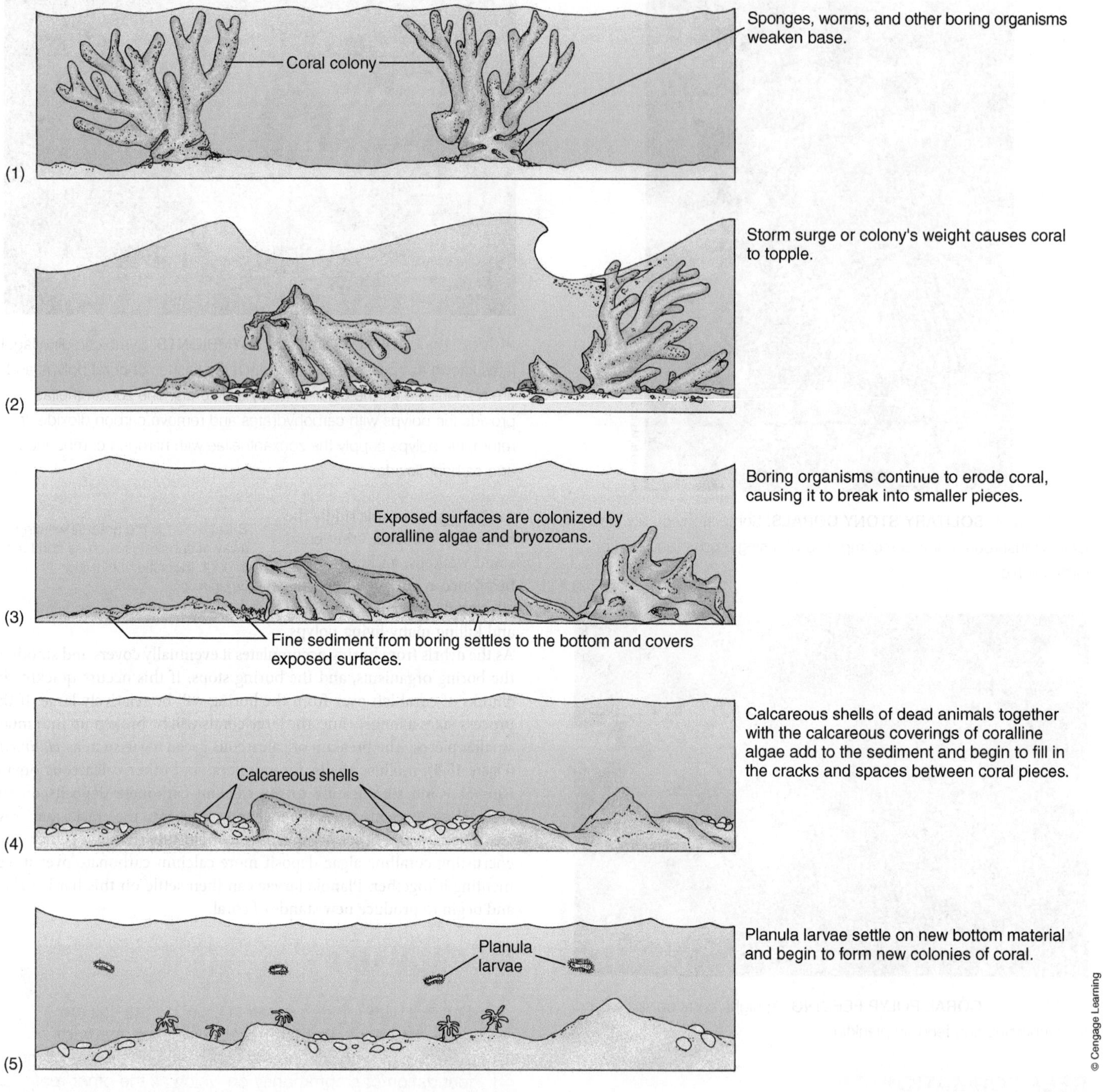

(1)   Coral colony

Sponges, worms, and other boring organisms weaken base.

(2)

Storm surge or colony's weight causes coral to topple.

(3)   Exposed surfaces are colonized by coralline algae and bryozoans.

Boring organisms continue to erode coral, causing it to break into smaller pieces.

Fine sediment from boring settles to the bottom and covers exposed surfaces.

(4)   Calcareous shells

Calcareous shells of dead animals together with the calcareous coverings of coralline algae add to the sediment and begin to fill in the cracks and spaces between coral pieces.

(5)   Planula larvae

Planula larvae settle on new bottom material and begin to form new colonies of coral.

© Cengage Learning

FIGURE 15-8 REEF CYCLE. (1) The destructive phase of a reef cycle begins when boring animals burrow through the bases of large corals, weakening them. (2) The weight of the coral or storm surges cause the coral to topple, killing the polyps. (3) The coral is broken into smaller pieces and bryozoans and coralline algae begin to colonize the exposed surfaces. (4) Shells of dead organisms contribute to filling in cracks and crevices. Coralline algae help cement the pieces together to form a new solid surface. (5) Planula larvae begin to settle and form new coral colonies.

IMPACT BUBBLE

DEFORESTATION AND BURNING of fossil fuel by humans have increased levels of atmospheric carbon dioxide by approximately one-third over the past few decades. Current projections suggest that, unless remedial steps are taken, CO_2 levels will more than double by the end of the century and potentially cause devastating changes to both terrestrial and aquatic systems. Rising atmospheric CO_2 levels cause global climate change by acting like a blanket that traps heat and prevents it from radiating to space (the greenhouse effect). In addition, about one-third of atmospheric CO_2 dissolves in the ocean, where it can detrimentally affect marine organisms through acidification.

When CO_2 dissolves in seawater, it reacts with water (H_2O) to form carbonic acid (H_2CO_3). Most of this carbonic acid dissociates into hydrogen ions (H^+), bicarbonate ions (HCO_3^-), and carbonate ions (CO_3^{2-}). The relative number of bicarbonate ions as opposed to carbonate ions depends on the pH of the water. By forming carbonic acid, dissolved CO_2 increases the number of hydrogen ions, thereby increasing the acidity and lowering the pH. At lower pH levels there are fewer carbonate ions available relative to bicarbonate ions. The deposition of calcium carbonate ($CaCO_3$) in the tissues of corals and many other marine organisms has been shown to be dependent on the concentration of dissolved carbonates. Hence, the increasing acidification of seawater has the potential to decrease the deposition rate of calcium carbonate in corals and other marine organisms. In addition, when carbonate ion concentrations fall too low, the calcium carbonate in coral skeletons and other organisms starts to dissolve. Researchers have suggested that these effects have the potential to cause a mass extinction of marine life rivaling that of 65 million years ago, when the dinosaurs disappeared.

Studies conducted on laboratory coral reef systems at the Biosphere-2 Ecosystem Center in Arizona, at the Hawaii Institute of Marine Biology, and at the University of Queensland's laboratory on Heron Island in Australia, among others, have demonstrated that increasing acidity can reduce coral deposition of carbonates by 20% or more. One of these studies even predicted a 40% decline from preindustrial levels by 2065. Acidity affects not only coral but also coralline algae and other calcifying organisms (e.g., echinoderms, crustaceans, and molluscs). Coralline algae is said to be the "glue" that holds coral reefs together and acts as a settlement site for coral larvae. Reduction of coralline algae, therefore, could adversely affect coral recruitment.

Fewer studies on acidification have been done in the field. A study of cores taken from massive *Porites* corals from the Great Barrier Reef found no change in calcification rates over the last century up to 1987. Since that time, however, there has been a 14% decline in calcification rate up to 2005. In a study conducted on patch reefs, bare sand, and coral rubble on the Molokai reef flat in Hawaii, K. K. Yates and R. B. Halley found that calcification rates were directly correlated with both the level of dissolved carbonates and atmospheric CO_2, as predicted in laboratory studies. They also found that fluctuating levels of atmospheric CO_2 (and resulting levels of dissolved CO_2) exceed the threshold value for decomposition of carbonates over their deposition about 18% of the time at the Molokai reef flat.

The oceans contain multitudes of calcifying organisms ranging from autotrophs to heterotrophs. The predicted doubling of CO_2 levels by 2065 from 1880 levels can therefore be expected to cause massive changes to both marine food webs and coral reef ecosystems.

Matthew Landau

FIGURE 15-9 CORAL REEF ALGAE. Calcareous green algae, such as this member of the genus *Halimeda*, are common inhabitants of coral reefs due to their ability to resist herbivory. The breakup of the thallus of these species contributes to calcium carbonate deposits on coral reefs.

15-2 Structure of Coral Reefs

Three main categories of coral reefs have been described: fringing reefs, barrier reefs, and atolls. Many reefs do not fit these categories exactly but are intermediate between them. There are many structural similarities among the different reef types, as detailed in the following sections.

REEF TYPES

Fringing reefs (Figure 15-10a) develop along shores of tropical or subtropical islands or continental landmasses—anywhere there is hard substrate. Because of their proximity to land, they are the most directly affected by human activity and freshwater runoff. The largest fringing reef in the world occurs along shores of the Red Sea, where desert conditions on land and the lack of freshwater runoff have helped to protect it. **Barrier reefs (Figure 15-10b)** are similar to fringing reefs but are separated from the landmass and fringing reef with which they are associated by lagoons or deepwater channels. The world's largest barrier reef is the Great Barrier Reef of Australia, which runs about 2,100 kilometers (1,260 miles)

Fringing reefs are formed close to the shoreline of an island or continent.

Barrier reefs are ridges of coral and rock separated from the land by lagoons that are too deep for coral growth.

FIGURE 15-10 **TYPES OF CORAL REEFS.** **(a)** A fringing reef is connected directly to the shore. **(b)** A barrier reef is separated from the shore by a lagoon. **(c)** The largest barrier reef in the world, the Great Barrier Reef, runs along the northeastern coast of Australia. **(d)** An atoll is a circular reef that encloses a lagoon. Atolls usually form on the cones of extinct volcanoes.

(c)

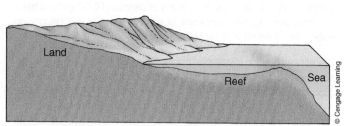

(a) Fringing reef

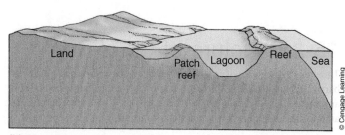

(b) Barrier reef

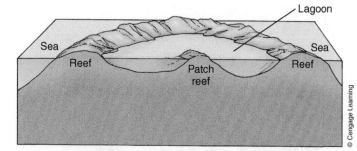

(d) Atoll

along the northeastern coast of Australia to New Guinea **(Figure 15-10c)**. This reef is so large that it is visible even to astronauts orbiting the earth. The second-largest barrier reef is off the coast of Belize, in the Caribbean Sea. **Atolls (Figure 15-10d)** are usually somewhat elliptical, arise out of deep water, and have a centrally located lagoon. They vary in size from less than 1 kilometer (about 0.6 miles) in diameter to over 130 kilometers (80 miles) long and 32 kilometers (20 miles) wide

Atolls are a form of coral reef that builds up around eroding volcanic islands. They enclose a shallow lagoon and are surrounded by a deep ocean.

Patch reefs are isolated coral formations within lagoons, barrier reefs, or atolls.

Bank reefs are isolated flat-topped reefs that are found growing on midshelf regions of island or continental shelves.

(Kwajalein in the Marshal Islands). More than 300 atolls are located in parts of the Indian and Pacific oceans, but only 10 atolls occur in the Atlantic Ocean. The Atlantic Ocean supports fewer reefs in general because the water is too cool and turbid for coral growth in many areas. The water in the lagoon of an atoll is not isolated but is connected to the open sea by gaps in the reef. Over time, parts of the reef may become exposed and eroded by the action of wind and waves. The sand formed by these physical processes can be the basis for the formation of an island. Sandy islands frequently surround the lagoon. Small **patch reefs** may occur within the lagoons associated with atolls and barrier reefs. Where patch reefs occur, they are usually numerous. The term **bank reef** is applied to larger, isolated, flat-topped reefs that are found growing on midshelf regions of island or continental shelves. They are usually larger than patch reefs.

1 Active volcano

2 Fringing reef

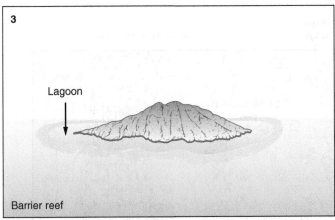

3 Lagoon

Barrier reef

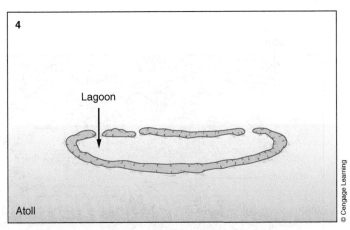

4 Lagoon

Atoll

© Cengage Learning

FIGURE 15-11 ATOLL FORMATION. According to Charles Darwin's theory, fringing reefs form in the shallow water around newly formed volcanic islands. Over time, the island erodes and becomes separated from the surrounding and continually growing reef. This reef is now termed a *barrier reef* because it is separated from the island by a lagoon. As the island completely erodes, an atoll is formed as the barrier reef continues to grow, making up for the subsidence of the island.

Charles Darwin proposed the currently accepted theory for the origin of atolls and their relationship to other coral reef types. According to Darwin's theory **(Figure 15-11)**, as newly formed volcanic islands arise from the sea, corals colonize the shallow areas around them and form a fringing reef. As these volcanic islands slowly erode and sink, the coral growth rate comes to equal the rate of sinking, and a barrier reef forms about the island. At this point, a lagoon separates this barrier reef from what remains of the central island. Finally, as the island completely erodes and sinks below the sea, corals continue to grow to the surface of the barrier reef, and an atoll is formed. Water flow is usually inadequate, and there is too much sediment where the island once existed for vigorous coral growth to occur, so a lagoon persists. Although Darwin's theory does link fringing reefs, barrier reefs, and atolls, it does not explain how *all* fringing and barrier reefs develop. Corals reefs probably occur around continents and islands simply because of suitable solid surfaces and conditions for their growth.

STRUCTURE OF REEFS

On the seaward side, coral reefs rise from the lower depths of the ocean to a level at or just below the surface of the water. This portion of the reef is called the **reef front (Figure 15-12a)**. The slope of the reef front can be gentle or quite steep. In some cases, the reef front forms a vertical wall referred to as a **drop-off**. More commonly, finger-like projections of the reef protrude seaward in a **spur-and-groove formation** (or buttress zone). This arrangement disperses wave energy and helps prevent damage to the reef and its inhabitants **(Figure 15-12b)**. Between some of these projections (spurs) lie sand-filled pockets (grooves) that channel sediments down and away from the living coral surface and provide habitat for many species of burrowing organisms. Coral growth is rich at the reef margin, on the spurs, and along the sides of the grooves, but declines with depth because of decreased light availability.

The **reef crest** is the highest point on the reef and the part that receives the full impact of wave energy. Where wave impact is very strong, the reef crest may consist of an **algal ridge**, composed of encrusting coralline algae and lacking most other organisms. Grooves of the spur-and-groove formation penetrate the algal ridge as

The **reef front** is the seaward-facing slope of a reef.

A **drop-off** is the steep downward slope of a reef front.

A **spur-and-groove formation (buttress zone)** consists of low ridges of corals (spurs) extending from the reef flat that are separated by sandy channels (grooves).

The **reef crest** is an erosion-resistant ridge composed of compacted coral and algal material lying on the seaward edge of a reef flat.

An **algal ridge** is the elevated platform of a windward coral reef built by encrusting coralline algae.

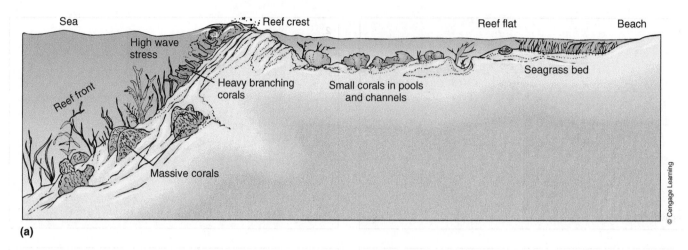

(a)

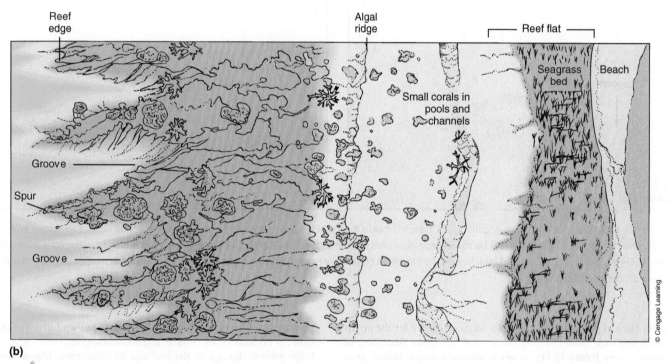

(b)

FIGURE 15-12 **REEF CHARACTERISTICS.** **(a)** The reef front rises from the ocean's depths to the reef crest. The area behind the reef is the reef flat. **(b)** A view of a reef from above looking down shows the characteristic spur-and-groove formation.

surge channels. Behind the reef crest lays the **reef flat** (back reef). The reef flat may be short or several hundred meters long. It may be shallow or cut through by surge channels several meters deep. The bottom of the flat consists of rock, sand, coral rubble, or some combination of these. Seagrass beds are common in the reef flat area. The reef flat of fringing reefs ends at the shoreline. The reef flat of atolls and barrier reefs descends into the lagoon.

Surge channels are channels in the windward side of a coral reef through which water flows both inward and outward.

The **reef flat** is the shallow flat expanse of dead reef rock, often exposed at low tide, that lies behind the reef crest.

Different areas of a reef support different species of coral and other organisms. Coral populations on the reef front are usually found at depths of 10 to 60 meters (33 to 200 feet). Massive dome-shaped brain corals (*Diploria*) and columnar pillar corals (*Dendrogyra*) are on intermediate slopes. Below this region, species that form plate-like formations such as lettuce leaf and elephant ear coral (*Pectinia, Pavona,* and *Agaricia* species; **Figure 15-13a**) predominate. Higher up on the reef, where wave stress is greatest, branching species of coral are found. Heavy spreading branches project toward the sea, where they break the force of incoming waves. In the Caribbean, this upper area is the habitat of elkhorn coral (**Figure 15-13b**). Wave stress is one of the most important factors in determining which species of coral and other organisms can occupy the reef crest.

In more protected areas behind the reef front, the deeper and less-turbulent water supports more delicate species of coral. In the Caribbean, this is frequently staghorn coral. In the Indo-Pacific region, species such as staghorn (*Acropora*), finger (*Stylophora*), and cluster and lace corals (*Pocillopora*) are prominent. Farther from the reef front in shallow

FIGURE 15-13 TYPES OF CORAL. (a) Corals that produce large plate-like formations, such as this lettuce leaf coral, are located farther down the reef front, where wave action is minimal. (b) Areas of a reef that receive heavy wave action contain thick branching corals such as this elkhorn coral, as well as large solid corals such as brain coral.

Jon L. Hawker

Richard N. Mariscal

(a)

(b)

calmer water are small species such as rose (*Meandrina* and *Manicina*), flower (*Mussa* and *Eusmilia*), and star (*Montastraea*) corals.

IN SUMMARY

- Three main types of coral reefs have been described, fringing reefs, barrier reefs and atolls. (Have You Wondered? #2)

- The reef front arises from the ocean's depths and is typically arranged in a spur-and-groove formation (or buttress zone) that disperses wave energy and prevents damage to the reef and its inhabitants.

- Because of differences in the physical environment, each area of the reef supports different combinations of coral and other reef organisms.

15-3 Distribution of Coral Reefs

Temperature, light availability, sediment accumulation, salinity, wave action, and duration of air exposure are the major factors determining the distribution of coral reefs. Tropical coral reefs do not develop in waters where the average annual minimum temperature is below 18°C, and the best development occurs where the average annual water temperature is between 23°C and 25°C. Almost all the world's coral reefs are therefore restricted to tropical seas **(Figure 15-14)**. Lack of light explains why coral reefs are rarely found below 60 meters (196 feet). Most grow in depths of 25 meters or less, and they are typically confined to the margins of continents or islands. Sedimentation from land-surface run-off can both reduce light available to zooxanthellae for photosynthesis and clog feeding structures, killing coral polyps. Many examples of the deleterious effect of sedimentation can be seen on South Florida reefs. Salinity level is also important, as demonstrated by the lack of coral reefs in such areas of massive freshwater outflow as the mouths of the Amazon and Orinoco rivers in South America. Moderate wave action

FIGURE 15-14 DISTRIBUTION OF CORAL REEFS. Coral reefs (brown areas) are restricted to the tropics and subtropics, the area within the surface 20°C isotherm.

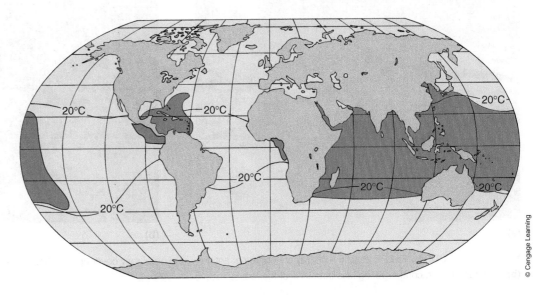

© Cengage Learning

is important for a healthy reef because it brings in fresh oxygenated seawater and removes sediment that may smother coral polyps. Heavy wave action during hurricanes, however, can physically damage reef structure, uprooting corals and destroying habitat. Although some coral species can sustain exposure to air for a few hours, most are killed by it. Therefore the lowest tide level limits upward coral growth.

IN SUMMARY

- Factors limiting reef distribution include temperature, light availability, salinity, sedimentation, wave action, and duration of exposure to air.

- Coral reefs are typically confined to shallow water at the margins of continents or islands where the average surface temperature is greater than 18°C.

- Coral reefs are rarely found below 60 meters due to low light levels. (Have You Wondered? #3)

15-4 Comparison of Atlantic and Indo-Pacific Reefs

Atlantic and Indo-Pacific reefs (**Figure 15-15**) differ in a number of ways. Pacific reefs are older than those in the geologically younger Atlantic Ocean and have a far greater depth of reef carbonates. Spur-and-groove formations are deeper on Atlantic reefs, and coral growth may extend down to 100 meters (328 feet). In contrast, coral growth in the Pacific seldom exceeds 60 meters (196 feet). The proportion of the reef covered by coral may approach 100% on some Pacific reefs but is usually less than 60% on those in the Atlantic. Algal ridges are more common in the Pacific than in the Atlantic because of wind and wave conditions. The hydrozoan *Millepora complanata* (fire coral) dominates on many Atlantic reefs, but similar species never dominate in the Pacific. Gorgonians (sea fans and sea whips) are much more abundant in the Atlantic, whereas Pacific reefs have a greater abundance of soft corals (subclass Alcyonaria). Most Atlantic corals are nocturnally active, but many Pacific species are diurnal. Atlantic corals often reproduce by fragmentation, whereas

Indo-Pacific forms most often propagate by sexual reproduction. Finally, the diversity of coral species is far greater in the Indo-Pacific region than in the Atlantic. The Indo-Pacific has 500 species of stony corals, whereas the Atlantic Ocean has only about 62 species. The dominant coral genus *Acropora* has about 200 species in the Indo-Pacific but only 3 in the Atlantic. Sponges on Atlantic reefs have far greater biomass than those of the Indo-Pacific, and most do not have phototrophic symbionts. Unlike the Pacific, the Atlantic Ocean has no giant clams or sea star species that are major coral predators.

The diversity of communities on Indo-Pacific reefs is much greater than that in the Atlantic. For example, there are more than 5,000 species of molluscs and about 2,200 species of ray-finned fish associated with reefs in the Indo-Pacific, but only about 1,200 molluscs and 550 fish species in the Atlantic. A possible reason for this difference is that more opportunities for species diversification existed due to the greater antiquity and size of the Indo-Pacific. A second possible reason is that Indo-Pacific coral reefs did not experience the severe temperature changes suffered by Caribbean reefs during the Pleistocene ice ages.

IN SUMMARY

- Indo-Pacific reefs are older than Atlantic reefs, have a greater diversity of organisms, have more extensive algal ridges, grow to 60 meters or less in depth, have a far greater depth of reef carbonates, and coral surface coverage can approach 100%. (Have You Wondered? #4)

- Atlantic reefs have deeper spur-and-groove formations, have coral growth extending to 100 meters, are commonly dominated by the hydrozoan *Millepora complanata* (fire coral), have a greater biomass of sponges, and have coral coverage rarely exceeding 60%. (Have You Wondered? #4)

- Most Atlantic corals are nocturnal, but many Pacific species are diurnal.

- Atlantic corals often reproduce by fragmentation, whereas most Indo-Pacific forms reproduce sexually.

(a)

(b)

FIGURE 15-15 CORAL REEFS. Pacific reefs **(a)** are of greater antiquity, and coral may cover up to 100% of the reef surface. Caribbean reefs **(b)** are younger, with coral covering less than 60% of the reef surface.

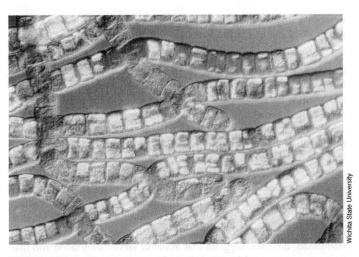

Wichita State University

FIGURE 15-16 **WIND-DRIVEN CYANOBACTERIA.** Cyanobacteria such as *Trichodesmium* are able to fix atmospheric nitrogen. Wind and currents deposit them on reefs, where their subsequent decomposition supplies important nutrients to the community.

15-5 Coral Reef Ecology

Coral reefs abound in life and are one of the most productive of all ecosystems. The net annual primary productivity on coral reefs is estimated to be 2,500 grams of carbon per square meter, a value comparable to tropical rainforests. The annual primary productivity of tropical oceans, in contrast, is estimated to be less than 50 grams of carbon per square meter. How is this possible in such nutrient-poor tropical waters?

SOURCE OF NUTRIENTS

The high level of productivity of coral reefs requires an abundant source of nutrients, particularly nitrogen and phosphorus. One can assume that reefs close to continents or islands receive nutrient runoff from the land, but the source of the nutrients supporting atolls is unclear. Although some researchers have suggested that upwelling or high flow rates of water across the reef are enough to supply the needed nutrients, many believe that, like rainforests, reefs have accumulated their nutrients over time. Reefs are probably able to maintain their high levels of productivity because they recycle nutrients so efficiently. Nutrients tied up in dissolved and particulate organic matter can be concentrated by reef bacteria and removed by the biofiltering activity of coral and other reef organisms. The important nutrient nitrate can also be supplied through nitrogen fixation by cyanobacteria and bacteria living within corals and other animals, in association with algae, in the soft sediments, as epiphytes on seaweeds, and free in the water column. The accumulation of wind-driven cyanobacteria such as *Trichodesmium rubrum* (**Figure 15-16**) from ocean waters and their subsequent decomposition can likewise supply needed nutrients to the reef. Many fish species feed away from the reef at night but return to it during the day. Their fecal matter can be an important source of nutrients to the reef. The reef, therefore, is a nutrient sink for material brought from the outside. Coral reef ecosystems are adapted to survive in low-nutrient waters by efficiently recycling these nutrients. Like tropical rain forests, nutrients are stored in the biomass of the community's inhabitants.

> **Gross primary production (P)** is the total production of organic compounds from carbon dioxide through the process of photosynthesis or chemosynthesis.
>
> **Community respiration (R)** is the total energy acquired from organic compounds that is used by all members of the community for their metabolism.

PHOTOSYNTHESIS ON REEFS

Reef photosynthetic organisms include the zooxanthellae of corals and other organisms, benthic algae, turf algae, sand algae, phytoplankton, and seagrasses. Their density is far greater than that in a similar area of open ocean, and their combined biomass is greater than that of the animals inhabiting the reef. The close association of zooxanthellae with the corals and other organisms, and a similar association of cyanobacteria with sponges, mean that nutrients are quickly recycled and not easily lost to the surrounding water. Most photosynthetic forms are firmly attached to the reef, and many free-living forms contain calcium carbonate, making them too heavy to be easily removed from the reef. Of all the photosynthetic organisms present on the reef, the turf algae seem to be the most important because they process the most organic carbon (**Table 15-1**).

REEF PRODUCTIVITY

The ratio of **gross primary production (P)** to **community respiration (R)**, called the P-R ratio, is sometimes used as a measure of the state of development of a biological community. When the P-R ratio is greater than 1, primary production exceeds the respiratory needs of the community, and excess biomass is available for community growth or harvesting. As communities mature, the P-R ratio approaches 1, little biomass remains available for growth or harvesting, and the community reaches a steady state or climax. P-R ratios measured for coral reefs are typically close to 1, indicating that the high productivity levels of coral reefs are balanced by high rates of community respiration, and that little excess is available for export or continued growth. Nutrient enrichment of coral reefs (eutrophication) typically manifests itself in a dramatic proliferation

TABLE 15-1 PRIMARY PRODUCTIVITY OF COMPONENT PRODUCER COMMUNITIES ON CORAL REEFS AND THEIR DISTRIBUTION

Producer Community	Productivity (g C m^{-2}d^{-1})	Approx. Cover (%)
Fleshy benthic algae	0.1–4.0	0.1–5.0
Crustose coralline algae	0.9–5.0	10–50
Turf algae	1.0–6.0	10–50
Zooxanthellae	0.6	10–50
Sand algae	0.1–0.5	10–50
Phytoplankton	0.1–0.5	10–50
Seagrasses	1.0–7.0	0–40

Sources: Larkum, A. W. D. 1983. Primary Productivity of Plant Communities. In D. J. Barnes, editor: *Perspectives on Coral Reefs*. Townsville, Australia: The Australian Institute of Marine Science, pp. 221–230; Chisholm, J. R. M. 2003. Primary Productivity of Reef-Building Crustose Coralline Algae, *Limnology and Oceanography* 48(4): 1376–1387.

of algae rather than increased coral growth. Unless increased grazing controls the algae, they will grow over the corals and smother them.

IN SUMMARY

- Despite existing in nutrient-poor tropical ocean water, coral reefs are one of the most productive of all ecosystems.

- It is thought that coral reefs have accumulated their nutrients over time and recycle them so efficiently that they are able to retain them in the biomass of the reef's organisms.

- Because community respiration is about the same as primary production, little excess biomass is available for export or community growth.

- Reef photosynthetic organisms include the zooxanthellae of the corals and other organisms, benthic algae, turf algae, sand algae, phytoplankton, and seagrasses. (Have You Wondered? #5)

15-6 The Coral Reef Community

The coral reef community is made up of an intricate array of producers and consumers. Because space and other resources are limited, various types of interactions have developed among the many species inhabiting the reef. In this section, we will explore some of these interactions.

COMPETITION AMONG CORALS

Because corals require both light and living space to survive, their many species compete for these resources. Fast-growing branching corals, for example, often grow over slower growing, encrusting, or massive corals and deny them light. In response, the slower-growing forms may extend mesenterial filaments from their digestive cavity and kill their competitor's polyps. Undamaged polyps on the fast-growing branching coral may respond by growing very long **sweeper tentacles** containing cnidocytes that kill polyps on the slower-growing form. The faster-growing form repairs the damage and continues to overgrow its competitor. In addition to sweeper tentacles and mesenterial filaments, corals have several other mechanisms available for attack or defense. In general, slower-growing corals are more aggressive than fast-growing species. In cases where a competitor cannot be overcome, however, corals may survive by taking advantage of differences in local habitats. Massive corals are generally more shade tolerant and able to survive at greater depths. Therefore, on many reefs, the fast-growing branching corals ultimately dominate at the upper, shallower portion of the reef, whereas more massive forms dominate in deeper areas.

COMPETITION AMONG REEF FISHES

The diversity and abundance of fishes associated with a coral reef is far greater than that found in any other marine habitat. A single patch of coral may provide habitat for 70 to 80 species. Fish also exploit just about every available energy source, from detritus to algae, sponges, coral, various invertebrates, and other fish. Although reef structure supplies an abundance of microhabitats for fish, habitat alone is not enough to explain their remarkable diversity. The *competitive exclusion principle* suggests that no two species can occupy the same ecological niche. Yet, 60% to 70% of reef fish can be

Sweeper tentacles are long stinging structures used by aggressive hard corals to sting other nearby corals in order to obtain growing space.

described as general carnivores, the rest being grazers on coral and algae (about 15%) or omnivorous.

Several hypotheses have been proposed to explain the great diversity of these carnivores. One hypothesis, the *competition model*, suggests that factors such as time, size of prey, position in the water column, and so on, partition the resource base to give each species a unique niche; hence there is no competition. In contrast, the *predation disturbance model* assumes competition among species but suggests that the effect of predation or other causes of mortality keep populations low enough to prevent competitive exclusion. The *lottery model* assumes competition occurs but suggests that it is unimportant:; Chance determines which species of larvae settling from the plankton colonize a particular area of reef. As each individual is lost over time, chance again determines which species will occupy the available space. All of these models assume that there is no limitation in the availability of fish larvae. The *resource limitation model*, however, suggests that available larvae are limited and that limitation prevents fish populations from ever reaching the carrying capacity of the habitat. Although different researchers have provided data that support one or more of these models, it remains unclear which, if any, best explains how so many species of fish can exist on coral reefs.

EFFECT OF GRAZING

Corals compete with many reef organisms, each with its own strategy for survival. Sponges, soft corals, and other sessile invertebrates, as well as fleshy algae, can overgrow stony corals and smother them. Algae are competitively superior to corals in shallow water but less so at depth. Survival of coral in shallow water depends on grazing by echinoderms and herbivorous fishes. In Jamaica, overfishing removed most of the herbivorous fish from coral reefs. Initially, increased populations of the grazing sea urchin *Diadema antillarum* **(Figure 15-17)** kept algal growth in check. An unknown pathogen reduced *Diadema* populations by 99% in 1982 and, as a result, coral cover declined from 52% to 3%, and algal cover increased from an average of 4% to 92%. Without grazers, fleshy algae were able to completely overgrow the coral. With grazers, however, the dominant algae on a healthy reef are usually fast-growing filamentous

De Agostini Picture Library/De Agostini/Getty Images

FIGURE 15-17 *DIADEMA.* The long-spined black sea urchin, *Diadema antillarum*, is considered a "keystone species" because of its importance in controlling macroalgae on reefs. In 1982, a pathogen reduced their population by 99%. The population has still not recovered, resulting in macroalgal overgrowth of some Caribbean reefs.

FIGURE 15-18 **THREESPOT DAMSELFISH.** The threespot damselfish, *Stegastes planifrons*, is highly territorial and will aggressively attack intruders regardless of their size. On Atlantic reefs, they act as "keystone species," allowing less competitive algae and other species to survive within their territories.

FIGURE 15-19 **REEF DESTRUCTION BY SEA STARS.** Population explosions of the coral-eating crown-of-thorns sea star (*Acanthaster planci*) have devastated many Pacific reefs. The crown-of-thorns sea star seen here is being attacked by a Triton's trumpet (*Charonia tritonis*), a major predator. Population explosions of the crown-of-thorns sea star seem to be a natural phenomenon, but the triggering mechanism is unknown.

forms or coralline algae, well protected by calcification and the production of noxious chemicals. Herbivory is greatest in the shallow part of the reef front but decreases with depth, where lower temperatures and light reduce algal growth. The reef is therefore a mosaic of microhabitats with different levels of grazing and different algal communities.

An additional complexity arises from the activity of damselfish (family Pomacentridae; **Figure 15-18**). Because they are territorial, many damselfish species exclude grazers and other species from certain areas of the reef. Algae grow rapidly in these territories, providing habitat for many small invertebrates but overgrowing the corals. Branching corals tend to dominate in damselfish territories because they are upright and faster growing than the more massive or encrusting forms. Damselfish also control algal diversity by weeding out undesirable species, making their territories, in many ways, "algal farms." Because damselfish territories can occupy up to 60% of the reef, they are very important in determining its structure.

EFFECT OF PREDATION

Although less studied than on rocky shores, predation has a significant influence on the community structure of coral reefs. Fish and other predators may preferentially prey on such competitors of corals as sponges, soft corals, and gorgonians, giving competitively inferior reef corals an advantage in securing space. Many species of fish, mollusc, annelid, crustacean, and echinoderm also feed directly on coral polyps or their mucus, limiting the growth of certain species. Several surgeonfish (family Acanthuridae) and parrotfish (family Scaridae) pass coral skeletons through their digestive tracts and add sediment to the reef. *Chaetodon unimaculatus*, the teardrop butterflyfish, slows the growth of the coral *Montipora verrucosa* in Hawaii through its feeding activity. Both fish and invertebrate corallivores (coral-feeding organisms) seem to attack faster-growing branching species preferentially, often preventing slower-growing forms from being overgrown. In general, it is rare for a corallivore to completely destroy a coral colony except in cases where tropical storms or humans have already done severe damage.

An exception to this generality is the infamous case of predation on Pacific corals by crown-of-thorns sea stars (*Acanthaster planci*; **Figure 15-19**). Since 1957, about one-third of the corals inhabiting Indo-Pacific reefs have been lost to predation by this species. Although small populations of crown-of-thorns sea stars can be selective in the coral species they eat, larger populations often form mass aggregations that destroy all corals in their path. Overpopulation of crown-of-thorns sea stars on the Great Barrier Reef in the 1960s caused so much damage that drastic, but mostly ineffective, measures were implemented for their control. Eventually, sea star population levels declined naturally.

Population explosions of crown-of-thorns sea stars seem to be a natural phenomenon, dating back thousands of years. What the triggering mechanisms are for the recent and more frequent outbreaks is still an active subject of debate. Some have suggested that the removal of sea star predators such as the Triton's trumpet (*Charonia tritonis*; see **Figure 15-19**) and several fish species by humans have left the crown-of-thorns sea star populations unchecked. Another suggestion was that nutrient runoff in wet years allowed larger phytoplankton blooms, providing increased food for *Acanthaster* larvae and increasing their numbers. Still another theory suggests that reef destruction by storms causes populations of sea stars to condense and attack the remaining coral reefs en masse. Whatever the cause, explosions of this species cause a dilemma for the managers of coral reef parks in Australia. Currently, sea stars are killed if they threaten particularly important parts of the reef, but otherwise a population explosion is allowed to run its course.

SYMBIOTIC RELATIONSHIPS

Symbiotic relationships help organisms exploit niches and better survive in the fiercely competitive world of the coral reef. These relationships can range from parasitism through commensalism to mutualism. The mutualistic relationship between coral polyps and zooxanthellae is only one of hundreds of examples of symbiotic relationships between reef

Richard N. Mariscal

FIGURE 15-20 GIANT CLAM. Giant clams (*Tridacna gigas*), such as this one on the Great Barrier Reef, contribute to the mass of coral reefs with their large heavy shells. They also maintain a symbiotic relationship with zooxanthellae similar to that of corals. Notice how the opening of the clam is facing upward. This allows the animal to drape its mantle over the edges of its shell so as to give the symbiotic zooxanthellae maximum exposure to sunlight.

organisms. The giant clam (*Tridacna gigas;* **Figure 15-20**) of Indo-Pacific reefs, for example, also hosts symbiotic zooxanthellae to supplement its energy needs. The clam provides carbon dioxide and nitrogen to the zooxanthellae in its mantle, which photosynthesize and return oxygen and carbohydrates to the clam. Sometimes the giant clam "harvests" some of its symbionts to supplement its diet. Like other clams, the giant clam is a filter feeder, but the quantity of plankton available on a coral reef is not great enough to sustain such a large animal. In many areas of the Pacific Ocean, this magnificent clam is overfished and endangered.

Cleaning Symbioses

At one time, marine biologists were puzzled by gatherings of many fish species at particular sites on a reef. It was later recognized that these fishes were waiting their turn to be cleaned by one of the reef's many cleaner organisms, such as the cleaner wrasse (*Labroides dimidiatus*). Cleaner wrasses, gobies, other small fish, and some invertebrates feed on the parasites of larger fishes **(Figure 15-21)**. These species set up a territory on the coral reef known as a cleaning station that "clients" frequently visit to have parasites and dead tissue removed. The cleaners benefit by obtaining a constant food supply, and the clients are kept healthy by the grooming. Wrasses and other species attract potential customers to their stations with their bright colors and a series of movements referred to as a "dance." Even aggressive predators such as the moray eel that usually eat small fish will submit to cleaning without harming the wrasse. Color changes or ritual postures by clients (also called hosts) may also be used to initiate and terminate the cleaning process.

Although the relationship between cleaners and their clients may superficially appear to be a mutualism, many cleaners often feed on their host's mucus, fins, and other healthy tissue in addition to removing parasites and diseased tissue. Indeed, removal of parasites may not be all that important. In most cases, experimental removal of cleaner fish from reefs had little effect on the overall health of the host's population. Hosts, therefore, may tolerate this behavior simply because they enjoy the tactile stimulation. Some predatory species such as the sabertooth blenny (*Aspidontus taeniatus*) take advantage of the cleaning relationship to

Chris Newbert/Minden Pictures

FIGURE 15-21 CLEANING SYMBIOSIS. A cleaner wrasse removes parasites from the gills of this mappa pufferfish (*Arothron mappa*).

attack their prey. This blenny resembles a cleaner wrasse both in body shape and color. It goes so far as to mimic the cleaner wrasse's dance that attracts other fish to its cleaning station. When the unsuspecting fish comes near to be cleaned, the blenny attacks, biting pieces of flesh from the unwary victim.

Other Symbiotic Relationships

Cleaning is not the only example of symbiosis in the coral reef community. The reef is full of examples of organisms that have evolved special relationships with other species. Anemonefishes (family Pomacentridae, subfamily Amphiprioninae), for example, seek shelter from their enemies in the tentacles of large anemones **(Figure 15-22)**. The mucus coating their body seems to protect them from the anemone's stings. It is unclear whether the fish or the anemone produces the mucus. The anemone

Courtesy of James W. Small

FIGURE 15-22 CLOWNFISH. Pacific anemonefishes, like these ocellaris clownfish (*Amphiprion ocellaris*), seek shelter from their enemies in the tentacles of large anemones. They seem to be protected from the anemone's stings by the mucus that coats their bodies.

benefits from the anemonefish's aggressive behavior toward potential predators and may gain some nutrition from food scraps dropped by the fish. The anemone's zooxanthellae may also benefit from the anemonefish's wastes. The anemone's stinging tentacles protect the anemonefish from predators.

Pearlfishes (family Carapidae) live in the hindgut of sea cucumbers, in the shells of oysters and clams, and in certain species of sea stars. Some species seem to enjoy a commensal relationship with their host, but others are parasites. The tiny conchfish (*Astrapogon stellatus*) is found only in the mantle cavity of the large queen conch (*Strombus gigas*). At night the conchfish emerges to feed on shrimp, sea lice, and other small crustaceans. On Indo-Pacific reefs certain species of gobies are found at the entrance to burrows dug by snapping shrimp. The shrimp digs the burrow and continuously clears it of debris. When predators appear, the gobies hide in the shrimp's burrow. The adult shrimp may benefit from the warning given by the gobies, but on the other hand, the gobies feed on young shrimp. Tiny shrimpfishes (family Centriscidae) hide between the spines of sea urchins, where they hover head down, their striped bodies looking very much like the urchin's spines. A species of cardinalfish seeks refuge in the spines of urchins and in return cleans its host's body.

Hermit crabs rely on the empty shells of dead snails to provide them with a protective covering. As the crabs grow, they exchange smaller shells for larger ones. Some species of hermit crab attach anemones to their shells for added protection. The anemones are carefully transferred to new shells when an exchange is made. The crab massages the base of the anemone until it releases its hold on the old shell. The crab then transfers the anemone to the new shell and holds it in place until the anemone attaches. The anemone benefits by being carried to different feeding grounds. Some species of hermit crabs even feed their anemones by dropping morsels of food into their mouths. The boxer crab (*Lybia tesselata*; **Figure 15-23**) attaches anemones to its claws, making them even more formidable weapons against potential predators.

These are only a few examples of the thousands of symbiotic relationships that occur in the coral reef community. In a sense, the entire coral reef community can be thought of as a huge interactive complex where all organisms are bound by interdependencies.

FIGURE 15-23 BOXER CRAB. The boxer crab (*Lybia tesselata*) attaches anemones to its chelipeds for use as defensive weapons.

IN SUMMARY

- Competition is very prevalent among coral species, as space and light are limited.

- Slower-growing corals extend mesenterial filaments from their digestive cavity and kill their competitor's polyps. In response, fast-growing corals may produce very long sweeper tentacles containing cnidocytes that kill their competitor's polyps.

- Fast-growing branching corals ultimately dominate at the upper shallower portion of the reef, whereas more massive forms, being more shade tolerant, dominate in deeper areas.

- Most reef fish are carnivores. How so many fish species coexist on coral reefs without being excluded by competition is a source of debate.

- Fish, sea urchins, and other predators prey on such competitors of corals as fleshy seaweeds, sponges, soft corals, and gorgonians, giving competitively inferior reef corals an advantage in securing space.

- Many species feed on corals, but it is rare for corallivores to completely destroy coral colonies.

- Many examples of symbioses, ranging from parasitism through commensalism to mutualism, occur on coral reefs.

15-7 Adaptations of Reef Dwellers

Organisms that inhabit coral reefs display an amazing array of adaptations that help them survive in this fiercely competitive environment. Many have evolved adaptive behaviors and specialized anatomical features that increase their chances of survival and reproduction.

ADAPTIVE BEHAVIORS TO AVOID PREDATION

A number of behavioral mechanisms have evolved among reef organisms to avoid predation. During the day, many invertebrates hide in the sand, in crevices, and beneath coral formations. At night, they emerge from their hiding place to feed. Sea cucumbers commonly shed their internal organs when stressed. When a predator sees these organs, it will typically eat them, giving the sea cucumber an opportunity to escape into a crevice or burrow into the sand. Later, the animal regenerates the lost organs.

When attacked, soapfishes (family Grammistidae; **Figure 15-24**) produce a sudsy coating of mucus that is poisonous and has an unappealing taste. Trunkfishes also produce a sudsy coating of poisonous mucus when disturbed. The pearly razor-fish (*Hemipteronotus novacula*), which feeds on small molluscs on the reef floor, dives headfirst into the sand and buries itself when disturbed. Pufferfishes (family Tetraodontidae) inflate themselves to look like too much of a mouthful for a potential predator.

A high proportion of predators are more active at night than at day. As a result, many fishes that are active in daylight such as parrotfishes (family Scaridae) and wrasses (family Labridae) seek places to hide as twilight approaches. These animals have well-developed eyesight that serves them

FIGURE 15-24 **PROTECTIVE BEHAVIORS.** This greater soapfish (*Rypticus saponaceus*) produces toxic mucus when it is disturbed.

FIGURE 15-26 **SEA ANEMONES.** The giant Caribbean anemone, *Condylactis gigantea*, is a common inhabitant of coral reefs in Florida and the Caribbean.

Radioles are heavily ciliated feather-like tentacles found in clusters on the crowns of fanworms and Christmas tree worms.

well during the day, but at night they cannot see well enough to avoid predators. As night approaches, they converge on the reefs and wedge themselves into crevices for the night. Some species of parrotfish take the extra precaution of secreting a mucous cocoon that surrounds their body and masks their scent from nocturnal predators **(Figure 15-25)**.

STRUCTURAL ADAPTATIONS FOR FEEDING

Sea anemones **(Figure 15-26)** and other cnidarians use stinging cells (cnidocytes) located on their tentacles to capture plankton, small crustaceans, and small fishes that pass by. Sea anemones usually remain fixed in one place, firmly attached to the bottom by means of their muscular mucus-secreting basal disk. If conditions become unfavorable, however, some species detach and, by undulating their basal disk, crawl snail-like

along the bottom at 10 centimeters (about 4 inches) per hour until they find a more suitable area. Other species turn themselves upside down and walk on the tips of their tentacles. Still others lie on their sides and ripple their bodies, moving along like oversized inchworms.

Among annelids found on coral reefs, Christmas tree worms (family Serpulidae; **Figure 15-27**) are sessile filter feeders that use **radioles**, hairlike appendages that circle outward from each of the organism's two "crowns," to catch phytoplankton from the water column. The related feather duster worms (family Sabellidae) are active suspension feeders, waving their "feathers" to create water vortices on the surface of the tentacles. Their tentacles have sticky mucus that traps any waterborne food particles.

Molluscs, with the exception of bivalves, use their radula to graze on algae, sponges, coral polyps, and other organisms. Cone snails (family Conidae) and some other groups use a modified radula to sting and paralyze such prey as molluscs and echinoderms. Also included among

FIGURE 15-25 **SLEEPING PARROTFISH.** Several species of parrotfish (family Scaridae) extrude mucous from their mouth to form a cocoon, presumably to hide their scent from predators. It may also act as an early warning system, allowing the parrotfish to flee when the cocoon is disturbed.

FIGURE 15-27 **REEF FILTER FEEDERS.** Reef dwellers such as these annelid Christmas tree worms filter their food from the surrounding water.

FIGURE 15-28 REEF OCTOPUS. This Caribbean reef octopus (*Octopus briareus*) can change color to match its background.

the molluscs are such active predators as the octopus **(Figure 15-28)** and squid. Adaptations that aid them in feeding include a specialized mantle for propulsion, tentacles with suckers, keen eyesight, a well-developed nervous system, and the ability to rapidly change color. They grasp their prey with tentacles and dismember it using a modified radula (squids) or a sharp beak (octopods). Squid feed primarily on fish, shrimp, and other cephalopods, whereas octopods prefer molluscs, polychaete worms, and crustaceans, especially crabs. In clams and other bivalves, water containing food particles is drawn into the body through an incurrent siphon by the action of cilia. Food is trapped by mucus covering the gills and transported along a groove to the palps, which push it into the mouth. An excurrent siphon carries the water away.

Among crustaceans, mantis shrimp (order Stomatopoda) have forward appendages shaped like the forelimbs of a praying mantis **(Figure 15-29)** so sharp they can cut another shrimp in half when they strike. They

FIGURE 15-29 MANTIS SHRIMP. Mantis shrimp, members of the crustacean order Stomatopoda, possess sharp powerful appendages that are used to stun and dismember prey. The appendages are so powerful that a single strike can break aquarium glass. Many divers refer to them as "thumb-splitters."

hide in crevices and under coral, where they wait for their prey to come close enough to attack. Snapping shrimp (*Alpheus*) use a modified cheliped (claw) to produce snapping sounds involved in territorial defense. They sometimes use the sound to stun the small fishes on which they feed. Feather stars (crinoids), an echinoderm, attach themselves to coral with feather-like appendages called cirri. They feed by forming a basket of mucus with their lacy cilia-covered tentacles and move them in such a way as to create a current that carries the food to the animal's mouth. Though generally sedentary, the feather star can move by creeping along on its cirri or using its arms in a swimming movement.

PROTECTIVE BODY COVERING

Many reef animals avoid predation by having tough defensive exteriors. Many species of algae, sponges, corals, and other reef organisms make themselves less attractive to predators by depositing calcium carbonate crystals in their tissues. Most molluscs produce calcium carbonate shells to protect themselves from predators, the thickness of their shells being related to the crushing ability of their predators. Caribbean gastropod shells, for example, are usually less thick than their Indo-Pacific counterparts, but their predators likewise have weaker crushing appendages. Crustaceans have hardened exoskeletons that offer protection from predators. Sea urchins have hardened tests and spines for protection. Trunkfishes (family Ostraciidae; **Figure 15-30**) have bony skins similar to the shell of a turtle. Triggerfishes (family Balistidae) also have tough leathery skins that protect them from some predators. Armor is not without its drawbacks, however, as some mobility and growth potential is sacrificed for this protective exterior.

ROLE OF COLOR

Coloration plays a vital role in survival on a coral reef. Many reef animals use color as camouflage to be less conspicuous to predators and prey. Color can also be used to communicate a warning, to signal a willingness to mate, or to indicate one's territory.

Use of Color for Concealment and Protection

Many invertebrates, including shrimp, crabs, molluscs, and worms, have colors and shapes that allow them to blend with their environment. Cryptically colored helmet snails (family Cassidae) partially bury

FIGURE 15-30 PROTECTIVE BODY COVERINGS. This trunkfish from the Caribbean has a tough exterior similar to that of a turtle, making it unappealing to many predators.

themselves in the sand pockets of the reef and emerge at night to feed on sea urchins and other echinoderms. The extremely venomous stone-fish (*Synanceja horrida*) of Indo-Pacific reefs can be found half buried in the sand using its dull coloration and body shape to resemble the rocky substrate, complete with algal growth. When a small fish swims too close, the stonefish opens its mouth and swallows its unsuspecting prey. Some fishes, such as flounders and certain groupers, conceal themselves by changing color to match their background. Many reef fishes, however, are brilliantly colored. Although in the open water such colorful fishes would stand out, among brilliantly colored corals, algae, and other colorful organisms, a brightly colored fish simply blends in. Some fishes sport bright colors during the day but take on duller, less conspicuous colors at night. The Spanish grunt (*Haemulon macrosto-mum*) is silver and yellow by day but becomes dull and blotched at night. The large yellow dorsal fin of the pork-fish (*Anisotremus virgini-cus*) turns black at night. Even in bright moonlight, only a dull outline is visible.

Other Roles of Color

Fishes that have an unpleasant taste or specific defensive weapons such as sharp venomous spines frequently display aposematic (warning) coloration. Lionfishes (*Pterois*), with their venomous dorsal spines, are an example of a strikingly colored reef fish displaying this type of coloration.

Coloration also plays a role in defending territories and in mating ritu-als. Some species such as the harlequin tusk wrasse (*Choerodon fascia-tus*) of Australia **(Figure 15-31)** display bright colors as they vigorously defend a territory. In essence, the bright coloration is a "no trespassing" sign to others of the same species. The bright colors of adult parrotfishes indicate the sex of an individual and help attract mates. Male gobies also display bright colors at mating time. The brightly colored male attracts the attention of a female, and she is lured to his lair by an elaborate courtship dance.

FIGURE 15-31 HARLEQUIN TUSK WRASSE. The harlequin tusk wrasse (*Choerodon fasciatus*) of Australia's Great Barrier Reef vigorously defends its territory while displaying bright colors that advertise its presence to others of the same species who might trespass.

Norbert Wu/Minden Pictures/National Geographic

IN SUMMARY

- Evolution has provided reef animals with a host of adaptations that help them feed, avoid predators, or become more efficient predators.

- Some animals use specialized body coverings to protect themselves from predators, and others use toxic substances or sharp spines.

- Many animals exhibit special behaviors that help them avoid predation or find prey.

- Some color patterns make animals difficult to see, allow-ing them to avoid predation or to sneak up on their prey.

- Color may be used as a warning to others that an animal is venomous or toxic, as an aid in defending territories, and in mating rituals.

15-8 Threats to Coral Reef Communities

The health of coral reefs is affected by both natural phenomena and human activities. Hurricanes and typhoons can cause massive destruc-tion by toppling coral formations and removing them from the reef. The global coupled ocean-atmosphere phenomenon known as the El Niño Southern Oscillation (ENSO) can change winds, ocean currents, temperatures, rainfall, and atmospheric pressure over large areas of tropical and subtropical ocean. The 1982–1983 ENSO raised water temperatures 2°C to 4°C above normal, lowered sea level by 44 centi-meters (17 inches), and spawned a number of massive storms, resulting in the destruction of 50% to 98% of the corals on some reefs of the East Pacific and significant destruction elsewhere. The destruction of corals also affected organisms associated with the reef, eliminating many spe-cies but encouraging increased populations of others. Because natural controls protecting them were lost, the surviving corals were subject to increased predation by urchins and sea stars On many reefs of the East-ern Pacific, little recovery has been seen. An equally severe ENSO oc-curred in 1997–1998, with devastating effects throughout the Pacific. Although the effects of these natural phenomena are significant, they pale in comparison to the destruction and degradation of reefs caused by human activity.

WHY ARE CORAL REEFS IMPORTANT?

The rapid decline of coral reefs worldwide brings up the question, why are they important? Coral reefs are important for a number of reasons. They protect coasts from high surf conditions generated by storms, slow-ing down the water before it hits the shore. They remove a tremendous amount of carbon dioxide from the water and air through their photo-synthetic activity. They provide habitat for a huge diversity of inverte-brates and fish, many of commercial importance. Coral reef fisheries provide both human food and pets for the aquarium trade. Indeed, many people earn their living through collecting and processing reef products. Reefs also provide a place of recreation for human divers and snorkelers, attracting millions of tourists each year. We are now just learning about the many significant pharmaceutical products that can be harvested from reefs. Drugs originating from inhabitants of coral reefs are already used to fight cancer and block ultraviolet rays. Finally, coral reefs may be sen-tinels, similar to the proverbial "canaries in the coal mine," providing a warning to humans that many of our practices threaten not only these beautiful ecosystems but also human health and welfare.

EFFECTS OF HUMAN ACTIVITIES

Over the past 50 years, approximately 20% of the planet's coral reefs have been lost and another 60% are in a state of decline, primarily due to the effects of human activity. Humans affect the health of coral reef communities through destructive fishing practices, coastal development, pollution, tourism, coral mining, and climate change resulting from deforestation and the burning of fossil fuels.

Destructive Fishing Practices

A top-down **trophic cascade** occurs when the removal of predators or grazing species causes multiple effects throughout an ecosystem. For example, a study in Kenya found that heavy fishing pressure increased sea urchin densities over 100 times that of protected reefs, thereby increasing the level of grazing and greatly impacting autotroph populations. We have already discussed how overfishing of Jamaican reefs after the 1982 demise of the grazing sea urchin, *Diadema antillarium*, allowed fleshy algae to overgrow and smother corals, changing the structure of the whole community. In Jamaica, even small reef fish were removed for commercial sale, meaning that few individuals remained to grow to reproductive age and replenish the population. The use of cyanide, a metabolic poison, to collect tropical fish for the aquarium trade or fresh fish for local markets not only removes great numbers of fish from the reef but also poisons corals and other invertebrates. Fishers squirt this chemical into coral crevices to stun the desired fish and make them easy to catch. Some fishers use dynamite or other explosives (blast fishing) to stun large numbers of fish, a practice that can cause massive damage to reef coral. Another destructive practice is bottom trawling, dragging a net along the ocean floor. This practice destroys coral structure and is currently one of the greatest threats to deep-water coral communities. Although we have only mentioned a few examples, the effects of overfishing on coral reefs are being felt worldwide.

Coastal Development

Coastal development produces runoff containing nutrients, pesticides, and other toxic wastes, increases sedimentation, and changes patterns of water flow. The effect of runoff on coral reefs has been well documented in long-term studies in Kaneohe Bay, Hawaii, a complex estuarine system with a barrier reef, fringing reefs, and numerous patch reefs. At 12.7 kilometers (7.6 miles) long and 4.3 kilometers (2.6 miles) wide, it is the largest sheltered body of water in Hawaii. Once renowned for its coral gardens, Kaneohe Bay was only sparsely populated by humans before 1950. The subsequent increase in development and population brought significant modification of the shoreline, increased runoff, and the release of up to 5 million gallons of sewage directly into the bay each day. The nutrient-rich sewage greatly increased phytoplankton levels and encouraged the growth of seaweeds, particularly the green bubble algae, *Dictyosphaeria cavernosa*. Seaweeds overgrew the corals, smothering them, and phytoplankton blooms made the water cloudy. In response to public outcry, sewage water was diverted offshore in the 1980s. The removal of the sewage effluent caused significant changes throughout the bay. Bubble algae initially declined and the reefs began to recover. Suburbanization of the watershed, however, continued to allow uncontrolled nutrient discharge to the bay from golf courses, lawn fertilizers, and road runoff. By the 1990s, recovery seemed to plateau and bubble algae began making a comeback, particularly in the southern part of the bay. A number of introduced red algal species, including several imported from the Philippines for an aquaculture experiment, began to

overgrow significant amounts of coral throughout the bay. Herbivorous fish failed to control these undesirable species either due to reduced populations (perhaps a result of overfishing) or the abundance and lack of palatability of these algae. The overgrowth of coral has become severe enough that the state has begun to remove these noxious seaweeds with a vacuuming device termed a "super sucker" **(Figure 15-32)**. In addition, native sea urchins, *Tripneustes gratilla*, are now being farmed and transplanted to reefs to aid in longterm algal control. Although early reports on seaweed removal by vacuuming and transplant of sea urchins seem promising, it is unknown how successful it will be in the long term.

Trophic cascades are effects that changes in the size of one population in a food web have on populations below it.

(a)

(b)

FIGURE 15-32 **REMOVING NOXIOUS ALGAE.** Scientists from the University of Hawaii are using this equipment, termed a "super sucker," to remove invasive red algae that are overgrowing coral in Kaneohe Bay. It has been suggested that the "super sucker" reduces algal populations to levels controllable by reef herbivores.

ECOLOGY & THE MARINE ENVIRONMENT

Is Atmospheric Dust Contributing to the Decline of Coral Reefs?

Scientists continue to debate the reasons for the marked decline in the world's coral reefs and their lack of recovery. Although the increasing incidence of coral diseases, overgrowth of coral by macroalgae, reef destruction by human activities, and mass mortalities of reef-building corals and associated organisms has been widely reported, in relatively few cases has the root cause of reef destruction and lack of recovery been identified. Virginia Garrison and her colleagues at the U.S. Geological Survey have proposed a hypothesis for this decline of coral reefs, suggesting that African and Asian dust air masses transport chemical and microbial contaminants to the Americas, the Caribbean, and the northern Pacific that adversely affect coral reef ecosystems (**Figure 15-A** (1)). Dust provides nutrients that are normally limited in these systems, encouraging overgrowth by macroalgae, and broadcasts bacterial, viral, or fungal spores that may cause diseases in reef organisms. In addition, industrial contaminants may weaken coral reef organisms and make them more susceptible to pathogens. This dust has been transported for millennia (Charles Darwin described it in 1846), but the hypothesis suggests that both the amount and the composition of dust have changed over the last 40 years due to changes in climate, drought, and such human activities as industrialization, the increased use of pharmaceuticals and pesticides, and the burning of forests and grasslands. African dust carries not only plant nutrients but also synthetic organic chemicals, viable microorganisms, trace metals, and other pollutants of human origin.

Increasing circumstantial evidence supports this hypothesis. Caribbean islands formed by coral (Barbados, for example) have red iron and clay rich soils that are known to be of Saharan dust origin. It is also known that the Amazon rain forest in South America derives a significant amount of essential nutrients (mainly phosphate) from Saharan dust. Studies in Barbados have shown that the average annual surface concentration of African dust has increased since 1965, and that the peak dust years of 1983 and 1987 correspond with massive changes in Caribbean reef systems (**Figure 15-A** (2)). Although we have only a very limited understanding of the impact of this dust on marine ecosystems, some researchers have suggested that the dust could introduce pollutants that detrimentally affect coral immune systems, interfere with reproduction, cause disease, and, by providing limiting nutrients, encourage overgrowth of reefs by macroalgae. Several scientists, for example, have suggested that the iron in African dust fuels blooms of the cyanobacterium *Trichodesmium*. *Trichodesmium* can make nitrogen available to other organisms, including dinoflagellates, the organisms causing red tide.

Air samples taken during dust events typically contain 20 or more times the number of microbes that are found in clear air. That a great many of these microbes are still viable after their multiday journey from Africa surprised many microbiologists. A terrestrial fungus (*Aspergillus sydowii*), identified as the pathogen causing aspergillosis in Caribbean sea fans, was isolated from dust samples taken in the Virgin Islands during a dust event in 1997. Garriet Smith, a researcher at the University of South Carolina, has been able to induce this disease by inoculating sea fans with *Aspergillus* spores taken from dust samples. The bacterium causing white plague in corals (*Aurantimonas coralicida*) may also be associated with dust, possibly explaining the spread of this disease far from human influence. Although the bacterial species causing black band disease may be naturally present on reefs, it has

The urbanization of the shoreline has exacerbated another problem, freshwater runoff. Corals are very sensitive both to changes in salinity and to sedimentation. Excess sediment can smother coral, and freshwater kills their polyps, as well as other marine organisms. Storm floods in 1987 reduced salinity by more than half in the surface waters of Kaneohe Bay and resulted in the massive death of coral and other shallow-water marine organisms. Nutrients transported into the bay by the flood caused a spectacular phytoplankton bloom in the following weeks. Although it is the best-studied example, the problems associated with coastal development discussed here are not confined to Kaneohe Bay's coral reefs but are serious and common threats to coral reefs worldwide.

Other Human Activities

In some countries, coral is directly mined for use as bricks, as road fill, or as a component of cement. It may also be removed for curios and jewelry that are sold to tourists. Snorkelers and divers can harm reefs by grabbing, kicking, walking on, and stirring up sediments. Tourist resorts in many countries contribute to reef degradation by emptying their sewage directly into the water surrounding coral reefs or allowing seepage from poorly maintained septic tanks. Boaters damage corals with their anchors and spillage of oil and gas. Trash thrown overboard can cover reefs, blocking sunlight that coral polyps need to survive. Because discarded plastic bags resemble jellyfish, turtles sometimes eat them. The bags block the digestive tract and cause the animal to starve to death. Lost or discarded fishing gear ("ghost nets") entangle thousands of fish, sea turtles, and marine mammals.

BLEACHING AND CORAL DISEASES

Coral bleaching (**Figure 15-33a**) results when corals expel their symbiotic zooxanthellae. Corals lose 60% to 90% of their zooxanthellae, and each zooxanthella may lose 50% to 80% of its photosynthetic pigments. Corals take on a pale, or "bleached," appearance because the calcareous skeleton shows through the now-translucent tissues. Elevated temperatures seem to be the primary trigger of coral bleaching, but low temperatures, an excessive amount of radiation, aerial exposure, sedimentation, bacterial infection, or a combination of these factors may also elicit this response. If the stress is not too prolonged or severe, many coral species

been postulated that the iron in dust may be the trigger that activates these bacteria and allows them to infect the coral. Although many questions remain, African dust is one mechanism by which pathogens can be distributed throughout the western Atlantic. Dust from Chinese deserts is transported in a similar fashion throughout the Pacific and into western North America. Associated pathogens, along with transported nutrients and various contaminants, may be another factor causing the deterioration of coral reefs.

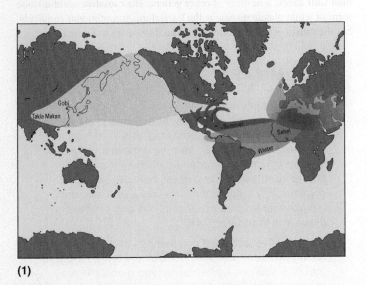

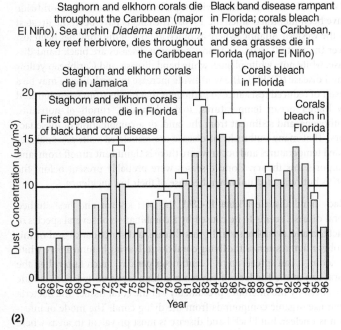

FIGURE 15-A **EFFECT OF DUST TRANSPORT. (1)** Dust travels from the Sahara and Sahal of Africa to the Americas, Europe, and Near East; and from the Takla Makan and Gobi deserts of China, across China, Korea, Japan, and the northern Pacific to North America. (*From Garrison et al. 2003. African and Asian Dust: From Desert Soils to Coral Reefs. Bioscience 53(5): 469–480; graphic by Betsy Boynton/USGS.*) **(2)** This figure describes the overall increase in African dust reaching Barbados, a Caribbean island, since 1965. Notice that the peak years for the dust deposition were 1983 and 1987, years of extensive environmental change on Caribbean coral reefs. (*Graph credit: Dr. Joe Prospero, University of Miami. Image Credit and Location: USGS.*)

FIGURE 15-33 **CORAL BLEACHING AND DISEASE. (a)** In coral bleaching, the zooxanthellae are expelled, possibly as a stress response to temperature. If they are not reacquired within a reasonable time, the coral colony dies. **(b)** Black band disease is caused by a consortium of bacteria and cyanobacteria. Because the bacterial species are naturally occurring on reefs, it is unclear what triggers the attack. **(c)** White pox disease is caused by the fecal bacterium *Serratia marcescens*. It has killed 85% of the elkhorn corals in the Florida Keys and an unknown number throughout the Caribbean.

regain their symbionts and recover. Some species may even be able to adapt to higher temperatures by acquiring symbionts that have differing physiological responses to both temperature and irradiance exposure. If the stress is prolonged, however, many corals fail to regain the zooxanthellae and eventually die. Although coral bleaching has been observed for more than 100 years, the incidence has greatly increased throughout the world since 1980. Bleaching associated with the 1997–1998 ENSO killed many corals in the Indian Ocean, although those in the Caribbean and on the Great Barrier Reef recovered when the waters cooled.

Over 20 diseases associated with significant coral mortality worldwide have been identified since the 1970s. The effects have been the greatest in the western Atlantic, where as much as 50% of the coral has been lost over the past 20 years. The most common diseases are black band disease, white pox, white plague, white band disease, and Caribbean yellow band disease (CYBD). The worldwide increase in coral disease may be a result of reduced resistance to pathogens that results from stress induced by higher ocean temperatures, toxic chemicals, runoff of nutrients, pathogens, and sediment from the land, or a combination of these factors. That many outbreaks of disease are associated with seasonal elevated temperatures and occur where there is significant runoff from land supports this theory. Coral diseases were probably present before the 1970s but remained unknown because of their low incidence.

Black band disease **(Figure 15-33b)**, the best-known and best-studied coral disease, seems to be caused by a combination of several species of bacteria and cyanobacteria. In this disease, a distinct dark band of microbes migrates across the living coral tissue, leaving behind a bare white skeleton. It can spread up to 1 centimeter per day, much faster than the coral can grow. The lack of oxygen and high levels of hydrogen sulfide produced by the microbes presumably kill the coral tissue. The bacteria then use organic compounds from the dying coral. The mode of infection is unclear, but black band disease is most prevalent in areas where corals experience stress from high temperature or pollution. Although only about 2% of coral is affected worldwide, this disorder is significant because it commonly attacks massive reef-building corals. First observed on Atlantic reefs, black band disease is now found worldwide.

Other coral diseases causing significant coral destruction include white pox, white band, white plague, and CYBD. White pox **(Figure 15-33c)** has killed 85% of the elkhorn corals in the Florida Keys and an unknown number throughout the Caribbean since it was first documented in 1996. The bacterium *Serratia marcescens*, found in human and animal feces, seems to be the cause of this disease. Laboratory studies have shown that the infection can be acquired within 5 days. White pox is highly contagious and spreads as rapidly as 2.5 square centimeters per day.

White band disease causes tissue to evenly peel off the skeletons of staghorn and elkhorn corals (*Acropora*). White plague is a similar-appearing disorder that attacks up to 33 species of stony corals. In CYBD, a spreading yellow patch appears on the coral, and the coral dies in the center of the patch. In the Caribbean, this disorder affects only certain members of the genus *Montastraea*, but a similar-appearing disorder in the Arabian Gulf affects a number of other genera. The causative agent for one form of white plague disease is the bacterium *Aurantimonas coralicida*, but the causes of CYBD and white band disease are unknown.

IN SUMMARY

- The health of coral reefs is affected by both natural phenomena and human activities. Corals reefs are important because they protect coasts from high surf conditions generated by storms, remove a tremendous amount of carbon dioxide from the water and air through photosynthetic activity, provide habitat for commercially important invertebrates and fish, and provide recreation for human divers and snorkelers. (Have You Wondered? #6)

- Humans affect coral reef communities through destructive fishing practices, coastal development and associated pollution, tourism, coral mining, and global climate change brought about by deforestation and the burning of fossil fuels.

- The incidence of coral bleaching and disease has greatly increased over the past few decades, probably as a stress response to environmental change.

KEY CONCEPTS

1. Scleractinian (stony) corals are responsible for the large colonial masses that make up the bulk of a coral reef. (15-1)

2. Corals obtain up to 90% of their energy from zooxanthellae, symbiotic dinoflagellates that use coral wastes, produce carbohydrates, and aid in calcium carbonate deposition. (15-1)

3. Coral reefs are constantly forming and breaking down, a process known as bioerosion. (15-1)

4. The major types of coral reefs are fringing reefs, barrier reefs, and atolls. (15-2)

5. The front of reefs is typically arranged in a spur-and-groove formation (or buttress zone) that disperses wave energy and

prevents damage to the reef and its inhabitants. (15-2)

6. Coral reefs are primarily found in tropical clear water, usually at depths of 60 meters or less. (15-3)

7. Indo-Pacific reefs differ from Atlantic reefs in a number of ways. Being older, they have a greater diversity of organisms, more extensive algal ridges, a far greater depth of reef carbonates, and coral surface coverage can approach 100%. (15-4)

8. The most important primary producers on coral reefs are turf algae and symbiotic zooxanthellae. (15-5)

9. Coral reefs are oases of high productivity in nutrient-poor tropical seas. Nutrients

are stored in reef biomass and efficiently recycled. (15-5)

10. Coral reefs are large interactive complexes full of intricate interdependencies among species. (15-6)

11. Inhabitants of coral reefs display many adaptations that help them avoid predation or be more efficient predators. (15-7)

12. Coral reefs are threatened by both natural phenomena such as the El Niño Southern Oscillation (ENSO) and human activities such as destructive fishing practices, coastal development, pollution, tourism, coral mining, and human-induced global climate change. (15-8)

1. Corals obtain up to 90% of their food energy from products produced by symbiotic zooxanthellae, primarily glucose, glycerol, and amino acids. A lesser amount comes from feeding on plankton, detritus, and dissolved organic matter.

2. Three main types of coral reefs are fringing reefs, barrier reefs and atolls. Fringing reefs surround islands or border continental landmasses; barrier reefs are separated from the nearest landmass by a lagoon; and atolls form on the tops of submerged volcanoes.

3. Low light levels prevent the development of coral reefs below about 60 meters.

4. Indo-Pacific reefs are older than Atlantic reefs, have a greater diversity of organisms, have more extensive algal ridges, grow to 60 meters or less in depth, have a far greater depth of reef carbonates, and coral surface coverage can approach 100%. Atlantic reefs have a deeper buttress zone, have coral growth extending down to 100 meters, are commonly dominated by the hydrozoan *Millepora complanata* (fire coral), have a greater biomass of sponges, and have coral coverage rarely exceeding 60%.

5. The most important primary producers on reefs are zooxanthellae of corals and other organisms, benthic algae, turf algae, sand algae, phytoplankton, and seagrasses, turf algae and zooxanthellae being the most important.

6. Corals reefs are important because they protect coasts from high surf conditions generated by storms, remove a tremendous amount of carbon dioxide from the water and air through photosynthetic activity, provide habitat for many important invertebrates and fish, and provide recreation for human divers and snorkelers.

QUESTIONS FOR REVIEW

Multiple Choice

1. The type of reef that is separated from its associated landmass by a lagoon is called
   a. a barrier reef
   b. an atoll
   c. a fringing reef
   d. a table reef
   e. a patch reef

2. The portion of a reef that rises from the ocean's depths is the
   a. reef crest
   b. reef front
   c. back reef
   d. reef flat
   e. spur-and-groove formation

3. The area of the reef that receives the full impact of wave energy is the
   a. reef front
   b. reef crest
   c. reef flat
   d. back reef
   e. drop-off

4. The corals that dominate areas of a reef that receive the most wave energy are
   a. fire corals
   b. elkhorn corals
   c. mushroom corals
   d. cluster corals
   e. flower corals

5. Corals supply their zooxanthellae with
   a. oxygen
   b. lipids
   c. sugars
   d. glycerol
   e. nitrogen

6. In the Caribbean, the most important reproductive mechanism of staghorn (*Acropora cervicornis*) and elkhorn (*Acropora palmata*) corals is
   a. budding
   b. sexual reproduction
   c. fragmentation
   d. broadcast spawning
   e. brooding

7. Which of the following is the major primary producer on coral reefs?
   a. turf algae
   b. coralline algae
   c. seagrasses
   d. zooxanthellae
   e. phytoplankton

8. An echinoderm that feeds on coral polyps is the
   a. sand dollar
   b. sea cucumber
   c. crown-of-thorns sea star
   d. feather star
   e. pencil urchin

9. Many reef organisms exhibit color patterns and body shapes that help them to
   a. blend in with their background
   b. defend a territory
   c. attract a mate
   d. be more efficient predators
   e. all of the above

10. An example of an animal that exhibits aposematic coloration is the
    a. stonefish
    b. lionfish
    c. cleaner wrasse
    d. harlequin tusk wrasse
    e. surgeonfish

11. Which of the following statements is *incorrect*?
    a. Corals on Pacific reefs seldom grow below 60 meters.
    b. Pacific coral reefs are of greater antiquity than Atlantic reefs.
    c. Corals of Pacific reefs seldom cover more than 60% of the surface.
    d. Sea fans (gorgonians) are much less abundant on Pacific reefs than Atlantic reefs.
    e. Many Pacific corals are diurnal, whereas most Atlantic corals are nocturnal.

12. An important predator of sea stars and sea urchins is the
    a. octopus
    b. squid
    c. shark
    d. Triton's trumpet snail
    e. sea cucumber

Matching

Match the following definitions with the appropriate term from the list below:

a. cocoon
b. fire coral (*Millepora*)
c. radioles
d. corallite
e. mesenterial filaments
f. brooders
g. dinoflagellates
h. bioerosion
i. patch reefs
j. mutualism
k. planula
l. coralline algae
m. bank reef
n. ahermatypic corals
o. soapfish

1. Group of organisms making up the zooxanthellae in corals.
2. Type of coral lacking zooxanthellae and not forming reefs.
3. Important noncoral contributor to the structure of Pacific reefs.
4. Noncoral species that is an important contributor to the structure of Atlantic reefs.
5. Type of larva produced by coral.
6. Calcium carbonate cup secreted by a single coral polyp.
7. Structures extruded from coral guts that secrete digestive enzymes and absorb digested organic matter.
8. Type of coral that broadcast their sperm into surrounding waters but retain eggs within the gastrovascular cavity.
9. The destructive phase of reef formation.
10. Type of reefs that occur within lagoons associated with barrier reefs or atolls.
11. Type of symbiosis between stony corals and zooxanthellae.
12. Large, isolated, flat-topped reefs found growing on midshelf regions of island or continental shelves.
13. Fish producing a sudsy coating of mucus that is poisonous and has an unappealing taste.
14. Secreted structure used by parrotfish to avoid predators.
15. Specialized appendages on Christmas tree worms used to trap phytoplankton.

Short Answer

1. What roles do coralline and calcareous algae play in reef formation?
2. Explain how the physical characteristics of the reef environment influence the species of corals that inhabit them.
3. Describe how a coral colony is formed.
4. Describe the process of reef formation.
5. Describe how an atoll is formed.
6. If the water surrounding coral reefs is nutrient poor, how can the coral reef support large numbers of organisms?
7. Describe what is meant by a cleaning symbiosis.
8. Why don't most coral species grow in the aphotic zone?
9. Why are coral reef fishes so brightly colored?
10. Why is moderate wave action important to the health of a reef?

Thinking Critically

1. How does the practice of using cyanide to collect aquarium fish affect corals and other reef organisms?
2. What impact does daily visits by hundreds of sport divers and snorkelers have on a reef community?
3. Why are coral reefs not found along warm coastal areas where large rivers such as the Mississippi, Amazon, and Congo empty into the sea?
4. Coral reefs are generally confined to the tropics, defined as the area between 23.4 degrees N and S latitudes. Yet coral reefs exist in Bermuda, which lies at 32 degrees N latitude. How is this possible?
5. What factors account for the small number of atolls in the Atlantic Ocean versus those in the Indo-Pacific?

SUGGESTIONS FOR FURTHER READING

Chaplin, Gordon. 2006. A Return to the Reefs, *Smithsonian* 36 (11): 40–48.

De'ath, G., J. M. Lough, and K. E. Fabricius. 2009. Declining Coral Calcification on the Great Barrier Reef, *Science* 323(5910): 116–119.

Dupree, J. 2007. Coral Crisis, *National Wildlife* 45(4): 22–30.

Dulvy, Nicholas K., Robert P. Freckleton, Nicholas V. C. Polunin. 2004. Coral Reef Cascades and the Indirect Effects of Predator Removal by Exploitation, *Ecology Letters* 7(5): 410–416.

Edwards, Helen J., I. A. Elliott, C. M. Eakin, A. Irikawa, J. S. Madin, M. McField, et al. 2011. How Much Time Can Herbivore Protection Buy for Coral Reefs Under Realistic Regimes of Hurricanes and Coral Bleaching? *Global Change Biology* 17(6): 2033–2048.

Lesser, M. P. 2007. Coral Reef Bleaching and Global Climate Change: Can Corals Survive the Next Century? *Proceedings of the National Academy of Sciences of the United States of America*, 104(13): 5259–5260.

Guest, J. 2008. Ecology: How Reefs Respond to Mass Coral Spawning, *Science* 320(5876): 621–623.

Henderson, C. 2006. Paradise Lost, *New Scientist* 191: 29–33.

Hoegh-Guldberg, O., P. J. Mumby, A. J. Hooten, R. S. Steneck, P. Greenfield, E. Gomez, et al. 2007. Coral Reefs under Rapid Climate Change and Ocean Acidification [Review], *Science* 318(5857): 1737–1742.

Jokiel, P. L., K. S. Rodgers, I. B. Kuffner, A. J. Andersson, E. F. Cox, and F. T. Mackenzie. 2008. Ocean Acidification and Calcifying Reef Organisms: A Mesocosm Investigation, *Coral Reefs* 27(3): 473–483.

Holland, Jennifer. 2011. A Fragile Empire, *National Geographic* 219(5): 35–36, 38, 41–46, 49–53.

Kolbert, Elizabeth. 2011. The Acid Sea, *National Geographic* 219 (4): 100–101, 103–109, 111–113, 115–117, 119–121.

Kuffner, Ilsa B., Andreas J. Andersson, Paul L. Jokiel, Kuʻulei S. Rodgers & Fred T. Mackenzie. 2008. Decreased abundance of crustose coralline algae due to ocean acidification, *Nature Geoscience* 1:114–117.

Mora, C. 2008. A Clear Human Footprint in the Coral Reefs of the Caribbean, *Proceedings of the Royal Society B: Biological Sciences* 275(1636): 767–773.

Pala, C. 2008. The Reel Downfall of Reefs, *E: The Environmental Magazine* 19(4): 16–18.

Prosek, James. 2010. Beautiful Friendship, *National Geographic* 217 (1): 121–122, 124–128, 130–131.

Prospero, J. M., and R. T. Nees. 1986. Impact of the North African Drought and El Niño on Mineral Dust in the Barbados Trade Winds, *Nature* 320: 735–738.

Sheppard, Charles R.C., Simon K. Davy, Graham M. Pilling. 2009. *The Biology of Coral Reefs*, Oxford University Press, New York, 352 pages.

Stone, Gregory. 2011. Phoenix Rising, *National Geographic* 219(1): 70–71, 73, 75–77, 79–80, 82.

Vroom, P., K. Page, J. Kenyon, and R. Brainard. 2006. Algae-Dominated Reefs, *American Scientist* 94(5): 430–437.

Warne, K., photos by B. Skerry. 2008. An Uneasy Eden, *National Geographic* 214(1): 144–157.

Yates, K. K., and R. B. Halley. 2006. CO_3^{2-} Concentration and pCO_2 Thresholds for Calcification and Dissolution on the Molokai Reef Flat, Hawaii. *Biogeosciences* 3: 357–369.

Have You Wondered?

1. Why waters over continental shelves are so productive? (16-1)

2. How the nature of bottom sediments affects benthic organisms? (16-1)

3. Why kelp communities are so important? (16-2)

4. Why coastal seas are so important to commercial fishing? (16-3)

16

Continental Shelves and Neritic Zone

IN THE RELATIVELY shallow waters above the continental shelves, plentiful sunlight and abundant nutrients combine to support enormous numbers of primary producers. Grazing on these producers are equally impressive numbers of consumers. In the water column, schools of fish feast on the abundance of food. On the bottom, large numbers of filter feeders, deposit feeders, and scavengers feast on the detritus that rains down from the sunlit waters above. In turn, bottom fishes prey on the abundance of benthic organisms or scavenge for food. In shallower waters with hard bottoms, forests and thickets of sizeable algae provide food and shelter for many animals.

The productive and bountiful coastal waters of the neritic zone are the source of most of the saltwater fish and shellfish that humans consume worldwide. Because of their high productivity, these areas are hotly contested by coastal countries, which vie for the exclusive rights to fish the waters. These areas are not infinitely productive, however, and overfishing in most parts of the world has started to affect the productivity of coastal seas.

16-1 Continental Shelves

Coastal seas lie over continental shelves. These shelves extend as little as 1.6 kilometers (1 mile) from the coastline of western North and South America to as far as 1,500 kilometers (900 miles) off the Arctic coast of Siberia (**Figure 16-1**). Most continental shelves, however, average about 67 kilometers (40 miles) wide. They descend gradually from the shore, reaching depths of 130 meters (430 feet) at their most distant edge. At this point, the gently sloping bottom may become a steep slope or a sheer drop-off. Off the coast of Florida, in the Gulf of Mexico, the continental shelf ends in a drop-off that runs for 800 kilometers (480 miles) and at some points has perpendicular drops of 1.6 kilometers (1 mile). By contrast, off the coast of Chile, the continental shelf ends in a steep incline that descends uninterrupted to a depth of more than 6 kilometers (3.6 miles).

Worldwide, rivers annually carry almost 750 million tons of sediment to coastal seas. Some ends up deposited on the bottom of the continental shelves, and the rest is dissolved in the seawater. The sediments contain nutrients such as nitrogen, phosphorus, calcium, and silica that are essential for life in the ocean. Coastal upwellings also supply nutrients to coastal seas. The heavy concentration of nutrients and the large amount of sunlight this area receives combine to make continental shelves and the seas above them highly productive areas.

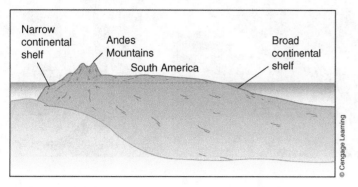

FIGURE 16-1 CONTINENTAL SHELVES. Continental shelves descend gradually from the shoreline out to sea, where they end at a steep slope or sheer drop-off. They can be broad, as they are along the Atlantic coasts of North and South America, or narrow, as they are along the Pacific coasts of the Americas.

WAVES, CURRENTS, AND LIGHT

One of the most important factors in determining the types of organism found in benthic communities is the stability of the environment. In shallower areas of the continental shelf, waves have an important effect. On open coasts, the effect of strong storm waves can reach to a depth of 80 meters (264 feet) **(Figure 16-2)**. In areas where waves cause the movement of sediments, they can be a major cause of mortality among benthic organisms. High wave action can also limit the body size of organisms that can survive in such a high-energy environment, because organisms with larger bodies are more susceptible to damage from wave shock.

Shear stress is force exerted by currents on an area of sea bottom.

A **plume** is a region of turbid water that spreads out from the mouth of a river or an estuary as it enters the sea.

The movement of water produces currents that can affect both shallow and deep portions of the continental shelf. The flow of water is extremely important because it affects the transport of organisms, their gametes, and their larvae. Water flow also affects the rate at which food is supplied to the shelf floor from the water above. The continental shelf closest to the shoreline is the area most affected by waves and currents. As one moves away from the shore, the effects of waves become progressively less. Further from shore the effects of currents on the seabed and its inhabitants tend to be more important. This shift is clearly reflected in the distribution of sessile filter feeders and deposit feeders, which increases with distance from the shore. Here the stress of waves and currents diminishes, and the bottom is subjected to less physical disturbance.

With the exception of the deep sea, currents at the seafloor are rarely as fast as surface currents. As water moves over the sea bottom it is slowed by friction produced as the water comes into contact with the uneven bottom. Currents and the force they exert on an area of sea bottom (**shear stress**) influence food availability for bottom dwellers. High levels of shear stress remove food from the seafloor and can injure epifauna. High current velocities also interfere with feeding activity. Organisms that live in such environments display characteristics or behaviors that help them cope with the extremes of this physical stress. Typically, attached organisms are either highly flexible (seaweeds) or encrusting (bryozoans). Mobile animals will often seek shelter from currents by burrowing within the sediments or hiding in crevices. On the other hand, areas of sea bottom with low current velocities will exhibit slow water movement, which will not replenish the food supply quickly enough. The slow input of food then becomes a limiting factor for growth in these regions.

Water depth and turbidity are important in determining the distribution of benthic algae in shallow waters close to the coast. At the mouths of rivers and openings of estuaries, the outflow of water carries sediments that spread out as they enter the sea. This flow pattern is called a **plume (Figure 16-3)**. Areas affected by estuarine plumes generally

FIGURE 16-2 EFFECTS OF WAVES. On exposed coastlines the effects of waves produced by storms can reach as deep as 80 meters. The wave force causes bottom sediments to shift and can increase the mortality rate in some benthic populations.

FIGURE 16-3 PLUMES. Sediments carried into the sea from rivers or estuaries tend to fan out, as shown in this photo, forming a plume. The plume increases the turbidity in these areas and limits the amount of sunlight that can reach the bottom. This in turn limits the size of populations of benthic algae.

(a)

(b)

FIGURE 16-4 BOTTOM COMMUNITIES. (a) Hard-bottom communities tend to be dominated by sedentary and sessile filter feeders, such as these sponges and tunicates. (b) Soft-bottom communities are home to many burrowing organisms, such as clams and worms.

receive minimal light due to high amounts of suspended sediment and detritus. These materials absorb light in the first few meters of the water column, leaving less to penetrate to the bottom. These areas tend to be dominated by animals, and any algae are restricted to very shallow water adjacent to the shore. Coastal areas that are open to the ocean and have limited discharge from rivers have much clearer waters. These areas support a large variety of algae, which can be the majority of biomass in these regions. The width of the zone dominated by algae depends on water clarity and slope of the seafloor, which influences the depth.

ROLE OF SEDIMENTS

The number and type of organisms that can live in and on the bottom of continental shelves are greatly influenced by the characteristics of the sediments that make up the bottom. In areas of a continental shelf where currents flow, the bottom is composed of coarse sediments. The moving water tends to carry away the fine light sediments such as silt, leaving behind larger sand, gravel, and rock particles. The bottom in these areas is constantly shifting and is not a good habitat for infauna, animals that burrow in the sediments, and **interstitial animals**, animals that live in the spaces between sediment particles. These animals cannot withstand abrasion or the constant shifting of the sediments. The flow of water, however, does carry with it a large supply of food. Epifauna that live on surface sediment, primarily sedentary or sessile filter feeders and suspension feeders such as sponges, anemones, and colonial cnidarians, are well adapted to this type of bottom **(Figure 16-4a)**. The size of the sediment particles provides them with a firm surface for attachment, and the large supply of suspended food provides a constant source of nutrients.

Where bottom currents are weak, the bottom is composed of fine sediments such as silt. The sediments here are more stable and support a variety of infauna that construct permanent burrows, such as polychaete worms, amphipods, and clams **(Figure 16-4b)**. Most organisms living in this habitat are deposit feeders, feeding on the organic material that settles from above and becomes trapped in the soft sediments. These bottoms do not support many filter feeders because of the scarcity of suspended food and the fine sediments that interfere with the animal's filtering structures.

IN SUMMARY

- Coastal seas lie over the continental shelves, a region known as the neritic zone.

- Continental shelves vary in their characteristics depending on the geographical and geological characteristics of the coastlines with which they are associated.

- The water column over the continental shelves is quite productive because of the abundance of sunlight and the high nutrient input from the adjacent coasts. (Have You Wondered? #1)

- The number and kinds of organisms that can live on the bottom of continental shelves is greatly influenced by waves, currents, light penetration, and sediment characteristics. (Have You Wondered? #2)

16-2 Benthic Communities

A variety of benthic communities can be found on the continental shelf from the shallow subtidal zone to the edge of the continental shelf. In some areas where the bottom is primarily composed of rock, large forests of seaweed form a habitat for a variety of animals that rely on these large algae for food and shelter. Where the bottom is soft, burrowing organisms of all kinds dominate. On continental shelves, food supply is the major limiting factor. In most areas of a continental shelf, large amounts of food in the form of detritus rain down from the sunlit waters above and fertilize the bottom sediments. The detritus consists of dead plankton, dead and dying larger animals, and organic debris, including human sewage, washed into the sea from land by rivers and runoff. This continuous food supply supports many filter feeders, suspension feeders, and detritus feeders, which in turn supply food for other animals. Filter feeders and suspension feeders such as sponges, tunicates, and bivalves feed directly on the detritus by filtering large volumes of water containing detritus

Interstitial animals are animals that live in the spaces between sediment particles.

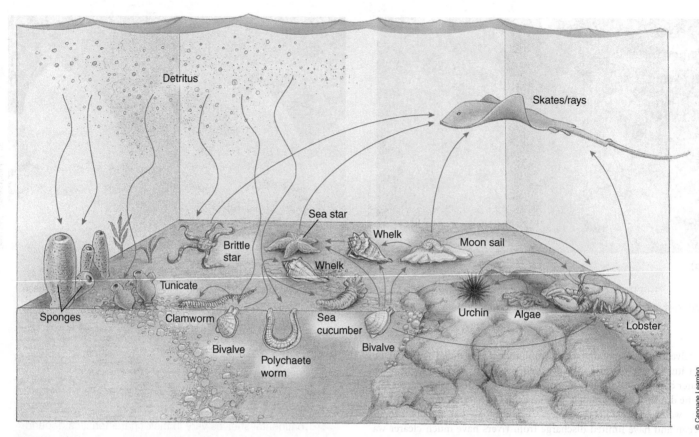

FIGURE 16-5 BENTHIC FOOD WEB. Detritus and plankton form the basis of the benthic food web in coastal seas.

through their bodies **(Figure 16-5)**. Several species of polychaete worm feed on trapped detritus as they consume bottom ooze in the process of making their burrows. Sea cucumbers use their sticky tentacles to gather food off the bottom, and their relatives the sea urchins move slowly over the bottom, chewing off bits of algae and detritus that have become attached to the firmer bottom sediments.

HARD-BOTTOM COMMUNITIES

The hard-bottom habitat consists of large sediments that cannot be pushed apart such as boulders, rock, and clam shells. These habitats are most often found off rocky coasts. Organisms that live in this type of habitat include sessile organisms that attach to the solid surface by cementing themselves in place or by attaching with strong, thread-like structures, and those that can freely move over the hard surfaces. Hard-bottom communities tend to be dominated by sessile colonial animals such as bryozoans, hydroids, sponges, and colonial tunicates. Other common animals in this habitat include anemones, bivalves, snails, crustaceans, and echinoderms. The large number of colonial animals in hard-bottom communities may reflect the inability of these organisms to deal effectively with the changes in temperature and salinity and the problem of desiccation encountered in shallower-water habitats such as the intertidal zone.

Epibenthic organisms, which live on the surface of the bottom sediments, are not evenly distributed on hard bottoms, and as a result there is an uneven distribution of benthic organisms. This characteristic distribution of organisms in a population or community is referred to as **patchiness (Figure 16-6)**. Groups of organisms are sometimes separated by a few centimeters and at other times by hundreds of meters. Exposure to sunlight is one factor that contributes to the patchy distribution. Vertical and near-vertical rock walls that receive less sunlight are dominated by sessile invertebrates, whereas a variety of algal species dominate the horizontal surfaces in water that is shallow enough to receive sufficient sunlight, except in areas that are heavily grazed by sea urchins.

Disturbance also plays a role in the distribution of benthic organisms. Landslides and shifting sediments have a greater impact on sessile invertebrates than on thick algal turf. On rock walls, large patches of diverse organisms dominated by bryozoans, sponges, and tunicates spread for hundreds of meters. Between these patches are patches of less diversity dominated by calcareous algae. Patches of low diversity are maintained by periodic landslides that clear away the more delicate invertebrates and allow the calcareous algae to dominate. The areas of high invertebrate diversity appear to be maintained by larval recruitment. These invertebrates produce larvae that spend a short time in the water column; thus they do not disperse too far from the population and can readily replenish it. Many invertebrate species in these communities also reproduce asexually. Species that reproduce asexually gain a competitive advantage by being able to either overgrow their neighbors or resist being crowded out by other species.

KELP COMMUNITIES

One of the most productive marine communities on some continental shelves is the kelp bed. Kelp is a type of brown algae that requires a rocky bottom, cold water, and a continuous supply of nutrients to support its high level of photosynthetic activity. Most kelp beds are found in water that is no more than 20 meters (66 feet) deep, but if the water is exceptionally clear, they may occur in water as deep as 30 meters (99 feet). Kelp are strictly cold-water organisms that rarely survive in areas where the average surface water temperature exceeds 20°C **(Figure 16-7)**.

Epibenthic organisms are organisms that live on the surface of bottom sediments.

Patchiness is an uneven distribution of benthic organisms.

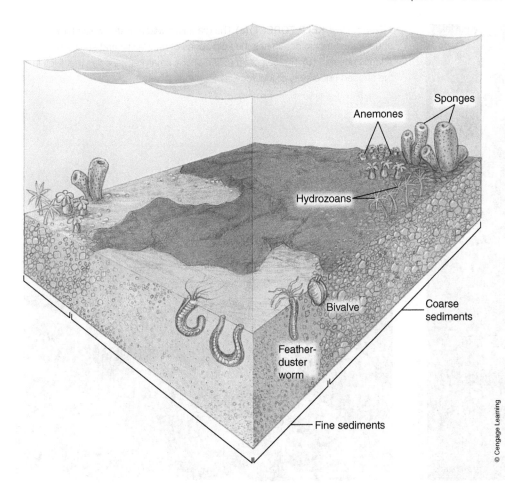

© Cengage Learning

FIGURE 16-6 PATCHINESS. Patchiness refers to the uneven distribution of benthic organisms that results from differences in the characteristics of bottom sediments. Areas of coarse sediments and fine sediments are randomly distributed on the continental shelf, and each supports different types of organisms.

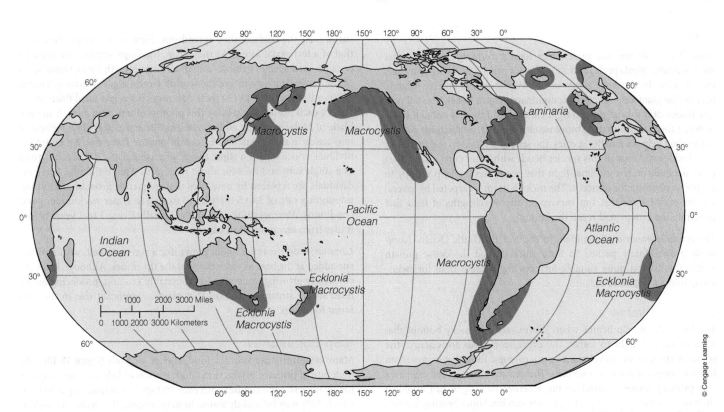

© Cengage Learning

FIGURE 16-7 KELP DISTRIBUTION. Kelps are found on continental shelves where the bottom is rocky and the water is cold. The kelp genera listed above are dominant in the indicated areas.

(a)

KIKE CALVO / Visual&Written SL / Alamy

(b)

Doug Sokell/Visuals Unlimited

FIGURE 16-8 KELP FOREST. (a) The giant kelp *Macrocystis*, which forms underwater forests, dominates the kelp zones of the Pacific and southern Atlantic oceans. **(b)** The sea palm (*Eisenia*) is a small kelp species that lives in the shaded understory of *Macrocystis*.

Kelp Beds

Some kelp beds are like underwater forests, forming a canopy that shades smaller algal species and an understory that is home to many animal species. In the kelp zones of the Pacific and southern Atlantic oceans, the giant kelp *Macrocystis* dominates **(Figure 16-8a)**. The tall narrow blades of this kelp grow in tight clusters, and there is enough space between them for animals to move about with ease. Beneath the canopy of the giant kelp is a smaller species, the sea palm (*Eisenia*) **(Figure 16-8b)**. The thick elastic stipe of this species bends with water currents. The sea palm depends on the minimal light that penetrates the kelp canopy to supply its photosynthesis needs. The rocky bottom is carpeted by several species of red algae that can survive on the wavelengths of light that penetrate the canopy and reach the bottom.

The genus *Laminaria* is dominant in the North Atlantic Ocean. *Laminaria* grows closely packed to form thickets, and the dense growth makes this habitat more suitable for crawling animals than for swimming animals **(Figure 16-9)**.

Kelp Life Cycles

The life cycle of kelp begins when spores settle on rocky bottom that receives enough light to satisfy the organism's needs for energy. The spores germinate and develop into a microscopic form that is preyed on by herbivores, especially sea urchins. The density of herbivores in an area is a primary factor in determining whether new algae can establish themselves. Once established, kelps grow quickly. Stipes bearing the flattened blades that carry out photosynthesis grow toward the light at the surface. At the surface, they spread out, forming a canopy similar to that of a terrestrial forest. As they grow, the kelps' appearance changes because each species exhibits a characteristic growth form. Giant kelps of the genus *Macrocystis* are the world's largest algae, reaching lengths of 20 to 40 meters (66 to 132 feet). *Macrocystis* has gas-filled floats at the base of each blade to help buoy this photosynthetic structure. A mature blade of *Macrocystis* can add as much as 50 centimeters (19.5 inches) of new tissue in a single day when conditions are favorable. An entire individual, consisting of a stipe, floats, and as many as 200 blades, grows as a single unit and lives for about 6 months. After 6 months, these individuals are replaced by new ones that frequently grow in pairs at the astounding rate of 3 to 5 meters (10 to 16.5 feet) per week under good conditions. *Macrocystis* is a perennial, usually living 3 to 7 years before it dies from any number of causes.

Laminaria, on the other hand, grows like a conveyor belt, with new tissue added at the base as older tissue at the tip erodes. Although kelps are constantly growing, they are also constantly eroding, producing an almost steady stream of detritus that plays a significant role in the kelp forest food web.

Kelp Community

Many organisms use kelp for food, shelter, or both **(Figure 16-10)**. The high rate of primary production and retention of kelp detritus supports highly diverse communities with over 200 species of large organisms, of which 36% may be mainly found in kelp forests. The dense network of kelp blades slows currents and decreases the force of all but the most

FIGURE 16-9 KELP BED. The kelp *Laminaria*, which forms dense thickets, is shown here growing along the base of a rock outcrop on the Atlantic coast.

Mark Conlin/Alamy

FIGURE 16-10 KELP FOREST COMMUNITY. Kelps provide both habitat and food for over 200 species of large marine organisms and countless smaller species.

energetic storm waves, thus providing convenient shelter for many fishes and mammals such as otters and seals. The large tree-sized algae greatly increase the amount of useful habitat in the area. In kelp forests off the western coast of the United States, echinoderms such as the omnivorous bat star (*Asterina miniata*) move slowly among the kelp, feeding on both animals and algae. Its relative, the many-rayed sunflower sea star (*Pycnopodia helianthoides*), feeds on urchins, other sea stars, and molluscs that it finds on the kelp forest floor. Fishes such as pipefish, blennies, eels, the Garibaldi (*Hypsypops rubicundus*), and many larger fish species find the kelp an ideal environment in which to live. Sheepshead (*Semicossyphus pulcher*) come to feed on many of the larger invertebrates, and several species of rockfish (*Sebastes*) feed on both invertebrates and other fishes. Without the kelp forests for habitat, these species would either disappear or be present in smaller numbers.

A species of mussel that lives in kelp beds is one of the few multicellular organisms that produces an enzyme that can digest cellulose. This species can feed not only on the bacteria that coat particles of kelp, as do other species, but also on the kelp fragments themselves. Other filter feeders attach directly to the surface of parts of the kelp, some growing over the surface of new blades almost as quickly as they are produced. Sponges grow at the base of kelp and, together with the kelp's holdfasts, form a habitat that shelters tiny shrimp, crabs, lobsters, brittle stars, and small fishes. The commercially important molluscs known as abalone (*Haliotis;* **Figure 16-11**) are common residents of kelp beds of the Pacific coast, and American lobsters (*Homarus americanus*) are a common resident of North Atlantic kelp beds **(Figure 16-12a)**. Lobsters inhabit holes and crevices in the bottom by day and come out at night to feed on molluscs, crustaceans, and urchins, as well as to scavenge. The American

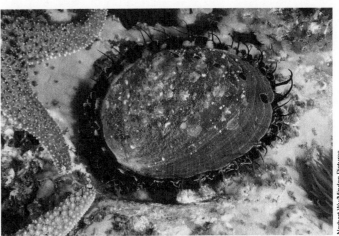

Norbert Wu/Minden Pictures

FIGURE 16-11 ABALONE. The red abalone (*Haliotis rufescens*) is one of several species of commercially important molluscs found in kelp forests. Not only are they harvested for human food but they are also a favorite food of sea otters.

FIGURE 16-12 LOBSTER. **(a)** The American lobster (*Homarus americanus*) is a commercially important crustacean that lives in the kelp beds off the New England coast. **(b)** Lobsters are caught in traps such as these that are placed on the bottom overnight. The lobsters enter the traps to feed on the bait they contain but can't get back out.

NOAA

(a)

William B. Folsom/NOAA

(b)

lobster is fished commercially using pots or traps. The animals enter these seeking shelter or attracted by bait and are unable to get out **(Figure 16-12b)**.

Although most of the organisms that live on kelp are filter feeders, some are herbivores that feed directly on the kelp itself. Snails crawl along the stipes, using their radula to scrape the outer layer of cells from the alga. Burrowing through the matted holdfasts are hordes of termite-like crustaceans called gribbles (*Phycolimnoria*; **Figure 16-13**). In some instances, the action of gribbles can be so destructive that the weakened holdfasts can no longer anchor the alga, and storm waves uproot and carry it away. It is thought that in a stable kelp forest gribbles may be the primary cause of mortality among adult kelp.

Impact of Sea Urchins on Kelp Communities

Kelps are also a favorite food of sea urchins. Wave action and predators that feed on sea urchins usually prevent urchins from doing significant damage to the upper portions of the kelp and the canopy, but there are few predators that prevent these animals from devouring young algae and damaging the important holdfasts. In the United States, a decline in the sea otter population along the West Coast and the lobster and cod populations along the North Atlantic coast has resulted in population explosions of urchins. The lobster population decreased by as much as 50%, the result of overfishing in the 1970s. During the same time period, an estimated 70% of the kelp beds in this area disappeared. Some researchers believe that increased numbers of urchins, along with naturally occurring

Courtesy of Pam Brown

FIGURE 16-13 GRIBBLES. The termite-like gribbles (*Phycolimnoria*) feed on the holdfasts of kelp and are a major cause of kelp mortality.

Geophoto/Natalia Chervyakova/imagebroker/Alamy

FIGURE 16-14 KELP HERBIVORES. The sea urchin, *Strongylocentrotus droebachiensis*, is a major kelp herbivore, primarily feeding on young kelp.

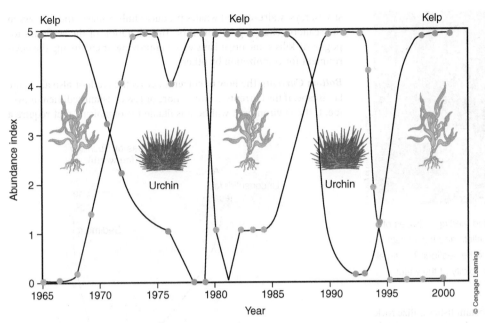

© Cengage Learning

FIGURE 16-15 EFFECTS OF SEA URCHIN GRAZING ON *LAMINARIA*. The depletion of cod and lobster due to overfishing over the last 40 years has allowed the populations of sea urchins to increase, resulting in the near decimation of kelp beds. Periodic episodes of sea urchin disease decrease the urchin population enough to allow the kelp to rebound to some extent. (Blue dots represent kelp abundance and yellow dots represent urchin abundance.)

events such as disease and climate change, may be primarily responsible for the decline in kelp beds in some areas over the last 50 years. Off the coast of Nova Scotia, the sea urchin *Strongylocentrotus droebachiensis* (**Figure 16-14**) is a major grazer of kelp. The overfishing of cod and lobster from inshore waters allowed the urchin population to increase, leading to the near elimination of kelp in the region. The increased size of the sea urchin population resulted in an increase in sea urchin disease and mortality. The decrease in the number of urchins allows the periodic reestablishment of kelp in the system. The current pattern is a boom-and-bust cycle (**Figure 16-15**) in which the urchins overgraze the kelp and, as their populations increase, disease spreads more readily, bringing the population down and allowing a partial recovery of the kelp. This is then followed by another cycle of increased urchin populations and overgrazing.

In areas where kelp has been completely removed, enough sea urchins remain to make it nearly impossible for new kelps to reestablish themselves. Although urchins may ultimately destroy the kelp bed, the animals continue to survive. Because they are primarily generalist feeders, sea urchins simply switch to feeding on other species of algae when kelps are not available. In southern California, sea urchins feed on the nutrients from treated sewage. Other animal species that depend on kelp for food or shelter are not as fortunate and soon disappear along with the kelp.

ROCK REEFS

Hard-bottom communities known as rock reefs contain several different microhabitats (**Figure 16-16**). The sides and overhangs of the reefs tend to be dominated by suspension feeders such as sponges, bryozoans, and tunicates. Many of these sessile organisms are adapted to living in environments where water currents tend to change. Current flow affects the delivery of food particles, and organisms that live in these habitats must be able to feed over a range of current flow rates. At low flow rates the delivery of

food particles may be so slow that feeding may cease, because it takes more energy to trap the minimal amount food than the food will supply. At high flow rates, feeding may not be physically possible as particles move so quickly that they are hard to trap or capture. The increased drag that occurs with faster currents may also increase the risk of body damage, so feeding structures are withdrawn. Many organisms feed only when current flow is optimal but they also exhibit patterns in feeding behavior that coincide with movements of zooplankton in the water column, so that they maximize feeding when food is most available.

The crevices and spaces in between rocks permit exchange of well-oxygenated water and provide organisms a safe haven from strong

Scott Leslie/Minden Pictures

FIGURE 16-16 ROCK REEF. Rock reefs offer many microhabitats for a variety of organisms that can live attached to the rock surface or in the many crevices and holes in the reef.

FIGURE 16-17 FECAL CASTS. The burrowing and feeding activities of deposit feeders produce large amounts of fecal pellets and fecal casts such as these that are deposited on the surface of the seafloor. The fecal matter breaks down and contributes to the local turbidity of the water.

currents and predators. Many commercially important fishes utilize rock reefs in their juvenile stages as protection from predators. The availability of suitable habitat is critical for commercially important animals such as lobsters (*Homarus*), which compete aggressively for the best refuges. When there is a shortage of suitable habitat, less competitive individuals are displaced and become vulnerable to predation.

SOFT-BOTTOM COMMUNITIES

Soft-bottom communities are found where the sediments are sand or mud or a mixture of both. Where currents are fairly strong, soft-bottom sediments are usually sand. Mud is common where currents are not very strong. Sandy bottoms are usually dominated by suspension feeders, whereas mud bottoms are generally dominated by deposit feeders. Suspension feeders and deposit feeders do not generally occur in the same area for several reasons. The water above muddy bottoms tends to be quite turbid because of the large number of fine particles suspended in the water. The turbidity of the water is increased by the burrowing and feeding activities of deposit feeders. These animals deposit large amounts of fecal pellets at the surface of the sediments **(Figure 16-17)**. The fecal pellets are easily broken down by bottom currents and increase the turbidity. The high concentration of suspended sediment particles clogs the feeding organs of suspension and filter-feeding animals, ultimately causing their death. The activities of deposit feeders also make bottom sediments more unstable, rendering the habitat unsuitable for suspension feeders because they do not have enough stable surface for attachment.

Patchiness in Soft-Bottom Communities

Like hard-bottom communities, the distribution of organisms in soft-bottom communities is patchy. The patchiness is the result of several factors, including changes in the bottom sediments, bottom currents, and patterns of larval settlement.

Changes in Sediment Distribution The animals that live in or visit soft sediments can modify what is originally a generally homogeneous area, changing it into an environment characterized by a discontinuous surface. For instance, burrowers, such as the sea cucumber *Molpadia oolitica* **(Figure 16-18a)** can produce fecal mounds that rise several centimeters above the bottom. These mounds are home to suspension feeders such as bivalves and polychaete worms. The areas between mounds are dominated by deposit feeders, whose activity makes the habitat inhospitable to suspension feeders. The feeding activity of predators such as

skates, rays, walruses, and whales produces hills, valleys, and furrows in the seafloor, changing its contours. This alteration of the seafloor's topography kills some organisms while improving or enhancing the environment for colonization by others.

Bottom Currents The flow of currents across the seafloor also alters the landscape. If the currents produce more or less constant sediment movement, the environment will be unsuitable for organisms that require a

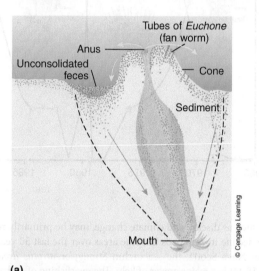

(a)

(b)

FIGURE 16-18 ALTERATIONS OF THE SEAFLOOR SURFACE.
(a) Living organisms can alter areas of the sea bottom by their activities. For instance, the burrowing sea cucumber *Molpadia oolitica* builds fecal mounds that provide a more stable surface for other bottom dwellers than the surrounding mud. **(b)** Bottom currents can produce a textured surface of ridges and troughs. The troughs contain large amounts of organic material that sinks to the bottom, providing a good habitat for deposit feeders. The ridges offer elevated surfaces for suspension feeders.

stable habitat. Currents may produce ripples on bottoms composed of fine sediments **(Figure 16-18b)**. The troughs between raised areas tend to collect more organic material and are good habitats for deposit feeders. On the other hand, suspension feeders are found on the more exposed crests. In areas where the currents are strong, the pattern of crests and troughs can change so quickly that they cannot be colonized by animals that require a more stable bottom.

Larval Settlement Many invertebrate animals have planktonic larval stages that may stay in the water column for weeks before settling. The pattern of settlement depends on currents, predation, and the type of bottom sediments. The uneven pattern of larval settlement results in a patchy distribution of organisms on the seafloor.

Soft-Bottom Food Chains

As in hard-bottom communities, the primary source of food for soft-bottom communities is detritus. Suspension feeders feed on detritus and plankton, and deposit feeders feed on detritus and the bacteria that decompose it. Burrowers, sedentary suspension feeders, and slow-moving grazers are preyed on by more-active animals. Polychaetes known as clamworms (*Nereis*), some as long as 0.5 meter (19.5 inches), feed on small bivalves that burrow in the soft bottom. Carnivorous snails such as members of the family Buccinidae (whelks) feed on both bivalves and other snails, usually by boring a small hole through the shell so that they can suck out the contents **(Figure 16-19)**. Sea stars move slowly along the bottom, feeding on mussels, oysters, and scallops. Brittle stars slither across the bottom powered by undulations of their serpentine rays. Some species burrow into the bottom, where they feed on decaying material, whereas others feed on detritus and small organisms that are passed along their tube feet toward their mouth. Brittle stars are perhaps the most abundant animals in soft-bottom communities. Some areas of soft bottom off the coast of Great Britain support as many as 80 million brittle stars per square kilometer (0.36 square mile).

Skates, rays, angel sharks (*Squatina*), and batfish (family Ogcocephalidae) are just a few of the fishes that forage along the bottom, feeding on molluscs and crustaceans that they crush with their powerful jaws **(Figure 16-20)**. These animals are well adapted to dwelling and feeding on the bottom. They usually have flattened or compressed bodies, retractable fishing lures, and specialized sense organs. They also have well-camouflaged dorsums that allow them to blend with the sea bottom.

FIGURE 16-20 COWNOSE RAY. This cownose ray (*Rhinoptera bonasus*) has powerful jaws to crush the hard shells of the molluscs and crustaceans on which it feeds.

A large number of fishes spend their lives on or near the bottom of the continental shelf. Fishes such as haddock (*Melanogrammus*), hake (*Merluccius*), pollock (*Pollachius*), cod (*Gadus*), and rockfish (*Sebastes*) are commercially valuable. They are caught in most of the coastal waters of the Northern Hemisphere. These fishes feed on molluscs, echinoderms, crabs, other fish, each other, and even their own young. At one time, it was estimated that as many as 10,000 cod lived above each acre of the continental shelf of the Grand Banks, off Nova Scotia, Newfoundland. As the result of overfishing, however, there are so few fish now that the area can no longer support a major commercial fishery. Another group of commercially valuable fish, the flatfishes (order Pleuronectiformes), live directly on the bottom. These include flounder (*Platyichthys*; **Figure 16-21**), halibut (*Hippoglossus*), turbot (*Rhombus*), and various species of sole (*Solea*).

Succession in Soft-Bottom Communities

When sediments are disturbed, by erosion or landslides, for instance, many animals are removed or killed. The disturbance also exposes deeper layers of sediments that are frequently anoxic because of the high levels of decomposition that occur there. After the disturbance

FIGURE 16-19 WHELKS. Whelks (Buccinidae) are carnivorous snails that feed on other molluscs, especially bivalves.

FIGURE 16-21 FLATFISHES. This flounder is just one of many important commercial species of flatfish that are harvested from coastal seas.

new organisms arrive by larval settlement to recolonize the area. Early colonizers are usually surface dwellers such as polychaete worms. These animals feed on the accumulating organic material. Their burrowing activities help aerate the surface sediments, and the increased oxygen content in the sediments allows other marine worms, along with bivalves and snails, to begin colonizing the area. The action of these animals continues to create tunnels in the sediments and improve aeration, making way for more permanent, deeper-burrowing organisms.

IN SUMMARY

- Much of the food that supports the primary consumers of bottom communities drifts down from the sunlit waters above and consists mostly of detritus and organic debris.

- The seabed of the continental shelf can be either hard bottom or soft bottom.

- Hard-bottom communities are dominated by epibenthic organisms that live on the surface of the hard sediments.

- Soft-bottom communities are dominated by burrowing organisms that are generally suspension feeders or deposit feeders.

- The distribution of organisms in benthic communities is uneven, or patchy due to several factors, including variations in environment, changes in sediment distribution, bottom currents, and larval settlement.

- Kelps form complex three-dimensional habitats for large numbers of invertebrate and vertebrate animals in areas of the continental shelf where the bottom is hard, the water cold, and nutrients are readily available. (Have You Wondered? #3)

- Kelps serve as an important food source, directly or indirectly, for many of the species that live in and around kelp beds.

- Rock reefs are hard-bottom habitats with many microhabitats.

16-3 Neritic Zone

The **neritic zone** comprises the coastal seas that lie along the edges of the continents and above the continental shelves. These waters contain an enormous number of phytoplankton, the microscopic photosynthetic organisms that form the basis of aquatic food webs. The density of plankton populations in coastal seas is astounding. A 10-milliliter (about 2 teaspoons) sample of surface water may contain thousands of planktonic organisms. Unlike the clear blue waters around coral reefs, coastal seas are green, a sign of high productivity. Grazing on these pastures of phytoplankton are hordes of tiny animals, the zooplankton. The abundant plankton provides food for large schools of fish **(Figure 16-22)**, and these in turn provide food for larger fishes, some marine mammals, seabirds, and humans. Although the neritic zone covers an area equivalent to only about the size of Asia and contains

The **neritic zone** is the zone of ocean that lies over the continental shelves.

Microphytoplankton are phytoplankton that measure between 20 and 200 micrometers.

Nanophytoplankton are phytoplankton that measure less than 20 micrometers.

JAMES FORTE/National Geographic Stock

FIGURE 16-22 COMMERCIAL FISHES. The large numbers of plankton in coastal seas support large schools of commercial fish.

only 10% of the ocean's expanse, it produces nearly 90% of the world's annual harvest of fish and shellfish. By comparison, the open ocean is a barren desert.

FOOD CHAINS IN THE NERITIC ZONE

The neritic zone receives freshwater runoff from the neighboring land. The runoff provides enough nutrients to support the growth of relatively large producers called **microphytoplankton** (20 to 200 micrometers), as well as smaller forms, the **nanophytoplankton** (less than 20 micrometers). In the colder waters around Antarctica and in the northern Pacific and Atlantic oceans, diatoms dominate the microphytoplankton. Planktonic organisms called coccolithophores are sometimes so numerous in areas of the North Sea that the water turns white. At certain times of the year, they even outnumber the diatoms. In warmer coastal waters, dinoflagellates are more numerous. Unicellular green algae occur in both tropical and polar seas. The composition of the phytoplankton varies from one region to another, from season to season, and sometimes even within a single season. Grazing by zooplankton and other animals also affects the composition of the phytoplankton. When the number of one type of phytoplankton decreases as the result of heavy grazing, another species will usually proliferate to take its place.

Drifting animals that make up the zooplankton **(Figure 16-23)** feed on the millions of diatoms, dinoflagellates, and other phytoplankton. The most abundant members of marine zooplankton are crustaceans called copepods. In coastal waters, the concentration of copepods can be as high as 100,000 individuals per cubic meter of water. The primary reasons for such large numbers are the tremendous reproductive capacity of these animals and the rich food supply. After being fertilized, a

IMPACT BUBBLE

THE WATERS OFF the New England coast once teemed with fish and supported a large varied fishing industry. For more than 400 years, commercial fishing has been an important part of the New England economy. An especially important fishing area is the Georges Bank, a huge shoal larger than the state of Massachusetts that extends for about 200 miles off the coast of southeastern New England. Historically, the bank has been a rich fishing area for groundfish, fish that swim close to the bottom such as cod, haddock, and yellowtail flounder. Unfortunately, fishers throughout the years, thinking there was no limit to the bounty of the sea, fished without regard to the fact that the numbers of fish were declining. The belief that the sea is a limitless resource for fish is not new:. That notion ultimately led to the overfishing and collapse of the New England halibut fishery in the late 1800s. Modern fishers did not learn from these past behaviors, and overfishing has pushed several populations of commercial fish in New England coastal seas to the brink of collapse.

In the past, some of the problems of overfishing were blamed on foreign fleets fishing in U.S. coastal waters. But with the passage of the Magnuson Act in 1976, which extended U.S. coastal waters to 200 miles offshore, the coastal waters of New England have been fished exclusively by domestic fishers. With the removal of foreign competition, the New England fishing fleet grew exponentially during the late 1970s and early 1980s. Fishing in the coastal zone was open access; anyone could fish anywhere without restrictions. New and better fishing boats were being built, with the latest in new electronic technology that increased the fish catch. With so much pressure being placed on limited numbers of fish, it was inevitable that a fishery collapse would occur.

By the mid-1990s, fish populations were in serious trouble. New England cod, once the most important commercial fish in the waters off southeast New England, reached their lowest levels ever recorded in the region in 1994. Alarmed by the significant decrease in the fish catch, government agencies implemented a new fishery management plan in 1995. Unfortunately, insufficient scientific data were available for estimating population sizes and analyzing trends in fish populations, and the new management plan did little to help the decreasing numbers of fish or to save jobs in the fishing industry. From 1995 to the present, fisheries managers have been struggling to balance fishing pressure with declining fish populations. Many fishers who entered the fishing business during the boom years of the 1980s and early 1990s were forced to fish for less desirable fish or went out of business, dealing a blow to some local economies. By 2005, the cod population on the Georges Bank was estimated to be 14% of what was considered a healthy population that could sustain commercial fishing. If current management practices are successful, it is estimated that the population of Atlantic cod on the Georges Bank might recover by 2026. Atlantic cod are not the only groundfish on the verge of commercial extinction due to overfishing. Some species of flounder are even more impacted than cod as the result of chronic pollution of their habitat compounding the effects of overfishing.

Recent action by fisheries managers to set catch limits based on solid scientific data and to develop accountability measures for fishers who exceed the limit will hopefully return fish populations along the New England coast to a more sustainable equilibrium. There are still problems to be resolved, such as how the rights to fish the recovered fish populations will be distributed and how a consolidation of the fishing fleet will affect jobs. With continued work on sustainability and sound management plans, however, fresh-caught fish will continue to support the region's fisheries and economy for years to come.

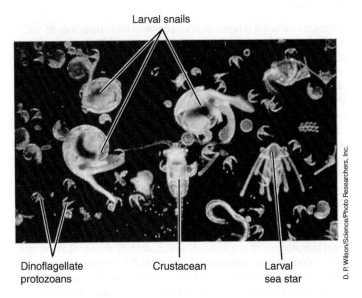

Larval snails

Dinoflagellate protozoans

Crustacean

Larval sea star

D. P. Wilson/Science/Photo Researchers, Inc.

FIGURE 16-23 ZOOPLANKTON. Members of the zooplankton include protists, adult crustaceans, jellyfish, and the larvae of both invertebrate and vertebrate animals.

female can produce as many as 100 eggs, and some species produce a new clutch of eggs every 4 or 5 days. Between 1 and 2 weeks after hatching, the new generation is mature and ready to reproduce. Copepods are the primary consumers of diatoms. A single copepod can consume as many as 120,000 diatoms per day.

Because tiny phytoplankton can be either eaten directly by small zooplankton or filtered from the water by benthic filter feeders such as clams and worms, food chains in coastal seas are frequently two or more steps shorter than those of the open sea. Small zooplankton are the preferred food of large fishes such as menhaden (*Brevoortia*) and alewives (*Alosa*), whereas large bottom fishes such as cod and haddock prey on filter feeders. In both cases there are only three trophic levels between primary producers and consumers of reasonable size **(Figure 16-24)**. Even larger animals such as tuna, sharks, and humans are only four trophic levels from producers. The combination of higher productivity and shorter food chains supports a larger number of higher-level consumers in coastal seas.

PRODUCTIVITY IN THE NERITIC ZONE

The most productive of the planktonic ecosystems are located in upwelling zones, where the combination of winds, ocean currents, and shape of the seafloor interact to bring nutrients into the photic zone from the

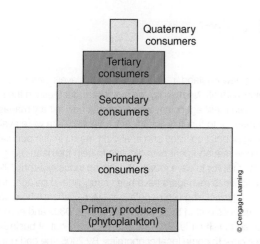

FIGURE 16-24 PRODUCTIVITY OF COASTAL SEAS. In this generalized pyramid, the size of each box indicates the relative amount of biomass at each trophic level. Because of the abundance of sunlight and nutrients, phytoplankton in coastal seas can reproduce at extremely rapid rates and can therefore support about five times their own biomass in primary consumers. These primary consumers, which include enormous amounts of zooplankton, are in turn a rich food supply for large populations of higher-order consumers. (D. P. Wilson, Science Source, Photo Researchers, Inc.)

ocean floor. The high productivity of these regions would not be possible without the activity of microorganisms. Bacteria break down the dead remains of plants and animals, releasing the nutrients to be used again. These nutrients are then returned by upwellings to the surface, where they are recycled by plankton. The shallow banks along the coasts of the North Atlantic and the Pacific coast of Peru are examples of such productive areas. These upwelling areas yield almost half the world's supply of commercial fishes. In these regions, the almost continuous supply of nutrients supports a phytoplankton community dominated by large chain-forming diatoms **(Figure 16-25)**. These chains, which consist of several cells linked together, are sizable enough to be eaten directly by large zooplankton such as the shrimp-like krill (*Euphausia*) of the Southern Ocean and small fish such as anchovies (family Engraulidae) of the Pacific Ocean off the coast of South America. Krill and anchovies, in turn, are large enough to be worthwhile prey for large fish, seabirds, seals, whales, and even humans. The high productivity and short food chains of upwelling areas support the greatest biomass of any planktonic system.

Biologists estimate that as much as one-seventh of the anchovies caught worldwide by humans and seabirds come from Peru's coastal seas in years when upwelling currents are strong. Smaller upwellings along the Pacific coast of the United States and islands such as Japan support locally productive fisheries. It is no wonder that so many animals, including humans, come to upwelling areas to feed. The primary production in these regions may be as much as six times higher than that of the open

ECOLOGY & THE MARINE ENVIRONMENT

The Bountiful Southern Ocean

At first, one might think that the frigid waters surrounding the continent of Antarctica wouldn't harbor many living organisms. In reality, though, the Southern Ocean, like other coastal seas, is an extremely productive area, supporting not only large numbers of aquatic organisms but also most of the animals that live in Antarctica. The basis of this rich food web is the numerous diatoms that dominate the phytoplankton. Dominant among the zooplankton that feed on these diatoms is krill (*Euphausia superba*). This herbivorous crustacean accounts for almost 50% of the zooplankton in the Southern Ocean. Although krill are found throughout the region, their distribution is not uniform. The greatest concentrations have been recorded in the Weddell Sea, where some swarms of krill have been estimated to be 2 million tons. It has been estimated that between 750 million and 1,300 million tons of krill are produced annually in the Southern Ocean.

During the summer, primary production in the Southern Ocean increases as the pack ice melts and the polar region receives more sunlight. Krill move to the surface to graze on the increased numbers of diatoms. During the winter, less light, increased pack ice cover, and increased turbulence lower primary production to near zero. Krill move to deeper water and apparently feed mostly on detritus.

As many as 20 species of squid, some relying heavily on krill as a food source, live in the Southern Ocean. The squid provide food for numerous species of toothed whales, seals, and birds. It is estimated that these animals consume 35 million tons of squid annually from this region. Many fish species are also found in the Southern Ocean. Some

species are permanent residents, and others are seasonal visitors, arriving in summer to feed on the large numbers of krill. It is estimated that all of the fish in the Southern Ocean consume 100 million tons of krill annually.

Although Antarctica is home to few species of bird, those that do live there are usually represented by large populations. Their diet consists of crustaceans, mainly krill and copepods, squid, fish, and carrion. Krill accounts for almost 78% of all the food they eat. The total annual direct and indirect consumption of krill by birds is approximately 115 million tons. Seven species of seal are found in the Southern Ocean. Of these, the crabeater seal's diet consists of 94% krill, the leopard seal's diet consists of 37% krill, and the fur seals of South Georgia Island feed almost exclusively on krill. The other seal species feed on fish and squid that in turn rely on krill. Larger baleen whales such as the blue whale feed mainly on krill, consuming about 43 million tons of it annually. Whaling has decreased the number and size of baleen whales in the Southern Ocean. It is estimated that 150 million tons of krill that was formerly consumed by baleen whales is now available to other krill-feeding species. Smaller whales, seals, and penguins appear to be the main beneficiaries of this change, as indicated by increases in the size of their populations, presumably because of the increased availability of food. Krill are also being harvested for human consumption. The krill being netted by fishers is assumed to be available because of the decrease in the size of whale populations. There is still some doubt, however, whether the productivity of krill is sufficient to feed humans without endangering other animal populations in the area.

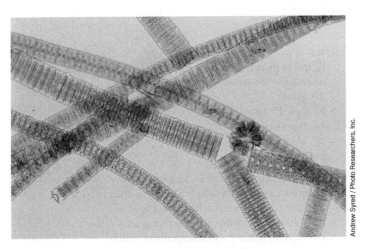

FIGURE 16-25 DIATOMS. The phytoplankton of upwelling areas are dominated by large chain-forming diatoms such as these.

sea, and they produce biomass at a rate more than 36,000 times as fast as that of the open sea. These limited areas represent valuable commercial resources that are protected by laws and require conservation.

OTHER ROLES OF PLANKTON IN COASTAL SEAS

The ecological importance of plankton goes beyond their role of supplying food for animals. An overwhelming number of animal species spend at least a part of their lives as members of the plankton. Sessile animals such as barnacles, mussels, and coral, which spend their adult lives fixed in one place, rely on their planktonic larval forms such as the nauplius, trochophore, veliger, and planula to colonize new territories. When spawning, these animals release hundreds to thousands of tiny eggs and sperm that unite in the water column and then scatter in the currents.

Almost all crabs, shrimps, and lobsters have larval stages that feed and grow as part of the plankton before settling down to lead their adult existence. Even free-swimming fishes such as herring and eels have tiny planktonic larvae and juvenile stages almost as small that join the zooplankton for the first weeks or months of their lives.

Only a small fraction of the larvae in plankton develop into adults. Odds are that only 1 in every 100,000 eggs found in the plankton will survive to adulthood. The health of plankton is important not only because of the role they play in supporting commercial fisheries but also because any disturbance to their community can cause unexpected changes in populations of animals whose adults are anything but planktonic.

IN SUMMARY

- The neritic zone comprises the water that lies above continental shelves. This area contains enormous numbers of phytoplankton, which form the basis of an extremely productive food web. (Have You Wondered? #4)

- The high productivity of these regions is due to a combination of physical and biological factors.

- Rivers and runoff deposit large amounts of nutrients directly into coastal seas, and the geography of continental shelves favors upwellings that bring nutrients from the bottom to the photic zone, where they can be used by plankton.

- Phytoplankton support large numbers of zooplankton and some large animals as well.

- Those large predators that do not prey directly on the plankton feed on smaller animals that are plankton feeders.

KEY CONCEPTS

1. The number and kinds of benthic organisms on continental shelves are influenced by waves, currents, light penetration, and sediment characteristics. (16-1)

2. Hard-bottom communities are dominated by epibenthic organisms. (16-1)

3. In areas north and south of the tropics, kelps (a type of brown algae) dominate the subtidal zone where the water is cold and the sediments are hard. (16-2)

4. Kelps are important primary producers and provide habitats for many animals. (16-2)

5. Soft-bottom communities are dominated by suspension feeders and deposit feeders. (16-2)

6. The distribution of organisms in benthic communities of the continental shelf is patchy. (16-2)

7. The neritic zone is the water column that lies above the continental shelves. (16-3)

8. The neritic zone receives high levels of nutrient input from rivers, coastal runoff, and upwellings. (16-3)

9. The neritic zone supports enormous amounts of phytoplankton. (16-3)

10. The high productivity of coastal seas supports large numbers of fish, birds, and marine mammals. (16-3)

ARE YOU STILL WONDERING?

1. Coastal waters receive nutrient runoff from the land and upwellings bring nutrients from deeper water to the surface. The heavy concentration of nutrients combined with large amounts of sunlight support large populations of phytoplankton which in turn support large numbers of zooplankton and fish.

2. Hard bottom sediments such as rock support seaweeds and a variety of sessile and sedentary filter feeders. Soft bottom sediments such as sand and mud support many burrowing organisms.

3. Kelp communities are important because they provide a complex three-dimensional habitat for hundreds of species as well as a food source for many species.

4. Although representing only 10% of the ocean, coastal seas provide 90% of the world's commercial catch. The high level of productivity found in coastal seas supports large populations of fish and shellfish, far more than the open ocean.

QUESTIONS FOR REVIEW

Multiple Choice

1. Most of the world's harvest of commercial fish comes from
   a. estuaries
   b. bays
   c. coastal seas
   d. the open ocean
   e. benthic regions

2. Burrowing organisms would more likely be found where
   a. sediments are predominantly gravel
   b. the sediments are sand, and the water currents are fast
   c. the sediments are silt, and the water currents are slow
   d. the sediments are silt, and the water currents are fast
   e. landslides frequently disrupt the bottom

3. Sedentary and sessile filter feeders and suspension feeders are better adapted to
   a. hard bottoms
   b. soft bottoms

4. The uneven distribution of bottom organisms and sediment type on the continental shelf is referred to as
   a. reticulation
   b. patchiness
   c. diversity
   d. sediment selection
   e. benthic orientation

5. In colder waters, ___ dominate the phytoplankton.
   a. krill
   b. copepods
   c. diatoms
   d. dinoflagellates
   e. kelp

6. The most abundant members of the zooplankton are
   a. krill
   b. copepods
   c. jellyfish
   d. diatoms
   e. dinoflagellates

7. Nutrients produced by bacteria in the bottom sediments are returned to the surface waters by
   a. gyres
   b. upwellings
   c. thermoclines
   d. Eckman spirals
   e. the Coriolis effect

8. In soft-bottom communities, suspension feeders and deposit feeders are usually not found in the same areas because
   a. they compete with each other for food
   b. they compete with each other for space
   c. deposit feeders prey on suspension feeders
   d. the suspended silt in regions inhabited by deposit feeders clogs the feeding structures of suspension feeders
   e. the type of sediments inhabited by deposit feeders will not support suspension feeders

9. The world's most productive upwelling area is located off the coast of
   a. California
   b. Peru
   c. China
   d. South Africa
   e. Europe

10. The dominant alga in southern California kelp forests is
    a. *Fucus*
    b. *Laminaria*
    c. *Eisenia*
    d. *Sargassum*
    e. *Macrocystis*

Matching

Match the following organisms with the appropriate description.

a. krill   1. A termite-like crustacean that damages large kelp by feeding on their holdfasts.

b. gribble   2. A type of kelp that grows in dense thickets along the Atlantic coast.

c. whelk   3. A crustacean member of the zooplankton that is the primary consumer of diatoms.

d. Laminaria   4. A carnivorous snail that feeds on bivalves and other molluscs.

e. copepod   5. A shrimplike crustacean that provides food for some seals, whales, and seabirds.

Short Answer

1. What are the main sources of nutrient input into coastal seas?

2. What factors affect the size of plankton populations?

3. How do waves and currents affect the types of benthic organisms found on the continental shelf?

4. Name two important functions of bottom currents in benthic communities.

5. Explain how the types of bottom sediments influence the diversity of life on the floor of the continental shelf.

6. List three explanations for the patchy distribution of organisms in soft-bottom communities.

7. Explain how decreases in the size of lobster and cod populations have affected North Atlantic kelp beds.

8. Why are kelp beds frequently compared to terrestrial rainforests?

9. Why is the neritic zone such a productive area?

10. Diagram a simple food web for the continental shelf.

11. Describe the process of succession that occurs in soft-bottom communities that are disturbed.

12. What are two ecological roles for plankton in the neritic zone?

Thinking Critically

1. Currently there is concern about climate change caused by the greenhouse effect. How might this affect the productivity of coastal seas?

2. Why do so many species of small fish found in coastal seas travel in large schools?

3. Why do disturbances such as landslides have a greater impact on invertebrate populations than on algal populations?

SUGGESTIONS FOR FURTHER READING

Ashjian, C. 2004. Life in the Arctic Ocean, *Oceanus* 43 (2): 20–23.

Epifanio, C. E., and R. W. Garvine. 2001. Larval Transport on the Atlantic Continental Shelf of North America: A Review, *Estuarine, Coastal and Shelf Science* 52: 51–77.

Steneck, R. S., M. H. Graham, B. J. Bourque, C. Corbett, J. M. Erlandson, J. A. Estes, and M. J. Tegener. 2002. Kelp Forest Ecosystems: Biodiversity, Stability, Resilience, and Future, *Environmental Conservation* 29: 436–459.

Walsh, R. 2000. The Lobster Pickle, *Natural History* 190(6).

Have You Wondered?

1. **What kinds of microbes and animals live in the ocean realm? (17-1)**

2. **What kinds of adaptations are needed to live in the open sea? (17-2)**

3. **If the ocean is so huge, why life within its waters is so limited in abundance? (17-2, 17-3)**

17

The Open Sea

BEYOND THE SHALLOW coastal seas over the continental shelves (neritic zone) lies the open ocean (oceanic zone). The ocean is the largest ecosystem on earth, yet it remains poorly known. We are familiar with shallow coastal regions, but we rarely venture out on the seas. Very few humans have explored the ocean beneath its surface, and then only within confining vessels that keep us from experiencing directly this mysterious realm. Most of what we know of the ocean comes from plumbing its depths with nets, dredges, scoops, and other gear and more recently with remote cameras. Our fascination with the creatures that are retrieved and photographed makes us wonder what else remains to be found within the vastness of its 1.4 billion cubic kilometers (327 million cubic miles) of salty water.

If we were to use our experiences as land-dwellers as a guide, we would guess wrongly about life in the sea. One might think that marine life extended only a short distance below the surface, just as trees, insects, and birds ascend no more than hundreds of feet above the land. A few kinds of organism might occasionally dive or sink deeply as migrating geese fly thousands of feet high or pollen grains are swept aloft on currents into the thin asphyxiating atmosphere. Indeed, early marine biologists and oceanographers did guess wrongly about the distribution of marine life, assuming that cold, darkness, pressure, and lack of oxygen would keep life close to the surface. But these factors have provided few obstacles to adaptation over eons. Virtually every gallon of seawater bears some living organisms in reasonable diversity.

When viewed from above, this open expanse of water appears monotonous. Even though it lacks obvious boundaries, the open sea can still be divided into regions based on the physical characteristics of the water and life forms in them. Vertical zonation depends on the depth to which sufficient light penetrates to support photosynthesis. As you may recall from Chapter 2, the photic zone is the layer that receives enough sunlight for phytoplankton to survive. In clear tropical waters, the photic zone extends down to a maximal depth of 200 meters (660 feet). The photic zone roughly corresponds to the **epipelagic zone**, a term used more in reference to the location of pelagic animals in the upper 200 meters of the ocean. Below this is the aphotic zone, which extends to the ocean bottom. In this zone,

The **epipelagic zone** is the upper 200 meters of ocean waters.

light rapidly disappears until the environment is totally dark. In this chapter we will be dealing with life in the photic and epipelagic zones. Life in the aphotic zone will be covered in the next chapter.

17-1 Life in the Open Sea

Two groups of organisms inhabit the oceanic zone: plankton and nekton. On the basis of production, biomass, numbers, and biodiversity, life in the oceanic zone consists far more of plankton than nekton. Because nekton are defined by their ability to swim strongly against ocean currents, they are restricted to the large, muscular, streamlined, higher consumers among the animals, such as tuna and other pelagic fishes and whales. Plankton, on the other hand, belong to all three domains of life and many phyla of microbes and animals.

CLASSIFICATION OF PLANKTON

Many different organisms make up the plankton of the open sea. To accommodate this great diversity, marine scientists use several different schemes to organize plankton into logical groups. Most often the oceanic plankton are classified according to taxonomic group, motility, size, life history, and spatial distribution in the sea.

Taxonomic Groups

Particles suspended in the sea are called **seston** and include the nonliving **tripton** (mineral particles, dead organisms, and decaying organic matter or detritus) and the plankton. The two most familiar kinds of plankton are phytoplankton and zooplankton. As you learned in Chapter 2, phytoplankton are primary producers that belong to many distantly related groups, most of which were described in Chapter 6. Zooplankton are the heterotrophic eukaryotic microbes and those animals that are so small or are such weak swimmers that their distribution in the sea is determined by ocean currents. **Bacterioplankton** generally include both archaeons and bacteria, many of which are phytoplankton. **Viriplankton** include free viruses, which are the most abundant plankton of all.

Motility

Some plankton such as viruses, diatoms, and foraminiferans do not move at all and are called **akinetic**. Most plankton, however, are **kinetic** and move by flagella, jet propulsion, undulation of the body, swimming appendages, or other means. Their movements orient them in the water column, help them avoid predators, or change their vertical position. Because of their small size, slow swimming speed, or both, movement of these planktonic organisms does not result in significant changes of location when compared with the effects of water currents and turbulent mixing in moving them.

Size

The classification originally applied to sizes of planktonic organisms depended on their visibility and methods of collecting them. **Macroplankton** were those organisms that were visible to the naked eye and generally exceeded 1 millimeter. **Microplankton** were small plankton that could be caught with standard plankton nets, whose mesh size became finer with advances in cloth fabrication technology. **Nanoplankton** (also called

centrifuge plankton to distinguish them from microplankton, or net plankton) passed through the mesh of plankton nets and could be concentrated best by centrifugation.

There has been a trend in recent years to develop a more consistent system for classifying plankton by size. This trend stems from the discovery of viruses and a variety of other kinds of centrifuge plankton of apparently great importance in pelagic ecosystems. One commonly used scheme of classification divides the full range of plankton into seven categories, each containing plankton of a size within one to two orders of magnitude based on the number 2 **(Figure 17-1)**. **Femtoplankton** and **picoplankton** include viruses and the smallest prokaryotes. Most prokaryotes and many kinds of eukaryotic phytoplankton (for example, diatoms and coccolithophores) compose the nanoplankton. The microplankton consist of a mixture of larger diatoms and dinoflagellates, invertebrate larvae, small crustaceans, and other adult invertebrates. Most other animal plankton, including larval fishes, fall into the larger categories of **mesoplankton**, macroplankton, and **megaplankton**, although the latter group also contains sargassum weed and detached rafts of benthic algae.

Life History

The many organisms that are planktonic throughout their lives are called **holoplankton**. These include the microbes and many groups of invertebrate animals, especially in the open sea. Arrowworms, salps, siphonophores, comb jellies, copepods, krill, and some specialized groups of gastropod molluscs are holoplanktonic **(Figure 17-2a)**. Many species of scyphozoans such as the sea wasps have evolved a life cycle that omits the benthic polyp stage, and these oceanic medusae are, therefore, holoplanktonic.

Not all plankton are planktonic in every phase of their life cycle. In the neritic zone, invertebrates that are benthic as adults often have larval stages in the plankton for feeding and dispersal. These planktonic larvae are called **meroplankton**. Meroplanktonic larvae of benthic invertebrates are less characteristic of the epipelagic zone of the open sea largely because the deep-sea floor is too far below the surface waters (4,000 to 6,000 meters, or 13,200 to 19,800 feet, deep) for epipelagic larvae to be an adaptive part of life cycles. On the other hand, the abundant larvae or juveniles of many nektonic fish and squid are small enough to be classified as meroplankton **(Figure 17-2b)**.

Spatial Distribution

Communities of plankton can be characterized as neritic or oceanic based on species composition. Neritic plankton communities are often distinguished by the presence of meroplankton such as the nauplius and cyprid larvae of barnacles. At low latitudes, diatoms are an abundant and diverse component of neritic plankton communities. Oceanic plankton communities are less diverse in diatoms and invertebrate meroplankton, but salps, larvaceans, arrowworms, and sea butterflies (a specialized group of snails) are among the more distinctive components of the epipelagic zone.

Plankton that live close to the water's surface are called **neuston** and represent a special case. These organisms often use the surface tension of water to remain at or near the surface, as does the purple sea snail (see **Figure 17-5c**). In addition to many microbes that are characteristically neustonic, about a dozen gastropod molluscs, several cnidarians, and a few seaweeds live here. Most of them are buoyed by gas bladders or bubbles and extend above the surface, where wind and water currents

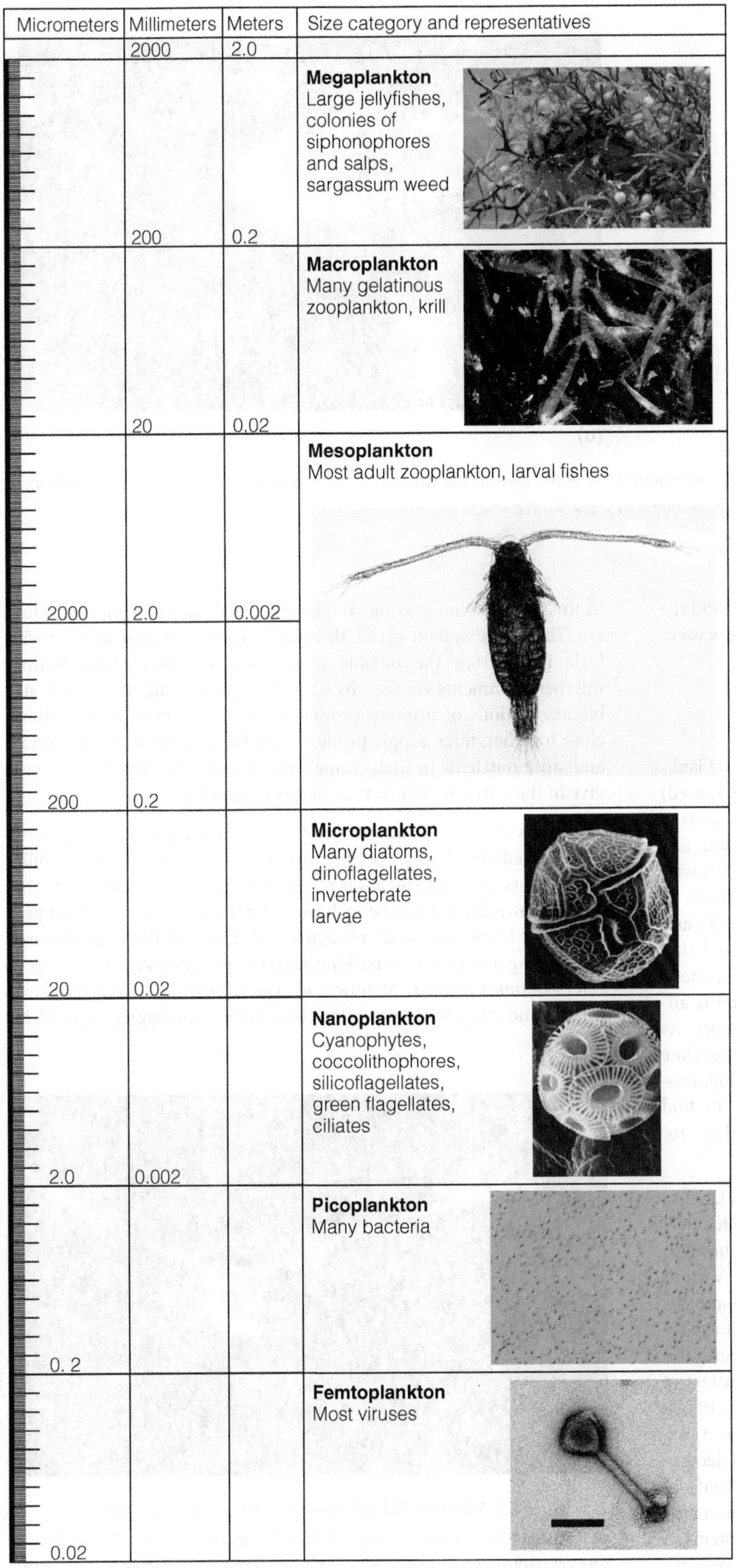

Micrometers	Millimeters	Meters	Size category and representatives
	2000	2.0	
			Megaplankton Large jellyfishes, colonies of siphonophores and salps, sargassum weed
	200	0.2	
			Macroplankton Many gelatinous zooplankton, krill
	20	0.02	
			Mesoplankton Most adult zooplankton, larval fishes
2000	2.0	0.002	
200	0.2		
			Microplankton Many diatoms, dinoflagellates, invertebrate larvae
20	0.02		
			Nanoplankton Cyanophytes, coccolithophores, silicoflagellates, green flagellates, ciliates
2.0	0.002		
			Picoplankton Many bacteria
0.2			
			Femtoplankton Most viruses
0.02			

FIGURE 17-1 CLASSIFICATION OF PLANKTON BY SIZE. This chart demonstrates a scheme of classification that divides the full range of plankton into seven categories, each containing plankton of a size within one to two orders of magnitude based on the number 2.

Photos: **Megaplankton:** Image Quest Marine; **Macroplankton:** Jamie Hall/NOAA; **Mesoplankton:** Courtesy of Angel Seery, Florida Institute of Technology; **Microplankton:** John D Dodge; **Nanoplankton:** Dr. Elizabeth Venrick/Scripps Institute of Oceanography; **Picoplankton:** Dennis Drenner; **Femtoplankton:** Bin Ni, Peter Weigele, Matt Sullivan, Sallie Chisholm, and MIT

Seston are particles, living or dead, suspended in seawater.

Tripton are particles of dead organic matter suspended in seawater.

Bacterioplankton are prokaryotic members of the plankton.

Viriplankton are planktonic virus particles.

Akinetic plankton are those that have no recognizable method of movement.

Kinetic plankton are those that are able to move by their own effort.

Macroplankton are 2 to 20 centimeters in size.

Microplankton are 20 to 200 micrometers in size.

Nanoplankton are 2 to 20 micrometers in size.

Centrifuge plankton include organisms that are so small that they are not retained by standard plankton nets and are best sampled by centrifugation of water samples.

Femtoplankton are 0.02 to 0.2 micrometers in size.

Picoplankton are 0.2 to 2.0 micrometers in size.

Mesoplankton are 0.2 to 20 millimeters in size.

Megaplankton are 20 to 200 centimeters in size.

Holoplankton are planktonic throughout their life cycles.

Meroplankton are planktonic stages of an organism that is benthic or nektonic at other stages in its life history.

Neuston are plankton that live at or near the surface of the ocean.

(a)

Courtesy of Angel Seery, Florida Institute of Technology

(b)

NOAA

FIGURE 17-2 PLANKTON. (a) Organisms such as this copepod that spend their entire lives in the water column as plankton are called holoplankton. **(b)** Meroplankton such as this larval fish spend only a part of their life cycle in a planktonic stage.

direct their paths, as in the by-the-wind sailor (see **Figure 17-5d**) and the water strider (see **Figure 4-2**). Neuston that break the surface of the water are called **pleuston**.

PATCHINESS IN THE OPEN SEA

The distribution of plankton in the sea is not uniform or random. Plankton occur in localized aggregations (that is, they are not evenly spaced) called patches, and many factors contribute to their formation. Areas of upwelling can cause large-scale patchiness in marine plankton and nekton communities by supporting bursts of primary production with the influx of nutrients. Factors such as highly localized variations in sea-surface conditions, vertical mixing of water, downwelling events, and meeting of waters of different density can cause more rapid population growth of phytoplankton in one place and diminished growth elsewhere.

Grazing by zooplankton is another major contributor: As phytoplankton densities decline with high rates of feeding, zooplankton move about to find adjacent denser patches for their next meal.

On a smaller scale, plankton biologists have found what they call **micropatchiness** throughout the photic zone. Various marine microbes become attached to particles of organic matter, particularly strands of mucus secreted by zooplankton, and form translucent cobwebby aggregates called **marine snow (Figure 17-3)**. Bits of marine snow and clusters of floating algal threads are home to populations of bacteria up to 10,000 times more concentrated than free bacteria in the open water. Along with these bacteria, which include both primary producers and decomposers, drift hordes of their microscopic predators. Further study of these drifting particles has led to discoveries indicating that the microscopic communities they harbor may in fact be complete ecosystems in miniature. On each of these floating islands of life, bacteria grow, respire, and are eaten. Primary producers reabsorb any nutrients released

Pleuston are neuston that breach the surface of the ocean and are carried by wind and water currents.

Micropatchiness is the small-scale clumped distribution of organisms in an ecosystem.

Marine snow is a loose network of live and dead particles in the plankton arising from webs of mucus released by plankton.

in this process, and grazing zooplankton feed on the primary producers. The whole system cycles through intense microbial activity with little input from the outside other than sunlight. These floating microenvironments are seen by some biologists as highly evolved, stable associations of primary producers and consumers. By remaining close together, microscopic producers and consumers can concentrate and store nutrients in their immediate vicinity, allowing them to survive in the nutrient-poor waters that surround them.

Patchiness of their planktonic prey may create a problem for grazing marine predators. Patches of prey vary in space and time, and the ability of predators to detect patches is reduced by the nature of seawater. That is, the knowledge that predators have about the distribution of their prey is limited. David Sims and colleagues have followed the vertical search and feeding movements of basking sharks (*Cetorhinus maximus*), bigeye tuna (*Thunnus obesus*), Atlantic cod (*Gadus morhua*), leatherback sea turtles, and Magellanic penguins fitted with data-logging tags. These

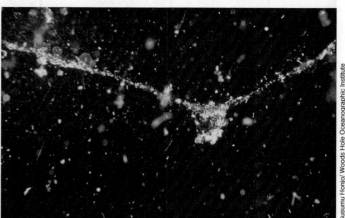

Susumu Honjo/ Woods Hole Oceanographic Institute

FIGURE 17-3 MARINE SNOW. Marine snow is an aggregation of marine microbes attached to particles of organic material such as the mucus secreted by zooplankton. These particles contain both producers and consumers and represent miniature ecosystems.

scientists have evidence that the movements of such predators occur in patchy fractal patterns called **Lévy walks** that lead to greater foraging success than would random feeding patterns. Moreover, there is some indication that the development of patchy feeding patterns is learned by juvenile predators.

PLANKTON MIGRATIONS

Many open-ocean zooplankton make daily migrations from the surface of the sea to depths of nearly 1.6 kilometers (1 mile). Some marine biologists believe that these movements take advantage of the benefits of feeding on the phytoplankton that live only in the photic zone and reduce to some extent the threat of predation by plankton-eating fishes, which are also more abundant in the epipelagic zone. These migratory zooplankton are often so densely packed that they form what is called a **deep scattering layer**, a mixed group of zooplankton and fishes that gives sonar systems a false image of a nearly solid surface hanging in midwater.

MEGAPLANKTON

Most of the organisms classified as megaplankton are animals. Among these the most prominent are the cnidarians, molluscs, and salps.

Cnidarian Zooplankton

The largest members of the plankton are jellyfish. One of the most common is the moon jellyfish (*Aurelia*). This animal's bell-shaped body measures from 15 to 45 centimeters (6 to 18 inches). A fringe of thin hair-like tentacles surrounds the bottom of the bell, and within the clear body are four oval, pigmented gonads. In addition to the tentacles, the outer surface of the bell is covered by bands of sticky mucus that ensnare any small organism that comes into contact with them. Cilia on the body surface constantly move the mucus and any trapped food to the edge of the bell. Fleshy projections on the underside of the bell move the large blobs of accumulated mucus and food into the mouth for digestion. The moon jellyfish feeds mainly on copepods and other small zooplankton.

Larger and more dangerous are the jellyfish known as lion's mane jellyfish (*Cyanea*; **Figure 17-4a**). One enormous species in the Arctic Ocean has a bell that averages 2.5 meters (8 feet) in diameter and has tentacles that extend downward for 30 meters (99 feet) or more, giving it the distinction of being the largest member of the zooplankton in the world. Species in temperate regions of the Atlantic and Pacific Oceans tend to be smaller. They feed on surface fishes, and their sting can easily kill a 30-centimeter (12-inch) fish.

One of the jellyfish species best adapted to life in the open ocean is *Pelagia noctiluca* (**Figure 17-4b**). This species has a beautiful, pastel body, and at night it is bioluminescent. Unlike most species of jellyfish that have an asexual polyp stage in their life cycle, *Pelagia* larvae mature in the open sea without going through a polyp stage; it is, therefore, holoplanktonic.

Molluscan Zooplankton

Even some molluscs have become adapted to life in the open sea. Pteropods (from *ptero*, meaning "wing," and *pod*, meaning "foot"), popularly known as sea butterflies (**Figure 17-5a**), are related to snails. The animal's foot has two large wing-like projections, and the shell is much reduced (**thecosome pteropods**) or absent (**gymnosome pteropods**). The projections on the foot propel the animal through the water. Cilia on the

> A **Lévy walk** is a random pattern of movement often used by predators in search of patches of prey.
>
> A **deep scattering layer** is a midwater false-bottom image on sonar produced by a mixed group of zooplankton and fishes.
>
> **Thecosome pteropods** are herbivorous oceanic gastropods with highly reduced shells that swim by flapping lateral extensions of the foot.
>
> **Gymnosome pteropods** are carnivorous shell-less oceanic gastropods that swim like thecosomes.

(a)

(b)

FIGURE 17-4 JELLYFISH OF THE OPEN SEA. (a) The lion's mane jellyfish, *Cyanea capillata*, can achieve bell sizes of 3 meters (10 feet) in diameter and can weigh 1 ton. **(b)** *Pelagia noctiluca* is bioluminescent and lacks an asexual polyp stage in its life cycle.

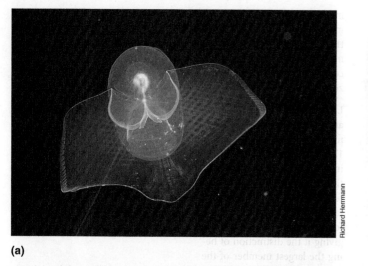

(a)

(b)

(c)

(d)

FIGURE 17-5 MOLLUSCS OF THE OPEN SEA. (a) Pteropods, or sea butterflies, have a foot that is modified to form a pair of winglike structures that animals use to propel themselves through the water column. **(b)** The thecosome pteropod *Hyalocylis striata* is so abundant in some areas of the South Atlantic Ocean that the ocean floor is covered with the shells of these dead organisms. **(c)** The purple sea snail keeps itself afloat by clinging to a bubble raft that it produces. This specimen seems to be eating a blue buttons, *Porpita porpita*. **(d)** The pelagic nudibranch (*Glaucus*) swims upside down just beneath the surface of the water. This specimen is feeding on a by-the-wind sailor.

surface of these projections on thecosome pteropods create a current that drives small plankton, mainly diatoms, toward the mouth. Some species secrete enormous mucous nets in which phytoplankton are trapped as food. These gastropods are abundant in the open ocean and sometimes form dense groups typically preyed on by fishes and whales.

In some areas of the South Atlantic Ocean, thecosome pteropods **(Figure 17-5b)** are so numerous that large areas of the bottom are covered with the shells of these dead animals, forming calcareous sediments called **pteropod ooze**. The less-abundant carnivorous gymnosome pteropods specialize in preying on the herbivorous thecosomes.

Pteropod ooze is a calcareous seafloor sediment formed by the accumulated shells of thecosome pteropods.

Another prominent mollusc in the open sea is the purple sea snail (*Janthina*; **Figure 17-5c**). Unlike pteropods, the purple sea snail has retained a relatively large, although light, shell. It keeps itself from sinking by producing a raft of bubbles surrounded by mucus. The snail clings

upside down to the underside of this bubble raft. This upside-down lifestyle has led to a reversal of the typical coloration of open-water organisms. What would normally be the animal's dark-colored upper surface is mainly white, whereas its under surface is dark purple rather than light colored. The purple sea snail is widely distributed in all tropical waters, where it feeds on larger members of the zooplankton such as copepods, jellyfish, and by-the-wind sailors (*Velella*), a siphonophore related to the Portuguese man-of-war.

Some species of nudibranch have also enjoyed success in the open sea. Among the most numerous are species in the genus *Glaucus* **(Figure 17-5d)**. These small blue nudibranchs glide along upside down beneath the surface of the water. Like the purple sea snail, *Glaucus* also displays a reverse color pattern with a dark ventral surface and a lighter-colored upper surface. *Glaucus* feeds on by-the-wind sailors, and like some other species of nudibranch, it retains the stinging cells from its prey for its own defense. *Glaucus* has been observed to leave clutches of eggs attached to the floats of by-the-wind sailors after having devoured all of the colony's polyps.

Urochordates

Collections and observations made by scuba-diving researchers have yielded a wealth of information about pelagic members of the subphylum Urochordata (tunicates) known as **salps** (thaliaceans) and **larvaceans**. Salps have barrel-shaped bodies that are open at both ends. Some species produce and maintain a mucous net inside their oral openings and use it to filter water that is pumped through their bodies. When filled, the net and its contents are digested. Salps eat bacteria too small for most sizable plankton to capture, making that source of organic material available to their own predators. On the other hand, salps produce fecal pellets from planktonic food in abundance, increasing the flow of nutrients downward and out of the photic zone.

Because their bodies are at least 95% water, these adaptable organisms can grow and reproduce very rapidly, creating swarms that can stretch over kilometers and contain up to 500 individuals in each cubic meter of water. **Pyrosomes (Figure 17-6a)**, close relatives of salps, produce colonies that measure as long as 14 meters (46 feet), although the usual range is from a few centimeters (1 inch) to 3 meters (10 feet). Each pyrosome comprises hundreds of individual animals that join to form a hollow cylinder that appears to be a single organism. Each member of the colony faces outward and sucks in water and small plankton. The incurrent water enters the center of the colony, flowing through and out the back, moving the colony through the water by means of slow ciliary propulsion. Pyrosomes occur worldwide but most commonly in tropical and subtropical seas.

Sometimes, a salp will be inhabited by an amphipod crustacean of the genus *Phronima* **(Figure 17-6b)**. This amphipod eats the internal organs of the salp and then uses the salp's exterior as a sort of mobile home in which it lives and broods its young. The salp's body continues to move through the water, propelled by the swimming action of the crustacean.

Larvaceans secrete mucous structures called houses. Water is drawn through the house when the larvacean undulates its tail. Inside the house, tiny plankton are trapped in a wing-shaped feeding filter and digested. Because the houses have no opening for the elimination of feces, larvaceans must abandon their houses several times a day as they become clogged with feces and large plankton; they then produce new ones. Interestingly, the discarded houses of larvaceans, which often contain up to 50,000 trapped living phytoplankton cells, immediately become homes for bacteria and end up as particles of marine snow. The types of feeding exhibited by salps, pyrosomes, and larvaceans allow these zooplankton to filter enormous volumes of water very efficiently.

NEKTON

Nekton are the actively swimming organisms whose movements are not governed by currents or tides. Included in this group of animals are some larger invertebrates, fishes, reptiles, birds, and mammals.

Invertebrates

The invertebrates that reign supreme in the open sea are the squids. Speed, keen eyesight, and intelligence make the squid a formidable predator. The animal's body is streamlined, and it has sucker-laden tentacles for grasping prey. By drawing water into the mantle cavity and forcefully expelling it through its siphon, squids can achieve bursts of speed that are faster than the fastest fishes. Although squids normally swim backward, they can reverse their direction by adjusting the position of their siphon. This is especially useful when maneuvering within a school of fish.

Fish

The nekton include many species of fish exhibiting a variety of adaptations that enable them to survive in this niche. The majority of bony fishes that inhabit the open sea have mouths at the front of the body with a lower jaw that protrudes farther forward than the upper jaw. This arrangement allows the fish to

> **Salps** are barrel-shaped, free-swimming tunicates of the class Thaliacea that move by muscular action of the body wall.
>
> **Larvaceans** are tadpole-like, free-swimming tunicates of the class Larvacea that move by undulations of the tail.
>
> **Pyrosomes** are colonial free-swimming tunicates of the class Thaliacea that move by cilia of the pharynx.

Norbert Wu/Minden Pictures/National Geographic

(a)

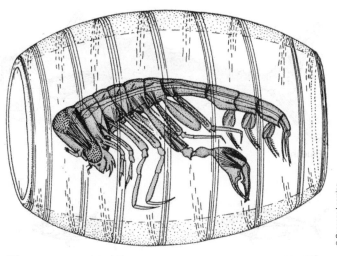

© Cengage Learning

(b)

FIGURE 17-6 PYROSOMES AND SALPS. (a) Large colonies of pyrosomes are formed by asexual reproduction from the original tunicate. **(b)** Amphipods of the genus *Phronima* feed on salps.

(a)

(b)

FIGURE 17-7 FISHES OF THE OPEN SEA. (a) The John Dory has jaws that protract for rapid grasping of prey. **(b)** The upper jaw of the sailfish is modified to form a bill with which it flails its prey.

grab prey from a variety of positions. Some species, such as the John Dory (*Zenopsis*; **Figure 17-7a**), have very extensible jaws. When the fish opens its mouth to seize its prey, the jaws project forward. They then retract quickly as the mouth closes.

Billfish The upper jaw of species such as marlin and sailfishes is greatly elongated, forming a bill **(Figure 17-7b)**. These fish lack the teeth characteristic of carnivores and use their bills to club their prey. The bill of a swordfish (*Xiphias gladius*) is broad and flat. As they swim into schools of mackerel, menhaden (*Brevoortia*), or squid, they flail their bill left and right, beating their prey and then swallowing their stunned victims whole.

Swordfish will attack virtually anything, often without any apparent provocation. The broken bills of swordfish have been found in the timbers of wooden ships, sometimes penetrating two layers of oak planks. Captured blue whales and fin whales have had broken bills embedded in their sides and backs, attesting to swordfish attacks.

Tuna The most wide-ranging fishes in the open ocean are tunas, which belong to the mackerel family **(Figure 17-8)**. The largest species of tuna are the bluefin tuna (*Thunnus thynnus*), with individuals achieving lengths of up to 4 meters (13 feet), and the yellowfin tuna (*Thunnus albacares*), which achieves lengths of 2 meters (7 feet) or more. Smaller relatives include the albacore (*Thunnus alalunga*), slightly more than 1 meter (3 feet) long, and the ocean bonito (*Sarda*), not quite 1 meter.

Tunas must swim constantly (some cruising at 27 kilometers, or 16 miles, per hour) or they will sink for lack of a swim bladder. This requires a large amount of energy and a good supply of oxygen. To supply the needed oxygen, these animals must swim swiftly and move large volumes of water past their gills. One adaptation for fast swimming in some tuna species is a body temperature 8°C to 10°C higher than the surrounding water. The higher temperature allows the fish to metabolize faster. Digestion is more rapid, nerve impulses travel more quickly, and

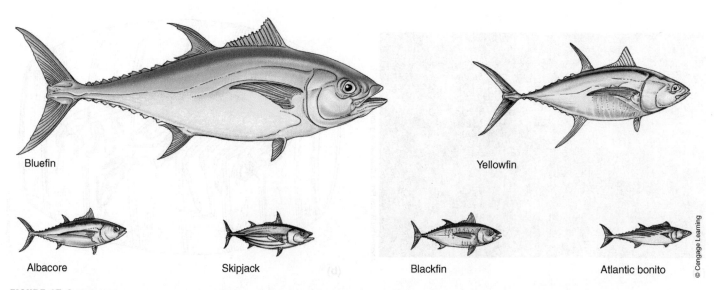

Bluefin

Yellowfin

Albacore

Skipjack

Blackfin

Atlantic bonito

FIGURE 17-8 TUNA. Tunas are fast swimmers that can be found throughout all seas. They range in size from the large bluefin tuna to the small Atlantic bonito.

FIGURE 17-9 OCEAN SUNFISH. Not much is known of the life history of this large, slow-moving fish.

the large skeletal muscles used for swimming contract and relax about three times faster than in other fishes. These adaptations help to account for the tuna's great speed and strength.

Many species of tuna exhibit bursts of speed that can exceed 54 kilometers (more than 33 miles) per hour. The main propulsive force comes from the high-keeled, sickle-shaped tail. To decrease resistance in the water, their bodies are streamlined and smooth. Their gill covers fit tightly against the sides of their body, and their pectoral fins lie retracted in grooves in their sides. Yellowfin tunas can reach speeds of 75 kilometers (45 miles) per hour, and one relative, the wahoo (*Acanthocybium solanderi*), has been clocked at 80 kilometers (48 miles) per hour. When they sense food, tunas quickly accelerate from cruising speed to top speed in less than 1 second. Tunas must consume large amounts of food to supply the energy they need for such vigorous swimming. Their diet is varied, consisting mainly of herring, anchovies, mackerel, sardines, flying fish, and squid.

Ocean Sunfish Quite different from the very active lifestyle of tunas is the more relaxed swimming of the ocean sunfish (*Mola mola*; **Figure 17-9**). Adults appear to spend most of their time lying on their side at the surface of the water, apparently basking in the sun; this behavior gives them their common name. The ocean sunfish has a highly modified caudal fin and prominent dorsal and anal fins. These fish are large, reaching lengths of more than 3 meters (10 feet) and weights exceeding 1 ton. They feed on larger zooplankton, especially jellyfish, and seem to have few natural predators.

The ocean sunfish is protected by a layer of cartilage 5 to 10 centimeters (2 to 4 inches) thick just beneath the surface of the skin. Although this thick layer offers protection against most predators, it is a haven for parasites. Almost every specimen examined has been infested with internal and external parasites, which may cause the death of many individuals. On the other hand, because parasite-infested animals are more easily caught, the data may show a bias.

Analysis of the stomach contents of some specimens reveals the remains of animals found only in deeper water, indicating that these large fish may not be as lazy as once perceived but may actually feed well below the surface. Some biologists suggest that the normal habitat for this fish is deep water, and only diseased individuals are found at the surface, where they come to die and are collected. In California, ocean sunfish regularly come near shore in summer to visit cleaning stations near kelp beds.

Sharks Some of the most efficient predators of the open ocean are sharks, whose streamlined bodies are well adapted to a pelagic lifestyle. Their reproductive behavior also reflects adaptations to life in the open sea. Unlike most bony fishes, which employ external fertilization, sharks copulate, with the male transferring sperm to the female. This increases the likelihood that the maximal number of eggs will be fertilized, and thus fewer eggs need to be produced. Many species are viviparous: The eggs are retained in the female's reproductive tract, where they hatch and for a time are nourished on a milky fluid until birth. This strategy helps to ensure better survival of the young when they enter the environment. Hammerhead sharks (*Sphyrna*) and blue sharks (*Prionace glauca*) not only hold their young within the body but also have a connection that supplies nutrients to the young from the mother's blood in a manner similar to that of mammals. Sharks produce fewer young each breeding season than other fishes because their reproductive strategies are more efficient, and the survival rate of their offspring is higher. For instance, whereas most bony fishes produce thousands of offspring each breeding season, the blue shark bears only about 30 young each season. Each pup is 60 to 70 centimeters (24 to 28 inches) long and fully independent when born, with a full set of teeth for feeding and defense.

Manta Rays A member of the nekton that is related to sharks is the manta ray (family Mobulidae), or devil-fish **(Figure 17-10)**. Fully adult mantas may measure 6 meters (20 feet) from one fin tip to the other and weigh as much as 1.5 tons. Mariners of old believed that manta rays could grab ships by their anchor chains and drag them to the ocean depths. It was also believed that these creatures would envelop swimmers in their large fins and devour them. Neither of these beliefs is true, of course, but injured or provoked manta rays have been known to reduce small wooden fishing boats to masses of floating splinters. Manta rays feed primarily on small fishes and plankton that they channel into their mouths with the large fleshy extensions called **labial flaps**.

> **Labial flaps** are fin-like extensions near the mouth that direct the flow of water during feeding by manta rays.

FIGURE 17-10 MANTA RAY. The manta ray, or devilfish, feeds on small fishes and plankton that it scoops up with its large mouth.

FIGURE 17-11 **SEA SNAKE**. The yellow-bellied sea snake, *Pelamis platurus*, is a member of the nekton that cruises the waters of the tropical Pacific in search of small fish on which to feed.

Tom McHugh / Photo Researchers, Inc.

Reptiles

Reptiles are not often thought of as oceanic nekton, although ichthyosaurs, other reptiles, and sharks ruled the seas during the Mesozoic era. In present-day seas, however, the yellow-bellied sea snake (*Pelamis platurus*; **Figure 17-11**) cruises the tropical waters of the Indian and Pacific Oceans along drift lines in search of small fish to capture as prey. Although the main predators of coastal sea snakes are sea eagles and sharks, the oceanic *Pelamis* seems to have no known enemies. Birds, fish, and sharks reject it as food. Laboratory experiments indicate that the

meat is distasteful, and large predators apparently learn to avoid it. Unlike its coastal relatives, the yellow-bellied sea snake cannot descend to the seafloor to rub its shedding skin against rocks or coral. Shedding allows the snake to rid itself of accumulated algae and other fouling organisms that increase frictional drag. Instead, this snake slips its body into and out of knots, abrading the skin against itself to remove the skin and its unwanted epizoics. Female yellow-bellied sea snakes retain their eggs within the reproductive tract until hatching, giving live birth at sea and relieving themselves of the need to crawl ashore for reproduction.

Another open-ocean reptile is the leatherback sea turtle (*Dermochelys coriacea*) (**Figure 17-12**). Its shell is much further reduced than that of other sea turtles, which spend much of their lives in coastal regions. Rather than feed on benthic prey, the leatherback sea turtle is one of the few nekton that specialize in consuming gelatinous zooplankton. Its large size (averaging 1.5 meters or about 60 inches in shell length and weighing up to 900 kilograms or 1 ton) enables it to dive to depths beyond 1,200 meters (4,000 feet), to inhabit cold waters of high latitudes while maintaining a warm body temperature, and to undergo extensive migrations between tropical nesting sites and their feeding grounds from the tropics to subpolar seas. Unlike the ichthyosaurs of ancient seas and the present-day yellow-bellied sea snake, female leatherbacks must return to land every 2 to 4 years to lay eggs.

Birds and Mammals

The only nektonic oceanic birds are the penguins of the southern oceans. The flightless penguins have adapted to life in the open sea except for their need to return to land to reproduce. When foraging at sea, penguins consume large numbers of krill, squid, and fish.

The largest nekton are the whales. Baleen whales filter krill, thecosome pteropods, and sometimes fish from the seawater. Toothed whales, such as sperm whales, feed on squid and fish.

FIGURE 17-12 **LEATHERBACK SEA TURTLE**. The largest sea turtle in modern seas, the leatherback is one of few nekton that must leave the open sea and return to land to reproduce.

Courtesy of Matthew Witt/Centre for Ecology and Conservation/University of Exeter Library

IN SUMMARY

- Although the open sea appears uniform and without divisions, it can be divided into regions based on the physical characteristics of the water and the kinds of organisms inhabiting it.

- The open sea can be divided into vertical zones according to the amount of sunlight available.

- The open sea lacks well-defined communities and is termed a pelagic ecosystem.

- Most members of this ecosystem are plankton, which vary widely in size, nutrition, style of locomotion, life history, and spatial distribution.

- The plankton belong to many different taxonomic groups of microbes and multicellular algae and animals. (Have You Wondered? #1)

- The nekton consist of free-swimming animals whose movements are not governed by currents and tides.

- Nekton are dominated by squid and a great diversity of fishes. (Have You Wondered? #1)

- Aside from fishes, few other vertebrates are found in the open sea, but the yellow-bellied sea snake, penguins and other seabirds, and whales are among them. (Have You Wondered? #1)

17-2 Survival in the Open Sea

Despite lower productivity compared with coastal seas, the epipelagic zone provides an environment that is generally more stable than others we have examined. Sudden fluctuations in salinity and temperature do not occur in the open seas, and even the most violent storms have little effect a few meters below the surface. Organisms that live in the well-lit surface waters, however, face other problems of survival. Low concentrations of mineral and dissolved organic nutrients can limit the populations of photosynthetic and osmotrophic microbes. Both large and small organisms must work to keep from sinking out of the photic zone, and with no place to hide, animals have to depend on other strategies for avoiding their predators.

COMPETING FOR LIMITED RESOURCES

Inorganic and organic nutrients that are so abundant in coastal zones and in areas of upwelling are far less concentrated in the open sea. Minerals are used by some plankton for protection and support of the cell, as in the silica of diatom frustules and in the skeletons of silicoflagellates. Nitrogen and phosphorus are critically needed for the production of amino acids and nucleic acids. Heterotrophic bacteria rely in part on the release of organic compounds by viral lysis of other microbes as well as from other causes of cell death. But the generally lower densities of organisms in the open sea limit the populations of osmotrophs.

Three features of marine microbes in the open sea reduce their dependence on nutrients. First, diatoms and other plankton that require silicon are less diverse in the open sea. Those that are present have lower concentrations in their cell walls. Second, the most abundant microbes in the open sea are very small, generally in the size range of picoplankton and nanoplankton. The most abundant primary producers in the oceans are species of the blue-green bacterial genera *Prochlorococcus* and *Synechococcus*, and the most abundant heterotroph is the bacterium *Pelagibacter ubique*. The spherical cells of *Prochlorococcus* and *Synechococcus* are 0.4–1.6 micrometers in diameter, and the rod-shaped cells of *Pelagibacter* are 0.2–0.5 micrometers long. Smaller size requires fewer resources. In addition, *Pelagibacter* uses absorbed organic matter solely for production of new molecules and uses sunlight captured by rhodopsin-like compounds for energy production in a way similar to the halobacteria, covered in Chapter 6. Third, these three genera are among the free-living cells with the fewest genes, minimizing the need of each cell for nitrogen and phosphorus.

REMAINING AFLOAT

Phytoplankton must remain in the photic zone to carry out photosynthesis, and the distribution of phytoplankton dictates the distribution of zooplankton that rely on these organisms for food. Nekton of the open sea have dense bodies, and remaining at or near the water surface requires mechanisms that do not interfere with their pursuit of food. Adaptations for staying afloat include swimming (locomotion) and reducing the sinking rate (**Figure 17-13**).

Swimming Methods

Some phytoplankton and many zooplankton and nekton remain afloat by actively swimming. Swimming may involve the use of flagella or

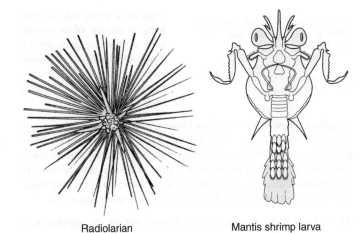

Radiolarian   Mantis shrimp larva

(a) Projections like the needles of the radiolarian and the flattened body of a mantis shrimp larva create friction and slow sinking.

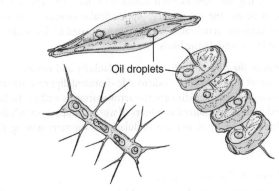

Oil droplets

(b) Diatoms store their food reserves as oil. The oil makes them more buoyant and offsets some of the weight of their shells.

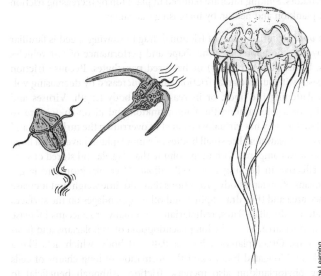

(c) Dinoflagellates and jellyfishes remain afloat by actively swimming.

FIGURE 17-13 HOW OPEN-WATER ORGANISMS REMAIN AFLOAT. Organisms that inhabit the water column of the open sea slow or prevent sinking by **(a)** increasing friction, **(b)** having adaptations that increase buoyancy, or **(c)** actively swimming.

An **appendicular swimmer** moves with legs, wings, or other extensions from the body.

Gelatinous zooplankton are pelagic animals that have a high content of body water.

cilia, as in many microorganisms, or the use of jet propulsion, as in some invertebrates. Many zooplankton and nekton swim with the aid of their appendages or by undulation of their bodies.

Flagella, Cilia, and Jet Propulsion Dinoflagellates, coccolithophores, and silicoflagellates use their whip-like flagella to swim, as does the blue-green bacterium *Synechococcus*. Tintinnids, other ciliates, and innumerable larval stages of invertebrates use cilia, as do the comb jellies with their ciliated ctenes and *Pyrosoma* with its ciliated pharynx. Jet propulsion is common among jellyfish, siphonophores, salps, and squid.

Appendages Many zooplankton and some nekton are **appendicular swimmers**, which use appendages to swim. The numerous copepods use their legs and antennae, stroking as fast as 600 times a second. All other planktonic crustaceans and their larvae are appendicular swimmers as well, but the appendages they use vary widely. The pteropod molluscs, or sea butterflies, flap the wing-like extensions of the foot. The legs of water striders scull across the sea surface. Appendicular swimming is very effective for nektonic vertebrates, which may use paired limbs, as do sea turtles, pinnipeds, and diving birds.

Undulations of the Body Finally, many zooplankton and nekton swim by undulation of the body, either side to side (horizontally) or vertically. Notable among the zooplankton are the arrowworms as well as the larvaceans and various specialized members of worm groups. Fish, whales, and dolphins of the open sea are undulatory swimmers among the nekton.

Reduction of Sinking Rates

Although swimming slows the rate at which organisms sink to greater depths, other methods that reduce sinking rates allow them to devote more of their swimming activity to orientation, feeding, and avoidance of predators. Sinking rates are reduced in plankton by increasing friction or in plankton and nekton by increasing buoyancy.

Frictional Drag The effect of frictional drag in slowing speed is familiar to anyone who appreciates the shape and performance of fast vehicles (cars, boats, planes, rockets) or has used parachutes. Because friction increases with surface area, friction can be increased by decreasing volume, flattening the body, or increasing the body length. Viruses and small prokaryotes among the femtoplankton and picoplankton are so small that gravity has virtually no effect in overriding the turbulence and viscosity of seawater. The small bodies of nanoplankton have such a large surface area compared with their volume that flagella and stored oils are very effective in keeping these cells afloat. Microplankton and larger organisms often have body projections that add little weight but increase surface area and thus drag. Spines and other appendages on the surfaces of diatoms, dinoflagellates, radiolarians, and many crustaceans increase their surface area, as do the long pseudopodia of radiolarians and foraminiferans. Other plankton have a flattened body, which acts like a parachute. Elongated bodies and the formation of long chains of cells among phytoplankton also increase friction. Although beneficial to plankton, frictional drag is a disadvantage to most nekton because it impedes achievement of high speeds.

Buoyancy Adaptations that increase friction do not prevent an organism from sinking; they merely slow it. Most members of the plankton must have other adaptations that allow them to counteract gravity and remain in the sunlit surface waters. In addition to locomotion, plankton often

use buoyancy mechanisms. In contrast to frictional drag, increasing buoyancy is a most effective adaptation for nekton to maintain their position in the sea. Buoyancy is increased by storage of oils, increasing the water content of the body, exchange of heavy ions for light ions, and use of gas spaces.

Because lipids have a density much lower than that of seawater, storage of oils, fatty acids, and other lipids is commonly used to compensate for the denser components of bodies. The products of photosynthesis may be stored as oils in the vacuoles of many phytoplankton, particularly diatoms and blue-green bacteria. Copepods, which graze heavily on diatoms, also store food as oil droplets. The use of fats in nekton is well known—whales, sharks, and codfish have long been harvested for whale oil, squalene, and cod-liver oil. Many fish deposit droplets of oils in their eggs to provide flotation.

The dilution of proteins and minerals with water immediately reduces the density of a body. Although such a mechanism is not suitable for nekton, which have a great need for muscle and skeleton for powerful locomotion, the bony fishes do have somewhat diluted body fluids. Their eggs also contain elevated water content, maintained by an impervious egg membrane until hatching. Many zooplankton, however, use this mechanism to an extreme and have bodies with a high water content.

Seawater is composed of many kinds of ions, some of which are heavier than others. Heavy ions such as magnesium, calcium, and sulfate ions frequently are replaced by the lighter ions ammonium and chloride in the vacuoles and body fluids of phytoplankton, zooplankton, and nekton. Although ammonium ion is toxic, its presence in confined parts of the bodies of many squids and crustaceans effectively controls buoyancy. In the body fluids of some squid, as much as four-fifths of the positively charged ions are ammonium.

Many groups of zooplankton combine watery bodies and ion exchange to achieve buoyancy. These animals, called **gelatinous zooplankton**, tend to be suspension feeders but run the gamut of locomotory types. The jellyfish are typical, as are their relatives the siphonophores and the medusae of other hydrozoans. These cnidarians move by jet propulsion. Exceptions are among the pleustonic cnidarians, such as the Portuguese man-of-war, by-the-wind sailor, and blue buttons (*Porpita porpita*), whose movements are driven by wind and water currents. Comb jellies are a common component of the gelatinous zooplankton, moving by their ciliary comb plates (ctenes). Pteropod molluscs are a group of pelagic snails that swim using wing-like extensions of their foot; thecosome pteropods have a reduced shell; and gymnosome pteropods lack a shell. Some squid have an enlarged body cavity filled with fluid of low density in which magnesium and other ions have been replaced by ammonium ions; they can "hang" motionless in the water column awaiting unsuspecting prey. A few species of gelatinous sea cucumbers swim by whole-body undulation or by peristaltic operation of an anterior hood or veil. All pelagic urochordates are gelatinous zooplankton, some moving by undulation of the tail (larvaceans), others by peristalsis of the barrel-shaped body (salps), and yet others by cilia on the pharynx (pyrosomes). Undulation is the mode of locomotion in the arrowworms and in the superficially similar leptocephalus larvae of eels and members of the tarpon family.

There is no more effective means to reduce density than gas bladders or evacuated chambers. Air and other gases have exceedingly low densities even at moderate depths in the sea. An organism needs only a small gas space to compensate for the density of well-developed muscle and skeleton. This mechanism is used by both plankton and nekton. Pleuston such as the Portuguese man-of-war and sargassum weed use gas-filled chambers to float at the surface of the ocean. Below the surface, siphonophores have gas bladders, and bubbles of gas occur in the cytoplasm of many blue-green bacteria, radiolarians, and foraminiferans. The rigid chambers of the shells of the nautilus and cuttlefish can be evacuated or filled with fluid to adjust buoyancy. Similarly, the flexible swim bladder of bony fishes can be inflated or deflated. At the greatest depths of the sea, compression of gas is so great that many deep-sea fishes have bladders filled with lipid rather than gas.

AVOIDING PREDATION

When threatened, animals of the open sea cannot quickly hide in a rock crevice, burrow into bottom sediments, or hide behind a stand of coral. Except for floating mats of sargassum weed, there are no large seaweeds, such as kelps, to provide a quick refuge. Pelagic organisms have evolved a variety of adaptations to make up for the lack of hiding places and increase their chances of survival. Larger members of the plankton such as jellyfish have stinging cells to deter potential enemies. Some animals avoid predators with speed; others such as flying fish and some species of squid take to the air. Some find safety in numbers, with many species of nektonic fish swimming in schools and some invertebrates forming colonies. Many animals of the epipelagic zone display adaptations that make them less visible or even invisible in the open water. Another ploy is possessing bodies of low nutritional value, which attract the attention of few predators.

Countershading is a camouflage pattern in which the upper surface of pelagic animals is dark and the lower surface is light.

Even with protective adaptations, the odds against survival in the open sea are tremendous. Many species compensate for the low survival rate by producing enormous numbers of offspring, thereby increasing the odds that a few will survive to propagate. The female ocean sunfish (*Mola mola*), for instance, produces as many as 30 million eggs in a single breeding season. Only two or three offspring need to survive to perpetuate the species.

Benefit of Being Less Conspicuous

In an environment that lacks hiding places, camouflage plays a central role in survival. Larger animals, from nudibranchs and pelagic snails to sharks and whales, exhibit **countershading (Figure 17-14)**. These animals have dorsal surfaces that are dark blue, gray, or green and ventral

FIGURE 17-14 COUNTERSHADING. (a) In a countershaded animal, the body surface that is usually exposed to sunlight is a dark color, and the opposite surface is a paler color or white. In the ocean environment, a uniformly colored fish **(b)** is highlighted from above and easy to see, whereas a countershaded fish **(c)** is effectively camouflaged.

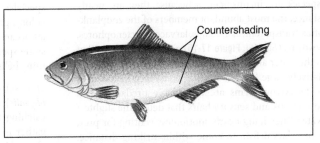

Countershading

(a) Countershaded fish

(b) Uniformly colored fish in natural lighting

(c) Countershaded fish in natural lighting

© Cengage Learning

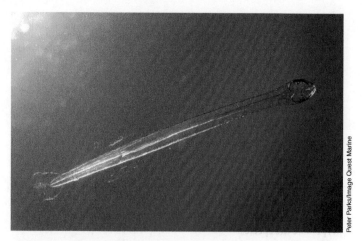

FIGURE 17-15 **ARROWWORMS.** Arrowworms are aggressive predators with transparent bodies. Transparency makes the arrowworm difficult to be seen by its own predators.

surfaces that are silvery or white. This arrangement of colors makes them difficult to see from above and below.

Many planktonic species are inconspicuous because they are nearly transparent. For instance, the most abundant members of the zooplankton, copepods, are often transparent, as are salps, larvaceans, ctenophores, and jellyfish. Arrowworms (*Sagitta*; **Figure 17-15**) are so transparent that they are invisible in the water and were not discovered until 1768, even though they are relatively large (up to 10 centimeters, or 4 inches) and distributed worldwide. Arrowworms are formidable predators with eyes that see in all directions and sensory hairs that detect the slightest movement in the water. They hang nearly motionless waiting for prey, such as a copepod or a larval fish, to come within striking distance. When the arrowworm senses prey, it darts toward it with amazing speed, grasping the prey with the hard pincer-like structures around its mouth.

Reduced Nutritional Quality

The watery bodies of gelatinous zooplankton that make them nearly transparent also make them unattractive as food for most predators. They convert otherwise rich sources of oils and proteins from their prey into the diluted mass of their own bodies. To acquire enough nutrients,

other zooplankton and nekton would have to fill their stomachs with large quantities of gelatinous zooplankton and deal with the water and ions consumed. Thus the possession of gelatinous bodies is an effective predator-deterrent mechanism, as well as a mechanism for buoyancy and to achieve transparency.

A few animals have become specialized predators of gelatinous zooplankton. Notable among them are the leatherback sea turtle and the ocean sunfish. In addition, some tubenosed sea birds and frigatebirds prey heavily on medusae. Some predators of gelatinous zooplankton are found within their own ranks. The comb jellies are sometimes eaten by cnidarian medusae, and some comb jellies eat medusae, salps, and even other comb jellies. There has been speculation recently that leptocephalus larvae, too, eat gelatinous zooplankton. Some of these specialist predators have become threatened in recent decades by the introduction of floating, transparent, plastic debris that may be mistaken for their natural prey **(Figure 17-16)**.

Safety in Numbers

Some animals increase their chances of survival in the open sea by forming colonies. One of the most successful of these is the cnidarian group Siphonophora. Although siphonophore colonies look like a single individual, they are actually made up of thousands of individuals, none of which can live independently of the colony. Some members of the colony specialize in capturing prey, whereas others digest food, maintain the colony, provide flotation, or reproduce. Siphonophores are common in all oceans, where they drift on or just beneath the surface of the water, some species propelled by the gentle pulsations of several tiny "swimming bells."

One well-known siphonophore, the Portuguese man-of-war (*Physalia physalis*), produces a gas-filled float that helps to propel the colony by catching the wind. Sometimes thousands of these colonies passively gather in drift lines caused by currents, forming a mass of siphonophores that stretches for miles across the open sea. The gas sac of a large man-of-war may be as long as 40 centimeters (16 inches), and the colony occasionally submerges the float to keep it moist. If the float dries out, it cracks, the gas escapes, and the colony sinks. The tentacles formed by feeding polyps in the colony can be as long as 8 meters (26 feet) in a large specimen.

IMPACT BUBBLE

ONE SIGNIFICANT PROBLEM that faces open-ocean food specialists is the increase in pelagic plastic debris. Sea birds, for example, evolved their feeding habits long before plastic bottles, bags, and other human products began floating on the seas. These items look like food to predators of gelatinous zooplankton. Consumption of pelagic debris fills the gut with indigestible material that may obstruct the digestive system and lead to wasting and death not only of adult birds but also the nestlings for which parents regurgitate their useless catch. One might expect innumerable generations to pass before the feeding behavior of sea birds is altered under natural selection to avoid ingestion of plastic waste.

FIGURE 17-16 **MISTAKEN AS FOOD.** This Laysan albatross probably died from starvation after having consumed a variety of floating plastic and other debris, mistaken as gelatinous zooplankton.

FIGURE 17-17 MAN-OF-WAR FISH. The small man-of-war fish gains protection by spending most of its time among the tentacles of the Portuguese man-of-war.

Although the stinging cells of the Portuguese man-of-war are capable of killing a large mackerel, one of the colony's main food items, the small man-of-war fish (*Nomeus gronovii*), lives symbiotically with the colony, spending most of its time swimming unharmed among the stinging tentacles (**Figure 17-17**). The little fish averages 8 centimeters (3 inches) long and is deep blue with vertical black stripes, blending quite well with the Portuguese man-of-war's tentacles. How the man-of-war fish avoids being killed by its host is not known. The fish has been observed to feed on some of the polyps, and it is possible that ingestion of the venom helps to immunize it against the stings. In laboratory experiments, the fish survived injections containing 10 times the venom that killed a different fish species of similar size.

IN SUMMARY

- One challenge of life in the open sea is remaining afloat. Plankton and nekton have adapted to pelagic life in many ways to keep from sinking to the deep sea, such as swimming (most pelagic organisms), increasing frictional drag (smaller organisms), and buoyancy control. (Have You Wondered? #2)

- Gelatinous zooplankton are efficient suspension feeders. The combination of their feeding efficiency and low food value greatly reduces the nutritional quality of pelagic communities. (Have You Wondered? #3)

- Animals of the open ocean have evolved ways of avoiding predation, including countershading, transparency, schooling, colony formation, toxins, and production of large numbers of offspring. (Have You Wondered? #2)

- Some plankton have evolved large size (megaplankton); among them are jellyfish, specialized kinds of gastropod molluscs, and colonies of pelagic tunicates.

17-3 Ecology of the Open Sea

The open sea represents a pelagic ecosystem in which the inhabitants live in the water column. In a pelagic ecosystem, the basis of food chains is the many species of small phytoplankton. These organisms derive the nutrients that they need from the seawater that surrounds them, but the low concentrations of nutrients limit the rate of production. The major herbivores in the open ocean are zooplankton, and these supply food for the nekton.

PRODUCTIVITY

In the great expanse of the open ocean, there are no vascular plants and few seaweeds that can serve as primary producers to support consumers. The seafloor to which plants and seaweeds might attach is deeply submerged and shrouded in darkness by the light-absorbing properties of the overlying water. In the open sea, all higher forms of life rely on the various species of plankton to supply food.

A dynamic floating ecosystem composed of plankton stretches across the entire surface of the sea and down into the depths for hundreds of meters. Pelagic ecosystems such as these include thousands of microbial and animal species, each with its own requirements for light, temperature, and nutrients, growth characteristics, and relationships with competitors and predators. Studying such a complex system and its inhabitants is no easy task, and our knowledge of open-water plankton communities continues to grow and change rapidly.

The blue of the open sea contrasts starkly with the green of coastal seas. Although surface waters of the open ocean receive large amounts of sunlight, they do not receive any of the nutrients that wash to sea from the land. This results in low levels of nutrients such as nitrogen and phosphorus that are necessary to support the life of phytoplankton. Waters just above the deep-sea floor contain higher concentrations of nutrients but do not receive enough sunlight to make them productive. The problem is compounded by the fact that there is very little mixing of deeper high-nutrient water with the low-nutrient water at the surface. Only in isolated parts of the open sea are nutrients brought to the surface by upwelling. Nutrient limitation makes nitrogen-fixing blue-green bacteria such as *Crocosphaera* and *Trichodesmium* critical to ocean productivity.

The low level of nutrients in the open ocean is most pronounced in tropical waters. These waters have relatively permanent layers that are separated by a thermocline with a warmer less-dense layer of water on top of colder denser water. This arrangement prevents significant exchange of nutrients from the deep water to the surface. Thus, even though tropical seas are the warmest and offer plankton large quantities of intense sunlight, the low concentrations of nutrients limit the size and composition of phytoplankton populations. At these latitudes, oceanic diatoms are low in abundance and diversity because silicon, needed for making the diatom frustule, is in limited supply. Phytoplankton production in the central South Pacific Ocean is only about half as much as the productivity off the California coast and only about one-sixth as much as in some coastal areas off Saudi Arabia or western Africa, where coastal upwelling brings nutrients to the surface. The low numbers of phytoplankton in tropical seas support even fewer numbers of the zooplankton that graze on the phytoplankton.

A **bacterial loop** is a subcomponent of a food web in which bacteria convert DOM and particulate organic matter into bacterial biomass.

Particulate organic matter (POM) consists of particles of organic matter suspended in seawater.

A **pyramid of production** is a diagram that illustrates the rate at which new biomass is produced at successive trophic levels.

Standing crop is the amount of biomass of organisms in a given area at a given time.

FOOD WEBS IN THE OPEN SEA

The base of food webs in the open sea is formed by the phytoplankton and heterotrophic bacteria (**Figure 17-18**). Most of the phytoplankton are nanoplankton, consisting of the smaller diatoms and dinoflagellates, as well as coccolithophores, silicoflagellates, and naked green flagellates. These photosynthetic nanoplankton dominate the phytoplankton in numbers and possibly in biomass, especially in tropical seas. The larger microphytoplankton might be grazed heavily by larger herbivorous zooplankton such as copepods, whereas nanophytoplankton are important food for invertebrate larvae, tintinnids, foraminiferans, radiolarians, and other smaller zooplankton. Although zooplankton normally serve as food for a variety of nekton, blooms of gelatinous zooplankton, which are highly efficient suspension feeders, can strongly depress population densities of nano-, micro-, meso-, and macroplankton, removing resources for nekton in space and time.

Dissolved and Particulate Organic Matter

In addition to providing food for herbivores, phytoplankton release some of their photosynthetic products into the surrounding seawater as dissolved organic matter (DOM). It is difficult for one-celled organisms to keep small molecules within their cell membranes and cell walls, and some leakage inevitably occurs. In a nutrient-poor environment, losses of DOM might limit the development of food webs. In recent decades, it has become apparent that heterotrophic bacteria play an important role in rapidly recycling DOM in the open sea. The production of bacterial biomass in plankton communities provides a second base to the food chain, but the bacteria are too small to be consumed by the familiar microzooplankton. Instead they are eaten by a poorly known group called heterotrophic nanoflagellates, which in turn are grazed by tintinnids and other ciliates. The ciliates then provide food for larger zooplankton. Bacteria can also metabolize some DOM and return it to the water in inorganic form, making the mineral nutrients again available to the phytoplankton. In this way, bacteria form a **bacterial loop** in returning nutrients rapidly to the phytoplankton in waters that are easily depleted of critical nutrients (see **Figure 17-18**).

Viriplankton, the most abundant plankton in the ocean, also play an important role at the base of ocean food webs. Lysis of bacterioplankton and phytoplankton by viruses releases DOM directly and produces **particulate organic matter** (**POM**) from the lysed cells. POM may slowly fall to greater depths, where other bacteria degrade it, but the bacterial biomass and recycled nutrients are then no longer available to the community in the photic zone. Although DOM released by viral lysis in surface waters will be absorbed by bacteria in the bacterial loop, the continuous destruction of bacteria by viruses short-circuits the loop, reducing the rate at which bacterial biomass is converted into biomass of higher consumers.

Efficiency of Open-Ocean Food Webs

The conversion of biomass from one level to the next in ocean food webs is surprisingly efficient. The phytoplankton or bacterial production of a given day may be entirely consumed by the next trophic level, leaving populations to begin the next day at densities of the previous day. Although not all food consumed can be converted into biomass at the next level because ecological efficiency is never 100%, conversion rates may be high. For example, bacteria convert DOM into bacterial biomass at 50% to 80% efficiency, and naked microflagellates consuming bacteria may have a conversion efficiency of 40%. Further reduction in conversion efficiency occurs at higher trophic levels, such as the copepods and krill that dominate the larger zooplankton.

The food webs in nutrient-poor open seas may have food chains with five or six links, leading to the nektonic top carnivores among the whales and sharks (**Figure 17-19**). Although the food chains may be long, the open sea can support few large animals for two reasons: the limited rate of primary production places a low ceiling on the total biomass that the area can sustain, and declining conversion efficiency along the food chain eventually exacts its toll. The outcome, measured as the rate of production of new biomass at various levels in the food chain, may be expressed in a **pyramid of production**, with a broad base of primary production and successive higher levels with decreasing rates of production. The base of the food chain in plankton communities is formed by microbes with short life cycles, and 100% of each day's production may be grazed entirely by zooplankton. As a result, the amount of biomass of phytoplankton at any given time (**standing crop**) might be very small, giving a pyramid of biomass that is partly inverted. It is only along the coastlines and at areas of upwelling that huge schools of tunas, flocks of seabirds, herds of seals, and pods of whales can form. The wealth of the open sea resides in its microscopic inhabitants.

IN SUMMARY

- The primary producers of the open sea are mainly phytoplankton.

- Even though the open sea receives large amounts of sunlight, it lacks the supply of nutrients that makes coastal seas so productive.

- The low level of nutrients is most pronounced in tropical oceans, where the water forms more or less permanent layers that are separated by thermoclines. (Have You Wondered? #3)

- Bacteria are important for nitrogen fixation and in the ability of the community to cycle nutrients efficiently, but viruses that infect bacteria short-circuit the bacterial loop and reduce the efficiency. (Have You Wondered? #3)

- The open sea does not support many large animals because of limited primary production and food webs involving several energy-wasting steps between primary producers and final consumers. (Have You Wondered? #3)

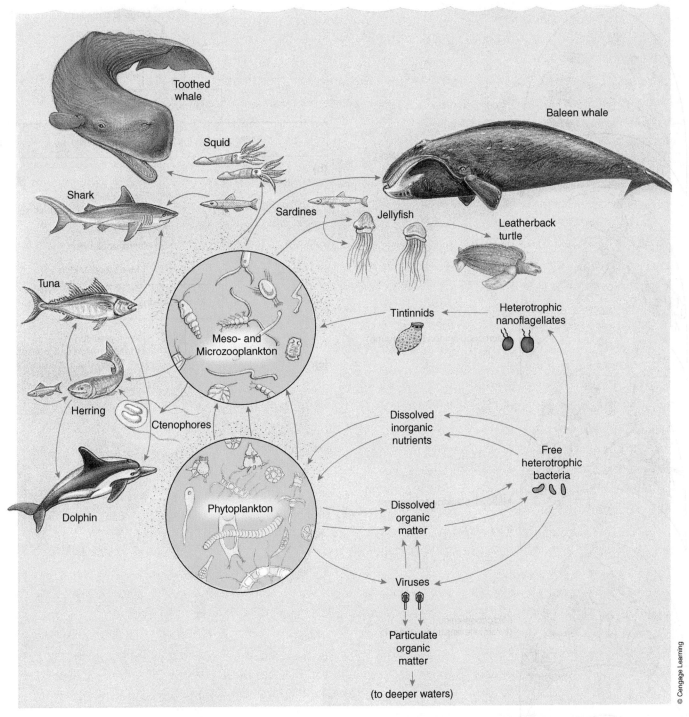

FIGURE 17-18 OPEN-OCEAN FOOD WEB. In open-ocean food webs, phytoplankton and heterotrophic bacteria form two foundations for a complex of food chains that form the pelagic food web. A common link in the chains is zooplankton, which are the primary food of many large animals.

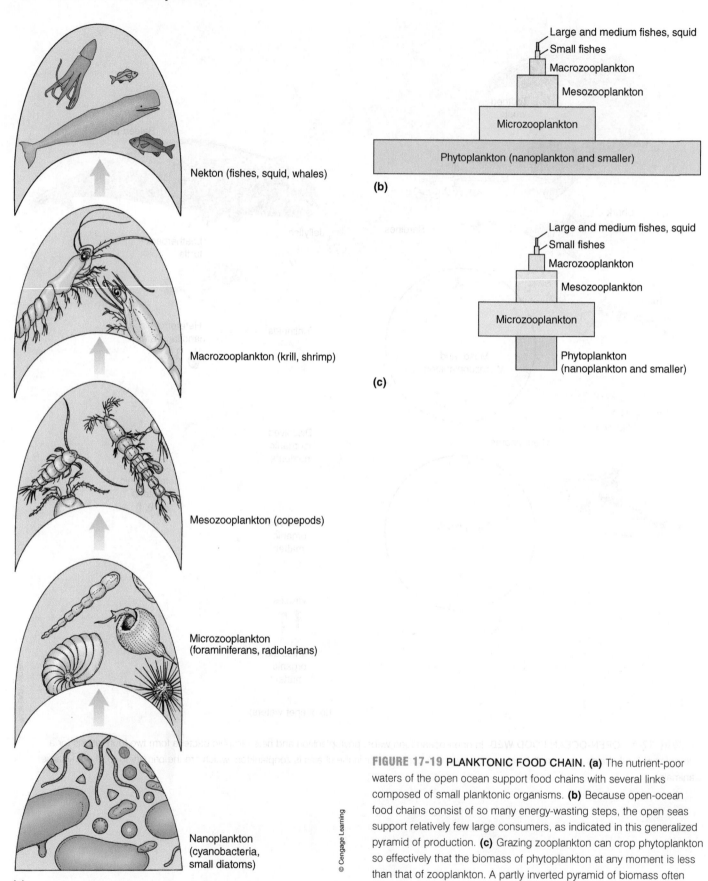

Nekton (fishes, squid, whales)

Macrozooplankton (krill, shrimp)

Mesozooplankton (copepods)

Microzooplankton
(foraminiferans, radiolarians)

Nanoplankton
(cyanobacteria,
small diatoms)

(a)

Large and medium fishes, squid
Small fishes
Macrozooplankton
Mesozooplankton
Microzooplankton
Phytoplankton (nanoplankton and smaller)

(b)

Large and medium fishes, squid
Small fishes
Macrozooplankton
Mesozooplankton
Microzooplankton
Phytoplankton
(nanoplankton and smaller)

(c)

© Cengage Learning

FIGURE 17-19 PLANKTONIC FOOD CHAIN. (a) The nutrient-poor waters of the open ocean support food chains with several links composed of small planktonic organisms. **(b)** Because open-ocean food chains consist of so many energy-wasting steps, the open seas support relatively few large consumers, as indicated in this generalized pyramid of production. **(c)** Grazing zooplankton can crop phytoplankton so effectively that the biomass of phytoplankton at any moment is less than that of zooplankton. A partly inverted pyramid of biomass often results.

KEY CONCEPTS

1. The open sea is a pelagic ecosystem, in which the living components are plankton and nekton.

2. Plankton range widely in size, taxonomic diversity, and lifestyle.

3. Phytoplankton are the primary producers in open-ocean food webs, and their productivity is limited by the scarcity of nutrients. (17-1, 17-3)

4. Heterotrophic bacteria provide a second base to open-ocean food webs, and they allow the scarce nutrients to be efficiently recycled. (17-1)

5. Limited primary production and food webs with several energy-wasting steps limit the number of large animals the open ocean can support. (17-3)

6. Gelatinous plankton such as salps and ctenophores play significant roles in open-ocean ecosystems because of their efficient feeding mechanisms, reduction of nutritional quality, and provision as prey for specialist carnivores. (17-1)

7. Several structural features and behaviors have evolved to keep afloat organisms that are not strong swimmers. (17-2)

8. Plankton display a number of interesting adaptations that help them avoid predation. (17-2)

9. Large zooplankton include jellyfish, gastropod molluscs, and colonial pelagic tunicates. (17-1)

10. Fish, squid, and mammals make up most of the nekton in the open sea. (17-1)

ARE YOU STILL WONDERING?

1. Although less diverse and abundant than in coastal waters, life in the open sea is remarkably diverse. Viruses, bacteria, and archaeons live there and play important roles in food webs and biogeochemical cycles. Eukaryotic phytoplankton primarily are microbes, but sargassum weeds are notable among the megaphytoplankton. Zooplankton include many kinds of microbes, but larvae of fish and several kinds of adult cnidarians, molluscs, and urochordates occur in the open sea. Nekton specialized for open-ocean life include billfish, sharks, yellow-bellied sea snake, leatherback sea turtle, penguins, and many species of whale.

2. Life in the open sea is challenging because of low levels of nutrients, lack of hiding places, and the need to remain afloat. Small cell size and reliance on a small number of genes reduce the dependence of primary producers and osmotrophs on mineral and organic resources. Patchiness, large numbers, transparency, and low-nutritive quality are several ways in which organisms in the open sea deal with predation. Many plankton and nekton swim by cilia, flagella, appendages, and undulatory tails, but mechanisms of buoyancy and increasing frictional drag also prevent pelagic organisms from sinking from the surface waters.

3. Life in the open sea is limited mainly by the low concentrations of nutrients, which keep primary production low. In addition, viral lysis of bacterioplankton slows the recycling of dissolved organic matter and the passage of production up the food web. Despite the high efficiency of trophic transfer, the nearly complete daily conversion of new biomass of primary producers to consumer levels keeps the standing crops of phytoplankton small.

QUESTIONS FOR REVIEW

Multiple Choice

1. The layer of ocean that receives enough sunlight to power photosynthesis is the
   a. photosynthetic zone
   b. photic zone
   c. aphotic zone
   d. abyss
   e. benthic zone

2. A collective term for animals whose distribution is not governed by waves or currents is
   a. zooplankton
   b. pelagic
   c. nekton
   d. akinetic
   e. pleuston

3. Productivity in the open ocean is limited in most cases by
   a. sunlight
   b. temperature
   c. space
   d. nutrients
   e. salinity

4. Food webs in the open sea are based on
   a. phytoplankton
   b. seaweeds
   c. dissolved organic matter
   d. viruses
   e. marine snow

5. Life in the open sea is characterized by
   a. uniform distribution
   b. low trophic efficiency
   c. dense populations
   d. food chains of 7 to 10 links
   e. patchiness

6. The rate of sinking can be slowed by anatomical adaptations that increase
   a. mass
   b. volume
   c. friction
   d. streamlining
   e. density

7. Which one of the following is not true of gelatinous zooplankton?
   a. Bodies are high in water content.
   b. Bodies are high in protein content.
   c. They reduce density by ion replacement.
   d. They have reduced skeletons.
   e. They are consumed by few kinds of predators.

8. The type of coloration exhibited by many large animals of the open sea is called
   a. countershading
   b. disruptive
   c. cryptic
   d. warning
   e. reticulated

9. The largest members of the zooplankton are
   a. copepods
   b. diatoms
   c. jellyfish
   d. sharks
   e. whales

10. The greatest diversity of nekton is among the
    a. squids
    b. fishes
    c. reptiles
    d. birds
    e. mammals

Matching

Match the following organisms with their method of locomotion:

a.	arrowworm	1.	Akinetic
b.	ctenophore	2.	Cilia
c.	dinoflagellate	3.	Flagella
d.	sargassum weed	4.	Jet propulsion
e.	sea turtle	5.	Limbs
f.	squid	6.	Undulation

Match the following organisms with their adaptations to life in the open sea:

a.	gymnosome pteropod	1.	Consumption of gelatinous zooplankton
b.	leatherback sea turtle	2.	Countershading
c.	*Pelagibacter ubique*	3.	Loss of shell
d.	salp	4.	Nitrogen fixation
e.	*Trichodesmium*	5.	Small size
f.	tuna	6.	Transparent body

Short Answer

1. What are some adaptations in open-ocean plankton that allow them to be productive despite the nutrient-poor environment?

2. What is marine snow, and what is its significance?

3. What are some adaptations that help organisms in the open ocean remain afloat?

4. What aspects of reproduction in pelagic sharks indicate adaptations to life in the open sea?

5. Describe some of the ways that animals living in the open sea hide from predators.

6. Describe some of the strategies animals use to increase their survival in the open sea.

7. Describe how the upside-down lifestyle of the purple sea snail and the nudibranch *Glaucus* has influenced their coloration.

8. Describe how a swordfish feeds.

9. Describe how tunas are adapted for fast swimming.

10. Why is the photic zone of the open ocean not as productive as the photic zone of coastal seas?

Thinking Critically

1. Why does it make good ecological sense for the largest sharks to be plankton feeders rather than predators of other nekton?

2. The water of the North Atlantic Ocean is colder than that of the tropical Atlantic Ocean. Do you think this difference might affect the shapes of phytoplankton living in these two areas? Explain your answer.

3. Why is it difficult to keep large tunas alive in public aquariums?

4. What problems in feeding mechanics, digestion, physiology, or nutrition might be encountered by predators that feed on gelatinous zooplankton? What additional problems might they face in today's seas with drifting bottles and plastic bags?

SUGGESTIONS FOR FURTHER READING

Anderson, N. M., and L. Cheng. 2004. The Marine Insect *Halobates* (Heteroptera: Gerridae): Biology, Adaptations, Distribution, and Phylogeny, *Oceanography and Marine Biology: An Annual Review* 42: 119–179.

Barton, A. D., S. Dutkiewicz, G. Fliert, J. Bragg, and M. J. Follows. 2010. Patterns of Diversity in Marine Phytoplankton, *Science* 327: 1509–1511.

Breitburg, D. L., B. C. Crump, J. O. Dabiri, and C. L. Gallegos. 2010. Ecosystem Engineers in the Pelagic Realm: Alteration of Habitat by Species Ranging from Microbes to Jellyfish, *Integrative and Comparative Biology* 50(2): 188–200.

DeLong, E. F. 2003. A Plenitude of Ocean Life, *Natural History* 112(4): 40–46.

Giovannoni, S. J. et al. 2005. Genome Streamlining in a Cosmopolitan Oceanic Bacterium, *Science* 309: 1242–1245.

Johnsen, S. 2001. Hidden in Plain Sight: The Ecology and Physiology of Organismal Transparency, *Biological Bulletin* 201(3): 301–318.

Klimley, A. P., J. E. Richert, and S. J. Jorgensen. 2005. The Home of Blue Water Fish, *American Scientist* 93(1): 42–49.

Moisander, P. H., R. A. Beinart, I Hewson, A. E. White, K. S. Johnson, C. A. Carlson, J. P. Montoya, and J. P. Zehr. 2010. Unicellular Cyanobacterial Distributions Broaden the Oceanic N_2 Fixation Domain, *Science* 327: 1512–1514.

Moore, C. 2003. Trashed: Across the Pacific Ocean, Plastics, Plastics, Everywhere, *Natural History* 112(9): 46–51.

Orr, J. C. et al. 2005. Anthropogenic Ocean Acidification over the Twenty-first Century and Its Impact on Calcifying Organisms, *Nature* 437: 681–686.

Prager, E. J., and S. A. Earle. 2000. *The Oceans*. New York: McGraw-Hill.

Sims, D. W. et al. 2008. Scaling Laws of Marine Predator Search Behaviour, *Nature* 451: 1098–1102.

Have You Wondered?

1. **What kinds of organisms live in the deep ocean? (18-2)**

2. **How organisms in the abyss deal with the cold temperatures and high pressures? (18-1)**

3. **How predators find prey in the dark? (18-2)**

4. **Why so many deep-sea fishes look so bizarre? (18-2)**

5. **How organisms can survive without sunlight? (18-2)**

18

Life in the Ocean's Depths

IN THE EARLY years of marine biology research, scientists believed that no animal life could exist in the dark abyss of the sea. There was no sunlight to power photosynthesis, so there would be no producers on which consumers could feed. The temperatures were too cold and the pressures too extreme to support life. As we noted in Chapter 1, retrieval of the transatlantic cable in the mid-1800s demonstrated that animal life could exist beneath the photic zone. On July 21, 1951, the Royal Danish research vessel *Galathea* recovered a variety of animals, including sea anemones, sea cucumbers, bivalves, amphipods, and bristle worms, from more than 10,000 meters (33,000 feet) deep in the Philippine Trench east of the Philippine Islands. Eight years later, Jacques Piccard and Don Walsh, using the bathyscaphe *Trieste*, discovered animal life at even greater depths at the bottom of the ocean's deepest trench, the Challenger Deep. These and other discoveries proved that animal life could not only survive in the dark recesses of the ocean but also thrive. In this chapter you will be introduced to the forbidding world of the ocean deep and the strange organisms that make it their home.

18-1 Survival in the Deep Sea

As late as 1843, the British naturalist Edward Forbes concluded that animal life could not exist in the sea below a depth of 55 meters (182 feet). At the time, his hypothesis was completely plausible, because what was known of the conditions of the deep ocean appeared to make it a most inhospitable place for life. As we have noted previously, sunlight rapidly fades with increasing water depth. Even in the clearest ocean water, there is generally not enough light to support photosynthesis below 200 meters (660 feet). Heat energy is also absorbed quickly by surface waters. Temperatures in the deep remain frigid throughout the year.

Adding to these forbidding conditions is the tremendous pressure of the ocean depths. The pressure at sea level is 1 atmosphere (760 mm Hg [millimeters of mercury] per square centimeter, or about 15 pounds per square inch). About every 10 meters (33 feet) below the surface of the ocean, the pressure increases by another atmosphere. At this rate the pressure exerted on an organism at 1,000 meters (3,300 feet) is 510 kilograms per square centimeter (1,500 pounds per square inch), or about 100 times greater than the pressure exerted at the surface. At a depth of 10,600 meters (35,000 feet), the pressure is 5,100 kilograms per square centimeter (7.5 tons per square inch).

Even if living organisms could survive in such an environment, they would have to find their food and their mates in total darkness. As adverse as these conditions may seem, they are at least stable, and, over millions of years of evolution, some organisms left the sunlit upper waters and adapted to life in the dark recesses of the sea (**Figure 18-1**).

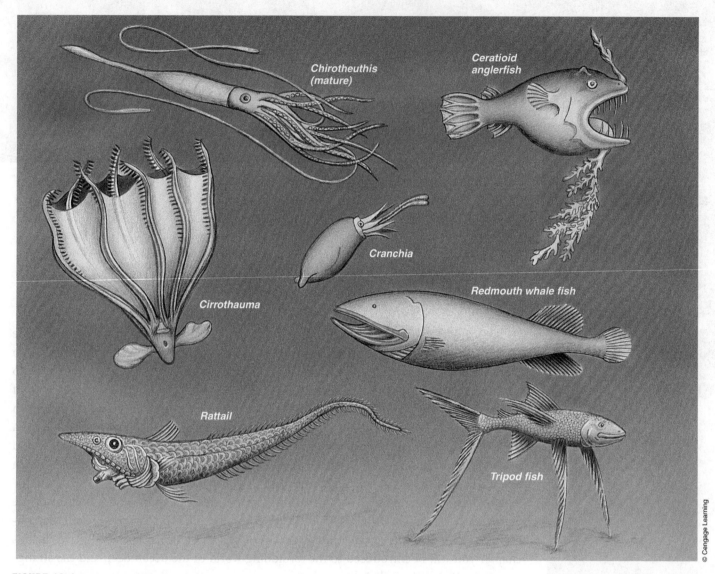

Chirotheuthis (mature)

Ceratioid anglerfish

Cranchia

Cirrothauma

Redmouth whale fish

Rattail

Tripod fish

© Cengage Learning

FIGURE 18-1 DEEP-SEA ANIMALS. Animals that live in the ocean's depths display a variety of intriguing adaptations that allow them to survive in a dark environment with cold temperatures and extreme pressures.

ADAPTATIONS TO PRESSURE

Early marine biologists believed that the enormous pressures encountered in the ocean depths would literally crush an organism. Interestingly enough, several thousand species of animal, especially echinoderms, annelids, and molluscs, tolerate the pressures. The fluid pressure in these animals' tissues matches the pressure of the surrounding water. The tissue fluid pressure pushes against the surrounding pressure with an equal but opposite force, preventing the animal from being crushed.

ADAPTATIONS TO COLD

Nearly all deep-sea animals have body temperatures that are close to the temperature of the surrounding water. Their body temperatures are so low that their metabolism is quite slow. As a result they move more slowly, grow more slowly, reproduce less frequently and later in life, and generally live longer than similar species from warmer surface waters. A lower metabolic rate also means that these animals require less food—scarce in the ocean deep—to stay alive.

One advantage of the cold is the increased density of the water. Animals that live in the deep have body densities very close to the density of their environment and do not have to expend energy to keep from sinking, as do animals in the photic zone.

IN SUMMARY

- Until the mid-1800s, biologists believed that life could not exist in the deepest parts of the ocean.

- The lack of sunlight to support photosynthesis, plus the cold temperatures and extreme pressures at great depths, were thought to limit, if not exclude, animal life.

- Deep-sea species have evolved several mechanisms for dealing with the high pressure and cold temperatures of the deep (Have You Wondered? #2)

- Because their metabolic rate is low, deep-sea animals do not have to eat as much or as often to survive.

- Cold water is denser than warm water, so deep-sea animals do not have the problem of remaining buoyant that their surface relatives have.

18-2 Life in the Dark

Of the three major factors that affect animals living in the ocean deep—light, temperature, and pressure—the lack of light has had the greatest evolutionary impact.

COLOR IN DEEP-SEA ORGANISMS

As we noted in the previous chapter, fishes that inhabit the well-illuminated surface waters exhibit countershading: they have dark dorsal surfaces and lighter ventral surfaces. From 150 to 450 meters (500 to 1,500 feet) below the surface, a region referred to as the **disphotic zone**, or **twilight zone**, there is still enough light to make countershading a useful means of camouflage. A common resident of these depths is the small hatchet fish (*Argyropelecus*; **Figure 18-2**). Its body is silvery or iridescent, with a dark dorsal surface and silvery underside. Like many deepwater fishes, hatchet fishes possess rows of light-producing organs called **photophores** along their bodies. These aid in species recognition, and the bioluminescence may make the ventral surface lighter, thus helping to camouflage.

FIGURE 18-2 HATCHET FISH. The silvery underside and dark dorsum of this hatchet fish are characteristic of the coloring of fish that inhabit the twilight zone.

NORBERT WU/MINDEN PICTURES/National Geographic Stock

Not all animals in this region, however, exhibit countershading. In the twilight zone, as well as the darker zones below, live animals that exhibit a variety of body colors. Black stomiatoid fishes (see **Figure 18-10**) have an iridescent sheen. Fishes known as gulper eels (see **Figure 18-9**) and swallowers (family Eurypharyngidae) are black or brown. These odd fish prowl the depths at 1,800 meters (6,000 feet), looking for prey. Whalefishes (suborder Cetomimoidei) resemble whales in their general appearance but are only a few centimeters in length. Their fins and jaws are bright orange and red. Deep-sea species of squid are permanently colored deep red, purple, or brown, and many species are bioluminescent. Benthic species of squid are often white. Shrimp that are blood red feed on red arrowworms and scarlet copepods. Interestingly, the bright red and orange colors of these animals appear as shades of black and gray in the depths at which they live. Studies by biologists at the Scripps Institution of Oceanography indicate that the red colors of deep-sea crustaceans vary according to the state of the molt cycle and may actually be involved in energy transfer during molting.

ROLES OF BIOLUMINESCENCE

Even though sunlight cannot penetrate to the depths of the ocean, many deep-sea animals produce their own light in the form of bioluminescence. This characteristic is especially common in animals that are found between 300 and 2,400 meters (1,000 and 8,000 feet). Some, such as certain species of squid, crustacean, and fish, have luminescent organs, whereas others harbor bioluminescent bacteria in species-specific locations. This symbiosis between bacteria and their host species is an example of mutualism. The host receives light that can be used to locate and recognize a mate or find prey, and the bacteria are given a place to live and supplied with food. Some species of fish are able to control the bioluminescence of their symbiotic bacteria by altering the flow of oxygenated blood to the regions where the bacteria reside. When oxygen levels are high, the bacteria luminesce; when oxygen levels are low, they do not.

Bioluminescence occurs when a protein called **luciferin** is combined with oxygen in the presence of an enzyme called **luciferase** and adenosine triphosphate (ATP). During the series of reactions that follows, the chemical energy of ATP is converted into light energy. The process is efficient, producing almost 100% light and little heat. Most bioluminescence associated with marine organisms is blue green, although red and yellow have also been reported.

Luminescent organs can be located at a variety of positions on an animal's body (**Figure 18-3**). Many deep-sea fishes display rows of photophores along their sides or on their bellies. Other species of fish may have depressions containing bioluminescent bacteria located on their heads or on fleshy growths attached to the head. Deep-sea squids exhibit bioluminescent spots on their tentacles and circling their eyes.

Camouflage

It would seem that such illuminated animals in a dark world would be easy marks for predators. In the twilight zone, where sunlight is fading into darkness, the presence of bioluminescence on the animal's ventral

---

The **disphotic zone (twilight zone)** is a region of ocean from 150 to 450 meters deep where sunlight is minimal.

Photophores are light-producing organs found on deep-sea fishes.

Luciferin is a protein that produces bioluminescence when combined with oxygen, luciferase, and ATP.

Luciferse is an enzyme that catalyzes the reactions that produce bioluminescence.

MARINE BIOLOGY & THE HUMAN CONNECTION

Exploring the Ocean's Depths

For centuries, humans dreamed of being able to explore the ocean's depths. Some tried to descend in containers such as wooden barrels equipped with an air hose, but the depths they could reach and the observations they could make were limited. In the last century, Dr. William Beebe pioneered underwater exploration. In 1934, he descended to a depth of 918 meters (3,028 feet) in a steel ball called a "diving bell" or "bathysphere." The diving bell had portholes and lights and was supplied with air and power by a mother ship on the surface.

The first self-contained craft that could operate free from a mother ship was the **bathyscaphe** (BATH-eh-skayf), invented in 1948 by Auguste Piccard. The vessel was basically a sphere that could accommodate two people. It was suspended from a huge float that contained gasoline to power the craft's systems. Auguste's son Jacques and Don Walsh (a lieutenant in the United States Navy) used the bathyscaphe *Trieste* in 1960 **(Figure 18-A)** to descend to the bottom of the Challenger Deep (11,020 meters, or 36,366 feet). Today's manned deep-sea submersibles are very sophisticated vessels equipped with the latest in photographic and robotic technology. Of the submersibles used by the United States, the best known is *Alvin*

Bettmann/CORBIS

FIGURE 18-A EARLY MANNED DIVING DEVICES. In 1960, the bathyscaphe *Trieste* was used by Jacques Piccard and Don Walsh to descend to the bottom of the Challenger Deep, the deepest point in the ocean.

NORBERT WU/MINDEN PICTURES/National Geographic Stock

FIGURE 18-3 BIOLUMINESCENT ORGANS. Luminescent organs (the row of small white dots on this fish's side) help species identify each other, locate prey, and avoid predation. They are frequently arranged in rows on the fish's body, as in this Pacific viperfish, or in depressions on the head.

surface may take the place of countershading in camouflaging the animal. If the intensity of the bioluminescence matches the intensity of sunlight at that level, the fish will not appear as a shadow or silhouette when seen from below.

Mating and Species Recognition

Bioluminescence can also play an important role in mating. During breeding times, the pattern of lights identifies an individual as being male or female. In lanternfishes (*Myctophum*), for example, males carry bright lights at the tops of their tails, whereas females have only weak lights on the underside of their tails **(Figure 18-4)**. Like fireflies, some species of deepwater fish signal their readiness to mate by a series of light flashes.

The pattern of lights also serves for species identification, especially among members of closely related species. For instance, one species of lantern fish has three rows of light spots along its ventral surface, whereas another closely related species displays only two. This would be of particular advantage in finding mates or in forming schools. The marine biologist William Beebe once remarked that the pattern of lights on deepwater fishes was so distinctive that he could recognize species in absolute darkness just by the light pattern that they presented.

(see photo in Chapter 1), which can carry a crew of three. *Alvin* averages 150 dives per year and was used by Robert Ballard in 1977 to first view the animals of hydrothermal vent communities. Ballard used *Alvin* again in 1985 to view the sunken passenger ship *Titanic*.

The United States Navy operates the *Sea Cliff,* a research-and-recovery vessel that can operate at depths of 6,100 meters (20,000 feet). In 1986, it was used by the U.S. Geological Survey to examine mineral deposits located off the coasts of California and Oregon. Two other submersibles owned by the Harbor Branch Oceanographic Institute in Florida, the *Johnson Sea Link 1* and *the Johnson Sea Link 2*, each can carry a crew of four. Both participated in the efforts to recover pieces of the space shuttle *Challenger* following its disastrous explosion shortly after takeoff in January 1986.

Manned submersibles have been used to photograph sharks, survey coral reefs, and produce motion-picture footage for documentaries on the sea and its creatures. They are also used commercially in offshore petroleum developments and other engineering projects.

The drawbacks of using manned submersibles are that they are expensive to operate and they expose passengers to risk. To deal with these problems, engineers have developed a set of unmanned units called remotely operated vehicles (ROVs; **Figure 18-B**). These carry photographic equipment, have robotic arms

FIGURE 18-B REMOTE OPERATING VEHICLES, OR ROVS. *Jason*, an ROV, is used by marine biologists to explore deep-sea ecosystems.

Woods Hole Oceanographic Institute

and other types of sampling equipment, and can be controlled from the surface, eliminating the need for an on-board operator. The main disadvantage of older ROVs is the need for a sizeable cable through which power is transmitted to the unit and data are transmitted back. However, this one disadvantage is far outweighed by lower construction and operating costs, not to mention the minimal risk to human life.

During the exploration of the *Titanic* in 1986, *Alvin* carried attached to it a small ROV called

Jason Jr., or *JJ. JJ* was used to explore the inside of the *Titanic*, something the larger, more cumbersome *Alvin* could not do. The newest generation of ROVs include *Jason II*, an ROV that can operate at depths of 6,500 meters (21,450 feet) and relay data to the surface via the Internet, and *Argo II*, a deep-sea imaging system that can be towed by a research vessel. Although ROVs will probably not replace the manned submersibles, they do offer the ability to examine larger areas of the sea for longer times at a greatly reduced cost and risk.

FIGURE 18-4 A LANTERNFISH. The bright, bioluminescent spots on the tail of this lanternfish identify it as a male. The pattern of bioluminescent spots helps the lanternfish recognize the opposite sex during the mating season.

NORBERT WU/MINDEN PICTURES/National Geographic Stock

Attracting Prey

Bioluminescence also functions to attract prey. Anglerfish (*Melanocetus johnsonii*; **Figure 18-5**) and stomiatoids attract prey with bioluminescent lures. Smaller fishes whose normal prey are luminescent are attracted by the light and, when close enough, become a meal for the larger predator. Dr. Beebe observed that a lanternfish apparently uses the light from its ventral surface to find prey. Copepods that were illuminated by the lights from the ventral surface of this fish were quickly eaten. Some species of stomiatoid fish have lights around their eyes that illuminate their field of vision. They have been observed to fix krill in the beam of light and then rapidly devour them. Deepwater krill also have illuminated spots around their eyes that may help them find smaller prey.

A **bathyscaphe** is a self-contained craft used in deep-sea exploration.

Defense

Some animals use bioluminescence for defense. In the dark waters of the deep, the inky cloud produced by surface squids would have no effect in deterring a predator. Some deepwater squid species, however, release a

467

FIGURE 18-5 ANGLERFISH. The fleshy projection on the head of this anglerfish contains a bioluminescent "lure" at the end. Smaller fish are attracted to the light and become an easy meal for the anglerfish.

NORBERT WU/MINDEN PICTURES/National Geographic Stock

bioluminescent fluid that clouds the water with light, confusing potential predators. This tactic is also used by some species of deepwater shrimp. The scarlet-colored opossum shrimp (family Mysidae; **Figure 18-6**) gets its name because the female carries her eggs in a pouch under the thorax. When these shrimp are threatened, they release a substance that bursts into a cloud of miniature light particles that look like stars in the night sky. The sudden burst of light frightens and confuses potential predators, allowing the shrimp to escape.

SEEING IN THE DARK

Do fishes in the aphotic zone of the sea perceive anything other than the light produced by other animals? To answer this question, the British biologist N. B. Marshall made an extensive study of the brains and eyes of deep-sea fishes, some species living as deep as 900 meters (3,000 feet) below the surface. He concluded that even though there is virtually no light at these depths, these fishes can probably see their food, their own species, and their predators, using what little light is available to stimulate vision.

In contrast to the typical spheroid vertebrate eye, the eyes of many deep-sea fishes are tubular and contain two retinas instead of one **(Figure 18-7)**. One retina is used to view distant objects, while the other sees things

closer in. Not only do fishes with tubular eyes see better in dim light, but they also have better depth perception. This allows them to judge the distance to their prey quite accurately, so that they are less likely to miss their catch—a definite advantage in an area where food is so scarce and where speed is not common.

The biologist G. L. Walls has noted that at depths between 900 and 1,500 meters (3,000 and 5,000 feet) the eyes of fishes become smaller and less functional. The anglerfish has well-developed eyes during its early life, which is spent in well-lit surface waters. As it grows into an adult, it begins to sink and eventually resides at depths of about 1,800 meters (6,000 feet). At this stage in its life, the eyes stop growing and sometimes begin to degenerate.

Deep-sea squid also have strange-looking eyes. They can be barrel-shaped, stalked, or unequal in size. It is thought that these eyes are probably adapted to the amount of light available in their habitat and to their method of locating and capturing prey.

Some animal species, both vertebrate and invertebrate, that occupy the deepest recesses of the sea have tiny eyes that are only slightly functional, or they are totally blind. These organisms rely more on tactile and chemical stimuli to find their prey and mates and to avoid predation.

FINDING MATES IN THE DARK

Some species of deepwater anglerfish have solved the problem of finding a mate in the dark in an unusual way. Because mates are hard to find, when a male encounters a female, he bites her and remains attached, sometimes for the rest of his life **(Figure 18-8)**. In some species of anglerfish, the male becomes a lifelong parasite. The skin around the male's mouth and jaws fuses with the female's body, and only a small opening remains on either side of the mouth for gas exchange. The eyes and most of the male's internal organs degenerate, and his circulatory system becomes connected to the female's. The male, in effect, becomes an external, sperm-producing appendage.

It is believed that the female anglerfish releases her eggs in deepwater, and fertilization is external. The fertilized eggs quickly float upward to hatch in surface waters, where the young feed on copepods and other small plankton. As the young grow, females develop fishing lures, and

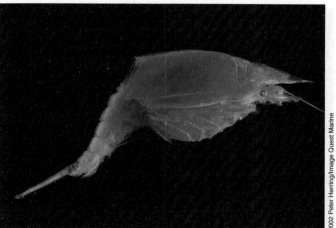

2002 Peter Herring/Image Quest Marine

FIGURE 18-6 SCARLET-COLORED OPPOSSUM SHRIMP. For protection, the scarlet-colored opossum shrimp releases a substance that produces flashes of light similar to fireworks. This sudden burst of light in an otherwise dark environment frightens potential predators.

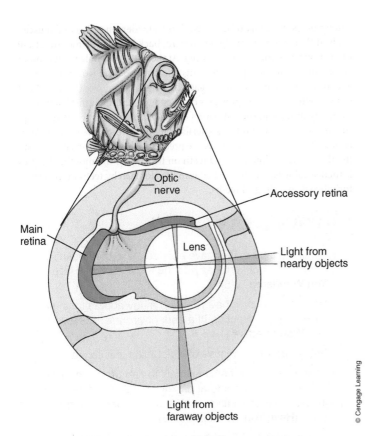

FIGURE 18-7 THE EYE OF DEEPWATER FISH. Fish that live in the ocean's depths have eyes that are adapted for seeing in dim light. They are tubular and contain two retinas. The shape of the eye gives them better depth perception, so they are less likely to miss their prey when they strike.

males develop gripping teeth on both their snouts and chins. When the fish mature to adults, they descend and begin their deepwater existence.

FINDING FOOD IN THE DARK

In the depths of the sea, food is scarce. There are no photosynthetic organisms to produce food, but organic wastes, scraps of food, and dead and dying organisms drift down slowly from above. Some of this food is consumed before it reaches the bottom. The rest reaches the floor of the ocean, where it supports a variety of benthic organisms and active deep-sea scavengers. Detritus feeders of the benthic community are, in turn, food for more active predators.

Not all deep-sea food chains rely on the rain of detritus from the surface. Many small fishes and invertebrates rise at night to feed in the rich surface waters, returning during the day to the deeper fathoms of the sea. These vertical migrations help to bring more food to the lower depths. Larger predators also make daily migrations to feed on smaller organisms.

Even at the best of times, food is scarce, and animals that survive in this environment exhibit some bizarre adaptations for feeding. Many deep-sea fishes appear to be nothing more than a huge, tooth-filled mouth that is attached to an expandable stomach with a small tail for swimming. Fish species called gulper eels **(Figure 18-9)** are common at depths below 1,500 meters (5,000 feet). They have jaws that are hinged like a

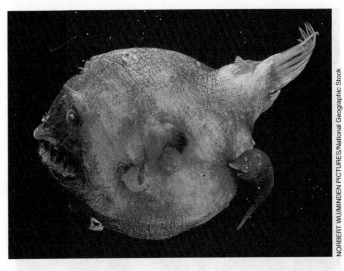

FIGURE 18-8 MATING IN ANGLERFISHES. Male anglerfishes attach themselves to a female, and in some species the relationship is permanent, with the male becoming a lifelong parasite of the female.

trapdoor and stomachs that can expand to several times their normal size to accommodate prey that is larger than the gulper eel itself. A research vessel captured a 15-centimeter (6-inch) gulper specimen with a 23-centimeter (9-inch) fish coiled in its stomach. Some species of gulper eel grow to 1.8 meters (6 feet) long. Most of this length is in their long, whiplike tail. The tip of the tail is luminescent and may play a role in luring potential prey close to the animal's mouth.

Stomiatoids **(Figure 18-10)** are also capable of ingesting prey larger than themselves. This gives them the ability to take advantage of rare large food items. Most species are 15 to 18 centimeters (6 to 7 inches) long. They have large heads with curved, fanglike teeth and elongated bodies that taper into a small tail. They are prominent residents at depths of 180 to 600 meters (600 to 2,000 feet).

> A **barbel** is a fleshy projection that dangles from the chin or throat of some fishes.

The typical stomiatoid has a fleshy projection called a **barbel** that dangles below its chin or throat (see **Figure 18-10**). The precise role of these

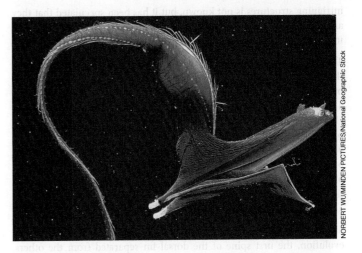

FIGURE 18-9 GULPER EEL. Many deepwater fish such as this gulper eel have huge mouths that can engulf prey as large as the predator.

NORBERT WU/MINDEN PICTURES/National Geographic Stock

FIGURE 18-10 STOMIATOID FISHES. Stomiatoid fishes, such as this black sea dragon, usually have a fleshy projection called a *barbel* dangling from their chin or throat. It is believed these structures function as lures to attract prey and as sensors for finding food in the bottom ooze.

intriguing structures is not known, but it has been speculated that they are lures or used to probe the bottom ooze for food. They may even serve in species recognition during mating.

The characteristics of the barbel vary from species to species. In some the barbel is a short hair, whereas in others it is a whip-like structure that may be 10 times longer than the animal's body. One species that is only 4 centimeters (1.5 inches) long has a barbel that is 40 centimeters (15 inches) long, and a 21-centimeter-long (8.5-inch-long) species has a barbel nearly 1 meter (approximately 3 feet) long. The shape of the barbel is as variable as the length. Some are single strands, whereas others branch repeatedly; still others resemble a strange flower or even a cluster of grapes. Many are also bioluminescent.

Deep-sea anglerfishes found at depths of 1,970 meters (6,500 feet) have a bony projection with a luminescent lure at the tip that is a modification of one of the spines of the fish's dorsal fin (see Figure 18-5). Through evolution, the first spine of the dorsal fin separated from the others, moved forward, and was modified into a fishing pole that, when not in use, lies in a groove on the top of the fish's head. Some species have short,

stubby poles, whereas others have long, slender ones. Sets of muscles evolved that can move the pole forward, back toward the mouth, or out of the way when the animal is eating. At the end of the pole is a luminous "lure." The anglerfish's prey may mistake the lure for a worm or shrimp, and as it moves closer to investigate, the anglerfish moves the lure closer to its mouth. When the prey gets close enough, the anglerfish's lower jaw suddenly drops and the gill covers expand, creating suction that sweeps the unwary victim into the anglerfish's mouth, which has become large enough to devour prey as big as the predator. The jaws snap shut, impaling the victim on long, curved teeth on the roof of the mouth. The prey is then swallowed whole, pushed into the stomach by teeth in the predator's throat (pharyngeal teeth).

IN SUMMARY

- Of all the factors that have shaped life in the ocean's depths, lack of light has had the most impact. (Have You Wondered? #5)
- Deep-sea animals have evolved many adaptations that allow them to function in a dark environment. (Have You Wondered? #1)
- Many deep-sea animals exhibit bioluminescence.
- Bioluminescence helps animals to find prey and mates in a world of almost total darkness. It also plays a role in species identification and, in some species, attracts prey. (Have You Wondered? #3)
- The shortage of food has led to many interesting adaptations in deep-sea fishes. (Have You Wondered? #4)

18-3 Giants of the Deep

Most deep-sea animals are small compared with those living in surface waters, but a few are giants compared with their shallow-water relatives. One possible reason for the large size of these organisms is that they live longer than their shallow-water relatives and thus have time to grow larger. Sea urchins that live on the ocean floor have bodies 30 centimeters (1 foot) in diameter, compared with shallow-water species only a few centimeters across. In the deep sea off the coast of Japan live hydroids that reach the amazing height of 2.5 meters (8 feet). Sea pens **(Figure 18-11a)** that are ordinarily 0.5 meter high grow as large as 2.5 meters in the ocean depths; isopod crustaceans (related to terrestrial pill bugs; **Figure 18-11b**) are as large as 20 centimeters (8 inches). Shrimp species that are bright red or bioluminescent reach lengths of nearly 30 centimeters (1 foot). Some of these species have antennae that are twice their body length, which may be used to capture small prey.

GIANT SQUIDS

Perhaps the most spectacular of the deep-sea giants is a mollusc, the giant squid (*Architeuthis dux*; **Figure 18-12**). The giant squid is the largest of all invertebrates, 9 to 16 meters (about 30 to 53 feet) long, including the tentacles. Its arms are as thick as a human thigh and covered with thousands of suckers. The two tentacles can be more than 12 meters (40 feet) long, and their flattened ends bear 100 or more suckers with serrated edges. As in other species of squid, the arms and tentacles capture prey and carry it to the animal's beak, where it is shredded into small pieces. The largest giant squid recorded washed up on a New Zealand beach in 1888. It measured 18 meters (59 feet) in total length, with the tentacles

FIGURE 18-11 GIANT DEEP-SEA ANIMALS. Some animals that live in the ocean depths such as this **(a)** sea pen and **(b)** isopod grow much larger than related species that live in the sunlit surface waters.

Robert Yin/Corbis RF/Alamy

(a)

David Wrobel/Visuals Unlimited, Inc.

(b)

accounting for 15 meters (49 feet) of the length. Although no giant squid has ever been accurately weighed, estimates run as high as 1 ton for a large specimen.

Little is known about the natural history of giant squids. They are found in all oceans, spending most of their lives at depths of 180 meters (600 feet) or more. They use the same method of swimming as other squids, a type of jet propulsion. Although there are reports of giant squids passing ships traveling 23 kilometers per hour (14 miles per hour), their anatomy suggests that they are weak swimmers that probably cannot capture active prey. No one knows for sure what these animals eat. It is assumed that they feed on a variety of small invertebrates

and fishes. A major predator of giant squids is the sperm whale. Sperm whales have been found with scars from the serrated suckers on their bodies. Analyses of the stomach contents of captured specimens have revealed the remains of giant squids.

NEW SPECIES OF DEEPWATER SQUID

In 1988, researchers making observations from a deep-sea submersible discovered a new and as yet unnamed species of large deepwater squid **(Figure 18-13)**. Since the original sighting, this animal has been observed at least eight times in water between 1,930 and 4,709 meters (6,367 and 15,580 feet) deep. These squid can reach lengths of 7 meters (23 feet) and have large, billowing fins. Arguably their most unusual feature is their arms, which are significantly longer than those of other known species of squid. The arms project from the squid's body for a short distance and then bend downward at sharp angles, possibly to

AP Photo/Tsunemi Kubodera of the National Science Museum

FIGURE 18-12 GIANT SQUID. Dr. Tsunemi Kubodera of the Japanese National Science Museum was the first to photograph a live giant squid in December 2006. This specimen was found 600 miles southeast of Tokyo and measured a little less than 8 meters (24 feet).

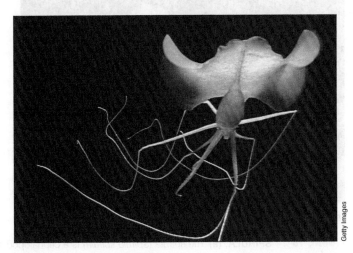

Getty Images

FIGURE 18-13 BIG FIN. This new species of giant squid was first observed in 1988 and may represent a new family of squid.

prevent them from becoming tangled. Researchers report that these squid don't seem to use their arms for grabbing prey but that they may be coated with a sticky substance that allows them to capture small crustaceans for food. Marine biologists speculate that this squid may be common in the deep sea because specimens have been observed in the Atlantic, Pacific, and Indian Oceans, as well as the Gulf of Mexico, all within a relatively short time. These squid exhibit behaviors different from their shallow-water relatives. For instance, when they release ink they hide in the ink cloud rather than swimming away, as squid do in sunlit waters. Pairs have been observed attached to each other, with one squid towing the other through the water, a behavior not observed in shallow-water squid. Many questions about this animal—not the least of which is whether it is a new kind of squid—will remain until a specimen can be captured and studied in detail.

IN SUMMARY

- Although most deep-sea species are small compared with similar shallow-water species, a few are giants.

- One possible explanation for the large size of some deep sea animals is that they live longer than shallow-water species and thus have more time to grow.

- One of the better-known deep-sea giants is the giant squid. Because live specimens have never been captured for study, very little is known about this intriguing animal.

18-4 Relicts from the Deep

Evolution by natural selection is stimulated by changes in the environment. Because environmental conditions of the deep sea are believed to have remained nearly stable for more than 100 million years, biologists theorize that the sea's depths hold many organisms that have undergone very little change from their early ancestors. Marine biologists

were encouraged to pursue this theory by the discovery in 1864 by Norwegian oceanographers of a large sea lily dredged from 540 meters (1,800 feet). Similar species were known from fossils 120 million years old, but until this discovery no living specimen had ever been discovered. In 1870, another expedition discovered a large, red sea urchin from the depths of the North Atlantic, a genus previously known only from 100-million-year-old fossils in the white chalk cliffs of Dover, England.

SPIRULA

The *Challenger* expedition of 1872 to 1876 netted a living specimen of *Spirula*, a mollusc named for its spiral-shaped internal shell **(Figure 18-14a)**. *Spirula* are small molluscs that resemble squids and octopods. They average 7.5 centimeters (3 inches) in length and have a barrel-shaped body with short thick arms. The animal swims with its head pointing downward and contains an internal shell divided into gas-filled chambers **(Figure 18-14b)**. Similar molluscs called belemnites were common in the seas 100 million years ago but disappeared about 50 million years ago, with the exception of the ancestors of *Spirula*.

VAMPIRE SQUID

In 1903, another strange animal was discovered. Named the vampire squid (family Vampyromorphidae; **Figure 18-15** because the webbing between the arms and dark color suggested the sinister figure of a mythical vampire, it appears to be the descendant of a group of molluscs that were intermediate between octopods and squid. These animals, some with webbing between the arms and others with paddle-like fins, disappeared from the fossil record about 100 million years ago. The vampire squid's ancestors may have avoided extinction by retreating to ocean depths of 900 to 2,700 meters (3,000 to 9,000 feet).

The vampire squid's muscles are soft and poorly developed, implying that it is not a very good swimmer. It probably drifts in the depths or moves feebly with its head down and the arms hanging limp. The arms are connected by a tissue that forms a webbing, and the combination forms a loose bag around the animal's mouth. Originally, the vampire squid was thought to be an octopus, but then researchers found two

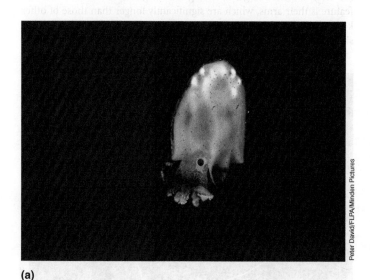
(a)

(b)

FIGURE 18-14 SPIRULA. (a) The spirula (*Spirula spirula*) is a cephalopod mollusc that is similar to extinct molluscs called belemnites that lived in the seas 100 million years ago. **(b)** The internal shell of spirula is divided into chambers like the shell of the nautilus. The gas content of the chambers helps to offset the animal's density and increases buoyancy.

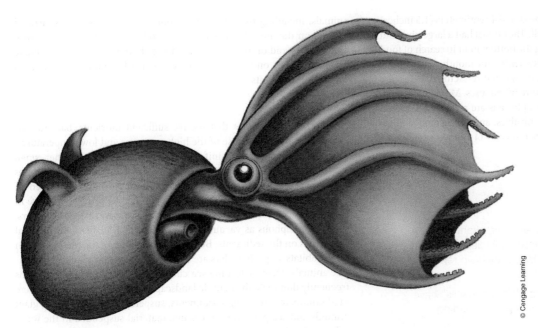

FIGURE 18-15 VAMPIRE SQUID.
This deepwater animal is not really a squid but an intermediate form between squids and octopods. It received its name because of the webbing between the arms.

© Cengage Learning

additional arms, bringing the total to ten, which is characteristic of squid. These two additional arms are long, lack suckers, and are quite different from the tentacles of other squid. They are thought to be feelers, and when not in use they are coiled into special pockets in the animal's web. Because of the animal's differences from squid and octopods, they are placed in a different order (Vampyromorpha).

Vampire squid apparently spend their entire lives in deepwater. Like other cephalopods, they have good eyesight and they possess several bioluminescent organs, including two particularly large ones that can be covered by flaps of skin. They are thought to feed on small animals that are not particularly active, possibly using their long feelers to find their prey.

Vampire squid mate in a fashion similar to octopods and squid, with the male using one of his arms to insert a packet of sperm into the female genital opening. Spawning and hatching occur at depths of 2,000 to 2,500 meters (6,500 to 8,500 feet). When the young first hatch, they have eight arms. The other two arms, the webbing, and light organs develop when they are about 2 centimeters (0.66 inch) long.

Vampire squid have been collected in several areas of the world, including off the coasts of South Africa, India, Indonesia, and New Zealand. One was taken from a submarine canyon north of New Zealand at a depth of 3,000 meters (9,850 feet). The largest vampire squid ever taken, a female, measured 20 centimeters (8.5 inches). No males larger than 12.5 centimeters (5 inches) have ever been recorded.

COELOCANTH

In 1938, fishers caught an unusual fish about 1.8 meters (6 feet) long off the coast of South Africa. The fish had large, thick scales and fleshy bundles between its body and fins. It was identified as the coelacanth (*Latimeria chalumnae*; see Figure 10-12 in Chapter 10), a species thought to have been extinct for 70 million years. This first known specimen was taken in relatively shallow water, 73 meters (240 feet). Subsequently, specimens have been caught at depths of 180 to 260 meters (600 to 1,200 feet). This discovery gave biologists essential insights into the evolution of tetrapods.

NEOPILINA

In 1952, researchers aboard the vessel *Galathea* discovered an unusual, limpet-like mollusc called *Neopilina* **(Figure 18-16)** at 3,600 meters (12,000 feet) in the muddy clay off the Pacific coast of Costa Rica. This mollusc belongs to the class Monoplacophora and was thought to have

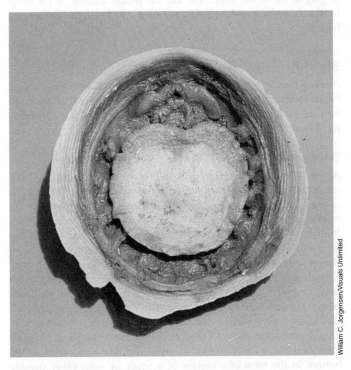

William C. Jorgensen/Visuals Unlimited

FIGURE 18-16 NEOPILINA. This limpet-like mollusc was known only from fossils until the discovery of a live specimen in 1957. Many researchers believe that this animal is a member of an ancient group of molluscs that were the ancestors of modern gastropods, bivalves, and cephalopods.

been extinct for 350 million years. It measured 4 centimeters (1.5 inches) long and had a lightweight conical shell. The animal had a large pink and blue foot, probably used to crawl along the bottom mud in search of food that it gathers with short fleshy tentacles behind the mouth. Several of its anatomical structures, such as gills, kidneys, and retractor muscles, are repeated several times, a primitive feature in molluscs. Many researchers who study molluscs believe monoplacophorans are an ancestral group that gave rise to modern gastropods, bivalves, and cephalopods. Since the discovery of *Neopilina*, many other species of monoplacophorans have been found in the deep sea.

IN SUMMARY

- Because the conditions of the deep have changed very little over the past 100 million years, biologists speculate that some animals living in the deep may have remained unchanged since they first evolved.

- Over the years, several examples of "living fossils" have been recovered from the ocean's depths, including invertebrates and fishes.

18-5 Life on the Sea Bottom

Marine biologists who have visited the sea bottom in submersibles describe a barren landscape with few living organisms compared with communities in sunlit surface waters. Life on the sea bottom is as challenging as life in the abyss, and like the pelagic animals of the deep, benthic fauna have evolved to fill the available niches in this challenging habitat.

BENTHIC COMMUNITIES

Animals that live on the sea bottom must deal with extreme pressure, low temperature, and darkness. The environmental factor that appears to have the greatest impact on benthic organisms, however, is the availability of food.

Sources of Food for Benthic Organisms

At the ocean floor there is no light to power photosynthesis, and cold temperatures slow the growth of bacteria and other microorganisms, so the usual bases of marine food chains are either nonexistent or severely limited. For the most part, food to sustain benthic flora comes from the sunlit surface waters. In some areas, potential food in the form of feces, decaying tissue, and other organic matter rains down almost constantly from above. What is not intercepted by pelagic animals accumulates and concentrates on the ocean floor. Although compared with bottom communities along the continental shelf this represents very little food, it does sustain a variety of meiofauna, infauna, suspension feeders, and omnivorous scavengers. Near the margins of continental shelves, turbidity currents deliver organic nutrients to abyssal plains and trenches. Occasionally, food may arrive on the bottom in the form of a carcass of a whale or some other sizeable organism that sank to the bottom with little interference from pelagic scavengers. The importance of such occasional large food items was demonstrated by two researchers, John Isaacs and Richard Schwartzlose. In experiments they lowered cameras mounted to buckets of dead fish to the seafloor. In as little as 30 minutes a large variety of benthic

animals, including fish, snails, echinoderms, and crustaceans, arrived to feast on the meal. Shortly after the food was consumed the bottom appeared devoid of life. The few predators found at these depths make a living preying on the small number of sizeable animals that live on top of the sediments.

Food Chains

In benthic communities that receive sufficient nutrients, bacteria can grow, albeit slowly because of the high pressures and low temperatures. Meiofauna such as foraminiferans and nematodes and other small worms feed on the bacteria, the limited organic matter, and sometimes each other. Larger worms and bivalves make up the infauna that feed on the meiofauna and available organic matter. Unlike bivalves that live in the sunlit waters near the surface, deep-sea bivalves are deposit feeders, using their siphons as vacuums to suck up the available food material that settles on the sediments. Because of the nature of the food source in these habitats, deposit feeders are usually the most abundant of the benthic animals. Deposit-feeding sea cucumbers, brittle stars, and urchins frequently dominate the benthic landscape (**Figure 18-17**). In areas where sand dominates the bottom sediments, suspension feeders such as giant crinoids and sea pens feed on organic material suspended in the water column. Predators in these communities are primarily fish, squid, and sea stars.

Farther away from the margins of the continental shelves, the availability of food diminishes even more. Although there is still enough food to support small fauna such as bacteria, meiofauna, and infauna, large deposit feeders are generally absent. In the deepest recesses of the ocean, the mid-ocean trenches, food is so scarce that even tiny organisms are rarely found.

Diversity of Benthic Organisms of the Deep

Although the density of benthic organisms in the deep ocean is only a fraction of that in the benthos of shallower seas, there is a surprising degree of diversity. One hypothesis for the large number of species is that they do not disperse their young as effectively as their relatives in shallower seas. This may contribute to isolation, which in turn is necessary for speciation. Another hypothesis is that the environmental conditions of the deep sea are quite stable and the benthic habitat may be much

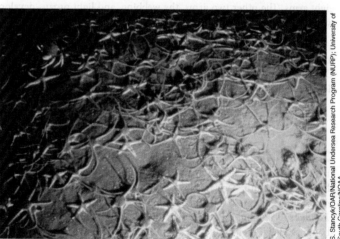

FIGURE 18-17 DEPOSIT FEEDERS. Because of the nature of the food source in deepwater benthic communities, the most numerous animals are deposit feeders, such these brittle stars.

S. Stancyk/OAR/National Undersea Research Program (NURP); University of South Carolina/NOAA

IMPACT BUBBLE

THE DEEP SEA REPRESENTS the largest ecosystem on earth and, because of its relative inaccessibility, one of the least studied. Even though the deep seafloor is a remote habitat, it is not immune to human activities. Disposal of wastes, mining, and climate change are having an impact on the deep-sea ecosystem.

As the human population increases, more litter is produced, a large amount of which eventually makes its way to the ocean and ultimately to the seafloor. For centuries all types of waste were dumped into deep water with mostly unknown and unstudied effects on deep-sea habitats and their inhabitants. Through the latter part of the 20th century, waste disposal and dumping represented the main human impact on the deep sea. Although ocean dumping is now legally banned, problems still persist due to accumulated waste and litter from the past. Samples from the seafloor indicate that the most common types of litter are plastic (both hard as in plastic bottles and soft as in plastic bags), glass, and metal. Large plastic items can degrade to small, microplastic particles that may be ingested by invertebrates. The toxic substances contained in some plastics could then move through food chains, accumulating to lethal levels in higher order predators. Sewage and wastes from dredging and mining have also been dumped into the deep sea. The organic material in sewage can be consumed by deposit feeders and may lead to changes in the size and distribution of benthic organisms which may alter food web dynamics. Dredging waste can alter the undersea landscape, and mining waste contains a variety of toxic substances that can kill organisms or accumulate in food chains. Waste from fishing vessels and factory ships adds large amounts of organic material to the water in the form of dead fish, fish parts, and other dead organisms. This may provide a food source for marine birds and some mammals, such as seals, but it also poses the risk of these animals becoming entangled in fish lines and nets and drowned. Organic matter discharged from fishing vessels that sinks to the bottom can create oxygen-depleted areas as it decomposes. Chemicals, pharmaceuticals, and radioactive materials have also been dumped into the sea in the past with little information on how these materials affect the deep sea environment.

As resources on land become depleted, interest focuses on the extraction of mineral resources from the seabed. Vent sites are currently being mined for sulfides, and there is an increased likelihood that hydrothermal vents may be mined in the future. Vent mining for sulfides produces large amounts of sediment that interfere with filter feeders. Vent mining also causes changes in hydrothermal circulation and adds chemical pollutants to the water. Another threat to the deep sea may come from climate change. Increased accumulation of carbon dioxide in the atmosphere not only contributes to global warming, but more carbon dioxide is dissolving in the ocean and causing an increase in acidity as well. The increased acid in the water leads to decreased amounts of calcium carbonate. Deep-sea animals such as echinoderms have high levels of carbonate in their skeletons and are likely to be the most affected by acidification of deep water. Molluscs will also be affected by the decrease in available calcium carbonate for their shells. Increases in the temperature of surface water may have an effect on the formation of cold deep water that could lead to changes in ocean circulation that supplies oxygen to deep water organisms. Changes in the temperature of seawater can lead to ocean stratification and decreased nutrient availability. This would have a negative impact on ocean productivity in surface waters and in turn decrease the amount of food that sinks to the seafloor. Changes in the amount of food from the sunlit surface waters could dramatically change the trophic structure of deep-sea communities.

As human endeavors continue to impact life in the deep, it is more important than ever to develop effective conservation measures. This will be challenging and will require continued research and monitoring of this ecosystem. Because much of the deep ocean is not under the jurisdiction of any one country, enforcement of conservation measures will be difficult. Improvements in technology not only allow individuals better access to the sea's resources, but also allow scientists to conduct more sophisticated studies. With time, advances in technology may give researchers and conservationists the tools they need to address these challenges and preserve the deep-sea ecosystem.

---

older than benthic habitats on continental shelves or in shallow seas. Because of the stability of the environment, fewer species may go extinct, and over time more species may accumulate. Until marine biologists are able to study this ecosystem in more detail, the cause of the diversity will remain a mystery.

VENT COMMUNITIES

Not all benthic communities in the deep ocean are populated by a relatively few small to medium-sized animals. In 1977, oceanographers discovered a thriving community off the Galápagos Islands along volcanic ridges in the ocean floor. Mineral-rich water, heated by the earth's core, raises the temperature of the water in these regions (normally 2°C to 8°C) to 16°C. Since this initial discovery, marine scientists have discovered several other vent communities in other parts of the world, including off the coast of Oregon, off the west coast of Florida, in the Gulf of Mexico, and in the central Gulf of California.

Despite the extremes of temperature and pressure, these self-contained vent communities are some of the most productive in the sea and demonstrate that communities can exist without solar energy. Not all vent communities surround hydrothermal vents. For instance, the Florida and Oregon vent communities are associated with cold-water seepage areas.

Vent Formation

You will recall from our discussion of geology in Chapter 3 that vents form at spreading centers and cold seawater seeps down near them through cracks and fissures in the ocean floor. The water comes into contact with hot, basaltic magma, where it loses some minerals to the magma and picks up others, notably sulfur, iron, copper, and zinc. The superheated water returns to the sea through chimney-like structures formed by minerals that have precipitated from the hot water. Some of these rise as high as 13 meters (43 feet) off the seafloor. The

FIGURE 18-18 BLACK SMOKER. The superheated water emitted from these chimneys is rich in sulfides. When the hot, sulfide-laden water mixes with the cold ocean water, the sulfides form black precipitates that give the appearance of smoke rising from the chimneys.

FIGURE 18-19 VENT COMMUNITY. Vestimentiferan worms and crabs are some of the larger animals found in this vent community in the Galápagos Rift.

chimneys are divided into two groups: white smokers and black smokers. White smokers produce a stream of milky fluid rich in zinc sulfide. The temperature of this water is normally less than 300°C. Black smokers are narrow chimneys that emit clear water, rich in copper sulfides, at temperatures of 300°C to 450°C. As the clear hot water encounters the cold ocean water, sulfide precipitates, producing the black color (Figure 18-18).

Vent Communities

Surrounding deep-sea vents for a few meters are rich marine communities (Figure 18-19) that include large clams, mussels, anemones, barnacles, limpets, crabs, worms, and fish. The clams (*Calyptogena*) are very large and exhibit the fastest known growth rate of any deep-sea animal (4 centimeters, or 1.5 inches, per year). Hydrothermal vent communities also contain large worms, called vestimentiferan worms (phylum Annelida) because of the covering that protects their bodies. Some of these worms may be as much as 3 meters (approximately 10 feet) long and several centimeters in diameter.

The primary producers in hydrothermal vent communities are chemosynthetic bacteria that oxidize compounds such as hydrogen sulfide (H_2S) to

provide the energy for producing organic compounds from carbon dioxide. Small animals such as crustaceans may feed directly on bacteria, whereas soft-bodied animals may be able to absorb organic molecules that are released by the bacteria when they die.

Clams, mussels, and worms are primary consumers, filtering the bacteria from the water or grazing on the bacterial film that covers the rocks. The clams *Calyptogena*, mussels *Bathymodiolus*, and vestimentiferan worms *Riftia* have symbiotic chemosynthetic bacteria in their tissues. *Riftia* has no mouth or digestive system; instead, the soft tissues of its body cavity are filled with symbiotic bacteria. The mussels have only a rudimentary digestive system, and both the clams and mussels harbor large numbers of bacteria in their gills.

The bivalves and worms have red flesh and red blood due to the presence of the oxygen-carrying protein hemoglobin. Hemoglobin supplies tissues with the large amounts of oxygen that are necessary for oxidation of molecules such as hydrogen sulfide. It also supplies the oxygen needs of the animal's tissues for growth and maintenance. Sulfides that the bacteria require for chemosynthesis are provided by the host's circulatory system. *Riftia*, for instance, has a sulfide-binding protein in its blood that allows it to concentrate sulfides from the environment and transport them to the bacteria. Most animals would be poisoned by the high concentrations of sulfide circulating in the worm's body, but the sulfide-binding protein has such a high affinity for free sulfide ions that it effectively binds them, thus preventing accumulation in the blood and poisoning of the worm's cells.

Light from Hydrothermal Vent Communities

While studying shrimp populations associated with Atlantic Ocean vents Cindy Lee Van Dover, a biologist from the Woods Hole Oceanographic Institution, made an intriguing discovery. She noticed that, even though these animals lack eyes, they have a reflective spot just behind the head containing a light-sensitive pigment similar to the visual pigment in the eyes of other animals. The evidence suggested that the spots might be modifications of the compound eyes in other species of crustacean. Because the organ lacks a lens, the animal cannot see images, but it can sense the presence of light or similar radiation. Analysis of the pigment suggests that it is sensitive to the longer wavelengths of visible light (red) and some of the infrared spectrum. Because these are the first wavelengths to be absorbed by water, they are the ones least available in the ocean's depths.

To determine the function of this light-sensitive organ, it was first necessary to determine if wavelengths of light in this range were present in hydrothermal vent communities. The submersible *Alvin* was used to examine the vent community associated with the Juan de Fuca ridge system off the northwestern coast of the United States. The vent was photographed with a digital camera under three different conditions: without illumination, illuminated by a halogen lamp mounted on *Alvin*, and finally with illumination from a flashlight shining through *Alvin's* port. When the photographs were analyzed, the data indicated that radiation coming from the vents was in the long wavelength range of the visible spectrum (800 to 900 nanometers). From these data, researchers concluded that the radiation came from the vent water or substances in it or from the vent chimney's internal wall.

The radiation associated with hot vents is an energy source that may be used in some way by bacteria that inhabit the area. Shrimp may use this energy to orient themselves relative to the vent or to locate their food. This discovery of light at the hot vents poses many new questions, and researchers are currently investigating how this energy is related to the vent's biological communities.

Rise and Fall of Vent Communities

It is generally believed that deepwater vents are not long-lived, forming and then going dormant. Shortly after the vents form, they are colonized by living organisms. These organisms thrive until geological changes deactivate the vent. Without a steady source of nutrients for chemosynthetic bacteria, the vent community dies. At some inactive vent sites, researchers have discovered pieces of clam shells. Because they know how long it takes the clams to grow and that it takes approximately 15 years for the shells to completely dissolve in the conditions around dormant vents, they estimate that active vents last 20 years. Radiometric dating of materials from vent sites supports this estimate.

For the species of vent communities to survive, they must colonize new vents. It has been demonstrated that, with their fast growth rate, large clams can mature in 4 to 6 years. It is believed that their large bodies not only allow them to harbor large numbers of symbiotic bacteria but also to produce large numbers of larvae. If the larvae remain suspended in the water for several weeks to several months, the deepwater currents could carry them over hundreds of kilometers, allowing them to disperse to other vent sites. Some biologists hypothesize that the larvae rise to the surface to be distributed more rapidly by surface currents before sinking back to the seafloor. Another hypothesis suggests that the larvae of vent animals use whale carcasses as an intermediate habitat where they can grow, and in some cases asexually reproduce, before colonizing new vents. More research still needs to be done, however, before any hypotheses are confirmed.

IN SUMMARY

- In 1977, oceanographers discovered thriving animal communities on the ocean floor.

- The basis of the food chain in these communities is chemosynthetic bacteria.

- These communities are in both the Pacific and Atlantic oceans and are the focus of much current research.

KEY CONCEPTS

1. Several thousand species have adaptations that allow them to survive in the deep-sea environment. (18-1)

2. The lack of light has had the most impact in shaping the organisms of the deep sea. (18-2)

3. Many deep-sea animals exhibit bioluminescence, which helps them find mates and prey in their dark environment. (18-2)

4. Deep-sea fish display a variety of adaptations such as sharp teeth, large mouths, and huge stomachs that help them survive in a habitat with limited food. (18-2)

5. Some deep sea organisms grow to great size, possibly because they live longer and thus have more time to grow. (18-3)

6. The environmental conditions of the deep sea have been relatively stable for more than 100 million years, and as a result several organisms have changed very little since they first evolved. (18-4)

7. Benthic communities consist of sparse populations that survive on the minimal food available in their environment. (18-5)

8. Thriving marine communities that depend on chemosynthetic bacteria for primary production exist on the ocean floor around hydrothermal vents. (18-5)

1. Organisms that live in the deep sea have adapted to deal with cold temperatures, extremely high pressures, and the absence of light. They exhibit adaptations and behaviors that help them find food and mates in a dark environment.

2. Organisms in the abyss deal with high pressure by having tissue fluid pressures that match the pressure of the surrounding water. They deal with their cold environment by having a very slow metabolism.

3. Predators find their prey in the dark using smell and other chemical senses or with the help of bioluminescence. Bioluminescence can illuminate prey or lure it close to the predator.

4. Many deep-sea fishes have adapted to efficiently catch large prey when it becomes available. These adaptations include wide mouths with many large sharp teeth, large stomachs, and small bodies. This disproportionate arrangement of body parts gives the animals a bizarre appearance.

5. Deep-sea organisms survive without sunlight by evolving symbiotic relationships with chemosynthetic bacteria, feeding on detritus and the occasional large food item that sinks from above, or using bioluminescence to find food. Bioluminescence is also used to identify potential mates and avoid predators.

QUESTIONS FOR REVIEW

Multiple Choice

1. Compared to animals with higher body temperatures, animals with lower body temperatures tend to
   a. consume more food
   b. move more slowly
   c. grow more quickly
   d. reproduce earlier in their life cycle
   e. all of the above

2. The environmental factor that has had the greatest evolutionary impact on animal life in the deep is
   a. temperature
   b. pressure
   c. light
   d. salinity
   e. oxygen levels

3. The eyes of many deep-sea fish are
   a. round
   b. spherical
   c. compound
   d. tubular
   e. identical to land vertebrates

4. Many deepwater fish use _____ to attract prey.
   a. color
   b. bioluminescence
   c. sound
   d. odor
   e. tactile sensors

5. In deepwater animals, bioluminescence functions in
   a. species identification
   b. locating mates
   c. locating prey
   d. defense
   e. all of the above

6. Worms known as *vestimentiferans* lack a _____ system.
   a. respiratory
   b. circulatory
   c. nervous
   d. digestive
   e. reproductive

7. The worm *Riftia* has a sulfur-binding protein in its blood that allows it to
   a. produce its own food
   b. photosynthesize
   c. feed on chemosynthetic bacteria
   d. concentrate high levels of sulfide in its blood
   e. carry oxygen for bacterial metabolism

Matching

Match each of the following organisms with the appropriate description:

a. spirula   1. A group of deepwater fishes that typically have a barbel that acts to lure prey.

b. *Neopilina*   2. A large worm that lacks a mouth and digestive system.

c. coelacanth   3. A small mollusc that resembles a squid or octopus and has an internal shell that contains multiple chambers.

d. stomiatoid   4. A fish that has fleshy bundles between its body and its fins.

e. vestimentiferan   5. A mollusc of the class Monoplacophora that retains many of the features of the ancestral mollusc.

Short Answer

1. What role does bioluminescence play in life in the deep?

2. What adaptations have evolved in deepwater fish to help them find and capture prey?

3. What role do vertical migrations play in bringing nutrients into deepwater?

4. Why did early biologists think that animal life could not survive in the ocean's depths?

5. Why are some deepwater animals bright red?

6. Why do some biologists believe that the sea's depths may harbor organisms that have changed very little from their early ancestors?

7. What food sources are available for animals that live on the ocean floor?

8. Some deep-sea fish have larvae that rise in the surface waters, where they grow before settling to the depths. What is the advantage of this strategy?

Thinking Critically

1. How might you determine whether an animal in a hydrothermal vent community derived its nutrition from symbionts or from feeding on other vent organisms?

2. The anglerfish spends the early part of its life in the upper regions of the ocean and its adult life in the abyss. What kinds of anatomical changes would you expect to observe as the animal matures?

3. Assuming deepwater fish could survive the physical environment of surface waters, do you think that they would be able to compete effectively with fish already adapted to that environment? Explain.

4. Do you think deepwater fish would have a swim bladder? Why or why not?

SUGGESTIONS FOR FURTHER READING

Dybas, C. L. 2004. Close Encounters of the Deep-Sea Kind, *Bioscience* 54 (10): 888–891.

Helmuth, L. 2007. Creatures of the Deep!, *Smithsonian* 38 (7): 68–75.

Kunzig, R. 2001. The Physics of Deep Sea Animals, *Discover* 22 (12): 40–47.

Lutz, R. A. 2003. Dawn in the Deep, *National Geographic* 203 (2): 92–103.

Morell, V. 2004. Way Down Deep, *National Geographic* 205 (6): 36–55.

Van Dover, C. L. 2000. *The Ecology of Deep-Sea Hydrothermal Vents*. New Jersey: Princeton University Press.

Yasuhara, M., T. M. Cronin, P. B. DeMenocal, H. Okahashi, and B. K. Linsley. 2008. Abrupt Climate Change and Collapse of Deep-Sea Ecosystems, *Proceedings of the National Academy of Sciences* 105(5): 1556–1560.

Have You Wondered?

1. If we are running out of fish to catch? (19-1)
2. Why it is important to manage marine fisheries? (19-1)
3. If fish can be farmed? (19-1)
4. If commercial fishing harms the marine environment? (19-1)
5. What other resources, besides food, are harvested from the ocean? (19-2)

Harvesting the Ocean's Resources

19

THE SEA HAS traditionally been a major source of food for human populations, and commercial fishing is important to the economies of coastal nations. In the waters of the northeastern United States there are 41 different invertebrate and fish species that each generate more than $1 million a year in fisheries revenue. Along the coast of the four western states, 55 species each yield at least $1 million in fisheries revenue annually. It was long believed that the oceans were an infinite source of food, but we now realize that we must conserve precious fishery resources if they are to last for our generation and future generations.

The sea is also seen as a potential new source of a variety of minerals and materials. The demand for natural resources has increased dramatically in the last 100 years, and the ability of onshore sources to meet the demand is being taxed. As terrestrial mineral resources are depleted, society will look with increasing interest toward the sea. Whether its minerals and materials will be tapped to the fullest extent depends on a number of factors, and many ecologists are concerned that the development of these resources could have devastating effects on an already stressed ecosystem.

19-1 Commercial Fishing

In November 2006, a team of 14 researchers from four countries published a paper in the journal *Science* in which they projected that by the year 2048 commercial fishers will have almost nothing left to catch. This shocking conclusion was based on an analysis of local experiments and regional and global fisheries data over the past 50 years. According to the study, at least 29% of fished species have collapsed, and several others are on the verge of collapse (**Table 19-1**). Not all fisheries scientists agree with this pessimistic outlook, but all agree that commercial marine fisheries are in serious trouble. Fisheries biologists and conservationists

TABLE 19-1 COMMERCIAL FISH SPECIES WITH POPULATION DECLINES OF 90% OR MORE

Sturgeon
Pacific salmon
Swordfish
Grouper
Red snapper
Bluefin tuna
Atlantic cod
Chilean sea bass
Most shark species

© Cengage Learning

around the globe agree that measures need to be taken immediately to better manage populations of commercially important fish and shellfish (crustaceans and molluscs) if this resource is to remain available for future generations.

Humans have harvested fish and shellfish from the ocean for thousands of years, but during the past 50 years there has been a dramatic increase in the amount of fish and shellfish taken from the sea. In 1950, the total world catch of marine fish and shellfish was approximately 21 million metric tons (1 metric ton is 1,000 kilograms, or 2,204.6 pounds). By 2000, it reached 86.7 million metric tons—the highest catch on record (**Figure 19-1a**). This increase in commercial fishing is the result of increased demand brought about by the growing number of humans, now more than 7 billion. Protein from fish and shellfish currently accounts for 16% of all animal protein consumed by the human population. Advances in fishing technology such as the use of giant factory ships to process large catches and the use of sonar and aircraft to locate schools of fish played a major role in increasing the size of the commercial catch. Recently, however, the world catch has not been increasing proportionately to the increased fishing effort and has been hovering around 80 million metric tons for the last 10 years. The total world catch of marine fish and shellfish in 2009 was 79.5 million metric tons. Unfortunately, most populations of ocean species peaked in the 1980s and are now declining.

Over the past 50 years, there has also been a change in the use of the commercial catch. In 1950, 90% of the world's catch was used for human consumption, and the remaining 10% was used for fish-meal products, mainly to feed the livestock of more developed countries. By 1988, 60% of the catch went to feeding people and 40% to feeding domestic animals, a trend that continues today (**Figure 19-1b**). Recall the "ten percent rule of ecology," which states that only about 10% of the energy available at one trophic level is passed on to the next. From this, it becomes apparent that feeding livestock with commercial catch is not a very efficient use of this resource. For each ton of commercial fish fed to animals, only about 0.1 ton of animal biomass would be produced. In addition, most of the commercial fish catch consists of fishes that are in higher trophic

Renewable resources are resources that can replenish themselves.

A **stock** is a separate population of fish that is assumed to be reproductively isolated from others in the range.

Tagging is a procedure for monitoring the distribution and movement of animals in the environment.

levels, such as tunas. This is a greater waste of energy than harvesting fishes at lower trophic levels.

Today such a large variety of food species are taken from the sea that whole ecosystems are being harvested rather than individual species. As the more traditional fisheries become depleted, fishers focus their efforts on species farther down the food chain. Current efforts in the field of fisheries management are attempting to correct some of the problems of recent years and maintain the health of commercial fisheries.

FISHERIES MANAGEMENT

Fish and shellfish are referred to as **renewable resources** because animals that aren't caught can continue to reproduce and replace those that are caught. The goal of fisheries management is to maintain these resources by enacting policies and setting catch limits that will prevent overfishing to the point of extinction and allow enough animals to survive and reproduce so that there will be fish and shellfish to catch in the future. To achieve these goals, fisheries biologists need detailed knowledge of a commercial species's life history. Ideally, this would include an understanding of the physical and biological factors that affect an organism's growth and survival, such as its optimal water temperature and salinity, where and on what it feeds, where and when it reproduces, and its migratory patterns. Until biologists better understand the life cycles, behavior, physiology, and ecology of commercial species, they will not be able to develop and implement the plans and conservation measures needed to properly manage these resources now and for future generations.

Managing marine fisheries is difficult because we know so little about the basic biology of commercial species. In addition to traditional experiments, fisheries biologists carefully and systematically monitor populations of commercially valuable species to learn more about them. They then use the information to develop plans to maintain adequate fish catches while not overexploiting the population.

Monitoring Fish Populations

To make informed decisions regarding the management of species, fisheries biologists need to know how populations are distributed in the ocean environment and their migrational movements. It is also necessary to have some idea of the population size and its age structure.

Determining Population Distribution and Movement Some commercial species can be found over a broad geographical range. For the purposes of management, the range is divided into separate populations known as **stocks**, and each stock is assumed to be reproductively isolated from other stocks. Fisheries biologists monitor the stocks by **tagging** or using molecular markers. In tagging, fishes are caught and marked with identification tags, usually made of plastic or metal, and then released (**Figure 19-2a**). When these fish are caught again by sport or commercial fishers, the tags are removed and sent, along with detailed information about where and how the animals were caught, to the biologists who inserted the tags. By mapping the locations where tagged fish are initially caught and released and where they are caught again, biologists track the movement and distribution of members of each stock. Other information such as how the fish were caught, what bait was used, or what they were feeding on help fisheries biologists better understand the natural history of the species. Because each stock is reproductively isolated from other stocks, individual members of the population possess DNA markers (unique DNA sequences) and proteins with amino acid sequences that are unique to the population.

FIGURE 19-1 WORLD POPULATION AND COMMERCIAL FISHING.
(a) Since 1940, as the population of the world has increased, so has the demand for fishery products. **(b)** Since 1950, a larger proportion of the world's fish catch has been going to feed livestock rather than humans. *[(a), Data are from the United Nations Food and Agriculture Organization (FAO) and the World Paper; (b), FAO.]*

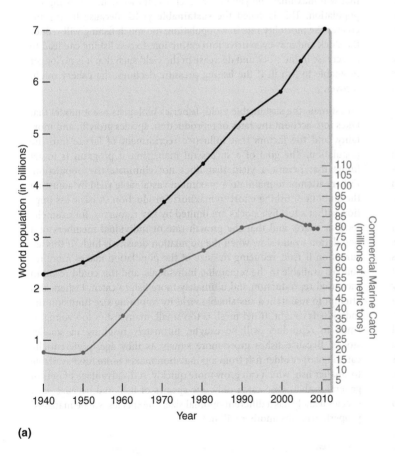

(a)

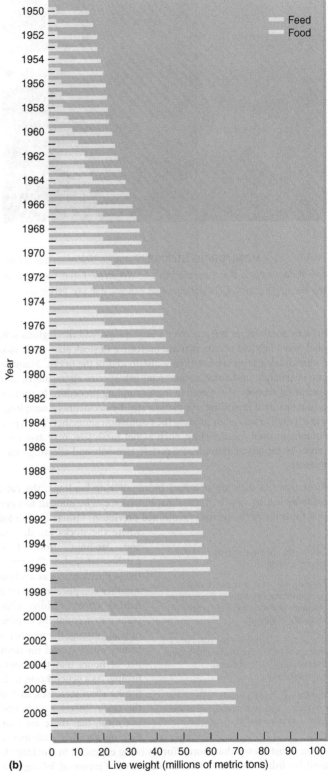

(b)

These molecules are markers that identify members of a specific stock. By analyzing these molecular markers, geographical distribution and other biological information can be determined for the species being studied **(Figure 19-2b)**.

Determining Population Size and Age Structure In addition to knowing the geographical range of a stock, it is necessary to estimate its size

to develop good management plans. Because fishes and shellfishes of only a certain size have economic value, it is also important to know the number of individuals of various ages and sizes in a given stock. To determine this information, fisheries biologists design sampling experiments that involve catching representative samples of the fishes or shellfishes from different areas of their range. This is not easy to do,

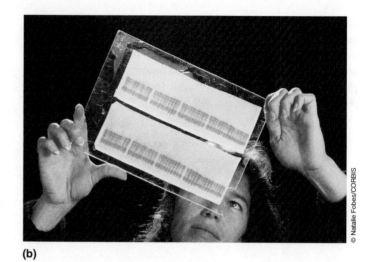

(a)

(b)

FIGURE 19-2 **MONITORING STOCKS.** Fisheries biologists can monitor fish stocks in several ways. **(a)** In tagging, a plastic or metal tag, such as the one shown here, is placed in the fish's fin. When the fish is caught again, the catch data and the tag are sent back to the laboratory that inserted it. **(b)** These gels show DNA patterns that indicate that two fishes are from the same stock.

because some fishes migrate over a large range, and the regions sampled may not yield results that are indicative of stock size. Also, most commercial species are not evenly distributed in the ocean. Many occur in patches, and some commercial fish species aggregate in large groups or schools. It is possible for a sampling procedure to produce erroneous data regarding population size by accidentally sampling a dense patch or school or missing the population altogether. The type of equipment used for sampling, such as net mesh size, may also cause errors by capturing animals of a certain size only and allowing others to escape.

Most fisheries data are obtained by monitoring **landings**, the catch made by fishing vessels. These data, however, can be difficult to interpret for some of the reasons mentioned previously. Fishers tend to fish where the stocks are densest, and landings from these patches are not indicative of the population size as a whole. Another complication involves **fishing effort**. Fishing effort takes into account the number of boats fishing, the number of fishers working, and the number of hours that they spend fishing. As a general rule, if the fishing effort increases, so does the catch. This, however, does not mean that there are more fish or shellfish available to catch; it just means that more effort is being expended in catching the available fish and shellfish. On the other hand, if increased fishing effort yields a smaller catch, it can be assumed that there are probably fewer animals to catch.

> **Landings** are the catch made by fishing vessels.
>
> **Fishing effort** is a measure of the number of vessels fishing, the number of fishers working, and the number of hours spent fishing.
>
> The **potential yield** is the number of pounds of fish or shellfish a stock can yield annually without being overexploited.
>
> The **sustainable yield** is the maximum yield over several years that will not stress a stock.

Fishery Yield

Using the population information, fisheries biologists try to estimate the number of pounds of fish or shellfish that the stock can yield per year without being overexploited. This is known as the **potential yield** of a stock. The information is then used to develop management plans that will maximize the yield over several years while not stressing the population. This is called the **sustainable yield**. Because fishing increases the mortality rate in a population, too much fishing will reduce the stock and may even drive it to extinction. Excess fishing can lead to a decrease in the stock and decrease in the yield such that it is no longer profitable to fish it. If the fishing pressure declines, the fishery might recover.

To estimate the sustainable yield, fisheries biologists use a model that takes into account the rates of reproduction, species growth, and mortality, and the factors that influence recruitment of larvae into the population. The goal of a successful management program is to set limits that permit a yield that does not eliminate the population. Fishers attempt to maintain a maximum sustainable yield by adjusting the amount of fishing effort to maximize production. A basic assumption is that adult fish stocks are limited by key resources, for example, food or space, and that the growth rate of individual members of a population would slow when the population density is high. If this assumption is true, reducing the size of the population makes more resources available to the remaining individuals, and this could result in increased reproduction and ultimately more fish to catch. Fishers also attempt to maintain a sustainable yield by imposing size limits on the individuals caught. If net mesh is too small, many fishes too young or small to reproduce will be caught, ultimately resulting in smaller stocks. Because fishes grow more slowly as they age, preferentially catching larger older fish from a population makes more food available to smaller fish, which can grow more quickly. A disadvantage of nets in general is that none catch only one species of fish, and because each species usually has a different optimal size, one species will be managed properly whereas another will not.

Problems in Managing Diverse Species

Fisheries diversity is another factor in fisheries management. Traditionally, fisheries biologists attempted to manage only one species at a time, but few single-species plans succeed with multiple species unless they are closely related. While implementing a management plan that meets the needs of one species, it is possible that other species will be negatively affected. For example, to better manage the cod (*Gadus morhua*; **Figure 19-3**) fishery in the North Atlantic, large areas in the Gulf of Maine

FIGURE 19-3 COD. Atlantic cod have been an important commercial fish for hundreds of years. Overfishing and destructive fishing practices have caused the population to fall to less than 10% of its original size.

were closed to bottom fishing for cod. Because the regulations were only for cod, dredging (a type of bottom fishing) for scallops (*Placopecten magellanicus*) was still permitted. The dredging operations not only disrupted benthic habitats but also killed and injured commercially valuable fishes, including the cod and yellowtail flounder (*Limanda ferruginei*). The attempt to properly manage one species conflicted with the management of other species. Some fisheries biologists suggest that the way to avoid problems such as this is to take an ecosystem-based or area-based management approach rather than trying to coordinate multiple single-species plans.

OVERFISHING

Overfishing occurs when fish are caught faster than they reproduce and replace themselves. At least 60% of the world's 200 most valuable fish species are either overfished or fished to the limit. In addition to depleting stocks, overfishing can adversely affect ecosystems by changing genetic and species diversity and damaging or destroying habitat.

Changes in Genetic Diversity

Overfishing can change the genetic characteristics of a population by selecting for or against certain traits, such as the size at which animals become sexually mature. This can occur when the larger mature fish in a population are preferentially caught. Constant removal of the larger fish over time tends to favor the survival of fish that mature at an earlier age and smaller size. If heavy fishing removes most fish early in their reproductive life, individuals that mature at a younger age or smaller size have a better chance of surviving and thus an evolutionary advantage. As a result, the genetic variability of the population will shift to a larger proportion of individuals that reproduce at an earlier age or smaller size. In this way, fishing can inadvertently exert a selective pressure for smaller animals. An example of this occurred in the Northeast Arctic cod (*Boreogadus saida*) fishery. Between the 1930s and 1950s the cod matured between the ages of 9 and 11 years, and individuals of breeding age had roughly a 40% chance of surviving and reaching their spawning grounds to breed. During the same period, the cod fishery began using **trawl nets** to increase the size of the catch. Trawls are large nets that are dragged along the ocean bottom, capturing virtually

everything that enters them **(Figure 19-4)**. The indiscriminate trawling resulted in large catches of fish and increased the mortality rate of older fish so much that it significantly reduced, to 2%, the chances of breeding-age cod reaching their spawning grounds. As a result, remaining cod that possessed genes allowing them to grow and mature faster dominated the breeding stock. This resulted in a gradual shift toward fish that mature at about 7 or 8 years of age, earlier and smaller than a half century earlier.

Overfishing can also affect genetic diversity by causing the population to become so small that it loses genetic variability because there are not enough individuals in the gene pool to carry the variety of traits that were once found in the population. An example of this is the orange roughy (*Hoplostethus atlanticus*; **Figure 19-5**), a commercial fish from the South Pacific. The orange roughy can live as long as 50 years and does not mature until it is 20 years old. In the 1980s, a large spawning population was discovered off New Zealand and fished intensely. By the early 1990s, the heavy fishing of the adults had caused the total biomass of orange roughy to decline by 60% to 70%. Fisheries biologists found that the genetic diversity within the orange roughy population had also decreased significantly during the same period, producing a genetically more homogeneous population that was less able to adapt to environmental changes.

> **Overfishing** is catching fish faster than they can reproduce and replace themselves.
>
> **Trawl nets** are large nets that are dragged along the sea bottom.

Changes in Species Diversity

Overfishing can affect the biological diversity of an area by reducing the number of species in an ecosystem. When a population declines to such low levels that the species no longer fulfills its role as prey, predator, or competitor in the ecosystem, it becomes ecologically extinct. This changes the relationships among species that had evolved naturally over time, causing shifts in the competitive and predator–prey relationships

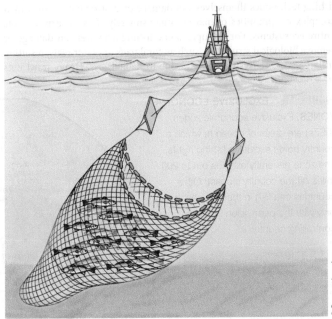

FIGURE 19-4 TRAWL FISHING. Trawls are large nets that are dragged along the bottom. The trawl captures virtually everything that enters it.

FIGURE 19-5 ORANGE ROUGHY. As the result of overfishing, not only have the numbers of this commercial fish decreased but so has the genetic diversity within the population. The loss of genetic diversity makes it more difficult for the surviving fish to adapt to changes in the environment.

in the ecosystem. Removing important species can restructure an entire ecosystem as predators adjust to new prey species and new predators fill niches vacated by overfished species. This in turn allows new species to become more dominant in the ecosystem. The restructuring of the ecosystem that results can change food chains and affect fisheries other than the one initially overfished.

Exclusive economic zone (EEZ) is the area of ocean that a coastal nation controls.

Surimi is a fishery product made from Alaskan Pollock.

Changes in Habitat

While overfishing severely depletes the stock of overfished species, the fishing techniques themselves can damage or destroy habitat on which complex communities of marine organisms rely. This in turn disrupts entire ecosystems. For example, some fishing activities can damage or remove biological structures, such as coral reefs or oyster reefs, resulting in large-scale ecosystem changes due to habitat loss. Heavy trawl nets

dragged along the ocean floor damage seagrass beds and rocky habitats and physically move or crush fish and shellfish, while changing the structural characteristics of the habitat. Trawling can disrupt feeding for animals such as shellfish and groundfish, which rely on food in undisturbed sediments.

Controlling Overfishing

In addition to managing fisheries, many fishing nations use other methods to take the pressure off heavily exploited stocks and to maintain sustainable yields. These include careful policing of fishing areas, development of new fisheries products from underused stocks, and campaigns to inform consumers about the problems of overfishing.

Coastal Zones In 1977, the United States, along with most other coastal nations, increased the area of ocean it controls out to 200 miles off the coast. These zones are called **exclusive economic zones** (EEZs) (**Figure 19-6**). Other countries can fish in a coastal zone only by specific agreement with the controlling government. What areas can be fished, what fish can be caught, and the limits are all negotiated with the government of the country in whose water the operator is fishing. In the United States, it is the job of the Coast Guard to enforce the agreements, and fisheries observers are placed on fishing vessels to monitor and record the catch. Unfortunately, coastal limits are often not enforced, especially in poor developing countries.

Developing New Fisheries Demand for fish has caused some significant decreases in traditional fisheries such as tuna, while stimulating the development of new fishery-related products. One such product is **surimi**, which is made from Alaskan pollock (*Theragra chalcogramma*), a bottom fish (**Figure 19-7a**). The flesh is processed to remove the fats and oils that give the fish its characteristic flavor, and a highly refined fish protein, surimi, is produced. The surimi can then be flavored to produce artificial crab, shrimp, lobster, or scallops (**Figure 19-7b**). Surimi is a major fish product in the Japanese market and is the product with the fastest-growing demand in the U.S. fish market. Unfortunately, the increased demand for this product in recent years is now starting to put pressure on pollock stocks.

Consumer Education Ultimately, consumer demand drives fisheries production. The more a species is purchased by consumers, the more intensely that species will be fished to supply the demand. It is hoped

FIGURE 19-6 EXCLUSIVE ECONOMIC ZONES. Exclusive economic zones (EEZs) are areas of ocean to which a country holds exclusive fishing rights. The zone generally extends out to 200 miles off the country's coast. Other countries can fish in these zones, but only with the permission of the controlling country.

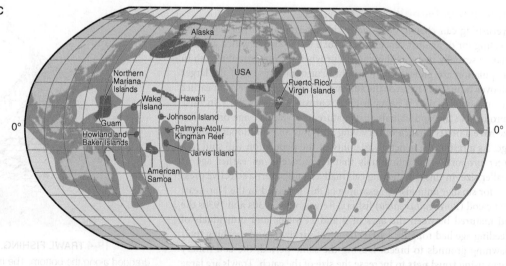

■ World Exclusive Economic Zones (EEZs)   ■ United States Exclusive Economic Zones (EEZs)

FIGURE 19-7 SURIMI PRODUCTS. **(a)** The Alaskan pollock is a bottom fish that is the source of refined fish protein called *surimi*. **(b)** Many processed seafood products, such as artificial crab, shrimp, and lobster, are made from surimi.

(a)

(b)

that educating consumers regarding the problems of overfishing will decrease the demand for endangered species, allowing the stocks to recuperate. **Figure 19-8** lists some commercial species of fish and shellfish and indicates which are best to purchase and which should be avoided if you want to support sustainable fisheries.

Good Fish
Fast-growing, abundant, sensibly managed, minimal bycatch and ecological impacts, minimally polluting farming methods
Anchovies
Bass, bluenosed
Bluefish, Atlantic
Catfish, farmed
Cod, Pacific
Crayfish (crawfish, crawdad)
Crab, Dungeness
Herring and sardines
Halibut, Pacific, Alaskan
Hoki
Mackerel
Mahi-mahi (dorado, dolphinfish)
Mussels, black and green-lipped, farmed
Oysters, farmed
Pollock, Pacific (surimi, krab)
Prawns, white-spotted
Salmon, wild, Alaskan and Californian
Scallops, farmed
Shrimp, pink
Squid (calamari)
Striped bass, farmed
Sturgeon, farmed
Tilapia, farmed
Trout, farmed
Tuna, Pacific albacore (tombo tuna)
Tuna, yellowfin (ahi)

Iffy Fish
Heavily fished or overfished, capture methods damage habitat and result in excessive bycatch
Crab, Alaskan King
Crab, Snow
Lobster, clawed, American, Maine
Snapper, tropical (huachinango)
Sole, petrale, English, Dover
Spiny lobsters (crayfish)

Bad Fish
Overfished and unmanaged, ecologically destructive capture destroys habitat and kills massive numbers of nontarget animals
Beluga sturgeon (beluga caviar)
Chilean seabass (Patagonian toothfish)
Clams, dredged
Grouper
Lingcod
Monkfish
Orange roughy (slimehead)
Oysters, dredged
Rockfish (Pacific red snapper, rock cod)
Salmon, Atlantic
Scallops, dredged
Shark (shark cartilage, shark fin)
Shrimp and prawns, farmed, trawled
Swordfish
Tuna, bluefin (maguro)

FIGURE 19-8 CHOICES. You can help stop overexploitation of marine fish and shellfish by choosing carefully in restaurants and supermarkets. This chart lists species that are abundant and those that should be avoided. (Source: California Academy of Sciences, San Francisco. Reprinted with permission.)

OTHER FACTORS AFFECTING MARINE FISHERIES

Although overfishing is a principal cause, it is not the only activity that causes stocks of marine fish and shellfish to decline. Destruction of coastal habitats, wasteful and destructive fishing practices, and even the practice of aquaculture, or fish farming, have all contributed to the alarming decline in commercial fisheries. These problems, combined with our lack of scientific information about most commercial species, make it difficult to develop adequate management policies and practices to reverse the damage.

Destruction of Coastal Habitats

Human destruction and development of coastal areas has resulted in a loss of feeding, breeding, and nursery grounds for marine fish and shellfish. For example, mangrove forests once covered 75% of tropical and subtropical coastlines. Today less than half remain (**Figure 19-9**).

FIGURE 19-9 HABITAT DESTRUCTION. The destruction of mangrove forests for development and aquaculture removes important nurseries for commercial fish and shellfish. The problems are compounded when the mangroves are burned, contributing greenhouse gases to the atmosphere.

The **incidental catch (bycatch)** is the noncommercial animals that are killed during fishing for commercial species.

Trash fish is the term used for incidental catch that is solely fish.

Drift nets are large nets that may stretch for as much as 60 kilometers.

Tans are individual sections of a drift net.

The destruction of mangrove forests contributes not only to a decline in local fish catches but also to the contamination of coastal waters. In addition, the destruction of mangroves leads to increased salinity (salinization) of coastal soils as seawater evaporates, leaving the salt behind. Without the root systems of mangroves to stabilize sediments, coastlines suffer severe erosion, and the loss of mangrove photosynthesis coupled with the burning of mangroves as they are cleared increases greenhouse gases such as carbon dioxide in the atmosphere. Draining estuaries for oceanfront development destroys seagrass beds and salt marshes that are spawning and nursery areas for many commercial species.

Wasteful and Destructive Fishing Practices

Biologists worldwide are concerned about the number of noncommercial marine animals killed each year during commercial fishing. These animals are referred to as **incidental catch, bycatch**, or, in the case of fish, "**trash fish**," and represent a terrible waste of marine resources.

Drift Nets Fishing with **drift nets** produces especially large numbers of by-catch. Drift nets are large nets composed of sections called **tans** (**Figure 19-10a**). Each tan is 40 to 50 meters (132 to 165 feet) long and 7 meters (23 feet) high. A catcher boat sets as much as 60 kilometers (36 miles) of drift net in the evening and retrieves the net in the morning. The commercial catch, usually squid or salmon, is taken to factory ships for processing. The by-catch is dumped overboard. In the North Pacific, drift-net fisheries resulted in the annual death of as many as 15,000 Dall's porpoises (*Phocaenoides dalli*) and 700,000 seabirds, as well as other species such as turtles, fur seals, and sharks

(**Figure 19-10b**). These animals become entangled in the drift nets when they try to feed on the catch. The use of drift nets for commercial fishing was banned by all commercial fishing nations in 1992, largely because of pressure from the United States. Unfortunately, outlaw vessels still use drift nets, and they tend to fish areas where they do the most damage.

Trawling Many fish, and shellfish such as shrimp and scallops, are caught in trawls or dredges that are dragged along the ocean floor. Although trawling is an efficient means of fishing, it damages benthic ecosystems and produces a large bycatch. Of the approximately 1.5 million tons of shrimp and 21 million tons of fish that are caught annually in U.S. waters, 3 to 5 million tons are discarded. The amount of the catch discarded varies from place to place. If the by-catch does not fetch a high enough price or processors are not available for the bycatch, it is discarded. In the past, shrimp fishing in the Gulf of Mexico usually produced 15 pounds of by-catch for each pound of shrimp (**Figure 19-11**). The figures are similar for the scallop industry. Almost all of the by-catch is returned, dead, to the ocean or carted off to landfills. Nets now used by shrimpers have a set of deflecting bars that push most of the by-catch below the net and not into it. This change has reduced by-catch to less than 5 pounds per pound of shrimp caught. By comparison, in Southeast Asia the demand for protein is so high that almost all of the incidental catch finds a local market (**Figure 19-12**) and very little is wasted, but overfishing is still a problem.

It has been estimated that shrimp trawlers catch as many as 45,000 sea turtles annually, with a mortality count of 12,000 or more. The introduction of turtle exclusion devices (TEDs) has greatly reduced the turtle mortality rate, but not every nation requires their shrimp fleet to use TEDs. Trawls for catching tuna can also catch and kill dolphins, although less so in recent years (see the case study of tuna fishing later in this chapter).

Inefficient Use of the Catch Some fisheries use only a portion of the catch and waste the rest. For example, some species of shark are caught

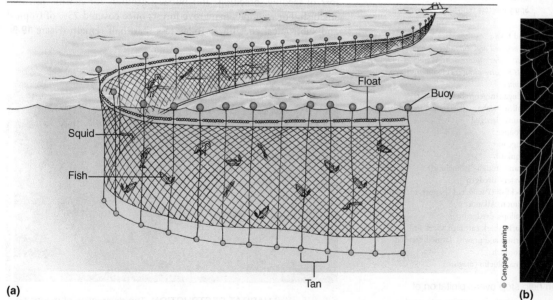

FIGURE 19-10 DRIFT NET FISHING. (a) Drift nets are made up of sections called tans. As much as 60 kilometers of net are set in the evening, and fish and squid get tangled as they swim into the net. The nets are then retrieved in the morning, and the catch is removed and refrigerated. **(b)** Other marine animals that are not being fished may also get tangled in the nets as they try to feed on the trapped fish.

FIGURE 19-11 **WASTEFUL FISHING.** Many fishing techniques capture large amounts of noncommercial species as well as commercial catch. The noncommercial species, or bycatch, is returned dead to the ocean and represents a waste of marine resources.

only for their fins. After the fins are removed, to make shark fin soup, a popular dish in some countries, the helpless shark is thrown overboard to die. In the United States, most commercial fishers are independents who sell their catch directly to processors. The biggest demand in the U.S. market is for fish fillets and fish steaks, and commercial fishers preferentially fish for higher-priced fishes such as swordfish, halibut, and salmon, because the filleting process uses only 20% to 50% of the fish's body weight. Compare this with the government-subsidized commercial fishing of some foreign countries, which processes the catch at sea at a lesser cost and uses substantially more of the animal's flesh. The net result is that it is cheaper to import some marine fisheries' products than to catch them domestically.

Aquaculture

Many fisheries biologists are looking into the possibility of increasing the commercial yield from the sea through **aquaculture** (also known as **mariculture**), the use of agricultural techniques to breed and raise

marine organisms. If only one species is raised, the process is called **monoculture**; if several species are raised together, it is called **polyculture**. The benefit of raising several species together in the same area is that it makes more efficient use of all the available resources. The Chinese were engaged in the aquaculture of oysters as far back as 1000 BC, and the peoples of Japan and Southeast Asia have used the practice for centuries. In these cultures, the methods have traditionally been labor intensive, and until recently most operations were small family-owned farms. Today aquaculture is becoming a big business, especially in low-income countries such as China, which alone accounts for a little more than 60% of the world's aquaculture production. In 1990, aquaculture of fishes and shellfishes accounted for 13% of the total world production. By 2009, it had risen to 38%, about 55.6 million metric tons (**Figure 19-13**). As the world fisheries catch continues to decline, aquaculture will become an increasingly important means of supplying commercial species to the world's markets.

Fish Aquaculture A good example of aquaculture today is in Israel, which has a large population concentrated in a small area. The Israelis have acres of fish ponds that produce an annual yield of 6,500 metric tons of carp and mullet in polyculture. In the United States, fish farming produces only 2% of the fisheries' total product, and most of that is catfish and trout, not marine species. Recently, there has been a worldwide increase in salmon farming. In 2009, world aquaculture of salmon produced 2.2 million metric tons, the greatest amounts being produced in Norway, Scotland, and Canada. The Canadian government has been encouraging salmon farming, with increased numbers of farms opening along the coast of British Columbia (**Figure 19-14**). Washington, California, and Maine allow salmon farming, but the procedure to obtain the proper permits is long and complicated. Texas and Florida have several farms that raise redfish. The initial costs, however, are quite high (approximately $400,000), which limits the number of people engaging in this fishery.

Aquaculture (mariculture) is the use of agricultural techniques to breed and raise marine organisms.

Monoculture is the process of raising a single species in aquaculture.

Polyculture is the process of raising several species together in aquaculture.

FIGURE 19-12 **BYCATCH.** In Asia, the demand for protein is so high that most of the catch is used, as can be seen by the variety of fish and shellfish in this fish market.

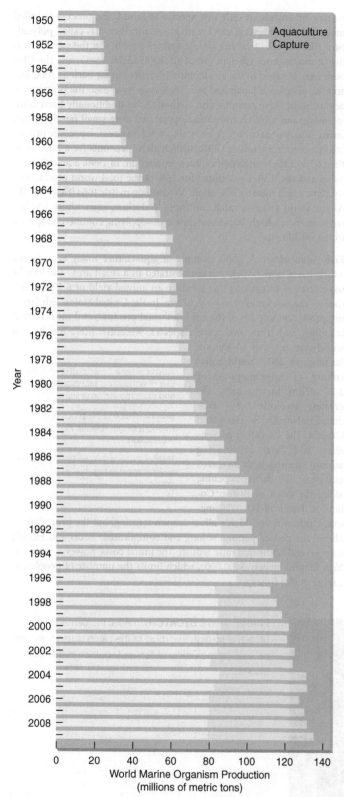

FIGURE 19-13 WORLD PRODUCTION FROM CAPTURE FISHERIES AND AQUACULTURE. Since 1950, more and more of the commercial catch is being produced by aquaculture. *[Data are from the United Nations Food and Agriculture Organization.]*

FIGURE 19-14 SALMON HATCHERIES. Salmon raised in hatcheries such as this one in Newfoundland, Canada, supplement the commercial salmon catch.

Raft Culture Commercially valuable molluscs are also being raised in aquaculture. Clams, mussels, and oysters are raised worldwide, but especially in Asia. In Europe and Asia a method of aquaculture known as **raft culture (Figure 19-15a)** is popular. In raft culture, juveniles of commercially valuable molluscs are collected from natural populations and attached to ropes suspended from rafts. This allows the farmer to keep the juveniles in a portion of the ocean where food supplies are high, while minimizing their exposure to natural predators. Raft culture of mussels in Spain produced 300,000 metric tons in 2009, and in the same year raft culture of oysters in Japan yielded 238,000 metric tons. The Japanese also culture scallops in net cages that hang from rafts, as well as the more conventional bottom methods. This has resulted in an annual scallop harvest of 280,000 metric tons in 2007. The United States, on the other hand, is just beginning to use raft culture.

Shrimp Farming Shrimp is another commercial shellfish produced in large quantities by aquaculture **(Figure 19-15b)**. In 2009, the world harvest of shrimp from aquaculture was 3 million metric tons. In Ecuador, shrimp farming is the second largest industry (oil production is the first). Some U.S. companies have shrimp farms in several Latin American countries, where production costs are lower. In Thailand the growing number of shrimp farms along coastal areas is beginning to negatively affect local mangrove communities.

Eco-Friendly Aquaculture At the Woods Hole Oceanographic Institution, biologist John Ryther has developed a model for using sewage waste in oyster aquaculture. Ryther used sewage nutrients from a town with a population of 50,000 to produce algae to feed oysters. His pilot plant produced an annual yield of 800 metric tons of oyster meat using this method. Because oysters produce large amounts of solid waste, Ryther introduced polychaete worms to feed on the oysters' waste. The worms could then be harvested and sold for fish bait.

In Long Island Sound, warm water from industrial cooling systems is used to increase oyster production. The average water temperature is 30°C (86°F), and it warms the animals enough to increase their metabolism and production. The success of the Long Island project has prompted some biologists to suggest that it would be possible to produce large amounts of oysters by aquaculture in the zones around the equator, where the water is warmer. Nutrients could be supplied from sewage or by pumping nutrient-rich deep water to the surface, producing an artificial upwelling.

(a)   **(b)**

FIGURE 19-15 MARICULTURE. (a) These oysters are being raised in raft cultures in Ireland. **(b)** This farm in Borneo raises large shrimp known as prawns. Although the prawns produced by this farm are an important food resource, the operation disrupts the ecology of the mangrove swamps where the farm is located.

Problems Associated with Aquaculture At one time, aquaculture was thought to be the solution to global fishing problems, but aquaculture is not always conducted in a sustainable way. Some types of aquaculture may, in fact, be causing some of the same problems it was intended to solve. For instance, shrimp from Ecuador and Asia are produced on coastal farm sites that were once natural mangrove forests. The mangrove ecosystem is destroyed to make room for the shrimp farms. To make matters worse, the farms quickly become so polluted from accumulated wastes and the chemicals added to the water to increase production that they are abandoned, and more mangroves are cleared to make room for more farms.

Some of the worst problems involve shrimp and salmon aquaculture. These animals require large quantities of fish meal and fish oil in their diets. To meet the nutritional demand of these aquacultures, large numbers of fish must be caught—fish that would normally support many other fish species in the wild. In addition, so many animals are crowded into a small area in these farms that antibiotics and pesticides must be constantly added to the water to keep them healthy. The excess of these chemicals along with excess nutrients are released into coastal waters, disrupting coastal ecosystems and polluting the ocean. Instead of becoming a substitute for ocean fishing and decreasing pressures on depleted natural stocks, this type of aquaculture adds more pressure and lowers yields in some natural fisheries.

CASE STUDIES

Each of the following examples of commercial fish and shellfish fisheries is a good lesson in how fishing and other human interventions have affected species, and what is being done to correct some of the problems.

Anchovies

Anchovies (family Engraulidae; **Figure 19-16a**) are small fast-growing fishes that feed on the bountiful phytoplankton in upwelling areas along the coast of Peru. Because anchovies travel in large dense schools, it is possible to catch great numbers with nets. Anchovies represent the world's largest fish catch for any single species. Most of the anchovies caught are not used for human consumption but end up as fish meal fed

to domestic animals. The commercial fishing of anchovies for fish meal began in 1950, with an annual harvest of 7,000 metric tons. By 1962, world demand for fish meal had so increased that the anchovy fishery yielded 6.5 million metric tons of fish. By 1972, this amount had increased to 12.3 million metric tons. As the tonnage taken increased over the years, the size of the fish being caught decreased—a common effect of overfishing. As a result, more fish needed to be caught to meet commercial demand.

In 1953, 1957 to 1958, 1965, and 1972 to 1973, the Peruvian coast was hit by major El Niños. During an El Niño, the change in wind patterns produces changes in surface currents and upwellings. This leads to a decrease in nutrients in the surface water, a lack of production by phytoplankton, and a massive die-off of zooplankton. The decomposing organic matter robs the surface water of oxygen and adds to the death toll. Seabirds and other animals that depend on anchovies for food die of starvation or migrate.

The intense fishing that produced the record catch of 1972, coupled with an El Niño event later that year, caused a dramatic decrease in the 1973 harvest to only 2 million metric tons **(Figure 19-16b)**. Alarmed by the situation, the Peruvian government set restrictions and placed quotas on the anchovy catch, and the fishery began a slow recovery in 1974 and 1975. In 1976, another El Niño caused another major decrease in the catch. From 1977 to 1982, catches again began to show a small increase, reaching 2 million metric tons by the end of 1982. An El Niño in 1982 and 1983 caused another decline, and from 1983 to 1985, the catch was below 150,000 metric tons. By 1997, the catch was back to 7.6 million metric tons, but a particularly bad El Niño in 1997 to 1998 caused the catch to drop to 1.7 million metric tons. By 2000, the catch had increased again to 11.2 million metric tons. El Niños in the early part of this century caused another decrease in the anchovy catch, but by 2004 it had rebounded to 10.6 million metric tons. The catch has been declining in recent years, and the 2008 anchovy catch was only 7.4 million

> **Raft culture** is a form of aquaculture in which the juveniles of commercially valuable molluscs are attached to ropes suspended from rafts.

IMPACT BUBBLE

THE MOST IMPORTANT marine fish raised in aquaculture is salmon. Many wild salmon stocks have been overfished, and aquaculture is seen as a way of relieving the fishing pressure while supplying market demand. Although aquaculture may benefit the species, problems associated with maintaining genetic diversity and hybridization could put even more pressure on already depressed stocks.

The goal of aquaculture is not only to supplement wild-caught fish but also to produce larger numbers of marketable fish in as short a time as possible. For thousands of years, humans have been altering the genetic makeup of their crops, livestock, and pets by selectively breeding individuals with the most desirable traits, and these techniques have been adapted to raising salmon as well as other marine species. It is important, however, that animals grown in aquaculture not become too genetically similar to avoid deformities, decreased disease resistance, and other genetic problems that might result from excessive inbreeding. There is a great deal of concern as to what would happen if inbred farmed animals escaped into the ocean. Jeff Hutchings, a fisheries biologist at Dalhousie University in Canada, found that mixing farmed Atlantic salmon with wild varieties usually has adverse effects on both the growth and survival rates of the wild fish. Since all species of wild Atlantic salmon are endangered, adding any new threat could push them closer to extinction.

Sometimes two different species produce hybrids that may grow to be larger or exhibit other beneficial characteristics. However, in the case of interspecific hybrids, any improvement in the animal may only occur in the first generation and not in subsequent breeding of the hybrid population, requiring continuous hybridization. By the 1990s, scientists had developed techniques for inserting genes from one species of organism into another, producing organisms with desirable traits they had not had before and providing an alternative to hybridization. New organisms produced by the transfer of genes are referred to in the marketplace as genetically modified organisms or GMOs. A company called AquaBounty has developed a genetically modified salmon it calls AquAdvantage, which has been modified with a growth hormone gene taken from the Chinook Salmon (*Oncorhynchus tshawytscha*) and regulatory genes taken from the ocean pout (*Zoarces americanus*). The new salmon, which is derived from the Atlantic salmon, grows to maturity in 16 to 18 months, twice as fast as wild salmon. The company is waiting for approval by the Food and Drug Administration, which may come by the end of 2011, to produce the fish on a large scale. Opponents to approval refer to the salmon as "Frankenfish" and question their safety both as food and for the environment. The company maintains that AquAdvantage Salmon is an environmentally sustainable alternative to current farmed salmon. The company has assured regulatory agencies that the salmon will be grown as sterile all-female populations in land-based facilities so they cannot escape or reproduce in the wild and therefore would pose no threat to wild salmon populations.

Scientists don't know what would happen if GMOs such as these escaped into the wild. As Hutchings has pointed out, genes that are inserted for one purpose sometimes have other unintended and unpredictable effects. It is difficult to make predictions but almost certainly the consequence of genetically modified salmon escaping and breeding with wild species would not be positive for the wild salmon, particularly given their poor conservation status. These potential problems are not just specific to salmon but apply to all species raised in aquaculture.

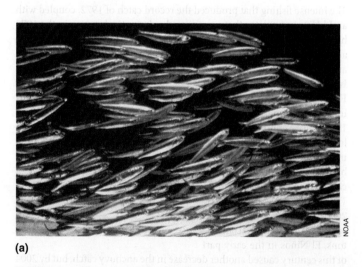

(a)

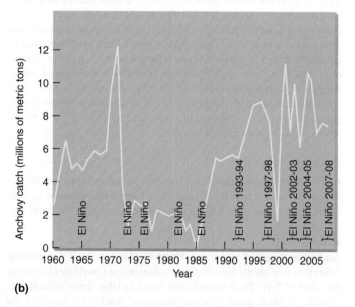

(b)

FIGURE 19-16 ANCHOVIES. (a) Anchovies swim in dense schools and are easily caught in large numbers. **(b)** The Peruvian anchovy catch since 1950 shows the combined impact of overfishing and natural phenomena such as El Niños. *[(b) Data from the United Nations Food and Agriculture Organization.]*

Hawaiian Aquaculture

Although Hawaii is better known for its warm coastal water, the deep waters (600 meters, or 2,000 feet) off the islands are cold and nutrient rich, providing a basis for a large and varied aquaculture. Engineers at the Natural Energy Laboratory of Hawaii (NELH), located at Kailua-Kona on the island of Hawaii **(Figure 19-A)**, have developed ways of using a combination of cold water from the deep and warm surface water to generate electricity, grow health foods, raise seafood and vegetables, and produce drinking water, all without pollution.

In addition to being cold, deep ocean water is rich in nutrients such as nitrates, phosphates, and silicates that are usually scarce in surface waters. NELH uses nine large plastic pipes to bring this water to the surface, creating an artificial upwelling. This technology can potentially enrich areas without natural upwellings. Another benefit of the cold seawater is its purity. It provides an essentially disease-free environment for the species being raised.

Phil and Joe Wilson, who run Aquaculture Enterprises, use the artificial upwelling to raise Maine lobsters in Hawaii without the need for boats or worries about bad weather. The cold water is actually a little too cold for Maine lobsters, but the Wilsons mix it with warmer surface water to produce a 21°C (70°F) bath that allows the lobsters to thrive. Lobsters in the wild usually require 7 years to grow to a market weight of 0.5 kilogram (a little more than 1 pound). The Wilson brothers can

accelerate growth by providing summertime water temperatures year round, thus avoiding the 3-month winter hibernation of animals in the wild. Under these conditions, lobsters reach market size in about 3 years.

Lobster fishers in Maine, however, do not have to worry about the Hawaiian competition. It is still cheaper to catch lobsters than to raise them in aquaculture, and the Wilsons do not have any noticeable impact on the northeastern market. The benefits of the Wilsons' operation is that they can deliver lobsters year round at a fixed price. Recently, the Wilsons have pioneered a lobster aquaculture that uses waste products of commercial fishing as a source of food for the lobsters.

Another group, Royal Hawaiian Sea Farms, grows a variety of seaweeds, such as nori (*Porphyra*), ogo (*Gracilaria*), and limu 'ele 'le (*Enteromorpha*). These "sea vegetables" are considered common foods by many Asians and Pacific Islanders. Royal Hawaiian's products are highly nutritious and quite popular with supermarkets and health food stores in Hawaii and California. The seaweeds are grown from spores attached to rope nets that sway in large water tanks or from vegetative fragments that grow unattached in large rotating drums (tumble culture) **(Figure 19-B)**. Nourished by the rich seawater and the abundant sunlight, the seaweed can be harvested once a week. Although many of these organisms would grow without the cold

nutrient-rich water, it contributes to more vigorous growth.

Another operation, Cyanotech, grows a microalga called *Spirulina* (a cyanobacterium). *Spirulina* is highly prized by health food enthusiasts as a source of B vitamins and beta carotene, and it contains up to 70% protein. Cyanotech produces 9 tons of *Spirulina* per month, which is sold to wholesalers for pills, dips, seasonings, and food additives. Cyanotech also grows algae for the production of phycobiloprotein. The protein is a pigment that fluoresces blue or red and is used in many areas of biomedical research for marking molecules, making them easier to locate and identify. Highly purified forms of this pigment sell for $10,000 per gram. Researchers at Cyanotech are investigating the production of a lower-grade phycobiloprotein that can be sold more cheaply and be used as a coloring agent for cosmetics and food.

Ocean Farms of Hawaii raises thousands of salmon in large ponds, where they are fed on krill. The cold-water zooplankton thrive in the cold water of the artificial upwelling. Ocean Farms also uses water from the artificial upwelling to raise abalone, sea urchins, and oysters. The facility can produce 1.8 million kilograms (4 million pounds) of salmon, 15 million oysters, and 1 million abalone and sea urchins per year. Similar operations are now functioning in St. Croix, U.S. Virgin Islands, and Muroto, Japan.

FIGURE 19-A NATIONAL ENERGY LABORATORY OF HAWAII.
At the National Energy Laboratory of Hawaii, an artificial upwelling is used to raise seafood.

FIGURE 19-B ALGAL AQUACULTURE. Seaweeds are grown from spores or vegetative fragments in large tanks such as the ones shown here.

metric tons. El Niños continue to appear and will continue to cause fluctuations in the anchovy catch. As the anchovy population declines, there is some evidence that sardines may be entering their niche, placing more pressure on the already stressed anchovy population.

There is a general lesson to be learned from these fluctuations in the anchovy fishery. Overfishing, especially when combined with unpredictable natural phenomena, can result in disastrous changes in key marine resources. Overfishing results in fewer reproductive adults in the population, which yields fewer young. Over time, the population continues to shrink. Without proper safeguards and conservation methods, important fisheries may collapse forever.

Tuna

Tuna fishing is big business for many countries, including the United States, which has an annual catch that averages 4.6 million metric tons.

Purse seines are huge nets that can be closed by pulling on a line, similar to the way a purse or bag of marbles is closed.

Backing down is a technique in which the skiff setting the purse seine backs up, causing the edge of the net to drop beneath the surface of the water, allowing dolphins to escape.

The fishing of tunas is a sophisticated and expensive operation, with a modern tuna boat costing in the neighborhood of $5.5 million. To meet the high demand for tuna, commercial fleets set out with state-of-the-art equipment, including purse seines. **Purse seines** are huge nets up to 1,100 meters (3,600 feet) long and 180 meters (595 feet) deep with bottoms that can be closed by pulling on a line, similar to the way a purse or bag of marbles is closed by pulling on the drawstrings **(Figure 19-17)**. These large nets take advantage of the schooling behavior of tuna. When a school of

fish is located, a powerful skiff launches from a large vessel called the purse seiner. The skiff hauls the purse seine in a circle around the school and back to the purse seiner. The bottom of the net is then closed off, and the closed seine containing the fish is hauled on board the purse seiner. This method of fishing is preferred because it yields a much larger catch with less effort than other techniques. A single haul can harvest as much as 150 tons of tuna. Although this method of tuna fishing is highly efficient, it has the drawback that dolphins are frequently trapped with the fish and drown. For unknown reasons, dolphins follow schools of tuna on their ocean migrations, swimming in the surface water just above the school. Tuna fishers locate schools of tuna by watching for dolphins. The total number of dolphins killed in tuna fishing is unknown, but it is estimated that the largest kill occurred in 1970, when more than 500,000 dolphins died.

In 1972, the Marine Mammal Protection Act was passed. It was enacted to reduce the number of dolphins killed in tuna fishing, among other things. The act allows the National Marine Fisheries Service (a division of the National Oceanic and Atmospheric Administration [NOAA]) to set a limit on the number of dolphins killed and to recall the fishing fleet if the limit is exceeded. The limit was set at 20,500 dolphins, and the only time the fleet was recalled for exceeding the limit was in 1977. In addition to setting limits, the law mandates that an observer from the Marine Mammal Protection Agency of NOAA be present on the tuna boats for at least part of the fishing season. These observers not only count the number of dolphins killed but also monitor fishing techniques and methods used to free the dolphins. Based on these observations, the observers attempt to develop new strategies and equipment for protecting the dolphins.

To save dolphins, purse seiners use a technique called **backing down**. In this procedure the skiff draws the purse seine halfway toward the purse

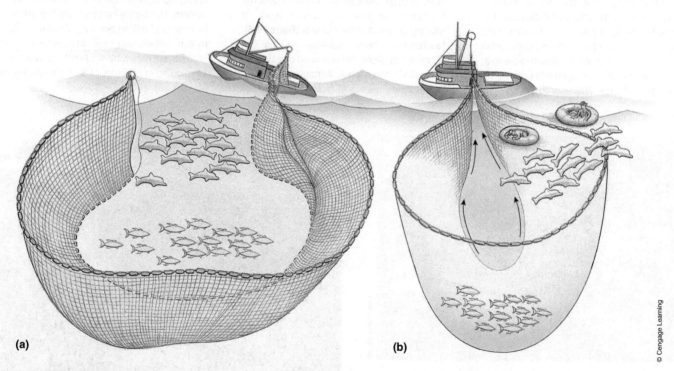

(a)

(b)

© Cengage Learning

FIGURE 19-17 PURSE SEINE NET. (a) The purse seine net is set in a large circle around a school of fish. The ends of the net are then drawn together and pulled closed, much like pulling on a drawstring bag. Unfortunately, when this technique is employed in tuna fishing, dolphins are also trapped in the net and drowned. **(b)** In the technique known as "backing down," fishers allow the edge of the purse seine to go slack, allowing dolphins to escape.

Bluefin Tuna

Bluefin tuna are the largest of the tuna and one of the largest fishes in the ocean, reaching lengths of 4.25 meters (14 feet) and weights in excess of 600 kilograms (1,320 pounds) (**Figure 19-Ca**). They are also arguably the most desirable food fishes in the world and the most endangered of all large fish species. Bluefin tuna are the fish of choice for sushi (rice topped or rolled with fish or vegetables) and sashimi (sliced raw fish) (**Figure 19-Cb**). The high demand for these fish has resulted in the severe overfishing of these species, pushing them to the brink of extinction.

In the early 1900s, bluefin tuna were known as horse mackerel, and their red strong-flavored meat was used for dog and cat food. Although they were not considered to be a good food fish, they were a favorite of sport fishers because they were a challenge to catch and put up a good fight. Bluefin tuna did not become a valuable food fish until later in the 20th century, when sushi became popular worldwide.

The widespread consumption of raw fish in sushi and sashimi is a relatively recent phenomenon that resulted from the widespread adoption of refrigeration in Japan following World War II. The ability to preserve fish for longer periods of time revolutionized Japanese eating habits. The Japanese have always depended on protein from the sea, but before refrigeration, fish were preserved by smoking or pickling, which preserved them for only short periods. With the advent of refrigeration, the fish samurai once referred to as unclean became a delicacy known as *maguro*. The popularity of the fish has made it highly valuable. In January 2011, a 341 kilogram (750 pound) bluefin tuna sold on the Tokyo fish market for $400,000. The high price that these fish would bring at market resulted in intense fishing. Like other tuna species, bluefin tuna are found in large schools and are relatively easy to catch with long lines or purse seines. Japanese fishing fleets initially fished for the Pacific bluefin (*Thunnus orientalis*) found off the shores of Japan and Asia. The North Atlantic bluefins (*Thunnus thynnus*), however, were larger and more plentiful, and Japanese fish buyers soon became regular customers at the fishing ports of the northeastern United States and Europe.

In 1969, a regulatory body, the International Commission for the Conservation of Atlantic Tunas (ICCAT), was established to manage the populations of bluefin tuna in the Atlantic. The ICCAT set limits for the western Atlantic population of bluefin but allowed large catches in the eastern Atlantic population. This has resulted in the near collapse of bluefin stocks in the eastern Atlantic, which includes the Mediterranean Sea. In the Mediterranean, fishers catch schools of half-grown tuna and haul them to offshore enclosures known as marine ranches, where they are fattened before being killed and shipped to market in Japan. Every country that borders on the Mediterranean, with the exception of Israel, maintains offshore tuna ranches. This practice of capturing large numbers of immature tuna is shortsighted because it accelerates the depletion of the Mediterranean stocks. In 2006, the World Wildlife Fund called for an end to all tuna fishing in the Mediterranean, but with such large financial rewards at stake, the plea was ignored. Unfortunately, ICCAT, at its November 2007 meeting, set quotas for the bluefin catch for 2008 at the same level as 2007. The organization did adopt a plan to scale back tuna fishing in the Mediterranean by 20% by 2010, with more reductions to come in later years. This may be too little, too late; even if ICCAT placed lower limits on the bluefin catch, stocks would continue to suffer as a result of widespread illegal fishing. As the tuna population continues to decline, the demand for bluefin continues to increase. This creates a vicious cycle of increased fishing effort producing fewer fish, leading to higher prices that drive more intense fishing.

Some researchers believe that the only hope for bluefin tuna and other commercially important species may be large-scale fish farming. Although many fish farms harm the environment by polluting coastal waters, destroying habitat, and putting pressure on wild populations, researchers such as Columbia University's John Marra suggest that bluefin tuna could be reared in large floating pens constructed on the edge of the continental shelf. These floating pens could be towed from place to place along the shelf to disperse pollutants generated in the process and mitigate the environmental impact. Scientists in the United States, Japan, Europe, and Australia are working on ways to breed bluefin in captivity, but with limited success at this time. For many, tuna aquaculture is the last hope to save this magnificent species.

All Canada Photos/Alamy

JTB Photo/JTB Photo Communications, Inc./Alamy

(a) (b)

FIGURE 19-C BLUEFIN TUNA (a) The Atlantic bluefin tuna is the largest of the tuna. **(b)** Bluefin tuna is the fish of choice for making the best sushi and sashimi.

seiner. When the tuna are close to the boat and the dolphins are at the edge of the net, the boat backs up, causing the edge of the net to go slack and drop beneath the surface of the water, allowing the dolphins to escape (see **Figure 19-17**). Other strategies that have been developed are not used: For instance, not setting the nets at night reduces dolphin kills, but it also reduces the size of the tuna catch and so is not practiced. Most U.S. tuna boats are now based in foreign countries, where they are free from U.S. fishing restrictions.

By 1990, the National Marine Fisheries Service estimated that the number of dolphins killed in tuna fishing by the U.S. fleet was zero. However, dolphins continue to be killed by foreign fishers who use older fishing methods. In 1990, the major tuna packers in the United States agreed that they would buy tuna only from fishers who did not harm or kill dolphins in the fishing process. To indicate this, their products carry a seal indicating that the tuna is dolphin free.

Salmon

The United States also plays a prominent role in the salmon fishery. Pacific salmon (*Oncorhynchus*; **Figure 19-18a**) are fished in the coastal waters of the Pacific Northwest and Alaska, and Atlantic salmon (*Salmo*) are fished in the north Atlantic. Salmon breed in freshwater but spend most of their adult lives in the ocean. Most of the commercial and sport salmon catch is made when the adults begin to migrate back to inland streams to reproduce. There is high demand for salmon, so they bring a high price per pound. As with other highly exploited fishes, salmon are becoming scarce, and their fishing is tightly regulated. Current regulations have shortened the fishing season and decreased the allowable catch, but fishers from several Asian countries still take these fish illegally on the open ocean.

Ocean ranching (sea ranching) is a process in which young fish are raised in pens and then returned to the sea to supplement natural populations.

Shellfish are invertebrate animals with a shell or hard external covering.

The problem of maintaining salmon populations involves not just overfishing but also protecting their spawning grounds. Salmon require shaded, clear, cool, fast-running, unpolluted streams for reproduction. Damming of rivers for flood control and hydroelectric power has prevented salmon access to many streams. Logging by the lumber industry in the Northwest removes shade and allows runoff of silt that clouds the water. Without shade, the water temperature rises. Pollutants such as human wastes have injured young fish and decreased water quality. Although these problems are being addressed, the measures are not implemented effectively or quickly enough, and the spawning population is quite small.

The price of salmon continues to increase not only because there are fewer fish but also because of the cost of cleaning up their native streams and constructing hatcheries to supplement natural populations. About half of the salmon now caught at sea are from hatchery-reared juveniles (**Figure 19-18b**). Hatchery methods, also known as **ocean ranching** or **sea ranching**, raise young fish and return them to the sea, where they develop into adults and increase the size of the population. Some commercial fishers oppose ocean ranching because they believe it will lead to a few large companies controlling the fishing industry and lowering the price of fish by increasing their availability.

Salmon are also raised in aquaculture, but diseases from farmed animals are spreading to wild populations and applying more pressure to already depressed stocks. For instance, fishes that are farmed in sea pens may become infested by parasites from wild fishes and in turn become sources of parasites for wild populations. Sea lice, copepods of the family Caligidae, are the most important disease-causing parasites in salmon aquaculture in the Americas and Europe. The damage they cause is estimated to cost the world industry $390 million per year. The proximity of coastal pens to the sea allows for free movement of the lice between farmed and wild fish. Farmed fish are held for months in the same location at high densities before harvesting—a situation that does not occur in nature. These conditions facilitate the spread of parasites and disease within the culture. When lice from wild fish infest a farm, their population may grow exponentially and then spread back into the environment. Studies present compelling evidence that salmon farms are the most significant source of sea lice on juvenile wild salmon in the Atlantic. Farms in coastal waters may contain millions of fishes almost year round, and unless the lice are controlled, may prove a continuous source of sea lice to wild populations. If farm fishes would escape and remain in coastal waters, they would provide an additional source of lice to wild populations. Studies have found that declines in salmon populations due to parasitic infestation correlated with the increase in salmon farming in both North America and Europe.

Not all salmon farms have sea lice problems, and local environmental conditions vary and influence the ability of parasites to spread. The wild fishery and aquaculture regulatory agencies have become increasingly involved in the problem of sea lice control on farms to minimize risks to wild fishes.

Shellfish

Although most people think of the commercial catch in terms of fish species, more and more of the most important marine fisheries species are **shellfish (Figure 19-19)**. In New England and along the West Coast of the United States, invertebrates, not fish, are the most valuable commercial fisheries. Crustaceans such as crabs, shrimp, prawns, and lobsters, and molluscs such as clams, mussels, oysters, scallops, octopus, and squid, are the most important. It is estimated that the world's shellfish catch is more than 7 million metric tons of molluscs and 5.5 million metric tons of crustaceans. Although the numbers, compared with fish, seem low, the high demand for shellfish places a high dollar value on each pound of catch. In the United States, oysters are harvested from New England, the Gulf of Mexico, Puget Sound, and Washington coastal estuaries. Lobsters are fished in New England (*Homarus americanus*) and Florida (*Panulirus* and *Scyllarides*). Shrimp are fished in the Gulf of Mexico, the Atlantic Ocean off the coasts of North and South Carolina and Georgia, and along the West Coast. Scallops are fished in the Gulf of Mexico and along the southern Atlantic states (*Argopecten*) and the northeastern states (*Placopecten magellanicus*). Crabs and clams are caught along virtually all coastal areas. Shellfish fisheries as a whole face the same problems as the rest of the fishing industry, primarily overfishing and lack of information regarding the basic biology of species. In addition, they are hard-hit by pollution that contaminates estuaries and near-shore waters and contributes to harmful algal blooms (red tides), which render some shellfish poisonous.

In the late 1970s, demand for king crab (*Paralithodes camtschatica*) increased dramatically (see **Figure 19-19c**). This animal is fished mainly in the Bering Sea, and as a result of the increased demand fishing pressure rose tremendously. Between 1979 and 1980, more than 84,000 metric tons of crab were harvested. By 1982, the catch had declined to 18,000 metric tons, less than one fourth the 1980 catch, and by 1985 the catch had declined to 7,300 metric tons. A combination of overfishing and lack of knowledge of the crab's natural history are blamed for the drop in

(a)

FIGURE 19-18 SALMON. **(a)** Salmon are popular with commercial fishers because the fish are large and bring a high price per pound. **(b)** Many salmon are now raised in hatcheries and released to the sea. This practice relieves some of the pressure on natural populations but may create new problems.

Sperm from male

Eggs from female

Salmon hatchery

Mature salmon captured from river

Eggs develop in the hatchery

Larval salmon

Juveniles released into river...

River

...and migrate downstream to Pacific Ocean

Salmon grow to maturity in 2–3 years

Mature salmon enter river and swim upstream toward spawning area

Pacific Ocean

(b)

© Cengagae Learning

(a)

(b)

(c)

FIGURE 19-19 SHELLFISH. Some commercially important invertebrates include **(a)** Maine lobster (*Homarus americanus*), **(b)** spiny lobster (*Panulirus*), and **(c)** Alaskan king crab (*Paralithodes camtschatica*).

commercial catch. King crabs migrate across the floor of the Bering Sea, but not enough is known about the migration stage in their life cycle, or how many migrate in any given direction, to implement appropriate protective measures. In 1986, the fishery came under tight government control. By 1987 the commercial catch rose to 13,800 tons. After several good years between 1987 and 1993, the catch reached an all-time low of 5,880 metric tons in 1994. With the exception of 1998, when the catch was 10,760 metric tons, the catch over the last 10 years has been averaging only 7,000 metric tons. Continued regulation of this fishery should help ensure the future catch, although a major problem, as with other fisheries, will still be the lack of enforcement of protective laws.

IN SUMMARY

- In the past 50 years, the demand for food from the sea has increased and so has commercial fishing.

- Natural populations of fish and shellfish have diminished, and as a result some fisheries have collapsed and others brought to the brink of collapse. (Have You Wondered? #1)

- Fisheries biologists study the size, distribution, and basic biology of commercial species of fish. They use the information to develop management plans that maximize yields while not overstressing the species. (Have You Wondered? #2)

- Overfishing can change the genetic diversity of a population and the species diversity of ecosystems.

- Some techniques used in fishing can decrease fish yields by damaging the habitat. (Have You Wondered? #4)

- The incidental catch, or by-catch, represents a large waste of marine resources.

- Aquaculture represents a way of increasing the commercial catch. (Have You Wondered? #3)

- Some aquaculture practices contribute to habitat destruction, coastal pollution, and increased fishing of noncommercial catch that supply food for many species.

19-2 Salt, Water, and Minerals

The most obvious nonfisheries products of the sea are its major constituents: salt and water. About 30% of the world's supply of salt (NaCl) comes from seawater. The remaining 70% comes from terrestrial salt deposits that were formed when water from ancient seas evaporated. In the south of France, Puerto Rico, central California, the Bahamas, Hawaii, and the Netherlands Antilles, sea salt is extracted and refined to produce table salt. To keep the cost of production low, as little industrial

FIGURE 19-20 SALT PONDS. A salt pond on the island of Bonaire in the Caribbean. Seawater is trapped in ponds such as this. When the water evaporates, it leaves behind deposits of salt, Bonaire's major export. The high-salinity water teems with brine shrimp and larval brine flies, the favorite foods of West Indian flamingos such as these.

energy as possible is used in the extraction process. The process begins by allowing seawater to enter shallow ponds **(Figure 19-20)**. Evaporation of the water produces a concentrated salt solution to which more seawater is added.

Finally, the water is allowed to totally evaporate, leaving a thick salt layer behind that can be processed commercially. In colder regions, seawater is allowed to enter shallow ponds and freeze. The ice layer that forms on top is nearly pure water, concentrating the salt in the water beneath. This highly concentrated saltwater is then removed and heated to drive off the remaining water.

In addition to salt, the sea provides a vast reservoir of water; however, the salts must be removed—a process known as **desalination**—to make the water fit for irrigation or drinking. The greatest problem with producing freshwater from seawater is the cost. The energy required to power desalination processes is expensive; in general, it costs more to produce freshwater by desalination than to obtain it from groundwater or surface water supplies. Whether desalination is used depends on how badly the water is needed, what other sources of water are available, the uses for the water produced, and the local cost of desalination.

In many areas of the world, the amount of freshwater is the major limiting factor for human populations and industrial expansion. These areas include Israel, Saudi Arabia, Morocco, Malta, Kuwait, and many Caribbean islands. In each of these areas, thousands of cubic meters of freshwater are produced daily by desalination. In Texas and California, desalination plants produce freshwater but at twice the cost of freshwater from other sources. The water produced by desalination in these states is primarily used for drinking and some light industry. It is too expensive, however, to be used for agriculture, which requires extremely large volumes of water.

MINERAL RESOURCES

The oceans contain large amounts of minerals, but not all of them can be easily obtained for commercial use. About 60% of the world's supply of magnesium and 70% of the world's supply of bromine come from seawater. Magnesium is combined with other metals to form alloys used in making lightweight portable tools and business machines and in the aerospace industry. Bromine is used as a disinfecting and bleaching agent and a reagent in numerous commercial chemical applications.

It is estimated that the seas contain as much as 10 million metric tons of dissolved gold and 4 billion metric tons of uranium. The concentrations,

however, are on the order of 1 part per billion or less, and no one has yet devised a commercially profitable extraction method.

SULFIDES

During the 1970s and 1980s, deposits of **sulfides** (mineral compounds containing sulfur) were found at several sites in the oceans. These deposits form when hot molten material beneath the earth's crust rises along rift valleys, heating the rocks and causing them to fracture, forming mineral-rich solutions. When the solutions come into contact with the colder seawater, the minerals precipitate and can form massive deposits on the seafloor more than 10 meters (33 feet) high and hundreds of meters long. Minerals in the form of sulfide ores in various regions of the world include zinc, iron, copper, gold, silver, platinum, molybdenum, lead, and chromium. At this time, however, the technology does not exist for selective sampling of these deposits or for mining them.

During the 1960s, muds containing metallic sulfides of iron, copper, zinc, and small amounts of gold and silver were discovered in the Red Sea. These deposits are 100 meters (330 feet) thick in small basins at depths of 1,900 to 2,200 meters (6,200 to 7,200 feet). The saltwater over these deposits contains hundreds of times the concentration of these minerals compared with normal seawater. Their estimated value is in the billions of dollars.

MANGANESE

Manganese is an element used in industry as a component of several alloys. In combination with iron, it strengthens steel and makes it easier to mold. In combination with copper, it forms alloys that are sensitive to temperature changes and that are used in temperature-activated switches such as thermostats. Manganese nodules are in many areas on the deep ocean floor. It is estimated that 16 million metric tons of these commercially valuable nodules accumulate on the ocean floor each year. Although first discovered in the 19th century by the *Challenger* expedition, only during the last 30 years has industry concentrated on mining and extracting them. Since the 1960s, several multinational groups have spent more than $600 million to locate the nodules and develop methods to collect them. Progress has been slow. By the 1980s, many groups had terminated or suspended their activities. The depressed market for metals and the question of ownership are two reasons for the decreased activity.

Desalination is the process of removing salt from seawater.

Sulfides are minerals containing sulfur.

IN SUMMARY

- Almost one-third of the world's supply of salt (NaCl) is produced from seawater by evaporation. (Have You Wondered? #5)

- Freshwater for drinking and irrigation can be obtained from desalination, removing the salt from seawater.

- The oceans contain vast amounts of minerals, but most of them are difficult to reach or expensive to extract. (Have You Wondered? #5)

19-3 Sand and Gravel

The most widespread seafloor mining operations extract sand and gravel used in cement and concrete and for building artificial beaches. The process of extraction is essentially the same as in land-based operations, and the major portion of the cost is related to transportation distance.

Approximately 112 billion metric tons are extracted worldwide annually. The removal of sand and gravel is the only major seabed mining done by the United States at this time. The

Fossil fuels are fuels formed from the remains of organisms that lived millions of years ago.

United States has an estimated 450 billion metric tons of sand reserves off the northeastern coast and large gravel deposits off the coast of New England in the area of the Georges Bank, as well as off the coast of New York City. The dredging of sand and gravel, however, severely damages benthic communities and their related ecosystems.

Deposits of calcium carbonate are found along the coastal areas of Texas, Louisiana, and Florida. The calcium carbonate from these deposits is used for lime and cement, as a source of calcium oxide for removing magnesium from seawater, and as a gravel substitute in road construction. In the Bahamas, sands are mined for calcium carbonate, and reef sands are mined in Florida, Hawaii, and Fiji.

Coastal sands in some areas of the world contain deposits of iron, tin, uranium, platinum, gold, and even diamonds. For hundreds of years, tin has been extracted from sand dredged from the coastal regions of Southeast Asia, from Thailand, along Malaysia, to Indonesia. Today, 1% of the world's tin is obtained from coastal sand in this region. In Thailand tin mining has resulted in heavy deposits of silt in the intertidal and subtidal regions—all but destroying the productivity of these coastal habitats. It is estimated that sands in the shallow coastal waters of Japan hold a reserve of 36 million metric tons of iron.

Since 1972, Russia and the former Soviet Union extracted uranium from bottom sediments of the Black Sea. The United States, Australia, and South Africa extract platinum from some coastal sands. In all of these instances, there is widespread concern about the effects of pollution and habitat destruction that accompany the extraction of minerals from the seabed.

IN SUMMARY

- The largest seafloor mining operations extract sand and gravel from the seafloor for use in making cement, concrete, roads, and artificial beaches.

19-4 Energy Sources: Coal, Oil, Natural Gas, and Methane Hydrate

Coal, oil, and natural gas are collectively referred to as **fossil fuels** because they are formed from the remains of plants and microorganisms that lived millions of years ago. Another source of fuel, methane hydrate, may be tapped in the future as reserves of fossil fuels are depleted.

COAL

Coal formed from the remains of plants such as ferns that lived in prehistoric swamps. When these plants died, they generally fell into the swamp and were covered with water. The fungi that play a major role in the decomposition of woody plant material could not survive in the anaerobic swamp water, and other decomposers, such as anaerobic bacteria, don't decompose wood very rapidly. As a result, very little decomposition took place, and over time, layers of sediment buried the plant material. The sediment's weight compressed the organic material, and over millions of years, the heat generated from the minimal decomposition and the pressure of the overlying sediment converted the plant material to coal. Some deposits that were eventually submerged under the sea are a source of coal today. In Japan coal is mined from under the sea using shafts that originate on land or that descend from artificial islands.

OIL AND NATURAL GAS

Ninety percent of the mineral value taken from the sea is in the form of oil and natural gas formed from the remains of microorganisms such as diatoms. When these organisms died, their remains settled to the bottom of the sea and were covered with sediments. The pressure and heat of being buried under tons of sediments converted the remains of these organisms to oil and natural gas over millions of years. Major offshore oil deposits are located in the Persian Gulf, North Sea, Gulf of Mexico, the northern coast of Australia, the southern coast of California, and the coastlines that border the Arctic Ocean. Many areas are still unexplored, such as the continental shelves of Asia, Africa, parts of South America, and Antarctica. Oil has also been discovered off the mouth of the Amazon river in South America and near the Philippine Islands in the Pacific Ocean. To extract the oil and gas, industry has had to develop huge drilling platforms and specialized equipment for drilling and developing wells at great depths **(Figure 19-21)**. This was a particularly challenging task for platforms in the North Sea, where heavy seas and storms are common.

The annual revenue worldwide from offshore oil and gas production is over $100 billion. The offshore reserves represent about one-third of the world's estimated total reserves. Although the cost of drilling and extracting offshore oil is about three to four times greater than the same development on land, the size of some deposits makes it financially worthwhile. Little is known about the oil reserves in the deep sea, because the deeper the sea, the more expensive the oil recovery. Methods and equipment developed for oceanographic research and deep-sea drilling have provided industry with models for developing the next generation of oil- and gas-drilling equipment. The development of various offshore sites proceeds slowly because of environmental concerns, legal restraints, and the uncertainty associated with global oil supplies. It appears that petroleum exploration and development will continue to be the main focus of ocean mining for the future.

FIGURE 19-21 OIL PLATFORM. Pumping equipment used to remove oil from the seabed is set up on large platforms such as this one in the Gulf of Mexico.

C. Lockwood/Earth Scenes/Animals Animals

crystals. They appear to be stable for long periods. When sediments containing methane hydrate are brought to the surface, where the temperature is warmer and the pressure is less than in the deep water where they are found, the methane gas rapidly escapes, like gas from a carbonated beverage that has been shaken. Because the methane hydrate contains methane, it can be ignited and will burn vigorously. In early 2002, an operation in the Canadian Arctic began to produce a controlled stream of gas from methane hydrate crystals. The results of the experimental attempt indicate that it is at least feasible to produce energy from these sources. Although this is exciting news, geologists point out that only a very small percentage of the known reserves have commercial potential. Most deposits are too thinly spread, and it is much too expensive to exploit them at this time.

Methane hydrate is ice that holds trapped methane gas.

Because methane is a greenhouse gas, some biologists speculate that the escape of methane from marine sediments may have contributed to ocean warming events that produced climate change in the past. They worry that continued global warming may cause methane to escape from sediments, increasing greenhouse gas levels and aggravating an already bad situation.

METHANE HYDRATE

Ice crystals that trap methane are known as **methane hydrate**, the ice that burns, and represent the world's largest known fuel reserve. Methane hydrate occurs as layers, nodules, and filler in and on ocean sediments of some continental slopes 200 to 500 meters (660 to 1,650 feet) deep worldwide. It forms when methane produced by decomposition of organic matter comes in contact with very cold water under very high pressure such that individual gas molecules are trapped in cages of ice

IN SUMMARY

- The ocean contains large reserves of energy in the form of coal, natural gas, oil, and methane hydrate.

- Offshore reserves of oil and natural gas represent about one-third of the world's total reserves.

KEY CONCEPTS

1. Fish and shellfish are renewable resources that must be properly managed to produce a sustainable yield. (19-1)

2. Increased demand for food from the sea has placed a great deal of pressure on natural fish and shellfish populations. (19-1)

3. The advent of mechanized fleets and better fishing techniques, coupled with natural phenomena, has caused a decrease in the size of commercial fish catches. (19-1)

4. Overfishing has brought some fisheries to the brink of collapse. (19-1)

5. Techniques such as aquaculture have helped relieve fishing pressure on natural populations, but not without new effects on natural environments. (19-1)

6. Large numbers of noncommercial animals are killed as a result of current mechanized fishing techniques. (19-1)

7. Our limited knowledge of the basic biology of many commercial species hampers our ability to properly manage and conserve these resources. (19-1)

8. The sea is an important source of minerals, including salt (NaCl) and manganese, and the sulfides of valuable metals such as gold and uranium. (19-2)

9. Fresh water for drinking and irrigation can be produced from seawater by removing the salt. (19-2)

10. The largest seafloor mining operations extract sand and gravel from the seafloor for use in making cement, concrete, roads, and artificial beaches. (19-3)

11. The oceans contain energy reserves in the form of fossil fuels and methane hydrate. (19-4)

1. Increased fishing effort has decreased many stocks of fish and shellfish to dangerously low levels. One recent report suggests that by the middle of this century there will be nothing left to catch.

2. Fisheries management is important to ensure that commercial fish stocks are not depleted and to maintain stable populations for future generations to fish.

3. Commercial fish and shellfish are raised using the same methods that are used in agriculture for terrestrial animals, a practice known as aquaculture or mariculture.

4. Many commercial fishing practices are harmful to the environment. Dredges and trawl nets disrupt benthic communities, destroy bottom habitat, and kill many noncommercial species. Since nets are not designed to catch only one type of fish, many noncommercial species (the by-catch) are caught and killed in fishing.

5. Other resources obtained from the sea include salt, water, gravel, sand, oil, natural gas, and a variety of minerals.

QUESTIONS FOR REVIEW

Multiple Choice

1. The largest seafloor mining operations extract
   a. salt
   b. manganese
   c. sulfides
   d. sand and gravel
   e. calcium

2. The most valuable substance (in dollar amount) that is removed from the sea is
   a. coal
   b. salt
   c. water
   d. oil
   e. sulfide

3. Many _____ are killed as a result of the methods of fishing for tuna.
   a. turtles
   b. seabirds
   c. dolphins
   d. noncommercial fishes
   e. molluscs

4. In the United States, most of the by-catch is
   a. used for feeding livestock
   b. consumed by humans
   c. dumped into landfills
   d. returned to the sea
   e. exported to other countries

5. A major problem associated with preferentially fishing for large fish such as tuna is that
   a. they are harder to catch in nets
   b. they feed at higher trophic levels
   c. fewer can be caught at one time
   d. they yield more waste when processed
   e. they do not swim in schools

6. By 1980, what percentage of the commercial catch was being converted into fish meal?
   a. 10%
   b. 20%
   c. 30%
   d. 40%
   e. 50%

7. Which of the following fishing techniques is most likely to damage habitat?
   a. purse seining
   b. drift netting
   c. gill fishing
   d. trawling
   e. pole fishing

Matching

Match the following terms with the appropriate definition or description:

a. landing	1.	A fishery product made from Alaskan pollock.
b. stock	2.	The catch made by fishing vessels.
c. tan	3.	Huge nets that can be closed by pulling on a line.
d. surimi	4.	A section of a drift net.
e. purse seine	5.	A separate population of fish that is assumed to be reproductively isolated from other populations in the range.

Short Answer

1. What is the goal of fisheries management?

2. In addition to increased mortality rates, what are two other problems associated with overfishing?

3. Why is it difficult to accurately determine the size of commercial stocks of fishes and shellfishes?

4. Compare potential yield with sustainable yield.

5. Besides overfishing, what other problems are contributing to the decline in fisheries production?

6. What are some benefits of aquaculture?

7. What are some ecological problems associated with aquaculture?

8. What is the relationship between the size of the world's population, fishing effort, and the size of the commercial catch of fish and shellfish?

9. How does the cost of fishing influence the kinds of fish caught and the methods used to process them?

10. Considering the vast mineral resources of the sea, why aren't more minerals mined from this rich area?

11. Why is it important for the fisheries industry to be regulated?

12. Why is it not ecologically sound to use anchovies for livestock feed?

13. Why are some commercial fishers opposed to ocean ranching?

Thinking Critically

1. From an ecological standpoint, would it make more sense for humans to eat anchovies or tuna? Explain.

2. When the size of the commercial catch increases, the size of individual fishes decreases. Why does this occur?

3. Suggest some ways that commercial fishers could decrease the size of their catch without becoming unemployed or going bankrupt.

SUGGESTIONS FOR FURTHER READING

Carrol, C. 2007. End of the Line, *National Geographic* 211(4): 90–99.

Haedrich, R. L. 2007. Deep Trouble. *Natural History* 116(8): 28–33.

Hutchings, J. A. and D. J. Fraser, 2008. The Nature of Fisheries and Farming-Induced Evolution, *Molecular Ecology* 17(1): 294–313.

Jeremy, B., C. Jackson, M. X. Kirby, W. H. Berger, K. A. Bjorndal, L. W. Botsford, et al. 2001. Historical Overfishing and the Recent Collapse of Coastal Ecosystems, *Science* 293(5530).

Montaigne, F. 2007. Still Water, *National Geographic* 211(4): 42–69.

Quammen, D. 2009. Where the Salmon Rule, *National Geographic* 218(2): 28–47, 54–55.

Rahel, F. J. 2008. Managing Fisheries and Conserving Fishes: A Difficult Balancing Act, *BioScience* 58(4): 354–356.

Safina, C. 2003. The Continued Danger of Overfishing, *Issues in Science and Technology* 19(4).

Warne, K. 2007. Blue Haven, *National Geographic* 211(4): 70–89.

Have You Wondered?

1. **If pollutants can accumulate in marine organisms? (20-1)**
2. **Why plastic is such a serious hazard for marine organisms? (20-1)**
3. **How climate change is affecting marine organisms? (20-2)**
4. **Why introduction of nonnative species upsets ecosystems? (20-3)**
5. **What you can do to help preserve the marine environment? (20-5)**

Oceans in Jeopardy

IN 1968, A GROUP led by Thor Heyerdahl crossed the southern Atlantic Ocean on a papyrus raft. After completing the trip, Heyerdahl reported that the ocean was polluted. His navigator, Norman Baker, noted that even though for days they would see no land, no ships, and no other humans, they would still see garbage and gobs of oil. In the 45 years since this expedition, the condition of the ocean has not improved. If anything, it has gotten worse. For years, humans have thought of the ocean as a huge waste receptacle, but now there is concern about what this waste is doing to the environment and what will become of the waste already dumped into the ocean. Once-magnificent recreational beaches are being rendered dangerous. Medical wastes such as syringes and other biological contaminants are brought in by the tide. Tar residues clot the sands.

Pollution is not the only factor disturbing the natural balance of the ocean ecosystem. Climate change is threatening coral reefs and the lives of other marine organisms. Introduced species are putting pressure on already stressed populations of marine organisms. Development of oceanfront property has resulted in erosion and damage to offshore habitat that may never be reversed **(Figure 20-1)**.

The United States is not the only country experiencing such problems. Nearly every nation on earth feels the effects of these disturbances. Without changes in federal and international policy, the situation will only become more serious.

20-1 Pollution

More so now than at any time in human history, the ocean is being contaminated by the products of human activity. Trash, sewage, chemicals (both toxic and nontoxic), and radioactive materials are just a few of the items polluting the sea. The effects of pollution on marine organisms and ecosystems are as varied as the pollutants themselves. Some injure or kill marine life or interfere with their ability to reproduce. Some so enhance productivity that they trigger algal blooms that can contribute to the death of marine organisms. Others interfere with productivity, resulting in long-term effects that can have a negative impact on humans as well as on the inhabitants of the sea.

FIGURE 20-1 THREATS TO OCEANS AND MARINE ORGANISMS. Marine ecosystems and organisms are threatened by many human activities, including pollution, overfishing, introduction of nonnative species, climate change, and habitat destruction. (Reprinted with permission of the *St. Louis Post-Dispatch*, © 1997.)

Major threats to oceans, wildlife

Pollution
Oil runoff form streets, driveways; agricultural runoff; cruise ship sewage

Overfishing
Populations of cod, haddock, yellowtail flounder at historic lows

Fishing damage
Longline nets catch about 2.3 billion pounds a year of unwanted fish, seabirds

Seafood farming
Nonnative species escape, pose threat to wild fish

Invasive species
Over 370 plant, animal invaders; 40% have appeared since 1970

Seafloor damage
Fishing gear drags along sea bottom, causing destruction

Climate change
Rising sea temperatures may destroy coral reef ecosystems

Coastal development
20,000 acres (8,000 hectares) of wetlands, estuaries lost each year

© Cengage Learning

SOLID WASTES

Almost from the beginning of human history, coastal countries around the world have used the sea as a dumping site for trash and garbage **(Figure 20-2)**. Over the years, domestic wastes, industrial wastes, and, more recently, radioactive wastes have been poured into the ocean. The dumping of human and industrial wastes into coastal waters is an inexpensive but shortsighted solution to eliminating these wastes.

In September 2010, the 25th annual Coastal Cleanup, coordinated by the Center for Marine Conservation, cleared 8.2 million pounds of debris from 14,520 miles (23,232 kilometers) of shoreline worldwide **(Figure 20-3)**. On average, volunteers collected 565 pounds (257 kilograms) of trash for every mile of shoreline. The most extensive cleanup was in the United States, where 245,447 volunteers picked up 3.9 million pounds of trash from 10,110 miles (16,176 kilometers) of coastline. This averages to 390 pounds (177 kilograms) of trash per mile. The effort in Florida is estimated to have saved the state over $500,000 in cleanup costs. The trash collected included 1.8 million cigarette butts and filters (enough to make 94,626 packs of cigarettes), 440,756 glass bottles, and 429,167 beverage cans. Medical wastes, including 14,555 syringes, were recovered, mostly on beaches of the Northeast and the Gulf Coast (for a more complete list see **Table 20-1**). Forty-eight percent of the trash collected along the beaches was plastic. This is of special concern because plastics do not degrade quickly and are hazardous to wildlife. A total of 488 animals were found trapped in the debris **(Figure 20-4)**. A project report released 10 months after the cleanup concluded that U.S. beaches are still being polluted by a wide variety of materials, especially plastic wastes. If more concerned citizens get involved with projects such as Coastal Cleanup, perhaps the state of our beaches will improve.

Plastic

It is estimated that naval and merchant ships dump 77 tons of plastic into the ocean annually. In addition to this, the National Academy of Sciences estimates that the fishing industry annually discards or loses 149,000 tons of fishing gear (nets, ropes, traps, and buoys, which are mostly plastic) and then dumps another 2,600 tons of plastic packaging material into the sea. To this, add the discarded plastic from pleasure boats, ocean

Images Etc. Ltd./Alamy

FIGURE 20-2 OCEAN DUMPING. Much of the trash on this recreational beach is the result of littering by thoughtless individuals.

Jeff Greenberg/Alamy

FIGURE 20-3 COASTAL CLEANUP. For over two decades the Ocean Conservancy has sponsored a day for coastal cleanup during which volunteers such as these collect millions of pounds of trash from the world's coastlines.

TABLE 20-1 TYPES OF DEBRIS POLLUTING THE MARINE ENVIRONMENT

Shoreline and Recreational Activities	Ocean and Waterway Activities	Smoking-Related Activities	Dumping-Related Activities	Medical and Personal Hygiene
Food wrappers/containers	Rope	Cigarettes/cigarette filters	Building materials	Tampons/tampon applicators
Caps, lids	Plastic sheeting/tarps	Cigar tips	Cars/car parts	Condoms
Cups, plates, forks, knives, spoons	Fishing line	Tobacco packaging/ wrappers	Tires	Diapers
Beverage cans	Bait containers/packaging	Cigarette lighters	Batteries	Syringes
Beverage bottles (glass)	Buoys/floats		Appliances (refrigerators, washers, etc.)	
Beverage bottles (plastic), 2 liters or less	Strapping bands		55-gallon drums	
Bags	Bleach/cleaner bottles			
Straws, stirrers	Oil/lube bottles			
Balloons	Fishing lures/light sticks			
Clothing, shoes	Fishing nets			
Pull tabs	Crab/lobster/fish traps			
Toys	Light bulbs/tubes			
Six-pack holders	Crates			
Shotgun shells/wadding	Pallets			

Medical/Personal Hygiene 1.5%

Dumping Activities 2.8%

Smoking-Related Activities 35.8%

Shoreline and Recreational Activities 52%

Ocean/Waterway Activities 7.9%

Source: Types of Debris Polluting the Marine Environment from "International Debris Breakdown," The 2011 Marine Debris Report by the Ocean Conservancy, pp. 50-79.

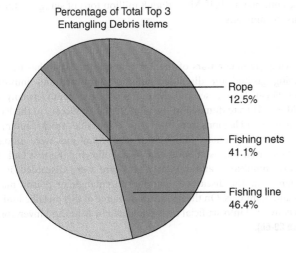

Percentage of Total Top 3 Entangling Debris Items

Rope 12.5%

Fishing nets 41.1%

Fishing line 46.4%

FIGURE 20-4 ENTANGLING DEBRIS ITEMS. Many marine organisms are injured or killed when they become entangled in trash along the shore and in the water. The top three entangling debris items are rope, fishing nets, and fishing line. (Source: Percentage of Total Top 3 Entangling Debris Items (International Coastal Cleanup Report 2011, p. 34.)

liners, oil platforms, and visitors to the beach. The Environmental Protection Agency estimates that somewhere between 500 billion and 1 trillion plastic bags are consumed worldwide each year. Of these, less than 1% are recycled. The majority end up in the ocean. Bags are blown into streams and rivers that empty into the sea, as well as into the sea itself. Plastic bags collect in drains and sewage pipes and are then carried into the ocean. In total, about 1 million tons of plastic waste enters the ocean annually. Carried by the currents, this refuse can appear on even the most remote beaches of the world. Plastic bags have been found floating north of the Arctic Circle near Spitzbergen, Norway, and as far south as the Falkland Islands. Some of the very characteristics that make plastic such a valuable component in manufacturing—its strength and durability—also make it one of the most hazardous materials dumped into the sea. Many marine biologists consider plastic trash to be as great a killer of marine life as oil spills and toxic chemicals. No one knows for sure how long plastic remains in the marine environment, but it is thought that a plastic six-pack ring could last as long as 450 years.

Large marine animals and seabirds face the greatest danger from plastic wastes. As many as 30,000 fur seals are killed annually when discarded fishing nets and cargo straps ensnare them **(Figure 20-5a)**. Discarded netting also traps fish, turtles, and other marine life. Shellfish traps are frequently made in part or entirely of plastic. When these traps are lost,

(a)

(b)

FIGURE 20-5 PLASTIC TRASH. (a) Plastic trash poses the single greatest threat to large marine animals, especially turtles, birds, and mammals. Animals such as this sea lion get tangled in the plastic and strangle, lose appendages, or drown. **(b)** Some consume the plastic, possibly mistaking it for a jellyfish, and die of intestinal blockage.

they continue to catch animals. Unable to escape, the animals ultimately die, frequently of starvation. On the western coast of Florida, more than 100,000 of these traps are set out each year, and approximately 25,000 are lost. Many seabirds die when they become entangled in six-pack rings or fishing line. Nearly 200 different species of marine life die because of plastic bags. Whales and dolphins have been found suffocated by plastic bags or sheets of plastic. Seabirds, sea turtles, and mammals all have a tendency to eat plastics, especially plastic bags that they mistake for food **(Figure 2-5b)**. The plastics form an indigestible mass that blocks the digestive tract and causes the animals to die of starvation. Plastics slowly **photodegrade** over time, as radiant energy breaks them down into microscopic particles composed of toxic chemicals that contaminate shores and the water column and eventually enter food chains.

> **Photodegrade** is the process of radiant energy breaking down a material.

The Marine Plastic Pollution Research and Control Act, passed in 1987 and effective as of 1988, prohibits the dumping of plastic in the ocean and requires ports and terminals to provide facilities for the disposal of this waste. The U.S. Navy, however, can still legally dump trash and plastic into the ocean. Similar laws have been passed by other nations, and an international agreement prohibits dumping plastic waste in the ocean. In the United States, the Coast Guard is mandated to enforce the law, but with all of its other duties and limited resources, violations still occur. Other nations are even less successful in enforcing dumping restrictions, and plastic continues to be a problem. Some manufacturers have started to produce plastics that are more biodegradable, with a shorter life span in the environment. Unfortunately, this will probably not do much to help. If consumers would use reusable cloth bags instead of plastic bags, it would decrease the number of plastic bags consumed by 288 bags per year per person. Only education and responsible action by those who use the sea for commerce and recreation, as well as all consumers, will reduce plastic in the ocean.

Commercial Dumping

Since 1890, garbage, sewage, and toxic chemicals have been dumped into the New York Bight, the area lying off the mouth of the Hudson River. Floating debris from the dumping ended up on beaches. By 1934,

the problem had reached such large proportions that legislation was passed prohibiting the dumping of floating waste. The legislation did not, however, limit the dumping of building wastes, storm and sewage wastes, or industrial wastes. It has been estimated that 1.4 million cubic meters of solid waste was dumped into the New York Bight between 1890 and 1971—enough to cover all of Manhattan Island to the height of a six-story building. Needless to say, this amount of dumping has greatly decreased water quality and the quality of bottom sediments. The materials dumped were both toxic and oxygen demanding (decomposition of these materials by microorganisms tends to deplete the water of oxygen). Occasionally, the contaminated water would upwell along the coast, killing shallow-water organisms. The offshore dump was finally closed in 1987 by the Environmental Protection Agency. The agency did, however, allow New York City, as well as several New Jersey communities, to continue dumping sewage sludge, a product of sewage treatment, at a site 171 kilometers (106 miles) out to sea at the edge of the continental shelf. More than 8 million tons of sludge is dumped at this site annually.

Military Refuse

Following the major wars of the last century, the ocean has been the dumping ground for discarded military hardware and munitions **(Figure 20-6a)**. Toxic gases and chemicals removed from Germany after World War II were dumped in the North Atlantic Ocean; in the Pacific Ocean, bays and lagoons were filled with jeeps, tanks, trucks, and munitions. After the Vietnam War, vehicles and munitions were unloaded into the coastal waters of Southeast Asia. Following the first Gulf War in 1991, munitions and military hardware were discarded in the Arabian Gulf. The discarded material can entangle or poison marine birds and mammals. On the positive side, some of this metallic junk has been converted into artificial reefs populated by fishes and invertebrates **(Figure 20-6b)**.

CHEMICAL POLLUTION

Throughout the world, major population centers have developed along coastlines, and around estuaries and the rivers that drain into them. In the United States, more than 50% of the population lives within 50 miles of a coastline (including the Great Lakes). Worldwide, the numbers are

(a)

(b)

FIGURE 20-6 ARTIFICIAL REEFS. (a) After military conflicts, such as World War II, military wastes are frequently dumped into the ocean. **(b)** Some, such as this sunken vessel, become artificial reefs, providing habitat for a large variety of marine organisms.

similar and sometimes higher. This dense population with its needs for energy, industry, and waste treatment, has placed a large burden on coastal seas and coastal habitats. Surface runoff from metropolitan areas contains pesticides, fertilizers from residential areas, chemicals from oil and gasoline, and residues from industry **(Figure 20-7)**. Pesticides used in agriculture can also enter the marine environment as they are rinsed into streams and rivers that flow to the sea. Toxic chemicals (including oil) from offshore oil rigs and the transport of petroleum products by tankers pose another threat to marine ecosystems.

Pesticides and Other Toxic Chemicals

Pesticides, including dichlorodiphenyltrichloroethane (commonly known as DDT); toxic organic compounds such as polychlorinated biphenyls (commonly known as PCBs); and heavy metals such as mercury, lead, zinc, and chromium, continue to enter the sea through various routes even though they are closely monitored and some no longer used. The use of DDT was suspended in the United States in the 1960s, and the United States ceased production of PCBs in the late 1970s. Both DDT and PCBs are, however, persistent toxic substances, remaining in the environment and ocean for long periods.

DDT is carried to the sea by runoff from the land. The year that DDT was banned, another pesticide made from DDT, dicofol, was approved for use. Hundreds of thousands of tons of this pesticide were used around the nation until 1988, when researchers found that dicofol contained 1% to 7% DDT contaminants. Seabirds that feed on fish contaminated with DDT lay thin-shelled eggs that break when the parents incubate them. This has led to high mortality rates of chicks and decreases in the size of some seabird populations, such as brown pelicans in California, already under pressure from other environmental changes. Even populations of the Bermuda petrel that live far from shores where the pollutant enters have been affected.

PCBs are mixtures of up to 209 individual chlorinated compounds that are either oily liquids or solids. PCBs were used as coolants and lubricants in transformers and in a number of consumer articles, including plastics, solvents, and electrical insulators. They are still present in many electrical devices that were manufactured before 1979. When these devices are burned, PCBs enter the atmosphere and then enter the water supply with precipitation. Fish absorb PCBs from contaminated sediments and their food. In January 2001, a British news program reported that salmon

raised in aquaculture contained high levels of contamination from PCBs as well as other pollutants. Subsequent studies found that the source of the PCBs in farmed fish was the fishmeal they were fed, which contains high levels of PCBs. PCBs are stored in fat, and because farmed salmon are generally larger and contain more fat than wild salmon, farmed salmon contain more PCBs. In July 2003 a report was released indicating that farmed salmon purchased in the United States has the highest level of PCBs in the food supply, 16 times the PCBs found in wild salmon and 3.4 times the levels in other seafood. A report in the journal *Science* (January 2004) indicated that farmed salmon contain 10 times more toxins than wild salmon and suggested eating no more than eight ounces of farmed salmon once every one or two months, as they may pose a risk of cancer to humans.

Phytoplankton are particularly sensitive to chronic pollution by toxic substances, as are many forms of zooplankton and larger crustaceans. Toxic pollutants inhibit photosynthesis, growth, and cell division in marine phytoplankton. They affect the growth and development of filter-feeding zooplankton, as well as the early developmental stages of other organisms, by interfering with cell division. The impact of marine pollution on plankton has barely been studied, but it is known that pollution in the North Atlantic Ocean has resulted in decreased numbers of species of both phytoplankton and zooplankton. This should serve as a warning. In the long term, pollution causes an overall reduction in marine primary productivity, altering interactions at every trophic level throughout the food web.

Chemical pollutants such as carbon dioxide and sulfur dioxide from the burning of some fossil fuels dissolve in water and enter the ocean directly, in precipitation, or indirectly, in river runoff. The water is acidic, which can lower the pH of seawater and interfere with the biological processes of marine organisms. Chlorine added to drinking water and then later to sewage during treatment forms chlorinated organic compounds in seawater that are toxic to some marine organisms.

Efforts to reduce the discharge of toxic materials into the marine environment are gradually succeeding. Since 1980, the increased use of unleaded fuels has significantly decreased the amount of organic lead entering the marine environment. The concentration of lead in surface waters of the Sargasso Sea dropped by 30% between 1980 and 1984. In the last 10 years, the amount of lead entering the Gulf of Mexico from the Mississippi River has decreased 40%.

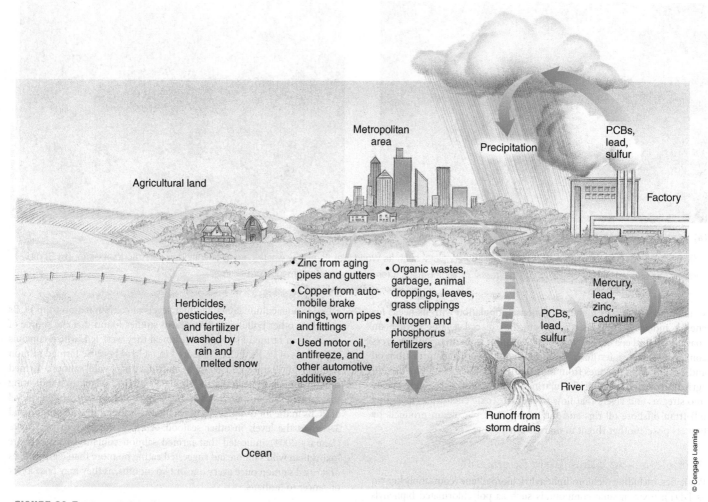

FIGURE 20-7 POLLUTION FROM LAND. Pollutants, including toxic organic compounds, pesticides, sewage, petroleum products, and heavy metals, are carried into the sea by rivers and washed into the sea by rainwater and melting snow.

Petroleum Pollution

The demand for oil to supply the needs of industry and motorists has exposed the ocean to one of its greatest threats. In addition to the petroleum and petroleum products released into the environment by runoff, industrial discharge, and other processes, each step in production and transport poses the risk of contamination. When petroleum is pumped from the seafloor by offshore drilling operations, there is the risk of well blowouts and leakage. When it is transported from production sites to processors and industrial areas in tankers, there is the risk of shipwreck. The tanker transport of oil and petroleum products poses the greatest risk of large spills.

Crude oil is a mixture containing nearly 10,000 different polluting chemical compounds. The majority of these chemicals are organic molecules composed primarily of carbon and hydrogen atoms. Crude oil contains two major types of these organic molecules: aromatic hydrocarbons and aliphatic hydrocarbons. **Aromatic hydrocarbons**

Aromatic hydrocarbons are compounds composed of carbon atoms that form ring structures.

Aliphatic hydrocarbons are compounds composed of carbon atoms that form straight-chain molecules.

such as benzene, naphthalene, and cyclohexane are made up of carbon atoms that form ring structures. **Aliphatic hydrocarbons,** on the other hand, are straight-chain molecules such as heptane, octane, and nonane.

When oil is released into water, it spreads in a film across the surface. Lighter components of the oil evaporate or are absorbed by particles of clay and other sediments in the water and sink to the bottom. Some inorganic components dissolve in seawater, whereas some organic components are degraded by bacteria and fungi. Many bacteria can digest the simpler organic molecules in oil but have difficulty digesting the more complex molecules. These persist in the sea for a long time as tarry chunks floating on the surface or lying on the bottom **(Figure 20-8)**. Little is known about the effect of oil on bottom-dwelling organisms of the deep. We do know that oil can suffocate organisms at the bottom of bays and harbors, and that some of the soluble components of oil are extremely toxic to many forms of marine life, especially eggs and young. Aromatic compounds are generally more toxic than aliphatic compounds, and the smaller the molecules, the more easily they can be taken up by marine organisms. Refined petroleum products such as gasoline and kerosene are even more toxic to marine life than the unrefined components of crude oil.

FIGURE 20-8 PETROLEUM POLLUTION. Complex molecules in oil can persist in the ocean for long periods of time and form tarry masses that contaminate water and wash up on beaches.

Oil Spills

At approximately 9:45 p.m. CDT, on April 20, 2010, British Petroleum's Deepwater Horizon oil rig, located 50 miles off the coast of Louisiana in the Gulf of Mexico, exploded and caught, fire killing 11 workers. Two days later, the oil platform sank and the uncapped well continued to spew oil into the Gulf. The blowout caused the biggest accidental release of oil into the ocean in history. By the time the well was finally capped on July 15, 2010, more than 4.9 million barrels of oil had poured into the ocean **(Figure 20-9a)**. Shortly after the well was capped, reports from government agencies indicated that about half the oil released had been removed from the environment. The other half still remained as small droplets in the water, in the form of tar balls, or buried in sediments. Nongovernment sources, however, indicated that the amount of oil remaining in the environment is much higher. Most studies thus far agree that a great deal of oil is likely still in the environment, and that the oil released caused significant harm to natural resources and the coastal economy. No one knows how much oil remains in the environment but past experience indicates there is the potential for it to persist for many years.

The spill threatened eight U.S. national parks and more than 400 species of wildlife. The toll on local wildlife was and continues to be significant. As of November 2010, 6,814 dead animals had been collected, mostly marine birds (6,104) and the remainder mainly sea turtles and dolphins.

Since the majority of dead animals are never found, the actual death toll is almost certainly significantly higher. In July 2010, scientists from Tulane University found that blue crab larvae in the Gulf showed signs of an oil-and-dispersant mixture under their carapaces. These findings indicated that dispersants used to treat the spill had broken oil into droplets small enough to easily enter the food chain. Due to possible contamination of fish and shellfish in the area, the waters around the spill were closed to fishing until November 15, 2010. A report of toxic chemicals in shrimp caused 4,200 square miles (2,700 square kilometers) to be closed again to shrimping on November 24, but the area was re-opened by early 2011.

Marine life continues to be affected in the aftermath of the oil spill. Since January 2011, 67 dead dolphins have been found in the area affected by the oil spill, 35 of them were premature or newborn calves **(Figure 20-9b)**. Although their deaths cannot be directly related to the oil spill, the cause is under investigation. In February 2011, the first birthing season for dolphins since the spill, the Institute for Marine Mammal Studies reported that dead baby dolphins were washing up along the shore at about 10 times the normal number for the first two months of the year. Clearly it will take several years to assess the full ecological impact of the Deepwater Horizon disaster.

Prior to the Deepwater Horizon spill, the largest oil spill in the United States occurred in March 1989, when the tanker *Exxon Valdez* ran onto a rocky reef 25 miles out of its home port of Valdez, Alaska. The accident tore five huge holes in the hull, and the ship spilled more than 240,000 barrels (10 million gallons) of oil into Alaska's pristine Prince William Sound **(Figure 20-10a)**. Several factors intensified the damage and hampered cleanup. Local plans for dealing with oil spills were not designed to handle an event of this magnitude. Cleanup equipment and personnel were not actively in service, and there was a delay while these were assembled. Rugged terrain and tidal currents of the enclosed area also aggravated the problem. The spill quickly spread over 2,610 square kilometers (900 square miles) of Alaska's most picturesque and biologically productive coastal areas. As the oil moved out of the sound, currents carried it down the Alaskan coast, where it coated rocky shores and damaged marine and terrestrial habitats **(Figure 20-10b)**. Thousands of birds, fish, and marine mammals died. The cold temperatures of this subarctic region slow the natural decomposition of oil by photochemical

(a)

(b)

FIGURE 20-9 THE *DEEPWATER HORIZON* SPILL. **(a)** For nearly 3 months, the *Deepwater Horizon* well poured over 4.9 million barrels of oil into the Gulf of Mexico, causing the largest oil spill disaster in history. **(b)** Thousands of marine animals, including this dolphin, were injured or killed by the oil and other toxins entering the marine environment from the *Deepwater Horizon* disaster.

(a)

(b)

FIGURE 20-10 OIL SPILLS. **(a)** Oil booms are used in an attempt to limit the spread of oil leaking from the damaged tanker *Exxon Valdez*. **(b)** Oil from the *Exxon Valdez* covers some of the rocky coast of Prince William Sound, Alaska, killing intertidal organisms.

reactions and bacteria. The marine life of the area also live longer and reproduce more slowly, so although the region is expected to recover, it will probably take longer than if it occurred in a temperate or tropical zone. Not all of the ecological damage was immediate: The herring fishery in Prince William Sound collapsed four years later due to the impact of the oil spill and other factors.

The longest-lasting oil spill in the world occurred in June 1979, when an offshore oil well in the Gulf of Mexico, the Ixtoc I, owned and operated by a Mexican company, blew out and caught fire. It required almost a year to cap the well, during which time more than 137 million gallons of petroleum poured into the Gulf of Mexico. Similar events have occurred in the North Sea and the Persian Gulf. During the Iran–Iraq war, the bombing of oil facilities resulted in large amounts of oil entering the Persian Gulf. One bombed oil rig pumped 50,000 gallons per day into the Gulf for almost 3 months. Miles of beaches on the west side of the Gulf are still covered with a hardened surface of sun-baked oil. Current reports indicate that the oil spills killed 40% of the coral reefs off the coast of Qatar and 30% of the reefs off the Bahrain coast. During the first Gulf War, oil facilities sabotaged when the Iraqi army retreated from Kuwait released thousands of gallons of oil into the Persian Gulf around Kuwait. The damage from this ecological disaster is still being assessed.

Hypothermia is the condition of having an abnormally low body temperature.

Although large oil spills receive the most media attention, more oil and petroleum products enter the sea by way of runoff from urban areas and from spills and leakage that occur in the loading and transfer of petroleum products.

When an oil spill occurs near a coastal area, damage to the environment and marine life can be substantial. Millions of seabirds such as cormorants, pelicans, and diving ducks and thousands of marine mammals, especially seals, sea otters, and dolphins, have fallen victim to oil spills. The effects on invertebrates and algae are just as damaging and have far-reaching ecological consequences. The immediate damage from oil spills is obvious, but the damage to the environment from gasoline, diesel fuel, and other more toxic petroleum products is more insidious. Many of these petroleum products evaporate and disperse quickly, and it is difficult to assess their immediate effects; however, because they are constantly released into the sea, they cause chronic problems. Most petroleum transported by sea is offloaded or transferred at major ports, and most major ports are located in the world's largest bays and estuaries, putting these delicate, productive, and valuable areas at risk of damage from oil and petroleum products. When oil spills occur in the open sea, the effects are difficult to assess. Because they occur far from land, they are difficult to monitor, and wind, waves, and currents dissipate the oil quickly. Presumably, the damage to marine life is the same as in coastal spills.

Effects on Birds and Mammals A heavy coating of oil impairs a bird's ability to fly and swim. In addition, large outer feathers cannot repel water, and down feathers cannot insulate. During cold weather, a spot of oil no bigger than a dime in a vital area can cause death by **hypothermia** (lowering the body temperature). Birds may ingest fatal amounts of oil as they preen to remove it from their feathers. Oil destroys the ability of an otter's fur to insulate the animal, causing it to die of hypothermia (**Figure 20-11** Oil clogs the ears and nostrils and irritates the eyes of otters and other mammals. Oil has also been shown to cause cancer in sea otters.

FIGURE 20-11 OIL-RELATED INJURY. This otter was a casualty of the *Exxon Valdez* oil spill. Even a small spot of oil on the feathers of a bird or on the fur of a marine mammal can cause the animal to lose body heat and die.

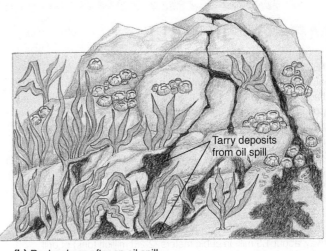

(a) Rocky shore before an oil spill

(b) Rocky shore after an oil spill

FIGURE 20-12 GENERALIZED CHANGES IN HABITAT THAT RESULT FROM OIL POLLUTION. **(a)** A generalized view of a temperate zone rocky shore before an oil spill. **(b)** The same rocky shore after an oil spill. Oil kills molluscs, allowing seaweeds to overgrow the intertidal rocks. Wave action will eventually remove much of the algae, resulting in a decrease in species diversity.

Effects on Invertebrates and Algae Less visible, but just as deadly, are the effects on invertebrate, algal, and plant life along the shore. On sandy shores, sand crabs and other organisms that live in spaces between sand particles are smothered by a coating of oil or killed by its toxic components.

Even after tides have washed the beaches clean, a residual layer of oil may persist several meters below the surface. Although some intertidal invertebrates such as bivalve molluscs seem to be resistant to oil pollution, their flesh becomes tainted with petroleum chemicals, which are passed along the food chain. On rocky shores, barnacles are fairly resistant to pollution, although in some instances they are smothered by the oil, but grazing molluscs such as limpets, periwinkles, and the carnivorous whelks are vulnerable. Toxic components of the oil are a narcotic, causing molluscs to lose their hold on the rocks and be washed away by the tides.

Community Effects The elimination of grazers allows seaweeds to colonize the vacated area. Eventually, the seaweeds become encrusted with oil and are torn away by the waves. The ultimate outcome of oil pollution in the intertidal zone is a decrease in species diversity, a simplification of the food web, and a disproportionate increase in the populations of resistant species **(Figure 20-12)**.

Oil Spill Cleanup

Current technology for dealing with ocean oil spills includes the use of oil booms and oil skimmers **(Figure 20-13)** that help confine the spill to a smaller area and recover some of the oil and chemical dispersants. These techniques are most efficient when the spill occurs in an area that is easily accessible and where equipment can be rapidly mobilized. In a spill such as the one involving the *Exxon Valdez*, where weather, sea conditions, rugged coastline, and distance from cleanup equipment are major factors, or the Deepwater Horizon, where huge amounts of oil are carried by currents over a large area, little is actually recovered. In some spills along an accessible coast, straw is used to soak up the oil, then is collected by hand and disposed of by burning—which transfers the pollution to the air. Microbiologists have developed a genetically engineered bacterium capable of degrading most of the organic compounds present in crude oil. It is being tested to determine its environmental impact. Current legislation prevents the use of such organisms in nature. Naturally occurring bacteria living near oil seeps on the ocean floor are used with some success to clean up oil spills.

NUTRIENT POLLUTION

It was once thought that coastal waters had an infinite capacity to absorb wastes from human populations. But too much waste entered too small an area in too short a time. This has resulted in a decrease in the economic and recreational value of coastal areas and, in some cases, endangered public health and safety.

Human Wastes

Cottages built along the shore use septic tanks to process sewage and drain wastes into the sandy soils. Some commercial developments intentionally and unintentionally drain human wastes into the ground, and coastal communities dump raw sewage into the shallow water off the coasts. On the East Coast, water from storm drains mixes with sewage and overflows sewage treatment facilities after heavy rains and when snow melts. The overflow enters the ocean. In addition to untreated runoff, coastal cities discharge treated sewage that adds its own array of contaminants to ocean water. This type of pollution damages recreational areas as well as coastal fisheries.

FIGURE 20-13 OIL SPILL CLEANUP. Absorbent booms (red) are used to contain and clean up oil spills.

IMPACT BUBBLE

IN THE PROCESS of drilling for oil and natural gas in the ocean, large amounts of material called cuttings are removed from the seafloor. The cuttings are contaminated with a number of toxins, and they are usually disposed of by dumping them back into the sea. Thousands of tons of cuttings, along with oil and other chemicals, are polluting the ocean. It is generally believed that the main toxic components in cuttings are oil and oil products. Regulatory agencies set the permissible amount of these substances in cuttings at no more than 100g /kg cuttings. Scientists point out that even if this requirement were met during actual drilling operations, the concentration is still 900 to1,000 times the amount necessary to produce toxic bottom sediments. Over the past 40 years, in an area of the North Sea bordered by Great Britain and Norway, about 1.3 million cubic meters (2-2.5 million tons) of drill cuttings and solid wastes have built up on the seafloor. The ecological damage can be detected up to 10 kilometers (6 miles) from the platforms producing the waste. These piles of cuttings smother benthic sea life and remain toxic for years.

Various mixtures of clay and liquids, known as muds, are used as lubricants and coolants in the drilling process. Oil-based muds (OBMs) contain oil and are the most toxic. Synthetic-based muds (SBMs) contain synthetic oil and are less toxic. Water-based mud (WBMs) uses water and is thought to be relatively harmless. The dumping of cuttings contaminated with OBMs and SBMs is banned in many areas of the world, but WBMs are permitted in most regions, including North America, as long as they are treated to prevent the formation of surface slicks from crude oil that mixes with the wastes. According to the U.S. Department of Energy, WBMs produce only short-term minor impacts on the seafloor, whereas OBM cuttings cause long-term severe impacts. Some studies, however, indicate that WBMs may not be as harmless as generally thought. Although the area of affected seafloor is much smaller where only water-based muds have been used, the ecological effects are still significant. WBMs may contain free oil, aromatic hydrocarbons, heavy metals, and radioactive materials. Thus WBM cuttings can poison marine life as well as smother it with artificial sediments or suffocate it with plumes of superfine suspended particles. Plumes of suspended particles from cuttings may contribute to the mass death of pelagic fishes, such as the one that occurred in June 1999 in Piltun Bay of Sakhalin Island off the East coast of Russia in the North Pacific.

Dealing with existing cuttings poses complex problems. For instance, if drilling platforms are removed, can this be accomplished without disturbing the cutting piles below them and releasing buried toxins into the water column? Can the cuttings themselves be removed, treated, and disposed of, without causing further ecological damage, or would it be better to leave them in place and let natural processes eventually remove them? It is estimated that removing all of the cuttings in the North Sea could impact the ecology of an area of 4,000 square kilometers (2,400 square miles) if the pollutants from the piles were released into the water column at a concentration of just 3%. At lower concentrations such as 5 parts per million the affected area could be 246,000 square kilometers (147,600 square miles). The costs of removing, treating, and disposing of the wastes in the North Sea are estimated at a minimum of $614 million. That would be in addition to the economic impact resulting from fishery losses if valuable fishing grounds were closed due to commercial fish stocks being contaminated by pollutants from the cutting piles.

Uncertainty over the potential ecological effects of disturbing the piles causes fundamental problems in developing solutions. Regardless of how existing wastes are handled, the fact remains that the expedient thing to do at this point is minimize all dumping and discharges from oil and gas production platforms.

Human wastes reduce water quality by adding harmful microorganisms and enriching the water with nutrients. Filter feeders such as clams, mussels, and oysters can concentrate in their tissues large amounts of harmful microbes such as hepatitis viruses and *Salmonella* bacteria from sewage, resulting in their being unfit for human consumption. In areas where these shellfish are eaten, diseases such as hepatitis and dysentery may occur. In many parts of the United States, bacteria from cities and individual septic systems have forced the closure of clamming beaches and oyster beds (Figure 20-14a). In some cases, the result has been economic hardship for clam and oyster fishers. Local environmental agencies monitor water quality by counting the number of **coliform bacteria** in the water. Coliform bacteria are found in the intestines of many animals, including humans, and are an indicator of the amount of animal wastes entering the water. Their presence indicates the potential for disease.

Coliform bacteria are bacteria found in the digestive tracts of animals.

Both raw (untreated) and treated sewage add large amounts of nutrients, such as ammonia and urea, to coastal waters. This leads to eutrophication, an increase in the amount of dissolved nutrients in the water. Eutrophication leads to blooms of phytoplankton and other marine microbes (Figure 20-14b). The large increase in phytoplankton populations frequently exceeds the ability of zooplankton to control it by their grazing. The increased amounts of phytoplankton also make the water turbid and interfere with primary production in deeper water. Eventually, the phytoplankton begin to die off and are decomposed by aerobic bacteria. As you learned in earlier chapters, excessive decomposition can rob the water of oxygen and result in the death of fish and other marine organisms, as well as the aerobic bacteria. Anaerobic bacteria continue the job of decomposition, removing even more oxygen from the water and in many cases producing hydrogen sulfide, a compound toxic to many marine organisms. The lack of oxygen and accumulation of hydrogen sulfide leads to the death of more marine organisms and their decomposition, and the vicious cycle continues.

In some instances, the addition of sewage and animal wastes can increase the productivity of an area. During the early 20th century, sewage from London entered the North Sea by way of the river Thames. A small region around the mouth of the river supported a significantly higher catch than other parts of the North Sea, presumably because of the increased nutrient input. Similar increases in coastal productivity have been noticed along the coast of southern California and the northeastern coast of the United States.

(a)

(b)

FIGURE 20-14 WATER QUALITY. (a) Sewage contamination of coastal waters can introduce disease-causing bacteria and lead to loss of recreational beaches due to beach closings. **(b)** The influx of nutrients from terrestrial sources can cause eutrophication of the water and algal blooms.

Agricultural Wastes

Fertilizers used on lawns, gardens, and crops wash to the sea during rainstorms and in snow melt. Surface runoff from agricultural lands can also contain wastes from livestock. Fertilizers and animal wastes affect marine communities in a manner similar to that of human wastes. As with human sewage, animal feces can add disease-causing microbes to the water. It is thought that the outbreak of *Pfiesteria*, a potentially dangerous dinoflagellate, in the Chesapeake Bay region may be related to the contamination of the water by feces from pig and chicken farms. Surface runoff containing nutrients is carried by streams and rivers to the sea, where it can poison and overfertilize the water. Nutrient addition from fertilizers causes eutrophication of coastal waters and supports phytoplankton blooms. The increase in algae outstrips the environment's ability to support it, and the algae die. The decomposition of the dead algae removes excessive amounts of oxygen from the water, causing more organisms to die and resulting in even lower levels of oxygen. Agricultural nutrients carried into the ocean can cause **hypoxic zones** (water with abnormally low oxygen content), with potentially devastating effects for local fisheries.

Although there has been increased input of nutrients from fertilizers used in agricultural areas, better sewage treatment has decreased the nutrient content of discharged sewage, so in some areas there has been no net increase in the level of nutrients reaching the sea. Unfortunately, this is not true everywhere.

RADIOACTIVE WASTES

Today's electronic, chemical, and defense industries and nuclear facilities produce large amounts of highly toxic or radioactive wastes that have been stored in landfills and in above-ground repositories. As concern for their effects on human health increases, it has been suggested that they be deposited on the ocean floor, for instance, in deep-sea trenches far from continental landmasses. Supporters of this proposal point out that because the trenches are subduction zones, the wastes would ultimately be taken into the mantle of the earth, eliminating the problem of long-term storage. Opponents point out that if the containers holding the waste rupture or decompose before subduction occurs, their contents could cause considerable habitat damage. Considering

our general ignorance of the deep-sea benthos and deep-sea food webs, no one really knows the long-range effects or dangers of this disposal method. Subduction-zone disposal is currently restricted by the Ocean Dumping Act of 1972, which requires an environmental impact statement and approval of both houses of Congress before radioactive waste can be dumped in the ocean.

> **Hypoxic zones** are areas of water with abnormally low oxygen content.

CONTROLLING POLLUTION

In the summer of 1988, many northeastern U.S. beaches closed because of polluted waters and dangerous wastes washing up on the beaches. In addition to gummy balls of sewage as much as 2 inches (5 centimeters) in diameter, medical wastes such as syringes, vials of blood, and medicine bottles were washing up on shore **(Figure 20-15)**.

FIGURE 20-15 MEDICAL WASTES. Toxic and medical wastes, such as this syringe, make recreational beaches unfit for visitors and kill marine organisms.

On a New Jersey beach just south of Long Island, New York, more than 100 vials of blood washed up. When tested, 5 of these vials were found to contain antibodies for human immunodeficiency virus (HIV). The ensuing public scare caused attendance at some beaches to drop by 85%.

In response, Congress passed legislation that prohibited dumping of sewage sludge or industrial wastes in the ocean after January 1, 1992. This legislation does not solve all of the problems, however. Much of the material that contaminated beaches in 1988 continues to contaminate beaches today. It is the product of overflowing storm sewers that empty into the ocean, as well as improperly dumped wastes.

The bigger threat to coastal water quality is not dumping of wastes but increasing coastal populations and improperly controlled residential and commercial development, discussed later in the chapter. In theory we could have both large populations and clean water if we spent the money to treat the sewage. Whether we decide to spend the money depends on ethical, economic, and political considerations.

BIOLOGICAL MAGNIFICATION

Not all pollutants that enter the marine environment dissolve in the water. Some are absorbed by suspended particles of clay and detritus. The particles clump and form denser masses that sink to the bottom. Toxic materials such as DDT, PCBs, petroleum products, and heavy metals always collect in higher quantities in sediments than in the overlying water. They continue seeping into the water for years. The tainted detritus and contaminated phytoplankton are used by various marine

organisms as a source of food. Because many of these organisms have no means of breaking down or excreting the toxins, the toxins concentrate in their body tissues. The toxins are passed along to predators, where they concentrate further, a process known as **biological magnification (Figure 20-16)**. At some level, the concentration of toxins can be so high that the next predator to feed on the contaminated flesh will become seriously ill or die. During the 1970s, this is exactly what occurred when mercury from industrial pollution accumulated in the food chain and contaminated tuna. Humans who consumed the contaminated tuna suffered from the effects of mercury toxicity, including neurological problems and death.

Between 1953 and 1960, an industrial plant in Minamata, Japan, dumped industrial wastes containing high levels of mercury into a local embayment. The mercury readily formed organic complexes that entered the food chain and were concentrated in the shellfish and fish that the people of the town used as a major food source. Many cases of severe mercury poisoning and even death resulted. The damage to the nervous system caused by the mercury was particularly noticeable in children whose mothers had consumed contaminated shellfish or fish while pregnant. Sadly, this is not an isolated incident; bottom fish in all of the estuaries of the United States have some amount of toxic materials. Shellfish concentrate toxic materials such as mercury, lead, and cadmium to several thousand times their concentrations in the surrounding water. Scallops concentrate cadmium to 2 million times that of its concentration in the water, and oysters concentrate DDT to more than 90,000 times that of its concentration in water. This contamination of the food supply puts large numbers of people at risk.

Biological magnification is the concentration of toxins in higher trophic levels of a food chain.

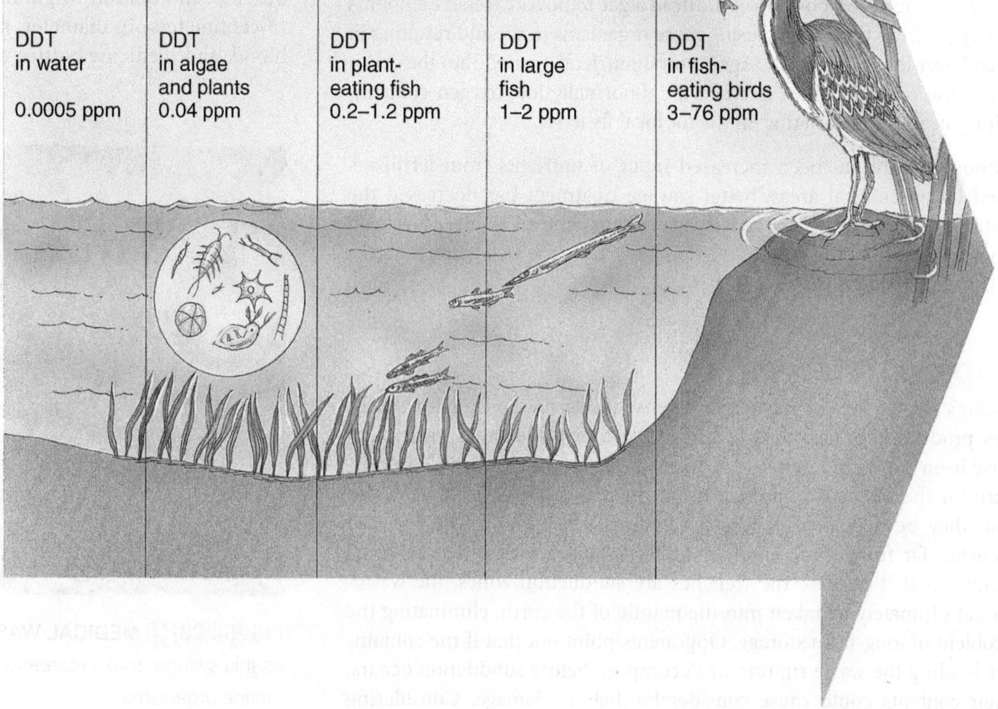

FIGURE 20-16 BIOLOGICAL MAGNIFICATION. Toxic substances can accumulate in the tissues of consumers as they feed on organisms that contain low levels of toxic material. Notice how the level of DDT in this example becomes more concentrated in the tissues of organisms as you move up the food chain. Ultimately, the toxic substance will reach a lethal level and cause the death of a higher-order consumer.

DDT in water	DDT in algae and plants	DDT in plant-eating fish	DDT in large fish	DDT in fish-eating birds
0.0005 ppm	0.04 ppm	0.2–1.2 ppm	1–2 ppm	3–76 ppm

© Cengage Learning

IN SUMMARY

- Almost from the beginning of human history, people have dumped pollutants into the ocean such as garbage, sewage, and industrial wastes.

- The indiscriminate dumping has decreased the economic and recreational value of some beaches and in some instances has posed health hazards.

- Agricultural and urban runoff contains a variety of pesticides, as well as human and industrial wastes.

- Some toxic materials can combine with bottom sediments, increasing the length of time that they remain in the environment and contaminate the water.

- Legislation and the use of less-toxic materials such as unleaded fuel are gradually decreasing the level of pollution in some regions.

- Not all pollutants that enter the marine environment dissolve in the water. Some are absorbed by particles, consumed by animals, and channeled through food chains. (Have You Wondered? #1)

- Large marine animals such as birds, reptiles, and mammals are most affected by plastic trash. (Have You Wondered? #2)

- Plastic traps for catching fish, crabs, and lobsters continue to trap and kill animals even after they are lost or discarded. (Have You Wondered? #2)

- There have been numerous examples of severe environmental damage due to oil spills and the damage can persist for many years.

20-2 Climate Change

Increased amounts of greenhouse gases in the atmosphere, especially carbon dioxide, during the last 100 years have caused the surface air temperature of the planet to increase about 0.74°C (1.33°F), and the temperature is continuing to rise. A similar increase in the surface temperature of the ocean has also been documented. Climate change resulting from the effects of this global warming affects both biological processes and the diversity and distribution of species in the ocean. The rise in sea temperature affects the distribution of marine species, the oxygen content of the water, the health of coral reefs, and wind and rainfall patterns. Such changes frequently have unpredictable effects on an ecosystem already under heavy pressure.

CHANGES IN SPECIES DISTRIBUTION

Water temperature in combination with other environmental factors plays a major role in determining the distribution of species in the ocean, and we are beginning to see changes in species distribution as a result of climate change. Rising temperatures are affecting the temperature gradient of seawater from the equator to the poles, with water at higher latitudes becoming increasingly warmer **(Figure 20-17)**. Scientists predict that this change in water temperature will tend to move temperate species ranges toward the poles. For example, as warmer water extends farther north, the range of warm temperate species such as the Venus clam (*Mercenaria mercenaria*) and eastern oyster (*Crossostrea virginica*) has begun to extend farther north as well. The range extension could result in increased competition for native species, especially those that are not able to tolerate warmer water temperatures. This could also spread disease, placing more pressure on indigenous populations **(Figure 20-18)**. Changes in range could lead to widespread species extinction. This is especially likely if a species has limited dispersal capabilities and is not able to relocate to regions of suitable temperature, if there is not sufficient habitat in the extended region to support native species and those increasing their range, or if native species succumb to introduced diseases. Temperature changes may also extend the range of nonnative species (see the following section) that, through competition or predation, may decrease native populations or drive them to extinction.

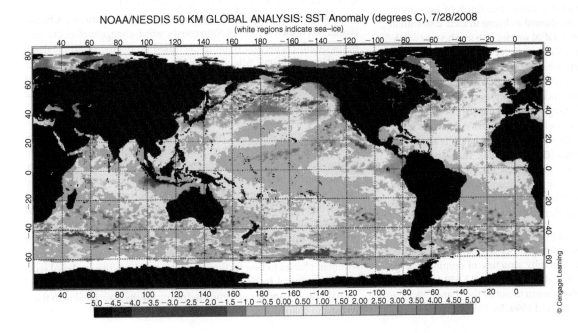

NOAA/NESDIS 50 KM GLOBAL ANALYSIS: SST Anomaly (degrees C), 7/28/2008
(white regions indicate sea–ice)

© Cengage Learning

FIGURE 20-17 CHANGING SEA TEMPERATURES. This map shows the increases in sea temperature that are occurring as a result of global warming. Notice how warm water is extending toward the poles.

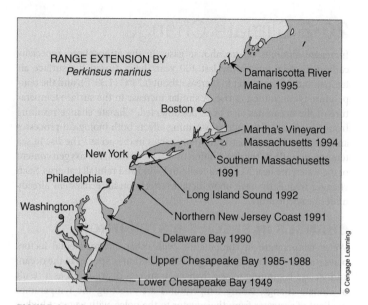

RANGE EXTENSION BY
Perkinsus marinus

Damariscotta River
Maine 1995

Boston

Martha's Vineyard
Massachusetts 1994

New York

Southern Massachusetts
1991

Philadelphia

Long Island Sound 1992

Washington

Northern New Jersey Coast 1991

Delaware Bay 1990

Upper Chesapeake Bay 1985-1988

Lower Chesapeake Bay 1949

© Cengage Learning

FIGURE 20-18 **RANGE EXTENSION CAUSED BY GLOBAL WARMING.** Warming of the ocean has allowed oysters (*Crassostrea virginica*) to move northward from their normal range around Chesapeake Bay. As the population has spread northward, it has also spread the oyster parasite *Perkinsus marinus*, a protist that causes a bivalve disease called *Dermo* in oyster populations.

Climate change can strongly influence the distribution and abundance of fishes, timing of spawning, growth rate, and larval survival. Increased water temperature can affect fish populations by producing changes in lower trophic levels, thus limiting the amount of food available to higher trophic levels. Oceanic "fronts" where cold and warm water masses collide concentrate food for large predators such as bluefin tuna and blue whales. As the ocean warms and water temperatures become more uniform, fewer of these fronts will occur. This will cause a shift in the food supply and decline in the species that rely on them. These changes may have significant effects on the nature and value of commercial fisheries, most of which consist of fishes at higher trophic levels.

A study conducted in 2005 found that almost two-thirds of fish species studied in the North Sea (21 of 36) showed a change in their distribution in response to warming of the water. Of 20 species with southern or northern range limits in the North Sea, the boundaries of the ranges of half of those species moved significantly with warming of the water. Previous studies off the eastern United States have shown that fishes with the most temperature-sensitive distributions included key prey species of predators that were not able to change their range. Since the majority of commercial fishes are predators, such a change could lead to dramatic decreases in the population, due to lack of food. Increased sea temperature is already affecting the beleaguered Atlantic cod (*Gadus morhua*). The warmer ocean temperature is causing the plankton on which the young cod feed to bloom too far to the north or too early in the season, when immature cod are not big enough to eat them. These changes, along with overfishing, have pushed the cod fishery in the North Sea to the brink of collapse.

Climate change is not just affecting pelagic ecosystems. The effects have also been documented in benthic communities at great depths. For instance, a major change in the community structure of dominant, large epibenthic organisms was observed at 4,100 meters (13,530 feet) depth in the northeast Pacific. Researchers began to notice the change after the El Niño that occurred between 1997 and 1999. By 2002, the populations

of two species of sea cucumber had decreased two to three orders of magnitude, whereas other echinoderm populations increased by one to two orders of magnitude.

CHANGES IN OXYGEN LEVELS

As you learned in Chapter 2, increased temperature decreases the amount of oxygen that will dissolve in water, thus lowering the oxygen-carrying capacity of the water. On the other hand, increased temperature increases the metabolism of most marine organisms, thus increasing their need for oxygen. This combination can lead to the death of marine organisms that are not able to move to locations with sufficient amounts of oxygen. Lower oxygen-carrying capacity combined with eutrophication will increase the likelihood of more oxygen-depleted dead zones, especially in semi-enclosed areas such as embayments. An increase in the number and size of dead zones will have an adverse affect on coastal fisheries. The effects of temperature on fisheries is of particular concern because stocks of many commercial species are already critically low as a result of overfishing. In the coastal waters of the United States, a reduction in the amount of oxygen-rich cool water is already beginning to change the distribution of striped bass (*Morone saxatalis*).

EFFECTS ON CORAL REEFS

Coral reefs provide ecosystem goods and services worth more than $375 billion to the global economy each year. But as you learned in Chapter 15, the world's coral reefs are in serious decline as a result of a number of environmental pressures, including climate change, which is thought to be a major cause of the coral bleaching that threatens their extinction. Another effect of climate change is melting of glaciers and polar ice. This raises the sea level and adds more freshwater to the sea. A rise in sea level could be devastating to corals that are already being stressed by other environmental changes and disease. On average, an upward growth of about 1 centimeter per year is enough to keep the coral within the sunlit waters they require. If water levels rise more quickly, the corals might not be able to keep pace. Without sufficient sunlight, the corals' zooxanthellae would decrease productivity or die, resulting in widespread death of coral reefs.

CHANGES IN RAINFALL AND WINDS

Increasing temperatures can affect global weather patterns. Climate change has been linked to El Niño events, which change surface currents in the South Pacific and result in decreased productivity and the death of marine organisms. Scientists predict that there will probably be an increase in El Niño–like conditions in the tropical Pacific as a result of climate change brought on by global warming. Worldwide changes in rainfall will also occur, with some areas receiving more annual rainfall and others less. Changes in rainfall will affect estuaries and coastal seas that receive river input and runoff from the land. An increase or a decrease in the influx of freshwater will alter the salinity of these important marine ecosystems, putting stress on estuarine organisms. In addition, the melting of polar ice and resulting increase in ocean levels could flood barriers that protect estuaries, increasing the input of saltwater and destroying these ecosystems. Increased rainfall combined with tidal surges can increase coastal flooding, affecting shorebirds and intertidal organisms. An increase in the rate and volume of water discharged from rivers may alter the position of density gradients and may cause coastal sediment plumes to flow further along the coastline. The increased sediment contributed by the plume will disrupt benthic communities and coastal fisheries and cause some marine organisms associated with these systems to change their distribution accordingly.

Another effect of climate change is an increase in the intensity of wind stress acting on the ocean surface along shorelines. This can lead to an increase in the intensity of coastal upwellings. Since the world's most productive fisheries occur in upwelling areas, it is highly likely that climate change will impact them. Along the West Coast of the United States, changes in the coastal upwelling have already supported higher than normal productivity levels that have led to an anoxic event resulting in mass kills of commercially important fish and shellfish. Changes in climate can lead to changes in the wind patterns that drive ocean currents. Ocean currents transport larvae, and changes in these currents can result in larvae no longer being transported to habitats suitable for further development and an overall decline in the species affected. Ocean currents also carry plankton and other important food organisms. A change in the strength and pattern of currents caused by changes in wind patterns can disrupt food webs, causing changes in the distribution of, and even death of, marine organisms.

IN SUMMARY

- During the past 100 years, surface air temperature and ocean temperatures have been rising faster than normal.

- The increase in temperature is mainly the result of human activities that increase the levels of greenhouse gases, especially carbon dioxide.

- The global warming and related climate change is having a negative impact on the biological processes and species diversity of the ocean. (Have You Wondered? #3)

- Decreased oxygen-carrying capacity caused by increased water temperature is driving some species to different habitats and creating dead zones along coastlines.

- Climate change is a major cause of coral bleaching, and the melting of glaciers and polar ice is raising sea levels, placing more stress on coral and coastal ecosystems. (Have You Wondered? #3)

- Changes in rainfall and winds are altering the nature of coastal ecosystems and the pattern of ocean currents that are so important in distributing larvae and carrying food.

20-3 Introduction of Nonnative Species

Nonnative species (also known as **alien** or **exotic species**) are species that have been introduced to a given geographical region. The introduction of nonnative species, accidentally or on purpose, frequently places pressure on native populations, driving some to the brink of extinction. Some introduced species are successful in their new environments because they are competitively superior or because they have no natural predators. Many introduced species are considered pests because they prey on commercially important species.

INTRODUCTION OF NONNATIVE SPECIES BY SHIPS' BALLAST

Nonnative species are frequently transported in a ship's ballast. In the days of sailing vessels, rocks were commonly used for ballast. The rocks taken from shorelines would contain the young and adults of many species.

When these rocks were replaced at distant ports, new species would be introduced. The European green crab (*Carcinus maenas*; **Figure 20-19a**) that has become established on the East Coast of the United States and in Australia is native to the North Atlantic coast of Europe and the coast of North Africa. It is believed to have been introduced in the early 19th century, most likely in the dry ballast of cargo ships. It is now found from Delaware to Nova Scotia and is the most common species of crab in many parts of its range. The European green crab is a voracious predator that feeds on clams, oysters, snails, and other crabs. It competes with native fishes, birds, and humans for food and is frequently cited as the cause of the collapse of Maine's soft shell clam industry. Modern vessels use water for ballast, and nonnative species are introduced when a ship's ballast water is discharged at port. The ballast water contains the larvae and planktonic adults of the area where the ballast was loaded and introduces these into new waters when the ballast is discharged. This is how the ctenophore *Mnemiopsis leydii* was introduced into the Caspian Sea and Black Sea.

> **Nonnative species (alien or exotic species)** are species that have been introduced to an area from another geographical region.

INTRODUCTION OF NONNATIVE SPECIES IN AQUACULTURE

Humans have deliberately introduced nonnative species into new geographical areas for many years in the process of aquaculture. Introduced species can have severe ecological consequences, and the financial implications for the aquaculture industry and commercial fisheries can be significant. They create problems by competing with native species, increasing predation pressure by preying on native organisms, and introducing diseases into native populations. In bivalve aquaculture, for instance, introduced species of bivalves may also introduce bivalve competitors such as slipper limpets; predators such as oyster drills; and diseases that could decimate native bivalves. In 2003, the United States considered introducing the Asian Suminoe oyster (*Crassostrea ariakensis*) to help revive the collapsed oyster fishery of the Chesapeake Bay. Scientists opposed this option, citing our limited knowledge of the species and the potential for irreversible consequences such as the introduction of disease.

In the 1970s, the alga *Eucheuma* (**Figure 20-19b**) from the Philippines was imported to Kaneohe Bay, Hawaii, for the purpose of studying its potential to be cultivated as a source of the food additive carrageenan. Methods developed during the study made *Eucheuma* the most widely farmed seaweed, cultivated in 23 countries. Since its introduction, the alga has spread across Kaneohe Bay and now covers half of the fringe reef, where it smothers and kills the coral. The state of Hawaii and several agencies have teamed up to use barge-mounted vacuum cleaners capable of clearing as much as 350 kilograms (159 pounds) of algae an hour to remove the introduced species. Scientists are hoping that this technique will buy them time until they can develop biological controls with the native sea urchin *Tripneustes gratilla*. Unfortunately, most *Eucheuma* farming takes place in remote areas where little is known of the extent of the damage. A similar situation has occurred in India, where the alga *Kappaphycus alvarezii*, introduced from the Philippines, has invaded coral reefs in a marine reserve in the Bay of Bengal. The red alga was imported in 1996 to help impoverished coastal farmers paid to cultivate it for the food additive carrageenan.

ACCIDENTAL INTRODUCTION OF NONNATIVE SPECIES

Some exotic species are released by hobbyists when they tire of their saltwater aquaria, or are introduced when accidents occur at private or public aquariums. Since 2000, lionfish (*Pterois volitans* and *Pterois miles*;

Figure 20-19c) from the Indo-Pacific region have been observed in coral, rocky, and artificial reefs along the southeastern coast of the United States and in the Bahamas. It is believed that they were introduced into the Atlantic in 1992, when Hurricane Andrew shattered a private aquarium and six fish spilled into Miami's Biscayne Bay. As of 2011, lionfish have been documented from Massachusetts to Florida and throughout the Caribbean as far south as the northern coast of South America. They do not appear to survive the cold winter temperatures to the North, and their continued presence is due to fish from the south moving north during the summer months. Scientists expect that the lionfish will continue to spread throughout the Caribbean Sea and Gulf of Mexico. Lionfish have no natural enemies in this region, and they compete with native species for food and shelter. They also prey on the young of many species, adding pressure to populations that are already stressed by pollution, climate change, and overfishing. This fish also has venomous spines and thus poses a risk to humans, especially anglers, snorkelers, and divers. Scientists at the National Oceanographic and Atmospheric Agency (NOAA) have been conducting research on the invasive lionfish to better understand its natural history, environmental tolerances, and genetics, that will be used to determine how to mitigate potential ecosystem and fisheries damage.

An **urchin barren** is a community dominated by sea urchins.

COMMUNITY CHANGES

Nonnative species can alter ecosystem structure and function and cause substantial economic losses. They are considered one of the largest threats to global biodiversity because of their negative impact on native populations. The introduction of nonnative species can change the structure of entire communities. Before the 1970s, the climax community in the shallow subtidal region of the Gulf of Maine was composed of *Laminaria* kelp beds with an understory of tree-like red algae. In the 1980s, a population explosion of the green sea urchin (*Strongylocentrotus droebachiensis*) decimated much of the kelp and created an alternate community, known as an **urchin barren,** dominated by urchins. Beginning in 2000, a new community has been observed in former urchin barrens and kelp beds. This community is principally composed of introduced species that include algae (*Codium fragile tomentosoides, Polysiphonia harveyi, Bonnemaisonia hamifera*), bryozoans (*Membranopora membranacea*), and tunicates (*Diplosoma listerianum, Botrylloides violaceus, Styela clava*), as well as opportunistic species such as blue mussels (*Mytilus edulis*) and the brown alga *Desmarestia aculeata*. Recent studies suggest that these introduced species are becoming the dominant organisms and altering the composition and structure of the subtidal community to the detriment of native species.

IN SUMMARY

- Nonnative species are species that have been introduced into a given geographical area where they did not previously exist.

- The introduction of nonnative species places pressure on native populations and can drive native species to extinction. (Have You Wondered? #4)

- Introduced species can thrive in new habitats because they are superior competitors or because they have no natural predators.

- Nonnative species can alter ecosystem structure and function and cause substantial economic losses. (Have You Wondered? #4)

(a)

(b)

(c)

FIGURE 20-19 NONNATIVE SPECIES. (a) The European green crab was introduced in the early 1800s and now occupies an extensive range in which it competes with other organisms, including humans, for food. **(b)** The alga *Eucheuma* was imported from the Philippines for aquaculture. It has now spread to neighboring reefs where it is killing the coral. **(c)** Lionfish were accidentally introduced into the Atlantic in 1992. They swiftly spread and now compete with native species for food and habitat.

20-4 Coastal Development

The marine environment is also being damaged by the destruction of habitats. As more and more coastal land is developed for housing and recreation, more habitats are lost and so are the species they support. Many of the most vulnerable habitats are nurseries for commercial species of fish and shellfish that are close to shore. Destruction of these areas not only threatens the survival of many ocean creatures but also aggravates an already bad situation by contributing to the decline in populations of commercially valuable species.

WETLANDS

Wetlands such as salt marshes provide nutrients, shelter, and spawning areas for a variety of marine organisms, including crabs, shrimp, oysters, and many commercial fishes. In the past, these areas have been drained, filled, and dredged to provide more ground for industry, channels into ports and harbors for large vessels, and beachfront real estate for the growing number of individuals who desire to live and vacation near the sea (Figure 20-20). It is estimated that in 1776 the United States had between 125 and 215 million acres of wetlands. By 1975, the amount of wetlands had decreased to 99 million acres. It is estimated that 18,000 acres of wetlands were lost annually between 1950 and 1970. The loss of estuarine wetlands has been greatest in Florida, California, Texas, New Jersey, and Louisiana. By 2000, more than half of the coastal wetlands in the contiguous 48 states had been destroyed, and the other half is threatened as demand increases for residential and recreational development along the coasts.

We now recognize that wetlands, including estuaries, are among the most productive of marine environments and that they play a major role as a nursery for a number of commercial fish and shellfish. In the 1970s and 1980s, increased public awareness led to the passage of new laws protecting wetlands in many states and passage of stricter federal regulations. Legislation closely regulates any development or modification of wetlands. Whenever possible, these valuable areas are being carefully managed and restored to their natural state. If estuarine or wetland areas must be altered, the current policies demand that other, damaged areas be restored to replace the lost wetlands. Although this practice sounds good in principle, in the few instances where it has actually been tried it has generally been unsuccessful because of loopholes in the legislation. To further complicate matters, the federal government, in response to pressure from special interest groups, continues to change its definition of what constitutes a protected wetland and what it means to restore an area. Changes such as these severely weaken the laws that were meant to protect these resources. A successful exception is the current restoration of wetlands in the San Francisco Bay area. Environmentalists are cooperating with federal and local governments, as well as local industries, to make their plan for restoring wetlands work. However, wetlands along the West Coast are still under heavy pressure for industrial and residential development, and net loss of these valuable areas continues.

BEACHES

Coastal areas are popular vacation and retirement destinations. Travel brochures display pictures of beautiful, immaculate, sandy beaches backed by vigorous shoreline vegetation (Figure 20-21a). Few show the reality—overcrowded beaches backed by hotels, resorts, and shops (Figure 20-21b). Development of the seashore for recreational and commercial use, combined with intensive seasonal use, has had a severe impact on intertidal areas.

FIGURE 20-20 WETLAND DESTRUCTION. This shoreline development at Barnegat Bay, New Jersey, was once a productive wetland that was drained and filled to allow for real estate development.

Direct Effects of Beach Use and Development on Marine Life

Heavy beach use has seriously affected intertidal wildlife, especially along sandy shores. Beach nesting birds such as the piping plover (*Charadrius melodus*) and the least tern (*Sterna albifrons*) are in danger of extinction along the U.S. Gulf Coast because of disturbances by bathers, pets, and dune buggies. Other shorebirds are subjected to competition for nesting sites and egg predation from growing populations of several species of large gulls. Gulls are highly tolerant of humans and can thrive on the garbage humans generate. Populations of sea turtles and horseshoe crabs are also declining because of a loss of nesting sites on sandy beaches and predation by domestic animals. The construction of beachfront houses, docks, and seawalls disrupts habitats and makes them unavailable for nesting sites, reproduction, and feeding. Some coastal communities are setting aside stretches of beach for bird and turtle nests, levying fines for littering beaches, and participating in projects such as the Coastal Cleanup project mentioned at the beginning of this chapter.

BEACH EROSION

Development of beaches for recreational use frequently destroys sand dunes and beach vegetation that prevents beach erosion, and along with too much human traffic, causes significant beach erosion. Ironically, vacation areas often spend hundreds of thousands of dollars each year to refortify beaches destroyed by the previous season's tourism.

> **Longshore currents** are currents moving parallel to shore in the surf zone.
>
> The **longshore transport process** is a process driven by longshore current that carries sediment along the shoreline.

Interfering with Natural Processes

Human interference with natural processes has a pronounced effect on beaches. When rivers are dammed to control floods or to produce power, sand and gravel that were once deposited on the coast is instead deposited in the lake behind the dam. The sand and gravel no longer replaces sediments removed by natural erosion processes.

Engineering projects such as breakwaters and jetties in coastal zones also produce changes in beaches. Breakwaters and jetties protect coastal areas from wave action. The areas behind these structures are quiet water. Sediments settle to the bottom in the quiet water rather than being carried by longshore currents to other beaches. **Longshore currents** are generated by waves that break at an angle to the beach. These currents move parallel to the shore in the surf zone, carrying sediments, in what is known as the **longshore transport process**.

FIGURE 20-21 BEACH DEVELOPMENT. (a) Travel brochures entice vacationers with pictures of beautiful sandy beaches and aquamarine water. **(b)** In reality, the beach is a small strip of sand crowded with tourists and surrounded by hotels, condominiums, shopping areas, and restaurants.

Historically, coastal engineering projects trigger chain reactions. Facilities constructed in one area create ecological problems in another that must be solved by other engineering projects, which create problems of their own, and so on. According to the U.S. Coastal Survey, 43% of our national shorelines, excluding Alaska, are losing more sediments than they receive. In other words, our beaches are disappearing into the sea. In some areas, eroding beaches are maintained by bringing in sand and gravel from inland areas or dredging it from offshore sandbars. These programs are expensive but necessary if the beaches are to be preserved.

IN SUMMARY

- The heavy development of coastal regions has severely damaged habitats, especially wetlands and sandy beaches.

- The large numbers of humans, along with their pets, that visit sandy beaches annually destroy habitats, kill marine organisms, and pollute the environment.

- Destruction of wetlands results in decreases in ocean productivity and in the populations of many important commercial species that rely on these areas as nursery grounds.

- Human interference with longshore transport processes are contributing to beach erosion.

20-5 Epilogue

The basic mechanisms of evolution and ecology are inextricably interrelated, and human activity changes those interactions. The scope and pace of the changes caused by humans are very different from those of nature. Nature works by means of small changes over long periods of time. Pollution, oil spills, climate change, introduced species, coastal development, and other human changes frequently involve entire biological communities and sometimes occur instantaneously. The result is widespread disruption to and damage of marine ecosystems. The environmental damage that results is significant not only because of the damage done this month or this year but because of the long-term changes in ecological relationships and disappearing niches that lead to long-term changes in evolutionary patterns and ultimately species extinction. Although all of the various regions of the sea are unique, the knowledge that we gain from studying one area helps us to better understand others. An understanding of the underlying patterns and processes in the sea will help us to judge the impact of our actions and help us determine just how much an area can be modified without jeopardizing its environmental or economic value. Humans will continue to interact with the sea, especially along coastal areas, and problems caused by this interaction will continue to occur. We owe it to ourselves and our children to make the most intelligent and knowledgeable use of these resources. With proper care, these resources can renew themselves and continue to be a source of delight, amazement, and sustenance.

Although the picture of our oceans presented in this chapter is not a pretty one, the situation is not hopeless. Many groups and individuals are working hard to preserve our marine resources. If you would like to know more about current efforts to save the ocean and its creatures, or how you as an individual can make a difference, here is a list of some organizations you can contact:

- Cousteau Society, 930 West 21st St., Norfolk, VA 23517. This group encourages protection and preservation of the oceans for future generations. Student volunteers are welcome to work at their Norfolk office. Students might wish to participate in Project Ocean Search, in which people from a variety of backgrounds spend 2 weeks studying an island ecosystem. Website: http://www.cousteau.org/?sPlug 1.

- Greenpeace USA, Inc., 1432 U St. NW, Suite 201-A, Washington, DC 20009. This international organization focuses on ocean ecology, hazardous wastes, disarmament, and atmospheric pollution issues. Website: http://www.greenpeace.org.

- League of Conservation Voters, 1150 Connecticut Ave. NW, Suite 201, Washington, DC 20036. This nonpartisan political action group works to elect pro-environment candidates to Congress. They also publish a scorecard that rates current members of Congress on their environmental votes. Website: http://www.lcv.org.

- National Wildlife Federation, 1400 16th St. NW, Washington, DC 20036. Conservation education is the primary mission of this group. Website: http://www.nwf.org/.

MARINE BIOLOGY & THE HUMAN CONNECTION

Effects of Artificial Processes on Beach Formation

An example of the effects of interfering with the natural processes of a coastal zone can be seen in the Santa Barbara Harbor project in California. In this section of the California coast, the longshore current transports sediment down the coast from north to south **(Figure 20-A)**. To supply the needs of recreational boaters, a jetty and breakwater were constructed to form a boat harbor. The jetty is on the north side of the harbor and continues out to sea before turning south to

form a breakwater. This structure not only shelters the area from waves but also blocks the longshore current. Thus more sand was deposited to the north of the structure than before the jetty was built, and the beach began to grow, whereas beaches south of the structure began to disappear. Eventually, the beach to the north grew until it reached the seaward limit of the jetty, and the longshore currents could again move sand to the south. Instead of carrying sand to more southerly

beaches, however, the current deposited sand at the end of the breakwater, forming a sand spit that began to fill in the harbor, and the southern beaches continued to erode. As a solution, a dredge pumps the sand deposited in the harbor back into the longshore current south of the harbor, which carries it to southern beaches. In this case, interference with a natural process resulted in the expenditure of large amounts of time, money, and energy to re-create a once natural process.

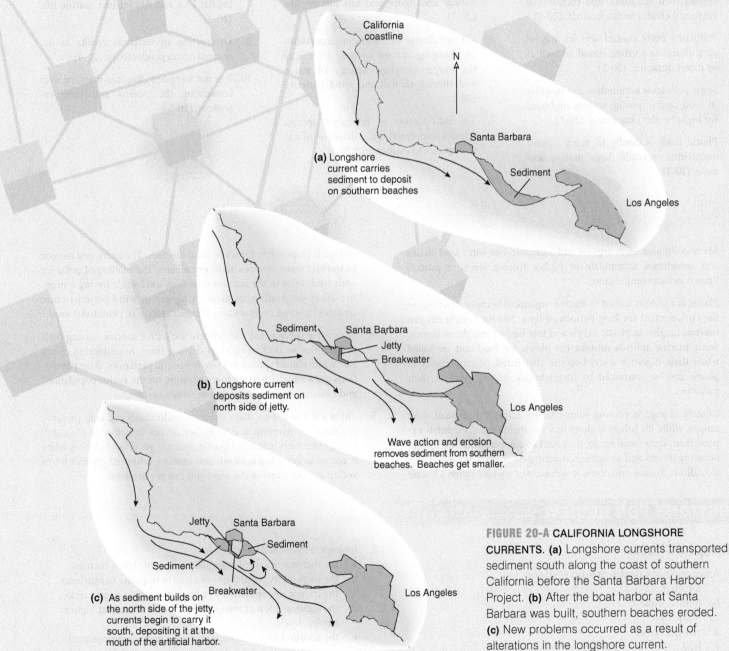

(a) Longshore current carries sediment to deposit on southern beaches

(b) Longshore current deposits sediment on north side of jetty.

Wave action and erosion removes sediment from southern beaches. Beaches get smaller.

(c) As sediment builds on the north side of the jetty, currents begin to carry it south, depositing it at the mouth of the artificial harbor.

FIGURE 20-A CALIFORNIA LONGSHORE CURRENTS. (a) Longshore currents transported sediment south along the coast of southern California before the Santa Barbara Harbor Project. **(b)** After the boat harbor at Santa Barbara was built, southern beaches eroded. **(c)** New problems occurred as a result of alterations in the longshore current.

- The Nature Conservancy, 1815 North Lynn St., Arlington, VA 22209. This group fosters global preservation of natural diversity by finding, protecting, and maintaining the best examples of communities, ecosystems, and endangered species in the natural world. Website: http://www.nature.org/.

- Ocean Conservancy, 2029 K St. NW, Washington, DC 20006. This group works to motivate individuals worldwide to work to restore the world's oceans. Website: http://www.oceanconservancy.org.

- World Wildlife Fund, 1250 24th St. NW, Washington, DC 20036. This group strives to save endangered species and acquire wildlife habitats. Website: http://www.panda.org/.

IN SUMMARY

- A better knowledge of the marine environment will help us to better preserve marine ecosystems.

- Concerned citizens can find numerous ways to become involved in protecting the marine environment. (Have You Wondered? #5)

KEY CONCEPTS

1. Dumping wastes into coastal seas decreases their economic and recreational value and creates health hazards. (20-1)

2. Pollutants enter coastal seas by way of agricultural and urban runoff, as well as by direct dumping. (20-1)

3. Some pollutants accumulate and magnify in food chains, posing serious problems for higher-order consumers. (20-1)

4. Plastic trash is deadly to many marine organisms, especially large marine animals. (20-1)

5. Oil spills damage significant amounts of habitat and injure and kill marine life. (20-1)

6. Climate change affects marine ecosystems by changing species ranges, decreasing the oxygen-carrying capacity of water, and altering rainfall and wind patterns. (20-2)

7. The introduction of nonnative species stresses communities and drives some native species to extinction. (20-3)

8. Development of coastal areas leads to habitat loss and diminished marine life. (20-4)

9. Destruction of wetlands results in decreased ocean productivity. (20-4)

10. It is not too late to become involved with conserving the oceans and their resources. (20-5)

ARE YOU STILL WONDERING?

1. Many pollutants, both organic and inorganic, can enter food chains and sometimes accumulate in higher trophic levels, a process known as biomagnification.

2. Plastic is a serious hazard to marine organisms because it persists in the environment for long periods of time. Marine vertebrates can become tangled in plastic debris and lose limbs, strangle, or drown. Some marine animals mistake the plastic for food and are killed when their digestive tracts become obstructed. Microparticles of plastic can be consumed by invertebrates, killing them or their predators.

3. Climate change is causing some marine animals to expand their ranges, while for others their ranges are shrinking. As animals expand from their usual range, they start to compete with the indigenous organisms and sometimes outcompete them. Climate change also allows disease organisms to spread over a wider range. Climate change is responsible for some coral diseases. The decreased oxygen in warmer water stresses many organisms. The melting of polar ice adds freshwater to the sea and raises sea level while having a negative effect on Arctic ecosystems. Changes in wind patterns cause changes in ocean currents and the distribution of planktonic food.

4. Nonnative species may outcompete the native species causing their populations to decline or even driving them to extinction. Since many introduced species are free of natural predators, their populations grow rapidly putting more pressure on the native populations and causing changes in ecosystem structures.

5. There are many ways that concerned individuals can help preserve the marine environment even if they do not live close to the coast. Using cloth bags instead of plastic, making good food choices when it comes to selecting seafood, and making informed choices when voting are just some of the ways you can get involved.

QUESTIONS FOR REVIEW

Multiple Choice

1. An example of a persistent toxin would be
   a. fertilizer
   b. human waste
   c. DDT
   d. red-tide toxin
   e. lead

2. Biological magnification refers to
   a. the increase in size of organisms as trophic levels increase
   b. the increase in populations of algae in response to nutrients
   c. the change in population size from generation to generation
   d. the accumulation of toxins in the flesh of animals at higher trophic levels
   e. the accumulation of pollutants in the marine environment

3. Sewage from beachfront cottages usually enters
   a. sewer system
   b. septic tanks
   c. the sea directly
   d. drainage ditches
   e. compost heaps

4. Damage to _____ has the greatest impact on many commercial species of fish and shellfish.
   a. sand beaches
   b. rocky beaches
   c. wetlands
   d. the benthos
   e. the open sea

Matching

Match each of the following pollutants with an associated problem.

a. DDT
b. PCB
c. oil
d. coliform bacteria
e. plastic

1. Hypothermia in marine birds and mammals.
2. Suffocation or loss of limbs.
3. Human disease.
4. Egg shells that easily break.
5. Dangerous levels stored in fish fat.

Short Answer

1. What environmental problems are associated with plastic trash?

2. What are the major sources of oil pollution?

3. What effect does an oil spill have on the ecology of a rocky shore?

4. What activities are most damaging to wetlands?

5. Describe how toxins and other pollutants can enter ocean food chains.

6. List some of the major problems that are associated with the agricultural runoff that enters the ocean.

7. Describe how an oil spill causes injury to birds and mammals.

8. How do changes in wind patterns caused by climate change affect marine organisms?

9. Why does the introduction of a nonnative species frequently result in a decrease in the number of native organisms?

10. What are some ways in which nonnative species are introduced into new environments?

11. Describe how recreational and commercial use of beaches affects beach ecology.

12. What are some causes and consequences of eutrophication?

Thinking Critically

1. Seashell collectors are frequently blamed for decreases in local mollusc populations in Florida and other areas of the world. The collectors state that development and pollution are more to blame than overcollecting. Do you think that the collectors' argument is valid? Explain.

2. What portion of the East Coast of the United States do you think would be more likely to experience algal blooms resulting from agricultural runoff?

3. An industrial waste is being dumped into the ocean off the coast of California, and you are asked to determine if it is accumulating in the aquatic food chains. How might you determine this?

4. You find that a snail from the Philippines has been accidentally introduced in the Bahamas and has spread to the reefs associated with most of the islands. What might be done to control the spread of the nonnative snail and eradicate it?

SUGGESTIONS FOR FURTHER READING

Harrigan, S. 2011. From Relics to Reefs, *National Geographic* 219(2): 84–103.

Harris, L. G., and M. C. Tyrrell. 2001. Changing Community States in the Gulf of Maine: Synergism between Invaders, Overfishing, and Climate Change, *Biological Invasions* 3: 9–21.

Kilbert, E. 2011. The Acid Sea, *National Geographic* 219(4): 82–121.

Rick, T. C, and J. M. Erlandson. 2008. *Human Impacts on Ancient Marine Ecosystems*. University of California Press.

Stap, D. 2002. Living on the Edge: With Wetlands Declining and Shorebirds in Trouble, the Question Is: How Do We Reclaim the Habitat the Birds Need to Survive?, *Audubon* 104(2).

Yasuhara, M., T. M. Cronin, P. B. deMenocal, H. Okahashi, and B. K. Linsley. 2008. Abrupt Climate Change and Collapse of Deep-Sea Ecosystems, *Proceedings of the National Academy of Sciences* 105(5): 1556–1560.

Glossary

abdomen A body region of an animal that corresponds to the belly. In arthropods this region is usually muscular and may contain gills.

abiotic factors (ay-by-AH-tik) The nonliving aspects of an organism's environment.

aboral surface The surface opposite the mouth of, for example, a sea star.

abyssal hill (uh-BIS-suhl) A hill on the ocean floor formed by volcanic activity.

abyssal plain A flat expanse at the bottom of an ocean basin.

abyssal zone The ocean bottom that extends from a depth of 4,000 to 6,000 meters.

acid A compound that releases hydrogen ions when added to water.

acrorhagi (a-kro-RAY-ji) Specialized tentacles found in some anemones that are used to prevent other anemones from getting too close.

actinopod A needle-like pseudopod typical of radiolarians.

active continental margin Location where the leading edge of a continent collides with an oceanic plate. Active continental margins are commonly the sites of such geologic activity as earthquakes, volcanoes, mountain building, and the formation of new rock.

adductor muscle (a-DUHK-tir) A large muscle that closes a bivalve's shell.

adenosine triphosphate (ATP) The major energy-carrying molecule in cells.

adhesion A property of a liquid by which the liquid is attracted to the surface of objects that carry an electrical charge.

adsorption The process by which ions adhere to the surface of an object.

aerenchyme (AIR-en-kime) A tissue in vascular plants that consists of spaces between the cell walls for carrying gases, usually to roots and stems in sediments that lack oxygen; also used to confer buoyancy.

aerial root A root that is exposed to the atmosphere rather than subterranean, such as drop roots, prop roots, and pneumatophores of mangroves.

aerobic organism An organism that requires oxygen to survive.

agar (AH-gar) A red algal phycocolloid that forms thick gels at very low concentrations. It is resistant to degradation by most microorganisms and is used to culture bacteria in microbiology laboratories.

ahermatypic coral A coral that lacks zooxanthellae in its tissues and does not form reefs.

akinetic Descriptive of planktonic organisms that have no recognizable method of movement.

alevins Newly hatched salmon that still have a yolk sac attached. They are known as *fry* after they lose their yolk sacs and begin feeding.

alga Any photosynthetic organism, unicellular or multicellular, in which all cells are photosynthetic.

algal bloom A dramatic increase in the population size of certain photosynthetic plankton; usually associated with the water being enriched with nutrients.

algal ridge The elevated margin of a windward coral reef built by actively growing coralline algae.

algal turf A dense carpet of low-growing seaweeds covering rock or sediment, similar to a closely cropped lawn.

alginate (AL-gen-ate) A class of phycocolloids found in the cell walls of brown algae; confers flexibility and strength to the thallus; harvested for commercial use as a gelling agent.

algologist (al-GAHL-o-jist) A scientist who studies seaweeds and phytoplankton; also referred to as a *phycologist*.

alien species Species that does not naturally occur in a given geographic area. Also known as *nonnative species* and *introduced species*.

aliphatic hydrocarbon Organic compound in which the carbon atoms form straight chains; examples are octane and heptane.

allantois (eh-LAN-tuh-wiss) An embryonic support membrane found in some vertebrates that functions in the elimination of wastes.

alleles Two or more forms of a single gene.

allopatric speciation (al-oh-PAT-rik) The formation of new species that occurs when two or more populations or parts of a large population become geographically isolated from each other.

alternation of generations The succession of two or more multicellular stages, usually sexual and asexual, in the life cycles of seaweeds, plants, and some animals.

alveolates A group of microbes with membranous sacs (alveoli) beneath the cell membrane.

alveolus (plural alveoli) (in animals) Tiny air sacs in the lungs of animals.

alveolus (in alveolates) A membranous sac beneath the cell membrane of a microbe belonging to the group Alveolata.

ambergris (AM-ber-gris) A digestive by-product of sperm whales that is used in making perfume.

ambulacral groove (am-byu-LAK-ruhl) A groove through which the tube feet of an echinoderm extend.

amino acids Molecules that are the building blocks of proteins.

ammocoetes The burrowing freshwater larvae of lampreys.

amnesic shellfish poisoning A human syndrome caused by consumption of shellfish that concentrate the toxic chemicals produced by diatom blooms.

amnion (AM-nee-uhn) A liquid-filled sac that contains the developing embryo of some vertebrate animals.

amniotic egg (am-nee-AH-tik) An egg covered by a protective shell and containing a liquid-filled sac called the *amnion*.

amphiblastula The planktonic larval stage of sponges.

amphipod A type of crustacean with a body similar to a shrimp belonging to the order Amphipoda.

ampulla An expanded area of a tubular structure; in echinoderms, a sac-like structure attached to the tube foot.

ampullae of Lorenzini (am-POOL-ee of lohren-ZEE-nee) Organs in sharks and other cartilaginous fishes that can detect electrical signals in the water.

anadromous (uh-NAD-ruh-muhs) A term applied to fishes that reproduce in freshwater and spend their adult lives in the marine environment.

anaerobic organism An organism that does not use oxygen in its metabolism.

anal fin The ventral unpaired fin of fish that lies behind the anus.

anatomical isolation The prevention of interbreeding between different species in nature due to incompatible copulatory organs.

anchor root A subterranean root that branches from a major root type to secure the plant in the sediment.

animal A multicellular organism that has cells lacking cell walls, that is an ingestive heterotroph, and that can usually move when adult.

annelid (AN-eh-lid) A segmented worm belonging to the phylum Annelida.

Antarctic Circle The latitude 66.5 degrees S of the equator, which marks the southernmost limit of sunlight at the June solstice.

Antarctic circumpolar current The largest of the ocean currents, it flows continuously eastward around the continent of Antarctica.

anthozoan (an-thuh-ZOH-uhn) An animal belonging to the cnidarian class Anthozoa, which includes corals and sea anemones.

aperture An opening, such as the opening into a gastropod's shell.

aphotic zone (ay-FOH-tik) The portion of the pelagic division where sunlight is absent.

aposematic coloration Bright and conspicuous coloration exhibited by animals that are distasteful or dangerous to their attackers. Such colors are presumed to warn away attackers. Also known as *warning coloration*.

appendicular swimmer An organism that swims with legs, fins, wings, or other extensions from the body.

aquaculture The farming of freshwater or marine organisms.

Archaea One of the three domains of life. It contains prokaryotes that lack molecules known as *peptidoglycans* in their cell walls and can tolerate extreme environmental conditions.

archaeocyte Cells in a sponge's body that resemble amoebas and that can form any cell type in the sponge.

archaeon Member of the domain Archaea, distinguished from bacteria by its biochemical and genetic makeup and especially noted for its production of methane and tolerance of extreme environmental conditions.

Arctic Circle The latitude 66.5 degrees N of the equator, which marks the northernmost limit of sunlight at the December solstice.

Aristotle's lantern A chewing structure composed of five teeth found in the mouths of sea urchins.

armored dinoflagellate Dinoflagellate with thick layers of cellulose in its alveoli, giving the appearance of having a protective cell covering.

aromatic hydrocarbon Organic compound in which the carbon atoms form ring structures; examples include benzene and naphthalene.

arrowworm A planktonic animal belonging to the phylum Chaetognatha.

arthropod (AR-thruh-pahd) An animal belonging to the phylum Arthropoda. Arthropods are characterized by jointed appendages and a hard exterior covering (exoskeleton).

artificial selection The process by which farmers and animal breeders select only animals and plants with certain desirable traits for breeding in an effort to produce more organisms with the same desirable traits.

ascocarp The fruiting (spore-producing) body of a member of the fungal group Ascomycota.

asconoid (AS-kuh-noyd) A type of sponge body that has only a single spongocoel that does not contain invaginations.

ascospores Haploid cells produced within an ascocarp that disperse and become new adult fungi.

ascus One component of a fungal ascocarp that produces four or eight spores.

asexual reproduction The process whereby offspring are produced from a single parent without the fusion of sex cells.

aspergillosis A fungal disease caused by the genus *Aspergillus*.

asthenosphere (as-THEN-uh-sfeer) The region of mantle that lies below the earth's crust.

atoll (a-TOHL) A coral reef somewhat circular in shape, with a centrally located lagoon.

ATP See *adenosine triphosphate*.

atrium One of the chambers of the vertebrate heart; it collects blood from the body and acts as a primer pump.

autotomize To purposely cast off a body part.

autotroph (AW-toh-trohf) An organism that is capable of producing its own food. Also known as a *producer*.

auxospore A growth stage in the life cycle of a diatom that serves to restore the maximal size of the cell.

avoidance A mechanism to reduce predation by placing an organism in a different dimension in time or space than the predator.

bacillus A bacterium with a cell shaped like a rod.

backing down A procedure used in tuna fishing to save dolphins, in which a skiff draws a purse seine halfway toward the purse seiner, then backs up when the tuna are close to the boat and the dolphins are at the edge of the net, causing the edge of the net to go slack and drop beneath the surface of the water, allowing the dolphins to escape.

backwash Water flowing down the beach after a wave breaks.

bacterial loop A subcomponent of a food web in which bacteria convert dissolved and particulate organic matter into bacterial biomass, available for consumption by suspension and deposit feeders; particularly applied to bacteria that use dissolved organic material (DOM) released by phytoplankton.

bacteriochlorophylls A class of primary photosynthetic pigments that do not release oxygen; found mostly in purple and green photosynthetic bacteria.

bacteriophage A virus, often called simply *phage*, that infects a bacterium.

bacterioplankton Prokaryotic members of the plankton.

bacteriorhodopsin A class of light-capturing proteins that serve in ATP production in the Halobacteria.

baleen (buh-LEEN) A proteinaceous structure used to strain food from the water; it takes the place of teeth in baleen whales.

bank reefs Isolated flat-topped reefs that are found growing on midshelf regions of islands or continental shelves.

barbel (BAHR-behl) A fleshy projection on the head of some fishes that usually fulfills a sensory role. Barbels typically contain taste buds.

bar-built estuary An estuary formed when islands or geographical barriers form a wall between freshwater from rivers and saltwater from the ocean.

barnacle A sessile crustacean whose body is usually covered by plates composed of calcium carbonate.

barrier reef A reef separated by a lagoon from the land mass with which it is associated.

base A compound that can remove hydrogen ions from solution.

bathyal zone (BATH-ee-uhl) The region of the ocean bottom that extends from the edge of the continental shelf to a depth of 4,000 meters.

bathygraphic features (bath-eh-GRAF-ik) The physical features of the ocean bottom.

bathymetric chart (bath-eh-MET-rik) Chart of the ocean that shows lines connecting points of similar depth.

bathyscaphe A submersible ship for deep-sea exploration that has a spherical watertight cabin attached to its underside.

bearers Fish that carry their offspring with them, either internally or externally.

behavioral isolation The prevention of interbreeding between different species in nature due to differences in mating behaviors.

the bends A condition that occurs when nitrogen gas dissolved in the blood comes out of solution and forms gas bubbles as the pressure decreases during an ascent from deep water. The bubbles can interfere with the circulation of blood to body tissues, causing tissue damage or even death.

benthic division (BEN-thik) The division of the ocean environment composed of the ocean bottom.

benthic spawners Fish that lay eggs that attach to some type of surface. There is no parental care after spawning, and the larvae may become pelagic or remain benthic.

beta-carotene A carotenoid pigment that confers a yellow or orange hue to a cell.

bilateral symmetry A type of symmetry in which body parts are arranged in such a way that only one plane through the midline of the central axis (the midsagittal plane) divides the organism into similar right and left halves.

binal virus A virus composed of an icosahedral head and a helical tail.

binary fission A type of asexual reproduction in which one cell splits into two after the original cell has duplicated its genetic material.

binomial nomenclature (by-NOH-mee-uhl NOH-men-klay-chur) The system of using two words (that is, a genus and species epithet) to name an organism.

biochemical isolation The prevention of interbreeding between different species in nature due to molecular differences at the cellular level.

bioerosion The gradual wearing away of substrates such as coral reefs through the actions of living organisms.

biogenous sediments (by-AH-gen-is) Sediments formed from the remains of living organisms.

biogeochemical cycles A combination of biological, chemical, and physical processes that recycle nutrients within the biosphere.

biological fitness The physical and behavioral characteristics of an organism that enable it to survive and reproduce in a particular environment.

biological magnification The concentration of pollutants or toxins in higher trophic levels of a food chain.

biosphere (BY-oh-sfeer) The collection of all of the earth's ecosystems together.

biotic factors The living aspects of an organisms environment.

bioturbation The mixing of sediments by the activities of benthic animals.

bivalve A type of mollusc whose body is covered by a two-piece hinged shell.

black zone The lowest zone of the supralittoral fringe of tropical rocky shores. The characteristic color of the rocks is due to the presence of certain species of algae.

bladder In seaweeds, an expanded part of an algal thallus that contains gas and is used for buoyancy.

blade In seaweeds, the large, flat, leaf-like part or *frond* of the thallus, lacking the vascular tissue of a plant leaf. In vascular plants such as grasses, the outer photosynthetic part of a leaf, usually flattened and free of the stem (compare with *sheath*).

blastocyst In mammals, the equivalent of the blastula, the hollow ball of cells that results from the cleavage stages of early embryonic development.

bloom A large increase in a population of phytoplankton.

blowhole A hole on the top of the head of a cetacean that serves as the opening to the animal's respiratory system.

blubber The fat layers of marine mammals.

bosses The large bumps on the snout of a humpback whale.

bottom-up factors Factors such as the availability of nutrients that affect the basal level of food chains and, thereby, the associated community.

brachiopod (BRAK-ee-uh-pawd) A type of generally benthic lophophorate with a bivalve shell belonging to the phylum Brachiopoda; commonly known as a lamp shell.

brackish water A mixture of freshwater and saltwater; water that is less saline than ocean water.

breaching A behavior thought to be a mating ritual of male humpback whales in which the animal jumps out of the water and comes crashing back down, creating a loud noise.

breaker A type of wave, the lower part of which is slowed by friction but whose crest continues to move toward the shore at a speed greater than that of the rest of the wave.

broadcast spawner Organism that directly releases eggs and sperm into open water and provides no parental care after spawning.

brood hiders Fish that hide their eggs but show no parental care after spawning.

brooders Organisms that broadcast their sperm into surrounding waters but retain the eggs within the body cavity.

brown alga A seaweed of the phylum Phaeophyta, characterized by possession of chlorophylls *a* and *c* and the pigment *fucoxanthin*, which gives them their olive-brown color.

bryozoan (bry-oh-ZOH-uhn) Tiny lophophorates belonging to the phylum Bryozoa. They are commonly called moss animals and form colonies on a wide variety of solid surfaces.

bubble net A ring of bubbles produced by humpback whales that is used to capture food.

budding A form of asexual reproduction in microbes and multicellular organisms in which unequal division of the adult produces two individuals.

buffer A substance that can maintain the pH of a solution at a relatively constant point.

bulb gland A gland found in molluscs of the family Conidae which produces the toxin that coats their radulae.

buttress zone The seaward-sloping portion of a coral reef that consists of alternating ridges and furrows. Also known as a *spur-and-groove formation.*

bycatch The noncommercial animals killed during fishing for commercial species. Also known as *incidental catch.*

byssal threads (BIS-suhl) Strong protein fibers secreted by mussels that fasten the animal to a rock or another solid surface.

cable root A subterranean root that arises from a stem or trunk, as in some mangroves, and extends horizontally through the sediment.

calcareous ooze Sediment composed of 30% or more of the remains of organisms that produce shells made of calcium carbonate.

calymma The vacuolated outermost cytoplasm of a radiolarian.

capillary action The ability of water to rise in narrow spaces.

capillary wave A wave for which the restoring force is the surface tension of the water.

capsid The outer coating of proteins of a mature virus (or virion).

capsule In radiolarians, an external organic layer that separates the inner nuclear region from the calymma.

carapace The hard dorsal covering of the bodies of animals such as arthropods and turtles.

carbohydrate An organic molecule composed of the elements carbon, hydrogen, and oxygen, frequently in a ratio of 1:2:1.

carbon fixation The process of converting inorganic carbon into an organic form.

carnivore An animal that feeds on other animals.

carotenoids (kuh-RAHT-in-noydz) A class of accessory pigments that absorb blue light and function in protection of primary photosynthetic pigments (chlorophylls) and to some degree in photosynthesis.

carpospore A nonmotile, diploid, dispersal stage released by the carposporophyte of a red alga.

carposporophyte A multicellular stage in the life cycle of red algae having a diploid thallus arising from the zygote, living parasitically on the female gametophyte, and producing carpospores for dispersal.

carrageenan (kar-uh-GEE-nuhn) A phycocolloid from red algae that is used commercially as a thickening and binding agent.

carrying capacity The maximum population of a species that a given environment can sustain for an extended period of time.

catadromous (kuh-TAD-ruh-muhs) A term applied to fishes that reproduce in the marine environment and spend their adult lives in freshwater.

caudal fin (KAW-duhl) The tail fin of a fish.

cell The basic unit of all living organisms.

cell membrane The plasma membrane (lipid bilayer) composed of phospholipids and proteins that surrounds a living cell.

cell wall A tough rigid structure that surrounds the cell membrane of most cells.

cellulose A polysaccharide found in the structural components of plants and algae and in the cell walls of many organisms.

centric diatom A shape of diatom with radially symmetrical valves.

centrifuge plankton Components of the plankton that are so small that they are not retained by standard plankton nets; best sampled by centrifugation of water samples; composed mainly of nanoplankton, picoplankton, and femtoplankton.

cephalization (sef-uh-luh-ZAY-shuhn) The concentration of sensory organs in the head region of an animal.

cephalochordate An animal belonging to the subphylum Cephalochordata. Also known as a *lancelet.*

cephalopod (SEF-uh-loh-pahd) An animal belonging to the molluscan class Cephalopoda, which includes squids, octopods, cuttlefish, and nautiloids.

cephalothorax (sef-uh-loh-THOR-aks) The combination of two body regions: the head and the thorax.

cerata (sir-AH-tuh) Projections on the body of a nudibranch that function in gas exchange.

cetacean (seh-TAY-shen) An animal belonging to the mammalian order Cetacea, which includes whales, dolphins, and porpoises.

chart A representation of the oceans and their features.

chart projection A map or chart made by projecting the features of the earth's surface together with reference lines onto a surface.

chelicera (keh-LI-suh-ruh) An appendage in chelicerates that is modified for feeding and takes the place of mouthparts.

chelicerate An animal belonging to the arthropod subphylum Chelicerata; they have a pair of oral appendages called *chelicerae* that are used in feeding.

chelipeds (KEE-leh-pehdz) Legs bearing claw-like structures (chela) in arthropods.

chemoreceptor (kee-moh-ree-SEP-tuhr) A sense organ capable of detecting changes in the chemical composition of an organism's environment.

chemosynthetic bacteria Bacteria that are able to form organic molecules from inorganic molecules using other chemicals rather than sunlight as a source of energy.

chitin (KY-tin) A polysaccharide found in the cell walls of fungi and in the exoskeletons of many arthropods.

chiton (KY-tuhn) A mollusc belonging to the class Polyplacophora. Chitons are characterized by a shell composed of eight separate plates bound together by a leathery girdle.

chloride cells Specialized cells located on marine fish gills that remove excess electrolytes from the body.

chlorophyll *a* A primary photosynthetic pigment of most autotrophic microbes, algae, and plants. It absorbs primarily violet and red light.

chlorophyll *b* A primary photosynthetic pigment in few microbes, in green algae, and in all plants. It absorbs primarily blue and red light.

chlorophyll *c* A primary photosynthetic pigment found in ochrophytes that is molecularly unrelated to true chlorophylls.

chloroplast Organelle found in eukaryotic photosynthetic organisms that contains the pigments necessary for photosynthesis.

choanocyte (koh-AN-oh-syt) A flagellated cell that lines cavities within the body of a sponge. Choanocytes are responsible for moving water through the body of a sponge. Also known as a *collar cell*.

choanoflagellate A group of microbes, similar to the choanocytes of sponges, that filter suspended particles through a collar of microvilli.

chordate An animal belonging to the phylum Chordata characterized by having a notochord, pharyngeal gill slits, postanal tail, and dorsal, hollow nerve tube at some time in its life cycle.

chorion (KOR-ee-uhn) An embryonic support membrane found in some vertebrates that functions in gas exchange.

chromatic adaptation The ability of photosynthetic organisms to alter the kind and quantity of their photosynthetic pigments in response to changes in the wavelengths and intensity of sunlight. The term *chromatic adaptation* was also used to refer to the now-rejected concept that the apparent vertical zonation of seaweeds can be explained by the way that seawater absorbs wavelengths of light with depth and by the variation in their complement of photosynthetic pigments, which differ in the optimal wavelengths of light used for photosynthesis.

chromatophore (kroh-MAT-uh-fohr) A special cell in an animal's skin that contains pigment molecules.

chromosome (KROH-muh-sohm) A structure consisting of DNA and protein located in the nucleus of eukaryotic cells.

chronometer (kroh-NAHM-i-ter) A clock that keeps accurate time and can be used to determine longitude.

ciguatera fish poisoning A human syndrome caused by consumption of fish that concentrate toxic chemicals produced by zooxanthellae.

cilia (SIL-ee-uh) Hair-like structures found on some cells that function in movement and feeding.

ciliate A group of alveolate microbes characterized by the presence of cilia for locomotion and feeding.

cingulum The groove around the middle of a dinoflagellate that holds one of its two flagella.

cirri Grasping structures found on some species of sea lily.

cirriped (SER-uh-ped) Feathery appendage used by barnacles to filter food from the surrounding water.

cladistics A taxonomic approach that classifies organisms according to the order in time at which branches arise along a phylogenetic tree without considering the degree of morphological divergence.

cladogram A diagram consisting of a series of branches where each branch point represents novel similarities that are unique to the species on that branch.

claspers Modified pelvic fins of male cartilaginous fishes that transfer sperm to the genital opening of a female.

cleaning station A territory established by one of several cleaner organisms that fishes visit at regular intervals to have parasites and dead tissue removed.

climax community A stable self-sustaining community that is the end result of ecological succession. In reality, climax communities may not exist due to the effects of frequent external disturbances and environmental change.

cloaca A common chamber and outlet for the digestive, urinary, and reproductive systems of certain animal species.

clutch size The number of offspring produced at each reproductive episode.

clumped pattern A dispersion pattern in which individuals are densely packed in patches.

cnida (NYD-uh) The stinging organelle in the cnidocyte.

cnidarian An animal that belongs to the phylum Cnidaria, which includes hydroids, jellyfish, and corals.

cnidocil (NYD-uh-sil) A bristle-like structure that extends from one end of a cnidocyte and functions as a trigger.

cnidocyte (NYD-uh-syt) The stinging cell found in members of the phylum Cnidaria.

coastal plain estuary An estuary formed between glacial periods, when water from melting glaciers raises the sea level and floods coastal plains and low-lying rivers. Also known as a *drowned river valley estuary*.

coccolith One of numerous disc-shaped calcareous plates that cover the surface of phytoplankton called coccolithophores.

coccolithophore (kahk-oh-LITH-oh-for) A photosynthetic haptophyte with the cell surface covered by coccoliths.

coccus A bacterium with a cell shaped like a sphere.

coenocytic (see-nuh-SIT-ik) The condition of having a body consisting of one giant cell or a few large cells containing more than one nucleus.

coleoid A cephalopod belonging to the subclass Coleoidea that does not have an external shell.

coliform bacteria Bacteria in the intestines of animals. Their number in a body of water indicates the amount of animal wastes entering it.

collar cell A flagellated cell, many of which line the cavities within sponges. They are responsible for producing the water current that flows through a sponge's body. Also known as a *choanocyte*.

colloblasts Specialized adhesive cells found on the tentacles of some ctenophores and used to capture prey.

commensalism A symbiotic relationship in which one organism benefits while the other is neither harmed nor benefited.

comb plate A row of cilia found on the surface of Ctenophores that functions in locomotion. Also known as a *ctene*.

community An assembly of populations of different species that occupy the same habitat at the same time.

community respiration (R) The total energy acquired from organic compounds that is used by all members of the community for their metabolism.

compensation depth The depth in the sea at which a primary producer can photosynthesize only enough to balance metabolism and not be able to grow; usually dependent on latitude, season, sea surface conditions, water clarity, and kind of primary producer.

competition model A hypothesis that each fish species inhabiting a coral reef has a unique niche and there is no competition between them.

competitive exclusion The process by which the less successful competitor for a limited resource is driven to extinction.

conceptacle In brown algae, a chamber within the receptacle that holds the gametangia.

conchiolin (kahn-KY-uh-lin; kahn-CHEE-uh-lin) The protein that makes up the periostracum of a molluscan shell.

conidiospore Asexually produced dispersal stage in the life cycle of a fungus.

conjugation A kind of sexual reproduction that involves the exchange of nuclei between two fused cells.

consolidation The formation of large aggregates by chemical or electrostatic attachment of particles to one another, as in the binding of loose sediments on a reef by seaweeds or the aggregation of suspended particles in seawater by bacteria.

continental drift The movement of continental masses as a result of seafloor spreading.

continental margin The region of ocean that lies beneath the neritic zone. It is composed of the continental shelf and the continental slope.

continental rise A gentle slope at the base of a steep continental slope.

continental shelf The edge of a continental landmass.

continental slope The transition between the continental shelf and the deep floor of the ocean.

control set The trial set in an experiment that does not contain the experimental variable.

convergent evolution An evolutionary process in which similar selective pressures bring about similar adaptations in unrelated groups of animals.

convergent plate boundary Regions of the ocean floor where the boundaries of lithospheric plates move toward each other and old lithosphere is destroyed.

copepod A crustacean belonging to the class Copepoda, usually the most abundant member of the marine zooplankton.

copulatory pleopod Specialized appendage on the abdomens of male decapods used for transferring sperm to females.

coral bleaching The whitening of coral colonies due to the loss of zooxanthellae from their polyps. Coral bleaching may be caused by such environmental stresses as increased temperature, sedimentation, and pollution.

coralline red alga A member of a group of red algae that deposit calcium carbonate in their cell walls; named for their resemblance to coral animals.

corallite The calcareous skeleton of a coral polyp.

corallivores Animals that eat coral polyps.

cordgrass A plant in the genus *Spartina*. It is the dominant vegetation in North American salt marsh communities.

Coriolis effect (kohr-ee-OH-lis) The apparent deflection of the path of winds and ocean currents that results from the rotation of the earth.

cosmogenous sediment Sediments formed from dust and debris from outer space.

countercurrent exchange system A biological mechanism involving fluids, with a concentration gradient between them, flowing in opposite directions in parallel channels. This arrangement maintains a stable gradient that allows maximum exchange of heat or dissolved substances between the two fluids.

countershading A camouflage pattern in which the upper surface of pelagic animals is dark and the lower surface light to blend in with the ocean depths or the surface, respectively, when observed by a potential predator. Also known as *obliterative countershading*.

craniates Members of the phylum Chordata that possess a cartilaginous or bony skull (cranium).

crinoid (KRY-noyd) An animal belonging to the echinoderm class Crinoidea. These generally sessile animals have bodies that resemble flowers and are commonly referred to as sea lilies.

crop A digestive organ found in some animals that stores food before it is processed.

crossing over The exchange of genetic material between adjacent chromosomes that occurs during meiosis.

crust (earth) The outermost layer of the earth.

crustacean An animal belonging to the arthropod class Crustacea, which includes crabs, lobsters, shrimp, and barnacles.

cryptic coloration Color patterns that camouflage animals, allowing them to blend with their environment to avoid predators or ambush prey.

ctene A row of cilia found on the surface of Ctenophores that functions in locomotion. Also known as a *comb plate*.

ctenoid scales (TEE-noyd) Thin overlapping scales on advanced bony fishes. Their exposed posterior edges have small spines.

ctenophore (TEEN-uh-fohr) An animal belonging to the phylum Ctenophora. Also known as a *comb jelly*.

cubozoans Cnidariabns known as boxy jellyfish belonging to the class Cubozoa.

culm The specialized kind of vertical stem typical of grasses, sedges, and rushes.

cuticle One of several terms used for a layer of an organism outside the most superficial layer of cells, or the outermost layer of a cell wall or exoskeleton of some animals, plants, and seaweeds.

Cuvierian tubule Small sticky tube expelled from the anus of some sea cucumbers to distract predators.

cyanobacteria (sy-AN-oh-bak-TEER-ee-uh) A group of photosynthetic prokaryotes that contain chlorophyll *a* and that release oxygen as a by-product of their photosynthesis. Cyanobacteria are sometimes referred to as *blue-green bacteria*.

cyanophycean starch The molecular form in which carbohydrates are stored in cyanobacteria.

cycloid scales (SY-kloyd) Thin overlapping scales with a smooth posterior edge that are found in less-advanced bony fishes.

cydippid larva The planktonic larva of ctenophores.

cyprid larva A planktonic larval stage of barnacles that develops from a nauplius larva and has compound eyes and a carapace composed of two shell plates.

cytokinesis The process of dividing a cell into two daughter cells.

cytoplasm The combination of cytosol and organelles.

cytosol The fluid contents of a cell.

cytostome An organelle in a eukaryotic microbe that serves as the site for ingestion of food.

decapod (DEK-uh-pahd) An animal with five pairs of walking legs that belongs to the arthropod order Decapoda, which includes crabs, lobsters, and shrimp.

dead zone An area of ocean that has become oxygen depleted.

decomposer An organism that breaks down the tissue of dead plants and animals, as well as animal wastes.

deductive reasoning A process of reasoning in which observations suggest a general principle from which a specific statement can be derived.

deep scattering layer A mixed group of zooplankton and fishes that causes sonar systems to generate a false image of a nearly solid surface hanging in midwater.

deep-sea vent community A community of marine organisms that depend on the specialized environment found at divergence zones in the ocean floor. Also known as a *rift community*.

deepwater wave A wave that occurs in water that is deeper than one half the wave's wavelength.

delphinid (del-FI-nid) A collective term referring to dolphins and porpoises.

density The mass of a substance in a given volume of that substance.

density-dependent factor A factor such as competition, predation, parasitism, etc., that influences population growth particularly strongly when population density is high.

density-independent factor A factor such as a change in weather or volcanic eruption that influences population growth to the same extent regardless of the population density.

deposit feeder An animal that feeds on bottom sediments.

derived characteristics Characteristics that evolved after a group diverged from its ancestral line.

desalination (dee-sal-uh-NAY-shun) The process of removing salt from seawater.

desiccation (dehs-ik-KAY-shun) The process of drying out or losing moisture.

deterrence A mechanism to reduce predation by repelling the predator or otherwise making the organism unattractive for consumption.

detritivore (deh-TRY-ti-vor) An organism that feeds on detritus.

detritus (deh-TRY-tuhs) Organic matter such as animal wastes and bits of decaying tissue.

diarrhetic shellfish poisoning A human syndrome caused by consumption of shellfish that concentrate the toxic chemicals produced by dinoflagellate blooms.

diatomaceous earth An industrial material harvested from exposed deposits of dead diatom frustules and used for filtration and other applications.

dinoflagellate An alveolate microbe with cellulose in its alveoli and with two heterokont flagella for locomotion.

dinosporin A chemical in the cellulose plates of dinoflagellates that increases their resistance to decay.

diploid number (2N) The number of paired chromosomes in a cell's nucleus.

disaccharide Carbohydrate that contains two bonded monosaccharides.

discrimination click High-frequency click used by toothed whales such as dolphins to get a precise picture of an object.

dispersion The pattern of spacing among individuals within a range.

disphotic zone (twilight zone) The region of ocean between 150 and 450 meters deep where there is not enough light to power photosynthesis. Also known as the *twilight zone*.

dissipative beach Type of beach on which wave energy is strong but is dissipated in a surf zone some distance from the beach face.

dissolved organic matter (DOM) Organic molecules dissolved in seawater that were lost or released from living organisms.

diurnal tide (dy-YUR-nuhl) The condition of having only one high tide and one low tide each day.

divergent plate boundary Region of the ocean floor where the boundaries of lithospheric plates move apart as new lithosphere is formed.

diversity (or biodiversity) The variation in numbers of living organisms within an ecosystem. Ecosystems with greater diversity (i.e., a greater number of species) are considered to be more stable than those with fewer species.

doldrums The area of rising air at the equator.

DOM See *dissolved organic matter*.

domain The taxonomic category that is higher than a kingdom.

dorsal fin A fin on the dorsal surface of a fish.

dorsal, hollow nerve tube A neural structure found in chordates some time in their life cycle.

downwelling The vertical movement of water in a downward direction.

drift alga A seaweed that breaks free from its holdfast and is carried by current or wind, sometimes accumulating in vast beds in quiet waters of bays and lagoons.

drift net A large net composed of sections called *tans* that may stretch as far as 60 kilometers. Drift nets entangle fish, squid, and other marine animals that swim into them.

drop root An aerial root of a mangrove that arises from a branch and descends to the sediment.

drop-off The vertical wall that is sometimes formed on a reef front.

drowned river valley estuary An estuary formed between glacial periods, when water from melting glaciers raises the sea level and floods coastal plains and low-lying rivers. Also known as a *coastal plain estuary*.

eastern-boundary current Current that flows along the eastern edge of an ocean basin carrying cold water toward the equator.

ebb tide A falling tide.

echinoderm (eh-KY-noh-derm) An animal belonging to the phylum Echinodermata, which includes sea stars, brittle stars, sea urchins, sea cucumbers, and crinoids.

echinoid An echinoderm belonging to the class Echinoidea, which includes sea urchins, heart urchins, sand dollars, and sea biscuits.

echiuran An animal that belongs to the class Echiura in the phylum Annelida. Also known as a *spoonworm*.

echolocation A process that allows some cetaceans to use sound waves to distinguish and home in on objects from distances of several meters.

ecological equivalent One of two different groups of animal that have evolved independently along the same lines in similar habitats and that therefore display similar adaptations.

ecological efficiency The percentage of energy that is taken in as food by one trophic level and then passed on as food to the next highest trophic level.

ecosystem A system that is composed of living organisms and their nonliving environment.

ectoparasite A parasite that attaches to the outer covering of its host.

ectotherm An animal that obtains most of its body heat from its surroundings.

Ekman spiral The spiral flow of water that results from the movement of water at the surface moving deeper layers of water due to friction. The movement of the successive deeper layers of water are deflected by the Coriolis effect, producing a downward spiral of water to a depth of about 100 meters.

Ekman transport The net movement of water to the 100-meter depth by the Ekman spiral.

El Niño Southern Oscillation (ENSO) Changes in oceanic and atmospheric current patterns that move warm Pacific Ocean surface water farther east than normal, resulting in worldwide changes in weather.

elasmobranch Group of cartilaginous fishes (class Chondrichthyes) containing the sharks, rays, and skates.

embayment Area of coastline where portions of the ocean are cut off from the rest of the sea.

elver A juvenile eel (Anguilliformes).

endoplasmic reticulum (ER) A series of membranes that wind through the cytoplasm of eukaryotic cells.

endoskeleton (EN-doh-SKEL-eh-tuhn) An internal skeleton.

endosymbiont A guest organism or organelle that lives within a host organism or cell.

endosymbiotic theory The body of evidence supporting the idea that some one-celled organisms have evolved by the incorporation of other one-celled organisms or their organelles into the host cell.

endotherm (EN-doh-therm) An animal that maintains a constant body temperature by generating heat internally.

energy pyramid A diagram that pictures the energy flow from one trophic level to the next. The diagram is shaped like a pyramid to indicate that there is successively less energy at each higher level.

envelope An outer covering of a virus (or virion) derived usually from the cell or nuclear membrane of the host.

environment All external factors that act on an organism.

enzyme A biological catalyst that increases the rate of the chemical reactions of metabolism, allowing them to be more efficient.

epibenthic organism An organism that lives on the surface of hard sediments.

epidermis (ehp-i-DER-mis) An outer layer of cells or the cellular covering of an organism.

epifauna (EP-i-faw-nuh) Benthic organisms that live on the ocean bottom.

epipelagic zone The waters over the continental shelf, or to 200 meters deep in the open sea, occupied by zooplankton and nekton.

epiphyte Any organism that grows on a multicellular primary producer (i.e., on a plant in the broadest sense).

epitoke The pelagic reproductive individual formed by some errant polychaetes.

epitoky (EP-i-toh-kee) A type of reproduction that occurs in some polychaetes that involves the production of a reproductive individual (epitoke) adapted for a pelagic existence from a nonreproductive individual adapted for a benthic existence.

epizoic Term describing any organism that grows on an animal.

equator A circle drawn around the center of the earth that is perpendicular to its axis of rotation.

equatorial upwelling Upwelling along the equator caused by water on either side of the equator being deflected toward the poles by the Coriolis effect. This pulls surface water away from the equator, and that water is replaced by deeper water, producing an upwelling.

errant polychaete (ER-ent PAHL-eh-keet) An actively mobile polychaete that has a mouth equipped with jaws or teeth.

escarpment A nearly continuous line of cliffs with sharp vertical drops formed by the movement of lithospheric plates along faults.

estuary A region where freshwater is mixed with saltwater.

Eubacteria The domain of life that contains prokaryotes with peptidoglycans in their cell walls.

Eukarya The domain of life that contains all of the eukaryotes.

eukaryote (yoo-KAR-ee-oht) An organism whose cells contain a nucleus and membrane-bound organelles.

eukaryotic cells Cells that contain a nucleus and membrane-bound organelles.

euryhaline Able to tolerate a broad range of salinities.

eutrophication Nutrient enrichment of an aquatic system that results in rapid algal growth.

evaporites Salt deposits produced when saltwater evaporates.

eviscerate The release of internal organs from the mouth or anus. This behavior is used by sea cucumbers to deter predation.

evolution The process whereby populations of organisms change over time.

exclusive economic zone (EEZ) Area of ocean exclusively controlled by a coastal nation.

exhalant opening The opening of a bivalve's or tunicate's siphon, through which water is expelled from the body.

exoenzyme Enzyme released from osmotrophic microbes for external digestion.

exoskeleton (EK-soh-SKEL-eh-tuhn) A hard protective exterior skeleton such as that found in arthropods.

experimental set The trial set in an experiment that contains the experimental variable.

exotic species Species that does not naturally occur in a given geographic area. Also known as *alien* or *nonnative species*.

experimental variable In an experiment, the factor that is altered in the experimental group but not in the control group.

exponential growth A pattern of population growth characterized by initial slow growth followed by increasingly rapid growth; when graphed the curve is typically a J-shape. Also known as *logarithmic growth*.

facultative anaerobe Organism that thrives in the presence or absence of oxygen.

facultative halophyte A plant that will thrive in the presence or absence of salt.

fault An area where the earth's plates move past each other.

fecal cast A pile of organic material and minerals defecated by deposit feeders. Also known as *casting*.

feeding polyp A type of polyp found in hydrozoan colonies that functions in capturing food and feeding the colony. Also known as a *gastrozooid*.

femtoplankton Plankton that range in size from 0.02 to 0.2 micrometers, composed primarily of viruses.

fertilizin A chemical released by some female epitokes that stimulates males to shed their sperm.

fetch The distance along the water over which the wind blows.

filamentous fungus A fungus that forms a mycelium of long thread-like masses.

filter-feeder An organism that filters its food from the water.

first-order consumer A consumer that feeds directly on a producer. Also known as a *primary consumer*.

fishing effort A measure of the number of boats fishing, the number of fishers working, and the number of hours that they spend fishing.

fission A type of asexual reproduction in which an adult organism literally tears itself into two pieces that eventually form two individuals.

fitness An organism's biological success as measured by the number of its own genes that are present in the next generation of a population.

fjord (fyord) A deep valley cut into the coastline by glaciers and filled with a mixture of freshwater and saltwater.

flagellum (plural, flagella) Long hair-like structure that functions in cellular locomotion.

flipper flapping A cetacean behavior in which a whale rolls over onto its back and flaps its flippers in the air.

flood tide A rising tide.

flowering plant A vascular plant that produces seeds in a fruit.

fluke A type of parasitic flatworm belonging to the class Trematoda. In sirenians and cetaceans the lobes of the animal's tail.

fluke down A cetacean behavior in which the tail is brought straight up such that the fluke clears the water but remains turned down.

fluke up A cetacean behavior in which the tail is brought straight up so that the ventral (bottom) surface is visible.

foliage leaf A leaf that produces a photosynthetic blade.

food chain A sequence of feeding relationships among a group of organisms that begins with producers and proceeds in a linear fashion to higher-level consumers.

food web A representation of the complex feeding networks that exist in an ecosystem.

foraminiferan (or foram) An amoeboid marine protozoan with reticulopods and usually bearing a calcareous test.

forced wave A wave produced by storms at sea.

forereef The outer seaward margin of a coral reef. Also known as the *reef front*.

fossil fuel Fuel—including coal, oil, and natural gas—formed from the remains of plants and microorganisms that lived millions of years ago.

fouling community An assembly of populations of many kinds of organisms growing on an intertidal or submerged artificial structure.

fracture zone Linear region of unusually irregular ocean bottom that runs perpendicular to ocean ridges.

fragmentation A type of asexual reproduction in which a part of the parent organism breaks off and forms a new individual.

free wave A wave that is no longer affected by the energy from a generating force.

frictional drag Resistance to movement that is generated by the passage of the medium (for example, seawater) over the surface of an object.

fringing reef A reef close to and surrounding newer volcanic islands or that borders continental landmasses.

frond The leaf-like part or *blade* of a large seaweed but lacking the vascular tissue of a plant leaf.

fruit A flowering plant structure formed after pollination from various parts of the flower and the embryo composed of several layers of tissue that surround the seed.

fruiting body The general term for a sexually reproductive structure of a fungus, including the stalk and cap of a mushroom.

frustule (FRUHS-tyool) The glassy structure, composed of silica, that covers diatom cells.

fucoxanthin An accessory carotenoid pigment in ochrophytes that is brown to golden-brown in color.

fundamental niche The broadest of all possible niches that an organism theoretically can occupy.

fungus (plural, fungi) An organism that belongs to the kingdom Fungi. Fungi are eukaryotic, have cell walls containing chitin, and are not photosynthetic.

fusiform body shape Spindle-shaped body of many fishes and other animals, characterized by a rounded middle portion and tapering toward each end.

gametangium The part of an algal thallus that produces gametes.

gamete (GAM-eet) A sex cell such as a sperm, egg, or pollen grain.

gametophyte (ga-MEE-toh-fyt) The stage in the life cycle of an alga or plant during which gametes are produced.

ganoid scales (GAN-oyd) Thick bony scales that do not overlap. Ganoid scales are found in some less-advanced bony fishes.

gas gland A gland that secretes gas into the swim bladder of fishes or into the float of some large pelagic cnidarians.

gastropod (GAS-troh-pahd) A member of the molluscan class Gastropoda, which includes snails, limpets, abalones, and nudibranchs.

gastrodermis The cellular lining of the gastrovascular cavity of cnidarians and ctenophores.

gastrovascular cavity (gas-troh-VAS-kyooler) A central cavity in the body of cnidarians, ctenophores, and flatworms that functions in digestion and in the movement of materials within the animal.

gastrozoid A type of polyp found in hydrozoan colonies that functions in capturing food and feeding the colony. Also known as a *feeding polyp*.

gelatinous zooplankton Components of the plankton that have a high content of body water.

gene pool All of the genes in a given population that exist at a given time.

gene A unit of hereditary information located on an organism's DNA.

generation time The average time between an individual's birth and the birth of its offspring.

generating force A force that disturbs the surface of water, producing a wave.

genetics The study of heredity.

genus (JEE-nuhs) The first name in the two-part Latin name for an organism. It is the taxonomic group that is one step above the species level.

geographical range The geographical area where a population lives.

gestation period The period of time that the young is carried in the uterus of a placental mammal.

gill filaments Thin rod-like structures that make up the gills found in some animals.

gill pump A series of muscles, found in cartilaginous fishes, that suck in water and push it past the gills.

gill rakers Finger-like projections of the gill arches of fish that can function in filter-feeding or in preventing solid substances from being carried into gill cavities.

global positioning system (GPS) A system of satellites that can be used to find an exact position anywhere on earth.

globigerina ooze A seafloor sediment consisting of the accumulated calcareous tests of foraminiferans.

glucose A monosaccharide that is the basic fuel for all living cells.

glycogen A polysaccharide found in some organisms that serves to store glucose reserves.

gnathopods Specialized appendages found on amphipods that are used for collecting food.

Golgi apparatus An organelle that functions in modifying molecules and forming membrane-bound sacs around them.

gonangium A type of polyp found in hydrozoan colonies that is specialized for the process of asexual reproduction. Also known as a *reproductive polyp*.

Gondwanaland The southern portion of the supercontinent Pangaea, composed of what became India, Africa, South America, Australia, and Antarctica.

gorgonian A type of octocorallian coral that belongs to the cnidarian order Gorgonacea, which includes sea fans, sea feathers, and whip corals.

grana (singular, granum) A stack of thylakoids within a chloroplast.

grasping spine A structure on the head of an arrowworm used to capture prey.

gravity wave A wave for which the restoring force is gravity.

gray zone The middle zone of the supralittoral fringe of tropical rocky shores. This zone is the farthest zone from the low tide line where macroscopic marine algae grow.

green alga Seaweeds of the phylum Chlorophyta, characterized by possession of chlorophylls *a* and *b*, which give them their green color.

greenhouse gas A gas such as carbon dioxide, methane, or a chlorofluorocarbon that collects in the atmosphere and prevents heat energy from radiating back to space.

Greenwich meridian See prime meridian.

guano (GWAH-noh) A phosphate-rich manure produced by birds.

gross primary production (P) The total production of organic compounds from carbon dioxide through the process of photosynthesis or chemosynthesis.

guarders Fish that guard their offspring until they hatch and, frequently, through the larval stage.

gular pouch (GYOO-ler; GUH-ler) A sac of skin that hangs between the flexible bones of a pelican's lower mandible.

gulf A small body of water that is mostly cut off from an ocean or sea by land formations.

Gulf Stream The largest of the western-boundary currents; it runs along the western edge of the North Atlantic gyre.

guyot A seamount with a flat top.

gymnosome pteropod A carnivorous oceanic gastropod that has no shell and swims by flapping lateral extensions of its foot.

gyre The circular flow pattern of water around the edge of an ocean basin.

habitat The specific place in the environment where an organism lives.

habitat isolation The prevention of interbreeding between different species in nature due to their occupying different habitats.

hadal zone (HAYD-uhl) The portion of the ocean bottom that lies at depths greater than 6,000 meters.

halobacteria A group of archaeons that require high concentrations of salt where they live.

halocline (HAL-oh-klyn) A zone in the ocean characterized by a rapid change in salinity with increasing depth.

halophyte (HAL-oh-fyt) A salt-tolerant flowering plant.

haploid number (*N*) The number of unpaired chromosomes in a cell's nucleus.

haptonema A rod-like organelle that projects between the two flagella of haptophytes and is used to capture prey.

haptophytes A group of eukaryotic microbes that possess a haptonema; includes the coccolithophores.

harmful algal bloom (HAB) The formation of a dense population of photosynthetic microbes and macroalgae that presents an environmental threat in natural and managed communities.

head-foot The region of the gastropod body that contains the head, with its mouth and sensory organs, and the foot, which is the animal's organ of locomotion.

helical virus A virus with a capsid of spirally arranged proteins.

hemichordate An animal in the phylum Hemichordata. Also known as an *acorn worm*.

hemipenes Paired penises that are characteristic of snakes and lizards.

herbivore An animal that eats only plants and algae.

hermaphrodite An animal that possesses both male and female sex organs.

hermatypic coral A colonial coral that maintains zooxanthellae in its tissues and forms reefs.

heterocercal tail (het-uh-roh-SIR-kuhl) The type of tail in some fishes in which the upper lobe is larger than the lower lobe and the vertebral column bends slightly upward into the upper lobe.

heterocyst A specialized cell of cyanobacteria in which conditions favorable for nitrogen fixation are maintained.

heterokont The condition of a cell that bears two different flagella, one simple and one with mastigonemes.

heterotroph (HET-uh-roh-trohf) An organism that relies on other organisms for food. Also known as a *consumer*.

high marsh The region of a salt marsh closest to shore.

high water The greatest height to which a high tide rises.

holdfast A branching system of fibers or root-like structures at the base of an algal thallus, securing the seaweed to the sea floor but lacking the vascular tissue of plant roots.

holocephalan A chimaera, a type of cartilaginous fish.

holoplankton Organisms that are planktonic throughout their life cycles.

holothurin A poison from the flesh of sea cucumbers that protects the animal from predation by fishes and that is used by Pacific islanders to poison fishes.

home scar A location on rock to which certain limpet species return as the tide recedes. The shape of the limpet shell often grows to precisely match the contours of this location. Returning to this location allows the animal to form a tighter seal to the rock, protecting it from predation and desiccation.

homeostasis (HOH-mee-oh-STAY-sis) The internal balance that living organisms must maintain to survive.

homeothermic Able to maintain a constant body temperature.

homocercal tail (hoh-moh-SIR-kuhl) A tail in which the upper and lower lobes are generally equal and into which the vertebral column does not extend.

horse latitudes The areas of descending air at 30 degrees north and south latitudes.

host In a parasitic relationship, the organism that supports the parasite.

hybrid The offspring produced by breeding between two different species.

hydrocoral A type of hydroid that secretes a calcareous skeleton around the polyps and resembles hard corals.

hydrogen bond Weak attractive force that occurs between slightly positive hydrogen atoms of one molecule and slightly negatively charged atoms such as oxygen or nitrogen on another molecule.

hydrogen hypothesis One concept on the origin of eukaryotic cells that postulates an endo-symbiotic relationship between a host archaeon that needed hydrogen for chemosynthesis and a guest bacterium that released hydrogen and became a mitochondrion.

hydrogen sulfide (H_2S) A colorless, flammable, foul-smelling gas produced by the anaerobic breakdown of sulfates in organic matter by bacteria.

hydrogenous sediment Sediments formed by the precipitation of dissolved minerals from seawater.

hydroid A colonial organism belonging to the cnidarian class Hydrozoa. Also known as a *hydrozoan*.

hydrophilous pollination The mechanism of dispersal of a pollen grain by water currents from the male flower to the female flower.

hydrophyte (HYD-roh-fyt) A flowering plant that generally lives submerged under water.

hydrostatic skeleton A body support that results from fluids under pressure in an animal's body cavity.

hydrozoan (hyd-roh-ZOH-en) A generally colonial organism that belongs to the cnidarian class Hydrozoa. Also known as a *hydroid*.

hyperthermophile Any microbe, usually an archaeon, that grows and reproduces best at temperatures exceeding 100° C.

hypha A thread-like filament that makes up a section of the mycelium of a nonyeast fungus.

hypocotyl The initial stem of a young or embryonic plant that arises below the embryonic leaves.

hypothermia Abnormally low body temperature.

hypothesis An explanation for observed events that can be tested by experiments.

hypoxic zone An area of ocean characterized by very low oxygen content due to eutrophication and excessive decomposition.

icosahedral virus A virus composed of a DNA or RNA core surrounded by a capsid with 20 sides.

incidental catch Noncommercial animals that are killed each year during fishing for commercial species. Also known as *bycatch*.

inductive reasoning A process of reasoning whereby a general explanation is derived from a series of observations.

infauna (IN-faw-nuh) Benthic organisms that live in the bottom sediments.

infralittoral fringe Zone of rocky shores that lies above the infralittoral zone and is exposed to air only at extreme low tides.

infralittoral zone Zone of rocky shores that lies below the extreme low water of spring tides.

inhalant opening The opening of a bivalve's or tunicate's siphon through which water is drawn into the body.

international date line The line that lies at approximately 180 degrees from the prime meridian that marks the beginning of each calendar date.

internode The part of a plant stem that lies between successive nodes, leaves, lateral stems, or roots.

interspecies competition (interspecific competition) Competition between similar species for a limited resource.

interstitial animal Animal that lives in the spaces between sediment particles.

intertidal zone The region of a shore that is covered by high tide and exposed at low tide.

intraspecies competition (intraspecific competition) Competition between members of a single species for a limited resource.

invertebrate An animal that lacks a vertebral column (backbone).

ion A particle that carries an electrical charge.

iridophore Type of chromatophore that produces iridescence from light striking crystals contained within the cell.

irregular echinoid An echinoid such as a heart urchin or sand dollar that is adapted to a burrowing lifestyle.

island arc A chain of volcanic islands usually found along deep-sea trenches.

isolating mechanism Mechanism that prevents members of different species from reproducing in nature.

isopycnal (eye-soh-PIK-nuhl) Having the same density at the top as at the bottom.

keratin A tough protein found in mammalian hair and nails and in the baleen of cetaceans.

keystone species An animal whose presence in a community makes it possible for many other species to live there. Also known as a *keystone predator*.

kinetic Descriptive of planktonic organisms that are able to move by flagella, limbs, jet propulsion, or other modes.

krill Pelagic shrimp-like creatures that belong to the arthropod order Euphausiacea.

K-strategist A species whose population is usually in equilibrium.

labial flaps Fin-like extensions near the mouth, which direct the flow of water during feeding by manta rays.

labyrinthomorph A group of stramenopiles with heterotrophic nutrition that are decomposers or pathogens.

labyrinthulid A labyrinthomorph that includes the pathogen responsible for wasting disease in eelgrass.

lactation period The period of time during which a female mammal nurses her young.

lactic acid A waste product produced in muscle tissue when there is not enough oxygen to support aerobic metabolism.

lactose A disaccharide composed of glucose and galactose. It is the sugar found in mammalian milk.

lacuna Large, open, gas space in the aerenchyme tissue of plants.

lagoon A relatively small area of the ocean that has been partially isolated by the development of a barrier.

laminarin The molecular form in which carbohydrates are stored in ochrophytes.

lancelet An animal belonging to the subphylum Cephalochordata. Also known as a *cephalochordate*.

landing The catch made by a fishing vessel.

larvacean A free-swimming tunicate that resembles a tadpole.

larval settlement The process by which planktonic larvae leave the water column and settle to the bottom.

latent heat of evaporation The amount of energy required to overcome the surface tension of a liquid before the molecules move into the vapor state.

lateral line system A system of canals running the length of a fish's body and over its head that functions in detecting movement in the water.

latitude The angular distance north and south of the equator measured in degrees. Also known as *parallel*.

Laurasia The northern portion of the supercontinent Pangaea composed of what became Europe, Asia, and North America.

lenticel A scar-like structure on the surface of a stem serving for passage of atmospheric gas through the cuticle and epidermis into the aerenchyme beneath.

leptocephalus larva Thin leaf-like larval form of eels, tarpon, bonefish, and their relatives.

leuconoid (LOO-keh-noyd) A type of sponge body containing many spongocoels and chambers leading to them.

Lévy walks A random pattern of movement defined by the lengths of the segments of travel, in which short or long lengths are disproportionately represented.

lichen Mutualistic association between a fungus and an alga, often found in dry habitats.

life history The events that characterize an organism's life, specifically birth, reproduction, and death.

light–dark-bottle method A procedure for measuring productivity that involves using two bottles: one that is clear and allows light to penetrate and one that is opaque and does not allow light to penetrate.

limiting nutrient A nutrient that limits the number or distribution of marine organisms.

lipid A macromolecule composed primarily of carbon and hydrogen. Also known as fat, oil, or wax.

lithification The conversion of an aggregate of particles into a solid mass with a mineral cement.

lithosphere (LITH-oh-sfeer) The part of the earth comprising the crust and upper mantle.

logarithmic growth A pattern of population growth characterized by initial slow growth followed by increasingly rapid growth; when graphed the curve is typically a J-shape. Also known as *exponential growth*.

logistic growth A pattern of population growth characterized by an initial exponential-growth phase that eventually levels off at zero growth as the population reaches an equilibrium point.

longitude The angular distance east and west of the prime (or Greenwich) meridian measured in degrees. Also known as *meridian*.

longshore current Current moving parallel to the shore in the surf zone that is generated by waves breaking at an angle to the beach.

longshore transport process A process driven by longshore currents that carries sediments along the shoreline.

lophophorate (lohf-uh-FOHR-ayt) A sessile animal that lacks a distinct head and possesses a feeding device called a *lophophore*.

lophophore (LOHF-uh-fohr) An arrangement of ciliated tentacles that surrounds the mouth of animals known as lophophorates. It functions in feeding and gas exchange.

lorica A loosely fitting external covering of a microbe.

lottery model A hypothesis that competition among fishes on coral reefs is unimportant; chance determines which species of larvae settling from the plankton colonize a particular area of reef. As each individual is lost over time, chance again determines which species will occupy the available space, and competitive exclusion does not occur.

low marsh The portion of a salt marsh that occupies the lower intertidal zone and that is covered by tidal water much of the day.

low water The lowest point to which a low tide falls.

luciferase An enzyme that catalyzes the reactions that produce bioluminescence.

luciferin A protein that produces bioluminescence when combined with oxygen in the presence of the enzyme luciferase and ATP.

lysis The rupture of a cell and release of its contents by its own enzymatic activity, viral action, osmotic shock, or other agent.

lysogenic cycle The life history of a virus that remains dormant within the host genome awhile before initiating viral replication.

lysosome A package of digestive enzymes found in eukaryotic cells.

lytic cycle The life history of a virus that has no dormant phase in the host before initiating viral replication.

macroalga Multicellular alga, such as a seaweed, visible to the naked eye.

macrofauna Organisms large enough to be seen by the naked eye.

macromolecule A large organic molecule such as a protein, carbohydrate, lipid, or nucleic acid.

macronucleus The larger nucleus of a ciliate that plays a role in metabolism of the cell.

macroplankton Plankton that range in size from 2 to 20 centimeters, composed primarily of animals.

maculae Sensory membranes within the inner ear containing neuromast-like sensory receptors that are used to detect fluid movement.

madreporite (mad-ruh-POHR-ryt) The structure at which water enters the water vascular system of an echinoderm.

magma Molten material located deep in the earth's mantle.

mammary gland Special gland in female mammals that produces milk.

mandible One component of the third pair of chewing or grinding appendages of crustaceans.

mandibulate A member of the arthropod subphylum Mandibulata, which includes the crustaceans.

mangal (MAN-guhl) Describes a community dominated by plants called *mangroves*.

mangrove Any of a variety of salt-tolerant trees and shrubs restricted to humid tropical coasts.

mannitol An alcohol and a product of photosynthesis in brown algae.

mantle (earth) The thickest layer of the earth and the one that contains the greatest mass of material.

mantle (mollusc) The part of a mollusc's body that secretes its shell.

mantle cavity The space between a mollusc's mantle and its body.

map A representation of the land features of the earth.

marine biology The study of the living organisms that inhabit the seas and of their interactions with each other and their environment.

marine snow A loose network of live and dead particles in the plankton arising from webs of mucus released by many kinds of microbes and zooplankton.

maritime zone Zone of rocky shores that lies above the extreme high water of spring tides. Also known as the *supralittoral zone*.

mark-recapture metod Method that estimates population size by capturing and marking or tagging individuals, the releasing them back into the wild and counting the number recaptured at a later time,

mastigoneme One of many hair-like filaments that extend from the shaft of some flagella and increase the effectiveness of locomotion.

medusa The free-floating form of cnidarian that resembles an umbrella or a bell.

megaplankton Plankton that range in size from 20 to 200 centimeters in size, composed primarily of the larger jellyfishes, siphonophores, salps, and floating mats of sargassum weed.

meiofauna (MY-oh-fawn-uh) The tiny organisms that are adapted to living in the spaces between sediment particles.

meiosis The process of nuclear division in which the number of chromosomes in the cell's nucleus is reduced to one half in the formation of gametes. Also known as *reduction division*.

melanin A brown or brown-black pigment found in many animals.

melon An oval mass of fatty, waxy material located between the blowhole and the end of the head in cetaceans capable of echolocation. The melon directs and focuses the sound waves produced by an animal.

membranelles Ribbon-shaped or tufted arrangements of cilia that increase the effectiveness of locomotion and feeding.

Mercator projection A type of modified cylindrical projection, used to produce maps and charts, that distorts the areas near the poles but offers the advantage that a straight line is a line of true direction.

meridian A line of longitude.

Mermaid's purse The egg case of skates, sharks, and rays.

meroplankton Planktonic stages of an organism that is benthic or nektonic at other stages in its life history.

mesenterial filaments Long, thin, tube-like structures, attached to the gut of cnidarians, that are involved in digestion and absorption.

mesoglea The layer between the epidermis and gastrodermis of cnidarians and ctenophores. It typically contains mostly water and protein.

mesoplankton Plankton that range in size from 0.2 to 20 millimeters, composed primarily of larger phytoplankton and larvae of zooplankton as well as many kinds of adult crustaceans and other zooplankton groups.

messenger RNA (mRNA) A type of RNA molecule that contains the instructions for synthesizing a protein.

metabolism The sum of all of the chemical reactions that occur within living cells.

methane hydrate Ice crystals containing trapped methane.

methanogen Archaeon with the ability to produce methane gas in its metabolism.

microbe A living organism too small to examine without the aid of a microscope.

microhabitat The smaller subdivisions of a habitat.

micronucleus The smaller nucleus of a ciliate that serves only to hold a complete diploid set of chromosomes for inheritance by the next generation.

micropatchiness The small-scale clumped distribution of organisms within an ecosystem, often explained by the similarly small-scale variation in the physical and chemical environment or disturbance effects of other organisms.

microphytoplankton (my-kroh-FYT-oh-plank-tuhn) Phytoplankton ranging from 20 to 200 micrometers.

microplankton Plankton ranging from 20 to 200 micrometers, composed of many kinds of phytoplankton, marine invertebrate larvae, and smaller adult zooplankton.

microvillus A short, hair-like cellular extension, usually one of many on the surface of a cell, that increases surface area for absorption or that filters out particles from a feeding current.

midlittoral zone Zone of rocky shores that lies between the supralittoral fringe and the infralittoral fringe. The midlittoral zone is the true intertidal, inhabited by both marine and amphibious organisms.

midocean ridge A long mountain range that forms along cracks on the ocean floor where erupting magma breaks through the earth's crust. Also known as *ridge system*.

midsagittal plane A plane through the center of the long axis of a bilaterally symmetrical animal that divides the animal into more or less identical right and left halves.

mitochondria (singular, mitochondrion) The membrane-bound organelles in which the energy of food molecules is used to produce ATP.

mitosis The process of nuclear division that occurs in eukaryotic cells and produces two identical nuclei.

mixed semidiurnal tide (mixed tide) A tide in which the high tide and the low tide are of different levels.

mixotrophic Nutrition of an organism by combining autotrophy and heterotrophy.

modern synthetic theory of evolution Darwin's theory of evolution by natural selection as modified by modern genetics.

molecular biology The study of the structure and function of macromolecules such as nucleic acids.

mollusc A soft-bodied animal that is a member of the phylum Mollusca. The bodies of most molluscs are covered by a shell.

molting In arthropods, the process by which an old exoskeleton is shed and a new one is formed.

monoculture The process of raising only one species in mariculture.

monophyletic Describes groups of organisms consisting of a common ancestor and all of its descendents. Another term that refers to this relationship is a *clade*.

monosaccharide A simple sugar that usually contains five or six carbon atoms.

morphology (mohr-FAHL-uh-jee) The structure or appearance of an organism.

mother-of-pearl layer Another term for the nacreous layer of pearl oysters.

mucilage (MYOO-suh-lij) The slimy gelatinous secretion covering algal cells for attachment of cells and for protection.

mutualism A symbiotic relationship in which both organisms benefit.

mycelium The term for the body of a fungus, whether a single cell or a large mass of hyphae.

mycologist A scientist who studies fungi.

mycology The scientific study of fungi.

myoglobin A pigmented protein in vertebrate muscle that stores oxygen.

nacreous layer (NAY-kree-uhs) The innermost layer of a molluscan shell. In oysters, it also is known as the *mother-of-pearl layer*.

nanoplankton Plankton ranging from 2 to 20 micrometers, composed primarily of small phytoplankton and larger bacteria. Also known as *centrifuge plankton*.

natural selection The mechanism that explains why organisms that possess variations best suited to their particular environments exhibit a better survival rate and reproductive capacity than do less well-suited organisms.

nauplius larva (NAW-plee-uhs) A larval stage in the life cycle of many crustaceans. Nauplius larvae are characterized by three pairs of appendages and a median eye.

nautical mile A unit of distance equal to 1 minute of latitude, 1.85 kilometers, and 1.15 land miles.

nautiloid A cephalopod that has an external shell belonging to the subclass Nautiloidea.

neap tide Tide that exhibits the smallest change between the high and low tide marks. Neap tides occur when the sun and the moon are at right angles to each other.

negative estuary An estuary in which the surface water flows toward the river and the water along the bottom moves out to sea.

nekton All actively swimming organisms whose movements are not governed by currents or tides.

nematocyst (neh-MAT-uh-sist) The stinging organelle found within the stinging cell of cnidarians.

nematode (NEM-uh-tohd) A round worm-like animal that belongs to the phylum Nematoda.

neritic province The water that overlies the continental shelves, same as neritic zone.

neritic zone (neh-RIT-ik) The zone of water that lies over the continental shelves, same as neritic province.

net plankton Components of the plankton that are large enough to be sampled with standard plankton nets; mainly composed of microplankton and larger organisms.

neuromast (NOO-roh-mast) A sense organ that can detect vibrations in the fluid that fills the canals of a fish's lateral line system.

neurotoxic shellfish poisoning A human syndrome caused by consumption of shellfish that concentrate toxic chemicals produced by blooms of certain species of dinoflagellate.

neuston Plankton that live at or near the surface of the ocean.

niche (nish; neesh) An organism's role in its environment.

nictitating membrane (NICK-ti-tay-ting) A clear membrane that covers and protects the eyes of some vertebrates.

nitrification (ny-truh-fi-KAY-shun) The process whereby ammonia from animal wastes and dead tissue is converted into nitrate ions.

nitrogen fixation The process by which some microorganisms are able to convert atmospheric nitrogen into a form that is usable by producer organisms.

nitrogenase The enzyme of nitrogen-fixing bacteria that is capable of breaking the strong bond of nitrogen gas for production of ammonia.

node The joint-like structure on a plant stem at which leaves and sometimes lateral stems and roots arise.

nonnative species Species that does not naturally occur in a given geographic area. Also known as *alien species* and *introduced species*.

nonselective deposit feeder An animal that ingests both organic and mineral particles and then digests the organic material, especially the bacteria that grow on the surface of the mineral particles.

nonsynchronous spawners Organisms that spawn at different times of the year.

nori An oriental food made from seaweed, often deriving from red algae of the genus *Porphyra*.

northeast trade winds The surface winds that occur between the equator and 30 degrees north latitude.

notochord A flexible supporting structure that runs the length of the body between the gut and the nerve cord. It is present in all chordates at some stage in their development.

nucleic acids Polymers composed of basic units called nucleotides.

nucleocapsid The combined core of nucleic acids and surrounding protein coat of a mature virus (or virion).

nucleolus The area of the nucleus where ribosomes are assembled.

nucleotide A molecule consisting of a five-carbon sugar, an organic base, and a phosphate group. Nucleotides are the building blocks of nucleic acids and are found in other important molecules as well.

nucleus A structure in eukaryotic cells that contains the cells' chromosomes and acts as the cellular control center.

nudibranch (NOO-di-brangk) A gastropod mollusc that does not have a shell and has many projections from its body, called *cerata*.

nutrient Any organic or inorganic material that an organism needs to metabolize, grow, and reproduce.

nutritive roots The finest division of roots for absorption of minerals.

obligate anaerobe Organism that thrives only in the absence of oxygen.

obliterative countershading A camouflage pattern in which the upper surface of pelagic animals is dark and the lower surface is light, helping the animal to blend in with the ocean depths or the surface, respectively, when observed by a potential predator. Also known as *countershading*.

observational science Science in which hypotheses are supported or denied on the basis of observations alone rather than controlled experiments.

ocean A continuous mass of water that covers nearly 70.8% of the earth's surface. The body of water that occupies an ocean basin.

ocean basin The floor of the ocean.

ocean productivity The production of organic matter by photosynthetic and chemosynthetic marine organisms.

ocean ranching The process of raising young fish in hatcheries and returning them to the sea when they have reached adulthood. Also known as *sea ranching*.

oceanic basaltic crust New earth crust formed when magma erupting from ridge systems cools.

oceanic province The water that covers the deep ocean basins.

oceanography The study of the oceans and their phenomena, such as waves, currents, and tides.

ochrophyte A photosynthetic stramenopile characterized by presence of chlorophylls *a* and *c*, the pigment *fucoxanthin*, and the starch *laminarin*; ochrophytes include diatoms, silicoflagellates, and brown algae.

octocoral A coral whose polyp has eight tentacles. These mostly colonial corals include soft corals such as sea fans and sea plumes and hard corals such as pipe-organ coral.

omnivore An animal that feeds on both producers and consumers.

ooze Sediment composed of fine biogenous particles.

operculum (oh-PER-kyoo-luhm) A hard or tough covering found in some molluscan gastropods that closes the opening to the shell when the animal retracts inside. In fishes, the protective covering of the animal's gills.

ophiuroid Echinoderm commonly referred to as brittle star or serpent star, belonging to the class Ophiuroidea.

optimal range The range of environmental factors to which an organism is best adapted.

oral tentacles Modified tube feet that are located around the mouth of a sea cucumber and used in acquiring food.

orca Another name for the killer whale, *Orcus orcina*.

organ A specialized structure made up of more than one tissue.

organelle Specialized structure in the cytoplasm of eukaryotic cells that performs a specific cellular function.

organ system A group of organs that function together for a specific purpose such as processing of food or gas exchange.

orientation click A low-frequency click produced by toothed whales such as dolphins that gives the animal a general idea of its surroundings.

osculum The opening through which water exits a sponge.

osmoconformer (ahz-moh-kuhn-FOHR-mer) An animal whose tissues and cells can tolerate dilution.

osmoregulator(ahz-moh-REG-yoo-lay-tir) An animal that can maintain an optimal salt concentration in its tissues regardless of the salt content of its environment.

osmosis (ahz-MOH-sis) The movement of water across a semipermeable barrier in response to differences in solute concentration on either side of the barrier.

osmotrophy Nutrition of an organism by absorption of small organic molecules from the external medium across the cell membrane.

ossicles Plates of calcium carbonate that make up the endoskeleton of echinoderms.

ostia The openings into the body of a sponge.

otoliths Structures in the inner ear of vertebrates that, together with sensory receptors, make up otolith organs, structures that are sensitive to gravity and linear acceleration.

overfishing The capture of fish faster than they can reproduce and replace themselves.

oviduct A tube that carries eggs to the outside of a female's body.

oviparity Reproductive mode characterized by eggs that hatch after leaving the body of the female.

ovoviviparity Reproductive mode characterized by females producing large, yolky eggs that hatch in the oviduct but are not nourished by the parent.

oxygen-minimum zone A region just below the sunlit surface waters in which oxygen is depleted by the resident animal life but not replaced by photosynthesis.

palp (bivalve) A pair of structures located near the mouth that forms filtered food into a mass and then moves it to the bivalve's mouth.

palp (sea spider) A pair of sensory structures.

Pangaea The single continental landmass or supercontinent that existed about 400 million years ago. It was composed of a northern portion, Laurasia, and a southern portion, Gondwanaland.

parallel The angular distance north or south of the equator measured in degrees. Also known as *latitude*.

paralytic shellfish poisoning A human syndrome caused by ingestion of saxitoxin in seafood.

parasite In a parasitic relationship, the organism that lives off its partner.

parasitism A symbiotic relationship in which one organism benefits and the other is harmed.

parr One- to five-year-old salmon that inhabit freshwater. As parr prepare to move into seawater they change into smolts.

partially mixed estuary An estuary that has a strong surface flow of freshwater and a strong influx of seawater.

particulate organic matter (POM) Particles of organic matter suspended in seawater; largely synonymous with *detritus* and *tripton*.

passive continental margins Location where sea and land meet that lies within a tectonic plate and on relatively wide continental shelves. As no plate collision or subduction is taking place, the dominant geologic processes are weathering, erosion, and the accumulation of sediments.

patch reef Small patch of reef located in a lagoon associated with an atoll or a barrier reef.

patchiness The uneven distribution of organisms in a population or a community.

pathogen An agent of disease or mortality such as a virus or bacterium.

pectoral fins The anterior paired fins of a fish.

pectoral stroking A cetacean behavior in which pectoral fins are used to stroke the body of another whale.

pedal disk The base of a sea anemone.

pedal laceration A type of asexual reproduction in which a portion of a sea anemone's basal disk is separated and left to form a new individual.

pedicellariae (ped-uh-suh-LEHR-ee-eh) Tiny pincer-like structures, found in some echinoderms, that keep the surface of the body clean and free of parasites and the settling larvae of fouling species. In some echinoderm species, they may also aid in obtaining food.

pedicle (PED-i-kuhl) A fleshy stalk found in some lamp shells that fastens the animal to a solid surface.

peduncle The part of a whale's body closest to the tail fluke.

pelagic division (pe-LAJ-ik) The division of the marine environment composed of the ocean's water.

pelagic ecosystem The animals, plants, and microbes of the water column and their realm of the open sea.

pelagic spawners Fish that release vast quantities of eggs into the water. The eggs are fertilized by the males and drift with the currents; there is no parental care.

pellicle The complex of alveoli and the cell membrane in alveolates.

pelvic fins The posterior pair of fins of a fish.

pen An internal strip of hard protein that helps support the mantle of a squid.

pennate diatom A shape of diatom with bilaterally symmetrical valves.

perennial A plant or alga that completes its life cycle in more than two growing seasons.

peridinin A kind of carotenoid pigment in dinoflagellates that gives them their golden-brown color and assists chlorophyll in gathering light energy for photosynthesis.

period The time required for one wavelength to pass a fixed point.

periostracum (per-eeAHS-treh-kuhm) The outermost layer of a molluscan shell. The periostracum is composed of the protein *conchiolin*.

periwinkle A mollusc in the family Littorinidae that inhabits the upper part of the intertidal zone.

pH scale A scale that indicates the acidity or basicity of a solution. The pH scale runs from 0 to 14, with 7 being the neutral point. The pH number reflects the concentration of hydrogen ions in a solution. Acids have a pH between 0 and 7, whereas bases have a pH between 7 and 14.

phage A virus that infects a bacterium; shortened form of *bacteriophage*.

phagocytosis The process by which a cell engulfs a particle by inward folding and separation of the cell membrane to form a vacuole.

phagotrophy A mode of nutrition by which one organism eats another, either by phagocytosis or by taking into the mouth.

pharyngeal gill slits Perforated slit-like openings that lead from the pharyngeal cavity to the outside. They are present in all chordates at some stage in their development.

pharynx (FA-ringks) A muscular tube that forms part of the digestive tract of an animal.

phenetics An approach to classification based on measureable similarities in physical characteristics.

pheromone (FAYR-eh-mohn) A hormone released into the environment by one individual that controls the behavior or development of other individuals of the same species.

phloem The kind of vascular tissue in plants that carries photosynthetic products from leaves to parts of the plant that need them as food or for storage.

phocid A seal that belongs to the family Phocidae. Also known as true seals, these animals lack external ears and are unable to rotate their flippers beneath their bodies.

phoronid (FOHR-oh-nid) A small worm-like lophophorate.

phospholipid A type of lipid that is made up of glycerol, two fatty acids, and a phosphate-containing molecule.

photic zone (FOH-tik) The portion of the pelagic division of the ocean that receives enough sunlight to support photosynthesis.

photodegrade The process of radiant energy breaking down a material.

photoperiod The amount of light and darkness in a 24-hour period.

photophore (FOH-toh-fohr) A specialized organ in some organisms that produces bioluminescence.

photoreceptor A sensory organ capable of responding to light.

photosynthesis The process by which some organisms use the energy of sunlight to produce organic molecules, usually from carbon dioxide and water.

phycobilins A class of accessory photosynthetic pigments in cyanobacteria and red algae; they capture wavelengths less used by chlorophylls and transfer the energy to them.

phycocolloid (fy-koh-KAHL-oyd) Any polysaccharide that has a gelling effect in water and that can be extracted from the cell walls of seaweeds; includes the commercially important agar, alginates, and carrageenan.

phycocyanin (fy-koh-SY-uh-nin) A blue phycobilin pigment that absorbs orange light.

phycoerythrin (fy-koh-e-RITH-rin) A red phycobilin pigment that absorbs green light.

phycologist (fy-KAHL-ah-jist) A scientist who studies seaweeds and phytoplankton. Also known as an *algologist*.

phylogenetic tree A branching diagram that shows lines of descent among a group of related species.

phylogeny The complete evolutionary history of a group of organisms.

physiographic chart (fiz-ee-oh-GRAF-ik) A chart of the ocean that uses perspective drawing, coloring, or shading to show the various depths.

phytoplankton (FY-toh-plank-tuhn) Tiny photosynthetic organisms that float in ocean currents.

picoplankton Plankton that range in size from 0.2 to 2.0 micrometers, composed primarily of larger viruses and many kinds of bacteria.

pinacocytes Flattened cells that form the outer body surface of a sponge.

pink zone The lower part of the midlittoral zone of tropical rocky shores characterized by the widespread encrustation of coralline algae.

pinniped (PIN-i-ped) An animal belonging to one of the following mammalian families: Otariidae, Phocidae, or Odobenidae. These animals are characterized by having flippers at the ends of their limbs and include seals, sea lions, elephant seals, and walruses.

placenta (plah-SEN-tuh) An organ present only during pregnancy in some female mammals that functions in maintaining the fetus until birth.

placental mammal Mammal that retains its young inside the body until birth.

placoid scale (PLAK-oyd) The type of scale in cartilaginous fishes that has a structure resembling the teeth of other vertebrates.

plankton Organisms that float or drift in the sea's currents.

plant Multicellular, photosynthetic organism whose cells are surrounded by a cell wall composed of cellulose.

planula larva The planktonic larval stage of cnidarians.

plastid An organelle in a eukaryotic cell derived originally by endosymbiosis with a cyanobacterium and having its own small number of genes.

plastid endosymbiosis The evolutionary process by which a heterotrophic host cell gains the ability to photosynthesize from a photoautotrophic guest cell.

plastron The ventral surface of a turtle's shell.

pleuston Components of the neuston that breach the surface of the ocean and are carried by wind as well as water currents.

plume A region of turbid water that spreads out from the mouth of a river or an estuary as it enters coastal seas.

plunger A type of wave in which the crest curls and curves over and outruns the rest of the wave.

pneumatophore An aerial root of a mangrove that arises from a cable root and grows out of the sediment, or the gas sac of a Portuguese man-of-war.

pod A group of related cetaceans.

podia The tube feet of sea stars.

polar easterlies Winds that occur in the Northern Hemisphere between 60 degrees north and the North Pole and blow from the north and east, and in the Southern Hemisphere between 60 degrees south and the South Pole and blow from the south and east.

polar molecule A molecule in which different parts of the molecule have different electrical charges.

pollen The male gametophyte of a seed-bearing plant; usually carried by wind, water, or pollinating animals to the female gametophyte within the cone or flower.

polychaete (PAHL-eh-keet) A type of annelid worm belonging to the class Polychaeta.

polyculture (PAHL-ee-kuhl-chur) The raising of more than one species in mariculture.

polygynous (pah-LIJ-eh-nehs) Relating to a male having more than one female mate.

polymer A large molecule that consists of multiple identical units linked together.

polyp (PAHL-uhp) A generally benthic form of cnidarian characterized by a cylindrical body that has an opening at one end that is usually surrounded by tentacles.

polypeptide A polymer of amino acids.

polysaccharide A type of carbohydrate that is a polymer of monosaccharides.

POM See *particulate organic matter*.

population A group of individuals of the same species that occupies a specified area.

positive estuary An estuary in which the surface water from the river flows out to sea and saltwater from the ocean moves into the estuary along the bottom.

postanal tail A tail that extends beyond an animal's digestive tract. It is present in all chordates at some stage in their development.

poster colors Bright, showy color patterns in fish and other animals that may advertise territorial ownership, help foraging individuals to keep in contact, or be important in sexual displays.

potential yield The number of pounds of fish or shellfish that a stock can yield per year without being overexploited.

P-R ratio The ratio of primary production (gross photosynthesis) to community respiration, sometimes used as a measure of the state of development (or succession) of a biological community.

precipitation nuclei Airborne particles that attract water droplets.

predation disturbance model A hypothesis that predation or other causes of mortality keep populations of coral reef fishes low enough to prevent competitive exclusion.

primary consumer A consumer that feeds directly on a producer. Also known as a *first-order consumer*.

primary plastid endosymbiosis The formation of an autotrophic eukaryote by endosymbiosis of a cyanobacterium within a heterotrophic eukaryotic cell.

primary productivity The process of producing energy-rich organic compounds from inorganic materials.

prime, or Greenwich, meridian The primary line of longitude that passes through the Royal Naval Observatory in Greenwich, England, and is designated 0 degrees longitude.

prismatic layer (priz-MAT-ik) The middle layer of a molluscan shell and the layer that contains most of the mass of the shell.

proboscis A tube that extends from the mouth of some animals.

producer An organism that can produce its own food. Also known as an *autotroph*.

progressive wave A wave generated by wind and restored by gravity that moves in a particular direction.

prokaryote (proh-KAR-ee-oht) A unicellular organism without a nucleus.

prokaryotic cell A cell that lacks a nucleus and membrane-bound organelles.

prop root An aerial root of a mangrove that arises from the trunk and descends to the sediment.

propagule A dispersal phase representing the next generation in a life cycle; specifically applied to viviparous mangroves, which do not produce dormant seeds.

protandry Condition in sequential hermaphroditic animals whereby sperm are produced in the life cycle before eggs.

protein synthesis The process by which ribosomes link amino acids into chains using the instructions provided by messenger RNA.

protein A polymer of amino acids.

protist A generic term used for eukaryotic organisms that are not classified as fungi, plants, or animals.

protogyny Condition in sequential hermaphroditic animals whereby eggs are produced in the life cycle before sperm.

pseudofeces (SOO-doh-fee-sees) Large semisolid particles produced by bivalves and consisting of phytoplankton and detritus that they filter but do not consume.

pseudopod (SOO-doh-pahd) A finger-like projection of cytoplasm and membrane that functions in locomotion and feeding in amoeboid protozoans.

pteropod ooze Calcareous sediments of the seafloor formed by the accumulation of shells of thecosome pteropods.

purse seine Huge net with a bottom that can be closed off by pulling on a line, similar to pulling the drawstring of a purse.

pycnocline (PIK-noh-klyn) A zone in the ocean that is characterized by a rapid change in density with depth.

pyramid of biomass A pyramid-shaped diagram that indicates the amount of biomass available at successive trophic levels.

pyramid of numbers A pyramid-shaped diagram that indicates the relative number of organisms a series of trophic levels can support.

pyramid of production A pyramid-shaped diagram that indicates the rate that organic matter is produced or the rate at which it is converted to the next level in the food chain.

pyrosome Large colony of pelagic tunicates that move by cilia of the pharynx. A colony may contain as many as 500 individuals per cubic meter of water and stretch over several meters.

radial symmetry The symmetrical organization of body parts around a central axis, similar to spokes on a wheel.

radiolarian An amoeboid marine protozoan with actinopods and usually producing a skeleton of silica.

radiolarian ooze A seafloor sediment consisting of the accumulated siliceous skeletons of radiolarians.

radioles Hair-like appendages found in certain Annelids that circle outward from the organism's central spine. They are used to catch phytoplankton from the water column.

radula (RAJ-oo-luh; RAD-yoo-luh) A ribbon of tissue that contains teeth. This structure is unique to molluscs.

raft culture A process in which juvenile commercial molluscs are attached to ropes that are suspended from rafts floating in regions of the ocean where food is plentiful and exposure to natural predators is minimized.

ram ventilation The production of a respiratory flow of water through the mouth and across the gills in some fish by swimming with the mouth open.

random pattern A dispersion pattern in which spacing among individuals varies in an unpredictable pattern.

raphe The slit along the valve by which some pennate diatoms move along surfaces.

realized niche The niche that an organism actually occupies. It is narrower than an organism's fundamental niche because of interspecific competition.

receptacle In brown algae, a swollen reproductive part of the thallus that holds chambers (*conceptacles*) for sexual reproduction.

reciprocal copulation The process in which two hermaphroditic animals fertilize each other.

recruitment The addition of new members to a population through reproduction or immigration.

rectal gland A gland that is associated with the rectal area of cartilaginous fishes and that functions in salt secretion.

red alga A member of the algal phylum Rhodophyta, with chlorophylls *a* and *d* and phycobilins.

redd The spawning area of salmon and trout, usually a cleared circular depression on the stream bottom.

reef crest The highest point on a coral reef.

reef flat The area opposite the reef front. Also known as the *back reef.*

reef front The outer, seaward margin of a coral reef. Also known as the *forereef.*

reflective beach Type of beach with a steep slope on which wave energy is directly dissipated.

regular echinoid An echinoid with a spherical body. Also known as a *sea urchin.*

renewable resource A resource such as a fishery that can replenish itself.

reproductive polyp A type of polyp found in hydrozoan colonies that is specialized for the process of asexual reproduction. Also known as a *gonangium.*

resource limitation model A hypothesis that available larvae are too limited for fish populations to ever reach the carrying capacity of the habitat and, therefore, competitive exclusion does not occur.

resource partitioning The process by which organisms share a resource.

respiratory tree A system of tubules found in most sea cucumbers that functions in gas exchange.

restoring force The force that causes the water in a wave to return to its undisturbed level.

reticulopod A pseudopod with branches that interconnect to form a net for the capture of particles.

rhizoid A root-like anchoring structure of a nonvascular plant.

rhizome A horizontal stem growing below ground or trailing on the surface.

rhizosphere The area below ground that is physically and chemically influenced by the complex of roots of a vascular plant.

ribbon worm An animal belonging to the phylum Nemertea. Also known as a *nemertean.*

ribonucleic acid (RNA) A usually single-stranded nucleic acid that functions in protein synthesis.

ribosomal RNA (rRNA) A type of structural RNA found in organelles called *ribosomes.*

ribosome An organelle consisting of protein and RNA. It is the site of protein synthesis.

ridge system A long mountain range that forms along cracks in the ocean floor where erupting magma breaks through the earth's crust. Also known as *midocean ridge.*

rift community A community of marine organisms that depend on the specialized environment found at divergence zones in the ocean floor. See also *deep-sea vent community.*

rift valley An area of high volcanic activity that runs along portions of a midocean ridge.

rift zone A region where the lithosphere splits, separates, and moves apart as new crust is formed.

roe An ovary with eggs.

root hair A hair-like extension of an epidermal cell of a root, for absorption of water and minerals.

rorqual A type of baleen whale in the family Balaenopteridae. These whales have a dorsal fin and ventral grooves on their throats.

rough endoplasmic reticulum (RER) Endoplasmic reticulum that has ribosomes attached to its surface.

r-strategist A species that maximizes its intrinsic rate of reproduction.

salinity A measure of the concentration of dissolved inorganic salts in water.

salp A free-swimming tunicate belonging to the class Thaliacea.

salt gland One or more cells of a plant or animal that concentrate salt from sap or blood and release a salty secretion outside the body to control the body's mineral balance.

salt wedge The angled boundary between saltwater and freshwater in an estuary that occurs when the rapid flow of river water prevents the saltwater from mixing with the freshwater.

salt-wedge estuary An estuary in which the seawater moves in and out along an angled boundary known as a *salt wedge.*

sampling methods Procedures used to estimate the size of populations.

saxitoxin A class of toxins produced by certain species of dinoflagellate that cause paralytic shellfish poisoning.

scale leaf A leaf that lacks a photosynthetic blade and is a protective sheath around the delicate growing tip of a stem.

scaphopod (SKA-foh-pahd) A mollusc belonging to the class Scaphopoda. Scaphopods are also known as *tusk shells* because their shells are shaped like the tusks of elephants.

school a grouping of individual fish that operates in a highly polarized and synchronized fashion.

scientific method The orderly pattern of gathering and analyzing information in science.

science An endeavor or study.

scleractinian coral Coral that forms hard skeletons of calcium carbonate around its polyps.

scyphozoan (sy-fuh-ZOH-uhn) An animal that belongs to the cnidarian class Scyphozoa. These are commonly called *jellyfish*.

sea A body of saltwater that is smaller than an ocean and is more or less landlocked.

sea anemone Large, heavy, complex polyps that belong to the cnidarian class Anthozoa.

seafloor spreading The process by which magma driven by convection currents is turned back by the lithosphere, moves laterally, and then descends, causing lateral movement of the earth's crust.

seagrass A lily-like flowering plant that grows and reproduces while submerged in seawater.

seamount A steep-sided formation that rises sharply from the ocean bottom.

seasonal delayed implantation A process by which a placental mammal can adjust her gestation period of less than one year into an annual time frame by delaying the implantation of the blastocyst in the uterus.

seaweed Any multicellular marine alga visible to the naked eye.

secondary consumer A carnivore that feeds on herbivores. Also known as a *second-order consumer*.

secondary metabolite A chemical made by cells for purposes other than routine metabolism.

secondary plastid endosymbiosis The condition of most eukaryotic photoautotrophs, in which an ancestor became host to a red algal cell that became the host's plastid.

second-order consumer A carnivore that feeds on herbivores. Also known as a *secondary consumer*.

sedentary polychaete (PAHL-eh-keet) A sessile polychaete that usually forms some sort of tube to cover its body.

sedimentation The settlement of particles from suspension in water, usually accumulating on the seafloor.

seed A specialized dormant stage in the life cycle of conifers and flowering plants. The seed contains the plant embryo and a supply of nutrients surrounded by a protective outer layer.

seismic sea wave A large wave that increases in amplitude as it approaches the shore. Sometimes referred to as a *tidal wave*. Also known as a *tsunami*.

selective deposit feeder An animal that separates organic material from minerals in sediments and ingests only the organic material.

selective forces The physical and biological characteristics of the environment, such as temperature, salinity, predation, and food availability, that favor the survival of one species over another.

semidiurnal tide The condition of having two high tides and two low tides each day.

senescence The process of aging.

sepia (SEE-pee-uh) A dark fluid produced in the ink glands of cephalopods.

septa (singular, septum) A divider or partition.

sequential hermaphrodite Individual possessing both functional male and female reproductive organs at different times in its life.

sessile A term applied to animals that spend most of their time fixed in one place.

seston Particles, living or dead, suspended in seawater.

setae (SEE-tee) Bristles on the bodies of annelid worms.

sextant A device used to measure the angle of a star with respect to the horizon.

sexual dimorphism (dy-MOR-fiz-uhm) The condition in which males and females look different from one another.

sexual reproduction The process by which two parent organisms produce an offspring by the fusion of sex cells produced by each parent.

shallow-water wave A deepwater wave that enters shallow water.

shear stress The force exerted by currents on an area of sea bottom.

sheath The basal, sometimes nonphotosynthetic part of the leaf of many vascular plants, usually wrapping around the stem.

shelf break An abrupt change in the underwater landscape that occurs where the continental shelf ends and the continental slope begins.

shelf zone The part of the ocean bottom that extends from the line of lowest tide to the edge of the continental shelf.

shellfish An invertebrate animal with a shell, usually molluscs and crustaceans.

shipworm A bivalve mollusc with a worm-like body belonging to the genus *Teredo* or *Bankia* that can burrow into wood.

shoaling An association of fish for various social reasons.

signature whistle A unique sound made by a dolphin that identifies the animal that is vocalizing, gives its location, and may relay other information as well.

siliceous ooze Sediment composed of 30% or more of the remains of organisms that produce shells made of silica.

silicoflagellate (sil-i-koh-FLAJ-uh-layt) A tiny photosynthetic protist that belongs to the phylum Chrysophyta. Silicoflagellates possess one or two flagella and internal shells composed of silica.

sinus venosus A thin-walled chamber that collects blood from the veins and conveys it to the atrium of the heart in fishes, amphibians, and reptiles.

siphon Tubular structure in some invertebrates that directs the flow of water in and out of the animal's body.

siphuncle (SY-fuhn-kuhl) A cord of tissue that runs through the chambers of a nautilus shell and functions in the removal of seawater from the chambers and in the regulation of the gas content of the chambers.

sipunculid Solitary worms belonging to the phylum Sipuncula.

sirenian (sy-REE-nee-uhn) An animal belonging to the mammalian order Sirenia, which includes manatees and dugongs.

slack water The period during a change in tide when tidal currents slow and then reverse.

smolts A stage in the life history of salmon characterized by a bright silvery color and loose scales. The smolt undergoes physiological changes that allow it to live in salt water.

smooth endoplasmic reticulum (SER) A type of endoplasmic reticulum that lacks ribosomes on its surface.

solute A dissolved substance.

solvent A medium for dissolving other substances.

southeast trade winds The surface winds between the equator and 30 degrees south latitude.

Southern Ocean The portion of the world ocean that surrounds the continent of Antarctica.

speciation (spee-see-AY-shun) The process by which new species are formed.

species (modern definition) One or more populations of potentially interbreeding organisms that are reproductively isolated from other such groups.

species (older definition) an organism with a definable set of characteristics that is visibly different from other similar organisms.

species epithet (EHP-i-thet) The second part of the two-part scientific name for an organism.

specific heat The amount of heat energy required to increase the temperature of 1 gram of a substance by 1°C. Also known as the *thermal capacity*.

spermaceti (spur-meh-SEH-tee) A thin, colorless, transparent oil that forms a waxy material when it comes into contact with air. Spermaceti is found in a cavity in the head of sperm whales.

spermatophore A package of sperm.

spicule (SPIK-yool) A support structure of sponges. A spicule can be composed of calcium carbonate, silica, or the protein *spongin*.

spiller A type of wave in which the energy is dissipated gradually as the wave moves over shallow bottom.

spiracle (SPEER-uh-kuhl) One of a pair of openings on the head of some cartilaginous fishes that functions in the passing of water to the gills.

spiral valve A valve shaped like a spiral staircase found in the intestine of sharks.

spirillum A bacterium with a cell shaped like a corkscrew.

splash zone The uppermost area of a rocky shore that is covered by only the highest tides and usually is just dampened by the spray of crashing waves. Also known as the *supralittoral fringe*.

spongin (SPUN-jin) A structural protein in some sponge spicules.

spongocoel (SPUN-joh-seel) A spacious water-filled cavity within the body of a sponge.

sporangium The part of a seaweed or plant that produces and usually releases spores for asexual reproduction.

sporophyte The multicellular stage in the life cycle of a seaweed or plant producing haploid or diploid spores that germinate into a subsequent stage without fertilization.

spring tide A tide that exhibits the greatest change between the high and low tide marks. Spring tide occurs when the earth, sun, and moon are in line with each other.

spur-and-groove formation The usual shape of a reef front, characterized by finger-like projections of coral that protrude seaward.

spy hopping A cetacean behavior in which a whale sticks its head straight up out of the water and surveys its surroundings.

squalene (SKWAY-leen) An oily material produced by the liver of sharks.

standing crop The amount of biomass in any trophic level in a food chain at a given time.

starch Polymer of glucose used by plants and some other organisms to store energy reserves.

statocyst An organ found in some animals that helps them maintain equilibrium.

steroid A type of lipid that contains complex ring structures. Many function as important chemical messengers in the bodies of animals.

stigma The part of the female flower that receives the pollen.

stilt root Any aerial root (drop or prop root) that holds up a mangrove tree.

stipe A stem-like structure in seaweeds, particularly brown algae, but lacking the vascular tissue of plants.

stock Separate population of commercial fishes or shellfishes within a species' geographic range that is assumed to be reproductively isolated from other stocks.

stoma (STOH-ma; plural, stomata, stoh-MAH-tuh) An opening in a plant leaf that allows water to escape and gases to enter.

stonewort A plant belonging to the phylum Charophyta that lives in fresh or brackish waters and is related to green algae and land plants.

stony (or true) corals The primary organisms depositing the calcium carbonate ($CaCO_3$) that builds the structure of coral reefs. Members of the phylum Cnidaria, class Anthozoa, order Scleractinia.

stramenopile A group of organisms, including diatoms, silicoflagellates, and brown algae, characterized by heterokont flagella with specialized mastigonemes.

stress zone A region above or below the optimal range which an organism must expend more energy than normal to maintain homeostasis.

stroma The liquid that contains enzymes for carbohydrate synthesis and surrounds the grana inside chloroplasts.

stromatolite Coral-like community of microbes that form a thin layer of living cells and filaments over an accumulated mass of dead and lithified material.

structural color Iridescent color produced by light reflecting from crystals located in specialized cells.

stylet A hard sharp point on the end of an organ.

subduction zone Region of the ocean floor where old crust sinks to the earth's core and is recycled.

submarine canyon Underwater canyon similar to canyons on land that are found on some continental slopes.

subtidal zone The region of the shore that is covered by water even during low tide. It is also known as the *infralittoral zone*.

succulent The condition of enlarged water-filled cells in the tissues of plants, making leaves or stems appear thicker than those of nonsucculent plants; a condition typical of plants in deserts or along marine shorelines.

sucrose A disaccharide composed of glucose and fructose.

sulfide A chemical compound that contains the element sulfur and one other element.

supralittoral fringe Uppermost area of a rocky shore that is partially covered by only the highest (spring) tides and is usually just dampened by the spray of crashing waves. Also known as the *splash zone*, it overlaps the boundary of the supralittoral and littoral zones.

supralittoral zone Zone of rocky shore that lies above the extreme high water of spring tides.

surf zone The area of a coast where waves slow down before striking the shore.

surface tension A property of liquids characterized by the clinging together of molecules at the exposed surface as the result of cohesive forces.

surge channel Groove in the buttress zone of a coral reef that penetrates the algal ridge.

surimi A product made from the flesh of the Alaskan pollock (*Theragra chalcogramma*). It is flavored to produce artificial crab, shrimp, and lobster.

survivorship The length of time, on average, an individual of a given age could be expected to live.

survivorship curve A graph that shows the number of individuals in a population still alive at each age from reproduction to death.

suspension feeder An organism that feeds on material suspended in the water.

sustainable yield The number of fishes and shellfishes that can be caught over several years without stressing the population.

sverdrup (sv) A measure of the volume of water transported in an ocean current. One sverdrup is equal to a flow of 1 million cubic meters of water per second.

swarm A large mass of krill.

swarming A behavior exhibited by some errant polychaetes in which the males and females congregate at the surface of the water to reproduce.

swash The water running up a beach after a wave breaks.

sweeper tentacles Very long, nematocyst-containing tentacles of certain corals that sweep over and kill polyps on competitive coral species.

swell A long-period, uniform wave that appears as a regular pattern of wave crests on the ocean's surface.

swim bladder A gas-filled sac in some bony fishes that allows them to maintain neutral buoyancy.

swimmeret An appendage modified for swimming in some crustaceans.

syconoid (SY-kuh-noyd) A type of sponge body with a single spongocoel that has many invaginations.

symbiosis (sim-by-OH-sis) An intimate living arrangement between two different species of organism.

synchronous hermaphrodite Individual possessing both functional male and female reproductive organs at the same time in its life.

synchronous spawners Organisms that spawn during a brief annual event.

T.E.D. See *turtle exclusion device*.

tagging A procedure for monitoring the distribution and movements of fisheries stocks in which fish that are caught are marked with identification tags and then released to be caught again.

tail fluke One of the two lobes of a whale's tail.

tail lobbing A cetacean behavior in which a whale lifts its tail high above the water and slaps it down on the surface with such force as to make a huge splash and produce a loud noise. Also known as *tail slapping*.

tail slapping A cetacean behavior in which a whale lifts its tail high above the water and slaps it down on the surface with such force as to make a huge splash and produce a loud noise. Also known as *tail lobbing*.

tan Individual section of a drift net.

tannin A class of compounds that are produced by plants and that have properties that reduce microbial infection and herbivory.

tapeworm A parasitic flatworm belonging to the class Cestoda. Tapeworms are found in the intestines of vertebrate animals.

taxonomy (tak-SAHN-uh-mee) The science of naming and classifying organisms.

tectonic estuary An estuary that forms when earthquakes cause the land at the mouth of a river to sink, allowing seawater to cover it.

telson A long spike attached to the posterior end of a horseshoe crab's abdomen; used for steering and defense.

temporal isolation The prevention of interbreeding between different species in nature due to their mating at different times of day or different times of the year.

ten percent rule A rule in ecology that states that, on average, only approximately 10% of the energy available at one trophic level is passed along to the next trophic level.

tentacle An arm-like or finger-like structure that projects from an animal's body, usually the head.

terrigenous sediment Sediments that originate on land and are washed into the sea.

tertiary consumer A carnivore that feeds on carnivores that eat herbivores. Also known as a *third-order consumer*.

test A term used for the external covering of foraminiferans; the hard skeleton of echinoid echinoderms such as sea urchins.

tetraspore The haploid, nonmotile, dispersal stage released by the tetrasporophyte of a red alga.

tetrasporophyte A multicellular diploid stage in the life cycle of a red alga growing independently and releasing haploid spores, which are produced by meiosis.

thallus (THAL-uhs) The body of an alga.

thecosome pteropod An herbivorous oceanic gastropod that has a highly reduced shell and swims by flapping lateral extensions of its foot.

theory of plate tectonics The theory stating that the movement of continental masses is the result of the movement of the rigid slabs, or plates, on which they rest.

thermal capacity The amount of heat energy required to increase the temperature of 1 gram of a substance by 1° C. Also known as *specific heat*.

theory A body of observations and their experimental supports that have stood the test of time.

thermocline (THER-moh-klyn) A zone in the ocean that is characterized by a rapid change in temperature with increasing depth.

thermohaline circulation Vertical mixing that results from density changes due to changes in temperature and salinity of the water.

third-order consumer A carnivore that feeds on carnivores that eat herbivores. Also known as a *tertiary consumer*.

thraustochytrids (thraw-stow-KY-trids) Labyrinthomorphs that are abundant decomposers in planktonic and benthic communities.

thylakoid A membrane-bound structure that contains chlorophyll and other photosynthetic pigments. They are found within chloroplasts.

tidal current Current that occurs as the result of changing tides.

tidal flat An area of estuary that is exposed at low tide and covered at high tide.

tidal overmixing A mixing action that occurs in some estuaries at high tide due to seawater at the surface moving upstream more quickly than the bottom water from the river, causing the denser surface water to sink to the bottom and displace lighter freshwater, which moves to the surface.

tidal wave A term sometimes used as a synonym for tsunami.

tide The changes in sea level that occur as the result of the gravitational pull of the moon and the sun on the water in the oceans.

tide pool A depression in intertidal rocks or in the intertidal zone of sandy beaches that continues to hold water during a low tide.

tiller A secondary stem that grows from the base of a main stem (culm) in a grass and gives the plant a tufted growth if arising in multiples.

tissue A group of similar cells that serve a specific function.

tolerance A mechanism to reduce the effects of predation by rapid recovery through regrowth or high reproductive output.

top-down factors Factors such as competition, herbivory, or predation that affect the structure of communities by acting on lower trophic levels.

trace elements Elements dissolved in seawater that are present in concentrations of less than one part per million.

transfer RNA (tRNA) A type of RNA that carries amino acids to ribosomes for their incorporation into protein.

transform fault A fault occurring in regions where lithospheric plates move past each other, particularly sections of the midocean ridge.

transverse currents Currents that connect eastern- and western-boundary currents within a gyre.

trash fish The noncommercial fishes that are killed during fishing for commercial species.

trawl A large net dragged along the bottom or in midwater, depending on the catch, by vessels called *trawlers*.

triglyceride A type of lipid composed of glycerol and three fatty acids.

tripton Particles of dead organic matter suspended in seawater; largely synonymous with *detritus* and *POM*.

trochophore larva (TROHK-eh-fohr) A free-swimming ciliated larva in the life cycles of many marine molluscs as well as some other marine organisms.

trophic cascades Effects that changes in the size of one population in a food web have on populations below it.

trophic level (TROH-fik) An energy-storing level in a food chain.

Tropic of Cancer The latitude 23.5 degrees north of the equator. It marks the northernmost extent of the tropics.

Tropic of Capricorn The latitude 23.5 degrees south of the equator. It marks the southernmost extent of the tropics.

trumpet cell A type of cell in the tissue of kelp carrying products of photosynthesis to parts of the thallus that lie in deeper water.

tsunami (soo-NAH-mee) A large, seismic, sea wave that increases in amplitude as it approaches a shore. Sometimes referred to as a *tidal wave*. Also known as a *seismic sea wave*.

tube feet Tubular structures in echinoderms that function in locomotion and feeding.

tunic The body covering of animals known as tunicates, phylum Urochordata, which is composed of a molecule similar to cellulose.

tunicate (TOO-ni-kayt) Animals belonging to the subphylum Urochordata. Tunicates are named for their body covering: a tunic composed of a substance similar to cellulose.

turbellarian flatworm A free-living (nonparasitic) type of flatworm (phylum Platyhelminthes).

turbidity The cloudy condition of seawater that results from the presence of living or dead suspended particles.

turbidity current A swift underwater avalanche of sediment and water.

turtle exclusion device (T.E.D.) A grate that is placed in shrimp trawls to direct sea turtles out of the net so they won't drown.

twilight zone The region of ocean between 150 and 450 meters deep where there is not enough light to power photosynthesis. Also known as the *disphotic zone*.

Type I survivorship curve A graph that indicates relatively low death rates during early and middle life, with increasing mortality among older individuals.

Type II survivorship curve A graph that indicates more-or-less constant mortality rates over time.

Type III survivorship curve A graph that indicates high mortality rates for the young, but a flat rate for individuals that survive to a certain critical age.

type specimen A museum specimen considered to be representative of the species it represents.

typological definition of species The defining of a species based on morphology or appearance.

umbo The area around the hinge of a bivalve's shell. This area represents the oldest part of the shell.

unarmored dinoflagellate Dinoflagellate with a few thin layers of cellulose in its alveoli, giving the appearance of having no protective cell covering.

understory The layer of low vegetation on the floor of a kelp forest.

uniform pattern Pattern wherein individuals are evenly spaced.

univalve A shell composed of one piece.

upwelling The process by which a combination of wind and ocean currents bring nutrient-laden material from the ocean bottom into the photic zone.

urchin barren A community dominated by sea urchins.

urea Waste product of protein metabolism that is stored in the bodies of cartilaginous fish and a few other animals to maintain osmotic balance. In vertebrates it is formed in the liver and excreted by the kidneys.

vacuole Structure surrounded by a plasma membrane that may contain food, wastes, or water.

valve (diatom) One half of the two-part cell wall, or frustule, of a diatom.

valve (mollusc and brachiopod) One of the two jointed halves of a shell that surround the bodies of bivalves and lamp shells.

vascular tissue Layers of cells that conduct water, minerals, and the products of photosynthesis to distant parts of a plant and also provide structural support for the plant body.

vegetative growth A form of asexual reproduction in seaweeds and plants in which growth regions produce additional units of the thallus or plant body of identical genetic makeup.

veliger larva (VEL-uh-jer) A free-swimming larval stage that develops from the trochophore larva of some molluscs.

vent community The populations of organisms that live around a deep-sea vent. Also known as a *rift community*.

ventricle Thick-walled and muscular pumping chamber of the vertebrate heart.

vertebra (plural, vertebrae) Segment of the spinal column of vertebrates.

vertebrate An animal that possesses an internal skeletal rod (commonly called a backbone) composed of units known as *vertebrae*.

vertical mixing The mixing of water in a water column that occurs when the surface water becomes more dense than the water beneath it.

vertical overturn The change-over in the water of an unstable water column.

vertical zonation The separation of organisms in a habitat into definite vertical zones or bands.

viral replication The process by which the parts of many new viruses are manufactured and assembled within a cell under the control of the infecting virus.

virion The mature infective viral particle released by a host cell.

viriplankton Planktonic virus particles.

virologist A scientist who studies viruses.

virology The scientific study of viruses.

visceral mass The dorsal region of the gastropod body that contains the circulatory, digestive, respiratory, excretory, and reproductive systems.

viviparity Reproductive mode characterized by females producing eggs that are retained and nourished in the reproductive system until the young are mature enough to be released to the outside.

warning coloration Bright and conspicuous coloration possessed by animals that are distasteful or dangerous to their attackers. Such colors are presumed to warn away attackers. Also known as *aposematic coloration*.

water column The ocean water in an ocean basin.

water table The level below which the ground remains saturated with water.

water vascular system A hydraulic system unique to echinoderms that functions in locomotion, feeding, gas exchange, and excretion.

wave refraction The horizontal bending of waves that enter shallow water at an angle to the shore caused by one end of the wave slowing as it drags on the bottom.

wave shock The force of waves as they crash against the coastline and the organisms living on it.

well-mixed estuary An estuary in which river flow is low and tidal currents play a major role in the circulation of the water, resulting in a seaward flow of water and a uniform salinity at all depths.

westerlies Winds that occur between 30 and 60 degrees north latitudes that blow from the south and the west and that occur between 30 and 60 degrees south that blow from the north and the west.

western-boundary currents The currents that flow along the western edge of ocean basins, moving warm water toward the poles.

wetland A basically terrestrial area that contains a large amount of water such as a salt marsh or swamp.

white zone The upper zone of the supralittoral fringe of tropical rocky shores. This zone is the true border between the land and the sea. No fully marine organisms occur here.

whorls The turns of a gastropod's spiral shell.

wind-induced vertical circulation The vertical movement of water that sometimes occurs when wind moves surface water horizontally, causing deeper water to move vertically.

world ocean A continuous mass of water that covers nearly 70.8% of the earth's surface.

xanthophylls (ZAN-thuh-filz) A subclass of carotenoid accessory pigments that confer a yellow or brown hue to some organisms.

xylem The kind of vascular tissue in plants that carries water and minerals from roots to parts of the plant that need them.

yeast A fungus that consists of a single cell.

yellow zone The upper part of the midlittoral zone of tropical rocky shores. The typical yellow or green color of the rock is due to microscopic boring algae.

yolk sac A sac-like structure in amniotic eggs that contains a supply of food in the form of yolk.

zoea larva The planktonic larval stage of decapod crustaceans.

zonation The separation of organisms in a habitat into definite zones or bands.

zone of drying sand The uppermost vertical zone in the high intertidal region of a sandy shore characterized by the presence of moisture only during the highest tides.

zone of intolerance A region that is so far removed from an organism's optimal range for an environmental variable that the organism cannot survive.

zone of resurgence A vertical zone of the middle and low intertidal regions of a sandy shore that retains water at low tide.

zone of retention A vertical zone in the high intertidal region of a sandy shore that retains water during low tide as the result of capillary action.

zone of saturation The vertical zone of the sandy intertidal region that is constantly moist.

zone of stress A region above or below an organism's optimal range for an environmental variable in which the organism must expend more energy than normal to maintain homeostasis.

zooid An individual member of a bryozoan colony.

zooplanktivorous The condition of filter feeding on small, nonphotosynthetic organisms that occur in the plankton.

zooplankton Components of the plankton that feed by ingestion of detritus or of other zooplankton.

zooxanthella (ZOH-oh-zan-THEL-ah) A dinoflagellate that is an important symbiont of diverse marine microbes and invertebrates.

zygote The diploid and genetically unique cell that is formed upon fusion of two haploid cells or gametes of different strains or, in the extreme case, when the egg is fertilized by a sperm.

Answers to Multiple-Choice Questions

Chapter 1

Multiple Choice
1. b 2. b 3. c 4. c 5. b

Matching
1. b 2. c 3. d 4. j 5. i 6. a 7. e 8. g
9. f 10. h

Chapter 2

Multiple Choice
1. e 2. a 3. d 4. b 5. c 6. c 7. c 8. b
9. c

Matching
1. b 2. d 3. a 4. e 5. c 6. c 7. e 8. a
9. b 10. d 11. c 12. e 13. b 14. d 15. a

Chapter 3

Multiple Choice
1. d 2. d 3. c 4. d 5. d 6. b 7. d 8. b
9. c 10. a

Matching
1. b 2. e 3. a 4. c 5. d 6. b 7. e 8. a
9. c 10. d 11. c 12. a 13. e 14. b 15. d
16. e 17. c 18. a 19. b 20. d

Chapter 4

Multiple Choice
1. b 2. a 3. c 4. d 5. c 6. b 7. e 8. c

Matching
1. c 2. d 3. b 4. e 5. a 6. c 7. e 8. a
9. b 10. d

Chapter 5

Multiple Choice
1. c 2. c 3. d 4. e 5. d

Matching
1. c 2. a 3. e 4. b 5. a 6. d 7. a 8. e
9. b 10. c

Chapter 6

Multiple Choice
1. b 2. b 3. a 4. b 5. d 6. a 7. c 8. a
9. c 10. a

Matching (Set #1)
1. c 2. e 3. f 4. b 5. a 6. d

Matching (Set #2)
1. f 2. c 3. a 4. g 5. e 6. h 7. d 8. b

Chapter 7

Multiple Choice
1. a 2. c 3. d 4. c 5. e 6. b 7. c 8. d
9. e

Matching (Set #1)
1. f 2. c 3. b 4. d 5. a 6. e

Matching (Set #2)
1. c 2. b 3. a 4. e 5. f 6. d

Chapter 8

Multiple Choice
1. c 2. c 3. a 4. d 5. e 6. d 7. c 8. b

Matching
1. b 2. e 3. a 4. c 5. d 6. c 7. d 8. b
9. e 10. a

Chapter 9

Multiple Choice
1. e 2. b 3. d 4. c 5. c 6. d 7. c 8. b
9. b 10. d

Matching
1. c 2. e 3. b 4. d 5. a 6. d 7. a 8. c
9. e 10. b

Chapter 10

Multiple Choice
1. e 2. b 3. d 4. b 5. c 6. d 7. d 8. d
9. e 10. a 11. b 12. c

Matching

1. g	2. h	3. l	4. d	5. n	6. c	7. b	8. o
9. a	10. f	11. m	12. j	13. k	14. e	15. i	

Chapter 11

Multiple Choice

1. c	2. b	3. d	4. e	5. c	6. b	7. e	8. c
9. d	10. d						

Matching

1. c	2. a	3. d	4. b

Chapter 12

Multiple Choice

1. c	2. d	3. c	4. d	5. c	6. e	7. b

Matching

1. c	2. e	3. d	4. a	5. b

Chapter 13

Multiple Choice

1. d	2. c	3. a	4. d	5. d	6. d	7. d	8. b
9. b	10. a						

Matching (Set #1)

1. a	2. e	3. b	4. d	5. c

Matching (Set #2)

1. e	2. d	3. c	4. b	5. a

Chapter 14

Multiple Choice

1. b	2. c	3. c	4. c	5. c	6. b	7. a	8. b
9. a	10. b						

Matching

1. d	2. e	3. a	4. c	5. b	6. e	7. c	8. a
9. b	10. d						

Chapter 15

Multiple Choice

1. a	2. b	3. b	4. b	5. e	6. c	7. a	8. c
9. e	10. b	11. c	12. d				

Matching

1. g	2. n	3. l	4. b	5. k	6. d	7. e	8. f
9. h	10. i	11. j	12. m	13. o	14. a	15. c	

Chapter 16

Multiple Choice

1. c	2. c	3. a	4. b	5. e	6. b	7. b	8. d
9. b	10. e						

Matching

1. b	2. d	3. e	4. c	5. a

Chapter 17

Multiple Choice

1. b	2. c	3. d	4. a	5. e	6. c	7. b	8. a
9. c	10. b						

Multiple Choice (Set #1)

1. d	2. b	3. c	4. f	5. e	6. a

Multiple Choice (Set #2)

1. b	2. f	3. a	4. e	5. c	6. d

Chapter 18

Multiple Choice

1. b	2. c	3. d	4. b	5. e	6. d	7. d

Matching

1. d	2. e	3. a	4. c	5. b

Chapter 19

Multiple Choice

1. d	2. d	3. c	4. d	5. b	6. d	7. d

Matching

1. d	2. a	3. e	4. c	5. b

Chapter 20

Multiple Choice

1. c	2. d	3. b	4. c

Matching

1. c	2. e	3. d	4. a	5. b

Index

Acidification

As the ocean absorbs carbon dioxide from the atmosphere, it becomes more acidic. This can damage corals and the shells of marine animals and make it difficult for new organisms to build calcium structures. The map shows the estimated decrease since 1870 in the aragonite saturation state, which is linked to an increase in ocean acidity.

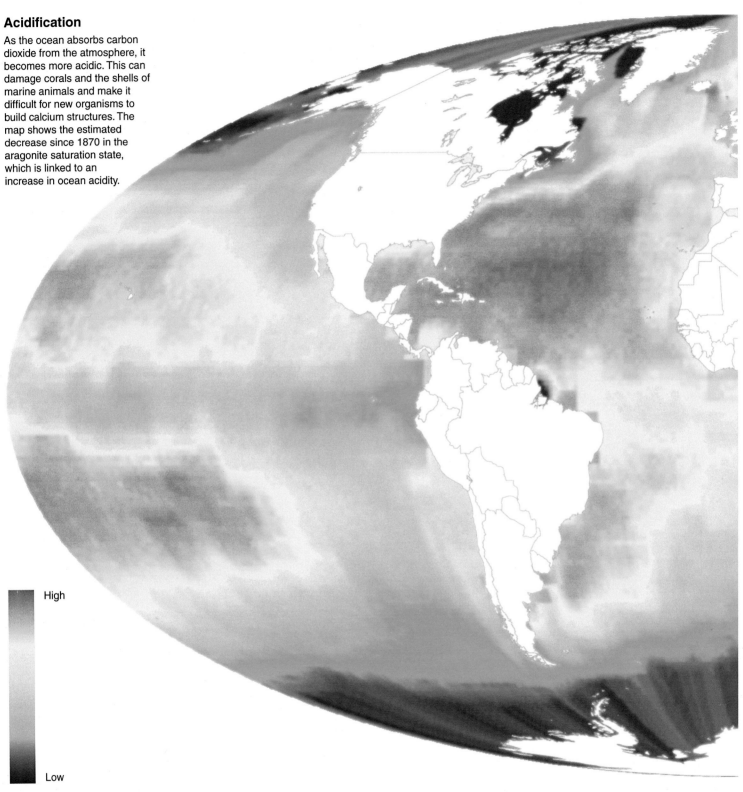

High

Low

Sources: Benjamin S. Halpern, National Center for Ecological Analysis and Synthesis; *Science* 319 (Feb. 2008):948–952.

You can learn more about the damaging effects of acidification of ocean water in Chapters 4, 15, and 20.

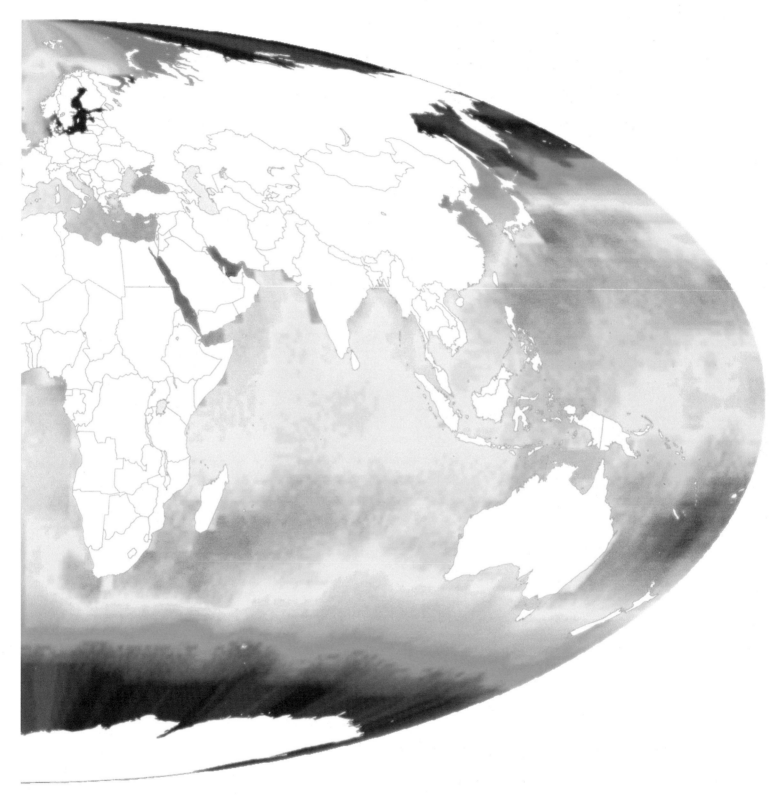